Model Reduction and Approximation

Computational Science & Engineering

The SIAM series on Computational Science and Engineering publishes research monographs, advanced undergraduate- or graduate-level textbooks, and other volumes of interest to an interdisciplinary CS&E community of computational mathematicians, computer scientists, scientists, and engineers. The series includes both introductory volumes aimed at a broad audience of mathematically motivated readers interested in understanding methods and applications within computational science and engineering and monographs reporting on the most recent developments in the field. The series also includes volumes addressed to specific groups of professionals whose work relies extensively on computational science and engineering.

SIAM created the CS&E series to support access to the rapid and far-ranging advances in computer modeling and simulation of complex problems in science and engineering, to promote the interdisciplinary culture required to meet these large-scale challenges, and to provide the means to the next generation of computational scientists and engineers.

Series Volumes

Benner, Peter, Cohen, Albert, Ohlberger, Mario, and Willcox, Karen, Editors, *Model Reduction and Approximation: Theory and Algorithms*

Kuzmin, Dmitri and Hämäläinen, Jari, *Finite Element Methods for Computational Fluid Dynamics: A Practical Guide*

Rostamian, Rouben, *Programming Projects in C for Students of Engineering, Science, and Mathematics*

Smith, Ralph C., *Uncertainty Quantification: Theory, Implementation, and Applications*

Dankowicz, Harry and Schilder, Frank, *Recipes for Continuation*

Mueller, Jennifer L. and Siltanen, Samuli, *Linear and Nonlinear Inverse Problems with Practical Applications*

Shapira, Yair, *Solving PDEs in C++: Numerical Methods in a Unified Object-Oriented Approach, Second Edition*

Borzì, Alfio and Schulz, Volker, *Computational Optimization of Systems Governed by Partial Differential Equations*

Ascher, Uri M. and Greif, Chen, *A First Course in Numerical Methods*

Layton, William, *Introduction to the Numerical Analysis of Incompressible Viscous Flows*

Ascher, Uri M., *Numerical Methods for Evolutionary Differential Equations*

Zohdi, T. I., *An Introduction to Modeling and Simulation of Particulate Flows*

Biegler, Lorenz T., Ghattas, Omar, Heinkenschloss, Matthias, Keyes, David, and van Bloemen Waanders, Bart, Editors, *Real-Time PDE-Constrained Optimization*

Chen, Zhangxin, Huan, Guanren, and Ma, Yuanle, *Computational Methods for Multiphase Flows in Porous Media*

Shapira, Yair, *Solving PDEs in C++: Numerical Methods in a Unified Object-Oriented Approach*

Edited by

PETER BENNER
Max Planck Institute for Dynamics
of Complex Technical Systems
Magdeburg, Germany

ALBERT COHEN
Université Pierre et Marie Curie
Paris, France

MARIO OHLBERGER
Universität Münster
Münster, Germany

KAREN WILLCOX
Massachusetts Institute of Technology
Cambridge, Massachusetts

Model Reduction and Approximation

Theory and Algorithms

Society for Industrial and Applied Mathematics
Philadelphia

10 9 8 7 6 5 4 3 2 1

Publisher	David Marshall
Executive Editor	Elizabeth Greenspan
Developmental Editor	Gina Rinelli Harris
Managing Editor	Kelly Thomas
Production Editor	Ann Manning Allen
Copy Editor	Julia Cochrane
Production Manager	Donna Witzleben
Production Coordinator	Cally Shrader
Compositor	Cheryl Hufnagle
Graphic Designer	Lois Sellers

Library of Congress Cataloging-in-Publication Data

Names: Benner, Peter, editor. | Cohen, Albert, 1965- editor. | Ohlberger, Mario, editor. | Willcox, Karen, editor.
Title: Model reduction and approximation : theory and algorithms / edited by Peter Benner, Max Planck Institute for Dynamics of Complex Technical Systems, Magdeburg, Germany, Albert Cohen, Université Pierre et Marie Curie, Paris, France, Mario Ohlberger, Universität Münster, Münster, Germany, Karen Willcox, Massachusetts Institute of Technology, Cambridge, Massachusetts.
Description: Philadelphia : Society for Industrial and Applied Mathematics, [2017] | Series: Computational science and engineering ; 15 | Includes bibliographical references and index.
Identifiers: LCCN 2017012782 (print) | LCCN 2017016389 (ebook) | ISBN 9781611974829 (ebook) | ISBN 9781611974812 (print)
Subjects: LCSH: Mathematical models. | Mathematical optimization. | Computational complexity.
Classification: LCC QA401 (ebook) | LCC QA401 .M3979 2017 (print) | DDC 511.3/4–dc23
LC record available at https://lccn.loc.gov/2017012782

List of Contributors

Athanasios C. Antoulas
Rice University
Houston, TX 77005
USA

Ulrike Baur
Max Planck Institute for Dynamics of Complex Technical Systems
Sandtorstr. 1
39106 Magdeburg
Germany

Christopher Beattie
Department of Mathematics
Virginia Tech
Blacksburg, VA 24061
USA

Peter Benner
Max Planck Institute for Dynamics of Complex Technical Systems
Sandtorstr. 1
39106 Magdeburg
Germany

Tobias Breiten
Karl-Franzens-Universität Graz
Institute for Mathematics and Scientific Computing
Heinrichstraße 36
8010 Graz
Austria

Martin Gubisch
University of Konstanz
Department of Mathematics and Statistics
Universitätsstr. 10
78457 Konstanz
Germany

Serkan Gugercin
Department of Mathematics
Virginia Tech
Blacksburg, VA 24061
USA

Bernard Haasdonk
University of Stuttgart
Institute of Applied Analysis and Numerical Simulation
Pfaffenwaldring 57
70569 Stuttgart
Germany

Christian Himpe
University of Münster
Institute for Computational and Applied Mathematics
Einsteinstr. 62
48149 Münster
Germany

A. Cosmin Ionita
The MathWorks, Inc.
Natick, MA 01760-2098
USA

Sanda Lefteriu
Ecole des Mines
59508 Douai
France

Immanuel Martini
University of Stuttgart
Institute of Applied Analysis and Numerical Simulation
Pfaffenwaldring 57
70569 Stuttgart
Germany

Anthony Nouy
Ecole Centrale de Nantes
Department of Computer Science and Mathematics, GeM
1 rue de la Noe
44321 Nantes
France

Mario Ohlberger
University of Münster
Institute for Computational and Applied Mathematics
Einsteinstr. 62
48149 Münster
Germany

Ivan Oseledets
Skolkovo Institute of Science and Technology
Moscow 143026
Russia

Ronald DeVore
Texas A&M University
College Station, TX 77843
USA

Stefan Volkwein
University of Konstanz
Department of Mathematics and Statistics
Universitätsstr. 10
78457 Konstanz
Germany

Contents

List of Figures

List of Tables

List of Algorithms

Preface

Many physical, chemical, biomedical, and technical processes can be described by means of partial differential equations (PDEs) or dynamical systems. If the underlying processes exhibit nonlinear dynamics, analysis and prediction of the complex behavior is often only possible by solving the (partial) differential equations numerically. For this reason, the design of efficient numerical schemes is a central research challenge. In spite of increasing computational capacity, many problems are of such high complexity that they are still only solvable with severe simplifications. In recent years, large-scale problems—often involving multiphysics, multiscale, or stochastic behavior—have become a particular focus of applied mathematical and engineering research. A numerical treatment of such problems is usually very time-consuming and thus requires the development of efficient discretization schemes that are often realized in large parallel computing environments. In addition, these problems often need to be solved repeatedly for many varying parameters, introducing a curse of dimensionality when the solution is also viewed as a function of these parameters. With this book we aim to introduce recent developments on complexity reduction of such problems, both from a theoretical and an algorithmic perspective.

This book is based partially on keynote lectures from the workshop "Model Reduction and Approximation for Complex Systems," held at the Centre International de Rencontres Mathématiques, June 10–14, 2013, in Luminy, Marseille, France. This workshop brought together some of the world's leading experts from mathematics and engineering sciences who are concerned with modeling, approximation, and model reduction of complex (parametrized) systems. In particular, the focus was on further developing and analyzing approaches for practically relevant problems that are modeled by PDEs or dynamical systems.

The workshop addressed complexity reduction of such problems in settings that include design, control, optimization, inverse problems, uncertainty analysis, and statistical sampling. In recent years, there has been a tremendous effort to develop efficient approaches to deal with such problems in various mathematical communities. In particular, relevant research areas are high-dimensional and sparse approximation, system-theoretic model order reduction (MOR) for dynamical systems, proper orthogonal decomposition (POD), reduced basis (RB) methods for parametrized PDEs, numerical multiscale methods, polynomial chaos approximations, stochastic finite elements (FEs), approximation by slow manifolds, multiresolution methods, and hierarchical dimension reduction techniques.

The major goal of the workshop was to bring together leading mathematicians and engineers from these research areas to foster collaboration and to stimulate an exchange of ideas among these communities, with the aim to act as a catalyst for new and innovative ideas in this challenging field of research.

The main purpose of the book resulting from the workshop is to extend the keynote lectures of the workshop to tutorials accessible to developers and users of mathematical methods for model reduction and approximation of complex systems. In addition to the keynote lecturers of the workshop, we also invited other experts to contribute chapters on methods not represented in the keynotes. The book thus contains tutorial-style introductions to several promising emerging fields in model reduction and approximation. It focuses in particular on sampling-based methods (Part I), such as the RB method and POD, approximation of high-dimensional problems by low-rank tensor techniques (Part II), and system-theoretic methods (Part III), such as balanced truncation (BT), interpolatory methods, and the Loewner framework.

Both application-driven aspects and fundamental points of view from approximation theory and information-based complexity are discussed. This reveals the great success of the proposed techniques for certain classes of applications but at the same time also shows their limitations. Real-life problems often pose major challenges that are currently covered by neither the mathematical theory nor the presented methods and thus constitute a driving force for future research.

We believe that the chapters collected in this book exhibit high tutorial value, so that in combination they can serve as the basis of graduate-level courses on the subject. Such courses could be valuable in any master's-level program ranging from applied and engineering mathematics to computational science and engineering. The tutorials could also be used as course material for integrated PhD programs or summer schools.

The second purpose of the book is to serve as a reference guide for examples and methods available to date, in particular for parametric MOR. It compares some of the methods for parametric model reduction using examples collected at the MOR Wiki.[1] Thus, it provides a first guide to the choice of suitable algorithms for model and complexity reduction of dynamical systems. This is merely a starting point, and further developments will be made available through the MOR Wiki.

December 2016 *Peter Benner*
Albert Cohen
Mario Ohlberger
Karen Willcox

[1]See `http://morwiki.mpi-magdeburg.mpg.de/`.

Part I

Sampling-Based Methods

Part I of this book is concerned with model order reduction (MOR) techniques for complex parametrized problems. The parameters may be of various type and dimension: the time variable in an evolution problem, the values of the diffusion, or other material quantities at different parts of the spatial domain. A common objective is to design numerical methods based on low-dimensional approximation spaces that are tailored to the considered problem at hand, as opposed to generic approximation spaces such as used in finite element (FE) or spectral methods. In particular, these methods are based on the evaluation of *snapshots*, that is, instances of the solution associated with particular values of the parameters.

In the first chapter, Martin Gubisch and Stefan Volkwein introduce the proper orthogonal decomposition (POD) method, which is based on applying singular value decomposition (SVD) to a family of representative snapshots. The POD spaces are generated by the leading eigenfunctions associated with the largest eigenvalues. The method is applied to the approximate resolution of quadratic optimal control problems governed by linear evolution equations.

In the second chapter, Bernard Haasdonk introduces the reduced basis (RB) method, in which the reduced spaces are generated by a relevant selection of snapshots during an *offline stage*. The selection strategies aim to guarantee that the RB approximation computed in the *online stage* has a certified accuracy for any parameter value. The method is applied to both stationary elliptic problems and time-dependent problems.

In the third chapter, Ron DeVore theoretically analyzes the performance of RB methods. Greedy strategies for the snapshot selection are emphasized. Such strategies allow the generation of RB spaces for which the approximation has the same convergence rate as the Kolmogorov *n-width*, which corresponds to the optimal, but not accessible, choice of n-dimensional spaces.

Chapter 1
Proper Orthogonal Decomposition for Linear-Quadratic Optimal Control

Martin Gubisch and Stefan Volkwein[2]

1.1 ▪ Introduction

Optimal control problems for partial differential equations (PDEs) are often hard to tackle numerically because their discretization leads to very large scale optimization problems. Therefore, different techniques of model reduction were developed to approximate these problems by smaller ones that are tractable with less effort.

Balanced truncation (BT) [2, 64, 79] is one well-studied model reduction technique for state-space systems. This method utilizes the solutions to two Lyapunov equations, the so-called controllability and observability Gramians. The BT method is based on transforming the state-space system into a balanced form so that its controllability and observability Gramians become diagonal and equal. Moreover, the states that are difficult to reach or to observe are truncated. The advantage of this method is that it preserves the asymptotic stability in the reduced-order system. Furthermore, a priori error bounds are available. Recently, the theory of BT model reduction was extended to descriptor systems; see, e.g., [48] and [21].

Recently the application of *reduced-order models* (ROMs) to linear time-varying and nonlinear systems, in particular to nonlinear control systems, has received an increasing amount of attention. The reduced-order approach is based on projecting the dynamical system onto subspaces consisting of basis elements that contain characteristics of the expected solution. This is in contrast to, e.g., finite element (FE) techniques (see, e.g., [7]), where the basis elements of the subspaces do not relate to the physical properties of the system that they approximate. The *reduced basis* (RB) method, as developed in [19, 54] and [31], is one such reduced-order method, where the basis elements correspond to the dynamics of expected control regimes. We refer the reader to [14, 23, 49, 53] for the successful use of RB methods in PDE-constrained optimization problems. Currently, *proper orthogonal decomposition* (POD) is probably

[2]The authors gratefully acknowledge support by the German Science Fund DFG grant VO 1658/2-1 *A-posteriori-POD Error Estimators for Nonlinear Optimal Control Problems Governed by Partial Differential Equations*. The first author is further supported by the Landesgraduiertenförderung of Baden-Württemberg.

the most used and most successful model reduction technique for nonlinear optimal control problems, where the basis functions contain information from the solutions of the dynamical system at prespecified time instances—so-called snapshots; see, e.g., [9, 29, 67, 75]. Due to a possible linear dependence or almost linear dependence, the snapshots themselves are not appropriate as a basis. Hence, a singular value decomposition (SVD) is carried out, and the leading generalized eigenfunctions are chosen as a basis, referred to as the POD basis. POD is successfully used in a variety of fields, including fluid dynamics, coherent structures [1, 3], and inverse problems [6]. Moreover, in [5] POD is successfully applied to compute reduced-order controllers. The relationship between POD and balancing was considered in [44, 61, 77]. An error analysis for nonlinear dynamical systems in finite dimensions was carried out in [58], and a missing point estimation in models described by POD was studied in [4].

ROMs are used in PDE-constrained optimization in various ways; see, e.g., [27, 63] for a survey. In optimal control problems, it is sometimes necessary to compute a feedback control law instead of a fixed optimal control. In the implementation of these feedback laws, models of reduced order can play an important and very useful role; see [5, 43, 46, 59]. Another useful application is in optimization problems, where a PDE solver is part of the function evaluation. Obviously, in the case of a gradient evaluation or even a step-size rule in the optimization algorithm, an expensive function evaluation leads to an enormous amount of computing time. Here, the ROM can replace the system given by a PDE in the objective function. It is quite common that a PDE can be replaced by a five- or ten-dimensional system of ordinary differential equations (ODEs). This results computationally in a very fast method for optimization compared to the effort for the computation of a single solution of a PDE. There is an extensive literature on engineering applications in this regard; we mention only the papers [47, 50]. Recent applications can also be found in finance using the RB model [56] and the POD model [62, 65] in the context of calibration for models in options pricing.

In the present work, we use POD to derive low-order models of dynamical systems. These low-order models then serve as surrogates for the dynamical system in the optimization process. We consider a linear-quadratic optimal control problem in an abstract setting and prove error estimates for the POD Galerkin approximations of the optimal control. This is achieved by combining techniques from [11, 12, 25] and [38, 40]. For nonlinear problems we refer the reader to [27, 55, 63]. However, unless the snapshots generate a sufficiently rich state space or are computed from the exact (unknown) optimal controls, it is not a priori clear how far the optimal solution of the POD problem is from the exact one. On the other hand, the POD method is a universal tool that is also applicable to problems with time-dependent coefficients or to nonlinear equations. Moreover, by generating snapshots from the real (large) model, a space is constructed that inhibits the main and relevant physical properties of the state system. This, and its ease of use, makes POD very competitive in practical use, despite a certain heuristic flavor. In this work, we review results for a POD a posteriori analysis; see, e.g., [71] and [20, 33, 34, 68, 69, 74, 76]. We use a fairly standard perturbation method to deduce how far the suboptimal control, computed on the basis of the POD model, is from the (unknown) exact one. This idea turned out to be very efficient in our examples. It compensates for the lack of a priori analysis for POD methods. We refer the reader to the papers [13, 18, 49], where a posteriori error bounds are computed for linear-quadratic optimal control problems approximated by the RB method.

This chapter is organized as follows. In Section 1.2 we introduce the method of POD in real, separable Hilbert spaces and discuss its relationship to SVD. We distinguish between two versions of the POD method: discrete and continuous. Reduced-order modeling with POD is carried out in Section 1.3. The error between the exact solution and its POD approximation is investigated by an a priori error analysis. In Section 1.4 we study quadratic optimal control problems governed by linear evolution problems and bilateral inequality constraints. These problems are infinite-dimensional convex optimization problems. Their optimal solutions are characterized by first-order optimality conditions. POD Galerkin discretizations of the optimality conditions are introduced and analyzed. By an a priori error analysis, the error of the exact optimal control and its POD suboptimal approximation are estimated. For the error control in the numerical realizations, we make use of an a posteriori error analysis, which turns out to be very efficient in our numerical examples, which are presented in Section 1.5.

1.2 ▪ The POD method

Throughout we suppose that X is a real Hilbert space endowed with the inner product $\langle\cdot,\cdot\rangle_X$ and the associated induced norm $\|\cdot\|_X = \langle\cdot,\cdot\rangle_X^{1/2}$. Furthermore, we assume that X is *separable*, i.e., X has a countable dense subset. This implies that X possesses a countable orthonormal basis; see, e.g., [60, p. 47]. For the POD method in complex Hilbert spaces we refer the reader to [73], for instance.

1.2.1 ▪ The discrete variant of the POD method

For fixed $n,\wp \in \mathbb{N}$, let the so-called snapshots $y_1^k,\dots,y_n^k \in X$ be given for $1 \le k \le \wp$. To avoid the trivial case we suppose that at least one of the y_j^k's is nonzero. Then, we introduce the finite-dimensional linear subspace

$$\mathscr{V}^n = \operatorname{span}\left\{y_j^k \,|\, 1 \le j \le n \text{ and } 1 \le k \le \wp\right\} \subset X \tag{1.1}$$

with dimension $d^n \in \{1,\dots,n\wp\} < \infty$. We call the set the $\mathscr{V}^n$ *snapshot subspace*.

Remark 1.1. Later we will focus on the following application: Let $0 \le t_1 < t_2 < \dots < t_n \le T$ be a given time grid in the interval $[0,T]$. To simplify the presentation, the time grid is assumed to be equidistant with step size $\Delta t = T/(n-1)$, i.e., $t_j = (j-1)\Delta t$. For nonequidistant grids, we refer the reader to [39, 40]. Suppose that we have trajectories $y^k \in C([0,T];X)$, $1 \le k \le \wp$. Here, the Banach space $C([0,T];X)$ contains all functions $\varphi : [0,T] \to X$ that are continuous on $[0,T]$; see, e.g., [70, p. 142]. Let the snapshots be given as $y_j^k = y^k(t_j) \in X$ or $y_j^k \approx y^k(t_j) \in X$. In Sections 1.3 and 1.4 we will choose trajectories as solutions to evolution problems.

In Section 1.2.3 we consider the case where the number n is varied. Therefore, we emphasize this dependence by using the superscript index n. We distinguish two cases:

1. The separable Hilbert space X has finite dimension m. Then, X is isomorphic to $\mathbb{R}^m$; see, e.g., [60, p. 47]. We set $\mathscr{I} = \{1,\dots,m\}$. Clearly, we have $d^n \le \min(n\wp, m)$.

2. Since X is separable, each orthonormal basis of X has countably many elements. In this case, X is isomorphic to the set ℓ_2 of sequences $\{x_i\}_{i\in\mathbb{N}}$ of real numbers that satisfy $\sum_{i=1}^\infty |x_i|^2 < \infty$; see [60, p. 47], for instance. Then, we define $\mathscr{I} = \mathbb{N}$.

The method of POD consists of choosing an orthonormal set $\{\psi_i\}_{i=1}^{\ell}$ in X such that for every $\ell \in \{1,\dots,d^n\}$ the mean square error between the $n\wp$ elements y_j^k and their corresponding ℓth partial Fourier sum is minimized on average:

$$\begin{cases} \min \displaystyle\sum_{k=1}^{\wp}\sum_{j=1}^{n}\alpha_j^n \Big\| y_j^k - \sum_{i=1}^{\ell}\langle y_j^k,\psi_i\rangle_X\,\psi_i\Big\|_X^2 \\ \text{subject to (s.t.) } \{\psi_i\}_{i=1}^{\ell} \subset X \text{ and } \langle\psi_i,\psi_j\rangle_X = \delta_{ij},\ 1\le i,j\le \ell, \end{cases} \tag{$\mathbf{P}_n^\ell$}$$

where the α_j^n's denote positive weighting parameters. Here, the symbol δ_{ij} denotes the Kronecker symbol satisfying $\delta_{ii}=1$ and $\delta_{ij}=0$ for $i\neq j$. An optimal solution $\{\bar\psi_i^n\}_{i=1}^{\ell}$ to ($\mathbf{P}_n^\ell$) is called a *POD basis* of *rank* ℓ, which can be extended to a complete orthonormal basis $\{\psi_i\}_{i\in\mathscr{I}}$ in the Hilbert space X. Notice that

$$\begin{aligned} &\Big\| y_j^k - \sum_{i=1}^{\ell}\langle y_j^k,\psi_i\rangle_X\,\psi_i\Big\|_X^2 \\ &= \Big\langle y_j^k - \sum_{i=1}^{\ell}\langle y_j^k,\psi_i\rangle_X\,\psi_i, y_j^k - \sum_{l=1}^{\ell}\langle y_j^k,\psi_l\rangle_X\,\psi_l\Big\rangle_X \\ &= \|y_j^k\|_X^2 - 2\sum_{i=1}^{\ell}\langle y_j^k,\psi_i\rangle_X^2 + \sum_{i=1}^{\ell}\sum_{l=1}^{\ell}\langle y_j^k,\psi_i\rangle_X\langle y_j^k,\psi_l\rangle_X\langle\psi_i,\psi_l\rangle_X \\ &= \|y_j^k\|_X^2 - \sum_{i=1}^{\ell}\langle y_j^k,\psi_i\rangle_X^2 \end{aligned} \tag{1.2}$$

for any set $\{\psi_i\}_{i=1}^{\ell}\subset X$ satisfying $\langle\psi_i,\psi_j\rangle_X=\delta_{ij}$. Thus, ($\mathbf{P}_n^\ell$) is equivalent to the maximization problem

$$\begin{cases} \max \displaystyle\sum_{k=1}^{\wp}\sum_{j=1}^{n}\alpha_j^n \sum_{i=1}^{\ell}\langle y_j^k,\psi_i\rangle_X^2 \\ \text{s.t. } \{\psi_i\}_{i=1}^{\ell} \subset X \text{ and } \langle\psi_i,\psi_j\rangle_X = \delta_{ij},\ 1\le i,j\le \ell. \end{cases} \tag{$\hat{\mathbf{P}}_n^\ell$}$$

Suppose that $\{\psi_i\}_{i\in\mathscr{I}}$ is a complete orthonormal basis in X. Since X is separable, any $y_j^k\in X$, $1\le j\le n$ and $1\le k\le\wp$, can be written as

$$y_j^k = \sum_{i\in\mathscr{I}}\langle y_j^k,\psi_i\rangle_X\,\psi_i, \tag{1.3}$$

and the (probably infinite) sum converges for all snapshots (even for all elements in X). Thus, the POD basis $\{\bar\psi_i^n\}_{i=1}^{\ell}$ of rank ℓ maximizes the absolute values of the first ℓ Fourier coefficients $\langle y_j^k,\psi_i\rangle_X$ for all $n\wp$ snapshots y_j^k in an average sense. Let us recall the following definition for linear operators in Banach spaces.

Definition 1.2. *Let $\mathscr{B}_1$, $\mathscr{B}_2$ be two real Banach spaces. The operator $\mathscr{T}:\mathscr{B}_1\to\mathscr{B}_2$ is called a* linear bounded operator *if these conditions are satisfied:*

1. $\mathscr{T}(\alpha u+\beta v)=\alpha\mathscr{T}u+\beta\mathscr{T}v$ *for all* $\alpha,\beta\in\mathbb{R}$ *and* $u,v\in\mathscr{B}_1$.
2. *There exists a constant* $c>0$ *such that* $\|\mathscr{T}u\|_{\mathscr{B}_2}\le c\,\|u\|_{\mathscr{B}_1}$ *for all* $u\in\mathscr{B}_1$.

The set of all linear bounded operators from $\mathscr{B}_1$ to $\mathscr{B}_2$ is denoted by $\mathscr{L}(\mathscr{B}_1, \mathscr{B}_2)$, which is a Banach space equipped with the operator norm [60, pp. 69–70]

$$\|\mathscr{T}\|_{\mathscr{L}(\mathscr{B}_1,\mathscr{B}_2)} = \sup_{\|u\|_{\mathscr{B}_1}=1} \|\mathscr{T} u\|_{\mathscr{B}_2} \quad \textit{for } \mathscr{T} \in \mathscr{L}(\mathscr{B}_1, \mathscr{B}_2).$$

If $\mathscr{B}_1 = \mathscr{B}_2$, we write $\mathscr{L}(\mathscr{B}_1)$ instead of $\mathscr{L}(\mathscr{B}_1, \mathscr{B}_2)$. The dual mapping $\mathscr{T}' : \mathscr{B}_2' \to \mathscr{B}_1'$ *of an operator $\mathscr{T} \in \mathscr{L}(\mathscr{B}_1, \mathscr{B}_2)$ is defined as*

$$\langle \mathscr{T}' f, u \rangle_{\mathscr{B}_1',\mathscr{B}_1} = \langle f, \mathscr{T} u \rangle_{\mathscr{B}_2',\mathscr{B}_2} \quad \textit{for all } (u,f) \in \mathscr{B}_1 \times \mathscr{B}_2',$$

where, for instance, $\langle \cdot,\cdot \rangle_{\mathscr{B}_1',\mathscr{B}_1}$ denotes the dual pairing of the space $\mathscr{B}_1$ with its dual space $\mathscr{B}_1' = \mathscr{L}(\mathscr{B}_1, \mathbb{R})$.

Let $\mathscr{H}_1$ and $\mathscr{H}_2$ denote two real Hilbert spaces. For a given $\mathscr{T} \in \mathscr{L}(\mathscr{H}_1, \mathscr{H}_2)$, the *adjoint operator* $\mathscr{T}^\star : \mathscr{H}_2 \to \mathscr{H}_1$ is uniquely defined by

$$\langle \mathscr{T}^\star v, u \rangle_{\mathscr{H}_1} = \langle v, \mathscr{T} u \rangle_{\mathscr{H}_2} = \langle \mathscr{T} u, v \rangle_{\mathscr{H}_2} \quad \text{for all } (u,v) \in \mathscr{H}_1 \times \mathscr{H}_2.$$

Let $\mathscr{J}_i : \mathscr{H}_i \to \mathscr{H}_i'$, $i = 1,2$, denote the Riesz isomorphisms satisfying

$$\langle u, v \rangle_{\mathscr{H}_i} = \langle \mathscr{J}_i u, v \rangle_{\mathscr{H}_i',\mathscr{H}_i} \quad \text{for all } v \in \mathscr{H}_i.$$

Then, we have the representation $\mathscr{T}^\star = \mathscr{J}_1^{-1} \mathscr{T}' \mathscr{J}_2$; see [70, p. 186]. Moreover, $(\mathscr{T}^\star)^\star = \mathscr{T}$ for every $\mathscr{T} \in \mathscr{L}(\mathscr{H}_1, \mathscr{H}_2)$. If $\mathscr{T} = \mathscr{T}^\star$ holds, $\mathscr{T}$ is said to be *self-adjoint*. The operator $\mathscr{T} \in \mathscr{L}(\mathscr{H}_1, \mathscr{H}_2)$ is called *nonnegative* if $\langle \mathscr{T} u, u \rangle_{\mathscr{H}_2} \geq 0$ for all $u \in \mathscr{H}_1$. Finally, $\mathscr{T} \in \mathscr{L}(\mathscr{H}_1, \mathscr{H}_2)$ is called *compact* if for every bounded sequence $\{u_n\}_{n\in\mathbb{N}} \subset \mathscr{H}_1$ the sequence $\{\mathscr{T} u_n\}_{n\in\mathbb{N}} \subset \mathscr{H}_2$ contains a convergent subsequence.

Now we turn to $(\mathbf{P}_n^\ell)$ and $(\hat{\mathbf{P}}_n^\ell)$. We make use of the following lemma.

Lemma 1.3. *Let X be a (separable) real Hilbert space, and let $y_1^k, \dots, y_n^k \in X$ be given snapshots for $1 \leq k \leq \wp$. Define the linear operator $\mathscr{R}^n : X \to X$ as follows:*

$$\mathscr{R}^n \psi = \sum_{k=1}^{\wp} \sum_{j=1}^{n} \alpha_j^n \, \langle \psi, y_j^k \rangle_X \, y_j^k \quad \textit{for } \psi \in X \tag{1.4}$$

with positive weights $\alpha_1^n, \dots, \alpha_n^n$. Then, $\mathscr{R}^n$ is a compact, nonnegative, and self-adjoint operator.

Proof. It is clear that $\mathscr{R}^n$ is a linear operator. From

$$\|\mathscr{R}^n \psi\|_X \leq \sum_{k=1}^{\wp} \sum_{j=1}^{n} \alpha_j^n |\langle \psi, y_j^k \rangle_X| \, \|y_j^k\|_X \quad \text{for } \psi \in X$$

and the Cauchy–Schwarz inequality [60, p. 38],

$$|\langle \varphi, \phi \rangle_X| \leq \|\varphi\|_X \|\phi\|_X \quad \text{for } \varphi, \phi \in X,$$

we conclude that $\mathscr{R}^n$ is bounded. Since $\mathscr{R}^n \psi \in \mathscr{V}^n$ for all $\psi \in X$, the range of $\mathscr{R}^n$ is finite dimensional. Thus, $\mathscr{R}^n$ is a finite rank operator that is compact; see [60, p. 199].

Next we show that $\mathscr{R}^n$ is nonnegative. For that purpose we choose an arbitrary element $\psi \in X$ and consider

$$\langle \mathscr{R}^n \psi, \psi \rangle_X = \sum_{k=1}^{\wp} \sum_{j=1}^{n} \alpha_j^n \langle \psi, y_j^k \rangle_X \langle y_j^k, \psi \rangle_X = \sum_{k=1}^{\wp} \sum_{j=1}^{n} \alpha_j^n \langle \psi, y_j^k \rangle_X^2 \geq 0.$$

Thus, $\mathscr{R}^n$ is nonnegative. For any $\psi, \tilde{\psi} \in X$, we derive

$$\begin{aligned}\langle \mathscr{R}^n \psi, \tilde{\psi} \rangle_X &= \sum_{k=1}^{\wp} \sum_{j=1}^{n} \alpha_j^n \langle \psi, y_j^k \rangle_X \langle y_j^k, \tilde{\psi} \rangle_X = \sum_{k=1}^{\wp} \sum_{j=1}^{n} \alpha_j^n \langle \tilde{\psi}, y_j^k \rangle_X \langle y_j^k, \psi \rangle_X \\ &= \langle \mathscr{R}^n \tilde{\psi}, \psi \rangle_X = \langle \psi, \mathscr{R}^n \tilde{\psi} \rangle_X.\end{aligned}$$

Thus, $\mathscr{R}^n$ is self-adjoint. □

Next we recall some important results from the spectral theory of operators (on infinite-dimensional spaces). We begin with the following definition; see [60, Section VI.3].

Definition 1.4. *Let $\mathscr{H}$ be a real Hilbert space and $\mathscr{T} \in \mathscr{L}(\mathscr{H})$.*

1. *A complex number λ belongs to the* resolvent set *$\rho(\mathscr{T})$ if $\lambda\mathscr{I} - \mathscr{T}$ is a bijection with a bounded inverse. Here, $\mathscr{I} \in \mathscr{L}(\mathscr{H})$ stands for the identity operator. If $\lambda \notin \rho(\mathscr{T})$, then λ is an element of the spectrum $\sigma(\mathscr{T})$ of $\mathscr{T}$.*
2. *Let $u \neq 0$ be a vector with $\mathscr{T} u = \lambda u$ for some $\lambda \in \mathbb{C}$. Then, u is said to be an* eigenvector *of $\mathscr{T}$. We call λ the corresponding* eigenvalue. *If λ is an eigenvalue, then $\lambda\mathscr{I} - \mathscr{T}$ is not injective. This implies $\lambda \in \sigma(\mathscr{T})$. The set of all eigenvalues is called the* point spectrum *of $\mathscr{T}$.*

We will make use of the next two essential theorems for compact operators; see [60, p. 203].

Theorem 1.5 (Riesz–Schauder). *Let $\mathscr{H}$ be a real Hilbert space and $\mathscr{T} : \mathscr{H} \to \mathscr{H}$ be a linear compact operator. Then, the spectrum $\sigma(\mathscr{T})$ is a discrete set with no limit points except perhaps zero. Furthermore, the space of eigenvectors corresponding to each nonzero $\lambda \in \sigma(\mathscr{T})$ is finite dimensional.*

Theorem 1.6 (Hilbert–Schmidt). *Let $\mathscr{H}$ be a real separable Hilbert space and $\mathscr{T} : \mathscr{H} \to \mathscr{H}$ be a linear, compact, self-adjoint operator. Then, there is a sequence of eigenvalues $\{\lambda_i\}_{i \in \mathscr{I}}$ and an associated complete orthonormal basis $\{\psi_i\}_{i \in \mathscr{I}} \subset X$ satisfying*

$$\mathscr{T} \psi_i = \lambda_i \psi_i \quad \textit{and} \quad \lambda_i \to 0 \textit{ as } i \to \infty.$$

Since X is a separable real Hilbert space and $\mathscr{R}^n : X \to X$ is a linear, compact, nonnegative, self-adjoint operator (see Lemma 1.3), we can utilize Theorems 1.5 and 1.6: there exist a complete countable orthonormal basis $\{\bar{\psi}_i^n\}_{i \in \mathscr{I}}$ and a corresponding sequence of real eigenvalues $\{\bar{\lambda}_i^n\}_{i \in \mathscr{I}}$ satisfying

$$\mathscr{R}^n \bar{\psi}_i^n = \bar{\lambda}_i^n \bar{\psi}_i^n, \quad \bar{\lambda}_1^n \geq \cdots \geq \bar{\lambda}_{d^n} > \bar{\lambda}_{d^n+1} = \cdots = 0. \tag{1.5}$$

The spectrum of $\mathscr{R}^n$ is a pure point spectrum except for possibly zero. Each nonzero eigenvalue of $\mathscr{R}^n$ has finite multiplicity, and zero is the only possible accumulation point of the spectrum of $\mathscr{R}^n$.

Remark 1.7. From (1.4), (1.5), and $\|\psi\|_X = 1$, we infer that

$$\begin{aligned}\sum_{k=1}^{\wp}\sum_{j=1}^{n}\alpha_j^n\,\langle y_j^k,\bar\psi_i^n\rangle_X^2 &= \Big\langle \sum_{k=1}^{\wp}\sum_{j=1}^{n}\alpha_j^n\,\langle y_j^k,\bar\psi_i^n\rangle_X y_j^k,\bar\psi_i^n\Big\rangle_X \\ &= \langle \mathscr{R}^n\bar\psi_i^n,\bar\psi_i^n\rangle_X = \bar\lambda_i^n \quad \text{for any } i\in\mathscr{I}.\end{aligned} \tag{1.6}$$

In particular, it follows that

$$\sum_{k=1}^{\wp}\sum_{j=1}^{n}\alpha_j^n\,\langle y_j^k,\bar\psi_i^n\rangle_X^2 = 0 \quad \text{for all } i > d^n. \tag{1.7}$$

Since $\{\bar\psi_i^n\}_{i\in\mathscr{I}}$ is a complete orthonormal basis and $\|y_j^k\|_X < \infty$ for $1 \le k \le \wp$, $1 \le j \le n$, we derive from (1.6) and (1.7) that

$$\begin{aligned}\sum_{k=1}^{\wp}\sum_{j=1}^{n}\alpha_j^n\,\|y_j^k\|_X^2 &= \sum_{k=1}^{\wp}\sum_{j=1}^{n}\alpha_j^n\sum_{\nu\in\mathscr{I}}\langle y_j^k,\bar\psi_\nu^n\rangle_X^2 \\ &= \sum_{\nu\in\mathscr{I}}\sum_{k=1}^{\wp}\sum_{j=1}^{n}\alpha_j^n\,\langle y_j^k,\bar\psi_\nu^n\rangle_X^2 = \sum_{i\in\mathscr{I}}\bar\lambda_i^n = \sum_{i=1}^{d^n}\bar\lambda_i^n.\end{aligned} \tag{1.8}$$

By (1.8) the (probably infinite) sum $\sum_{i\in\mathscr{I}}\bar\lambda_i^n$ is bounded. It follows from (1.2) that the objective of $(\mathbf{P}_n^\ell)$ can be written as

$$\begin{aligned}&\sum_{k=1}^{\wp}\sum_{j=1}^{n}\alpha_j^n\Big\|y_j^k-\sum_{i=1}^{\ell}\langle y_j^k,\bar\psi_i^n\rangle_X\,\bar\psi_i^n\Big\|_X^2 \\ &\quad= \sum_{i=1}^{d^n}\bar\lambda_i^n - \sum_{k=1}^{\wp}\sum_{j=1}^{n}\alpha_j^n\sum_{i=1}^{\ell}\langle y_j^k,\bar\psi_i^n\rangle_X^2,\end{aligned} \tag{1.9}$$

which we will use in the proof of Theorem 1.8.

Now we can formulate the main result for $(\mathbf{P}_n^\ell)$ and $(\hat{\mathbf{P}}_n^\ell)$.

Theorem 1.8. *Let X be a separable real Hilbert space, $y_1^k,\dots,y_n^k \in X$ for $1 \le k \le \wp$, and $\mathscr{R}^n : X \to X$ be defined by* (1.4). *Suppose that $\{\bar\lambda_i^n\}_{i\in\mathscr{I}}$ and $\{\bar\psi_i^n\}_{i\in\mathscr{I}}$ denote the nonnegative eigenvalues and associated orthonormal eigenfunctions of $\mathscr{R}^n$ satisfying* (1.5). *Then, for every $\ell \in \{1,\dots,d^n\}$, the first ℓ eigenfunctions $\{\bar\psi_i^n\}_{i=1}^{\ell}$ solve $(\mathbf{P}_n^\ell)$ and $(\hat{\mathbf{P}}_n^\ell)$. Moreover, the value of the cost evaluated at the optimal solution $\{\bar\psi_i^n\}_{i=1}^{\ell}$ satisfies*

$$\sum_{k=1}^{\wp}\sum_{j=1}^{n}\alpha_j^n\Big\|y_j^k-\sum_{i=1}^{\ell}\langle y_j^k,\bar\psi_i^n\rangle_X\,\bar\psi_i^n\Big\|_X^2 = \sum_{i=\ell+1}^{d^n}\bar\lambda_i^n \tag{1.10}$$

and

$$\sum_{k=1}^{\wp}\sum_{j=1}^{n}\alpha_j^n\sum_{i=1}^{\ell}\langle y_j^k,\bar\psi_i^n\rangle_X^2 = \sum_{i=1}^{\ell}\bar\lambda_i^n. \tag{1.11}$$

Proof. We prove the claim for ($\hat{\mathbf{P}}_n^\ell$) by finite induction over $\ell \in \{1,\ldots,d^n\}$.

1. The base case: Let $\ell = 1$ and $\psi \in X$ with $\|\psi\|_X = 1$. Since $\{\bar\psi_\nu^n\}_{\nu\in\mathscr{I}}$ is a complete orthonormal basis in X, we have the representation

$$\psi = \sum_{\nu\in\mathscr{I}} \langle \psi, \bar\psi_\nu^n\rangle_X \,\bar\psi_\nu^n. \tag{1.12}$$

Inserting this expression for ψ in the objective of ($\hat{\mathbf{P}}_n^\ell$), we find that

$$\begin{aligned}
\sum_{k=1}^{\wp}\sum_{j=1}^{n}\alpha_j^n \,\langle y_j^k,\psi\rangle_X^2 &= \sum_{k=1}^{\wp}\sum_{j=1}^{n}\alpha_j^n \Big\langle y_j^k, \sum_{\nu\in\mathscr{I}}\langle\psi,\bar\psi_\nu^n\rangle_X\,\bar\psi_\nu^n\Big\rangle_X^2\\
&= \sum_{k=1}^{\wp}\sum_{j=1}^{n}\alpha_j^n \sum_{\nu\in\mathscr{I}}\sum_{\mu\in\mathscr{I}}\Big(\big\langle y_j^k,\langle\psi,\bar\psi_\nu^n\rangle_X\,\bar\psi_\nu^n\big\rangle_X\big\langle y_j^k,\langle\psi,\bar\psi_\mu^n\rangle_X\,\bar\psi_\mu^n\big\rangle_X\Big)\\
&= \sum_{k=1}^{\wp}\sum_{j=1}^{n}\alpha_j^n \sum_{\nu\in\mathscr{I}}\sum_{\mu\in\mathscr{I}}\Big(\langle y_j^k,\bar\psi_\nu^n\rangle_X\langle y_j^k,\bar\psi_\mu^n\rangle_X\langle\psi,\bar\psi_\nu^n\rangle_X\langle\psi,\bar\psi_\mu^n\rangle_X\Big)\\
&= \sum_{\nu\in\mathscr{I}}\sum_{\mu\in\mathscr{I}}\bigg(\Big\langle\sum_{k=1}^{\wp}\sum_{j=1}^{n}\alpha_j^n\langle y_j^k,\bar\psi_\nu^n\rangle_X\,y_j^k,\bar\psi_\mu^n\Big\rangle_X\langle\psi,\bar\psi_\nu^n\rangle_X\langle\psi,\bar\psi_\mu^n\rangle_X\bigg).
\end{aligned}$$

Utilizing (1.4), (1.5), and $\|\bar\psi_\nu^n\|_X = 1$, we find that

$$\begin{aligned}
\sum_{k=1}^{\wp}\sum_{j=1}^{n}\alpha_j^n \,\langle y_j^k,\psi\rangle_X^2 &= \sum_{\nu\in\mathscr{I}}\sum_{\mu\in\mathscr{I}}\Big(\langle\bar\lambda_\nu^n\bar\psi_\nu^n,\bar\psi_\mu^n\rangle_X\langle\psi,\bar\psi_\nu^n\rangle_X\langle\psi,\bar\psi_\mu^n\rangle_X\Big)\\
&= \sum_{\nu\in\mathscr{I}}\bar\lambda_\nu^n\,\langle\psi,\bar\psi_\nu^n\rangle_X^2.
\end{aligned}$$

From $\bar\lambda_1^n \ge \bar\lambda_\nu^n$ for all $\nu\in\mathscr{I}$ and (1.6), we infer that

$$\begin{aligned}
\sum_{\nu\in\mathscr{I}}\bar\lambda_\nu^n\,\langle\psi,\bar\psi_\nu^n\rangle_X^2 \le \bar\lambda_1^n\sum_{\nu\in\mathscr{I}}\langle\psi,\bar\psi_\nu^n\rangle_X^2 &= \bar\lambda_1^n\,\|\psi\|_X^2 = \bar\lambda_1^n\\
&= \sum_{k=1}^{\wp}\sum_{j=1}^{n}\alpha_j^n\,\langle y_j^k,\bar\psi_1^n\rangle_X^2,
\end{aligned}$$

i.e., $\bar\psi_1^n$ solves ($\hat{\mathbf{P}}_n^\ell$) for $\ell = 1$ and (1.11) holds. This gives the base case. Notice that (1.9) and (1.11) imply (1.10).

2. The induction hypothesis: Now we suppose that

$$\left\{\begin{aligned}
&\text{for any } \ell\in\{1,\ldots,d^n-1\} \text{ the set } \{\bar\psi_i^n\}_{i=1}^{\ell}\subset X \text{ solves } (\hat{\mathbf{P}}_n^\ell)\\
&\text{and } \sum_{k=1}^{\wp}\sum_{j=1}^{n}\alpha_j^n\sum_{i=1}^{\ell}\langle y_j^k,\bar\psi_i^n\rangle_X^2 = \sum_{i=1}^{\ell}\bar\lambda_i^n.
\end{aligned}\right. \tag{1.13}$$

3. The induction step: We consider

$$\left\{\begin{aligned}
&\max\sum_{k=1}^{\wp}\sum_{j=1}^{n}\alpha_j^n\sum_{i=1}^{\ell+1}\langle y_j^k,\psi_i\rangle_X^2\\
&\text{s.t. } \{\psi_i\}_{i=1}^{\ell+1}\subset X \text{ and } \langle\psi_i,\psi_j\rangle_X = \delta_{ij},\ 1\le i,j\le\ell+1.
\end{aligned}\right. \tag{$\hat{\mathbf{P}}_n^{\ell+1}$}$$

By (1.13), the elements $\{\bar{\psi}_i^n\}_{i=1}^{\ell}$ maximize the term

$$\sum_{k=1}^{\wp}\sum_{j=1}^{n}\alpha_j^n\sum_{i=1}^{\ell}\langle y_j^k,\psi_i\rangle_X^2.$$

Thus, $(\hat{\mathbf{P}}_n^{\ell+1})$ is equivalent to

$$\begin{cases} \max\displaystyle\sum_{k=1}^{\wp}\sum_{j=1}^{n}\alpha_j^n\langle y_j^k,\psi\rangle_X^2 \\ \text{s.t. } \psi\in X \text{ and } \|\psi\|_X=1,\ \langle\psi,\bar{\psi}_i^n\rangle_X=0,\ 1\le i\le\ell. \end{cases} \tag{1.14}$$

Let $\psi\in X$ be given satisfying $\|\psi\|_X=1$ and $\langle\psi,\bar{\psi}_i^n\rangle_X=0$ for $i=1,\dots,\ell$. Then, using the representation (1.12) and $\langle\psi,\bar{\psi}_i^n\rangle_X=0$ for $i=1,\dots,\ell$, we derive, as above,

$$\sum_{k=1}^{\wp}\sum_{j=1}^{n}\alpha_j^n\langle y_j^k,\psi\rangle_X^2=\sum_{\nu\in\mathscr{I}}\bar{\lambda}_\nu^n\langle\psi,\bar{\psi}_\nu^n\rangle_X^2=\sum_{\nu>\ell}\bar{\lambda}_\nu^n\langle\psi,\bar{\psi}_\nu^n\rangle_X^2.$$

From $\bar{\lambda}_{\ell+1}^n\ge\bar{\lambda}_\nu^n$ for all $\nu\ge\ell+1$ and (1.6), we conclude that

$$\begin{aligned}\sum_{k=1}^{\wp}\sum_{j=1}^{n}\alpha_j^n\langle y_j^k,\psi\rangle_X^2&\le\bar{\lambda}_{\ell+1}^n\sum_{\nu>\ell}\langle\psi,\bar{\psi}_\nu^n\rangle_X^2\le\bar{\lambda}_{\ell+1}^n\sum_{\nu\in\mathscr{I}}\langle\psi,\bar{\psi}_\nu^n\rangle_X^2\\&=\bar{\lambda}_{\ell+1}^n\|\psi\|_X^2=\bar{\lambda}_{\ell+1}^n=\sum_{k=1}^{\wp}\sum_{j=1}^{n}\alpha_j^n\langle y_j^k,\bar{\psi}_{\ell+1}^n\rangle_X^2.\end{aligned}$$

Thus, $\bar{\psi}_{\ell+1}^n$ solves (1.14), which implies that $\{\bar{\psi}_i^n\}_{i=1}^{\ell+1}$ is a solution to $(\hat{\mathbf{P}}_n^{\ell+1})$ and

$$\sum_{k=1}^{\wp}\sum_{j=1}^{n}\alpha_j^n\sum_{i=1}^{\ell+1}\langle y_j^k,\bar{\psi}_i^n\rangle_X^2=\sum_{i=1}^{\ell+1}\bar{\lambda}_i^n.$$

Again, (1.9) and (1.11) imply (1.10).

It follows that the claim is proved. □

Remark 1.9. Theorem 1.8 can also be proved by using the theory of nonlinear programming; see [29, 73], for instance. In this case, $(\hat{\mathbf{P}}_n^{\ell})$ is considered as an equality-constrained optimization problem. Applying a Lagrangian framework it turns out that (1.5) are first-order necessary optimality conditions for $(\hat{\mathbf{P}}_n^{\ell})$.

For the application of POD to concrete problems, the choice of ℓ is certainly of central importance. It appears that no general a priori rules are available. Rather, the choice of ℓ is based on heuristic considerations combined with observing the ratio of the modeled to the "total" energy contained in the snapshots $y_1^k,\dots,y_n^k$, $1\le k\le\wp$, which is expressed by

$$\mathscr{E}(\ell)=\frac{\sum_{i=1}^{\ell}\bar{\lambda}_i^n}{\sum_{i=1}^{d^n}\bar{\lambda}_i^n}\in[0,1].$$

Utilizing (1.8), we have

$$\mathscr{E}(\ell) = \frac{\sum_{i=1}^{\ell} \bar{\lambda}_i^n}{\sum_{k=1}^{\wp} \sum_{j=1}^{n} \alpha_j^n \|y_j^k\|_X^2},$$

i.e., it is not necessary to compute the eigenvalues $\{\bar{\lambda}_i\}_{i=\ell+1}^{d}$. This is utilized in numerical implementations when iterative eigenvalue solvers, such as the Lanczos method, are applied; see [2, Chapter 10], for instance.

In the following we will discuss three examples that illustrate that POD is strongly related to the SVD of matrices.

Remark 1.10 (*POD in Euclidean space $\mathbb{R}^m$; see* [37]). Suppose that $X = \mathbb{R}^m$ with $m \in \mathbb{N}$ and $\wp = 1$. Then, we have n snapshot vectors $y_1,\ldots,y_n$, and we introduce the rectangular matrix $Y = [y_1|\ldots|y_n] \in \mathbb{R}^{m\times n}$ with rank $d^n \le \min(m,n)$. Choosing $\alpha_j^n = 1$ for $1 \le j \le n$, problem $(\mathbf{P}_n^\ell)$ has the form

$$\begin{cases} \min \displaystyle\sum_{j=1}^{n} \Big\| y_j - \sum_{i=1}^{\ell} (y_j^\top \psi_i)\psi_i \Big\|_{\mathbb{R}^m}^2 \\ \text{s.t. } \{\psi_i\}_{i=1}^{\ell} \subset \mathbb{R}^m \text{ and } \psi_i^\top \psi_j = \delta_{ij},\ 1 \le i,j \le \ell, \end{cases} \tag{1.15}$$

where $\|\cdot\|_{\mathbb{R}^m}$ stands for the Euclidean norm in $\mathbb{R}^m$ and $\top$ denotes the transpose of a given vector (or matrix). From

$$(\mathscr{R}^n \psi)_i = \Big(\sum_{j=1}^{n} (y_j^\top \psi) y_j \Big)_i = \sum_{j=1}^{n} \sum_{l=1}^{m} Y_{lj} \psi_l Y_{ij} = (YY^\top \psi)_i, \quad \psi \in \mathbb{R}^m,$$

for each component $1 \le i \le m$ we infer that (1.5) leads to the symmetric $m \times m$ eigenvalue problem

$$YY^\top \bar{\psi}_i^n = \bar{\lambda}_i^n \bar{\psi}_i^n, \quad \bar{\lambda}_1^n \ge \cdots \ge \bar{\lambda}_{d^n}^n > \bar{\lambda}_{d^n+1}^n = \cdots = \bar{\lambda}_m^n = 0. \tag{1.16}$$

Recall that (1.16) can be solved by utilizing the SVD [51]: there exist real numbers $\bar{\sigma}_1^n \ge \bar{\sigma}_2^n \ge \cdots \ge \bar{\sigma}_{d^n}^n > 0$ and orthogonal matrices $\Psi \in \mathbb{R}^{m\times m}$ with column vectors $\{\bar{\psi}_i^n\}_{i=1}^m$ and $\Phi \in \mathbb{R}^{n\times n}$ with column vectors $\{\bar{\phi}_i^n\}_{i=1}^n$ such that

$$\Psi^\top Y \Phi = \begin{pmatrix} D & 0 \\ 0 & 0 \end{pmatrix} =: \Sigma \in \mathbb{R}^{m\times n}, \tag{1.17}$$

where $D = \operatorname{diag}(\bar{\sigma}_1^n,\ldots,\bar{\sigma}_{d^n}^n) \in \mathbb{R}^{d\times d}$ and the zeros in (1.17) denote matrices of appropriate dimensions. Moreover, the vectors $\{\bar{\psi}_i^n\}_{i=1}^d$ and $\{\bar{\phi}_i^n\}_{i=1}^d$ satisfy

$$Y\bar{\phi}_i^n = \bar{\sigma}_i^n \bar{\psi}_i^n \quad \text{and} \quad Y^\top \bar{\psi}_i^n = \bar{\sigma}_i^n \bar{\phi}_i^n \quad \text{for } i = 1,\ldots,d^n. \tag{1.18}$$

They are eigenvectors of $YY^\top$ and $Y^\top Y$, respectively, with eigenvalues $\bar{\lambda}_i^n = (\bar{\sigma}_i^n)^2 > 0$, $i = 1,\ldots,d^n$. The vectors $\{\bar{\psi}_i^n\}_{i=d^n+1}^m$ and $\{\bar{\phi}_i^n\}_{i=d^n+1}^n$ (if $d^n < m$, respectively, $d^n < n$) are eigenvectors of $YY^\top$ and $Y^\top Y$ with eigenvalue zero. Consequently, in the case $n < m$ one can determine the POD basis of rank ℓ as follows: compute the eigenvectors $\bar{\phi}_1^n,\ldots,\bar{\phi}_\ell^n \in \mathbb{R}^n$ by solving the symmetric $n \times n$ eigenvalue problem

$$Y^\top Y \bar{\phi}_i^n = \bar{\lambda}_i^n \bar{\phi}_i^n \quad \text{for } i = 1,\ldots,\ell$$

and set, by (1.18),

$$\bar{\psi}_i^n = \frac{1}{(\bar{\lambda}_i^n)^{1/2}} Y \bar{\phi}_i^n \quad \text{for } i = 1,\ldots,\ell.$$

For historical reasons this method of determining the POD basis is sometimes called the *method of snapshots*; see [67]. On the other hand, if $m < n$ holds, we can obtain the POD basis by solving the $m \times m$ eigenvalue problem (1.16). If the matrix Y is badly scaled, we should avoid building the matrix product $YY^\top$ (or $Y^\top Y$). In this case the SVD turns out to be more stable for the numerical computation of the POD basis of rank ℓ.

Remark 1.11 (*POD in $\mathbb{R}^m$ with weighted inner product*). As in Remark 1.10, we choose $X = \mathbb{R}^m$ with $m \in \mathbb{R}^m$ and $\wp = 1$. Let $W \in \mathbb{R}^{m\times m}$ be a given symmetric positive definite matrix. We supply $\mathbb{R}^m$ with the weighted inner product

$$\langle \psi, \tilde{\psi} \rangle_W = \psi^\top W \tilde{\psi} = \langle \psi, W\tilde{\psi} \rangle_{\mathbb{R}^m} = \langle W\psi, \tilde{\psi} \rangle_{\mathbb{R}^m} \quad \text{for } \psi, \tilde{\psi} \in \mathbb{R}^m.$$

Then, problem $(\mathbf{P}_n^\ell)$ has the form

$$\begin{cases} \min \displaystyle\sum_{j=1}^n \alpha_j^n \Big\| y_j - \sum_{i=1}^\ell \langle y_j, \psi_i \rangle_W \psi_i \Big\|_W^2 \\ \text{s.t. } \{\psi_i\}_{i=1}^\ell \subset \mathbb{R}^m \text{ and } \langle \psi_i, \psi_j \rangle_W = \delta_{ij},\ 1 \le i,j \le \ell. \end{cases}$$

As in Remark 1.10, we introduce the matrix $Y = [y_1 | \ldots | y_n] \in \mathbb{R}^{m\times n}$ with rank $d^n \le \min(m,n)$. Moreover, we define the diagonal matrix $D = \operatorname{diag}(\alpha_1^n, \ldots, \alpha_n^n) \in \mathbb{R}^{n\times n}$. We find that

$$\begin{aligned} (\mathscr{R}^n \psi)_i &= \Big(\sum_{j=1}^n \alpha_j^n \langle y_j, \psi \rangle_W y_j \Big)_i = \sum_{j=1}^n \sum_{l=1}^m \sum_{\nu=1}^m \alpha_j^n Y_{lj} W_{l\nu} \psi_\nu Y_{ij} \\ &= \big(YDY^\top W\psi \big)_i \quad \text{for } \psi \in \mathbb{R}^m \end{aligned}$$

for each component $1 \le i \le m$. Consequently, (1.5) leads to the eigenvalue problem

$$YDY^\top W \bar{\psi}_i^n = \bar{\lambda}_i^n \bar{\psi}_i^n, \quad \bar{\lambda}_1^n \ge \cdots \ge \bar{\lambda}_{d^n}^n > \bar{\lambda}_{d^n+1}^n = \cdots = \bar{\lambda}_m^n = 0. \tag{1.19}$$

Since W is symmetric and positive definite, W possesses an eigenvalue decomposition of the form $W = QBQ^\top$, where $B = \operatorname{diag}(\beta_1,\ldots,\beta_m)$ contains the eigenvalues $\beta_1 \ge \cdots \ge \beta_m > 0$ of W and $Q \in \mathbb{R}^{m\times m}$ is an orthogonal matrix. We define

$$W^r = Q\operatorname{diag}(\beta_1^r,\ldots,\beta_m^r)Q^\top \quad \text{for } r \in \mathbb{R}.$$

Note that $(W^r)^{-1} = W^{-r}$ and $W^{r+s} = W^r W^s$ for $r, s \in \mathbb{R}$. Moreover, we have

$$\langle \psi, \tilde{\psi} \rangle_W = \langle W^{1/2}\psi, W^{1/2}\tilde{\psi} \rangle_{\mathbb{R}^m} \quad \text{for } \psi, \tilde{\psi} \in \mathbb{R}^m$$

and $\|\psi\|_W = \|W^{1/2}\psi\|_{\mathbb{R}^m}$ for $\psi \in \mathbb{R}^m$. Analogously, the matrix $D^{1/2}$ is defined. Inserting $\psi_i^n = W^{1/2}\bar{\psi}_i^n$ into (1.19), multiplying (1.19) by $W^{1/2}$ from the left, and setting $\hat{Y} = W^{1/2}YD^{1/2}$ yields the symmetric $m \times m$ eigenvalue problem

$$\hat{Y}\hat{Y}^\top \psi_i^n = \bar{\lambda}_i^n \psi_i^n, \quad 1 \le i \le \ell.$$

Note that

$$\hat{Y}^\top \hat{Y} = D^{1/2}Y^\top WYD^{1/2} \in \mathbb{R}^{n\times n}. \tag{1.20}$$

Thus, the POD basis $\{\bar{\psi}_i^n\}_{i=1}^\ell$ of rank ℓ can also be computed by the method of snapshots as follows. First, we solve the symmetric $n \times n$ eigenvalue problem

$$\hat{Y}^\top \hat{Y}\phi_i^n = \bar{\lambda}_i^n \phi_i^n,\ 1 \le i \le \ell \quad \text{and} \quad \langle \phi_i^n, \phi_j^n\rangle_{\mathbb{R}^n} = \delta_{ij},\ 1 \le i,j \le \ell.$$

Then, we set (by using the SVD of $\hat{Y}$)

$$\bar{\psi}_i^n = W^{-1/2}\psi_i^n = \frac{1}{\bar{\sigma}_i^n} W^{-1/2}\hat{Y}\phi_i^n = \frac{1}{\bar{\sigma}_i^n} YD^{1/2}\phi_i^n, \quad 1 \le i \le \ell. \tag{1.21}$$

Note that

$$\langle \bar{\psi}_i^n, \bar{\psi}_j^n\rangle_W = (\bar{\psi}_i^n)^\top W \bar{\psi}_j^n = \frac{1}{\bar{\sigma}_i^n \bar{\sigma}_j^n} (\phi_i^n)^\top \underbrace{D^{1/2}Y^\top WYD^{1/2}}_{=\hat{Y}^\top \hat{Y}} \phi_j^n = \delta_{ij}$$

for $1 \le i,j \le \ell$. Thus, the POD basis $\{\bar{\psi}_i^n\}_{i=1}^\ell$ of rank ℓ is orthonormal in $\mathbb{R}^m$ with respect to the inner product $\langle\cdot,\cdot\rangle_W$. We observe from (1.20) and (1.21) that computing $W^{1/2}$ and $W^{-1/2}$ is not required. For applications, where W is not just a diagonal matrix, the method of snapshots turns out to be more attractive with respect to the computational costs even if $m > n$ holds.

Remark 1.12 (*POD in $\mathbb{R}^m$ with multiple snapshots*). Let us discuss the more general case $\wp = 2$ in the setting of Remark 1.11. The extension for $\wp > 2$ is straightforward. We introduce the matrix $Y = [y_1^1 | \dots | y_n^1 | y_1^2 | \dots | y_n^2] \in \mathbb{R}^{m\times(n\wp)}$ with rank $d^n \le \min(m, n\wp)$. Then, we find

$$\begin{aligned}\mathcal{R}^n\psi &= \sum_{j=1}^n \Big(\alpha_j^n \langle y_j^1,\psi\rangle_W\, y_j^1 + \alpha_j^n \langle y_j^2,\psi\rangle_W\, y_j^2\Big)\\ &= Y\underbrace{\begin{pmatrix} D & 0\\ 0 & D\end{pmatrix}}_{=:\tilde{D}\in\mathbb{R}^{(n\wp)\times(n\wp)}} Y^\top W\psi = Y\tilde{D}Y^\top W\psi \quad \text{for } \psi \in \mathbb{R}^m.\end{aligned}$$

Hence, (1.5) corresponds to the eigenvalue problem

$$Y\tilde{D}Y^\top W\bar{\psi}_i^n = \bar{\lambda}_i^n \bar{\psi}_i^n, \quad \bar{\lambda}_1^n \ge \dots \ge \bar{\lambda}_{d^n}^n > \bar{\lambda}_{d^n+1}^n = \dots = \bar{\lambda}_m^n = 0. \tag{1.22}$$

Setting $\psi_i^n = W^{1/2}\bar{\psi}_i^n$ in (1.22) and multiplying by $W^{1/2}$ from the left yields

$$W^{1/2}Y\tilde{D}Y^\top W^{1/2}\psi_i^n = \bar{\lambda}_i^n \psi_i^n. \tag{1.23}$$

Let $\hat{Y} = W^{1/2}Y\tilde{D}^{1/2} \in \mathbb{R}^{m\times(n\wp)}$. Using $W^\top = W$ as well as $\tilde{D}^\top = \tilde{D}$, we infer from (1.23) that the POD basis $\{\bar{\psi}_i^n\}_{i=1}^\ell$ of rank ℓ is given by the symmetric $m\times m$ eigenvalue problem

$$\hat{Y}\hat{Y}^\top\psi_i^n = \bar{\lambda}_i^n\psi_i^n,\ 1\le i\le\ell, \qquad \langle\psi_i^n,\psi_j^n\rangle_{\mathbb{R}^m} = \delta_{ij},\ 1\le i,j\le\ell,$$

and $\bar{\psi}_i^n = W^{-1/2}\psi_i^n$. Note that

$$\hat{Y}^\top\hat{Y} = \tilde{D}^{1/2}Y^\top WY\tilde{D}^{1/2} \in \mathbb{R}^{(n\wp)\times(n\wp)}.$$

Thus, the POD basis of rank ℓ can also be computed by the method of snapshots as follows. First, we solve the symmetric $(n\wp)\times(n\wp)$ eigenvalue problem

$$\hat{Y}^\top\hat{Y}\phi_i^n = \bar{\lambda}_i^n\phi_i,\ 1\le i\le\ell \quad\text{and}\quad \langle\phi_i^n,\phi_j^n\rangle_{\mathbb{R}^{n\wp}} = \delta_{ij},\ 1\le i,j\le\ell.$$

Then, we set (by SVD)

$$\bar{\psi}_i^n = W^{-1/2}\psi_i^n = \frac{1}{\bar{\sigma}_i^n}W^{-1/2}\hat{Y}\phi_i^n = \frac{1}{\bar{\sigma}_i^n}Y\tilde{D}^{1/2}\phi_i^n$$

for $1\le i\le\ell$.

1.2.2 ▪ The continuous variant of the POD method

As in Remark 1.1, let $0\le t_1<t_2<\cdots<t_n\le T$ be a given time grid in the interval $[0,T]$ with step size $\Delta t = T/(n-1)$, i.e., $t_j = (j-1)\Delta t$. Suppose that we have trajectories $y^k\in C([0,T];X)$, $1\le k\le\wp$. Let the snapshots be given as $y_j^k = y^k(t_j)\in X$ or $y_j^k\approx y^k(t_j)\in X$. Then, the snapshot subspace $\mathscr{V}^n$ introduced in (1.1) depends on the chosen time instances $\{t_j\}_{j=1}^n$. Consequently, the POD basis $\{\bar{\psi}_i^n\}_{i=1}^\ell$ of rank ℓ as well as the corresponding eigenvalues $\{\bar{\lambda}_i^n\}_{i=1}^\ell$ also depend on the time instances (which has already been indicated by the superscript n). Moreover, we have not discussed so far the motivation for introducing the positive weights $\{\alpha_j^n\}_{j=1}^n$ in $(\mathbf{P}_n^\ell)$. For this reason we proceed by investigating the following two questions:

- How should we choose good time instances for the snapshots?
- What are appropriate positive weights $\{\alpha_j^n\}_{j=1}^n$?

To address these two questions, we will introduce a *continuous version* of the POD. In Section 1.2.1 we introduced the operator $\mathscr{R}^n$ in (1.4). By $\{\bar{\psi}_i^n\}_{i\in\mathscr{I}}$ and $\{\bar{\lambda}_i^n\}_{i\in\mathscr{I}}$ we denoted the eigenfunctions and eigenvalues for $\mathscr{R}^n$ satisfying (1.5). Moreover, we set $d^n = \dim\mathscr{V}^n$ for the dimension of the snapshot set. Let us now introduce the snapshot set by

$$\mathscr{V} = \text{span}\Big\{y^k(t)\,|\,t\in[0,T]\text{ and }1\le k\le\wp\Big\}\subset X$$

with dimension $d\le\infty$. For any $\ell\le d$, we are interested in determining a POD basis of rank ℓ that minimizes the mean square error between the trajectories y^k and the corresponding ℓth partial Fourier sums on average in the time interval $[0,T]$:

$$\begin{cases} \min\displaystyle\sum_{k=1}^{\wp}\int_0^T\Big\|y^k(t)-\sum_{i=1}^{\ell}\langle y^k(t),\psi_i\rangle_X\,\psi_i\Big\|_X^2\,\mathrm{d}t \\ \text{s.t. } \{\psi_i\}_{i=1}^\ell\subset X \text{ and } \langle\psi_i,\psi_j\rangle_X = \delta_{ij},\ 1\le i,j\le\ell. \end{cases} \tag{$\mathbf{P}^\ell$}$$

An optimal solution $\{\bar\psi_i\}_{i=1}^\ell$ to $(\mathbf{P}^\ell)$ is called a *POD basis of rank* ℓ. Analogous to $(\hat{\mathbf{P}}_n^\ell)$ we can—instead of $(\mathbf{P}^\ell)$—consider the problem

$$\begin{cases} \max \displaystyle\sum_{k=1}^{\wp}\int_0^T \sum_{i=1}^{\ell} \langle y^k(t),\psi_i\rangle_X^2\,\mathrm{d}t \\ \text{s.t. } \{\psi_i\}_{i=1}^\ell \subset X \text{ and } \langle \psi_i,\psi_j\rangle_X=\delta_{ij},\ 1\le i,j\le \ell. \end{cases} \tag{$\hat{\mathbf{P}}^\ell$}$$

A solution to $(\mathbf{P}^\ell)$ and to $(\hat{\mathbf{P}}^\ell)$ can be characterized by an eigenvalue problem for the linear integral operator $\mathcal{R}: X\to X$ given as

$$\mathcal{R}\psi=\sum_{k=1}^{\wp}\int_0^T \langle y^k(t),\psi\rangle_X\, y^k(t)\,\mathrm{d}t \quad \text{for } \psi\in X. \tag{1.24}$$

For the given real Hilbert space X, we denote by $L^2(0,T;X)$ the Hilbert space of square integrable functions $t\mapsto\varphi(t)\in X$ so that [70, p. 143]

- the mapping $t\mapsto\varphi(t)$ is measurable for $t\in[0,T]$ and
- $\|\varphi\|_{L^2(0,T;X)}=\Big(\int_0^T \|\varphi(t)\|_X^2\,\mathrm{d}t\Big)^{1/2}<\infty.$

Recall that $\varphi:[0,T]\to X$ is called *measurable* if there exists a sequence $\{\varphi_n\}_{n\in\mathbb{N}}$ of step functions $\varphi_n:[0,T]\to X$ satisfying $\varphi(t)=\lim_{n\to\infty}\varphi_n(t)$ for almost all $t\in[0,T]$. The standard inner product on $L^2(0,T;X)$ is given by

$$\langle\varphi,\psi\rangle_{L^2(0,T;X)}=\int_0^T\langle\varphi(t),\phi(t)\rangle_X\,\mathrm{d}t \quad \text{for } \varphi,\phi\in L^2(0,T;X).$$

Lemma 1.13. *Let X be a (separable) real Hilbert space and $y^k\in L^2(0,T;X)$, $1\le k\le \wp$, be given snapshot trajectories. Then, the operator $\mathcal{R}$ introduced in* (1.24) *is compact, nonnegative, and self-adjoint.*

Proof. First, we write $\mathcal{R}$ as a product of an operator and its Hilbert space adjoint. For that purpose, let us define the linear operator $\mathcal{Y}:L^2(0,T;\mathbb{R}^\wp)\to X$ by

$$\mathcal{Y}\phi=\sum_{k=1}^{\wp}\int_0^T \phi^k(t)y^k(t)\,\mathrm{d}t \quad \text{for } \phi=(\phi^1,\dots,\phi^\wp)\in L^2(0,T;\mathbb{R}^\wp). \tag{1.25}$$

Utilizing the Cauchy-Schwarz inequality [60, p. 38] and $y^k\in L^2(0,T;X)$ for $1\le k\le \wp$, we infer that

$$\begin{aligned}\|\mathcal{Y}\phi\|_X &\le \sum_{k=1}^{\wp}\int_0^T|\phi^k(t)|\|y^k(t)\|_X\,\mathrm{d}t\le\sum_{k=1}^{\wp}\|\phi^k\|_{L^2(0,T)}\|y^k\|_{L^2(0,T;X)}\\ &\le\Big(\sum_{k=1}^{\wp}\|\phi^k\|^2_{L^2(0,T)}\Big)^{1/2}\Big(\sum_{k=1}^{\wp}\|y^k(t)\|_X^2\Big)^{1/2}\\ &=C_{\mathcal{Y}}\|\phi\|_{L^2(0,T;\mathbb{R}^\wp)} \quad \text{for any } \phi\in L^2(0,T;\mathbb{R}^\wp),\end{aligned}$$

where we set $C_{\mathcal{Y}}=(\sum_{k=1}^{\wp}\|y^k(t)\|_X^2)^{1/2}<\infty$. Hence, the operator $\mathcal{Y}$ is bounded. Its Hilbert space adjoint $\mathcal{Y}^\star:X\to L^2(0,T;\mathbb{R}^\wp)$ satisfies

$$\langle\mathcal{Y}^\star\psi,\phi\rangle_{L^2(0,T;\mathbb{R}^\wp)}=\langle\psi,\mathcal{Y}\phi\rangle_X \quad \text{for } \psi\in X \text{ and } \phi\in L^2(0,T;\mathbb{R}^\wp).$$

Since we derive

$$\begin{aligned}\langle \mathscr{Y}^\star \psi, \phi\rangle_{L^2(0,T;\mathbb{R}^\wp)} &= \langle \psi, \mathscr{Y}\phi\rangle_X = \Big\langle \psi, \sum_{k=1}^{\wp}\int_0^T \phi^k(t) y^k(t)\,\mathrm{d}t\Big\rangle_X \\ &= \sum_{k=1}^{\wp}\int_0^T \langle \psi, y^k(t)\rangle_X \phi^k(t)\,\mathrm{d}t = \Big\langle \big(\langle \psi, y^k(\cdot)\rangle_X\big)_{1\le k\le \wp}, \phi\Big\rangle_{L^2(0,T;\mathbb{R}^\wp)}\end{aligned}$$

for $\psi \in X$ and $\phi \in L^2(0,T;\mathbb{R}^\wp)$, the adjoint operator is given by

$$(\mathscr{Y}^\star \psi)(t) = \begin{pmatrix} \langle \psi, y^1(t)\rangle_X \\ \vdots \\ \langle \psi, y^\wp(t)\rangle_X \end{pmatrix} \quad \text{for } \psi \in X \text{ and } t \in [0,T] \text{ a.e.},$$

where a.e. stands for almost everywhere. From (1.4) and

$$\big(\mathscr{Y}\mathscr{Y}^\star\big)\psi = \mathscr{Y}\begin{pmatrix} \langle \psi, y^1(\cdot)\rangle_X \\ \vdots \\ \langle \psi, y^\wp(\cdot)\rangle_X \end{pmatrix} = \sum_{k=1}^{\wp}\int_0^T \langle \psi, y^k(t)\rangle_X y^k(t)\,\mathrm{d}t \quad \text{for } \psi \in X,$$

we infer that $\mathscr{R} = \mathscr{Y}\mathscr{Y}^\star$ holds. Since the operator $\mathscr{Y}$ is bounded, its adjoint and therefore $\mathscr{R} = \mathscr{Y}\mathscr{Y}^\star$ are bounded operators. To prove that $\mathscr{R}$ is compact, we show that $\mathscr{Y}^\star$ is compact. Let $\{\chi_n\}_{n\in\mathbb{N}} \subset X$ be a sequence converging weakly to an element $\chi \in X$, i.e.,

$$\lim_{n\to\infty} \langle \chi_n, \psi\rangle_X = \langle \chi, \psi\rangle_X \quad \text{for all } \psi \in X.$$

This implies that

$$\lim_{n\to\infty} (\mathscr{Y}^\star \chi_n)(t) = \begin{pmatrix} \langle \chi_n, y^1(t)\rangle_X \\ \vdots \\ \langle \chi_n, y^\wp(t)\rangle_X \end{pmatrix} = \begin{pmatrix} \langle \chi_n, y^1(t)\rangle_X \\ \vdots \\ \langle \chi_n, y^\wp(t)\rangle_X \end{pmatrix} = \big(\mathscr{Y}^\star \chi\big)(t)$$

for $t \in [0,T]$ a.e. Thus, the sequence $\{\mathscr{Y}^\star \chi_n\}_{n\in\mathbb{N}}$ converges weakly to $\mathscr{Y}^\star\chi$ in $L^2(0,T;\mathbb{R}^\wp)$. Consequently, $\mathscr{R} = \mathscr{Y}\mathscr{Y}^\star$ is compact. From

$$\begin{aligned}\langle \mathscr{R}\psi, \psi\rangle_X &= \Big\langle \sum_{k=1}^{\wp}\int_0^T \langle \psi, y^k(t)\rangle_X\, y^k(t)\,\mathrm{d}t, \psi\Big\rangle_X \\ &= \sum_{k=1}^{\wp}\int_0^T \big|\langle \psi, y^k(t)\rangle_X\big|^2\,\mathrm{d}t \ge 0 \quad \text{for all } \psi \in X,\end{aligned}$$

we infer that $\mathscr{R}$ is nonnegative. Finally, we have $\mathscr{R}^\star = (\mathscr{Y}\mathscr{Y}^\star)^\star = \mathscr{R}$, i.e., the operator $\mathscr{R}$ is self-adjoint. □

Remark 1.14. It follows from the proof of Lemma 1.13 that $\mathscr{K} = \mathscr{Y}^\star\mathscr{Y} : L^2(0,T;\mathbb{R}^\wp) \to L^2(0,T;\mathbb{R}^\wp)$ is compact as well. We find that

$$\big(\mathscr{K}\phi\big)(t) = \begin{pmatrix} \sum\limits_{k=1}^{\wp}\int_0^T \langle y^k(s), y^1(t)\rangle_X \phi^k(s)\,\mathrm{d}s \\ \vdots \\ \sum\limits_{k=1}^{\wp}\int_0^T \langle y^k(s), y^\wp(t)\rangle_X \phi^k(s)\,\mathrm{d}s \end{pmatrix}, \quad \phi \in L^2(0,T;\mathbb{R}^\wp).$$

The compactness of $\mathscr{K}$ can also be shown as follows. Notice that the kernel function

$$r_{ik}(s,t)=\langle y^k(s),y^i(t)\rangle_X,\quad (s,t)\in[0,T]\times[0,T] \text{ and } 1\le i,k\le\wp,$$

belongs to $L^2(0,T)\times L^2(0,T)$. Here, we write $L^2(0,T)$ for $L^2(0,T;\mathbb{R})$. Then, it follows from [78, pp. 197 and 277] that the linear integral operator $\mathscr{K}_{ik}: L^2(0,T)\to L^2(0,T)$ defined by

$$\mathscr{K}_{ik}(t)=\int_0^T r_{ik}(s,t)\phi(s)\,\mathrm{d}s,\quad \phi\in L^2(0,T),$$

is a compact operator. This implies that the operator $\sum_{k=1}^{\wp}\mathscr{K}_{ik}$ is compact for $1\le i\le\wp$ as well.

In the next theorem, we formulate how the solution to $(\mathbf{P}^\ell)$ and $(\hat{\mathbf{P}}^\ell)$ can be found.

Theorem 1.15. *Let X be a separable real Hilbert space and $y^k\in L^2(0,T;X)$ be given trajectories for $1\le k\le\wp$. Suppose that the linear operator $\mathscr{R}$ is defined by* (1.24). *Then, there exist nonnegative eigenvalues $\{\bar\lambda_i\}_{i\in\mathscr{I}}$ and associated orthonormal eigenfunctions $\{\bar\psi_i\}_{i\in\mathscr{I}}$ satisfying*

$$\mathscr{R}\bar\psi_i=\bar\lambda_i\bar\psi_i,\quad \bar\lambda_1\ge\dots\ge\bar\lambda_d>\bar\lambda_{d+1}=\dots=0. \tag{1.26}$$

For every $\ell\in\{1,\dots,d\}$, the first ℓ eigenfunctions $\{\bar\psi_i\}_{i=1}^\ell$ solve $(\mathbf{P}^\ell)$ and $(\hat{\mathbf{P}}^\ell)$. Moreover, the values of the objectives evaluated at the optimal solution $\{\bar\psi_i\}_{i=1}^\ell$ satisfy

$$\sum_{k=1}^{\wp}\int_0^T\Big\|y^k(t)-\sum_{i=1}^{\ell}\langle y^k(t),\bar\psi_i\rangle_X\,\bar\psi_i\Big\|_X^2\,\mathrm{d}t=\sum_{i=\ell+1}^{d}\bar\lambda_i \tag{1.27}$$

and

$$\sum_{k=1}^{\wp}\int_0^T\sum_{i=1}^{\ell}\langle y^k(t),\bar\psi_i\rangle_X^2\,\mathrm{d}t=\sum_{i=1}^{\ell}\bar\lambda_i, \tag{1.28}$$

respectively.

Proof. The existence of sequences $\{\bar\lambda_i\}_{i\in\mathscr{I}}$ of eigenvalues and $\{\bar\psi_i\}_{i\in\mathscr{I}}$ of associated eigenfunctions satisfying (1.26) follows from Lemma 1.13, Theorem 1.5, and Theorem 1.6. Analogous to the proof of Theorem 1.8 in Section 1.2.1, one can show that $\{\bar\psi_i\}_{i=1}^\ell$ solves $(\mathbf{P}^\ell)$ as well as $(\hat{\mathbf{P}}^\ell)$ and that (1.27) (respectively (1.28)) is valid. □

Remark 1.16. Similar to (1.6), we have

$$\sum_{k=1}^{\wp}\int_0^T\|y^k(t)\|_X^2\,\mathrm{d}t=\sum_{i=1}^{d}\bar\lambda_i. \tag{1.29}$$

In fact,

$$\mathscr{R}\bar\psi_i=\sum_{k=1}^{\wp}\int_0^T\langle y^k(t),\bar\psi_i\rangle_X\,y^k(t)\,\mathrm{d}t\quad\text{for every } i\in\mathscr{I}.$$

Taking the inner product with $\bar{\psi}_i$, using (1.26), and summing over i, we get

$$\sum_{i=1}^{d}\sum_{k=1}^{\wp}\int_0^T \langle y^k(t),\bar{\psi}_i\rangle_X^2\,\mathrm{d}t=\sum_{i=1}^{d}\langle\mathscr{R}\bar{\psi}_i,\bar{\psi}_i\rangle_X=\sum_{i=1}^{d}\bar{\lambda}_i.$$

Expanding each $y^k(t)\in X$ in terms of $\{\bar{\psi}_i\}_{i\in\mathscr{I}}$, for each $1\le k\le\wp$, we have

$$y^k(t)=\sum_{i=1}^{d}\langle y^k(t),\bar{\psi}_i\rangle_X\,\bar{\psi}_i$$

and hence

$$\sum_{k=1}^{\wp}\int_0^T \|y^k(t)\|_X^2\,\mathrm{d}t=\sum_{k=1}^{\wp}\sum_{i=1}^{d}\int_0^T \langle y^k(t),\bar{\psi}_i\rangle_X^2\,\mathrm{d}t=\sum_{i=1}^{d}\bar{\lambda}_i,$$

which is (1.29).

Remark 1.17 (*SVD*). Suppose that $y^k\in L^2(0,T;X)$. By Theorem 1.15, there exist nonnegative eigenvalues $\{\bar{\lambda}_i\}_{i\in\mathscr{I}}$ and associated orthonormal eigenfunctions $\{\bar{\psi}_i\}_{i\in\mathscr{I}}$ satisfying (1.26). From $\mathscr{K}=\mathscr{Y}^\star\mathscr{Y}$ it follows that there is a sequence $\{\bar{\phi}_i\}_{i\in\mathscr{I}}$ such that

$$\mathscr{K}\bar{\phi}_i=\bar{\lambda}_i\bar{\phi}_i,\quad 1,\ldots,\ell.$$

We set $\mathbb{R}_0^+=\{s\in\mathbb{R}\,|\,s\ge 0\}$ and $\bar{\sigma}_i=\bar{\lambda}_i^{1/2}$. The sequence $\{\bar{\sigma}_i,\bar{\phi}_i,\bar{\psi}_i\}_{i\in\mathscr{I}}$ in $\mathbb{R}_0^+\times L^2(0,T;\mathbb{R}^\wp)\times X$ can be interpreted as an SVD of the mapping $\mathscr{Y}:L^2(0,T;\mathbb{R}^\wp)\to X$ introduced in (1.25). In fact, we have

$$\mathscr{Y}\bar{\phi}_i=\bar{\sigma}_i\bar{\psi}_i,\quad \mathscr{Y}^\star\bar{\psi}_i=\bar{\sigma}_i\bar{\phi}_i,\quad i\in\mathscr{I}.$$

Since $\bar{\sigma}_i>0$ holds for $1=1,\ldots,d$, we have $\bar{\psi}_i=\mathscr{Y}\bar{\phi}_i/\sigma_i$ for $i=1,\ldots,d$.

1.2.3 ▪ Perturbation analysis for the POD basis

The eigenvalues $\{\bar{\lambda}_i^n\}_{i\in\mathscr{I}}$ satisfying (1.5) depend on the time grid $\{t_j\}_{j=1}^n$. In this section we investigate the sum $\sum_{i=\ell+1}^{d^n}\bar{\lambda}_i^n$, the value of the cost in $(\mathbf{P}_n^\ell)$ evaluated at the solution $\{\bar{\psi}_i^n\}_{i=1}^\ell$ for $n\to\infty$. Clearly, $n\to\infty$ is equivalent to $\Delta t=T/(n-1)\to 0$.

In general, the spectrum $\sigma(\mathscr{T})$ of an operator $\mathscr{T}\in\mathscr{L}(X)$ does not depend continuously on $\mathscr{T}$. This is an essential difference from the finite-dimensional case. For the compact and self-adjoint operator $\mathscr{R}$, we have $\sigma(\mathscr{R})=\{\bar{\lambda}_i\}_{i\in\mathscr{I}}$. Suppose that for $\ell\in\mathbb{N}$ we have $\bar{\lambda}_\ell>\bar{\lambda}_{\ell+1}$ so that we can separate the spectrum as follows: $\sigma(\mathscr{R})=\mathscr{S}_\ell\cup\mathscr{S}_\ell'$ with $\mathscr{S}_\ell=\{\bar{\lambda}_1,\ldots,\bar{\lambda}_\ell\}$ and $\mathscr{S}_\ell'=\sigma(\mathscr{R})\setminus\mathscr{S}_\ell$. Then, $\mathscr{S}_\ell\cap\mathscr{S}_\ell'=\emptyset$. Moreover, setting $V^\ell=\mathrm{span}\,\{\bar{\psi}_1,\ldots,\bar{\psi}_\ell\}$, we have $X=V^\ell\oplus(V^\ell)^\perp$, where the linear space $(V^\ell)^\perp$ stands for the X-orthogonal complement of V^ℓ. Let us assume that

$$\lim_{n\to\infty}\|\mathscr{R}^n-\mathscr{R}\|_{\mathscr{L}(X)}=0. \tag{1.30}$$

Then it follows from the perturbation theory of the spectrum of linear operators [35, pp. 212–214] that the space $V_n^\ell=\mathrm{span}\,\{\bar{\psi}_1^n,\ldots,\bar{\psi}_\ell^n\}$ is isomorphic to V^ℓ if n is sufficiently large. Furthermore, the change in a finite set of eigenvalues of $\mathscr{R}$ is small provided $\|\mathscr{R}^n-\mathscr{R}\|_{\mathscr{L}(X)}$ is sufficiently small. Summarizing, the behavior of the spectrum is much the same as in the finite-dimensional case if we can ensure (1.30). Therefore, we start this section by investigating the convergence of $\mathscr{R}^n - \mathscr{R}$ in the operator norm.

Recall that the Sobolev space $H^1(0,T;X)$ is given by

$$H^1(0,T;X) = \{\varphi \in L^2(0,T;X) \,|\, \varphi_t \in L^2(0,T;X)\},$$

where φ_t denotes the weak derivative of φ. The space $H^1(0,T;X)$ is a Hilbert space with inner product

$$\langle \varphi, \phi \rangle_{H^1(0,T;X)} = \int_0^T \langle \varphi(t), \phi(t) \rangle_X + \langle \varphi_t(t), \phi_t(t) \rangle_X \,\mathrm{d}t \text{ for } \varphi, \phi \in H^1(0,T;X)$$

and induced norm $\|\varphi\|_{H^1(0,T;X)} = \langle \varphi, \varphi \rangle^{1/2}_{H^1(0,T;X)}$.

Let us choose the trapezoidal weights

$$\alpha_1^n = \frac{T}{2(n-1)},\ \alpha_j^n = \frac{T}{n-1} \text{ for } 2 \le j \le n-1,\ \alpha_n^n = \frac{T}{2(n-1)}. \tag{1.31}$$

For this choice we observe that for every $\psi \in X$ the element $\mathscr{R}^n \psi$ is a trapezoidal approximation of $\mathscr{R}\psi$. We will make use of the following lemma.

Lemma 1.18. *Suppose that X is a (separable) real Hilbert space and that the snapshot trajectories y^k belong to $H^1(0,T;X)$ for $1 \le k \le \wp$. Then,* (1.30) *holds true.*

Proof. For an arbitrary $\psi \in X$ with $\|\psi\|_X = 1$, we define $F : [0,T] \to X$ by

$$F(t) = \sum_{k=1}^{\wp} \langle y^k(t), \psi \rangle_X \, y^k(t) \quad \text{for } t \in [0,T].$$

It follows that

$$\begin{aligned} \mathscr{R}\psi &= \int_0^T F(t)\,\mathrm{d}t = \sum_{j=1}^{n-1} \int_{t_j}^{t_{j+1}} F(t)\,\mathrm{d}t, \\ \mathscr{R}^n \psi &= \sum_{j=1}^{n} \alpha_j F(t_j) = \frac{\Delta t}{2} \sum_{j=1}^{n-1} \big(F(t_j) + F(t_{j+1})\big). \end{aligned} \tag{1.32}$$

Then, we infer from $\|\psi\|_X = 1$ that

$$\|F(t)\|_X^2 \le \Big(\sum_{k=1}^{\wp} \|y^k(t)\|_X^2 \Big)^2. \tag{1.33}$$

Now we show that F belongs to $H^1(0,T;X)$ and its norm is bounded independently of ψ. Recall that $y^k \in H^1(0,T;X)$ implies that $y^k \in C([0,T];X)$ for $1 \le k \le \wp$. Using (1.33), we have

$$\|F\|^2_{L^2(0,T;X)} \le \int_0^T \Big(\sum_{k=1}^{\wp} \|y^k\|^2_{C([0,T];X)} \Big)^2 \mathrm{d}t \le C_1,$$

with $C_1 = T(\sum_{k=1}^{\wp} \|y^k\|^2_{C([0,T];X)})^2$. Moreover, $F \in H^1(0,T;X)$, with

$$F_t(t) = \sum_{k=1}^{\wp} \langle y_t^k(t), \psi \rangle_X \, y^k(t) + \langle y^k(t), \psi \rangle_X \, y_t^k(t) \quad \text{for almost all (f.a.a.) } t \in [0,T].$$

Thus, we derive

$$\|F_t\|^2_{L^2(0,T;X)} \le 4\int_0^T \Big(\sum_{k=1}^{\wp} \|y^k(t)\|_X \|y_t^k(t)\|_X\Big)^2 \mathrm{d}t \le C_2,$$

with $C_2 = 4\sum_{k=1}^{\wp} \|y^k\|^2_{C([0,T];X)} \sum_{l=1}^{\wp} \|y_t^l\|^2_{L^2(0,T;X)} < \infty$. Consequently,

$$\|F\|_{H^1(0,T;X)} = \Big(\int_0^T \|F(t)\|_X^2 + \|F_t(t)\|_X^2 \,\mathrm{d}t\Big)^{1/2} \le C_3, \tag{1.34}$$

with the constant $C_3 = (C_1 + C_2)^{1/2}$, which is independent of ψ. To evaluate $\mathscr{R}^n\psi$, we notice that

$$\begin{aligned}
\int_{t_j}^{t_{j+1}} F(t)\,\mathrm{d}t &= \frac{1}{2}\int_{t_j}^{t_{j+1}} \Big(F(t_j) + \int_{t_j}^{t} F_t(s)\,\mathrm{d}s\Big)\mathrm{d}t \\
&\quad + \frac{1}{2}\int_{t_j}^{t_{j+1}} \Big(F(t_{j+1}) + \int_{t_{j+1}}^{t} F_t(s)\,\mathrm{d}s\Big)\mathrm{d}t \\
&= \frac{\Delta t}{2}\big(F(t_j) + F(t_{j+1})\big) \\
&\quad + \frac{1}{2}\int_{t_j}^{t_{j+1}} \Big(\int_{t_{j+1}}^{t} F_t(s)\,\mathrm{d}s + \int_{t_{j+1}}^{t} F_t(s)\,\mathrm{d}s\Big)\mathrm{d}t.
\end{aligned} \tag{1.35}$$

Utilizing (1.32) and (1.35), we obtain

$$\begin{aligned}
\|\mathscr{R}^n\psi - \mathscr{R}\psi\|_X &= \Big\|\sum_{j=1}^{n-1}\Big(\frac{\Delta t}{2}\big(F(t_j) + F(t_{j+1})\big) - \int_{t_j}^{t_{j+1}} F(t)\,\mathrm{d}t\Big)\Big\|_X \\
&\le \frac{1}{2}\sum_{j=1}^{n-1}\Big\|\int_{t_j}^{t_{j+1}}\int_{t_j}^{t} F_t(s)\,\mathrm{d}s\mathrm{d}t\Big\|_X + \frac{1}{2}\sum_{j=1}^{n-1}\Big\|\int_{t_j}^{t_{j+1}}\int_{t_{j+1}}^{t} F_t(s)\,\mathrm{d}s\mathrm{d}t\Big\|_X.
\end{aligned}$$

From the Cauchy–Schwarz inequality [60, p. 38], we deduce that

$$\begin{aligned}
&\sum_{j=1}^{n-1}\Big\|\int_{t_j}^{t_{j+1}}\int_{t_j}^{t} F_t(s)\,\mathrm{d}s\mathrm{d}t\Big\|_X \le \sum_{j=1}^{n-1}\int_{t_j}^{t_{j+1}}\Big\|\int_{t_j}^{t} F_t(s)\,\mathrm{d}s\Big\|_X \mathrm{d}t \\
&\le \sqrt{\Delta t}\sum_{j=1}^{n-1}\Big(\int_{t_j}^{t_{j+1}}\Big\|\int_{t_j}^{t} F_t(s)\,\mathrm{d}s\Big\|_X^2 \mathrm{d}t\Big)^{1/2} \\
&\le \sqrt{\Delta t}\sum_{j=1}^{n-1}\Big(\int_{t_j}^{t_{j+1}}\Big(\int_{t_j}^{t} \|F_t(s)\|_X\,\mathrm{d}s\Big)^2 \mathrm{d}t\Big)^{1/2} \\
&\le \Delta t\sum_{j=1}^{n-1}\Big(\int_{t_j}^{t_{j+1}}\int_{t_j}^{t} \|F_t(s)\|_X^2\,\mathrm{d}s\mathrm{d}t\Big)^{1/2} \le T\sqrt{\Delta t}\,\|F\|_{H^1(0,T;X)}.
\end{aligned} \tag{1.36}$$

Analogously, we derive

$$\sum_{j=1}^{n-1}\Big\|\int_{t_j}^{t_{j+1}}\int_{t_{j+1}}^{t} F_t(s)\,\mathrm{d}s\mathrm{d}t\Big\|_X \le T\sqrt{\Delta t}\,\|F\|_{H^1(0,T;X)}. \tag{1.37}$$

From (1.34), (1.36), and (1.37), it follows that

$$\|\mathscr{R}^n\psi - \mathscr{R}\psi\|_X \le \frac{C_4}{\sqrt{n}},$$

where $C_4 = C_3 T^{3/2}$ is independent of n and ψ. Consequently,

$$\|\mathscr{R}^n - \mathscr{R}\|_{\mathscr{L}(X)} = \sup_{\|\psi\|_X=1} \|\mathscr{R}^n\psi - \mathscr{R}\psi\|_X \xrightarrow{n\to\infty} 0$$

which gives the claim. □

Now we follow [40, Section 3.2]. We suppose that $y^k \in H^1(0,T;X)$ for $1 \le k \le \wp$. Thus, $y^k \in C([0,T];X)$, which implies that

$$\sum_{k=1}^{\wp}\sum_{j=1}^{n} \alpha_j^n \|y^k(t_j)\|_X^2 \to \sum_{k=1}^{\wp}\int_0^T \|y^k(t)\|_X^2\,\mathrm{d}t \quad \text{as } n \to \infty. \tag{1.38}$$

Combining (1.38) with (1.8) and (1.29), we find

$$\sum_{i=1}^{d^n} \bar\lambda_i^n \to \sum_{i=1}^{d} \bar\lambda_i \quad \text{as } n \to \infty. \tag{1.39}$$

Now choose and fix

$$\ell \quad \text{such that} \quad \bar\lambda_\ell \ne \bar\lambda_{\ell+1}. \tag{1.40}$$

Then, by spectral analysis of compact operators and Lemma 1.18, it follows that

$$\bar\lambda_i^n \to \bar\lambda_i \quad \text{for } 1 \le i \le \ell \text{ as } n \to \infty. \tag{1.41}$$

Combining (1.39) and (1.41), we derive

$$\sum_{i=\ell+1}^{d^n} \bar\lambda_i^n \to \sum_{i=\ell+1}^{d} \bar\lambda_i \quad \text{as } n \to \infty.$$

In particular, if $\lambda_1 > \lambda_2 > \cdots > \lambda_\ell$ is satisfied, we conclude from (1.40) and Lemma 1.18 that $\lim_{n\to\infty} \|\bar\psi_i^n - \bar\psi_i\|_X = 0$ for $i = 1,\dots,\ell$. In summary, the following theorem has been shown.

Theorem 1.19. *Let X be a separable real Hilbert space, the weighting parameters $\{\alpha_j^n\}_{j=1}^n$ be given by* (1.31), *and $y^k \in H^1(0,T;X)$ for $1 \le k \le \wp$. Let $\{(\bar\psi_i^n, \bar\lambda_i^n)\}_{i\in\mathscr{I}}$ and $\{(\bar\psi_i, \bar\lambda_i)\}_{i\in\mathscr{I}}$ be eigenvector-eigenvalue pairs satisfying* (1.5) *and* (1.26), *respectively. Suppose that $\ell \in \mathbb{N}$ is fixed such that* (1.40) *holds. Then, we have*

$$\lim_{n\to\infty} |\bar\lambda_i^n - \bar\lambda_i| = 0 \quad \text{for } 1 \le i \le \ell$$

and

$$\lim_{n\to\infty} \sum_{i=\ell+1}^{d^n} \bar\lambda_i^n = \sum_{i=\ell+1}^{d} \bar\lambda_i.$$

In particular, if $\lambda_1 > \lambda_2 > \cdots > \lambda_\ell$, then we even have

$$\lim_{n\to\infty} \|\bar{\psi}_i^n - \bar{\psi}_i\|_X = 0 \qquad \text{for } 1 \le i \le \ell.$$

Remark 1.20. Theorem 1.19 gives an answer to the two questions posed at the beginning of Section 1.2.2: the time instances $\{t_j\}_{j=1}^n$ and the associated positive weights $\{\alpha_j^n\}_{j=1}^n$ should be chosen such that $\mathscr{R}^n$ is a quadrature approximation of $\mathscr{R}$ and $\|\mathscr{R}^n - \mathscr{R}\|_{\mathscr{L}(X)}$ is small (for reasonable n). A different strategy is applied in [42], where the time instances $\{t_j\}_{j=1}^n$ are chosen by an optimization approach. Clearly, other choices for the weights $\{\alpha_j^n\}_{j=1}^n$ are also possible, provided (1.30) is guaranteed. For instance, we can choose the Simpson weights.

1.3 ▪ Reduced-order modeling for evolution problems

In this section we utilize the POD method to derive low-dimensional models for evolution problems. For that purpose the POD basis of rank ℓ serves as test and ansatz functions in a POD Galerkin approximation. The a priori error of the POD Galerkin scheme is investigated. It turns out that the resulting error bounds depend on the number of POD basis functions. We refer the reader to, e.g., [20, 22, 30, 38–40, 62] and [32], where POD Galerkin schemes for parabolic equations and elliptic equations are studied. Moreover, we would like to mention the recent papers [8] and [66], where improved rate of convergence results are derived.

1.3.1 ▪ The abstract evolution problem

In this subsection, we introduce the abstract evolution problem that we will consider in Sections 1.3 and 1.4. Let V and H be real, separable Hilbert spaces, and suppose that V is dense in H with compact embedding. By $\langle\cdot,\cdot\rangle_H$ and $\langle\cdot,\cdot\rangle_V$ we denote the inner products in H and V, respectively. In particular, there exists a constant $c_V > 0$ such that

$$\|\varphi\|_H \le c_V \|\varphi\|_V \quad \text{for all } \varphi \in V. \tag{1.42}$$

Let $T > 0$ be the final time. For $t \in [0,T]$ we define a time-dependent symmetric bilinear form $a(t;\cdot,\cdot) : V \times V \to \mathbb{R}$ satisfying

$$|a(t;\varphi,\phi)| \le \gamma \|\varphi\|_V \|\phi\|_V \qquad \forall \varphi \in V \text{ a.e. in } [0,T], \tag{1.43a}$$

$$a(t;\varphi,\varphi) \ge \gamma_1 \|\varphi\|_V^2 - \gamma_2 \|\varphi\|_H^2 \qquad \forall \varphi \in V \text{ a.e. in } [0,T] \tag{1.43b}$$

for constants $\gamma, \gamma_1 > 0$ and $\gamma_2 \ge 0$ that do not depend on t. Identifying H with its dual H', it follows that $V \hookrightarrow H = H' \hookrightarrow V'$, with each embedding being continuous and dense. Here, V' denotes the dual space of V. Recall that the function space (see [10, pp. 472–479] and [70, pp. 146–148], for instance)

$$W(0,T) = \left\{\varphi \in L^2(0,T;V) \,\middle|\, \varphi_t \in L^2(0,T;V')\right\}$$

is a Hilbert space endowed with the inner product

$$\langle \varphi, \phi \rangle_{W(0,T)} = \int_0^T \langle \varphi(t), \phi(t) \rangle_V + \langle \varphi_t(t), \phi_t(t) \rangle_{V'} \,\mathrm{d}t \text{ for } \varphi, \phi \in W(0,T)$$

and the induced norm $\|\varphi\|_{W(0,T)} = \langle \varphi, \varphi \rangle_{W(0,T)}^{1/2}$. Furthermore, $W(0,T)$ is continuously embedded into the space $C([0,T];H)$. Hence, $\varphi(0)$ and $\varphi(T)$ are meaningful in H for an element $\varphi \in W(0,T)$. The integration by parts formula reads

$$\int_0^T \langle \varphi_t(t), \phi(t) \rangle_{V',V} \,\mathrm{d}t + \int_0^T \langle \phi_t(t), \varphi(t) \rangle_{V',V} \,\mathrm{d}t = \frac{\mathrm{d}}{\mathrm{d}t} \int_0^T \langle \varphi(t), \psi(t) \rangle_H \,\mathrm{d}t$$
$$= \varphi(T)\phi(T) - \varphi(0)\phi(0)$$

for φ, $\phi \in W(0,T)$, where $\langle \cdot, \cdot \rangle_{V',V}$ stands for the dual pairing between V and its dual space V'. Moreover, we have the formula

$$\langle \varphi_t(t), \phi \rangle_{V',V} = \frac{\mathrm{d}}{\mathrm{d}t} \langle \varphi(t), \phi \rangle_H \quad \text{for } (\varphi, \phi) \in W(0,T) \times V \text{ and f.a.a. } t \in [0,T].$$

Since we will consider optimal control problems in Section 1.4, we introduce the evolution problem with an input term here. We suppose that for $N_u \in \mathbb{N}$ the input space $U = L^2(0,T;\mathbb{R}^{N_u})$ is chosen. In particular, we identify U with its dual space U'. For $u \in U$, $y_\circ \in H$, and $f \in L^2(0,T;V')$, we consider the linear evolution problem

$$\begin{aligned} \frac{\mathrm{d}}{\mathrm{d}t} \langle y(t), \varphi \rangle_H + a(t; y(t), \varphi) &= \langle (f + \mathcal{B}u)(t), \varphi \rangle_{V',V} \\ &\qquad \forall \varphi \in V \text{ a.e. in } (0,T], \\ \langle y(0), \varphi \rangle_H = \langle y_\circ, \varphi \rangle_H &\qquad \forall \varphi \in H, \end{aligned} \tag{1.44}$$

where $\mathcal{B} : U \to L^2(0,T;V')$ is a continuous linear (control or input) operator.

Remark 1.21. Notice that the techniques presented in this chapter can be adapted for problems where the input space U is given by $L^2(0,T;L^2(\mathcal{D}))$ for some open and bounded domain $\mathcal{D} \subset \mathbb{R}^{\tilde{N}_u}$ for an $\tilde{N}_u \in \mathbb{N}$.

Theorem 1.22. *For $t \in [0,T]$ let $a(t;\cdot,\cdot) : V \times V \to \mathbb{R}$ be a time-dependent symmetric bilinear form satisfying* (1.43). *Then, for every $u \in U$, $f \in L^2(0,T;V')$, and $y_\circ \in H$, there is a unique weak solution $y \in W(0,T)$ satisfying* (1.44) *and*

$$\|y\|_{W(0,T)} \le C \left(\|y_\circ\|_H + \|f\|_{L^2(0,T;V')} + \|u\|_U \right) \tag{1.45}$$

for a constant $C > 0$ that is independent of u, $y_\circ$, and f. If $f \in L^2(0,T;H)$, $a(t;\cdot,\cdot) = a(\cdot,\cdot)$ (independent of t), and $y_\circ \in V$, we even have $y \in L^\infty(0,T;V) \cap H^1(0,T;H)$. Here, $L^\infty(0,T;V)$ stands for the Banach space of all measurable functions $\varphi : [0,T] \to V$ with $\operatorname{esssup}_{t\in[0,T]} \|\varphi(t)\|_V < \infty$ (see [70, p. 143], *for instance).*

Proof. For a proof of the existence of a unique solution we refer to [10, pp. 512–520]. The a priori error estimate follows from standard variational techniques and energy estimates. The regularity result follows from [10, pp. 532–533] and [17, pp. 360–364]. ☐

Remark 1.23. We split the solution to (1.44) into two parts, one that depends on the fixed initial condition $y_\circ$ and right-hand side f, and the other depending linearly on the input variable u. Let $\hat{y} \in W(0,T)$ be the unique solution to

$$\begin{aligned} \frac{\mathrm{d}}{\mathrm{d}t}\langle \hat{y}(t),\varphi\rangle_H + a(t;\hat{y}(t),\varphi) &= \langle f(t),\varphi\rangle_{V',V} && \forall \varphi \in V \text{ a.e. in } (0,T],\\ \hat{y}(0) &= y_\circ && \text{in } H. \end{aligned}$$

We define the subspace

$$W_0(0,T) = \big\{\varphi \in W(0,T)\,\big|\, \varphi(0) = 0 \text{ in } H\big\}$$

endowed with the topology of $W(0,T)$. Let us now introduce the linear solution operator $\mathscr{S} : U \to W_0(0,T)$: for $u \in U$ the function $y = \mathscr{S}u \in W_0(0,T)$ is the unique solution to

$$\frac{\mathrm{d}}{\mathrm{d}t}\langle y(t),\varphi\rangle_H + a(t;y(t),\varphi) = \langle (\mathscr{B}u)(t),\varphi\rangle_{V',V} \quad \forall \varphi \in V \text{ a.e. in } (0,T].$$

From $y \in W_0(0,T)$ we infer $y(0) = 0$ in H. The boundedness of $\mathscr{S}$ follows from (1.45). Now, the solution to (1.44) can be expressed as $y = \hat{y} + \mathscr{S}u$.

1.3.2 ▪ The POD method for the evolution problem

Let $u \in U$, $f \in L^2(0,T;V')$, and $y_\circ \in H$ be given and $y = \hat{y} + \mathscr{S}u$. To keep the notation simple we apply only a spatial discretization with POD basis functions, but no time integration by, e.g., the implicit Euler method. Therefore, we utilize the continuous version of the POD method introduced in Section 1.2.2. In this section we distinguish two choices for X: $X = H$ and $X = V$. It turns out that the choice for X leads to different rate of convergence results. We suppose that the snapshots y^k, $k = 1,\ldots,\wp$, belong to $L^2(0,T;V)$. Later, we will present different rate of convergence results for appropriate choices of the y^k's. Let us introduce the following notation:

$$\begin{aligned} \mathscr{R}_V\psi &= \sum_{k=1}^{\wp}\int_0^T \langle \psi, y^k(t)\rangle_V\, y^k(t)\,\mathrm{d}t && \text{for } \psi \in V,\\ \mathscr{R}_H\psi &= \sum_{k=1}^{\wp}\int_0^T \langle \psi, y^k(t)\rangle_H\, y^k(t)\,\mathrm{d}t && \text{for } \psi \in H. \end{aligned} \tag{1.46}$$

Moreover, we set $\mathscr{K}_V = \mathscr{R}_V^\star$ and $\mathscr{K}_H = \mathscr{R}_H^\star$. In Remark 1.17 we introduced the SVD of the operator $\mathscr{Y}$ defined by (1.25). To distinguish the two choices for the Hilbert space X we denote by the sequence $\{(\sigma_i^V,\psi_i^V,\phi_i^V)\}_{i\in\mathscr{I}}^{\ell} \subset \mathbb{R}_0^+ \times V \times L^2(0,T;\mathbb{R}^\wp)$ of triples the SVD for $X = V$, i.e., we have that

$$\mathscr{R}_V\psi_i^V = \lambda_i^V\psi_i^V, \quad \mathscr{K}_V\phi_i^V = \lambda_i^V\phi_i^V, \quad \sigma_i^V = \sqrt{\lambda_i^V}, \quad i \in \mathscr{I}.$$

Furthermore, let the sequence $\{(\sigma_i^H,\psi_i^H,\phi_i^H)\}_{i\in\mathscr{I}}^{\ell} \subset \mathbb{R}_0^+ \times H \times L^2(0,T;\mathbb{R}^\wp)$ satisfy

$$\mathscr{R}_H\psi_i^H = \lambda_i^H\psi_i^H, \quad \mathscr{K}_H\phi_i^H = \lambda_i^H\phi_i^H, \quad \sigma_i^H = \sqrt{\lambda_i^H}, \quad i \in \mathscr{I}. \tag{1.47}$$

The relationship between the singular values σ_i^H and σ_i^V is investigated in the next lemma, which is taken from [66].

Lemma 1.24. *Suppose that the snapshots $y^k \in L^2(0,T;V)$, $k = 1,\ldots,\wp$. Then we have the following:*

1. *For all $i \in \mathcal{I}$ with $\sigma_i^H > 0$, $\psi_i^H \in V$.*
2. *$\sigma_i^V = 0$ for all $i > d$ with some $d \in \mathbb{N}$ if and only if $\sigma_i^H = 0$ for all $i > d$, i.e., we have $d_H = d_V$ if the rank of $\mathcal{R}_V$ is finite.*
3. *$\sigma_i^V > 0$ for all $i \in \mathcal{I}$ if and only if $\sigma_i^H > 0$ for all $i \in \mathcal{I}$.*

Proof. We argue similarly as in the proof of Lemma 3.1 in [66].

1. Let $\sigma_i^H > 0$. Then, it follows that $\lambda_i^H > 0$. We infer from $y^k \in L^2(0,T;V)$ that $\mathcal{R}_H\psi \in V$ for any $\psi \in H$. Hence, we infer from (1.47) that $\psi_i^H = \mathcal{R}_H\psi_i^H/\lambda_i^H \in V$.
2. Assume that $\sigma_i^V = 0$ for all $i > d$ with some $d \in \mathbb{N}$. Then, we deduce from (1.27) that
$$y^k(t) = \sum_{i=1}^{d} \langle y^k(t), \psi_i^V\rangle_V \psi_i^V \quad \text{for every } k = 1,\ldots,\wp. \tag{1.48}$$
From
$$\begin{aligned}\mathcal{R}_H\psi_j^H &= \sum_{k=1}^{\wp}\int_0^T \langle \psi_j^H, y^k(t)\rangle_H y^k(t)\,\mathrm{d}t\\ &= \sum_{i=1}^{d}\Big(\sum_{k=1}^{\wp}\int_0^T \langle \psi_j^H, y^k(t)\rangle_H \langle y^k(t),\psi_i^V\rangle_V \,\mathrm{d}t\Big)\psi_i^V, \quad j\in\mathcal{I},\end{aligned}$$
we conclude that the range of $\mathcal{R}_H$ is at most d-dimensional, which implies that $\lambda_i^H = 0$ for all $i > d$. Analogously, we deduce from $\sigma_i^H = 0$ for all $i > d$ that the range of $\mathcal{R}_V$ is at most d.
3. The claim follows directly from part 2.

Thus, Lemma 1.24 is proved. □

Let us define the two POD subspaces
$$V^\ell = \operatorname{span}\{\psi_1^V,\ldots,\psi_\ell^V\} \subset V, \quad H^\ell = \operatorname{span}\{\psi_1^H,\ldots,\psi_\ell^H\} \subset V \subset H,$$
where $H^\ell \subset V$ follows from part 1 of Lemma 1.24. Moreover, we introduce the orthogonal projection operators $\mathcal{P}_H^\ell : V \to H^\ell \subset V$ and $\mathcal{P}_V^\ell : V \to V^\ell \subset V$ as follows:
$$\begin{aligned} v^\ell &= \mathcal{P}_H^\ell\varphi \text{ for any } \varphi\in V \quad \text{iff } v^\ell \text{ solves } \min_{w^\ell\in H^\ell}\|\varphi - w^\ell\|_V,\\ v^\ell &= \mathcal{P}_V^\ell\varphi \text{ for any } \varphi\in V \quad \text{iff } v^\ell \text{ solves } \min_{w^\ell\in V^\ell}\|\varphi - w^\ell\|_V.\end{aligned} \tag{1.49}$$
It follows from the first-order optimality conditions that $v^\ell = \mathcal{P}_H^\ell\varphi$ satisfies
$$\langle v^\ell, \psi_i^H\rangle_V = \langle \varphi, \psi_i^H\rangle_V, \quad 1\le i\le \ell. \tag{1.50}$$
Writing $v^\ell \in H^\ell$ in the form $v^\ell = \sum_{j=1}^\ell \mathrm{v}_j^\ell \psi_j^H$, we derive from (1.50) that the vector $\mathrm{v}^\ell = (\mathrm{v}_1^\ell,\ldots,\mathrm{v}_\ell^\ell)^\top \in \mathbb{R}^\ell$ satisfies the linear system
$$\sum_{j=1}^{\ell} \langle \psi_j^H, \psi_i^H\rangle_V \mathrm{v}_j^\ell = \langle \varphi, \psi_i^H\rangle_V, \quad 1\le i\le\ell.$$

For the operator $\mathscr{P}_V^\ell$ we have the explicit representation

$$\mathscr{P}_V^\ell \varphi = \sum_{i=1}^{\ell} \langle \varphi, \psi_i^V \rangle_V \, \psi_i^V \quad \text{for } \varphi \in V.$$

Since the linear operators $\mathscr{P}_V^\ell$ and $\mathscr{P}_H^\ell$ are orthogonal projections, we have $\|\mathscr{P}_V^\ell\|_{\mathscr{L}(V)} = \|\mathscr{P}_H^\ell\|_{\mathscr{L}(V)} = 1$. As $\{\psi_i^V\}_{i\in\mathscr{I}}$ is a complete orthonormal basis in V, we have

$$\lim_{\ell\to\infty} \int_0^T \|w(t) - \mathscr{P}_V^\ell w(t)\|_V^2 \, \mathrm{d}t = 0 \quad \text{for all } w \in L^2(0,T;V). \tag{1.51}$$

Next we review an essential result from [66, Theorem 5.2], which we will use in our a priori error analysis for the choice $X = H$. Recall that $\psi_i^H \in V$ for $1 \le i \le d_H$ and that the image of $\mathscr{P}_H^\ell$ belongs to V. Consequently, $\|\psi_i^H - \mathscr{P}_H^\ell \psi_i^H\|_V$ is well defined for $1 \le i \le d_H$.

Theorem 1.25. *Suppose that $y^k \in L^2(0,T;V)$ for $1 \le k \le \wp$. Then,*

$$\sum_{k=1}^{\wp} \int_0^T \|y^k(t) - \mathscr{P}_H^\ell y^k(t)\|_V^2 \, \mathrm{d}t = \sum_{i=\ell+1}^{d_H} \lambda_i^H \, \|\psi_i^H - \mathscr{P}_H^\ell \psi_i^H\|_V^2. \tag{1.52}$$

Here, d_H is the rank of the operator $\mathscr{R}_H$, which may be infinite. Moreover, $\mathscr{P}_H^\ell y^k$ converges to y^k in $L^2(0,T;V)$ as ℓ tends to ∞ for each $k \in \{1,\dots,\wp\}$.

Proof. Suppose that $1 \le \ell \le d_H$ and $1 \le \ell_\circ < \infty$. Then, $\lambda_i^H > 0$ for $1 \le i \le \ell$. Let $\mathscr{I} \in \mathscr{L}(V)$ denote the identity operator. As $\mathscr{I} - \mathscr{P}_H^\ell$ is an orthonormal projection on V, we conclude that $\|\mathscr{I} - \mathscr{P}_H^\ell\|_{\mathscr{L}(V)} = 1$. Furthermore, $y^k \in L^2(0,T;V)$ for each $k \in \{1,\dots,\wp\}$. Thus, (1.51) implies that $\mathscr{P}_V^{\ell_\circ} y^k \to y^k$ in $L^2(0,T;V)$ as $\ell_\circ \to \infty$ for each k. The proof of (1.52) is essentially based on Hilbert–Schmidt theory and on the following result [66, Lemma 5.1]:

$$\begin{aligned}
&\sum_{k=1}^{\wp} \int_0^T \|(\mathscr{I} - \mathscr{P}_H^\ell)\mathscr{P}_V^{\ell_\circ} y^k(t))\|_V^2 \, \mathrm{d}t \\
&\quad = \sum_{i=1}^{\ell_\circ} \lambda_i^V \, \|\psi_i^V - \mathscr{P}_H^\ell \psi_i^V\|_V^2 \le \sum_{i:\lambda_i^V>0} \lambda_i^V \, \|\psi_i^V - \mathscr{P}_H^\ell \psi_i^V\|_V^2 < \infty
\end{aligned} \tag{1.53}$$

for any $\ell_\circ \in \mathbb{N}$. To prove that $\mathscr{P}_H^\ell y^k$ converges to y^k in $L^2(0,T;V)$ as ℓ tends to ∞ for each $k \in \{1,\dots,\wp\}$, we observe that

$$\begin{aligned}
\sum_{i=\ell+1}^{d_H} \lambda_i^H \, \|\psi_i^H - \mathscr{P}_H^\ell \psi_i^H\|_V^2 &\le \sum_{i=\ell+1}^{d_H} \lambda_i^H \, \|\mathscr{I} - \mathscr{P}_H^\ell\|_{\mathscr{L}(V)} \|\psi_i^H\|_V^2 \\
&= \sum_{i=\ell+1}^{d_H} \lambda_i^H \, \|\psi_i^H\|_V^2.
\end{aligned}$$

Utilizing SVD (see Remark 1.17), it is shown in [66, Theorem 5.2] that $\sum_{i=\ell+1}^{d_H} \lambda_i^H \, \|\psi_i^H\|_V^2 < \infty$. Therefore,

$$\lim_{\ell_\circ\to\infty} \sum_{k=1}^{\wp} \int_0^T \|(\mathscr{I} - \mathscr{P}_H^\ell)\mathscr{P}_V^{\ell_\circ} y^k(t))\|_V^2 \, \mathrm{d}t = 0,$$

which gives the claim. $\square$

We will also need the following result, which follows from the continuous embedding $V \hookrightarrow H$. For a proof we refer to [66, Proposition 5.5].

Lemma 1.26. *Let $y^k \in L^2(0,T;V)$ for each $k \in \{1,\dots,\wp\}$ and $\lambda_i^H > 0$ for all $i \in \mathcal{I}$. Then,*

$$\lim_{\ell\to\infty} \|\varphi - \mathcal{P}_H^\ell \varphi\|_V = 0 \quad \text{for all } \varphi \in V.$$

1.3.3 ▪ The POD Galerkin approximation

After computing a POD basis of rank ℓ, we are interested in deriving a low-dimensional approximation for the evolution problem (1.44). In the context of Section 1.2.2, we choose $\wp = 1$, $y^1 = \mathcal{S}u$ and compute a POD basis $\{\psi_i\}_{i=1}^\ell$ of rank ℓ by solving $(\mathbf{P}^\ell)$ with $\psi_i = \psi_i^V$ for $X = V$ and $\psi_i = \psi_i^H$ for $X = H$. Then, we define the subspace $X^\ell = \operatorname{span}\{\psi_1,\dots,\psi_\ell\}$, i.e., $X^\ell = V^\ell$ for $X = V$ and $X^\ell = H^\ell$ for $X = H$. Now we approximate the state variable y by the Galerkin expansion

$$y^\ell(t) = \hat{y}(t) + \sum_{i=1}^\ell \mathrm{y}_i^\ell(t)\psi_i \in V \quad \text{a.e. in } [0,T] \tag{1.54}$$

with coefficient functions $\mathrm{y}_i^\ell : [0,T] \to \mathbb{R}$. We introduce the vector-valued coefficient function

$$\mathrm{y}^\ell = \left(\mathrm{y}_1^\ell,\dots,\mathrm{y}_\ell^\ell\right) : [0,T] \to \mathbb{R}^\ell.$$

Since $\hat{y}(0) = y_\circ$, we suppose that $\mathrm{y}^\ell(0) = 0$. Then, $y^\ell(0) = y_\circ$, i.e., the POD state matches the initial condition exactly. Inserting (1.54) into (1.44) and using the test space in V^ℓ for $1 \le i \le \ell$, we obtain the following POD Galerkin scheme for (1.44): $y^\ell \in W(0,T)$ solves

$$\begin{aligned} &\frac{\mathrm{d}}{\mathrm{d}t}\langle y^\ell(t),\psi\rangle_H + a(t;y^\ell(t),\psi) = \langle (f+\mathcal{B}u)(t),\psi\rangle_{V',V} \quad \forall \psi \in X^\ell \text{ a.e.,} \\ &\mathrm{y}^\ell(0) = 0. \end{aligned} \tag{1.55}$$

We call (1.55) a *low-dimensional* or *reduced-order model* for (1.44).

Proposition 1.27. *Let all assumptions of Theorem 1.22 be satisfied and the POD basis of rank ℓ be computed as described at the beginning of Section 1.3.2. Then, there exists a unique solution $y^\ell \in H^1(0,T;V) \hookrightarrow W(0,T)$ solving (1.55).*

Proof. Choosing $\psi = \psi_i$, $1 \le i \le \ell$, and applying (1.54) we infer from (1.55) that the coefficient vector y^ℓ satisfies

$$\mathrm{M}^\ell \dot{\mathrm{y}}^\ell(t) + \mathrm{A}^\ell(t)\mathrm{y}(t) = \hat{\mathrm{F}}^\ell(t) \text{ a.e. in } [0,T], \quad \mathrm{y}^\ell(0) = 0, \tag{1.56}$$

where we have set

$$\begin{aligned} &\mathrm{M}^\ell = \left(\left(\langle \psi_i,\psi_j\rangle_H\right)\right) \in \mathbb{R}^{\ell\times\ell}, \quad \mathrm{A}^\ell(t) = \left(\left(a(t;\psi_i,\psi_j)\right)\right) \in \mathbb{R}^{\ell\times\ell}, \\ &\hat{\mathrm{F}}^\ell(t) = \left(\langle (f+\mathcal{B}u)(t) - \hat{y}_t(t),\psi_i\rangle_{V',V} - a(t;\hat{y}(t),\psi_i)\right) \in \mathbb{R}^\ell, \end{aligned} \tag{1.57}$$

with $\psi_i = \psi_i^V$ for $X = V$ and $\psi_i = \psi_i^H$ for $X = H$. Since (1.56) is a linear ODE system, the existence of a unique $\mathrm{y}^\ell \in H^1(0,T;\mathbb{R}^\ell)$ follows by standard arguments. □

Remark 1.28.

1. Suppose $\hat{y} \neq 0$. In contrast to [28, 71], for instance, the POD approximation does admit values $y^\ell(t)$ in X^ℓ, but $(y^\ell - \hat{y})(t) \in X^\ell$. The benefit of this approach is that $y^\ell(0) = y_\circ$—and not $y^\ell(0) = \mathscr{P}_H^\ell y_\circ$ or $y^\ell(0) = \mathscr{P}_V^\ell y_\circ$. This improves the approximation quality of the POD basis, which is illustrated in our numerical tests.

2. We proceed analogously to Remark 1.23 and introduce the linear and bounded solution operator $\mathscr{S}^\ell : U \to W_0(0,T)$: for $u \in U$ the function $w^\ell = \mathscr{S}^\ell u \in W(0,T)$ satisfies $w^\ell(0) = 0$ and

$$\frac{\mathrm{d}}{\mathrm{d}t}\langle w^\ell(t), \psi\rangle_H + a(t; w^\ell(t), \psi) = \langle(\mathscr{B}u)(t), \psi\rangle_{V',V} \quad \forall \psi \in X^\ell \text{ a.e.}$$

Then, the solution to (1.55) is given by $y^\ell = \hat{y} + \mathscr{S}^\ell u$. Analogous to the proof of (1.45), we derive that there exists a positive constant C_2 that does not depend on ℓ or u so that

$$\|\mathscr{S}^\ell u\|_{W(0,T)} \leq C\,\|u\|_U.$$

Thus, $\mathscr{S}^\ell$ is bounded uniformly with respect to ℓ.

To investigate the convergence of the error $y - y^\ell$, we make use of the following two inequalities:

1. *Gronwall's inequality*: For $T > 0$ let $v : [0,T] \to \mathbb{R}$ be a nonnegative differentiable function satisfying

$$v'(t) \leq \varphi(t)v(t) + \chi(t) \quad \text{for all } t \in [0,T],$$

where φ and χ are real-valued, nonnegative, integrable functions on $[0,T]$. Then,

$$v(t) \leq \exp\left(\int_0^t \varphi(s)\,\mathrm{d}s\right)\left(v(0) + \int_0^t \chi(s)\,\mathrm{d}s\right) \quad \text{for all } t \in [0,T]. \tag{1.58}$$

In particular, if

$$v' \leq \varphi v \text{ in } [0,T] \quad \text{and} \quad v(0) = 0,$$

then $v = 0$ in $[0,T]$.

2. *Young's inequality*: For every $a, b \in \mathbb{R}$ and for every $\varepsilon > 0$ we have

$$ab \leq \frac{\varepsilon a^2}{2} + \frac{b^2}{2\varepsilon}.$$

Theorem 1.29. *Let $u \in U$ be chosen arbitrarily so that $0 \neq \mathscr{S}u \in H^1(0,T;V)$.*

1. *To compute a POD basis $\{\psi_i\}_{i=1}^\ell$ of rank ℓ we choose $\wp = 1$ and $y^1 = \mathscr{S}u$. Then,*

$y = \hat{y} + \mathscr{S} u$ and $y^\ell = \hat{y} + \mathscr{S}^\ell u$ satisfy the a priori error estimate

$$\|y^\ell - y\|^2_{H^1(0,T;V)} \le C_1 \cdot \begin{cases} \displaystyle\sum_{i=\ell+1}^{d_V} \lambda_i^V + \|y_t^1 - \mathscr{P}_V^\ell y_t^1\|^2_{L^2(0,T;V)} & \text{if } X = V, \\ \displaystyle\sum_{i=\ell+1}^{d_H} \lambda_i^H \|\psi_i^H - \mathscr{P}_H^\ell \psi_i^H\|_V^2 \\ \qquad + \|y_t^1 - \mathscr{P}_H^\ell y_t^1\|^2_{L^2(0,T;V)} & \text{if } X = H, \end{cases} \tag{1.59}$$

where the constant C_1 depends on the terminal time T and the constants γ, γ_1, γ_2 introduced in (1.43).

2. *If we set $\wp = 2$ and compute a POD basis of rank ℓ using the trajectories $y^1 = \mathscr{S} u$ and $y^2 = (\mathscr{S} u)_t$, it follows that*

$$\|y^\ell - y\|^2_{H^1(0,T;V)} \le C_3 \cdot \begin{cases} \displaystyle\sum_{i=\ell+1}^{d_V} \lambda_i^V & \text{for } X = V, \\ \displaystyle\sum_{i=\ell+1}^{d_H} \lambda_i^H \|\psi_i^H - \mathscr{P}_H^\ell \psi_i^H\|_V^2 & \text{for } X = H \end{cases} \tag{1.60}$$

for a constant C_3 that depends on γ, γ_1, γ_2, and T.

3. *If $\mathscr{S} \tilde{u}$ belongs to $H^1(0,T;V)$ for every $\tilde{u} \in U$ and if $\lambda_i^H > 0$ for all $i \in \mathscr{I}$, then we have*

$$\lim_{\ell \to \infty} \|\mathscr{S} - \mathscr{S}^\ell\|_{\mathscr{L}(U,W(0,T))} = 0. \tag{1.61}$$

Proof.

1. For almost all $t \in [0,T]$ we make use of the decomposition

$$\begin{aligned} y^\ell(t) - y(t) &= \hat{y}(t) + (\mathscr{S}^\ell u)(t) - \hat{y}(t) - (\mathscr{S} u)(t) \\ &= (\mathscr{S}^\ell u)(t) - \mathscr{P}^\ell\big((\mathscr{S} u)(t)\big) + \mathscr{P}^\ell\big((\mathscr{S} u)(t)\big) - (\mathscr{S} u)(t) \\ &= \vartheta^\ell(t) + \varrho^\ell(t), \end{aligned} \tag{1.62}$$

where $\vartheta^\ell = \mathscr{S}^\ell u - \mathscr{P}^\ell(\mathscr{S} u) \in X^\ell$ and $\varrho^\ell = \mathscr{P}^\ell(\mathscr{S} u) - \mathscr{S} u$. In (1.62) we will consider the two choices $\mathscr{P}^\ell = \mathscr{P}_H^\ell$ for $X = H$ and $\mathscr{P}^\ell = \mathscr{P}_V^\ell$ for $X = V$. From $y^1 = \mathscr{S} u$ and (1.27) we infer that

$$\|\varrho^\ell\|^2_{H^1(0,T;V)} = \sum_{i=\ell+1}^{d_V} \lambda_i^V + \|y_t^1 - \mathscr{P}_V^\ell y_t^1\|^2_{L^2(0,T;V)} \tag{1.63}$$

when $X = V$, where d_V stands for the rank of $\mathscr{R}_V$. For the choice $X = H$ we derive from $y^1 = \mathscr{S} u$ and Theorem 1.25 that

$$\|\varrho^\ell\|^2_{H^1(0,T;V)} = \sum_{i=\ell+1}^{d_H} \lambda_i^H \|\psi_i^H - \mathscr{P}_H^\ell \psi_i^H\|_V^2 + \|y_t^1 - \mathscr{P}_H^\ell y_t^1\|^2_{L^2(0,T;V)}. \tag{1.64}$$

Here, d_H denotes the rank of $\mathscr{R}_H$. Using $\vartheta_t^\ell(t) \in H$ for almost all $t \in [0,T]$, (1.44), (1.55), and (1.43a) we derive that

$$\begin{aligned}
&\frac{\mathrm{d}}{\mathrm{d}t}\langle \vartheta^\ell(t), \psi\rangle_H + a(t;\vartheta^\ell(t),\psi) \\
&\quad = \langle y_t^1(t) - \mathscr{P}^\ell y_t^1(t), \psi\rangle_H + a(t; y^1(t) - \mathscr{P}^\ell y^1(t), \psi) \\
&\quad \leq \|y_t^1(t) - \mathscr{P}^\ell y_t^1(t)\|_H \|\psi\|_H + \gamma\, \|y^1(t) - \mathscr{P}^\ell y^1(t)\|_V \|\psi\|_V
\end{aligned} \tag{1.65}$$

for all $\psi \in X^\ell$ and for almost all $t \in [0,T]$. From choosing $\psi = \vartheta^\ell(t)$, (1.43b), and (1.65) we find

$$\begin{aligned}
&\frac{\mathrm{d}}{\mathrm{d}t}\|\vartheta^\ell(t)\|_H^2 + \gamma_1 \|\vartheta^\ell(t)\|_V^2 - 3\gamma_2 \|\vartheta^\ell(t)\|_H^2 \\
&\quad \leq \frac{1}{\gamma_2}\|y_t^1(t) - \mathscr{P}^\ell y_t^1(t)\|_H^2 + \frac{\gamma^2}{\gamma_1}\|y^1(t) - \mathscr{P}^\ell y^1(t)\|_V^2.
\end{aligned}$$

From (1.58)—setting $v(t) = \|\vartheta^\ell(t)\|_H^2 \geq 0$,

$$\chi(t) = \frac{1}{\gamma_2}\|y_t^1(t) - \mathscr{P}^\ell y_t^1(t)\|_H^2 + \frac{\gamma^2}{\gamma_1}\|y^1(t) - \mathscr{P}^\ell y^1(t)\|_V^2 \geq 0,$$

and $\varphi(t) = 3\gamma_2 > 0$—and $\vartheta^\ell(0) = 0$ it follows that

$$\|\vartheta^\ell(t)\|_H^2 \leq c_1 \left(\|y_t^1 - \mathscr{P}^\ell y_t^1\|_{L^2(0,T;H)}^2 + \|y^1 - \mathscr{P}^\ell y^1\|_{L^2(0,T;V)}^2\right)$$

for almost all $t \in [0,T]$, with the constants $c_1 = c_2 \exp(3\gamma_2 T)$ and $c_2 = \max(1/\gamma_2, \gamma^2/\gamma_1)$, so that

$$\begin{aligned}
\|\vartheta^\ell\|_{L^2(0,T;V)}^2 &\leq c_3\Big(\|\vartheta^\ell\|_{L^2(0,T;H)}^2 + \|y_t^1 - \mathscr{P}^\ell y_t^1\|_{L^2(0,T;H)}^2\Big) \\
&\quad + c_3\,\|y^1 - \mathscr{P}^\ell y^1\|_{L^2(0,T;V)}^2 \\
&\leq c_4\Big(\|y_t^1 - \mathscr{P}^\ell y_t^1\|_{L^2(0,T;H)}^2 + \|\varrho^\ell(t)\|_{L^2(0,T;V)}^2\Big),
\end{aligned} \tag{1.66}$$

with $c_3 = \max(3\gamma_2, c_2)/\gamma_1$ and $c_4 = c_3(1 + c_1 T)$. We conclude from (1.43a), (1.59), and (1.66) that

$$\begin{aligned}
&\|\vartheta_t^\ell\|_{L^2(0,T;(V^\ell)')} \\
&\quad = \sup\left\{\int_0^T \langle \vartheta_t^\ell(t), \psi(t)\rangle_{V',V} \,\Big|\, \|\psi\|_{L^2(0,T;V)} = 1,\ \psi(t) \in V^\ell\right\} \\
&\quad \leq \gamma\,\|\vartheta^\ell\|_{L^2(0,T;V)} + \|y_t^1 - \mathscr{P}^\ell y_t^1\|_{L^2(0,T;H)} \\
&\quad \leq c_5\Big(\|y_t^1 - \mathscr{P}^\ell y_t^1\|_{L^2(0,T;H)} + \|y^1 - \mathscr{P}^\ell y^1\|_{L^2(0,T;V)}\Big),
\end{aligned} \tag{1.67}$$

with $c_5 = 1 + c_4\gamma$. By assumption we have $\vartheta_t^\ell \in L^2(0,T;V)$. Then, by the Riesz theorem [60, p. 43] we have

$$\|\vartheta_t^\ell\|_{L^2(0,T;(V^\ell)')} = \|\vartheta_t^\ell\|_{L^2(0,T;V^\ell)} = \|\vartheta_t^\ell\|_{L^2(0,T;V)},$$

so that (1.62)–(1.64), (1.66), and (1.67) imply (1.59).

2. The claim follows directly from

$$\|y_t^1-\mathcal{P}^\ell y_t^1\|^2_{L^2(0,T;V)}=\|y^2-\mathcal{P}^\ell y^2\|^2_{L^2(0,T;V)},$$

(1.27), and Theorem 1.25.

3. Recall that the space $H^1(0,T;V)$ is continuously embedded in $W(0,T)$. Therefore, there is a constant $c_W>0$ satisfying

$$\|\varphi\|_{W(0,T)}\le c_W\,\|\varphi\|_{H^1(0,T;V)}\quad\text{for all } \varphi\in H^1(0,T;V).$$

Using $\mathcal{S}\tilde{u}\in H^1(0,T;V)$ for any $\tilde{u}\in U$, referring to Remark 1.9, and applying the arguments as in part 1, we infer that there exists a constant c_6 that is independent of ℓ satisfying

$$\begin{aligned}
&\|\mathcal{S}-\mathcal{S}^\ell\|_{\mathcal{L}(U,W(0,T))}\\
&\quad=\sup_{\|\tilde{u}\|_U=1}\|(\mathcal{S}-\mathcal{S}^\ell)\tilde{u}\|_{W(0,T)}\le c_W\sup_{\|\tilde{u}\|_U=1}\|(\mathcal{S}-\mathcal{S}^\ell)\tilde{u}\|_{H^1(0,T;V)}\\
&\quad\le c_6\sup_{\|\tilde{u}\|_U=1}\int_0^T\|\tilde{y}(t)-\mathcal{P}^\ell\tilde{y}(t)\|_V^2+\|\tilde{y}_t(t)-\mathcal{P}^\ell\tilde{y}_t(t)\|_V^2\,\mathrm{d}t,
\end{aligned}$$

with $\tilde{y}=\mathcal{S}\tilde{u}$. By assumption, the elements $\tilde{y}(t)$ and $\tilde{y}_t(t)$ belong to the space $L^2(0,T;V)$. Therefore, the claim follows for $X=V$ from (1.51) and for $X=H$ from Lemma 1.26.

Thus, Theorem 1.29 is proved. □

Remark 1.30.

1. Note that the a priori error estimates (1.59) and (1.60) depend on the arbitrarily chosen, but fixed, control $u\in U$, which is also utilized to compute the POD basis. Moreover, these a priori estimates do not involve errors by the POD discretization of the initial condition $y_\circ$—in contrast to the error analysis presented in [28, 38, 40, 62, 71], for instance. Further, let us mention that the a priori error analysis holds for $T<\infty$.

2. From (1.61) we infer

$$\|\hat{y}+\mathcal{S}^\ell\tilde{u}-\hat{y}-\mathcal{S}\tilde{u}\|_{W(0,T)}\le\|\mathcal{S}-\mathcal{S}^\ell\|_{\mathcal{L}(U,W(0,T))}\|\tilde{u}\|_U\overset{\ell\to\infty}{\longrightarrow}0$$

for any $\tilde{u}\in U$.

3. For the numerical realization, we also have to utilize a time integration method such as the implicit Euler or the Crank–Nicolson method. We refer the reader to [38–40], where different time discretization schemes are considered. Moreover, in [45, 62] a finite element discretization of the ansatz space V is incorporated in the a priori error analysis.

Example 1.31. Accurate approximation results are achieved if the subspace spanned by the snapshots is (approximately) of low dimension. Let $T>0$, $\Omega=(0,2)\subset\mathbb{R}$, and $Q=(0,T)\times\Omega$. We set $f(t,x)=e^{-t}(\pi^2-1)\sin(\pi x)$ for $(t,x)\in Q$ and $y_\circ(x)=\sin(\pi x)$ for $x\in\Omega$. Let $H=L^2(\Omega)$, $V=H_0^1(\Omega)$, and

$$a(t;\varphi,\phi)=\int_\Omega \varphi'(x)\phi'(x)\,\mathrm{d}x \quad \text{for } \varphi,\phi\in V,$$

i.e., the bilinear form a is independent of t. Finally, we choose $u=0$. Then, the exact solution to (1.44) is given by $y(t,x)=e^{-t}\sin(\pi x)$. Thus, the snapshot space $\mathscr{V}$ is the one-dimensional space $\{\alpha\psi\mid\alpha\in\mathbb{R}\}$ with $\psi(x)=\sin(\pi x)$. Choosing the space $X=H$, this implies that all eigenvalues of the operator $\mathscr{R}_H$ introduced in (1.46) except the first one are zero and $\psi_1=\psi\in V$ is the single POD element corresponding to a nontrivial eigenvalue of $\mathscr{R}_H$. Further, the ROM of the rank-one POD Galerkin ansatz

$$\begin{aligned}\dot{\mathrm{y}}^1(t)+\|\psi_1'\|_H^2\mathrm{y}^1(t)&=\langle f(t),\psi_1\rangle_H \quad \text{for } t\in(0,T],\\ \mathrm{y}^1(0)&=\langle y_\circ,\psi_1\rangle_H\end{aligned}$$

has the solution $\mathrm{y}^1(t)=e^{-t}$, so both the projection

$$(\mathscr{P}^1y)(t,x)=\langle y(t),\psi_1\rangle_X\psi_1(x),\quad (t,x)\in\overline{Q},$$

of the state y on the POD Galerkin space and the reduced-order solution $y^1(t)=\mathrm{y}^1(t)\psi_1$ coincide with the exact solution y. In the latter case, this is because the data functions f and $y_\circ$ as well as all time derivative snapshots $\dot{y}(t)$ are already elements of $\mathrm{span}(\psi_1)$, so no projection error occurs here—compare the a priori error bounds given in (1.60). In the case $X=V$, we get the same results with $\psi_1(x)=\sin(\pi x)/\sqrt{1+\pi^2}$ and $\mathrm{y}^1(t)=\sqrt{1+\pi^2}e^{-t}$. ■

Utilizing the techniques in the proof of Theorem 6.5 in [66] one can derive an a priori error bound without including the time derivatives in the snapshot subspace. In the next proposition we formulate the a priori error estimate.

Proposition 1.32. *Let $y_\circ\in V$ and $u\in U$ be chosen arbitrarily so that $\mathscr{S}u\neq 0$. To compute a POD basis $\{\psi_i\}_{i=1}^\ell$ of rank ℓ we choose $\wp=1$ and $y^1=\mathscr{S}u$. Then, $y=\hat{y}+\mathscr{S}u$ and $y^\ell=\hat{y}+\mathscr{S}^\ell u$ satisfy the a priori error estimate*

$$\|y^\ell-y\|_{L^2(0,T;V)}^2\le C\cdot\begin{cases}\displaystyle\sum_{i=\ell+1}^{d_V}\lambda_i^V\,\|\psi_i^V-\mathscr{P}_{H,V^\ell}^\ell\psi_i^V\|_V^2 & \text{if } X=V,\\ \displaystyle\sum_{i=\ell+1}^{d_H}\lambda_i^H\,\|\psi_i^H\|_V^2 & \text{if } X=H,\end{cases}\tag{1.68}$$

where the constant C depends on the terminal time T and the constants γ, γ_1, γ_2 introduced in (1.43). *Moreover, $\mathscr{P}_{H,V^\ell}^\ell : H\to V^\ell$ is the H-orthogonal projection given as follows:*

$$v^\ell=\mathscr{P}_{H,V^\ell}^\ell\varphi \text{ for any } \varphi\in H \text{ iff } v^\ell \text{ solves } \min_{w^\ell\in V^\ell}\|\varphi-w^\ell\|_H.$$

In particular, we have $y^\ell\to y$ in $L^2(0,T;V)$ as $\ell\to\infty$.

1.4 ▪ The linear-quadratic optimal control problem

In this section we apply a POD Galerkin approximation to linear-quadratic optimal control problems. Linear-quadratic problems are interesting in several respects: In particular, they occur at each level of sequential quadratic programming (SQP) methods; see, e.g., [52]. In contrast to methods of BT type (see, e.g., Chapter 6), the POD method is somehow lacking a reliable a priori error analysis. Unless its snapshots are generating a sufficiently rich state space, it is not a priori clear how far the optimal solution of the POD problem is from the exact one. On the other hand, the POD method is a universal tool that is also applicable to problems with time-dependent coefficients or to nonlinear equations. By generating snapshots from the real (large) model, a space is constructed that contains the main and relevant physical properties of the state system. This, and its ease of use, make POD very competitive in practical use, despite a certain heuristic.

Here we prove convergence and derive a priori error estimates for the optimal control problem. The error estimates rely on the (unrealistic) assumption that the POD basis is computed from the (exact) optimal solution. However, these estimates are utilized to develop an a posteriori error analysis for the POD Galerkin approximation of the optimal control problem. Using a perturbation method [16], we deduce how far the suboptimal control, computed by the POD Galerkin approximation, is from the (unknown) exact one. This idea turned out to be very efficient in our numerical examples. Thus, we can compensate for the lack of an a priori error analysis for the POD method.

1.4.1 ▪ Problem formulation

In this section we introduce our optimal control problem, which is a constrained optimization problem in a Hilbert space. The objective is a quadratic function. The evolution problem (1.44) serves as an equality constraint. Moreover, bilateral control bounds lead to inequality constraints in the minimization. For the reader's convenience, we recall (1.44) here. Let $U = L^2(0,T;\mathbb{R}^{N_u})$ denote the control space with $N_u \in \mathbb{N}$. For $u \in U$, $y_\circ \in H$, and $f \in L^2(0,T;V')$ we consider the state equation

$$\begin{aligned} \frac{\mathrm{d}}{\mathrm{d}t}\langle y(t),\varphi\rangle_H + a(t;y(t),\varphi) &= \langle (f+\mathscr{B}u)(t),\varphi\rangle_{V',V} \\ &\qquad \forall \varphi\in V \text{ a.e. in } (0,T], \\ \langle y(0),\varphi\rangle_H = \langle y_\circ,\varphi\rangle_H &\qquad \forall\varphi\in H, \end{aligned} \tag{1.69}$$

where $\mathscr{B} : U \to L^2(0,T;V')$ is a continuous linear operator. Due to Theorem 1.22, there exists a unique solution $y \in W(0,T)$ to (1.69).

We introduce the Hilbert space

$$X = W(0,T) \times U$$

endowed with the natural product topology, i.e., with the inner product

$$\langle x,\tilde{x}\rangle_X = \langle y,\tilde{y}\rangle_{W(0,T)} + \langle u,\tilde{u}\rangle_U \quad \text{for } x=(y,u),\ \tilde{x}=(\tilde{y},\tilde{u})\in X$$

and the norm $\|x\|_X = (\|y\|^2_{W(0,T)} + \|u\|^2_U)^{1/2}$ for $x=(y,u)\in X$.

Assumption 1. *For $t \in [0,T]$, let $a(t;\cdot,\cdot) : V \times V \to \mathbb{R}$ be a time-dependent symmetric bilinear form satisfying* (1.43). *Moreover, $f \in L^2(0,T;V')$, $y_\circ \in H$, and $\mathscr{B} \in \mathscr{L}(U,L^2(0,T;V'))$.*

In Remark 1.23 we introduced the particular solution $\hat{y} \in W(0,T)$ as well as the linear, bounded solution operator $\mathscr{S}$. Then, the solution to (1.69) can be expressed as $y = \hat{y} + \mathscr{S}u$. By X_{ad} we denote the closed, convex, and bounded set of admissible solutions for the optimization problem as

$$X_{\mathsf{ad}} = \big\{(\hat{y} + \mathscr{S}u, u) \in X \,\big|\, u_a \le u \le u_b \text{ in } \mathbb{R}^{N_u} \text{ a.e. in } [0,T]\big\},$$

where $u_a = (u_{a,1},\ldots,u_{a,N_u}), u_b = (u_{b,1},\ldots,u_{b,N_u}) \in U$ satisfy $u_{a,i} \le u_{b,i}$ for $1 \le i \le N_u$ a.e. in $[0,T]$. Since $u_{a,i} \le u_{b,i}$ for $1 \le i \le N_u$, we infer from Theorem 1.22 that the set X_{ad} is nonempty.

The quadratic objective $J : X \to \mathbb{R}$ is given by

$$J(x) = \frac{\sigma_Q}{2}\int_0^T \|y(t) - y_Q(t)\|_H^2 \,\mathrm{d}t + \frac{\sigma_\Omega}{2}\|y(T) - y_\Omega\|_H^2 + \frac{\sigma}{2}\|u\|_U^2 \tag{1.70}$$

for $x = (y,u) \in X$, where $(y_Q, y_\Omega) \in L^2(0,T;H) \times H$ are given desired states. Furthermore, σ_Q, $\sigma_\Omega \ge 0$ and $\sigma > 0$. Of course, more general cost functionals can be treated analogously.

Now the quadratic programming problem is given by

$$\min J(x) \quad \text{s.t.} \quad x \in X_{\mathsf{ad}}. \tag{$\mathbf{P}$}$$

From $x = (y,u) \in X_{\mathsf{ad}}$ we infer that $y = \hat{y} + \mathscr{S}u$. Hence, y is a dependent variable. We call u the *control* and y the *state*. In this way, (**P**) becomes an *optimal control problem*. Utilizing the relationship $y = \hat{y} + \mathscr{S}u$, we define a so-called reduced cost functional $\hat{J} : U \to \mathbb{R}$ by

$$\hat{J}(u) = J(\hat{y} + \mathscr{S}u, u) \quad \text{for } u \in U.$$

Moreover, the set of admissible controls is given as

$$U_{\mathsf{ad}} = \big\{u \in U \,\big|\, u_a \le u \le u_b \text{ in } \mathbb{R}^{N_u} \text{ a.e. in } [0,T]\big\},$$

which is convex, closed, and bounded in U. Then, we consider the reduced optimal control problem

$$\min \hat{J}(u) \quad \text{s.t.} \quad u \in U_{\mathsf{ad}}. \tag{$\hat{\mathbf{P}}$}$$

Clearly, if $\bar{u}$ is the optimal solution to ($\hat{\mathbf{P}}$), then $\bar{x} = (\hat{y} + \mathscr{S}\bar{u}, \bar{u})$ is the optimal solution to (**P**). On the other hand, if $\bar{x} = (\bar{y}, \bar{u})$ is the solution to (**P**), then $\bar{u}$ solves ($\hat{\mathbf{P}}$).

Example 1.33. We introduce an example for (**P**) and discuss the presented theory for this application. Let $\Omega \subset \mathbb{R}^d$, $d \in \{1,2,3\}$, be an open and bounded domain with Lipschitz-continuous boundary $\Gamma = \partial\Omega$. For $T > 0$ we set $Q = (0,T) \times \Omega$ and $\Sigma = (0,T) \times \Gamma$. We choose $H = L^2(\Omega)$ and $V = H_0^1(\Omega)$ endowed with the usual inner products,

$$\langle \varphi, \psi \rangle_H = \int_\Omega \varphi\psi \,\mathrm{d}\boldsymbol{x}, \quad \langle \varphi, \psi \rangle_V = \int_\Omega \varphi\psi + \nabla\varphi \cdot \nabla\psi \,\mathrm{d}\boldsymbol{x},$$

and their induced norms, respectively. Let $\chi_i \in H$, $1 \le i \le N_u$, denote given control shape functions. Then, for given control $u \in U$, initial condition $y_\circ \in H$, and inhomogeneity $f \in L^2(0,T;H)$, we consider the linear heat equation

$$\begin{aligned}
y_t(t,\boldsymbol{x}) - \Delta y(t,\boldsymbol{x}) &= f(t,\boldsymbol{x}) + \sum_{i=1}^{N_u} u_i(t)\chi_i(\boldsymbol{x}), && \text{a.e. in } Q,\\
y(t,\boldsymbol{x}) &= 0, && \text{a.e. in } \Sigma,\\
y(0,\boldsymbol{x}) &= y_\circ(\boldsymbol{x}), && \text{a.e. in } \Omega.
\end{aligned} \tag{1.71}$$

We introduce the time-independent symmetric bilinear form

$$a(\varphi,\psi)=\int_\Omega \nabla\varphi\cdot\nabla\psi\,\mathrm{d}x \quad \text{for } \varphi,\psi\in V$$

and the bounded linear operator $\mathscr{B}: U\to L^2(0,T;H)\hookrightarrow L^2(0,T;V')$ as

$$(\mathscr{B}u)(t,x)=\sum_{i=1}^{m}u_i(t)\chi_i(x) \quad \text{for } (t,x)\in Q \text{ a.e. and } u\in U.$$

Hence, we have $\gamma=\gamma_1=\gamma_2=1$ in (1.43). It follows that the weak formulation of (1.71) can be expressed in the form (1.44). Moreover, the unique weak solution to (1.71) belongs to the space $L^\infty(0,T;V)$ provided $y_\circ\in V$. ∎

1.4.2 ▪ Existence of a unique optimal solution

We suppose the following hypothesis for the objective.

Assumption 2. *In* (1.70) *the desired states* (y_Q,y_Ω) *belong to* $L^2(0,T;H)\times H$. *Furthermore,* $\sigma_Q,\sigma_\Omega\geq 0$ *and* $\sigma>0$ *are satisfied.*

Let us review the following result for quadratic optimization problems in Hilbert spaces; see [70, pp. 50–51].

Theorem 1.34. *Suppose that* $\mathscr{U}$ *and* $\mathscr{H}$ *are given Hilbert spaces with norms* $\|\cdot\|_{\mathscr{U}}$ *and* $\|\cdot\|_{\mathscr{H}}$*, respectively. Furthermore, let* $\mathscr{U}_{\mathsf{ad}}\subset\mathscr{U}$ *be nonempty, bounded, closed, and convex and* $z_d\in\mathscr{H}$, $\varkappa\geq 0$. *The mapping* $\mathscr{G}:\mathscr{U}\to\mathscr{H}$ *is assumed to be a linear and continuous operator. Then, there exists an optimal control* $\bar u$ *solving*

$$\min_{u\in\mathscr{U}_{\mathsf{ad}}}\mathscr{J}(u):=\frac{1}{2}\|\mathscr{G}u-z_d\|_{\mathscr{H}}^2+\frac{\varkappa}{2}\|u\|_{\mathscr{U}}^2. \tag{1.72}$$

If $\varkappa>0$ *or if* $\mathscr{G}$ *is injective, then* $\bar u$ *is uniquely determined.*

Remark 1.35. In the proof of Theorem 1.34 we only used the fact that $\mathscr{J}$ is continuous and convex. Therefore, the existence of an optimal control follows for general convex continuous cost functionals $\mathscr{J}:\mathscr{U}\to\mathbb{R}$ with a Hilbert space $\mathscr{U}$.

Next we can use Theorem 1.34 to obtain an existence result for the optimal control problem ($\hat{\mathbf{P}}$), which implies the existence of an optimal solution to (**P**).

Theorem 1.36. *Let Assumptions* 1 *and* 2 *be valid. Moreover, let the bilateral control constraints* $u_a,u_b\in U$ *satisfy* $u_a\leq u_b$ *componentwise in* $\mathbb{R}^{N_u}$ *a.e. in* $[0,T]$. *Then,* ($\hat{\mathbf{P}}$) *has a unique optimal solution* $\bar u$.

Proof. Let us choose the Hilbert spaces $\mathscr{H}=L^2(0,T;H)\times H$ and $\mathscr{U}=U$. Moreover, $\mathscr{E}:W(0,T)\to L^2(0,T;H)$ is the canonical embedding operator, which is linear and bounded. We define the operator $\mathscr{E}_2:W(0,T)\to H$ by $\mathscr{E}_2\varphi=\varphi(T)$ for $\varphi\in W(0,T)$. Since $W(0,T)$ is continuously embedded in $C([0,T];H)$, the linear operator $\mathscr{E}_2$ is continuous. Finally, we set

$$\mathscr{G} = \begin{pmatrix} \sqrt{\sigma_Q}\,\mathscr{E}_1\mathscr{S} \\ \sqrt{\sigma_\Omega}\,\mathscr{E}_2\mathscr{S} \end{pmatrix} \in \mathscr{L}(\mathscr{U},\mathscr{H}), \quad z_d = \begin{pmatrix} \sqrt{\sigma_Q}(y_Q - \hat{y}) \\ \sqrt{\sigma_\Omega}\big(y_\Omega - \hat{y}(T)\big) \end{pmatrix} \in \mathscr{H}, \tag{1.73}$$

and $\mathscr{U}_{\mathsf{ad}} = U_{\mathsf{ad}}$. Then, ($\hat{\mathbf{P}}$) and (1.72) coincide. Consequently, the claim follows from Theorem 1.34 and $\sigma > 0$. □

Next we consider the case that $u_a = -\infty$ and/or $u_b = +\infty$. In this case, U_{ad} is not bounded. However, we have the following result [70, p. 52].

Theorem 1.37. *Let Assumptions* 1 *and* 2 *be satisfied. If* $u_a = -\infty$ *and/or* $u_b = +\infty$, *problem* ($\hat{\mathbf{P}}$) *admits a unique solution.*

Proof. We utilize the setting of the proof of Theorem 1.36. By assumption there exists an element $u_0 \in U_{\mathsf{ad}}$. For $u \in U$ with $\|u\|_U^2 > 2\hat{J}(u_0)/\sigma$ we have

$$\hat{J}(u) = \mathscr{J}(u) = \frac{1}{2}\|\mathscr{G}u - z_d\|_{\mathscr{H}}^2 + \frac{\sigma}{2}\|u\|_U^2 \geq \frac{\sigma}{2}\|u\|_U^2 > \hat{J}(u_0).$$

Thus, the minimization of $\hat{J}$ over U_{ad} is equivalent to the minimization of $\hat{J}$ over the bounded, convex, and closed set

$$U_{\mathsf{ad}} \cap \left\{ u \in U \,\Big|\, \|u\|_U^2 \leq \frac{2\hat{J}(u_0)}{\sigma} \right\}.$$

Now the claim follows from Theorem 1.34. □

1.4.3 ▪ First-order necessary optimality conditions

In (1.72) we introduced the quadratic programming problem

$$\min_{u \in \mathscr{U}_{\mathsf{ad}}} \mathscr{J}(u) = \frac{1}{2}\|\mathscr{G}u - z_d\|_{\mathscr{H}}^2 + \frac{\sigma}{2}\|u\|_{\mathscr{U}}^2. \tag{1.74}$$

Existence of a unique solution was investigated in Section 1.4.2. In this section we characterize the solution to (1.74) by first-order optimality conditions, which are essential to prove convergence and rate of convergence results for the POD approximations in Section 1.4.4. To derive first-order conditions, we require the notion of derivatives in function spaces. Therefore, we recall the following definition [70, pp. 56–57].

Definition 1.38. *Suppose that* $\mathscr{B}_1$ *and* $\mathscr{B}_2$ *are real Banach spaces,* $\mathscr{U} \subset \mathscr{B}_1$ *is an open subset, and* $\mathscr{F} : \mathscr{U} \supset \mathscr{B}_1 \to \mathscr{B}_2$ *is a given mapping. The derivative of* $\mathscr{F}$ *at a point* $u \in \mathscr{U}$ *in the direction* $h \in \mathscr{B}_2$ *is defined by*

$$D\mathscr{F}(u;h) := \lim_{\varepsilon \searrow 0} \frac{1}{\varepsilon}\big(\mathscr{F}(u + \varepsilon h) - \mathscr{F}(u)\big)$$

provided the limit exists in $\mathscr{B}_2$. *Suppose that the directional derivative exists for all* $h \in \mathscr{B}_1$ *and there is a linear continuous operator* $\mathscr{T} : \mathscr{U} \to \mathscr{B}_2$ *satisfying*

$$D\mathscr{F}(u;h) = \mathscr{T}h \quad \textit{for all } h \in \mathscr{U}.$$

Then, $\mathscr{F}$ is said to be Gâteaux-differentiable *at u and $\mathscr{T}$ is the* Gâteaux derivative *of $\mathscr{F}$ at u. We write $\mathscr{T} = \mathscr{F}'(u)$.*

Remark 1.39. Let $\mathscr{H}$ be a real Hilbert space and $\mathscr{F} : \mathscr{H} \to \mathbb{R}$ be *Gâteaux-differentiable* at $u \in \mathscr{H}$. Then, its Gâteaux derivative $\mathscr{F}'(u)$ at u belongs to $\mathscr{H}' = \mathscr{L}(\mathscr{H}, \mathbb{R})$. Due to the Riesz theorem [60, p. 43], there exists a unique element $\nabla\mathscr{F}(u) \in \mathscr{H}$ satisfying

$$\langle \nabla\mathscr{F}(u), v \rangle_{\mathscr{H}} = \langle \mathscr{F}'(u), v \rangle_{\mathscr{H}',\mathscr{H}} \quad \text{for all } v \in \mathscr{H}.$$

We call $\nabla\mathscr{F}(u)$ the *(Gâteaux) gradient* of $\mathscr{F}$ at u.

Theorem 1.40. *Let $\mathscr{U}$ be a real Hilbert space and $\mathscr{U}_{\mathsf{ad}}$ be a convex subset. Suppose that $\bar{u} \in \mathscr{U}_{\mathsf{ad}}$ is a solution to (1.74):*

$$\min_{u \in \mathscr{U}_{\mathsf{ad}}} \mathscr{J}(u).$$

Then, the following variational inequality holds:

$$\langle \nabla\mathscr{J}(\bar{u}), u - \bar{u} \rangle_{\mathscr{U}} \geq 0 \quad \textit{for all } u \in \mathscr{U}_{\mathsf{ad}}, \tag{1.75}$$

where the gradient of $\mathscr{J}$ is given by

$$\nabla\mathscr{J}(u) = \mathscr{G}^\star(\mathscr{G}u - z_d) + \sigma u \quad \textit{for } u \in \mathscr{U}.$$

If $\bar{u} \in \mathscr{U}_{\mathsf{ad}}$ solves (1.75), then $\bar{u}$ is a solution to (1.74).

Proof. Since $\mathscr{J}$ is Gâteaux-differentiable and convex in $\mathscr{U}$, the result follows directly from [70, p. 63]. ☐

Inequality (1.75) is a first-order necessary and sufficient condition for (1.74), which can be expressed as

$$\langle \mathscr{G}\bar{u} - z_d, \mathscr{G}u - \mathscr{G}\bar{u} \rangle_{\mathscr{H}} + \langle \sigma\bar{u}, u - \bar{u} \rangle_{\mathscr{U}} \geq 0 \quad \text{for all } u \in \mathscr{U}_{\mathsf{ad}}. \tag{1.76}$$

Next we study (1.76) for $(\hat{\mathbf{P}})$. Utilizing the setting from (1.73), we obtain

$$\begin{aligned}
&\langle \mathscr{G}\bar{u} - z_d, \mathscr{G}(u - \bar{u}) \rangle_{\mathscr{H}} \\
&\quad = \sigma_Q \langle \mathscr{S}\bar{u} - (y_Q - \hat{y}), \mathscr{S}(u - \bar{u}) \rangle_{L^2(0,T;H)} \\
&\qquad + \sigma_\Omega \langle (\mathscr{S}\bar{u})(T) - (y_\Omega - \hat{y}(T)), (\mathscr{S}(u - \bar{u}))(T) \rangle_H \\
&\quad = \sigma_Q \langle \mathscr{S}\bar{u}, \mathscr{S}(u - \bar{u}) \rangle_{L^2(0,T;H)} + \sigma_\Omega \langle (\mathscr{S}\bar{u})(T), (\mathscr{S}(u - \bar{u}))(T) \rangle_H \\
&\qquad - \sigma_Q \langle y_Q - \hat{y}, \mathscr{S}(u - \bar{u}) \rangle_{L^2(0,T;H)} - \sigma_\Omega \langle y_\Omega - \hat{y}(T), (\mathscr{S}(u - \bar{u}))(T) \rangle_H.
\end{aligned}$$

Let us define the two linear bounded operators $\Theta : W_0(0,T) \to W_0(0,T)'$ and $\Xi : L^2(0,T;H) \times H \to W_0(0,T)'$ by

$$\begin{aligned}
\langle \Theta\varphi, \phi \rangle_{W_0(0,T)', W_0(0,T)} &= \int_0^T \langle \sigma_Q \varphi(t), \phi(t) \rangle_H \, \mathrm{d}t + \langle \sigma_\Omega \varphi(T), \phi(T) \rangle_H, \\
\langle \Xi z, \phi \rangle_{W_0(0,T)', W_0(0,T)} &= \int_0^T \langle \sigma_Q z_Q(t), \phi(t) \rangle_H \, \mathrm{d}t + \langle \sigma_\Omega z_\Omega, \phi(T) \rangle_H
\end{aligned} \tag{1.77}$$

for φ, $\phi \in W_0(0,T)$ and $z=(z_Q,z_\Omega)\in L^2(0,T;H)\times H$. Then, we find

$$\begin{aligned}&\langle \mathscr{G}\bar u - z_d, \mathscr{G}(u-\bar u)\rangle_{\mathscr{H}}\\ &= \langle \Theta(\mathscr{S}\bar u) - \Xi(y_Q-\hat y, y_\Omega - \hat y(T)), \mathscr{S}(u-\bar u)\rangle_{W_0(0,T)', W_0(0,T)}\\ &= \langle \mathscr{S}'\Theta\mathscr{S}\bar u, u-\bar u\rangle_U - \langle \mathscr{S}'\Xi(y_Q-\hat y, y_\Omega-\hat y(T)), u-\bar u\rangle_U.\end{aligned} \tag{1.78}$$

Let us define the linear operator $\mathscr{A} : U \to W(0,T)$ as follows: for given $u \in U$ the function $p = \mathscr{A}u \in W(0,T)$ is the unique solution to

$$\begin{aligned}-\frac{\mathrm{d}}{\mathrm{d}t}\langle p(t),\varphi\rangle_H + a(t;p(t),\varphi) &= -\sigma_Q\langle(\mathscr{S}u)(t),\varphi\rangle_H \quad \forall\varphi\in V \text{ a.e.},\\ p(T) &= -\sigma_\Omega(\mathscr{S}u)(T) \quad \text{in } H.\end{aligned} \tag{1.79}$$

It follows from (1.43) and $\mathscr{S}u \in W(0,T)$ that the operator $\mathscr{A}$ is well defined and bounded.

Lemma 1.41. *Let Assumption* 1 *be satisfied and $u, v \in U$. We set $y = \mathscr{S}u \in W_0(0,T)$, $w = \mathscr{S}v \in W_0(0,T)$, and $p = \mathscr{A}v \in W(0,T)$. Then,*

$$\int_0^T \langle(\mathscr{B}u)(t),p(t)\rangle_{V',V}\,\mathrm{d}t = -\int_0^T \sigma_Q\langle w(t),y(t)\rangle_H\,\mathrm{d}t - \sigma_\Omega\langle w(T),y(T)\rangle_H.$$

Proof. We derive from $y = \mathscr{S}u$, $p = \mathscr{A}u$, $y \in W_0(0,T)$, and integration by parts that

$$\begin{aligned}\int_0^T \langle(\mathscr{B}u)(t),p(t)\rangle_{V',V}\,\mathrm{d}t &= \int_0^T \langle y_t(t),p(t)\rangle_{V',V} + a(t;y(t),p(t))\,\mathrm{d}t\\ &= \int_0^T -\langle p_t(t),y(t)\rangle_{V',V} + a(t;p(t),y(t))\,\mathrm{d}t + \langle p(T),y(T)\rangle_H\\ &= -\int_0^T \sigma_Q\langle w(t),y(t)\rangle_H\,\mathrm{d}t - \sigma_\Omega\langle w(T),y(T)\rangle_H,\end{aligned}$$

which is the claim. □

We define $\hat p \in W(0,T)$ as the unique solution to

$$\begin{aligned}-\frac{\mathrm{d}}{\mathrm{d}t}\langle \hat p(t),\varphi\rangle_H + a(t;\hat p(t),\varphi) &= \sigma_Q\langle y_Q(t)-\hat y(t),\varphi\rangle_H \quad \forall\varphi\in V \text{ a.e.},\\ p(T) &= \sigma_\Omega(y_\Omega-\hat y(T)) \quad \text{in } H.\end{aligned} \tag{1.80}$$

Then, for every $u \in U$ the function $p = \hat p + \mathscr{A}u$ is the unique solution to

$$\begin{aligned}-\frac{\mathrm{d}}{\mathrm{d}t}\langle p(t),\varphi\rangle_H + a(t;p(t),\varphi) &= \sigma_Q\langle y_Q(t)-y(t),\varphi\rangle_H \quad \forall\varphi\in V \text{ a.e.},\\ p(T) &= \sigma_\Omega(y_\Omega-y(T)) \quad \text{in } H,\end{aligned}$$

with $y = \hat y + \mathscr{S}u$. Moreover, we have the following result.

Lemma 1.42. *Let Assumption* 1 *be satisfied. Then, $\mathscr{B}'\mathscr{A} = -\mathscr{S}'\Theta\mathscr{S} \in \mathscr{L}(U)$, where linear and bounded operator Θ is defined in* (1.77). *Moreover, $\mathscr{B}'\hat p = \mathscr{S}'\Xi(y_Q-\hat y, y_\Omega-\hat y(T))$, where $\hat p$ is the solution to* (1.80).

Proof. Let $u, v \in \mathscr{U}$ be chosen arbitrarily. We set $y = \mathscr{S} u \in W_0(0,T)$ and $w = \mathscr{S} v \in W_0(0,T)$. Recall that we identify U with its dual space U'. From the integration by parts formula and Lemma 1.41 we infer that

$$\begin{aligned}
&\langle \mathscr{S}'\Theta\mathscr{S} v, u\rangle_U = \langle \Theta\mathscr{S} v, \mathscr{S} u\rangle_{W_0(0,T)',W_0(0,T)} = \langle \Theta w, y\rangle_{W_0(0,T)',W_0(0,T)}\\
&= \int_0^T \sigma_Q \langle w(t), y(t)\rangle_H \,\mathrm{d}t + \sigma_\Omega \langle w(T), y(T)\rangle_H\\
&= -\langle \mathscr{B} u, p\rangle_{L^2(0,T;V'),L^2(0,T;V)} = -\langle u, \mathscr{B}' p\rangle_U = -\langle \mathscr{B}'\mathscr{A} v, u\rangle_U.
\end{aligned}$$

Since $u, v \in U$ are chosen arbitrarily, we have $\mathscr{B}'\mathscr{A} = \mathscr{S}'\Theta\mathscr{S}$. Further, we find

$$\begin{aligned}
&\langle \mathscr{S}'\Xi(y_Q - \hat{y}, y_\Omega - \hat{y}(T)), u\rangle_U = \langle \Xi(y_Q - \hat{y}), y_\Omega - \hat{y}(T)), \mathscr{S} u\rangle_{W_0(0,T)',W_0(0,T)}\\
&= \int_0^T \sigma_Q \langle y_Q(t) - \hat{y}(t), y(t)\rangle_H \,\mathrm{d}t + \sigma_\Omega \langle y_\Omega - \hat{y}(T), y(T)\rangle_H\\
&= \int_0^T -\langle \hat{p}_t(t), y(t)\rangle_H + a(t; \hat{p}(t), y(t))\,\mathrm{d}t + \langle \hat{p}(T), y(T)\rangle_H\\
&= \int_0^T \langle y_t(t), \hat{p}(t)\rangle_H + a(t; y(t), \hat{p}(t))\,\mathrm{d}t = \int_0^T \langle (\mathscr{B} u)(t), \hat{p}(t)\rangle_{V',V} \,\mathrm{d}t\\
&= \langle \mathscr{B}'\hat{p}, u\rangle_U,
\end{aligned}$$

which gives the claim. □

We infer from (1.78) and Lemma 1.42 that

$$\langle \mathscr{G}\bar{u} - z_d, \mathscr{G} u\rangle_{\mathscr{H}} = -\langle \mathscr{B}'(\hat{p} + \mathscr{A}\bar{u}), u - \bar{u}\rangle_U.$$

This implies the following variational inequality for $(\hat{\mathbf{P}})$:

$$\begin{aligned}
&\langle \mathscr{G}\bar{u} - z_d, \mathscr{G} u - \mathscr{G}\bar{u}\rangle_{\mathscr{H}} + \sigma \langle \bar{u}, u - \bar{u}\rangle_{\mathscr{U}}\\
&\quad = \langle \sigma\bar{u} - \mathscr{B}'(\hat{p} + \mathscr{A}\bar{u}), u - \bar{u}\rangle_U \geq 0 \quad \text{for all } u \in U_{\mathsf{ad}}.
\end{aligned}$$

In summary, we have proved the following result.

Theorem 1.43. *Suppose that Assumptions* 1 *and* 2 *hold. Then,* $(\bar{y}, \bar{u})$ *is a solution to* (**P**) *if and only if* $(\bar{y}, \bar{u})$ *satisfy together with the adjoint variable* $\bar{p}$ *the first-order optimality system*

$$\bar{y} = \hat{y} + \mathscr{S}\bar{u}, \quad \bar{p} = \hat{p} + \mathscr{A}\bar{u}, \quad u_a \leq \bar{u} \leq u_b, \tag{1.81a}$$

$$\langle \sigma\bar{u} - \mathscr{B}'\bar{p}, u - \bar{u}\rangle_U \geq 0 \quad \textit{for all } u \in U_{\mathsf{ad}}. \tag{1.81b}$$

Remark 1.44. Using a Lagrangian framework, it follows from Theorem 1.43 and [70] that the variational inequality (1.81b) is equivalent to the existence of two functions $\bar{\mu}_a, \bar{\mu}_b \in U$ satisfying $\bar{\mu}_a, \bar{\mu}_b \geq 0$,

$$\sigma\bar{u} - \mathscr{B}'\bar{p} + \bar{\mu}_b - \bar{\mu}_a = 0,$$

and the complementarity condition

$$\bar{\mu}_a(t)^\top (u_a(t) - \bar{u}(t)) = \bar{\mu}_b(t)^\top (\bar{u}(t) - u_b(t)) = 0 \quad \text{f.a.a. } t \in [0,T].$$

Thus, (1.81) is equivalent to the system

$$\begin{aligned} &\bar{y} = \hat{y} + \mathscr{S}\bar{u}, \quad \bar{p} = \hat{p} + \mathscr{A}\bar{u}, \quad \sigma\bar{u} - \mathscr{B}'\bar{p} + \bar{\mu}_b - \bar{\mu}_a = 0, \\ &u_a \leq \bar{u} \leq u_b, \quad 0 \leq \bar{\mu}_a, \quad 0 \leq \bar{\mu}_b, \\ &\bar{\mu}_a(t)^\top(u_a(t) - \bar{u}(t)) = \bar{\mu}_b(t)^\top(\bar{u}(t) - u_b(t)) = 0 \text{ a.e. in } [0,T]. \end{aligned} \tag{1.82}$$

Utilizing a complementarity function, it can be shown that (1.82) is equivalent to

$$\begin{aligned} &\bar{y} = \hat{y} + \mathscr{S}\bar{u}, \quad \bar{p} = \hat{p} + \mathscr{A}\bar{u}, \quad \sigma\bar{u} - \mathscr{B}'\bar{p} + \bar{\mu}_b - \bar{\mu}_a = 0, \quad u_a \leq \bar{u} \leq u_b, \\ &\bar{\mu}_a = \max\big(0, \bar{\mu}_a + \eta(\bar{u} - u_a)\big), \ \bar{\mu}_b = \max\big(0, \bar{\mu}_b + \eta(\bar{u} - u_b)\big), \end{aligned} \tag{1.83}$$

where $\eta > 0$ is an arbitrary real number. The max and min operations are interpreted componentwise in the pointwise-everywhere sense.

The gradient $\nabla\hat{J} : U \to U$ of the reduced cost functional $\hat{J}$ is given by

$$\nabla J(u) = \sigma u - \mathscr{B}^\star p, \quad u \in U,$$

where $p = \hat{p} + \mathscr{A} u$; see, e.g., [26]. Thus, a first-order sufficient optimality condition for ($\hat{\mathbf{P}}$) is given by the variational inequality

$$\langle \sigma\bar{u} - \mathscr{B}'\bar{p}, u - \bar{u}\rangle_U \geq 0 \quad \text{for all } u \in U_{\mathsf{ad}}, \tag{1.84}$$

with $\bar{p} = \hat{p} + \mathscr{A}\bar{u}$.

Problem ($\hat{\mathbf{P}}$) can be solved numerically by a primal-dual active set strategy with the choice $\eta = \sigma$. In this case the method is equivalent to a locally superlinearly convergent semismooth Newton algorithm applied to (1.83); see [24, 26, 72]. In Algorithm 1.1 we formulate the method in the context of our application. In Section 1.5 we compare Algorithm 1.1 with the Banach fixed-point iteration as well as with the projected gradient method [36, 52].

***ALGORITHM* 1.1. Primal-dual active set strategy.**

Require: Starting value (u^0, λ^0) and maximal iteration number k_{max}.

1: Set $k = 0$. For $i = 1, \ldots, m$ determine the active sets

$$\mathscr{A}_{ai}^k = \big\{t \in [0,T] \,\big|\, \sigma u_i^k + \lambda_i^k < u_{ai} \text{ a.e.}\big\},$$
$$\mathscr{A}_{bi}^k = \big\{t \in [0,T] \,\big|\, \sigma u_i^k + \lambda_i^k > u_{bi} \text{ a.e.}\big\}$$

and the inactive set $\mathscr{I}_i^k = [0,T] \setminus \mathscr{A}_i^k$ with $\mathscr{A}_i^k = \mathscr{A}_{ai}^k \cup \mathscr{A}_{bi}^k$.

2: **repeat**

3: Compute the solution (y, p, u) to the optimality system

$$y = \hat{y} + \mathscr{S}u, \quad p = \hat{p} + \mathscr{A}u, \quad u_i = \begin{cases} u_{ai} & \text{on } \mathscr{A}_{ia}^k, \\ u_{bi} & \text{on } \mathscr{A}_{ib}^k, \\ (\mathscr{B}'p)_i/\sigma & \text{on } \mathscr{I}_i^k, \end{cases} \quad 1 \leq i \leq m.$$

4: Set $(y^{k+1}, u^{k+1}, p^{k+1}) = (y, u, p)$, $\lambda^{k+1} = \mathscr{B}'p^{k+1} - \sigma u^{k+1}$, and $k = k+1$.

5: Compute the active and inactive sets according to step 1.

6: **until** ($\mathscr{A}_{ai}^k = \mathscr{A}_{ai}^{k-1}$ **and** $\mathscr{A}_{bi}^k = \mathscr{A}_{bi}^{k-1}$) **or** $k = k_{\mathsf{max}}$.

1.4.4 ▪ The POD Galerkin approximation

In this subsection we introduce the POD Galerkin schemes for the variational inequality (1.84) using a POD Galerkin approximation for the state and dual variables. Moreover, we study the convergence of the POD discretizations, where we make use of the analysis in [28, 38–40, 66, 71]. For a general introduction we also refer the reader to the survey paper [27].

In (1.55) we have introduced a POD Galerkin scheme for the state equation (1.69). Suppose that $\{\psi_i\}_{i=1}^\ell$ is a POD basis of rank ℓ computed from $(\mathbf{P}^\ell)$ with $\psi_i = \psi_i^V$ for $X = V$ and $\psi_i = \psi_i^H$ for $X = H$. We set $X^\ell = \operatorname{span}\{\psi_1, \ldots, \psi_\ell\} \subset V$. Let the linear and bounded projection operator $\mathscr{P}^\ell$ be denoted as $\mathscr{P}_V^\ell$ for $X = V$ and $\mathscr{P}_H^\ell$ for $X = H$; see (1.49).

Recall the POD Galerkin ansatz (1.54) for the state variable. Analogously, we approximate the adjoint variable $p = \hat{p} + \mathscr{A} u$ by the Galerkin expansion

$$p^\ell(t) = \hat{p}(t) + \sum_{i=1}^{\ell} \mathrm{p}_i^\ell(t)\psi_i \in V \quad \text{for } t \in [0,T], \tag{1.85}$$

with coefficient functions $\mathrm{p}_i^\ell : [0,T] \to \mathbb{R}$ and with $\hat{p}$ from (1.80). Let the vector-valued coefficient function be given by

$$\mathrm{p}^\ell = \left(\mathrm{p}_1^\ell, \ldots, \mathrm{p}_\ell^\ell\right) : [0,T] \to \mathbb{R}^\ell.$$

If we assume that $\mathrm{p}^\ell(T) = -\sigma_\Omega \mathrm{y}^\ell(T)$, then we infer from $\hat{p}(T) = \sigma_\Omega(y_\Omega - \hat{y}(T))$ and (1.85) that

$$p^\ell(T) = \hat{p}(T) - \sigma_\Omega \sum_{i=1}^{\ell} \mathrm{y}_i^\ell(t)\psi_i = \sigma_\Omega\left(y_\Omega - y^\ell(T)\right).$$

This motivates the following POD scheme to approximate $p = \hat{p} + \mathscr{A} u$: $p^\ell \in W(0,T)$ satisfies

$$\begin{aligned} -\frac{\mathrm{d}}{\mathrm{d}t}\langle p^\ell(t), \psi\rangle_H + a(t; p^\ell(t), \psi) &= \sigma_Q \left\langle (y_Q - y^\ell)(t), \psi\right\rangle_H \quad \forall \psi \in X^\ell \text{ a.e.}, \\ \mathrm{p}^\ell(T) &= -\sigma_\Omega \mathrm{y}^\ell(T). \end{aligned} \tag{1.86}$$

It follows by similar arguments as for (1.55) that there is a unique solution $p^\ell \in W(0,T)$.

Remark 1.45. Recall that we introduced the linear and bounded solution operator $\mathscr{S}^\ell : U \to W(0,T)$ as an approximation for the state solution operator $\mathscr{S}$; see Remark 1.28, part 2). Analogously, we define an approximation of the adjoint solution operator $\mathscr{A}$ as follows. Let $\mathscr{A}^\ell : U \to W(0,T)$ denote the solution operator to

$$\begin{aligned} -\frac{\mathrm{d}}{\mathrm{d}t}\langle w^\ell(t), \psi\rangle_H + a(t; w^\ell(t), \psi) &= -\sigma_Q \left\langle (\mathscr{S}^\ell u)(t), \psi\right\rangle_H \qquad \forall \psi \in X^\ell \text{ a.e.}, \\ w^\ell(T) &= -\sigma_\Omega(\mathscr{S}^\ell u)(T). \end{aligned}$$

Then, $p^\ell = \hat{p} + \mathscr{A}^\ell u$ is the unique solution to (1.86).

Lemma 1.46. *Let Assumption* 1 *be satisfied and* $u, v \in U$. *We set* $y^\ell = \mathscr{S}^\ell u \in W_0(0,T)$, $w^\ell = \mathscr{S}^\ell v \in W_0(0,T)$, *and* $p^\ell = \mathscr{A}^\ell v \in W(0,T)$. *Then,*

$$\int_0^T \left\langle (\mathscr{B} u)(t), p^\ell(t)\right\rangle_{V',V} \mathrm{d}t = -\int_0^T \sigma_Q \langle w^\ell(t), y^\ell(t)\rangle_H \,\mathrm{d}t - \sigma_\Omega \langle w^\ell(T), y^\ell(T)\rangle_H.$$

Moreover, $\mathscr{B}'\mathscr{A}^\ell = -(\mathscr{S}^\ell)'\Theta\mathscr{S}^\ell \in \mathscr{L}(U)$*, where the linear and bounded operator* Θ *is defined in* (1.77).

Proof. Since the POD bases for the state and adjoint coincide, the claim follows by the same arguments used to prove Lemmas 1.41 and 1.42. □

Theorem 1.47. *Suppose that Assumptions* 1 *and* 2 *hold. Let* $u \in U$ *be arbitrarily given so that* $\mathscr{S}u, \mathscr{A}u \in H^1(0,T;V)\setminus\{0\}$.

1. *To compute a POD basis* $\{\psi_i\}_{i=1}^\ell$ *of rank* ℓ *we choose* $\wp = 4$, $y^1 = \mathscr{S}u$, $y^2 = (\mathscr{S}u)_t$, $y^3 = \mathscr{A}u$, *and* $y^4 = (\mathscr{A}u)_t$. *Then,* $p = \hat{p} + \mathscr{A}u$ *and* $p^\ell = \hat{p} + \mathscr{A}^\ell u$ *satisfy the a priori error estimate*

$$\|p^\ell - p\|^2_{H^1(0,T;V)} \le \begin{cases} C \sum\limits_{i=\ell+1}^{d_V} \lambda_i^V & \text{if } X = V, \\ C \sum\limits_{i=\ell+1}^{d_H} \lambda_i^H \|\psi_i^H - \mathscr{P}_H^\ell \psi_i^H\|_V^2 & \text{if } X = H \end{cases} \tag{1.87}$$

for a constant C *that depends on* γ, γ_1, γ_2, T, σ_Ω, *and* σ_Q.

2. *If* $\mathscr{S}\tilde{u}$ *and* $\mathscr{A}\tilde{u}$ *belong to* $H^1(0,T;V)$ *for every* $\tilde{u} \in U$ *and if* $\lambda_i^H > 0$ *for all* $i \in \mathscr{I}$, *then we have*

$$\lim_{\ell\to\infty} \|\mathscr{A} - \mathscr{A}^\ell\|_{\mathscr{L}(U,W(0,T))} = 0. \tag{1.88}$$

Proof. Analogous to (1.62), we have $p^\ell(t) - p(t) = \theta^\ell(t) + \rho^\ell(t)$ for almost all $t \in [0,T]$ with $\theta^\ell = \mathscr{A}^\ell u - \mathscr{P}^\ell(\mathscr{A}u)$ and $\rho^\ell = \mathscr{P}^\ell(\mathscr{A}u) - \mathscr{A}u$. Here, $\mathscr{P}^\ell = \mathscr{P}_V^\ell$ for $X = V$ and $\mathscr{P}^\ell = \mathscr{P}_H^\ell$ for $X = H$. Now, the proof of the claims follows by similar arguments as the proofs of Theorem 1.29, Proposition 4.7 in [28], Proposition 4.6 in [71], and Theorem 6.3 in [66]. To estimate the terminal term $\theta^\ell(T)$ we observe that

$$\begin{aligned}
\|\theta^\ell(T)\|_H &= \Big\|\mathscr{P}^\ell((\mathscr{A}u)(T)) - (\mathscr{A}^\ell u)(T)\Big\|_H \\
&\le \sigma_\Omega\Big(\Big\|\mathscr{P}^\ell((\mathscr{S}u)(T)) - (\mathscr{S}u)(T)\Big\|_H + \Big\|(\mathscr{S}u)(T) - (\mathscr{S}^\ell u)(T)\Big\|_H\Big) \\
&\le \sigma_\Omega\Big(\Big\|\mathscr{P}^\ell(\mathscr{S}u) - \mathscr{S}u\Big\|_{C([0,T];H)} + \Big\|\mathscr{S}u - \mathscr{S}^\ell u\Big\|_{C([0,T];H)}\Big) \\
&\le \sigma_\Omega c_E\Big(\Big\|\mathscr{P}^\ell(\mathscr{S}u) - \mathscr{S}u\Big\|_{H^1(0,T;V)} + \Big\|\mathscr{S}u - \mathscr{S}^\ell u\Big\|_{H^1(0,T;V)}\Big),
\end{aligned}$$

with an embedding constant c_E. The first term on the right-hand side can be handled by (1.27); the second term is estimated in Theorem 1.29. Finally, (1.88) follows from (1.61) and the fact that the operator $\mathscr{S}^\ell$ is bounded uniformly with respect to ℓ. □

Remark 1.48.

1. The inclusion of adjoint information in the snapshot ensemble also improves the approximation quality for nonlinear problems; see [15].

2. Analogous to Remark 1.30, part 2, the a priori estimate (1.87) holds for an arbitrarily chosen but fixed control $u \in U$. Furthermore, (1.88) implies that
$$\lim_{\ell \to \infty} \|\hat{p} + \mathscr{A}^\ell \tilde{u} - \hat{p} - \mathscr{A}\tilde{u}\|_{W(0,T)} = 0$$
for any $\tilde{u} \in U$.

3. We can also extend the results in Proposition 1.32 for the adjoint equation and get an a priori error estimate choosing $\wp = 2$, $y^1 = \mathscr{S}u$, and $y^2 = \mathscr{A}u$.

The POD Galerkin approximation for $(\hat{\mathbf{P}})$ is as follows:
$$\min \hat{J}^\ell(u) \quad \text{s.t.} \quad u \in U_{\mathsf{ad}}, \tag{$\hat{\mathbf{P}}^\ell$}$$
where the cost is defined by $\hat{J}^\ell(u) = J(\hat{y} + \mathscr{S}^\ell u, u)$ for $u \in U$. Let $\bar{u}^\ell$ be the solution to $(\hat{\mathbf{P}}^\ell)$. Then, a first-order sufficient optimality condition is given by the variational inequality
$$\langle \sigma \bar{u}^\ell - \mathscr{B}'\bar{p}^\ell, u - \bar{u}^\ell \rangle_U \geq 0 \quad \text{for all } u \in U_{\mathsf{ad}}, \tag{1.89}$$
where $\bar{p}^\ell = \hat{p}^\ell + \mathscr{A}^\ell \bar{u}^\ell$.

Theorem 1.49. *Suppose that Assumptions* 1 *and* 2 *hold. Let $u \in U$ be arbitrarily given so that $\mathscr{S}u, \mathscr{A}u \in H^1(0,T;V) \setminus \{0\}$.*

1. *To compute a POD basis $\{\psi_i\}_{i=1}^\ell$ of rank ℓ, we choose $\wp = 4$, $y^1 = \mathscr{S}u$, $y^2 = (\mathscr{S}u)_t$, $y^3 = \mathscr{A}u$, and $y^4 = (\mathscr{A}u)_t$. Then, the optimal solution $\bar{u}$ to $(\hat{\mathbf{P}})$ and the associated POD suboptimal solution $\bar{u}^\ell$ to $(\hat{\mathbf{P}}^\ell)$ satisfy*
$$\lim_{\ell \to \infty} \|\bar{u}^\ell - \bar{u}\|_U = 0 \tag{1.90}$$
for $X = V$ and $X = H$.

2. *If an optimal POD basis of rank ℓ is computed by choosing $\wp = 4$, $y^1 = \mathscr{S}\bar{u}$, $y^2 = (\mathscr{S}\bar{u})_t$, $y^3 = \mathscr{A}\bar{u}$, and $y^4 = (\mathscr{A}\bar{u})_t$, then we have*
$$\|\bar{u}^\ell - \bar{u}\|_U \leq \begin{cases} \dfrac{C}{\sigma} \displaystyle\sum_{i=\ell+1}^{d_V} \lambda_i^V & \text{if } X = V, \\ \dfrac{C}{\sigma} \displaystyle\sum_{i=\ell+1}^{d_H} \lambda_i^H \|\psi_i^H - \mathscr{P}_H^\ell \psi_i^H\|_V^2 & \text{if } X = H, \end{cases} \tag{1.91}$$
where the constant C depends on $\gamma, \gamma_1, \gamma_2, T, \sigma_\Omega, \sigma_Q$, and the norm $\|\mathscr{B}'\|_{\mathscr{L}(L^2(0,T;V),U)}$.

Proof. Choosing $u = \bar{u}^\ell$ in (1.84) and $u = \bar{u}$ in (1.89), we get the variational inequality
$$0 \leq \langle \sigma(\bar{u} - \bar{u}^\ell) - \mathscr{B}'(\bar{p} - \bar{p}^\ell), \bar{u}^\ell - \bar{u} \rangle_U. \tag{1.92}$$

Utilizing Lemma 1.46 and $\langle \Theta\varphi, \varphi \rangle_{W_0(0,T)',W_0(0,T)} \geq 0$ for all $\varphi \in W_0(0,T)$, we infer from (1.92) that

$$\begin{aligned}0 &\leq \langle \mathscr{B}'\mathscr{A}^\ell \bar{u}^\ell - \mathscr{B}'\mathscr{A}\bar{u}, \bar{u}^\ell - \bar{u}\rangle_U - \sigma\|\bar{u} - \bar{u}^\ell\|_U^2 \\ &= \langle \mathscr{B}'\mathscr{A}^\ell(\bar{u}^\ell - \bar{u}) + \mathscr{B}'(\mathscr{A}^\ell - \mathscr{A})\bar{u}, \bar{u}^\ell - \bar{u}\rangle_U - \sigma\|\bar{u} - \bar{u}^\ell\|_U^2 \\ &\leq \langle \Theta\mathscr{S}^\ell(\bar{u} - \bar{u}^\ell), \mathscr{S}^\ell(\bar{u}^\ell - \bar{u})\rangle_U + \|\mathscr{B}'(\mathscr{A}^\ell - \mathscr{A})\bar{u}\|_U\|\bar{u}^\ell - \bar{u}\|_U - \sigma\|\bar{u} - \bar{u}^\ell\|_U^2 \\ &\leq \|\mathscr{B}'(\mathscr{A}^\ell - \mathscr{A})\bar{u}\|_U\|\bar{u}^\ell - \bar{u}\|_U - \sigma\|\bar{u} - \bar{u}^\ell\|_U^2.\end{aligned}$$

Consequently,

$$\|\bar{u} - \bar{u}^\ell\|_U \leq \frac{1}{\sigma}\|\mathscr{B}'(\mathscr{A}^\ell - \mathscr{A})\bar{u}\|_U.$$

Now (1.90) and (1.91) follow from (1.88) and (1.87), respectively. □

In Algorithm 1.2 we formulate a discrete version of the primal-dual active set method (see Algorithm 1.1), which is utilized to solve ($\hat{\mathbf{P}}^\ell$) in Section 1.4.3.

ALGORITHM **1.2. POD discretized primal-dual active set strategy.**

Require: POD basis $\{\psi_i\}_{i=1}^\ell$, starting value $(u^{\ell 0}, \lambda^{\ell 0})$, maximal iteration number $k_{\max}$.

1: Set $k = 0$; determine the active sets

$$\mathscr{A}_{ai}^{\ell k} = \{t \in [0,T] \,|\, \sigma u_i^{k\ell} + \lambda_i^{k\ell} < u_{ai} \text{ a.e.}\},$$
$$\mathscr{A}_{bi}^{\ell k} = \{t \in [0,T] \,|\, \sigma u_i^{k\ell} + \lambda_i^{k\ell}(t) > u_{bi}(t)\}$$

and the inactive sets $\mathscr{I}_i^{\ell k} = [0,T] \setminus \mathscr{A}_i^{\ell k}$ with $\mathscr{A}_i^{\ell k} = \mathscr{A}_{ai}^{\ell k} \cup \mathscr{A}_{bi}^{\ell k}$.

2: **repeat**

3: Determine the solution (y^ℓ, u^ℓ, p^ℓ) to the optimality system

$$y^\ell = \hat{y} + \mathscr{S}^\ell u^\ell, \quad p^\ell = \hat{p} + \mathscr{A}^\ell u^\ell, \quad u^\ell = \begin{cases} u_a & \text{on } \mathscr{A}_a^{k\ell}, \\ u_b & \text{on } \mathscr{A}_b^{k\ell}, \\ \mathscr{B}'p^\ell/\sigma & \text{on } \mathscr{I}^{k\ell}. \end{cases}$$

4: Set $(y^{\ell,k+1}, u^{\ell,k+1}, p^{\ell,k+1}) = (y^\ell, u^\ell, p^\ell)$, $\lambda^{\ell,k+1} = \mathscr{B}'p^{\ell,k+1} - \sigma u^{\ell,k+1}$, and $k = k+1$.

5: Compute the active and inactive sets according to step 1.

6: **until** $(\mathscr{A}_a^{\ell k} = \mathscr{A}_a^{\ell,k-1}$ **and** $\mathscr{A}_b^{\ell k} = \mathscr{A}_b^{\ell,k-1})$ **or** $k = k_{\max}$.

1.4.5 ▪ POD a posteriori error analysis

In [71] POD a posteriori error estimates are presented that can also be applied to our optimal control problem. Based on a perturbation method [16], it is deduced how far the suboptimal control $\bar{u}^\ell$ is from the (unknown) exact optimal control $\bar{u}$. Thus, our goal is to estimate the norm $\|\bar{u} - \bar{u}^\ell\|_U$ without knowing the optimal solution $\bar{u}$. In general, $\bar{u}^\ell \neq \bar{u}$, so that $\bar{u}^\ell$ does not satisfy the variational inequality (1.84). However, there exists a function $\zeta^\ell \in U$ such that

$$\langle \sigma\bar{u}^\ell - \mathscr{B}'\tilde{p}^\ell + \zeta^\ell, u - \bar{u}^\ell\rangle_U \geq 0 \quad \forall v \in U_{\text{ad}}, \tag{1.93}$$

with $\tilde{p}^\ell = \hat{p} + \mathscr{A}\bar{u}^\ell$. Therefore, $\bar{u}^\ell$ satisfies the optimality condition of the perturbed parabolic optimal control problem

$$\min_{u\in U_{\text{ad}}} \tilde{J}(u) = J(\hat{y} + \mathscr{S}u, u) + \langle \zeta^\ell, u\rangle_U$$

with "perturbation" ζ^ℓ. The smaller ζ^ℓ is, the closer $\bar{u}^\ell$ is to $\bar{u}$. Next we estimate $\|\bar{u} - \bar{u}^\ell\|_U$ in terms of $\|\zeta^\ell\|_U$. By Lemma 1.42 we have

$$\mathscr{B}'\big(\bar{p} - \tilde{p}^\ell\big) = \mathscr{B}'\mathscr{A}\big(\bar{u} - \bar{u}^\ell\big) = -\mathscr{S}'\Theta\mathscr{S}\big(\bar{u} - \bar{u}^\ell\big) = \mathscr{S}'\Theta\big(\tilde{y}^\ell - \bar{y}\big), \tag{1.94}$$

with $\tilde{y}^\ell = \hat{y} + \mathscr{S}\bar{u}^\ell$. Choosing $u = \bar{u}^\ell$ in (1.84) and $u = \bar{u}$ in (1.93) and using (1.94), we obtain

$$\begin{aligned}
0 &\le \langle -\sigma(\bar{u} - \bar{u}^\ell) + \mathscr{B}'(\bar{p} - \tilde{p}^\ell) + \zeta^\ell, \bar{u} - \bar{u}^\ell\rangle_U\\
&= -\sigma\,\|\bar{u} - \bar{u}^\ell\|_U^2 + \langle \mathscr{S}'\Theta(\tilde{y}^\ell - \bar{y}), \bar{u} - \bar{u}^\ell\rangle_U + \langle \zeta^\ell, \bar{u} - \bar{u}^\ell\rangle_U\\
&= -\sigma\,\|\bar{u} - u_p\|_U^2 - \langle \Theta(\bar{y} - \tilde{y}^\ell), \bar{y} - \tilde{y}^\ell\rangle_{W_0(0,T)',W_0(0,T)} + \langle \zeta^\ell, \bar{u} - \bar{u}^\ell\rangle_U\\
&= -\sigma\,\|\bar{u} - \bar{u}^\ell\|_U^2 + \langle \zeta^\ell, \bar{u}^\ell - \bar{u}^\ell\rangle_U \le -\sigma\,\|\bar{u} - \bar{u}^\ell\|_U^2 + \|\zeta^\ell\|_U\|\bar{u} - \bar{u}^\ell\|_U.
\end{aligned}$$

Hence, we get the a posteriori error estimation

$$\|\bar{u} - \bar{u}^\ell\|_U \le \frac{1}{\sigma}\,\|\zeta^\ell\|_U.$$

Theorem 1.50. *Suppose that Assumptions* 1 *and* 2 *hold. Let $u \in U$ be arbitrarily given so that $\mathscr{S}u, \mathscr{A}u \in H^1(0,T;V)\setminus\{0\}$. To compute a POD basis $\{\psi_i\}_{i=1}^\ell$ of rank ℓ, we choose $\wp = 4$, $y^1 = \mathscr{S}u$, $y^2 = (\mathscr{S}u)_t$, $y^3 = \mathscr{A}u$, and $y^4 = (\mathscr{A}u)_t$. Define the function $\zeta^\ell \in U$ by*

$$\zeta_i^\ell(t) = \begin{cases} -\min(0, \xi_i^\ell(t)) & \text{a.e. in } \mathscr{A}_{ai}^\ell = \big\{t \in [0,T]\,|\,\bar{u}_i^\ell(t) = u_{ai}(t)\big\},\\ -\max(0, \xi_i^\ell(t)) & \text{a.e. in } \mathscr{A}_{bi}^\ell = \big\{t \in [0,T]\,|\,\bar{u}_i^\ell(t) = u_{bi}(t)\big\},\\ -\xi_i^\ell(t) & \text{a.e. in } [0,T]\setminus\big(\mathscr{A}_{ai}^\ell \cup \mathscr{A}_{bi}^\ell\big), \end{cases}$$

where $\xi^\ell = \sigma\bar{u}^\ell - \mathscr{B}'(\hat{p} + \mathscr{A}\bar{u}^\ell)$ in U. Then, the a posteriori error estimate is

$$\|\bar{u} - \bar{u}^\ell\|_U \le \frac{1}{\sigma}\,\|\zeta^\ell\|_U. \tag{1.95}$$

In particular, $\lim_{\ell\to\infty} \|\zeta^\ell\|_U = 0$.

Proof. Estimate (1.95) has already been shown. We proceed by constructing the function ζ^ℓ. Here we adapt the lines of the proof of Proposition 3.2 in [71] to our optimal control problem. Suppose that we know $\bar{u}^\ell$, and $\tilde{p}^\ell = \hat{p} + \mathscr{A}\bar{u}^\ell$. The goal is to determine $\zeta^\ell \in U$ satisfying (1.93). We distinguish three different cases.

- $\bar{u}_i^\ell(t) = u_{ai}(t)$ for fixed $t \in [0,T]$ and $i \in \{1,\dots,N_u\}$: Then, $u_i(t) - \bar{u}_i^\ell(t) = u_i(t) - u_{ai}(t) \ge 0$ for all $u \in U_{ad}$. Hence, $\zeta_i^\ell(t)$ has to satisfy

$$\big(\sigma\bar{u}^\ell - \mathscr{B}'\tilde{p}^\ell\big)_i(t) + \zeta_i^\ell(t) \ge 0. \tag{1.96}$$

  Setting $\zeta_i^\ell(t) = -\min(0, (\sigma\bar{u}^\ell - \mathscr{B}'\tilde{p}^\ell)_i(t))$, the value $\zeta_i^\ell(t)$ satisfies (1.96).

- $\bar{u}_i^\ell(t) = u_{bi}(t)$ for fixed $t \in [0,T]$ and $i \in \{1,\dots,N_u\}$: Now, $u_i(t) - \bar{u}_i^\ell(t) = u(t) - u_{bi}(t) \leq 0$ for all $u \in U_{ad}$. Analogous to the first case, we define $\zeta_i^\ell(t) = -\max(0, (\sigma \bar{u}^\ell - \mathscr{B}'\tilde{p}^\ell)_i(t))$ to ensure (1.96).
- $u_{ai}(t) < \bar{u}_i^\ell(t) < u_{bi}(t)$ for fixed $t \in [0,T]$ and $i \in \{1,\dots,N_u\}$: Consequently, $(\sigma \bar{u}^\ell - \mathscr{B}'\tilde{p}^\ell)_i(t) + \zeta_i^\ell(t) = 0$, so that $\zeta_i^\ell(t) = -(\sigma \bar{u}^\ell - \mathscr{B}'\tilde{p}^\ell)_i(t)$ guarantees (1.96).

It remains to prove that ζ^ℓ tends to zero for $\ell \to \infty$. Here we adapt the proof of Theorem 4.11 in [71]. By Theorem 1.49, part 1, the sequence $\{\bar{u}^\ell\}_{\ell \in \mathbb{N}}$ converges to $\bar{u}$ in U. Since the linear operator $\mathscr{B}'\mathscr{A}$ is bounded and $\tilde{p}^\ell = \hat{p} + \mathscr{A}\bar{u}^\ell$, $\{\mathscr{B}'\tilde{p}^\ell\}_{\ell \in \mathbb{N}}$ tends to $\mathscr{B}'\bar{p} = \mathscr{B}'\mathscr{A}\bar{u}$ as well. Hence, there exist subsequences $\{\bar{u}^{\ell_k}\}_{k \in \mathbb{N}}$ and $\{\mathscr{B}'\tilde{p}^{\ell_k}\}_{k \in \mathbb{N}}$ satisfying

$$\lim_{k\to\infty} \bar{u}_i^{\ell_k}(t) = \bar{u}_i(t) \quad \text{and} \quad \lim_{k\to\infty} (\mathscr{B}'\tilde{p}^{\ell_k})_i(t) = (\mathscr{B}'\bar{p})_i(t)$$

f.a.a. $t \in [0,T]$ and for $1 \leq i \leq N_u$. Next we consider the active and inactive sets for $\bar{u}$.

- Let $t \in \mathscr{I}_i = \{t \in [0,T] \,|\, u_{ai}(t) < \bar{u}_i(t) < u_{bi}(t)\}$ for $i \in \{1,\dots,N_u\}$. For $k_\circ = k_\circ(t) \in \mathbb{N}$ sufficiently large, $\bar{u}_i^{\ell_k}(t) \in (u_{ai}(t), u_{bi}(t))$ for all $k \geq k_\circ$ and f.a.a. $t \in \mathscr{I}_i$. Thus, $(\sigma \bar{u}^{\ell_k} - \mathscr{B}'\tilde{p}^{\ell_k})_i(t) = 0$ for all $k \geq k_\circ(t)$ in $\mathscr{I}_i$ a.e. This implies
$$\zeta_i^{\ell_k}(t) = 0 \quad \forall k \geq k_\circ \text{ and f.a.a. } t \in \mathscr{I}_i. \tag{1.97}$$
- Suppose that $t \in \mathscr{A}_{ai} = \{t \in [0,T] \,|\, u_{ai}(t) = \bar{u}_i(t)\}$ for $i \in \{1,\dots,N_u\}$. From $(\sigma \bar{u}_i - \mathscr{B}'\bar{p})_i(t) \geq 0$ in $\mathscr{A}_{ai}$ a.e., we deduce
$$\lim_{k\to\infty} \zeta_i^{\ell_k}(t) = -\lim_{k\to\infty} \min(0, (\sigma \bar{u}^{\ell_k} - \mathscr{B}'\tilde{p}^{\ell_k})_i(t)) = 0 \quad \text{f.a.a. } t \in \mathscr{A}_{ai}.$$
- Suppose that $t \in \mathscr{A}_{bi} = \{t \in [0,T] \,|\, u_{bi}(t) = \bar{u}_i(t)\}$. Analogous to the second case, we find
$$\lim_{k\to\infty} \zeta_i^{\ell_k}(t) = 0 \quad \text{f.a.a. } t \in \mathscr{A}_{bi}. \tag{1.98}$$

Combining (1.97) and (1.98), we conclude that $\lim_{k\to\infty} \zeta_i^{\ell_k} = 0$ a.e. in $[0,T]$ and for $1 \leq i \leq N_u$. Utilizing the dominated convergence theorem [60, p. 24], we have

$$\lim_{k\to\infty} \|\zeta^{\ell_k}\|_U = 0.$$

Since all subsequences contain a subsequence converging to zero, the claim follows from a standard argument. $\square$

Remark 1.51.

1. Theorem 1.50 shows that $\|\zeta^\ell\|_U$ tends to zero as ℓ goes to infinity. Thus, $\|\zeta^\ell\|_U$ is smaller than any tolerance $\epsilon > 0$ provided that ℓ is taken sufficiently large. Motivated by this result, we set up Algorithm 1.3. Note that the approximation quality of the POD Galerkin scheme is improved only by increasing the

number of POD basis elements: a rank-ℓ POD basis can be extended to a rank-$(\ell+1)$ POD basis by adding a new eigenfunction and keeping all the old ones. In particular, the system matrices and projected data functions can be extended by the new element; they do not have to be reconstructed in all components. Another approach is to update the POD basis in the optimization process; see, e.g., [1, 3, 41].

2. We infer from Proposition 1.32 and Remark 1.48, part 3, that Theorem 1.50 still holds true if we take $\wp = 2$, $y^1 = \mathscr{S}u$, and $y^2 = \mathscr{A}u$.

3. In [68], POD a posteriori error estimates are tested numerically for a linear-quadratic optimal control problem. It turns out that in certain cases a change of the POD basis is required to improve the approximation quality of the POD scheme; see also [41, 74], for instance.

4. Let us refer to [33], where POD a posteriori error estimates are combined with an SQP method to solve a nonlinear PDE constrained optimal control problem. Furthermore, the presented analysis for linear-quadratic problems can be extended to semilinear optimal control problems by a second-order analysis; see [34].

ALGORITHM **1.3. POD reduced-order method with a posteriori estimator.**

Require: Initial control $u^{0\ell} \in U$, initial number ℓ for the POD ansatz functions, a maximal number $\ell_{\mathsf{max}} > \ell$ of POD ansatz functions, and a stopping tolerance $\epsilon > 0$.

Determine $\hat{y}$, $\hat{p}$, $y^1 = \mathscr{S}u^{0\ell}$, $y^2 = \mathscr{A}u^{0\ell}$.
Compute a POD basis $\{\psi_i\}_{i=1}^{\ell_{\mathsf{max}}}$ choosing y^1 and y^2. Set $\ell = 1$.
repeat
    Establish the POD Galerkin discretization using $\{\psi_i\}_{i=1}^{\ell}$.
    Call Algorithm 1.2 to compute suboptimal control $\bar{u}^\ell$.
    Determine ζ^ℓ according to Theorem 1.47, and compute $\epsilon_{\mathsf{ape}} = \|\zeta^\ell\|_U/\sigma$.
    if $\epsilon_{\mathsf{ape}} < \epsilon$ **or** $\ell = \ell_{\mathsf{max}}$ **then**
        Return ℓ and suboptimal control $\bar{u}^\ell$ and STOP.
    end if
    Set $\ell = \ell + 1$.
until $\ell > \ell_{\mathsf{max}}$

1.5 ▪ Numerical experiments

In this section we present numerical test examples to illustrate our theoretical findings. The programs are written in MATLAB utilizing the Partial Differential Equation Toolbox for the computation of the FE discretization. For the temporal integration, the implicit Euler method is applied based on the equidistant time grid $t_j = (j-1)\Delta t$, $j = 1, \ldots, n$ and $\Delta t = T/(n-1)$.

Run 1 (POD for the heat equation). Let us apply the setting of Example 1.33. We choose the final time $T = 3$; the spatial domain $\Omega = (0,2) \subset \mathbb{R}$; the Hilbert spaces $H = L^2(\Omega)$, $V = H_0^1(\Omega)$; the source term $f(t,x) = t^3 - x^2$ for $(t,x) \in Q$; and the discontinuous initial value $y_\circ(x) = \chi_{(0.5,1.0)} - \chi_{(1,1.5)}$ for $x \in \Omega$, where, e.g., $\chi_{(0.5,1)}$

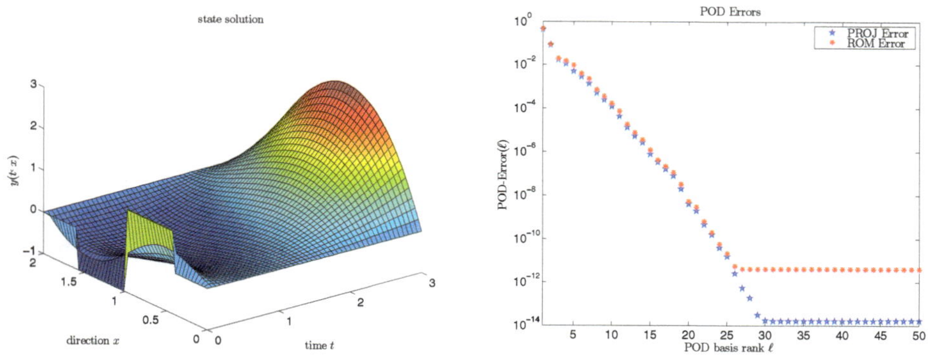

Figure 1.1. *Run* 1*: The FE solution y^h (left) and the residuals corresponding to the POD basis of rank ℓ (right).*

denotes the characteristic function on the subdomain $(0.5, 1) \subset \Omega$, $\chi_{(0.5,1)}(x) = 1$ for $x \in (0.5, 1)$ and $\chi_{(0.5,1)}(x) = 0$ otherwise. We consider a discretization of the controlled linear heat equation

$$\begin{aligned} y_t(t,x) - \Delta y(t,x) &= f(t,x) + \sum_{i=1}^{m} u_i(t)\chi_i(x), && \text{a.e. in } Q, \\ y(t,x) &= 0, && \text{a.e. in } \Sigma, \\ y(0,x) &= y_\circ(x), && \text{a.e. in } \Omega. \end{aligned} \tag{1.99}$$

To obtain an accurate approximation of the exact solution, we choose $n = 4000$ so that $\Delta t \approx 7.5 \cdot 10^{-4}$ holds. For the FE discretization, we choose $m = 500$ spatial grid points and the equidistant mesh size $h = 2/(m+1) \approx 4 \cdot 10^{-3}$. Thus, the FE error—measured in the H-norm—is of the order 10^{-4}. In the left graphic of Figure 1.1, the FE solution y^h to the state equation (1.71) is visualized. To compute a POD basis $\{\psi_i\}_{i=1}^{\ell}$ of rank ℓ, we utilize the multiple discrete snapshots $y_j^1 = y^h(t_j)$ for $1 \le j \le n$ as well as $y_1^2 = 0$ and $y_j^2 = (y^h(t_j) - y^h(t_{j-1}))/\Delta t$, $j = 2, \ldots, n$, i.e., we include the temporal difference quotients. We choose $X = H$ and utilize the (stable) SVD to determine the POD basis of rank ℓ; compare Remark 1.12. We address this issue in more detail in Run 4. Since the snapshots are FE functions, the POD basis elements are also FE functions. In the right plot of Figure 1.1, the projection and reduced-order error given by

$$\begin{aligned} \text{PROJ Error}(\ell) &= \Bigg(\sum_{j=1}^{n} \alpha_j \, \Big\| y^h(t_j) - \sum_{i=1}^{\ell} \langle y^h(t_j), \psi_i \rangle_H \, \psi_i \Big\|_H^2 \Bigg)^{1/2}, \\ \text{ROM Error}(\ell) &= \Bigg(\sum_{j=1}^{n} \alpha_j \, \| y^h(t_j) - y^\ell(t_j) \|_H^2 \Bigg)^{1/2} \end{aligned}$$

are plotted for different POD basis ranks ℓ. The chosen trapezoidal weights α_j were introduced in (1.31). We observe that both errors decay rapidly and coincide until the accuracy 10^{-12}, which is already significantly smaller than the FE discretization error. This numerical result reflects the a priori error estimates of Theorem 1.29.

Run 2 (POD for a convection-dominated parabolic problem). To present a more challenging situation, we study a convection-reaction-diffusion equation with a source

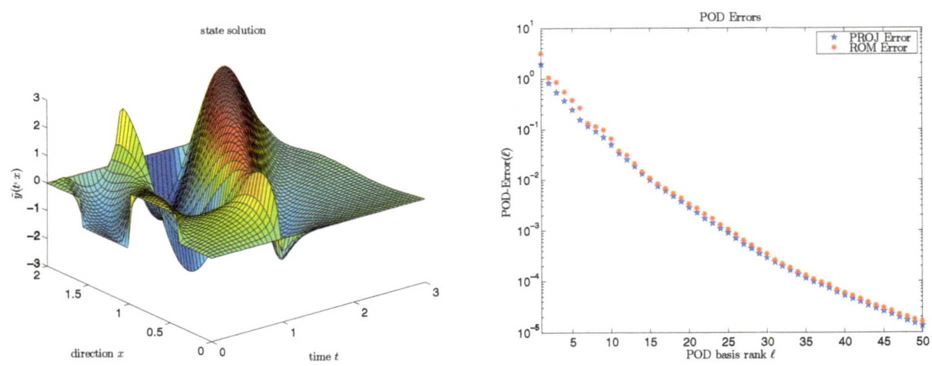

Figure 1.2. *Run 2: The FE solution y^h (left) and the residuals corresponding to the POD basis of rank ℓ (right).*

term that is close to singular: let T, Ω, $y_\circ$, H, and V be given as in Run 1. The time-independent bilinear form a corresponding to

$$\begin{aligned} y_t(t,x) - \eta_2 y_{xx}(t,x) & \\ +\eta_1 y_x(t,x) + \eta_0 y(t,x) &= f(t,x) + (\mathscr{B}u)(t,x), && \text{a.e. in } Q, \\ y(t,x) &= 0, && \text{a.e. in } \Sigma, \\ y(0,x) &= y_\circ(x) && \text{a.e. in } \Omega. \end{aligned} \tag{1.100}$$

is given by

$$a(\phi,\varphi) = \eta_2 \langle \phi', \varphi' \rangle_H + \eta_1 \langle \phi', \varphi \rangle_H + \eta_0 \langle \phi, \varphi \rangle_H \quad \text{for } \varphi, \phi \in V.$$

We choose the diffusivity $\eta_2 = 0.025$, the velocity $\eta = 1.0$ that determines the speed at which the initial profile $y_\circ$ is shifted to the boundary, and the reaction rate $\eta_0 = -0.001$. Finally, $f(t,x) = \mathbb{P}(\frac{1}{1-t})\cos(\pi x)$ for $(t,x) \in Q$, where $(\mathbb{P}z)(t) = \min(+l, \max(-l, z(t)))$ restricts the image of z on a bounded interval. In this situation, the state solution y develops a jump at $t = 1$ for $l \to \infty$; see the left plot of Figure 1.2. The right plot of Figure 1.2 demonstrates that in this case, the decay of the reconstruction residuals and the decay of the errors are much slower. The manifold dynamics of the state solution require an inconveniently large number of POD basis elements. Since the supports of these ansatz functions in general cover the whole domain Ω, the corresponding system matrices M^ℓ and A^ℓ of the reduced model (compare (1.57)) are not sparse in contrast to the matrices arising in the FE Galerkin framework, so the model order reduction (MOR) cannot be provided efficiently for this example if good accuracy of the solution function y^ℓ is required.

Run 3 (True and exact approximation error). Let us consider the setting of Run 1 again. The exact solution to (1.71) does not possess a representation by elementary functions. Hence, the presented reconstruction and reduction errors are actually the residuals with respect to a high-order FE solution y^h. To compute an approximation y of the exact solution y_{ex}, we apply a Crank–Nicolson method (with Rannacher smoothing [57]), ensuring $\|y - y_{ex}\|_{L^2(0,T;H)} = \mathcal{O}(\Delta t^2 + h^2) \approx 10^{-5}$. In the context of model reduction, such a state is sometimes called the "true" solution. To compute the FE state y^h, we apply the implicit Euler method. In the left plot of Figure 1.3 we compare the true solution with the associated POD approximation for different values

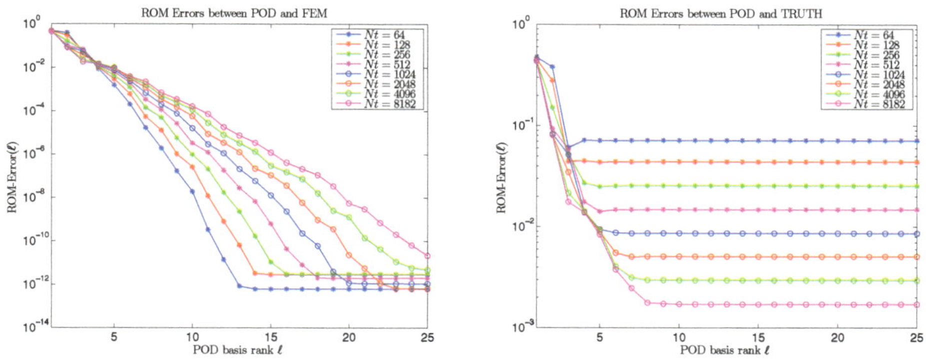

Figure 1.3. *Run 3: The ROM errors with respect to the true solution (left) and the exact one (right).*

$n = Nt \in \{64, 128, 256, \ldots, 8192\}$ of the time integration and for the spatial mesh size $h = 4 \cdot 10^{-3}$. For the norm we apply a discrete $L^2(0, T; H)$-norm as in Run 1. Let us mention that we compute for every n a corresponding FE solution y^h. We observe that the residuals ignore the errors arising by the application of time and space discretization schemes for the full-order model. The errors decay below the discretization error 10^{-5}. If these discretization errors are taken into account, the residuals stagnate at the level of the full-order model accuracy instead of decaying to zero; see the right plot of Figure 1.3. Due to the implicit Euler method, we have $\|y^h - y_{ex}\|_{L^2(0,T;H)} = \mathcal{O}(\Delta t + h^2)$, with $h = 4 \cdot 10^{-3}$. In particular, from $n \in \{64, 128, 256, \ldots, 8192\}$, it follows that $\Delta t > 3 \cdot 10^{-4} > h^2 = 1.6 \cdot 10^{-5}$. Therefore, the spatial error is dominated by the time error for all values of n. We can observe that the exact residuals do not decay below a limit of order Δt. One can observe that for fixed POD basis rank ℓ, the residuals with respect to the true solution increase if the high-order accuracy is improved by enlarging n, since the reduced-order model has to approximate a more complex system in this case, where the residuals with respect to the exact solution decrease due to the lower limit of stagnation $\Delta t = 3/(n-1)$.

Run 4 (Different strategies for the POD basis computation). Let $Y \in \mathbb{R}^{m \times n}$ denote the matrix of snapshots in the discrete setting, $W = (\langle \varphi_i, \varphi_j \rangle_X) \in \mathbb{R}^{m \times m}$ be the (sparse) spatial weight matrix arising from the FE basis $\{\varphi_i\}_{i=1}^m$, and $D = \Delta t \operatorname{diag}(\frac{1}{2}, 1, \ldots, 1, \frac{1}{2}) \in \mathbb{R}^{n \times n}$ be the trapezoidal time integration matrix fitting to implicit Euler discretization. As stated in Remark 1.11, the POD basis $\{\psi_i\}_{i=1}^\ell$ of rank ℓ can be determined by providing an eigenvalue decomposition of the matrix $\hat{Y}\hat{Y}^\top = W^{1/2} Y D Y^\top W^{1/2} \in \mathbb{R}^{m \times m}$, one of $\hat{Y}^\top \hat{Y} = D^{1/2} Y^\top W Y D^{1/2} \in \mathbb{R}^{n \times n}$, or an SVD of $\hat{Y} = W^{1/2} Y D^{1/2} \in \mathbb{R}^{m \times n}$. Since $n \gg m$ in Runs 1 to 3, the first variant is the cheapest one from a computational point of view. In the case of multiple space dimensions or if a second-order time integration scheme such as a Crank–Nicolson technique is applied, the situation is the reverse. On the other hand, SVD is more accurate than eigenvalue decomposition if the POD elements corresponding to eigenvalues/singular values that are close to zero are taken into account: since $\lambda_i = \sigma_i^2$ holds for all eigenvalues λ_i and singular values σ_i, the singular values decay to machine precision, where the eigenvalues stagnate significantly above. This is illustrated in the left graphic of Figure 1.4. Indeed, for $\ell > 20$ the EIG-ROM system matrices become singular due to the numerical errors in the eigenfunctions, and the reduced-order system is ill posed in this case, while the

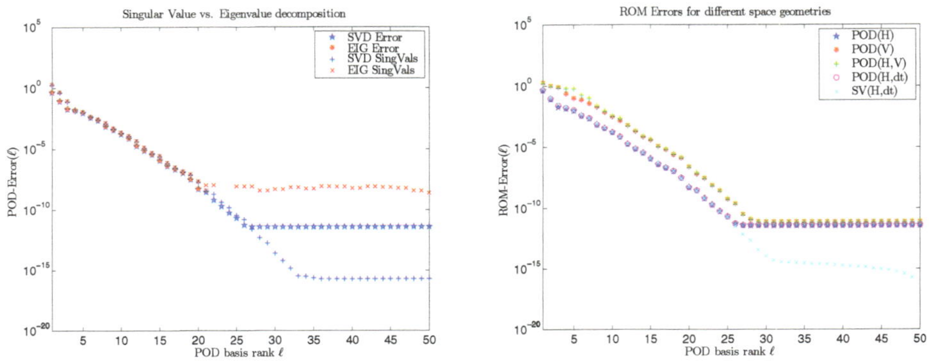

Figure 1.4. *Run 4: Singular values σ_i using the SVD (SVD Vals) or the eigenvalue decomposition (EIG Vals) and the associated ROM errors (SVD error and EIG Error, respectively) (left); ROM errors for different choices for X, the error norm, and the snapshot ensembles (right).*

SVD-ROM model remains stable. In the right plot of Figure 1.4, POD elements are constructed with respect to different scalar products, and the resulting ROM errors are compared: $\|\cdot\|_H$-residuals for $X = H$ (denoted by POD(H)), $\|\cdot\|_V$-residuals for $X = V$ (denoted by POD(V)), and $\|\cdot\|_V$-residuals for $X = H$ (denoted by POD(H,V)). This also works quite well, considering time derivatives in the snapshot sample (denoted by POD(H,dt)), allowing us to apply the a priori error estimate given in (1.60) and the corresponding sums of singular values (denoted by SV(H,dt)) corresponding to the unused eigenfunctions in the latter case, which indeed nearly coincide with the ROM errors. In many applications, the quality of the ROM does not vary significantly if the weights matrix W refers to the space $X = H$ or $X = V$ and if time derivatives of the used snapshots are taken into account or not. In particular, the ROM residual decays with the same order as the sum over the remaining singular values, $\|y - y^\ell\|_{W(0,T)} \sim \sum_{i=\ell+1}^{\infty} \sigma_i$, independent of the chosen geometrical framework.

Run 5 (Iterative methods for the optimal control problem). In this numerical test, we consider solution techniques for the linear-quadratic optimal control problem (**P**). We define the weights $\sigma_Q = 1$, $\sigma_\Omega = 0$; the desired state $y_Q(t,x) = t(1-(x-1)^2)$ for $(t,x) \in Q$; the desired final state $y_\Omega = 0$ (which is redundant due to $\sigma_\Omega = 0$, of course); the upper and lower bounds $u_a = 0.25$, $u_b = 0.75$; the control operator $(\mathcal{B}u)(t,x) = u_1(t)\chi_{\Omega_1}(x) + \cdots + u_{10}(t)\chi_{\Omega_{10}}(x)$, where $\{\Omega_i \mid i = 1,\dots,10\}$ is a uniform partition of Ω (in particular, $(\mathcal{B}^\star p)_i(t) = \int_\Omega \chi_i(x) p(t,x)\,\mathrm{d}x$); and an initial control $u_\circ(t) \equiv 1$.

1. *Banach fixed-point method*: The first-order necessary and sufficient optimality conditions (1.81) can be reformulated as the equivalent fixed-point problem

$$u = \mathbb{P}\Big(\frac{1}{\sigma}\Big(\mathcal{B}'\mathcal{A}u - \mathcal{B}'\hat{p}\Big)\Big) =: F(u),$$

   where $\mathbb{P}(u) = \min(\max(u, u_a), u_b)$ is the orthogonal projection on the set of admissible points U_{ad}. The optimal control $\bar{u} \in U$ can therefore be determined by the Banach fixed-point iteration $u_{k+1} = F(u_k)$ $(k > 0)$ with arbitrary initialization $u_0 \in U_{\mathrm{ad}}$ provided that F is a contraction. Since $\mathbb{P}$ is Lipschitz continuous

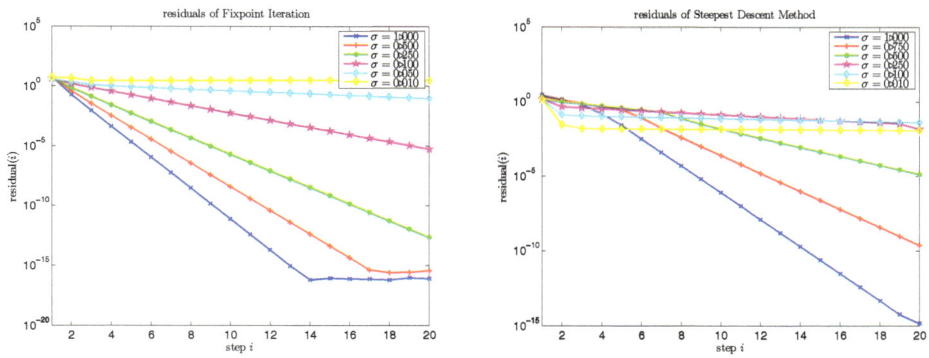

Figure 1.5. *Residuals of the Banach fixed-point iteration (left) and the projected gradient method (right) for different regularization parameters σ.*

with respect to the Lipschitz constant one, we get

$$||F(u)-F(v)||_U \le \frac{||\mathscr{B}'\mathscr{A}||_{\mathscr{L}(U)}}{\sigma}||u-v||_U \quad \text{for all } u,v\in U,$$

so the contraction of F is guaranteed if the regularization parameter σ is sufficiently large. Except for matrix multiplications, each iteration step requires forward solving the state equation for $\tilde{y}(u)=\mathscr{S}u$ and backward solving the adjoint equation for $\tilde{p}(\tilde{y})=\mathscr{A}u$. As can be observed in the left plot of Figure 1.5, the iteration indeed does not converge if σ is smaller than some critical value $\sigma_\circ \approx 0.02$. Furthermore, the convergence speed of the iteration loop tends to zero for $\sigma \downarrow \sigma_\circ$. We can therefore make use of this method if the control term $||u||_U^2/2$ in the objective functional J models a control cost such as the required energy and hence is small. On the other hand, if we just penalize the objective functional to enforce the strict convexity property and are interested in the case $\sigma \to 0$ (the resulting controls are usually of "bang-bang" type in this case, i.e., $u(t)\in\{u_a,u_b\}$ for almost all $t\in[0,T]$), we apply some other optimization technique.

2. *Projected gradient method*: A suitable steepest descent method for the control-constrained optimization problem is the projected gradient algorithm; see [36], for instance. Here, the next iteration point is given by the formula $u_{k+1} = \mathbb{P}(u_k+s_kd_k)$, where $d_k=-\nabla J(u_k)=-\sigma u_k+\mathscr{B}'(\mathscr{A}u_k+\hat{p})$ is the direction of the steepest descent of J in the current iteration point u_k, and $s_k>0$ is chosen by Algorithm 1.4. This procedure is globally convergent. However, as before, the convergence speed becomes extremely slow for $\sigma\to 0$. In addition, if the step-size condition $\hat{J}(u_k+s^{(j)}d_k)\le\hat{J}(u_k)-c/s^{(j)}\,||d_k||_U$ is just fulfilled for very small step sizes $s^{(j)}$, many evaluations of the reduced objective functional are required to test whether $\hat{J}(u_k+s^{(j)}d_k)\le\hat{J}(u_k)-cs^{(j)}\,||d_k||_U$ is satisfied. Here, each evaluation requires solving the state equation. Therefore, the single iteration steps may become quite expensive. The right plot of Figure 1.5 demonstrates that the projected gradient method cannot deal with small regularizations. In contrast to the Banach iteration, the residuals decay for arbitrarily small values of σ, but the numerical effort explodes if σ tends to zero.

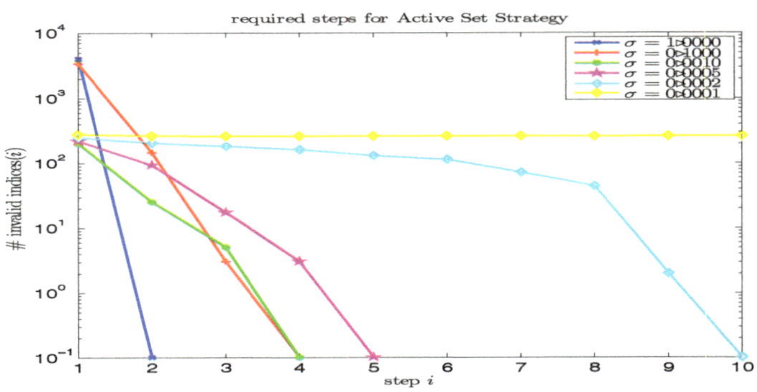

Figure 1.6. *Run 5: The number of grid points, where the previous and the actual active sets differ for different regularization parameters σ.*

3. *Primal-dual active set strategy*: This method—see Algorithm 1.1 for the infinite-dimensional case and Algorithm 1.2 for the POD discretization—solves the state and the adjoint equation simultaneously within the implicit linear scheme

$$u_{k+1}(t) = \chi_{\mathcal{A}_a^k}(t)u_a + \chi_{\mathcal{A}_b^k}(t)u_b(t) + \chi_{\mathcal{I}^k}(t)\frac{1}{\sigma}(\mathcal{B}'\mathcal{A}\, u_{k+1})(t)$$

f.a.a. $t \in [0,T]$. Since this technique is equivalent to a semismooth Newton procedure [24], locally superlinear convergence rates are provided. Further, the algorithm can deal with smaller regularizations than the other two methods presented: reasonable computation times are provided for all $\sigma > \sigma_\circ \approx 0.0002$; see Figure 1.6. For parameters below this critical value, the bang-bang control u oscillates between u_a and u_b at the boundary grid points of the active sets. Notice that both the critical $\sigma_\circ$ and the error between the exact solution and the suboptimal final iteration depend on the number of discretization points. The numerical effort of the simultaneous solving operations in each iteration step is significantly larger than the single solution since the initial condition for the state and the final condition for the adjoint state prevent us from iteratively solving a system of dimension $2m$ n times; instead, all time and space values $(y(t_i,\boldsymbol{x}_j), p(t_i,\boldsymbol{x}_j))$ are determined by solving a linear system of dimension $2nm$. Here, the MOR techniques come into play, which will lead to formidable calculation time reductions (or even make an execution of the primal-dual active set strategy possible). In the following, we will make use of this optimization procedure.

ALGORITHM **1.4. Backtracking strategy.**

Require: Maximal number $j_{\max}$ of iterations and parameter $c \in (0,1)$.
1: Set $s^{(0)} = 1$ and $j = 1$.
2: **while** $\hat{J}(u_k + s^{(j)} d_k) > \hat{J}(u_k) - c/s^{(j)}\, \|d_k\|_U$ **and** $j < j_{\max}$ **do**
3: Set $s^{(j+1)} = s^{(j)}/2$ and $j = j + 1$.
4: **end while**
5: **return** $s_k = s^{(j)}$

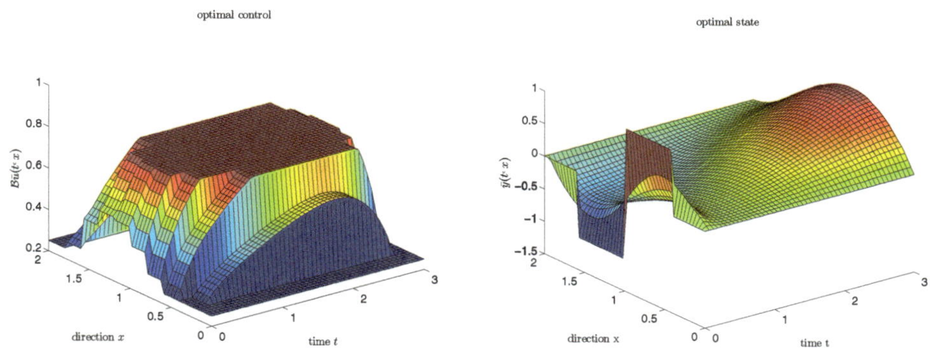

Figure 1.7. *Run 6: The optimal FE control $\mathcal{B}\bar{u}^h$ (left) and the optimal FE state $\bar{y}^h$ (right).*

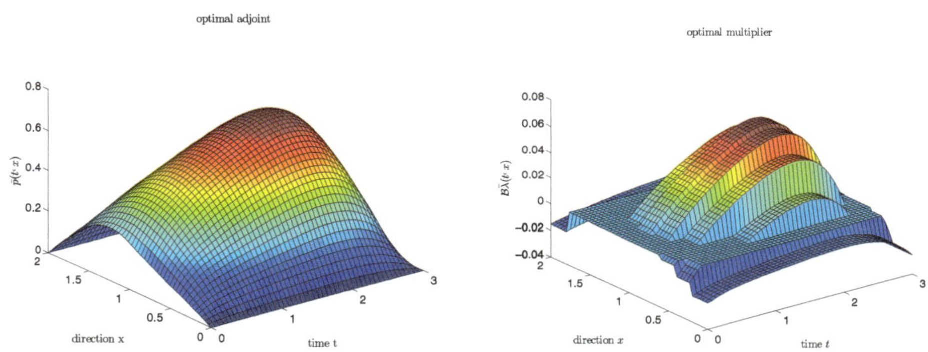

Figure 1.8. *Run 6: The optimal FE adjoint state $\bar{p}^h$ and the optimal FE Lagrange multiplier $\mathcal{B}\bar{\lambda}^h$.*

Run 6 (Different Galerkin expansions). In this run we compare the *modified* POD Galerkin expansions (1.54) for the state variable and (1.85) for the dual variable with the *standard* Galerkin approximations

$$y^\ell(t) = \sum_{i=1}^{\ell} \mathrm{y}_i^\ell(t)\psi_i, \quad p^\ell(t) = \sum_{i=1}^{\ell} \mathrm{p}_i^\ell(t)\psi_i \quad \text{for } t \in [0,T]. \tag{1.101}$$

We choose the same setting as in Run 5. Let $\sigma = 0.1$. In Figures 1.7 and 1.8 we plot the optimal FE solution components $(\bar{y}^h, \bar{u}^h, \bar{p}^h, \bar{\lambda}^h)$ obtained by using the primal-dual active set strategy. We observe that the support of the multiplier $\mathcal{B}\bar{\lambda}^h$ coincides with the active set for the control variable $\mathcal{B}\bar{u}^h$. Further, the relation $\bar{u}^h = \mathbb{P}(\mathcal{B}'\bar{p}^h/\sigma)$ can be observed. As is stated in Remark 1.28, part 1, the advantage of the modified Galerkin ansatz is that the ROM errors do not include the projection of the initial value on the POD space. Figure 1.9 illustrates the impact of homogenization, where we plot not only the ROM errors but also the a posteriori error estimates for different ℓ; compare Section 1.4.5. First we see that the ROM errors and the a posteriori error estimate nearly coincide in all scenarios. In the left plot of Figure 1.9 the POD basis is computed from snapshots of the state equation taking the control guess $u_0 \equiv 1$. One observes that the dynamics of the corresponding homogeneous snapshots in the

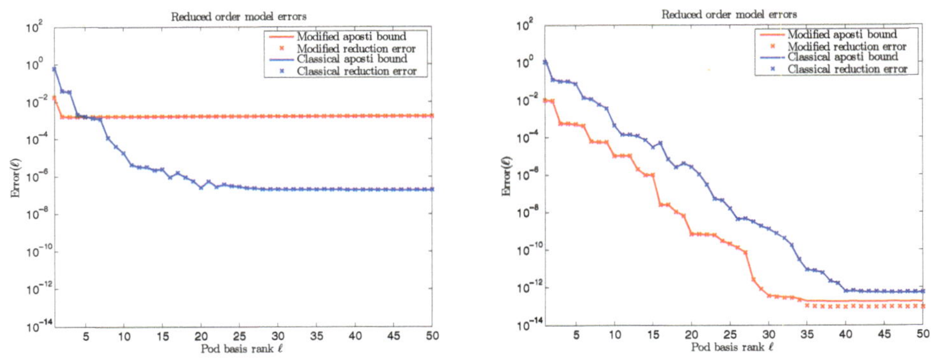

Figure 1.9. *Run 6: The ROM errors for the standard and the modified POD ansatz for initial control guesses $u_\circ = 1$ (left) and $u_\circ = \bar{u}^h$ (right).*

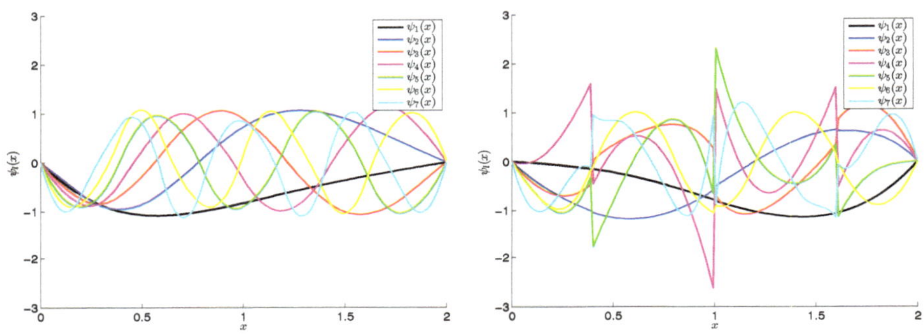

Figure 1.10. *Run 6: The first POD basis elements for the modified (left) and the standard (right) Galerkin expansion.*

modified ansatz are not sufficient to decrease the control error below a level of 10^{-3}, while the standard Galerkin ansatz, exploiting the dynamics of the initial value and the inhomogeneity, induces a higher-dimensional POD space and leads to an error order below 10^{-6}. In the right plot of Figure 1.9, the optimal FE control $\bar{u}^h$ creates the snapshots. Here, the modified Galerkin ansatz pays off: the approximation error in the standard Galerkin ansatz is dominated by the projection error of the initial value $y_\circ$ on the POD Galerkin ansatz space. This example also shows that good approximations of the ROM are only guaranteed when the snapshots that build up the POD basis include the dynamics of the optimal state solution; otherwise, enlarging the POD basis rank does not necessarily improve the accuracy of the results. Algorithm 1.3 proposes a solution for this problem: here, basis updates are provided if the a posteriori error estimator presented in Theorem 1.50 indicates that the control error does not decay in the current POD model. Figure 1.9 shows that these error bounds are sharp. Indeed, if the algorithm is initialized with the control guess $u_\circ \equiv 1$ and a single basis update is provided, a new POD basis is calculated with respect to the achieved suboptimal POD control $u_1^{\ell_{\max}}$. This new POD basis coincides with the POD basis associated with the best (but usually unknown) control guess $\bar{u}^h$. Thus, the resulting error decay by enlarging ℓ is the same as in the right graphic of Figure 1.9. In Figure 1.10, the

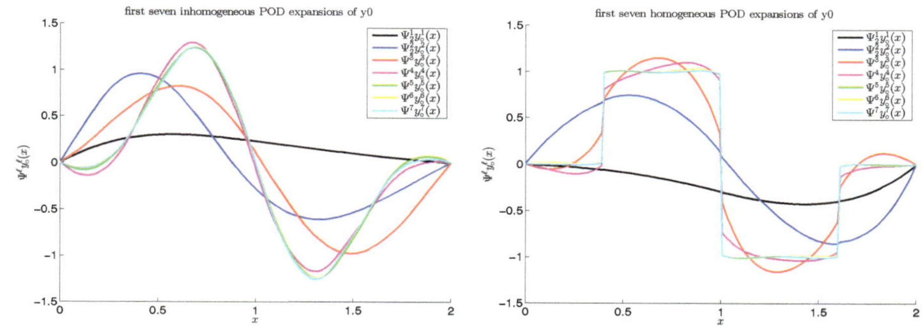

Figure 1.11. *Run 6: The reconstruction error* $\Psi^\ell y_\circ = \sum_{i=1}^{\ell} \langle y_\circ, \psi_i \rangle_H \psi_i$ *for the initial condition* $y_\circ$ *for the modified (left) and the standard (right) POD Galerkin expansions.*

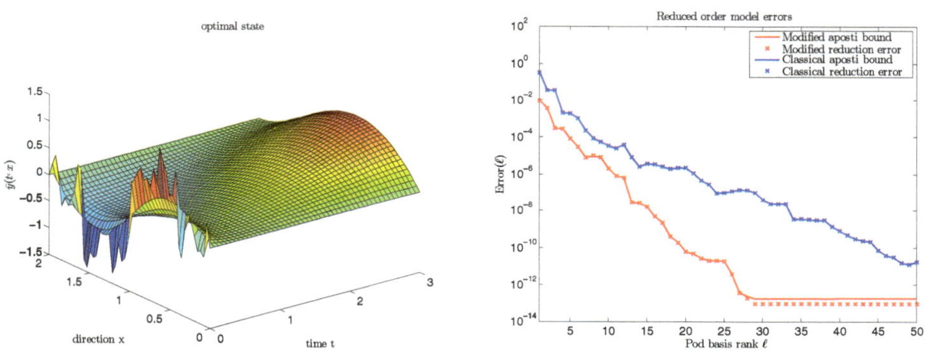

Figure 1.12. *Run 6: The optimal state solution for perturbed initial data (left) and the ROM errors for the two POD ansatzes (right).*

first POD basis functions are presented for the modified and standard Galerkin expansions. Consequently, the reconstruction of the initial condition $y_\circ$ with the standard Galerkin ansatz works quite well, as shown in Figure 1.11—in particular, due to the shape of the POD basis functions, no oscillations at the jump points occur, as can be observed by trigonometric Fourier approximations, for instance. For the modified POD Galerkin ansatz, it is neither required nor possible to build up the initial value $y_\circ$ accurately. But this is not needed because the initial condition is explicitly included in the initial condition; see (1.54). If the model data are perturbed by noise, the improvement of homogenization is significantly stronger. For the following simulation, we add random data to the initial value $y_\circ$. The controls gained by the modified model then reach the optimal precision 10^{-13} with 29 POD basis functions, where even 50 basis elements are not sufficient in the standard ansatz to decrease the error below a level of 10^{-11}; see Figure 1.12. We observe that the noise in the initial value is inherited by the POD basis elements of the modified Galerkin ansatz; despite this perturbation, their shape does not differ much from those of the POD basis for the unperturbed initial conditions of the standard Galerkin ansatz. This is different for the standard POD Galerkin ansatz; compare Figures 1.10 and 1.13.

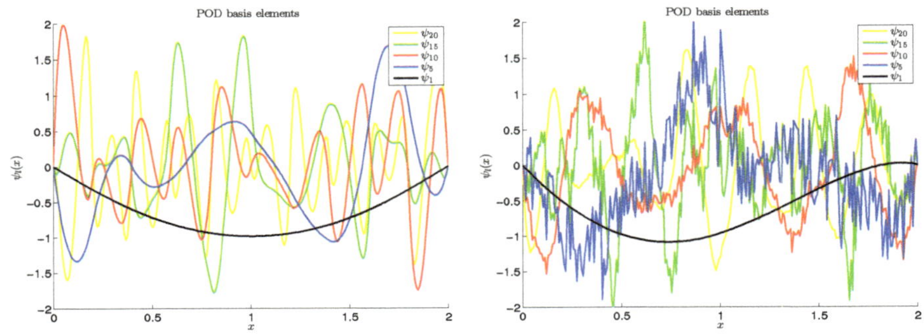

Figure 1.13. *Run 6: The first POD basis elements for the modified (left) and the standard (right) Galerkin expansion in the case of the perturbed initial condition.*

Bibliography

[1] K. AFANASIEV AND M. HINZE, *Adaptive control of a wake flow using proper orthogonal decomposition*, Lecture Notes in Pure and Applied Mathematics, 216 (2001), pp. 317–332.

[2] A.C. ANTOULAS, *Approximation of Large-Scale Dynamical Systems*, Advances in Design and Control 6, SIAM, Philadelphia, 2005.

[3] E. ARIAN, M. FAHL, AND E.W. SACHS, *Trust-Region Proper Orthogonal Decomposition for Flow Control*, Tech. Report 2000-25, ICASE, 2000.

[4] P. ASTRID, S. WEILAND, K. WILLCOX, AND T. BACKX, *Missing point estimation in models described by proper orthogonal decomposition*, IEEE Transactions on Automatic Control, 53 (2008), pp. 2237–2251.

[5] J.A. ATWELL, J.T. BORGGAARD, AND B.B. KING, *Reduced-order controllers for Burgers' equation with a nonlinear observer*, International Journal of Applied Mathematics and Computer Science, 11 (2001), pp. 1311–1330.

[6] H.T. BANKS, M.L. JOYNER, B. WINCHESKY, AND W.P. WINFREE, *Nondestructive evaluation using a reduced-order computational methodology*, Inverse Problems, 16 (2000), pp. 1–17.

[7] S.C. BRENNER AND L.R. SCOTT, *The Mathematical Theory of Finite Element Methods*, Texts in Applied Mathematics, Springer, New York, Berlin, Paris, 2008.

[8] D. CHAPELLE, A. GARIAH, AND J. SAINT-MARIE, *Galerkin approximation with proper orthogonal decomposition: New error estimates and illustrative examples*, Mathematical Modelling and Numerical Analysis, 46 (2012), pp. 731–757.

[9] A. CHATTERJEE, *An introduction to the proper orthogonal decomposition*, Current Science, 78 (2000), pp. 539–575.

[10] R. DAUTRAY AND J.-L. LIONS, *Mathematical Analysis and Numerical Methods for Science and Technology. Volume 5: Evolution Problems I*, Springer, Berlin, Heidelberg, New York, 2000.

[11] K. Deckelnick and M. Hinze, *Error estimates in space and time for tracking-type control of the instationary Stokes system*, in Control and Estimation of Distributed Parameter Systems, W. Desch, F. Kappel, and K. Kunisch, eds., International Series of Numerical Mathematics 143, Birkhäuser, Basel, 2002, pp. 87–103.

[12] ——, *Semidiscretization and error estimates for distributed control of the instationary Navier-Stokes equations*, Numerische Mathematik, 97 (2004), pp. 297–320.

[13] L. Dedè, *Reduced basis method and a posteriori error estimation for parametrized linear-quadratic optimal control problems*, SIAM Journal on Scientific Computing, 32 (2010), pp. 997–1019.

[14] M. Dihlmann and B. Haasdonk, *Certified nonlinear parameter optimization with reduced basis surrogate models*, Proceedings in Applied Mathematics and Mechanics, 13 (2013), pp. 3–6.

[15] F. Diwoky and S. Volkwein, *Nonlinear boundary control for the heat equation utilizing proper orthogonal decomposition*, in Fast Solution of Discretized Optimization Problems, K.-H. Hoffmann, R.H.W. Hoppe, and V. Schulz, eds., International Series of Numerical Mathematics 138, Birkhäuser, Basel, 2001, pp. 73–87.

[16] A.L. Dontchev, W.W. Hager, A.B. Poore, and B. Yang, *Optimality, stability, and convergence in nonlinear control*, Applied Mathematics and Optimization, 31 (1995), pp. 297–326.

[17] L.C. Evans, *Partial Differential Equations*, Graduate Studies in Mathematics, American Mathematical Society, Providence, RI, 2008.

[18] M. Grepl and M. Kärcher, *A posteriori error estimation for reduced order solutions of parametrized parabolic optimal control problems*, Mathematical Modelling and Numerical Analysis, 48 (2014), pp. 1615–1638.

[19] M.A. Grepl, Y. Maday, N.C. Nguyen, and A.T. Patera, *Efficient reduced-basis treatment of nonaffine and nonlinear partial differential equations*, Mathematical Modelling and Numerical Analysis, 41 (2007), pp. 575–605.

[20] M. Gubisch and S. Volkwein, *POD a-posteriori error analysis for optimal control problems with mixed control-state constraints*, Computational Optimization and Applications, 58 (2014), pp. 619–644.

[21] M. Heinkenschloss, D.C. Sorensen, and K. Sun, *Balanced truncation model reduction for a class of descriptor systems with application to the Oseen equations*, SIAM Journal on Scientific Computing, 30 (2008), pp. 1038–1063.

[22] S. Herkt, M. Hinze, and R. Pinnau, *Convergence analysis of Galerkin POD for linear second order evolution equations*, Electronic Transactions on Numerical Analysis, 40 (2013), pp. 321–337.

[23] C. Himpe and M. Ohlberger, *Cross-Gramian-based combined state and parameter reduction for large-scale control systems*, Mathematical Problems in Engineering, 2014 (2014), pp. 1–13.

[24] M. HINTERMÜLLER, K. ITO, AND K. KUNISCH, *The primal-dual active set strategy as a semismooth Newton method*, SIAM Journal on Optimization, 13 (2003), pp. 865–888.

[25] M. HINZE, *A variational discretization concept in control constrained optimization: The linear-quadratic case*, Computational Optimization and Applications, 30 (2005), pp. 45–61.

[26] M. HINZE, R. PINNAU, M. ULBRICH, AND S. ULBRICH, *Optimization with PDE Constraints*, Mathematical Modelling: Theory and Applications, Springer Netherlands, 2009.

[27] M. HINZE AND S. VOLKWEIN, *Proper orthogonal decomposition surrogate models for nonlinear dynamical systems: Error estimates and suboptimal control*, in Dimension Reduction of Large-Scale Systems, Lecture Notes in Computational Science and Engineering 45, Springer-Verlag, Berlin, Heidelberg, Germany, 2005, pp. 261–306.

[28] ——, *Error estimates for abstract linear-quadratic optimal control problems using proper orthogonal decomposition*, Computational Optimization and Applications, 39 (2008), pp. 319–345.

[29] P. HOLMES, J.L. LUMLEY, G. BERKOOZ, AND C.W. ROWLEY, *Turbulence, Coherent Structures, Dynamical Systems and Symmetry*, Cambridge Monographs on Mechanics, Cambridge University Press, 2012.

[30] D. HÖMBERG AND S. VOLKWEIN, *Control of laser surface hardening by a reduced-order approach using proper orthogonal decomposition*, Mathematical and Computer Modelling, 38 (2003), pp. 1003–1028.

[31] K. ITO AND S.S. RAVINDRAN, *A reduced basis method for control problems governed by PDEs*, in Control and Estimation of Distributed Parameter Systems, W. Desch, F. Kappel, and K. Kunisch, eds., International Series of Numerical Mathematics 126, Birkhäuser, Basel, 1998, pp. 153–168. Proceedings of the International Conference in Vorau, 1996.

[32] M. KAHLBACHER AND S. VOLKWEIN, *Galerkin proper orthogonal decomposition methods for parameter dependent elliptic systems*, Discussiones Mathematicae: Differential Inclusions, Control and Optimization, 27 (2007), pp. 95–117.

[33] ——, *POD a-posteriori error based inexact SQP method for bilinear elliptic optimal control problems*, Mathematical Modelling and Numerical Analysis, 46 (2012), pp. 491–511.

[34] E. KAMMANN, F. TRÖLTZSCH, AND S. VOLKWEIN, *A method of a-posteriori error estimation with application to proper orthogonal decomposition*, Mathematical Modelling and Numerical Analysis, 47 (2013), pp. 555–581.

[35] T. KATO, *Perturbation Theory for Linear Operators*, Grundlehren der mathematischen Wissenschaften, Springer-Verlag, Berlin, Heidelberg, New York, 1980.

[36] C.T. KELLEY, *Iterative Methods for Optimization*, Frontiers in Applied Mathematics 18, SIAM, Philadelphia, 1999.

[37] K. KUNISCH AND S. VOLKWEIN, *Control of Burgers' equation by a reduced order approach using proper orthogonal decomposition*, Journal on Optimization Theory and Applications, 102 (1999), pp. 345–371.

[38] ——, *Galerkin proper orthogonal decomposition methods for parabolic problems*, Numerische Mathematik, 90 (2001), pp. 117–148.

[39] ——, *Crank-Nicolson Galerkin proper orthogonal decomposition approximations for a general equation in fluid dynamics*, in Proceedings of the 18th GAMM Seminar on Multigrid and Related Methods for Optimization Problems, 2002, pp. 97–114.

[40] ——, *Galerkin proper orthogonal decomposition methods for a general equation in fluid dynamics*, SIAM Journal on Numerical Analysis, 40 (2002), pp. 492–515.

[41] ——, *Proper orthogonal decomposition for optimality systems*, Mathematical Modelling and Numerical Analysis, 42 (2002), pp. 1–23.

[42] ——, *Optimal snapshot location for computing POD basis functions*, Mathematical Modelling and Numerical Analysis, 44 (2010), pp. 509–529.

[43] K. KUNISCH, S. VOLKWEIN, AND L. XIE, *HJB-POD-based feedback design for the optimal control of evolution problems*, SIAM Journal on Applied Dynamical Systems, 3 (2004), pp. 701–722.

[44] S. LALL, J.E. MARSDEN, AND S. GLAVASKI, *A subspace approach to balanced truncation for model reduction of nonlinear control systems*, International Journal on Robust and Nonlinear Control, 12 (2002), pp. 519–535.

[45] O. LASS AND S. VOLKWEIN, *POD Galerkin schemes for nonlinear elliptic-parabolic systems*, SIAM Journal on Scientific Computing, 35 (2013), pp. A1271–A1298.

[46] F. LEIBFRITZ AND S. VOLKWEIN, *Reduced order output feedback control design for PDE systems using proper orthogonal decomposition and nonlinear semidefinite programming*, Linear Algebra and Its Applications, 415 (2006), pp. 542–575.

[47] H.V. LY AND H.T. TRAN, *Modeling and control of physical processes using proper orthogonal decomposition*, Mathematical and Computer Modelling, 33 (2001), pp. 223–236.

[48] V. MEHRMANN AND T. STYKEL, *Balanced truncation model reduction for large-scale systems in descriptor form*, in Dimension Reduction of Large-Scale Systems, Lecture Notes in Computational Science and Engineering 45, Springer-Verlag, Berlin, Heidelberg, Germany, 2005, pp. 83–115.

[49] F. NEGRI, G. ROZZA, A. MANZONI, AND A. QUATERONI, *Reduced basis method for parametrized elliptic optimal control problems*, SIAM Journal on Scientific Computing, 35 (2013), pp. A2316–A2340.

[50] B. NOACK, K. AFANASIEV, M. MORZYNSKI, G. TADMOR, AND F. THIELE, *A hierarchy of low-dimensional models for the transient and post-transient cylinder wake*, Journal of Fluid Mechanics, 497 (2003), pp. 335–363.

[51] B. NOBLE AND J.W. DANIEL, *Applied Linear Algebra*, Prentice Hall, Upper Saddle River, 3rd ed., 1998.

[52] J. NOCEDAL AND S.J. WRIGHT, *Numerical Optimization*, Springer Series in Operation Research, Springer, Berlin, 2006.

[53] M. OHLBERGER AND M. SCHAEFER, *Error control based model reduction for parameter optimization of elliptic homogenization problems*, in 1st IFAC Workshop on Control of Systems Governed by Partial Differential Equations, CPDE 2013; Paris, France, 25 September 2013 through 27 September 2013; Code 103235, Yann Le Gorrec, ed., vol. 1, International Federation of Automatic Control (IFAC), 2013, pp. 251–256.

[54] A.T. PATERA AND G. ROZZA, *Reduced Basis Approximation and a Posteriori Error Estimation for Parametrized Partial Differential Equations*, MIT-Pappalardo Graduate Monographs in Mechanical Engineering, © MIT, Massachusetts Institute of Technology, Cambridge, MA, 2007.

[55] R. PINNAU, *Model reduction via proper orthogonal decomposition*, in Model Order Reduction: Theory, Research Aspects and Applications, W.H.A. Schilders, H.A. van der Vorst, and J. Rommes, eds., vol. 13 of Mathematics in Industry, Springer, Berlin, Heidelberg, 2008, pp. 95–109.

[56] O. PIRONNEAU, *Calibration of options on a reduced basis*, Journal of Computational and Applied Mathematics, 232 (2009), pp. 139–147.

[57] R. RANNACHER, *Finite element solution of diffusion problems with irregular data*, Numerische Mathematik, 43 (1984), pp. 309–327.

[58] M. RATHINAM AND L. PETZOLD, *Dynamic iteration using reduced order models: A method for simulation of large scale modular systems*, SIAM Journal on Numerical Analysis, 40 (2002), pp. 1446–1474.

[59] S.S. RAVINDRAN, *Reduced-order adaptive controllers for fluid flows using POD*, Journal on Scientific Computing, 15 (2000), pp. 457–478.

[60] M. REED AND B. SIMON, *Methods of Modern Mathematical Physics* I*: Functional Analysis*, Academic Press, New York, 1980.

[61] C.W. ROWLEY, *Model reduction for fluids, using balanced proper orthogonal decomposition*, International Journal of Bifurcation and Chaos, 15 (2005), pp. 997–1013.

[62] E.W. SACHS AND M. SCHU, *A-priori error estimates for reduced order models in finance*, Mathematical Modelling and Numerical Analysis, 47 (2013), pp. 449–469.

[63] E.W. SACHS AND S. VOLKWEIN, *POD Galerkin approximations in PDE-constrained optimization*, GAMM-Mitteilungen, 33 (2010), pp. 194–208.

[64] W.H.A. SCHILDERS, H.A. VAN DER VORST, AND J. ROMMES, *Model Order Reduction: Theory, Research Aspects and Applications*, Mathematics in Industry, Springer, 2008.

[65] M. SCHU, *Adaptive Trust-Region POD Methods and Their Application in Finance*, PhD thesis, University of Trier, 2013.

[66] J.R. SINGLER, *New POD expressions, error bounds, and asymptotic results for reduced order models of parabolic PDEs*, SIAM Journal on Numerical Analysis, 52 (2014), pp. 852–876.

[67] L. SIROVICH, *Turbulence and the dynamics of coherent structures. Parts* I–II, Quarterly of Applied Mathematics, XVL (1987), pp. 561–590.

[68] A. STUDINGER AND S. VOLKWEIN, *Numerical analysis of POD a-posteriori error estimation for optimal control*, in Control and Optimization with PDE Constraints, K. Bredies, C. Clason, K. Kunisch, and G. von Winckel, eds., International Series of Numerical Mathematics 164, Springer, Basel, 2013, pp. 137–158.

[69] T. TONN, K. URBAN, AND S. VOLKWEIN, *Comparison of the reduced-basis and POD a-posteriori error estimators for an elliptic linear quadratic optimal control problem*, Mathematical and Computer Modelling of Dynamical Systems, 17 (2011), pp. 355–369.

[70] F. TRÖLTZSCH, *Optimal Control of Partial Differential Equations. Theory, Methods and Applications*, vol. 112, American Mathematical Society, Providence, RI, 2010.

[71] F. TRÖLTZSCH AND S. VOLKWEIN, *POD a-posteriori error estimates for linear-quadratic optimal control problems*, Computational Optimization and Applications, 44 (2009), pp. 83–115.

[72] M. ULBRICH, *Semismooth Newton Methods for Variational Inequalities and Constrained Optimization Problems in Function Spaces*, MOS-SIAM Series on Optimization 11, SIAM, Philadelphia, 2011.

[73] S. VOLKWEIN, *Optimal control of a phase-field model using proper orthogonal decomposition*, Zeitschrift für Angewandte Mathematik und Mechanik, 81 (2001), pp. 83–97.

[74] ——, *Optimality system POD and a-posteriori error analysis for linear-quadratic problems*, Control and Cybernetics, 40 (2011), pp. 1109–1125.

[75] ——, *Proper Orthogonal Decomposition: Theory and Reduced-Order Modelling*. http://www.math.uni-konstanz.de/numerik/personen/volkwein/teaching/POD-Book.pdf, 2013. Lecture Notes, University of Konstanz.

[76] G. VOSSEN AND S. VOLKWEIN, *Model reduction techniques with a-posteriori error analysis for linear-quadratic optimal control problems*, Numerical Algebra, Control and Optimization, 2 (2012), pp. 465–485.

[77] K. WILLCOX AND J. PERAIRE, *Balanced model reduction via the proper orthogonal decomposition*, American Institute of Aeronautics and Astronautics, 40 (2002), pp. 2323–2330.

[78] K. YOSIDA, *Functional Analysis*, vol. 123 of Classics in Mathematics, Reprint of the 1980 edition, Springer-Verlag, Berlin, Heidelberg, 1995.

[79] K. ZHOU, J.C. DOYLE, AND K. GLOVER, *Robust and Optimal Control*, Prentice-Hall, Upper Saddle River, NJ, 1996.

Chapter 2

Reduced Basis Methods for Parametrized PDEs—A Tutorial Introduction for Stationary and Instationary Problems

Bernard Haasdonk[3]

2.1 ▪ Abstract

In this part we are concerned with a class of model reduction techniques for parametric partial differential equations (PDEs), the so-called reduced basis (RB) methods. These allow us to obtain low-dimensional parametric models for various complex applications, enabling accurate and rapid numerical simulations. Important aspects are basis generation and certification of the simulation results by suitable a posteriori error control. The main terminology, ideas, and assumptions will be explained for the case of linear stationary elliptic, as well as parabolic or hyperbolic, instationary problems. Reproducible experiments will illustrate the theoretical findings. We close with a discussion of further recent developments.

2.2 ▪ Introduction

Discretization techniques for PDEs frequently lead to very high-dimensional numerical models with corresponding high demand for hardware and computation time. This is the case for various discretization types, such as finite element (FE), finite volume (FV), and discontinuous Galerkin (DG) methods. These high computational costs pose a serious problem in the context of multiquery, real-time, or slim computing scenarios. *Multiquery* scenarios comprise problems whose setting varies and where multiple simulations are requested. Such situations can be observed in parameter studies, design, optimization, inverse problems, and statistical analysis. *Real-time* scenarios consist of problems where the simulation result is required very quickly. This can be the case for simulation-based interaction with real processes, e.g., control or prediction, or interaction with humans, e.g., a development engineer working with simulation software and requiring rapid answers. *Slim computing* scenarios denote settings

[3]The author wants to acknowledge the Baden-Württemberg Stiftung gGmbH for funding as well as the German Research Foundation (DFG) for financial support within the Cluster of Excellence in Simulation Technology (EXC 310/1) at the University of Stuttgart.

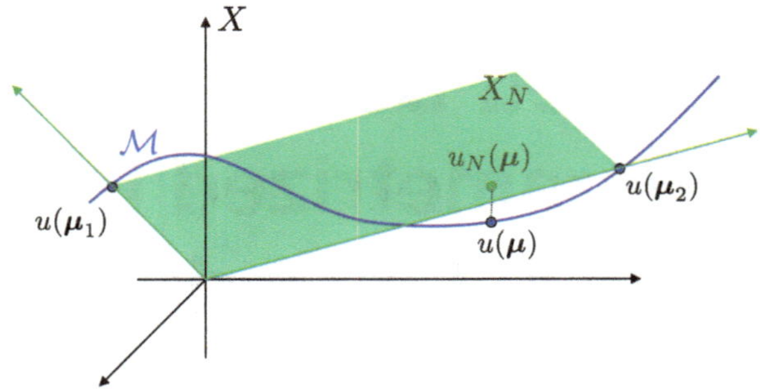

Figure 2.1. *Illustration of the solution manifold and the RB approximation.*

where computational capabilities are very limited with respect to speed or memory but accurate simulation answers are still required. This can comprise simple technical controllers, smartphone apps, etc.

In the above scenarios, the "varying" quantities that describe the problem will be denoted as *parameters* and are collected in a parameter vector $\mu \in \mathcal{P}$. Here we assume that $\mathcal{P} \subset \mathbb{R}^p$ is a set of possible/admitted parameters of low dimension p. The parametric solution will then be denoted $u(\mu)$ and typically stems from a solution space X that can be infinite or at least very high dimensional. Frequently, not the solution itself but rather a quantity of interest $s(\mu)$ depending on the solution is desired. So, the standard computational procedure is to start with a low-dimensional parameter, compute a typically high-dimensional solution $u(\mu)$, and then derive a low-dimensional output quantity. Clearly, the computationally intensive part in this chain is the computation of the solution $u(\mu)$. Therefore, the aim of *model reduction* is to develop techniques that provide low-dimensional and hence rapidly computable approximations for the solution $u(\mu)$ and possible outputs. RB methods focus on a certain problem class, namely parametric PDEs. The crucial insight enabling simplification of parametric problems is the fact that the solution manifold $\mathcal{M}$, i.e., the set of parametric solutions, can often be well approximated by a low-dimensional subspace $X_N \subset X$. One popular RB method is to construct this subspace by *snapshots*, i.e., X_N is spanned by solutions $u(\mu^{(i)})$ for suitable parameters $\mu^{(i)}, i = 1, \dots, N$. A crucial requirement for constructing a good approximating space is a careful choice of these parameters. After construction of the space, the reduced model is obtained, e.g., by Galerkin projection, and provides an approximation $u_N(\mu) \in X_N$ of the solution and an approximation $s_N(\mu)$ of the output quantity of interest. See Figure 2.1 for an illustration of the RB approximation scenario. In addition to the pure reduction, error control is also desired, i.e., availability of computable and *rigorous*, i.e., provable, upper bounds for the state or output error. Additionally, these error bounds should be *effective*, i.e., not arbitrarily overestimate the error. The computational procedure is ideally decomposed into an offline and an online phase. During the *offline* phase, performed once, an RB is generated and further auxiliary quantities are precomputed. Then, in the *online* phase, for varying parameters μ, the approximate solution, output, and error bounds can be provided rapidly. The computational complexity of the online phase will usually not depend on the dimension of the full space X; hence, the space X can be assumed to be arbitrarily accurate. Instead, the computational complexity of the online phase will typically be only polynomial in N, the dimension of the reduced space. The runtime advantage of an RB model in the context of a multiquery scenario is illustrated in

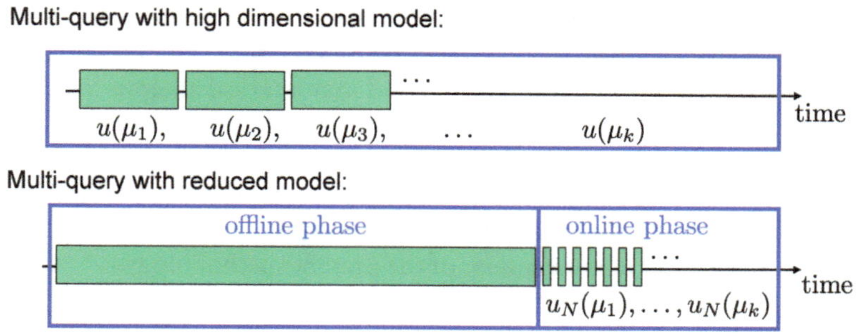

Figure 2.2. *Runtime advantage of RB model in multiquery scenarios.*

Figure 2.2: the offline phase is typically much more expensive than several full simulations. However, if a sufficient number of reduced solutions are required in the online phase, the overall computation time will be decreased in contrast to many full simulations. We collect some of the motivating questions that will be addressed in this chapter:

- How can we construct good spaces X_N? Can such procedures be provably "good"?
- How can we obtain a good approximation $u_N(\mu) \in X_N$?
- How can $u_N(\mu)$ be determined rapidly, i.e., computationally efficiently?
- Can stability or convergence with growing N be obtained?
- Can the RB error be rigorously bounded? Are the error bounds fully computable?
- Do the error bounds largely overestimate the error, or can the "effectivity" be quantified?
- For which problem classes can we expect low-dimensional approximation to be successful?

RB methods can be traced back to the last century [21, 55, 58, 60] but have received plenty of attention in the last decade in terms of *certification*, i.e., providing error and accuracy control, leading to applicable and efficient procedures for various problem types. Examples are stationary linear elliptic problems; stationary noncoercive problems; saddle point problems, in particular Stokes flow; instationary parabolic and hyperbolic problems; nonlinear problems; and geometry parametrizations. Various papers and PhD theses are devoted to the topic. For the moment, we refer to the electronic book [57], the overview article [63], and references therein. Also, we want to refer to the excellent collection of papers and theses at `http://augustine.mit.edu`. Concerning software, different packages have also been developed that address RB methods. Apart from our package *RBmatlab*, available for download at the website `www.morepas.org/software`, we mention the packages *rbMIT* and *rbAPPmit*, available at `augustine.mit.edu/methodology`, and *pymor*, publically accessible at the website `github.com/pymor`. Further references to papers and other electronic resources can be found at `www.morepas.org` and `www.modelreduction.org`. We postpone giving individual further references to the concluding Section 2.5. The purpose of the present chapter is to provide a tutorial introduction to RB methods. It may serve as material for an introductory course on the topic. The suggested audience is students or researchers that have a background in numerical analysis for PDEs and elementary discretization techniques. We aim at a self-contained document, collecting the central statements and providing elementary proofs or explicit references to corresponding literature. We include experiments that can be reproduced with the software package *RBmatlab*. We also provide plenty of exercises that are recommended for deepening the theoretical understanding of the present methodology.

The document is structured into two main parts. The first part consists of Section 2.3, where we consider elliptic coercive problems. This material is largely based on existing work, in particular [57] and lecture notes [27]. We devote the second part, Section 2.4, to the time-dependent case and give corresponding RB formulations. We close this chapter by providing an outlook and references on further topics and recent developments in Section 2.5. A selection of accompanying exercises is given in Section 2.6.

2.3 ▪ Stationary problems

We start with stationary problems and focus on symmetric or nonsymmetric elliptic PDEs. Overall, the collection of results in this chapter serves as a *general RB pattern* for new problem classes. This means that the current sequence of results/procedures can be used as a schedule for other problems. One can try to sequentially obtain analogous results for a new problem along the lines of this section.

Note that the results and procedures of this section are mostly well known and can be considered to be standard. Hence, we do not claim any (major) novelty in the current section but rather see it as a collection and reformulation of existing results and methodology. We introduce slight extensions or intermediate results at some points. Some references that must be attributed are [57, 63] and references therein, but similar formulations can also be found in further publications.

2.3.1 ▪ Model problem

A very elegant model problem is given in [57], which we also want to adopt (with minor modification) as a driving model example for the methodology in this section. It is an example of a parametrized PDE modeling the heat transport through a block of solid material that is assembled by subblocks of different heat conductivities. The values of the piecewise-constant heat conductivities are considered as parameters in the problem. Consequently, the example is called a thermal block. The block is heated on a part of its boundary, insulated on other parts, and cooled to a reference temperature on the remaining boundary part. We are interested in the average temperature on the heating boundary part.

Figure 2.3(a) explains the geometry and the notation. Let $\Omega = (0,1)^2$ be the unit square and $B_1, B_2 \in \mathbb{N}$ the number of subblocks per dimension. The subblocks are denoted $\Omega_i, i = 1, \dots, p$, for $p := B_1 B_2$ counted rowwise starting from the bottom left. The bottom boundary is denoted by $\Gamma_{N,1}$, with unit outward normal $n(x)$, where we will prescribe a unit flux into the domain. The left and right boundaries are insulated no-flux boundaries denoted by $\Gamma_{N,0}$. The upper boundary Γ_D is a homogeneous Dirichlet boundary, where we assign zero as the temperature. The heat conductivities are defined as parameters $\mu_i, i = 1, \dots, p$. We prescribe a suitable parameter domain for the parameter vector $\mu = (\mu_1, \dots, \mu_p)^T \in \mathcal{P} := [\mu_{\min}, \mu_{\max}]^p$, namely logarithmi-

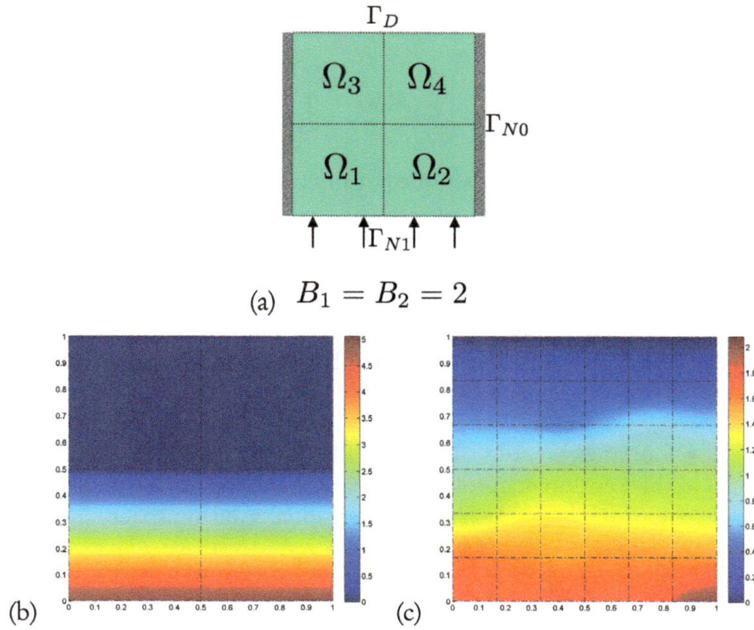

Figure 2.3. *Thermal block:* (a) *geometry and notation,* (b) *solution sample for* $B_1 = B_2 = 2$ *and* $\mu = (1,1,1,1)^T$*, and* (c) *solution sample for* $B_1 = B_2 = 6$ *and random parameter* μ*.*

cally symmetric around one, i.e., $\mu_{\min} = 1/\mu_{\max}$ for $\mu_{\max} > 1$. Note that the model in [57] further assumes the first parameter to be normalized to one; hence that formulation essentially has one parameter less than the current formulation. The space- and parameter-dependent heat-conductivity function is then expressed as a piecewise-constant function via indicator functions χ_{Ω_q}:

$$\kappa(x;\mu) := \sum_{q=1}^{p} \mu_q \chi_{\Omega_q}(x).$$

The parametrized PDE that needs to be solved for the parametric solution $u(x;\mu)$ is the elliptic problem

$$\begin{aligned} -\nabla\cdot(\kappa(x;\mu)\nabla u(x;\mu)) &= 0, \quad x\in\Omega, \\ u(x;\mu) &= 0, \quad x\in\Gamma_D, \\ (\kappa(x;\mu)\nabla u(x;\mu))\cdot n(x) &= i, \quad x\in\Gamma_{N,i},\ i=0,1. \end{aligned}$$

The weak form is based on the solution space $H^1_{\Gamma_D}(\Omega) := \{u \in H^1(\Omega)|\ u|_{\Gamma_D} = 0\}$ of functions vanishing on Γ_D (in the trace sense). Here $H^1(\Omega)$ denotes the standard Sobolev space of square integrable functions, which have square integrable derivatives. Then, for given $\mu\in\mathcal{P}$ we are interested in the solution $u(\cdot;\mu)\in H^1_{\Gamma_D}(\Omega)$ such that

$$\sum_{q=1}^{p}\int_{\Omega_q} \mu_q \nabla u(x;\mu)\cdot\nabla v(x)dx = \int_{\Gamma_{N,1}} v(x)dx$$

for all test functions $v \in H^1_{\Gamma_D}(\Omega)$. Then, we evaluate a scalar output value, e.g., the average temperature at the bottom

$$s(\mu) := \int_{\Gamma_{N,1}} u(x;\mu)dx.$$

This model allows some simple but interesting insights into the structure of the solution manifold for varying μ.

- Simple solution structure: In the case of $B_1 = 1$ (or $B_1 > 1$ but identical parameters in each row), the solution exhibits horizontal symmetry; see Figure 2.3(b). One can easily show that the solution for all $\mu \in \mathcal{P}$ is piecewise linear and contained in a B_2-dimensional linear subspace of the infinite-dimensional space $H^1_{\Gamma_D}(\Omega)$; see Exercise 2.97.
- Complex solution structure: Plot (c) indicates a more complex example, where the solution manifold for $B_1 > 1$ with independent parameters cannot be exactly approximated by a finite-dimensional solution space.
- Parameter redundancy: The solution manifold is invariant with respect to scaling of the parameter vector, i.e., if $u(\mu)$ is a given solution, then $u(c\mu) = \frac{1}{c}u(\mu)$ for $c > 0$ is the solution for the parameter $c\mu$. This is an important insight for parametric models: more parameters do not necessarily increase the parametric complexity of the solution manifold.

The thermal block is an excellent motivation and opportunity for RB methods: solution manifolds may possess structural simplicity or redundancy although the solution space is very high or even infinite dimensional. Identifying such structures in the solution manifold may offer chances for low-dimensional accurate approximation. Based on this example of a parametrized PDE, we can formulate the abstract setting.

2.3.2 ▪ Full problem

An abstract formulation for a large class of linear stationary problems will be given. This will be the basis for the exposition in the subsequent sections. We assume X to be a real separable Hilbert space with inner product $\langle \cdot,\cdot \rangle$, norm $||\cdot||$, and dual space X' with norm $||\cdot||_{X'}$. We assume a parameter domain $\mathcal{P} \subset \mathbb{R}^p$, a parameter-dependent bilinear form $a(\cdot,\cdot;\mu)$, and linear forms $l(\cdot;\mu), f(\cdot;\mu) \in X'$ for all $\mu \in \mathcal{P}$. We do not require symmetry for $a(\cdot,\cdot;\mu)$. We assume the bilinear form and the linear forms to be uniformly continuous and the bilinear form to be uniformly coercive in the following sense.

Definition 2.1 (Uniform continuity and coercivity). *The parametric bilinear form $a(\cdot,\cdot;\mu)$ is assumed to be* continuous, *i.e., there exists $\gamma(\mu) \in \mathbb{R}$ with*

$$\gamma(\mu) := \sup_{u,v \in X\setminus\{0\}} \frac{a(u,v;\mu)}{||u||\,||v||} < \infty,$$

and the continuity is uniform with respect to μ *in the sense that for some $\bar{\gamma} < \infty$, $\gamma(\mu) \leq \bar{\gamma}$ for all $\mu \in \mathcal{P}$. Further, $a(\cdot,\cdot;\mu)$ is assumed to be* coercive, *i.e., there exists $\alpha(\mu)$ with*

$$\alpha(\mu) := \inf_{u \in X\setminus\{0\}} \frac{a(u,u;\mu)}{||u||^2} > 0,$$

and the coercivity is uniform with respect to μ *in the sense that for some* $\bar{\alpha} > 0$, $\alpha(\mu) \geq \bar{\alpha}$ *for all* $\mu \in \mathcal{P}$. *Similarly, we assume the parametric linear forms* $f(\cdot;\mu), l(\cdot;\mu)$ *to be* uniformly continuous, *i.e., there exist constants* $\bar{\gamma}_f, \bar{\gamma}_l < \infty$ *such that for all* $\mu \in \mathcal{P}$,

$$\|l(\cdot;\mu)\|_{X'} \leq \bar{\gamma}_l, \quad \|f(\cdot;\mu)\|_{X'} \leq \bar{\gamma}_f.$$

Example 2.2 (Possible discontinuity with respect to μ). Note that continuity of a, f, and l with respect to u, v does not imply continuity with respect to μ. A simple counterexample can be formulated by $X = \mathbb{R}$, $\mathcal{P} := [0,2]$, $l : X \times \mathcal{P} \to \mathbb{R}$ defined as $l(x;\mu) := x\chi_{[1,2]}(\mu)$ with $\chi_{[1,2]}$ denoting the indicator function of the specified interval. Obviously, l is continuous with respect to x for all μ but discontinuous with respect to μ. ■

With these assumptions, we can define the full problem that is to be approximated by the subsequent RB scheme. The full problem can comprise both a continuous PDE in infinite-dimensional function spaces and an FE discretization of a PDE. The former is interesting from a theoretical point of view (how well can solution manifolds in function spaces be approximated); the latter is important from a practical point of view, as highly resolved discretized PDEs will serve as snapshot suppliers for the RB generation and as the reference solution to compare with.

Definition 2.3 (Full problem ($P(\mu)$)). *For* $\mu \in \mathcal{P}$, *find a solution* $u(\mu) \in X$ *and output* $s(\mu) \in \mathbb{R}$ *satisfying*

$$a(u(\mu), v;\mu) = f(v;\mu), \quad v \in X,$$
$$s(\mu) = l(u(\mu);\mu).$$

Under the above conditions, one obtains the well-posedness and stability of $(P(\mu))$.

Proposition 2.4 (Well-posedness and stability of ($P(\mu)$)). *The problem* $(P(\mu))$ *admits a unique solution satisfying*

$$\|u(\mu)\| \leq \frac{\|f(\cdot;\mu)\|_{X'}}{\alpha(\mu)} \leq \frac{\bar{\gamma}_f}{\bar{\alpha}}, \quad |s(\mu)| \leq \|l(\cdot;\mu)\|_{X'} \|u(\mu)\| \leq \frac{\bar{\gamma}_l \bar{\gamma}_f}{\bar{\alpha}}.$$

Proof. The existence, uniqueness, and bound for $u(\mu)$ follow from the Lax–Milgram theorem; see for instance [8]. Uniform continuity and coercivity then give the parameter-independent bound for $u(\mu)$. The definition of the output functional gives uniqueness for $s(\mu)$, and uniform continuity of l yields the second bound for $s(\mu)$. □

After having ensured solvability, it makes sense to introduce the solution manifold.

Definition 2.5 (Solution manifold). *We introduce the* solution manifold $\mathcal{M}$ *of the full problem* $(P(\mu))$ *as*

$$\mathcal{M} := \{u(\mu) | u(\mu) \textit{ solves } (P(\mu)) \textit{ for } \mu \in \mathcal{P}\} \subset X.$$

We use the notion "manifold," although strictly speaking, it may not be a manifold in the differential geometry sense, as we do not assume continuity/differentiability of $\mathcal{M}$.

A crucial property for efficient implementation of RB methods is parameter separability of all (bi)linear forms.

Definition 2.6 (Parameter separability). *We assume the forms a, f, l to be* parameter separable, *i.e., there exist coefficient functions $\theta_q^a(\mu) : \mathcal{P} \to \mathbb{R}$ for $q = 1, \ldots, Q_a$ with $Q_a \in \mathbb{N}$ and parameter-independent continuous bilinear forms $a_q(\cdot,\cdot) : X \times X \to \mathbb{R}$ such that*

$$a(u, v; \mu) = \sum_{q=1}^{Q_a} \theta_q^a(\mu) a_q(u, v), \qquad \mu \in \mathcal{P}, u, v \in X,$$

and similar definitions for l and f with corresponding continuous linear forms l_q, f_q; coefficient functions θ_q^l, θ_q^f; and numbers of components Q_l, Q_f.

Note that in the literature this property is commonly called *affine parameter dependence*. However, this notion is slightly misleading, as the decomposition can be arbitrarily nonlinear (and hence nonaffine) with respect to the parameter μ. Therefore, we instead denote it *parameter separability*.

In the above definition, the constants Q_a, Q_f, Q_l are assumed to be preferably small (e.g., 1–100), as the complexity of the RB model will explicitly depend on these. We further assume that the coefficient functions $\theta_q^a, \theta_q^f, \theta_q^l$ can be evaluated rapidly.

If such a representation does not exist for a given linear or bilinear form, a parameter-separable approximation can be constructed by the empirical interpolation (EI) method; we comment on this in Section 2.3.7.

Obviously, boundedness of the coefficient functions θ_a^q and continuity of the components a_q imply uniform continuity of $a(\cdot, \cdot; \mu)$, and similarly for f, l. However, coercivity of components a_q only transfers to coercivity of a under additional assumptions; see Exercise 2.99.

Example 2.7 (Thermal block as instantiation of $(P(\mu))$). It can easily be verified that the thermal block model satisfies the above assumptions; see Exercise 2.100. ∎

Example 2.8 ($(P(\mu))$ for matrix equations). The problem $(P(\mu))$ can also be applied to model order reduction (MOR) for parametric matrix equations, i.e., solving and reducing systems

$$\mathbf{A}(\mu)u = \mathbf{b}(\mu)$$

for $\mathbf{A}(\mu) \in \mathbb{R}^{\mathcal{N} \times \mathcal{N}}$, $\mathbf{b}(\mu) \in \mathbb{R}^{\mathcal{N}}$ with $\mathcal{N} \in \mathbb{N}$. This can simply be obtained by considering $X = \mathbb{R}^{\mathcal{N}}$, $a(u, v; \mu) := u^T \mathbf{A}(\mu) v$, and $f(v; \mu) := v^T \mathbf{b}(\mu)$ (and ignoring or choosing arbitrary l). ∎

Example 2.9 ($(P(\mu))$ with given solution). One can prescribe any arbitrarily complicated parametric function $u : \mathcal{P} \to X$ that defines a corresponding manifold $\mathcal{M} := \{u(\mu)\}$. Then, one can construct an instantiation of $(P(\mu))$ that has this solution. For this we only need to set $a(u, v) := \langle u, v \rangle$ and $f(v; \mu) := \langle u(\mu), v \rangle$ and immediately verify that $u(\mu)$ solves the corresponding $(P(\mu))$. This means that the class $(P(\mu))$ may have arbitrarily complex, nonsmooth, or even discontinuous solution manifolds. ∎

Example 2.10 ($(P(\mu))$ for $Q_a = 1$). If $a(\cdot,\cdot;\mu)$ consists of a single component, one can show that $\mathcal{M}$ is contained in an (at most) Q_f-dimensional linear space; see Exercise 2.101. Hence, a finite-dimensional RB space will provide an exact approximation. ∎

We state the following remark on the general misunderstanding of complexity in parametric problems.

Remark 2.11 (Parameter number and complexity). Frequently, problems with many parameters are concluded to have high solution manifold complexity. This is in general wrong. First, as already seen in the parameter redundancy of the thermal block, having many parameters does not necessarily imply a complex solution structure. In the extreme case, one can devise models with an arbitrary number of parameters but one-dimensional solution set; see Exercise 2.98. On the other extreme, one can devise models where one parameter induces arbitrary complex solution behavior. In view of Example 2.9, one can choose an arbitrary connected irregular manifold and assign a one-parameter "space-filling-curve" as solution trajectory on this manifold. The misconception that the number of parameters is directly related to the manifold complexity is very common, even in the MOR community. Nevertheless, certainly in specific examples, the parameter number may influence the solution complexity, as exemplified for the thermal block later on.

It is of interest to state properties of $\mathcal{M}$ that may give information on its complexity/approximability. The first insight is that Proposition 2.4 obviously implies the boundedness of the manifold.

It is possible to show a certain regularity of the manifold. First, one can prove Lipschitz continuity of the manifold along the lines of [20]; see Exercise 2.102.

Proposition 2.12 (Lipschitz continuity). *If the coefficient functions $\theta_q^a, \theta_q^f, \theta_q^l$ are Lipschitz continuous with respect to μ, then the forms a, f, l and the solution $u(\mu)$ and $s(\mu)$ are Lipschitz continuous with respect to μ.*

Further, if the data functions are differentiable, one can even conclude differentiability of the solution manifold by "formally" differentiating $(P(\mu))$. We leave the proof of the following as Exercise 2.103.

Proposition 2.13 (Differentiability, sensitivity problem). *If the coefficient functions θ_q^a, θ_q^f are differentiable with respect to μ, then the solution $u : \mathcal{P} \to X$ is differentiable with respect to μ and the partial derivative* (sensitivity derivative) $\partial_{\mu_i} u(\mu) \in X$ *for $i = 1, \ldots, p$ satisfies the* sensitivity problem

$$a(\partial_{\mu_i} u(\mu), v; \mu) = \tilde{f}_i(v; u(\mu), \mu)$$

for right-hand side $\tilde{f}_i(\cdot; u(\mu), \mu) \in X'$,

$$\tilde{f}_i(\cdot; u(\mu), \mu) := \sum_{q=1}^{Q_f} (\partial_{\mu_i} \theta_q^f(\mu)) f_q(\cdot) - \sum_{q=1}^{Q_a} (\partial_{\mu_i} \theta_q^a(\mu)) a_q(u(\mu), \cdot; \mu).$$

Similar statements hold for higher-order differentiability. So, the partial derivatives of $u(\mu)$ satisfy a similar problem as $(P(\mu))$, in particular involving the identical bilinear form, but a right-hand side that depends on the lower-order derivatives. So we conclude that smoothness of coefficient functions transfers to corresponding smoothness of the solution manifold. Then, the smoother the manifold, the better approximability by low-dimensional spaces may be expected.

2.3.3 ▪ Primal RB approach

We now formulate two RB approaches for the above problem class, which were similarly introduced in [59]. For the moment, we assume we have a low-dimensional space

$$X_N := \mathrm{span}(\Phi_N) = \mathrm{span}\{u(\mu^{(1)}), \dots, u(\mu^{(N)})\} \subset X \tag{2.1}$$

with a basis $\Phi_N = \{\varphi_1, \dots, \varphi_N\}$ available, which will be called the RB space in the following. The functions $u(\mu^{(i)})$ are suitably chosen *snapshots* of the full problem at parameter samples $\mu^{(i)} \in \mathcal{P}$. We will give details on procedures for their choice in Section 2.3.6. At the moment we only assume that the $\{u(\mu^{(i)})\}_{i=1}^N$ are linearly independent. The first RB formulation is a straightforward Galerkin projection. It is denoted "primal," as we will later add a "dual" problem.

Definition 2.14 (Primal RB-problem ($P_N(\mu)$)). *For $\mu \in \mathcal{P}$ find a solution $u_N(\mu) \in X_N$ and output $s_N(\mu) \in \mathbb{R}$ satisfying*

$$\begin{aligned} a(u_N(\mu), v; \mu) &= f(v; \mu), \quad v \in X_N, \\ s_N(\mu) &= l(u_N(\mu)). \end{aligned} \tag{2.2}$$

Again, well-posedness and stability of the reduced solution follow by Lax–Milgram, even using the same constants as for the full problem, as continuity and coercivity are inherited by subspaces.

Proposition 2.15 (Well-posedness and stability of ($P_N(\mu)$)). *The problem ($P_N(\mu)$) admits a unique solution satisfying*

$$\|u_N(\mu)\| \leq \frac{\|f(\mu)\|_{X'}}{\alpha(\mu)} \leq \frac{\bar{\gamma}_f}{\bar{\alpha}}, \quad |s_N(\mu)| \leq \|l(\cdot;\mu)\|_{X'} \|u_N(\mu)\| \leq \frac{\bar{\gamma}_l \bar{\gamma}_f}{\bar{\alpha}}.$$

Proof. We verify the applicability of the Lax–Milgram theorem with the same constants as the full problem using $X_N \subset X$:

$$\begin{aligned} \sup_{u,v \in X_N \setminus \{0\}} \frac{a(u,v;\mu)}{\|u\|\|v\|} &\leq \sup_{u,v \in X \setminus \{0\}} \frac{a(u,v;\mu)}{\|u\|\|v\|} = \gamma(\mu), \\ \inf_{u \in X_N \setminus \{0\}} \frac{a(u,u;\mu)}{\|u\|^2} &\geq \inf_{u \in X \setminus \{0\}} \frac{a(u,u;\mu)}{\|u\|^2} = \alpha(\mu). \end{aligned}$$

Then the argument of Proposition 2.4 applies. □

From a computational viewpoint, the problem ($P_N(\mu)$) is solved by a simple linear equation system.

Proposition 2.16 (Discrete RB problem). *For $\mu \in \mathcal{P}$ and a given RB $\Phi_N = \{\varphi_1, \ldots, \varphi_N\}$, define the matrix, right-hand side, and output vector as*

$$\mathbf{A}_N(\mu) := (a(\varphi_j, \varphi_i; \mu))_{i,j=1}^N \in \mathbb{R}^{N \times N},$$

$$\mathbf{f}_N(\mu) := (f(\varphi_i; \mu))_{i=1}^N \in \mathbb{R}^N, \quad \mathbf{l}_N(\mu) := (l(\varphi_i; \mu))_{i=1}^N \in \mathbb{R}^N.$$

Solve the following linear system for $\mathbf{u}_N(\mu) = (u_{N,i})_{i=1}^N \in \mathbb{R}^N$:

$$\mathbf{A}_N(\mu)\mathbf{u}_N(\mu) = \mathbf{f}_N(\mu).$$

Then, the solution of $(P_N(\mu))$ is obtained by

$$u_N(\mu) = \sum_{j=1}^N u_{N,j}\varphi_j, \quad s_N(\mu) = \mathbf{l}_N^T(\mu)\mathbf{u}_N(\mu). \tag{2.3}$$

Proof. Using linearity, it directly follows that $u_N(\mu)$ and $s_N(\mu)$ from (2.3) satisfy $(P_N(\mu))$. □

Interestingly, in addition to analytical stability from Proposition 2.15, we can also guarantee algebraic stability by using an orthonormal RB. This means we do not have to use snapshots directly as RB vectors, but Φ_N can be a postprocessed set of snapshots, as long as (2.1) holds. For example, a standard Gram–Schmidt orthonormalization can be performed to obtain an orthonormal RB Φ_N.

Proposition 2.17 (Algebraic stability for orthonormal basis). *If $a(\cdot,\cdot;\mu)$ is symmetric and Φ_N is orthonormal, then the condition number of $\mathbf{A}_N(\mu)$ is bounded (independently of N) by*

$$\mathrm{cond}_2(\mathbf{A}_N(\mu)) = \|\mathbf{A}_N(\mu)\| \left\|\mathbf{A}_N(\mu)^{-1}\right\| \leq \frac{\gamma(\mu)}{\alpha(\mu)}.$$

Proof. As $\mathbf{A}_N$ is symmetric and positive definite, we have $\mathrm{cond}_2(\mathbf{A}_N) = \lambda_{\max}/\lambda_{\min}$ with largest/smallest-magnitude eigenvalue of $\mathbf{A}_N$. Let $\mathbf{u} = (u_i)_{i=1}^N \in \mathbb{R}^N$ be an eigenvector of $\mathbf{A}_N$ for eigenvalue $\lambda_{\max}$, and set $u := \sum_{i=1}^N u_i\varphi_i \in X$. Then, due to orthonormality, we obtain

$$\|u\|^2 = \left\langle \sum_{i=1}^N u_i\varphi_i, \sum_{j=1}^N u_j\varphi_j \right\rangle = \sum_{i,j=1}^N u_i u_j \left\langle \varphi_i, \varphi_j \right\rangle = \sum_{i=1}^N u_i^2 = \|\mathbf{u}\|^2.$$

By definition of $\mathbf{A}_N$ and continuity we get

$$\lambda_{\max}\|\mathbf{u}\|^2 = \mathbf{u}^T\mathbf{A}_N\mathbf{u} = a\left(\sum_{i=1}^N u_i\varphi_i, \sum_{j=1}^N u_j\varphi_j\right) = a(u,u) \leq \gamma\|u\|^2,$$

and we conclude that $\lambda_{\max} \leq \gamma$. Similarly, one can show that $\lambda_{\min} \geq \alpha$, which then gives the desired statement. □

This uniform stability bound is a relevant advantage over a nonorthonormal snapshot basis. In particular one can easily realize that snapshots of "close" parameters will result in almost collinear snapshots leading to similar columns and therefore ill-conditioning of the reduced system matrix. But still, for small RBs (e.g., of size 1–10), the orthonormalization can be omitted to prevent additional numerical errors of the orthonormalization procedure.

Remark 2.18 (Difference of FE method from RB). At this point we can note a few distinct differences between the reduced problem and a discretized full problem. For this we denote $\mathbf{A} \in \mathbb{R}^{\mathcal{N}\times\mathcal{N}}$ for some large $\mathcal{N} \in \mathbb{N}$ the FE (or FV, FG, etc.) matrix of the linear system for $(P(\mu))$. Then,

- the RB matrix $\mathbf{A}_N \in \mathbb{R}^{N\times N}$ is small, but typically dense, in contrast to $\mathbf{A}$, which is large but typically sparse, and
- the condition of the matrix $\mathbf{A}_N$ does not deteriorate with growing N if an orthonormal basis is used, in contrast to the high-dimensional $\mathbf{A}$, whose condition number typically grows polynomially in $\mathcal{N}$.

The RB approximation will always be as good as the best approximation, up to a constant. This is a simple version of the lemma of Céa.

Proposition 2.19 (Céa, relation to best approximation). *For all* $\mu \in \mathcal{P}$,

$$\|u(\mu) - u_N(\mu)\| \le \frac{\gamma(\mu)}{\alpha(\mu)} \inf_{v\in X_N} \|u(\mu) - v\|. \tag{2.4}$$

If additionally $a(\cdot,\cdot;\mu)$ *is symmetric, we have the sharpened bound*

$$\|u(\mu) - u_N(\mu)\| \le \sqrt{\frac{\gamma(\mu)}{\alpha(\mu)}} \inf_{v\in X_N} \|u(\mu) - v\|. \tag{2.5}$$

Proof. For all $v \in X_N$, continuity and coercivity result in

$$\begin{aligned}\alpha\,\|u-u_N\|^2 &\le a(u-u_N, u-u_N) = a(u-u_N, u-v) + a(u-u_N, v-u_N)\\ &= a(u-u_N, u-v) \le \gamma\,\|u-u_N\|\,\|u-v\|,\end{aligned}$$

where we used Galerkin orthogonality $a(u-u_N, v-u_N) = 0$, which follows from $(P(\mu))$ and $(P_N(\mu))$ as $v - u_N \in X_N$. For the sharpened bound (2.5) we refer to [57] or Exercise 2.104. □

Similar best-approximation statements are known for interpolation techniques, but the corresponding constants mostly diverge to infinity as the dimension of the approximating space N grows. For the RB approximation, the constant does not grow with N. This is the conceptional advantage of RB approximation by Galerkin projection rather than some other interpolation techniques.

For error analysis, the following error residual relation is important, which states that the error satisfies a variational problem with the same bilinear form, but the residual as right-hand side.

Proposition 2.20 (Error residual relation). *For* $\mu \in \mathcal{P}$ *we define the residual* $r(\cdot;\mu) \in X'$ *via*

$$r(v;\mu) := f(v;\mu) - a(u_N(\mu), v;\mu), \quad v \in X. \tag{2.6}$$

Then, the error $e(\mu) := u(\mu) - u_N(\mu) \in X$ *satisfies*

$$a(e, v; \mu) = r(v; \mu), \quad v \in X. \tag{2.7}$$

Proof. $a(e,v;\mu) = a(u,v;\mu) - a(u_N,v;\mu) = f(v) - a(u_N,v;\mu) = r(v;\mu)$. □

Hence, the residual in particular vanishes on X_N as $X_N \subset \ker(r(\cdot;\mu))$.

A basic consistency property for an RB scheme is reproduction of solutions. The following statement follows trivially from the error estimators, which will be introduced soon. But in case of the absence of error estimators for an RB scheme, this reproduction property can still be investigated. It states that if a full solution happens to be in the reduced space, then the RB scheme will identify this full solution as the reduced solution, giving zero error.

Proposition 2.21 (Reproduction of solutions). *If* $u(\mu) \in X_N$ *for some* $\mu \in \mathcal{P}$*, then* $u_N(\mu) = u(\mu)$.

Proof. If $u(\mu) \in X_N$, then $e = u(\mu) - u_N(\mu) \in X_N$ and we obtain by coercivity and $(P(\mu))$ and $(P_N(\mu))$

$$\alpha(\mu)\|e\|^2 \leq a(e,e;\mu) = a(u(\mu),e;\mu) - a(u_N(\mu),e;\mu) = f(e;\mu) - f(e;\mu) = 0;$$

hence $e = 0$. □

This is a trivial, but useful, statement for at least two reasons, which we state as remarks.

Remark 2.22 (Validation of RB scheme). On the practical side, the reproduction property is useful to validate the implementation of an RB scheme. Choose Φ_N directly as a snapshot basis, i.e., $\varphi_i = u(\mu^{(i)})$ without orthonormalization, and set $\mu = \mu^{(i)}$; then the RB scheme must return $\mathbf{u}_N(\mu) = \mathbf{e}_i$, the ith unit vector, because $u_N(\mu) = \sum_{n=1}^{N} \delta_{ni}\varphi_n$ (with δ_{ni} denoting the Kronecker δ) is obviously the solution expansion.

Remark 2.23 (Uniform convergence of RB approximation). From a theoretical viewpoint we can conclude convergence of the RB approximation to the full continuous problem. We see that the RB solution $u_N : \mathcal{P} \to X$ interpolates the manifold $\mathcal{M}$ at the snapshot parameters $\mu^{(i)}$. Assume that $\mathcal{P}$ is compact and snapshot parameter samples are chosen such that the sets $S_N := \{\mu^{(1)}, \ldots, \mu^{(N)}\} \subset \mathcal{P}$ get dense in $\mathcal{P}$ for $N \to \infty$, i.e., the so-called fill distance h_N tends to zero:

$$h_N := \sup_{\mu \in \mathcal{P}} \operatorname{dist}(\mu, S_N), \quad \lim_{N\to\infty} h_N = 0.$$

Here, $\operatorname{dist}(\mu, S_N) := \min_{\mu' \in S_N} \|\mu - \mu'\|$ denotes the distance of the point μ from the set S_N. If the data functions are Lipschitz continuous, one can show as in Proposition 2.12 that $u_N : \mathcal{P} \to X_N$ is Lipschitz continuous with Lipschitz constant L_u independent of N. Then, obviously, for all N, μ and "closest" $\mu^* = \arg\min_{\mu' \in S_N} \|\mu - \mu'\|$,

$$\begin{aligned}\|u(\mu) - u_N(\mu)\| &\leq \|u(\mu) - u(\mu^*)\| + \|u(\mu^*) - u_N(\mu^*)\| + \|u_N(\mu) - u_N(\mu^*)\| \\ &\leq L_u\|\mu - \mu^*\| + 0 + L_u\|\mu - \mu^*\| \leq 2h_N L_u.\end{aligned}$$

Therefore, we obtain uniform convergence:

$$\lim_{N\to\infty} \sup_{\mu\in\mathcal{P}} ||u(\mu)-u_N(\mu)|| = 0.$$

Note, however, that this convergence rate is linear in h_N and thus is of no practical value, as h_N decays much too slowly with N, and N must be very large to guarantee a small error. In Section 2.3.6 we will see that a more clever choice of $\mu^{(i)}$ can even result in exponential convergence.

We now turn to an important topic in RB methods, namely *certification* by a posteriori error control. This is also based on the residual. We assume we have a rapidly computable lower bound $\alpha_{\mathrm{LB}}(\mu)$ for the coercivity constant available and that α_{LB} is still large in the sense that it is bounded away from zero:

$$0 < \bar{\alpha} \leq \alpha_{\mathrm{LB}}(\mu).$$

This can be assumed without loss of generality, as in the case of $\alpha_{\mathrm{LB}}(\mu) < \bar{\alpha}$ we should better choose $\alpha_{\mathrm{LB}}(\mu) = \bar{\alpha}$ and obtain a larger lower bound constant (assuming $\bar{\alpha}$ to be computable).

Proposition 2.24 (A posteriori error bounds). *Let $\alpha_{\mathrm{LB}}(\mu) > 0$ be a computable lower bound for $\alpha(\mu)$. Then, we have for all $\mu \in \mathcal{P}$*

$$||u(\mu)-u_N(\mu)|| \leq \Delta_u(\mu) := \frac{||r(\cdot;\mu)||_{X'}}{\alpha_{\mathrm{LB}}(\mu)}, \tag{2.8}$$

$$|s(\mu)-s_N(\mu)| \leq \Delta_s(\mu) := ||l(\cdot;\mu)||_{X'}\, \Delta_u(\mu). \tag{2.9}$$

Proof. The case $e = 0$ is trivial, so we assume nonzero error. Testing the error residual equation with e yields

$$\alpha(\mu)\, ||e||^2 \leq a(e,e;\mu) = r(e;\mu) \leq ||r(\cdot;\mu)||_{X'}\, ||e||\,.$$

Division by $||e||$ and α yields the bound for $||e||$. The bound for the output error follows by continuity from

$$|s(\mu)-s_N(\mu)| = |l(u(\mu);\mu)-l(u_N(\mu);\mu)| \leq ||l(\cdot;\mu)||_{X'}\, ||u(\mu)-u_N(\mu)||,$$

which concludes the proof. □

Note that bounding the error by the residual is a well-known technique in FE method analysis for comparing an FE method solution to the analytical solution. However, in that case X is infinite dimensional and the norm $||r||_{X'}$ is not available analytically. In our case, by using X to be a fine discrete FE space, the residual norm becomes a computable quantity, which can be computed *after* the reduced solution $u_N(\mu)$ is available; hence it is an a posteriori bound.

The above technique is an example of a general procedure for obtaining error bounds for RB methods. Show that the RB error satisfies a problem similar to the original problem, but with a residual as inhomogeneity. Then, apply an a priori stability estimate to get an error bound in terms of a residual norm, which is computable in the RB setting.

As the bounds are provable upper bounds to the error, they are denoted *rigorous* error bounds. The availability of a posteriori error bounds is the motivation to denote the approach a *certified* RB method, as we obtain not only an RB approximation but simultaneously a certification by a guaranteed error bound.

Having a bound, we wonder how tight this bound is. The first desirable property of an error bound is that it should be zero if the error is zero; hence we can a posteriori identify an exact approximation.

Corollary 2.25 (Vanishing error bound). *If $u(\mu) = u_N(\mu)$, then*

$$\Delta_u(\mu) = \Delta_s(\mu) = 0.$$

Proof. As $0 = a(0, v; \mu) = a(e, v) = r(v; \mu)$, we see that

$$\|r(\cdot; \mu)\|_{X'} := \sup_u \|r(u; \mu)\| / \|u\| = 0.$$

This implies that $\Delta_u(\mu) = 0$ and $\Delta_s(\mu) = 0$. □

This may give hope that the quotient of error bounds and true error behaves well. In particular, the factor of overestimation can be investigated and, ideally, be bounded by a small constant. The error bounds are then called *effective*. This is possible for $\Delta_u(\mu)$ in our scenario, and the so-called effectivity can be bounded by the continuity and coercivity constants. Thanks to the uniform continuity and coercivity, this is even parameter independent.

Proposition 2.26 (Effectivity bound). *The* effectivity $\eta_u(\mu)$ *is defined and bounded by*

$$\eta_u(\mu) := \frac{\Delta_u(\mu)}{\|u(\mu) - u_N(\mu)\|} \leq \frac{\gamma(\mu)}{\alpha_{\mathrm{LB}}(\mu)} \leq \frac{\bar{\gamma}}{\bar{\alpha}}. \tag{2.10}$$

Proof. Let $v_r \in X$ denote the Riesz representative of $r(\cdot; \mu)$, i.e., we have

$$\langle v_r, v \rangle = r(v; \mu), \quad v \in X, \quad \|v_r\| = \|r(\cdot; \mu)\|_{X'}.$$

Then, we obtain via the error residual equation (2.7) and continuity

$$\|v_r\|^2 = \langle v_r, v_r \rangle = r(v_r; \mu) = a(e, v_r; \mu) \leq \gamma(\mu)\|e\|\|v_r\|.$$

Hence, $\frac{\|v_r\|}{\|e\|} \leq \gamma(\mu)$. We then conclude

$$\eta_u(\mu) = \frac{\Delta_u(\mu)}{\|e\|} = \frac{\|v_r\|}{\alpha_{\mathrm{LB}}(\mu)\|e\|} \leq \frac{\gamma(\mu)}{\alpha_{\mathrm{LB}}(\mu)}$$

and obtain the parameter-independent bound via uniform continuity and coercivity. □

Note that in view of this statement, Corollary 2.25 is trivial. Still, the property stated in Corollary 2.25 has a value of its own, and in more complex RB scenarios without effectivity bounds it may be all one can get. Due to the proven reliability and effectivity of the error bounds, these are also denoted *error estimators*, as they are obviously equivalent to the error up to suitable constants.

In addition to absolute error bounds, it is also possible to derive relative error and effectivity bounds. We again refer to [57] for corresponding proofs and similar statements for other error measures. See also [27] or Exercise 2.105.

Proposition 2.27 (Relative error bound and effectivity). *We have for all $\mu \in \mathcal{P}$*

$$\begin{aligned} \frac{\|u(\mu)-u_N(\mu)\|}{\|u(\mu)\|} \leq \Delta_u^{\mathrm{rel}}(\mu) &:= 2\cdot\frac{\|r(\cdot;\mu)\|_{X'}}{\alpha_{\mathrm{LB}}(\mu)}\cdot\frac{1}{\|u_N(\mu)\|}, \\ \eta_u^{\mathrm{rel}}(\mu) &:= \frac{\Delta_u^{\mathrm{rel}}}{\|e(\mu)\|/\|u(\mu)\|} \leq 3\cdot\frac{\gamma(\mu)}{\alpha_{\mathrm{LB}}(\mu)} \end{aligned} \tag{2.11}$$

if $\Delta_u^{\mathrm{rel}}(\mu)\leq 1$.

Hence, these relative bounds are valid if the error estimator is sufficiently small.

One can put on different "glasses" when analyzing an error, i.e., use different norms, and perhaps obtain sharper bounds. This is possible for the current case by using the (parameter-dependent) energy norm. For this we assume that a is symmetric and define

$$\langle u,v\rangle_\mu := a(u,v;\mu).$$

This form is positive definite by coercivity of a. Hence, $\langle\cdot,\cdot\rangle_\mu$ is a scalar product and induces the *energy norm*

$$\|u\|_\mu := \sqrt{\langle u,u\rangle_\mu}.$$

By coercivity and continuity of a, one can easily see that the energy norm is equivalent to the norm on X by

$$\sqrt{\alpha(\mu)}\,\|u\| \leq \|u\|_\mu \leq \sqrt{\gamma(\mu)}\,\|u\|, \quad u\in X. \tag{2.12}$$

With respect to this norm, one can derive an improved error bound and effectivity. We omit the proof and refer to [57] or [27] and Exercise 2.106.

Proposition 2.28 (Energy norm error bound and effectivity). *For $\mu \in \mathcal{P}$ with symmetric $a(\cdot,\cdot;\mu)$, we have*

$$\begin{aligned} \|u(\mu)-u_N(\mu)\|_\mu \leq \Delta_u^{\mathrm{en}}(\mu) &:= \frac{\|r(\cdot;\mu)\|_{X'}}{\sqrt{\alpha_{\mathrm{LB}}(\mu)}}, \\ \eta_u^{\mathrm{en}}(\mu) := \frac{\Delta_u^{\mathrm{en}}}{\|e\|_\mu} &\leq \sqrt{\frac{\gamma(\mu)}{\alpha_{\mathrm{LB}}(\mu)}}. \end{aligned} \tag{2.13}$$

As $\gamma(\mu)/\alpha_{\mathrm{LB}}(\mu)\geq 1$, this is an improvement by a square root compared to (2.10).

The energy norm allows another improvement in the RB methodology. By choosing a specific $\bar\mu \in \mathcal{P}$, one can choose $\|\cdot\| := \|\cdot\|_{\bar\mu}$ as norm on X. Then, by definition, one obtains $\gamma(\bar\mu) = 1 = \alpha(\bar\mu)$. This means that for the selected parameter the effectivity is $\eta_u(\bar{\mu}) = 1$; hence the error bound exactly corresponds to the error norm. In this sense, the error bound is optimal. Assuming continuity of $\alpha(\mu), \gamma(\mu)$, one can therefore expect that choosing this norm on X will also give highly effective RB error bounds in an environment of $\bar{\mu}$.

We continue with further specialization. For the special case of a *compliant* problem, the above RB scheme $(P_N(\mu))$ turns out to be very good; we obtain effectivities and an output bound that is quadratic in $\Delta_u(\mu)$ instead of only linear. The proof of this statement can be found in [57] or follows as a special instance from Proposition 2.32 in the next section; see Remark 2.34. Still, as this bound can be seen as a central statement, we give the proof.

Proposition 2.29 (Output error bound and effectivity for "compliant" case). *If $a(\cdot,\cdot;\mu)$ is symmetric and $l = f$ (the so-called compliant case), we obtain the improved output bound*

$$0 \leq s(\mu) - s_N(\mu) \leq \Delta_s'(\mu) := \frac{\|r(\cdot;\mu)\|_{X'}^2}{\alpha_{\mathrm{LB}}(\mu)} = \alpha_{\mathrm{LB}}(\mu)\Delta_u(\mu)^2 \tag{2.14}$$

and effectivity bound

$$\eta_s'(\mu) := \frac{\Delta_s'(\mu)}{s(\mu) - s_N(\mu)} \leq \frac{\gamma(\mu)}{\alpha_{\mathrm{LB}}(\mu)} \leq \frac{\bar{\gamma}}{\bar{\alpha}}. \tag{2.15}$$

Proof. Using $a(u_N, e) = 0$, due to Galerkin orthogonality we obtain (omitting μ for brevity)

$$s - s_N = l(u) - l(u_N) = l(e) = f(e) \tag{2.16}$$

$$= f(e) - a(u_N, e) = r(e) = a(e, e). \tag{2.17}$$

Coercivity then implies the first inequality of (2.14). The second inequality and the last equality of (2.14) follow from the error residual relation and the bound for u:

$$a(e,e) = r(e) \leq \|r\|\|e\| \leq \|r\|\Delta_u = \|r\|\frac{\|r\|}{\alpha_{\mathrm{LB}}} = \alpha_{\mathrm{LB}}\Delta_u^2. \tag{2.18}$$

For the effectivity bound (2.15) we first note that with Cauchy–Schwarz and norm equivalence (2.12) the Riesz representative v_r satisfies

$$\|v_r\|^2 = \langle v_r, v_r\rangle = r(v_r) = a(e, v_r) = \langle e, v_r\rangle_\mu \leq \|e\|_\mu \|v_r\|_\mu \leq \|e\|_\mu \sqrt{\gamma}\|v_r\|.$$

Assuming $v_r \neq 0$, division by $\|v_r\|$ yields

$$\|r\|_{X'} = \|v_r\| \leq \|e\|_\mu \sqrt{\gamma}.$$

For $v_r = 0$ this inequality is trivially satisfied. This allows us to conclude using the definitions and (2.17) that

$$\eta_s'(\mu) = \frac{\Delta_s}{s - s_N} = \frac{\|r\|_{X'}^2/\alpha}{a(e,e)} = \frac{\|r\|_{X'}^2}{\alpha\|e\|_\mu^2} \leq \frac{\gamma\|e\|_\mu^2}{\alpha\|e\|_\mu^2} \leq \frac{\bar{\gamma}}{\bar{\alpha}}. \qquad \square$$

Note that the proposition gives a definite sign on the output error, i.e., we always have $s_N(\mu) \leq s(\mu)$.

We will conclude this section with some experimental results to illustrate the theoretical findings. These results can be reproduced via the package *RBmatlab*, which is available for download at `www.morepas.org`. It is a package providing different grid types for spatial discretization, different discretization schemes for PDEs, and various models and implementations of RB schemes. One example is the thermal block model, which is also realized in that package; in particular the plots in Figure 2.3 and the subsequent experiments can be reproduced by the program `rb_tutorial.m`. In the following we recommend that the reader inspect the source code of that program and verify the following results by running different parts of the script. If the reader does not want to install the complete package, `rb_tutorial_standalone.m`, the standalone script using some precomputed data files, offers the same functionality. These files are also accessible via `www.morepas.org`.

We consider the thermal block with $B_1 = B_2 = 2$, $\mu_{\min} = 1/\mu_{\max} = 0.1$; choose five sampling points $\mu^{(j)} = (0.1 + 0.5(j-1), c, c, c)^T, j = 1, \ldots, 5$, with $c = 0.1$; and plot the error estimator $\Delta_u(\mu)$ and true error $\|u(\mu) - u_N(\mu)\|$ for $\mu = (\mu_1, c, c, c)$ over μ_1. The results are depicted in Figure 2.4(a). We can see that the error estimator is finely resolved, as a parameter sweep is computationally cheap thanks to the reduced model. The true error has been sampled more coarsely, as solving the full problem is more tedious. We see that the error bound is indeed an upper bound for the error, confirming the rigor. Further, we see that the true error is indeed (numerically) zero for the chosen sampling points, due to the reproduction of solutions; see Proposition 2.21. Also, the error bound is zero in these points, as is expected by the vanishing error bound property of Corollary 2.25. Finally, the error between sampling points is growing for low-value intervals. This reflects the requirement that for small diffusivity coefficients denser sampling is necessary for uniform error distribution. This fact will be supported later by some a priori analysis.

If we look at the effectivities in Figure 2.4(b), we indeed see that $\eta_u(\mu)$ is only well defined for parameters with nonzero error and it is bounded from above by $\gamma(\mu)/\alpha(\mu)$, in accordance with Proposition 2.26. The values of the effectivities are only on the order of 10, which is considered quite good.

2.3.4 ▪ Primal-dual RB approach

As seen in the previous section, the output error bound $\Delta_s(\mu)$ scales linearly with Δ_u for the general case, Proposition 2.24, and quadratically for the compliant case, Proposition 2.29. By involving a goal-oriented strategy [3] via a corresponding dual problem, one can improve the output and output error estimation in the sense that the output error bounds will also show this "quadratic" behavior for noncompliant problems. For an early reference on the presented RB approach we refer to [59]. We first define the full dual problem.

Definition 2.30 (Full dual problem ($P'(\mu)$)). *For $\mu \in \mathcal{P}$ find a solution $u^{\text{du}}(\mu) \in X$ of*

$$a(v, u^{\text{du}}(\mu); \mu) = -l(v; \mu), \quad v \in X.$$

Again, well-posedness and stability are guaranteed due to coercivity and continuity.

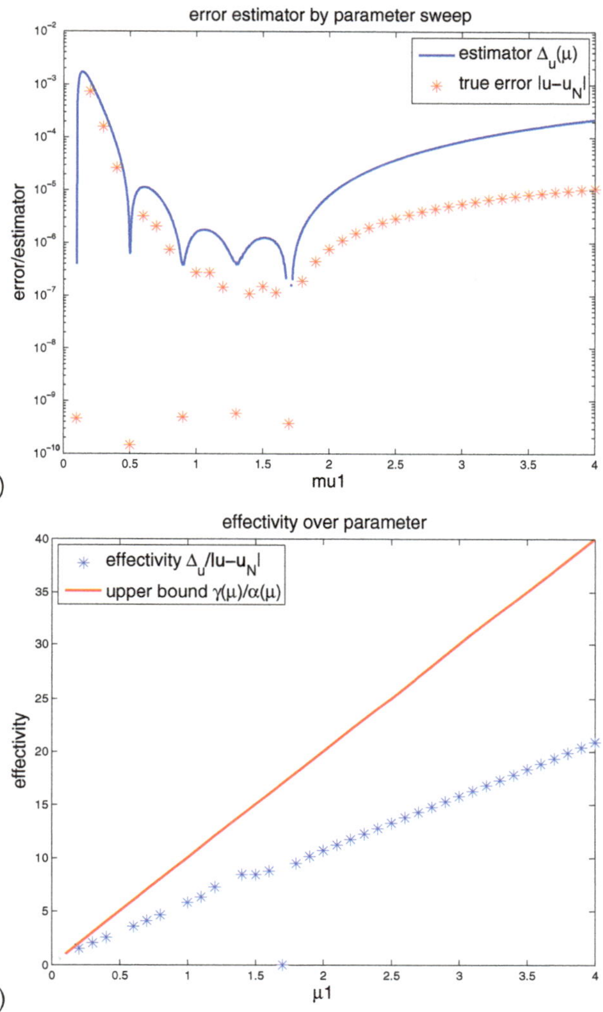

(a) (b)

Figure 2.4. *Illustration of* (a) *error and error bound and* (b) *effectivity and effectivity bound over parameter.*

We assume that an RB space $X_N^{\mathrm{du}} \subset X$ with dimension N^{du}, not necessarily equal to N, is available; see Section 2.3.6 for comments on corresponding basis generation strategies. Then we can define the primal-dual RB approach.

Definition 2.31 (Primal-dual RB problem $(P'_N(\mu))$). *For $\mu \in \mathcal{P}$ let $u_N(\mu) \in X_N$ be the solution of $(P_N(\mu))$. Then, find a solution $u_N^{du}(\mu) \in X_N^{\mathrm{du}}$ and output $s'_N(\mu) \in \mathbb{R}$ satisfying*

$$
\begin{aligned}
a(v, u_N^{\mathrm{du}}(\mu); \mu) &= -l(v; \mu), \quad v \in X_N^{\mathrm{du}}, \\
s'_N(\mu) &= l(u_N(\mu)) - r(u_N^{\mathrm{du}}; \mu).
\end{aligned}
$$

Again, well-posedness and stability follow by coercivity and continuity.

We observe that compared to the primal output $s_N(\mu)$ from (2.2) we have an output estimate $s'_N(\mu)$ using a "correction term" given by the primal residual evaluated at the dual solution. This "correction" allows us to derive sharper error bounds.

For the primal solution $u_N(\mu)$, the error bound (2.8) and effectivity (2.10) are still valid. For the dual variable and the corrected output we obtain the following.

Proposition 2.32 (A posteriori error and effectivity bounds). *For all $\mu \in \mathcal{P}$ we introduce the dual residual*

$$r^{\mathrm{du}}(v;\mu) := -l(v;\mu) - a(v, u_N^{\mathrm{du}}(\mu);\mu), \quad v \in X,$$

and obtain a posteriori error bounds

$$\left\| u^{\mathrm{du}}(\mu) - u_N^{\mathrm{du}}(\mu) \right\| \leq \Delta_u^{\mathrm{du}}(\mu) := \frac{\left\| r^{\mathrm{du}}(\cdot;\mu) \right\|_{X'}}{\alpha_{\mathrm{LB}}(\mu)},$$

$$|s(\mu) - s'_N(\mu)| \leq \Delta'_s(\mu) := \frac{\left\| r^{\mathrm{du}}(\cdot;\mu) \right\|_{X'} \| r(\cdot;\mu) \|_{X'}}{\alpha_{\mathrm{LB}}(\mu)} \tag{2.19}$$

$$= \alpha_{\mathrm{LB}}(\mu) \Delta_u(\mu) \Delta_u^{\mathrm{du}}(\mu) \tag{2.20}$$

and effectivity bound

$$\eta_u^{\mathrm{du}}(\mu) := \frac{\Delta_u^{\mathrm{du}}(\mu)}{\left\| u^{\mathrm{du}}(\mu) - u_N^{\mathrm{du}}(\mu) \right\|} \leq \frac{\gamma(\mu)}{\alpha_{\mathrm{LB}}(\mu)} \leq \frac{\bar{\gamma}}{\bar{\alpha}}.$$

Proof. The error and effectivity bound for the dual solution $u_N^{\mathrm{du}}(\mu)$ follow with identical arguments as for the primal solution. For the improved output bound we first note

$$\begin{aligned} s - s'_N &= l(u) - l(u_N) + r(u_N^{\mathrm{du}}) = l(u - u_N) + r(u_N^{\mathrm{du}}) \\ &= -a(u - u_N, u^{\mathrm{du}}) + f(u_N^{\mathrm{du}}) - a(u_N, u_N^{\mathrm{du}}) \\ &= -a(u - u_N, u^{\mathrm{du}}) + a(u, u_N^{\mathrm{du}}) - a(u_N, u_N^{\mathrm{du}}) \\ &= -a(u - u_N, u^{\mathrm{du}} - u_N^{\mathrm{du}}). \end{aligned} \tag{2.21}$$

Therefore, with the dual error $e^{\mathrm{du}} := u^{\mathrm{du}} - u_N^{\mathrm{du}}$ we obtain

$$\begin{aligned} |s - s'_N| &\leq |a(e, e^{\mathrm{du}})| = |r(e^{\mathrm{du}})| \leq \|r\|_{X'} \left\| e^{\mathrm{du}} \right\| \\ &\leq \|r\|_{X'} \cdot \Delta_u^{\mathrm{du}} \leq \|r\|_{X'} \left\| r^{\mathrm{du}} \right\|_{X'} / \alpha_{\mathrm{LB}} = \alpha_{\mathrm{LB}} \Delta_u \Delta_u^{\mathrm{du}}. \end{aligned}$$ □

Assuming $\Delta_u(\mu) \approx \Delta_u^{\mathrm{du}}(\mu) \approx h \ll 1$, we see a quadratic dependence of Δ'_s on h, in contrast to the simple linear dependence in the output estimate and bound of Proposition 2.24. Hence, the primal-dual approach is expected to give much better output error bounds.

Example 2.33 (Missing effectivity bounds for $\Delta_s(\mu), \Delta'_s(\mu)$). Note that in the general noncompliant case we cannot hope to obtain effectivity bounds for the output error estimators. This is because $s(\mu) - s_N(\mu)$ or $s(\mu) - s'_N(\mu)$ may be zero, while $\Delta_s(\mu), \Delta'_s(\mu)$ are not. Hence, the effectivity as the quotient of these quantities is not well defined. Choose vectors $v_l \perp v_f \in X$ and a subspace $X_N^{\text{du}} = X_N \perp \{v_l, v_f\}$. For $a(u,v) := \langle u, v \rangle, f(v) := \langle v_f, v \rangle$, and $l(v) := -\langle v_l, v \rangle$, we obtain $u = v_f$ as primal solution, $u^{\text{du}} = v_l$ as dual solution, and $u_N = 0, u_N^{\text{du}} = 0$ as RB solutions. Hence, $e = v_f, e^{\text{du}} = v_l$. This yields $s = s_N = 0$, but $r \neq 0$ and $r^{\text{du}} \neq 0$, so $\Delta_s(\mu) > 0$. Similarly, from (2.21) we obtain $s - s'_N = -a(e, e^{\text{du}}) = \langle v_f, v_l \rangle = 0$. But $r \neq 0$ and $r^{\text{du}} \neq 0$, so $\Delta'_s(\mu) > 0$. So, further assumptions, such as in the compliant case, are required to derive output error bound effectivities. ∎

Remark 2.34 (Equivalence of $(P_N(\mu))$ and $(P'_N(\mu))$ for compliant case). In Proposition 2.29 we gave a quadratic output bound statement for the compliant case. In fact, that bound is a simple corollary of Proposition 2.32: We first note that $(P_N(\mu))$ and $(P'_N(\mu))$ are equivalent for the compliant case, assuming $X_N = X_N^{\text{du}}$. Because $f = l$ and a is symmetric, $u_N^{\text{du}}(\mu) = -u_N(\mu)$ solves the dual problem, and $\Delta_u(\mu) = \Delta_u^{\text{du}}(\mu)$. The residual correction term vanishes—$r(u_N^{\text{du}}) = 0$ as $X_N \in \ker(r)$—and therefore $s'_N(\mu) = s_N(\mu)$. Similarly, equivalence of $(P(\mu))$ and $(P'(\mu))$ holds, so $e^{\text{du}} = -e$. Then, from (2.21) we conclude that $s - s_N = -a(e, e^{\text{du}}) = a(e,e) \geq 0$, and the second inequality in (2.14) follows from (2.20). Therefore, $(P_N(\mu))$ is fully sufficient in these cases, and the additional technical burden of the primal-dual approach can be circumvented.

2.3.5 ▪ Offline/online decomposition

We now address computational aspects of the RB methodology. We will restrict ourselves to the primal RB problem; the primal-dual approach can be treated similarly. As the computational procedure will assume that $(P(\mu))$ is a high-dimensional discrete problem, we will first introduce the corresponding notation. We assume that $X = \text{span}(\psi_i)_{i=1}^{\mathcal{N}}$ is spanned by a large number of basis functions ψ_i. We introduce the system matrix, inner product matrix, and functional vectors as

$$\mathbf{A}(\mu) := (a(\psi_j, \psi_i; \mu))_{i,j=1}^{\mathcal{N}} \in \mathbb{R}^{\mathcal{N} \times \mathcal{N}}, \quad \mathbf{K} := (\langle \psi_i, \psi_j \rangle)_{i,j=1}^{\mathcal{N}} \in \mathbb{R}^{\mathcal{N} \times \mathcal{N}}, \tag{2.22}$$

$$\mathbf{f}(\mu) := (f(\psi_i; \mu))_{i=1}^{\mathcal{N}} \in \mathbb{R}^{\mathcal{N}}, \qquad \mathbf{l}(\mu) := (l(\psi_i; \mu))_{i=1}^{\mathcal{N}} \in \mathbb{R}^{\mathcal{N}}. \tag{2.23}$$

Then, the full problem $(P(\mu))$ can be solved by determining the coefficient vector $\mathbf{u} = (u_i)_{i=1}^{\mathcal{N}} \in \mathbb{R}^{\mathcal{N}}$ for $u(\mu) = \sum_{j=1}^{\mathcal{N}} u_j \psi_j$ and output from

$$\mathbf{A}(\mu)\mathbf{u}(\mu) = \mathbf{f}(\mu), \quad s(\mu) = \mathbf{l}(\mu)^T \mathbf{u}(\mu). \tag{2.24}$$

We do not further limit the type of discretization. The system matrix may be obtained from an FE, an FV, or a DG discretization. Typically, $\mathbf{A}(\mu)$ is a sparse matrix, which is always obtained if local differential operators of the PDE are discretized with basis functions of local support. However, the RB methodology can in principle also be applied to discretizations resulting in full system matrices, e.g., integral equations or equations with nonlocal differential terms.

Let us first start with a rough complexity consideration for computing a full and reduced solution. We assume that a single solution of $(P(\mu))$ via (2.24) requires $\mathcal{O}(\mathcal{N}^2)$

operations (e.g., resulting from $\mathcal{N}$ steps of an iterative solver based on $\mathcal{O}(\mathcal{N})$ for each sparse matrix-vector multiplication). In contrast, the dense reduced problem in Proposition 2.16 is solvable in $\mathcal{O}(N^3)$ (assuming direct inversion of $\mathbf{A}_N$ or N steps of an iterative solver based on $\mathcal{O}(N^2)$ for each matrix-vector multiplication). Hence, we clearly see that the RB approach requires $N \ll \mathcal{N}$ to realize a computational advantage.

Let us collect the relevant steps for the computation of an RB solution (and not consider orthonormalization or the error estimators for the moment):

1. N snapshot computations via $(P(\mu^{(i)}))$: $\mathcal{O}(N\mathcal{N}^2)$;
2. N^2 evaluations of $a(\varphi_j, \varphi_i; \mu)$: $\mathcal{O}(N^2\mathcal{N})$;
3. N evaluations of $f(\varphi_i; \mu)$: $\mathcal{O}(N\mathcal{N})$;
4. solution of the $N \times N$ system $(P_N(\mu))$: $\mathcal{O}(N^3)$.

So, RB procedures clearly *do not pay off* if a solution for a single parameter μ is required. But in the case of multiple solution queries, the RB approach will pay off due to so-called offline/online decomposition, as already mentioned in the introduction. During the *offline phase*, *μ-independent*, high-dimensional quantities are precomputed. The operation count typically depends on $\mathcal{N}$; hence this phase is *expensive* but is only performed *once*. During the *online phase*, which is performed for *many* parameters $\mu \in \mathcal{P}$, the offline data are combined to give the small μ-dependent discretized reduced system, and the reduced solution $u_N(\mu)$ and $s_N(\mu)$ is computed rapidly. The operation count of the online phase is ideally completely independent of $\mathcal{N}$ and typically scales polynomially in N.

In view of this desired computational splitting, we see that step 1 above clearly belongs to the offline phase, while step 4 is part of the online phase. But steps 2 and 3 cannot clearly be assigned to either of the two phases as they require expensive as well as parameter-dependent operations. This is where the parameter separability of Definition 2.6 comes into play by suitably dividing steps 2 and 3. The crucial insight is that due to the linearity of the problem, parameter separability of a, f, l transfers to parameter separability of $\mathbf{A}_N, \mathbf{f}_N, \mathbf{l}_N$.

Corollary 2.35 (Offline/online decomposition of $(P_N(\mu))$).
Offline phase: After computation of an RB $\Phi_N = \{\varphi_1, \ldots, \varphi_N\}$, construct the parameter-independent component matrices and vectors

$$\begin{aligned}
\mathbf{A}_{N,q} &:= (a_q(\varphi_j, \varphi_i))_{i,j=1}^N \in \mathbb{R}^{N\times N}, \quad q = 1, \ldots, Q_A,\\
\mathbf{f}_{N,q} &:= (f_q(\varphi_i))_{i=1}^N \in \mathbb{R}^N, \quad q = 1, \ldots, Q_f,\\
\mathbf{l}_{N,q} &:= (l_q(\varphi_i))_{i=1}^N \in \mathbb{R}^N, \quad q = 1, \ldots, Q_l.
\end{aligned}$$

Online phase: For a given $\mu \in \mathcal{P}$, evaluate the coefficient functions $\theta_q^a(\mu)$, $\theta_q^f(\mu)$, $\theta_q^l(\mu)$ for q in suitable ranges and assemble the matrix and vectors

$$\mathbf{A}_N(\mu) = \sum_{q=1}^{Q_a} \theta_q^a(\mu)\mathbf{A}_{N,q}, \quad \mathbf{f}_N(\mu) = \sum_{q=1}^{Q_f} \theta_q^f(\mu)\mathbf{f}_{N,q}, \quad \mathbf{l}_N(\mu) = \sum_{q=1}^{Q_l} \theta_q^l(\mu)\mathbf{l}_{N,q},$$

which exactly results in the discrete system of Proposition 2.16, which can then be solved for $u_N(\mu)$ and $s_N(\mu)$.

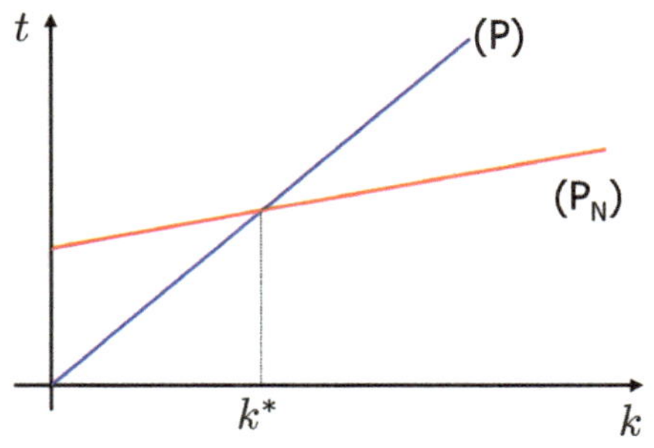

Figure 2.5. *Runtime behavior of the full and the reduced model with increasing number k of simulations.*

Note that the computation of $\mathbf{A}_{N,q}$ can be realized in a very simple way. Let the RB vectors φ_j be expanded in the basis $\{\psi_i\}_{i=1}^{\mathcal{N}}$ of the discrete full problem by $\varphi_j = \sum_{i=1}^{\mathcal{N}} \varphi_{i,j}\psi_i$ with coefficient matrix

$$\mathbf{\Phi}_N := (\varphi_{i,j})_{i,j=1}^{\mathcal{N},N} \in \mathbb{R}^{\mathcal{N}\times N}. \tag{2.25}$$

If the component matrices $\mathbf{A}_q := (a_q(\psi_j, \psi_i))_{i,j=1}^{\mathcal{N}}$ and the component vectors $\mathbf{l}_q := (l_q(\psi_i))_{i=1}^{\mathcal{N}}$, $\mathbf{f}_q := (f_q(\psi_i))_{i=1}^{\mathcal{N}}$ from the full problem are available, the computations for the reduced matrices and vectors reduce to matrix-vector operations:

$$\mathbf{A}_{N,q} = \mathbf{\Phi}_N^T \mathbf{A}_q \mathbf{\Phi}_N, \quad \mathbf{f}_{N,q} = \mathbf{\Phi}_N^T \mathbf{f}_q, \quad \mathbf{l}_{N,q} = \mathbf{\Phi}_N^T \mathbf{l}_q.$$

Concerning complexities, we realize that the offline phase scales on the order of $\mathcal{O}(N\mathcal{N}^2 + N\mathcal{N}(Q_f + Q_l) + N^2\mathcal{N}Q_a)$, the dominating part being the snapshot computation. We see from Corollary 2.35 that the online phase is computable in $\mathcal{O}(N^2 \cdot Q_a + N\cdot(Q_f + Q_l) + N^3)$, in particular completely independent of $\mathcal{N}$. This is why we notationally do not discriminate between the analytical and FE solutions: $\mathcal{N}$ can be chosen arbitrarily high, i.e., the discrete full solution can be chosen arbitrarily accurate, without affecting the computational complexity of the online phase. Certainly, in practice, a certain finite $\mathcal{N}$ must be chosen and then the reduced dimension N must be adapted. Here it is important to note that the RB approximation and error estimation procedure is only informative as long as N is not too large and the reduction error dominates the FE error. A too-large choice of N for a fixed given $\mathcal{N}$ does not make sense. In the limit case $N = \mathcal{N}$ we would have RB errors and estimators being zero, hence exactly reproducing the discrete solution but not the analytical (Sobolev space) solution.

The computational separation can also be illustrated by a runtime diagram; see Figure 2.5. Let $t_{\text{full}}, t_{\text{offline}}, t_{\text{online}}$ denote the computational times for the single computation of a solution of $(P(\mu))$ and the single computations of the offline and online phases of $(P_N(\mu))$. Assuming that these times are constant for all parameters, we obtain linear/affine relations of the overall computation time on the number of simulation requests k: the overall time for k full solutions is $t(k) := k \cdot t_{\text{full}}$, while the reduced model (including offline phase) requires $t_N(k) := t_{\text{offline}} + k \cdot t_{\text{online}}$. As noted earlier, the reduced model pays off as soon as sufficiently many, i.e., $k > k^* := \frac{t_{\text{offline}}}{t_{\text{full}} - t_{\text{online}}}$, simulation requests are expected.

As noted earlier, *certification* by a posteriori error bounds is an important topic. Hence, we will next address the offline/online decomposition of the a posteriori error estimators. The crucial insight is that parameter separability also holds for the residuals and hence for the residual norms.

Proposition 2.36 (Parameter separability of the residual). *Set $Q_r := Q_f + NQ_a$ and define $r_q \in X', q = 1, \ldots, Q_r$ via*

$$(r_1, \ldots, r_{Q_r}) := (f_1, \ldots, f_{Q_f}, a_1(\varphi_1, \cdot), \ldots, a_{Q_a}(\varphi_1, \cdot), \\ \ldots, a_1(\varphi_N, \cdot), \ldots, a_{Q_a}(\varphi_N, \cdot)).$$

Let $u_N(\mu) = \sum_{i=1}^N u_{N,i}\varphi_i$ be the solution of $(P_N(\mu))$ and define $\theta_q^r(\mu), q = 1, \ldots, Q_r$ by

$$(\theta_1^r, \ldots, \theta_{Q_r}^r) := (\theta_1^f, \ldots, \theta_{Q_f}^f, -\theta_1^a \cdot u_{N,1}, \ldots, -\theta_{Q_a}^a \cdot u_{N,1}, \\ \ldots, -\theta_1^a \cdot u_{N,N}, \ldots, -\theta_{Q_a}^a \cdot u_{N,N}).$$

Let $v_r, v_{r,q} \in X$ denote the Riesz representatives of r, r_q, i.e., $r(v) = \langle v_r, v\rangle$ and $r_q(v) = \langle v_{q,r}, v\rangle, v \in X$. Then the residual and its Riesz representatives are parameter separable via

$$r(v;\mu) = \sum_{q=1}^{Q_r} \theta_q^r(\mu) r_q(v), \quad v_r(\mu) = \sum_{q=1}^{Q_r} \theta_q^r(\mu) v_{r,q}, \quad \mu \in \mathcal{P}, v \in X. \quad (2.26)$$

Proof. Linearity of $a(\cdot,\cdot;\mu)$ directly implies that the first equation in (2.26) is a reformulation of r defined in (2.6). Linearity of the Riesz map gives the second statement in (2.26) for the Riesz representative. □

In the error estimation procedure, it is necessary to compute Riesz representatives of linear functionals. We briefly want to comment on how this can be realized.

Lemma 2.37 (Computation of Riesz representatives). *Let $g \in X'$ and $X = \mathrm{span}(\psi_i)_{i=1}^{\mathcal{N}}$ with basis functions ψ_i. We introduce the coefficient vector $\mathbf{v} = (v_i)_{i=1}^{\mathcal{N}} \in \mathbb{R}^{\mathcal{N}}$ of the Riesz representative $v_g = \sum_{i=1}^{\mathcal{N}} v_i \psi_i \in X$. Then, $\mathbf{v}$ can simply be obtained by solving the linear system*

$$\mathbf{K}\mathbf{v} = \mathbf{g}$$

with vector $\mathbf{g} = (g(\psi_i))_{i=1}^{\mathcal{N}} \in \mathbb{R}^{\mathcal{N}}$ and (typically sparse) inner product matrix $\mathbf{K}$ given in (2.22).

Proof. We verify for any test function $u = \sum_{i=1}^{\mathcal{N}} u_i \psi_i$ with coefficient vector $\mathbf{u} = (u_i)_{i=1}^{\mathcal{N}} \in \mathbb{R}^{\mathcal{N}}$ that

$$g(u) = \sum_{i=1}^{\mathcal{N}} u_i g(\psi_i) = \mathbf{u}^T \mathbf{g} = \mathbf{u}^T \mathbf{K}\mathbf{v} = \left\langle \sum_{i=1}^{\mathcal{N}} u_i \psi_i, \sum_{j=1}^{\mathcal{N}} v_j \psi_j \right\rangle = \left\langle v_g, u \right\rangle.$$

□

The parameter separability of the residual allows us to compute the norm of the residual in an offline/online decomposition.

Proposition 2.38 (Offline/online decomposition of the residual norm).
Offline phase: After the offline phase of the RB model, according to Corollary 2.35 we define the matrix

$$\mathbf{G}_r := (r_q(v_{r,q'}))_{q,q'=1}^{Q_r} \in \mathbb{R}^{Q_r \times Q_r}$$

via the residual components r_q *and their Riesz-representatives* $v_{r,q}$.
Online phase: For given μ *and RB solution* $u_N(\mu)$, *we compute the residual coefficient vector* $\theta_r(\mu) := (\theta_1^r(\mu), \ldots, \theta_{Q_r}^r(\mu))^T \in \mathbb{R}^{Q_r}$ *and obtain*

$$\|r(\cdot;\mu)\|_{X'} = \sqrt{\theta_r(\mu)^T \mathbf{G}_r \theta_r(\mu)}.$$

Proof. First we realize that $\mathbf{G}_r = (\langle v_{r,q}, v_{r,q'} \rangle)_{q,q'=1}^{Q_r}$ due to definition of the Riesz representatives. Isometry of the Riesz map and parameter separability (2.26) yield

$$\|r(\mu)\|_{X'}^2 = \|v_r(\mu)\|^2 = \left\langle \sum_{q=1}^{Q_r} \theta_q^r(\mu) v_{r,q}, \sum_{q'=1}^{Q_e} \theta_{q'}^r(\mu) v_{r,q'} \right\rangle = \theta_r(\mu)^T \mathbf{G}_r \theta_r(\mu).$$

□

Again, the online phase is independent of $\mathcal{N}$ as it has complexity $\mathcal{O}(Q_r^2)$. In a completely analogous way, we can compute the dual norm of the output functional used in the output bound (2.9); we omit the proof.

Proposition 2.39 (Offline/online decomposition of $\|l(\cdot;\mu)\|_{X'}$).
Offline phase: We compute the Riesz representatives $v_{l,q} \in X$ *of the output functional components, i.e.,* $\langle v_{l,q}, v \rangle = l_q(v), v \in X$, *and define the matrix*

$$\mathbf{G}_l := (l_q(v_{l,q'}))_{q,q'=1}^{Q_l} \in \mathbb{R}^{Q_l \times Q_l}.$$

Online phase: For given μ, *compute the functional coefficient vector*

$$\theta_l(\mu) := (\theta_1^l(\mu), \ldots, \theta_{Q_l}^l(\mu))^T \in \mathbb{R}^{Q_l}$$

and obtain

$$\|l(\cdot;\mu)\|_{X'} = \sqrt{\theta_l(\mu)^T \mathbf{G}_l \theta_l(\mu)}.$$

A further quantity appearing in relative a posteriori error estimates is the norm of u_N, which can be similarly decomposed.

Proposition 2.40 (Offline/online decomposition of $\|u_N(\mu)\|$).
Offline phase: After the offline phase of the RB model according to Corollary 2.35, we define the reduced inner product matrix

$$\mathbf{K}_N := (\langle \varphi_i, \varphi_j \rangle)_{i,j=1}^N \in \mathbb{R}^{N \times N}. \tag{2.27}$$

Online phase: For given μ and $u_N(\mu) \in \mathbb{R}^N$ computed in the online phase according to Corollary 2.35, we compute

$$\|u_N(\mu)\| = \sqrt{\mathbf{u}_N^T(\mu)\mathbf{K}_N\mathbf{u}_N(\mu)}.$$

Proof. We directly verify that

$$\|u_N\|^2 = \left\langle \sum_i u_{N,i}\varphi_i, \sum_j u_{N,j}\varphi_j \right\rangle = \sum_{i,j=1}^{N} u_{N,i}u_{N,j}\left\langle \varphi_i, \varphi_j \right\rangle = \mathbf{u}_N^T\mathbf{K}_N\mathbf{u}_N.$$

□

Here, too, the online phase is independent of $\mathcal{N}$ as it has complexity $\mathcal{O}(N^2)$. Again, $\mathbf{K}_N$ can be easily obtained by matrix operations. With the full inner product matrix $\mathbf{K}$ defined in (2.22) and the RB coefficient matrix $\mathbf{\Phi}_N$ from (2.25), we compute

$$\mathbf{K}_N = \mathbf{\Phi}_N^T\mathbf{K}\mathbf{\Phi}_N.$$

Analogously, for the relative energy norm bound in the symmetric case, the energy norm can be computed as

$$\|u_N\|_\mu^2 = \mathbf{u}_N^T\mathbf{A}_N\mathbf{u}_N.$$

The remaining ingredient of the a posteriori error estimators is the computation of lower bounds $\alpha_{\mathrm{LB}}(\mu)$ of the coercivity constant. We note that using the uniform lower bound is a viable choice, i.e., $\alpha_{\mathrm{LB}}(\mu) := \bar{\alpha}$, if the latter is available/computable a priori. Further, for some model problems, $\alpha(\mu)$ may be exactly and rapidly computable; hence $\alpha_{\mathrm{LB}}(\mu) := \alpha(\mu)$ may be a valid choice. For example, this is indeed available for the thermal block model; see Exercise 2.107.

A more general approach, the so-called min-theta procedure [57], can be applied under certain assumptions. It makes use of the parameter separability and the (expensive, offline) computation of a single coercivity constant for the full problem. Then, this lower bound can be evaluated rapidly in the online phase.

Proposition 2.41 (Min-theta approach for computing $\alpha_{\mathrm{LB}}(\mu)$). *We assume that the components of $a(\cdot,\cdot;\mu)$ satisfy $a_q(u,u) \geq 0, q = 1,\ldots,Q_a, u \in X$, and the coefficient functions fulfill $\theta_q^a(\mu) > 0, \mu \in \mathcal{P}$. Let $\bar{\mu} \in \mathcal{P}$ such that $\alpha(\bar{\mu})$ is available. Then, we have*

$$0 < \alpha_{\mathrm{LB}}(\mu) \leq \alpha(\mu), \quad \mu \in \mathcal{P},$$

with the lower bound

$$\alpha_{\mathrm{LB}}(\mu) := \alpha(\bar{\mu}) \cdot \min_{q=1,\ldots,Q_a} \frac{\theta_q^a(\mu)}{\theta_q^a(\bar{\mu})}.$$

Proof. Because $0 < \alpha(\bar{\mu})$ and $0 < C(\mu) := \min_q \theta_q^a(\mu)/\theta_q^a(\bar{\mu})$, we have $0 < \alpha(\bar{\mu})C(\mu) =$

$\alpha_{\mathrm{LB}}(\mu)$. For all $u \in X$,

$$\begin{aligned} a(u,u;\mu) &= \sum_{q=1}^{Q_a} \theta_q^a(\mu) a_q(u,u) = \sum_{q=1}^{Q_a} \frac{\theta_q^a(\mu)}{\theta_q^a(\bar{\mu})} \theta_q^a(\bar{\mu}) a_q(u,u) \\ &\geq \sum_{q=1}^{Q_a} \left(\min_{q'=1,\dots Q_a} \frac{\theta_{q'}^a(\mu)}{\theta_{q'}^a(\bar{\mu})} \right) \theta_q^a(\bar{\mu}) a_q(u,u) \\ &= C(\mu) a(u,u;\bar{\mu}) \geq C(\mu)\alpha(\bar{\mu}) \|u\|^2 = \alpha_{\mathrm{LB}}(\mu)\|u\|^2. \end{aligned}$$

In particular, $\alpha(\mu) = \inf_{u \in X\setminus\{0\}}(a(u,u;\mu)/\|u\|^2) \geq \alpha_{\mathrm{LB}}(\mu)$. □

This lower bound is obviously computable in $\mathcal{O}(Q_a)$, hence fast, if we assume a small/decent number of components Q_a.

For the min-theta approach, we require a single evaluation of $\alpha(\mu)$ for $\mu = \bar{\mu}$ in the offline phase. This can be obtained via solving a high-dimensional, hence expensive, eigenvalue problem; see [57] for the continuous formulation.

Proposition 2.42 (Computation of $\alpha(\mu)$ of discretized full problem). *Let* $\mathbf{A}(\mu), \mathbf{K} \in \mathbb{R}^{\mathcal{N}\times\mathcal{N}}$ *denote the high-dimensional discrete system and inner product matrix as given in* (2.22). *Define* $\mathbf{A}_s(\mu) := \frac{1}{2}(\mathbf{A}(\mu) + \mathbf{A}^T(\mu))$ *as the symmetric part of* $\mathbf{A}(\mu)$. *Then,*

$$\alpha(\mu) = \lambda_{\min}(\mathbf{K}^{-1}\mathbf{A}_s(\mu)), \tag{2.28}$$

where $\lambda_{\min}$ *denotes the smallest eigenvalue.*

Proof. We make use of a decomposition $\mathbf{K} = \mathbf{L}\mathbf{L}^T$, e.g., Cholesky or matrix square root, and a substitution $\mathbf{v} := \mathbf{L}^T\mathbf{u}$, and we omit the parameter μ for notational simplicity:

$$\alpha = \inf_{u \in X} \frac{a(u,u)}{\|u\|^2} = \inf_{\mathbf{u} \in \mathbb{R}^{\mathcal{N}}} \frac{\mathbf{u}^T\mathbf{A}\mathbf{u}}{\mathbf{u}^T\mathbf{K}\mathbf{u}} = \inf_{\mathbf{u} \in \mathbb{R}^{\mathcal{N}}} \frac{\mathbf{u}^T\mathbf{A}_s\mathbf{u}}{\mathbf{u}^T\mathbf{K}\mathbf{u}} = \inf_{\mathbf{v} \in \mathbb{R}^{\mathcal{N}}} \frac{\mathbf{v}^T\mathbf{L}^{-1}\mathbf{A}_s\mathbf{L}^{-T}\mathbf{v}}{\mathbf{v}^T\mathbf{v}}.$$

Thus, α is a minimum of a Rayleigh quotient; hence it is the smallest eigenvalue of the symmetric matrix $\tilde{\mathbf{A}}_s := \mathbf{L}^{-1}\mathbf{A}_s\mathbf{L}^{-T}$. The matrices $\tilde{\mathbf{A}}_s$ and $\mathbf{K}^{-1}\mathbf{A}_s$ are similar because

$$\mathbf{L}^T(\mathbf{K}^{-1}\mathbf{A}_s)\mathbf{L}^{-T} = \mathbf{L}^T\mathbf{L}^{-T}\mathbf{L}^{-1}\mathbf{A}_s\mathbf{L}^{-T} = \tilde{\mathbf{A}}_s;$$

hence they have identical eigenvalues, which proves (2.28). □

Certainly, an inversion of $\mathbf{K}$ needs to be prevented; hence in practice one can use an eigenvalue solver that only requires matrix-vector products: as soon as a product $\mathbf{y} = \mathbf{K}^{-1}\mathbf{A}_s\mathbf{x}$ is required, one solves the system $\mathbf{K}\mathbf{y} = \mathbf{A}_s\mathbf{x}$. Alternatively, one can make use of solvers for generalized eigenvalue problems of the form $\mathbf{A}_s\mathbf{u} = \lambda\mathbf{K}\mathbf{u}$ and determine the smallest generalized eigenvalue.

Concerning computational complexity for the a posteriori error estimators $\Delta_u(\mu)$ and $\Delta_s(\mu)$, we obtain as offline complexity $\mathcal{O}(\mathcal{N}^3 + \mathcal{N}^2(Q_f + Q_l + NQ_a) + \mathcal{N}Q_l^2)$. The dominating part corresponds to an eigenvalue problem in cubic complexity for the computation of $\alpha(\bar{\mu})$ for the min-theta procedure. Then, the online phase merely scales as $\mathcal{O}((Q_f + NQ_a)^2 + Q_l^2 + Q_a)$, again fully independent of the high dimension $\mathcal{N}$.

There also exist techniques for computing upper bounds $\gamma_{\mathrm{UB}}(\mu)$ of continuity constants $\gamma(\mu)$; see Exercise 2.108 for a *max-theta* approach. This allows us to evaluate effectivity bounds according to (2.10), etc., online.

For problems where the min-theta approach cannot be applied, the so-called successive constraint method (SCM) [40] is an option. Based on precomputation of many $\alpha(\mu^{(i)})$ in the offline phase, in the online phase a small linear optimization problem is solved for any new $\mu \in \mathcal{P}$, which gives a rigorous lower bound $\alpha_{\mathrm{LB}}(\mu)$.

Definition 2.43 (SCM). *Let $a(\cdot,\cdot;\mu)$ be uniformly coercive with respect to μ and parameter separable with $Q := Q_a$ components. Let $C, D \subset \mathcal{P}$ be finite subsets and $M_\alpha, M_+ \in \mathbb{N}$. Define*

$$Y := \left\{ y = (y_1,\ldots,y_Q) \in \mathbb{R}^Q | \exists u \in X \text{ with } y_q = a_q(u,u)/\|u\|^2, q = 1,\ldots,Q \right\}.$$

We define a target function $J : \mathcal{P} \times \mathbb{R}^Q \to \mathbb{R}$ by

$$J(\mu,y) := \sum_{q=1}^{Q} \theta_q^a(\mu) y_q$$

and a polytope B_Q by

$$\sigma_q^- := \inf_{u\in X} \frac{a_q(u,u)}{\|u\|^2}, \quad \sigma_q^+ := \sup_{u\in X} \frac{a_q(u,u)}{\|u\|^2}, \tag{2.29}$$

$$B_Q := \prod_{q=1}^{Q} [\sigma_q^-, \sigma_q^+] \subset \mathbb{R}^Q. \tag{2.30}$$

For $M \in \mathbb{N}, \mu \in \mathcal{P}$ define $P_M(\mu, C) \subset C$ by

$$P_M(\mu,C) := \begin{cases} M\text{-nearest neighbors of } \mu \text{ in } C & \text{if } \quad 1 \le M \le |C| \\ C & \text{if } \quad |C| \le M \\ \emptyset & \text{if } \quad M = 0. \end{cases}$$

Then, we define for $\mu \in \mathcal{P}$ the sets

$$\begin{aligned} Y_{\mathrm{LB}}(\mu) := \quad & \{ y \in B_Q | J(\mu', y) \ge \alpha(\mu') \forall \mu' \in P_{M_\alpha}(\mu, C) \text{ and} \\ & J(\mu', y) \ge 0 \forall \mu' \in P_{M_+}(\mu, D) \}, \\ Y_{\mathrm{UB}} := & \{ y^*(\mu') | \mu' \in C \} \quad \text{with} \quad y^*(\mu') := \arg\min_{y\in Y} J(\mu', y) \end{aligned}$$

and herewith the quantities

$$\alpha_{\mathrm{LB}}(\mu) := \min_{y \in Y_{\mathrm{LB}}(\mu)} J(\mu, y), \quad \alpha_{\mathrm{UB}}(\mu) := \min_{y \in Y_{\mathrm{UB}}} J(\mu, y). \tag{2.31}$$

With the above definitions, it is easy to show the following bounding property.

Proposition 2.44 (Coercivity constant bounds by SCM). *For all $\mu \in \mathcal{P}$,*

$$\alpha_{\mathrm{LB}}(\mu) \le \alpha(\mu) \le \alpha_{\mathrm{UB}}(\mu). \tag{2.32}$$

Proof. First we see that

$$\alpha(\mu) = \inf_{u \in X \setminus \{0\}} \frac{\sum_{q=1}^{Q} \theta_q^a(\mu) a_q(u,u)}{\|u\|^2} = \min_{y \in Y} J(\mu, y). \tag{2.33}$$

Further, we see that $Y_{\mathrm{UB}} \subset Y \subset Y_{\mathrm{LB}}(\mu)$. For the first inclusion we verify for any $y \in Y_{\mathrm{UB}}$ that there exists $\mu' \in C$ with

$$y = y^*(\mu') = \arg\min_{\bar{y} \in Y} J(\mu', \bar{y});$$

thus $y \in Y$. For the second inclusion we choose $y \in Y$; thus $y \in B_Q$ and we see for any $\mu' \in C$ that

$$\alpha(\mu') = \min_{\bar{y} \in Y} J(\mu', \bar{y}) \leq J(\mu', y).$$

Analogously, for any $\mu' \in D$,

$$0 < \alpha(\mu') \leq J(\mu', y).$$

Hence, $y \in Y_{\mathrm{LB}}(\mu)$. The nestedness of the sets then yields

$$\min_{y \in Y_{\mathrm{LB}}(\mu)} J(\mu, y) \leq \min_{y \in Y} J(\mu, y) \leq \min_{y \in Y_{\mathrm{UB}}} J(\mu, y),$$

which directly implies (2.32) with the definition of the bounds (2.31) and (2.33). □

For a fixed μ, the function J is obviously indeed linear in y, which implies the necessity to solve a small linear optimization problem in the online phase. For further details on the SCM, we refer to [40, 63].

This concludes the section, as we have provided offline/online computational procedures to evaluate all ingredients required for efficient computation of the reduced solution, a posteriori error estimates, and effectivity bounds.

2.3.6 ▪ Basis generation

In this section, we address the topic of basis generation. While this point can be seen as part of the offline phase for generation of the reduced model, it is such a central issue that we devote this separate section to it. One may ask why basis generation was not presented prior to the RB methodology of the previous sections. The reason is that basis generation will make full use of the presented tools of Sections 2.3.3 and 2.3.5 for constructively generating a problem-dependent RB.

We already used the most simple RB type earlier.

Definition 2.45 (Lagrangian RB). *Let $S_N := \{\mu^{(1)}, \ldots, \mu^{(N)}\} \subset \mathcal{P}$ be such that the* snapshots $\{u(\mu^{(i)})\}_{i=1}^N \subset X$ *are linearly independent. We then call*

$$\Phi_N := \{u(\mu^{(1)}), \ldots, u(\mu^{(N)})\}$$

a Lagrangian RB.

An alternative to a Lagrangian RB may be seen in a Taylor RB [21], where one includes sensitivity derivatives of the solution around a certain parameter, e.g., a first-order Taylor RB

$$\Phi_N := \{u(\mu^{(0)}), \partial_{\mu_1} u(\mu^{(0)}), \ldots, \partial_{\mu_p} u(\mu^{(0)})\}.$$

While this basis is expected to give rather good approximation locally around $\mu^{(0)}$, a Lagrangian RB can provide globally well approximating models if S_N is suitably chosen; see the interpolation argument of Remark 2.23.

As we are aiming at global approximation of the manifold $\mathcal{M}$, we define a corresponding error measure. We are interested in finding a space X_N of dimension N minimizing

$$E_N := \sup_{\mu \in \mathcal{P}} \|u(\mu) - u_N(\mu)\|. \tag{2.34}$$

This optimization over all subspaces of a given dimension N is very complex. Hence, a practical relaxation is an incremental procedure. We construct an approximating subspace by iteratively adding new basis vectors. The choice of each new basis vector is led by the aim of minimizing E_N. This is the rationale behind the *greedy procedure*, which was first used in an RB-context in [71] and is standard for stationary problems. It incrementally constructs both the sample set S_N and the basis Φ_N. We formulate the abstract algorithm and comment on practical aspects and choices for its realization. As the main ingredient we require an error indicator $\Delta(Y,\mu) \in \mathbb{R}^+$ that predicts the expected approximation error for the parameter μ when using $X_N = Y$ as the approximation space.

Definition 2.46 (Greedy procedure). *Let $S_{\text{train}} \subset \mathcal{P}$ be a given training set of parameters and $\varepsilon_{\text{tol}} > 0$ a given error tolerance. Set $X_0 := \{0\}, S_0 = \emptyset, \Phi_0 := \emptyset, n := 0$ and define iteratively*

$$\begin{aligned}
\textit{while } \varepsilon_n &:= \max_{\mu \in S_{\text{train}}} \Delta(X_n,\mu) > \varepsilon_{\text{tol}}, \\
\mu^{(n+1)} &:= \arg\max_{\mu \in S_{\text{train}}} \Delta(X_n,\mu), \\
S_{n+1} &:= S_n \cup \{\mu^{(n+1)}\}, \\
\varphi_{n+1} &:= u(\mu^{(n+1)}), \\
\Phi_{n+1} &:= \Phi_n \cup \{\varphi_{n+1}\}, \\
X_{n+1} &:= X_n \oplus \operatorname{span}(\varphi_{n+1}), \\
n &\leftarrow n+1, \\
&\textit{end while.}
\end{aligned} \tag{2.35}$$

The algorithm produces the desired RB space X_N and basis Φ_N by setting $N := n+1$ as soon as (2.35) is false. We can state a simple termination criterion for the above algorithm: if for all $\mu \in \mathcal{P}$ and subspaces $Y \subset X$ we have

$$u(\mu) \in Y \Rightarrow \Delta(Y,\mu) = 0, \tag{2.36}$$

then the above algorithm terminates in at most $N \leq |S_{\text{train}}|$ steps, where $|\cdot|$ indicates the cardinality of a given set. The reason is that with (2.36) no sample in S_{train} will be selected twice. This criterion is easily satisfied by reasonable indicators.

Alternatively, the first iteration, i.e., determination of $\mu^{(1)}$, is frequently skipped by choosing a random initial parameter vector.

The greedy procedure generates a Lagrangian RB with a carefully selected sample set. The basis is *hierarchical* in the sense that $\Phi_n \subset \Phi_m$ for $n \leq m$. This allows us to adjust the accuracy of the reduced model online by varying its dimension.

The training set S_{train} is mostly chosen as a (random or structured) finite subset. The maximization is then a linear search. The training set must represent $\mathcal{P}$ well in order to not "miss" relevant parts of the parameter domain. In practice it should be taken as large as possible.

Remark 2.47 (Choice of error indicator $\Delta(Y,\mu)$). There are different options for the choice of the error indicator $\Delta(Y,\mu)$ in the greedy procedure, each with advantages, disadvantages, and requirements.

(i) Projection error as indicator: In some cases, it can be recommended to use the above algorithm with the best-approximation error (which is an orthogonal projection error)

$$\Delta(Y,\mu) := \inf_{v\in Y} \|u(\mu)-v\| = \|u(\mu)-P_Y u(\mu)\|.$$

Here, P_Y denotes the orthogonal projection onto Y. In this version of the greedy procedure, the error indicator is expensive to evaluate because high-dimensional operations are required; hence S_{train} must be of moderate size. Also, all snapshots $u(\mu)$ must be available, which possibly limits the size of S_{train} due to memory constraints. Still, the advantage of this approach is that the RB model is decoupled from the basis generation: no RB model or a posteriori error estimators are required; the algorithm purely constructs a good approximation space. By statements such as the best-approximation relation of Proposition 2.19, one can then be sure that the corresponding RB model using the constructed X_N will be good. This version of the greedy algorithm will be denoted as the *strong greedy* procedure.

(ii) True RB error as indicator: If one has an RB model available, but no a posteriori error bounds, one can use

$$\Delta(Y,\mu) := \|u(\mu)-u_N(\mu)\|.$$

Again, in this version of the greedy procedure, the error indicator is expensive to evaluate; hence S_{train} must be of moderate size. And again, all snapshots $u(\mu)$ must be available, limiting the size of S_{train}. The advantage of this approach is that the error criterion that is minimized exactly is the measure used in E_N in (2.34).

(iii) A posteriori error estimator as indicator: This is the recommended approach if one has both an RB model and an a posteriori error estimator available. Hence, we choose

$$\Delta(Y,\mu) := \Delta_u(\mu).$$

The evaluation of $\Delta(Y,\mu)=\Delta_u(\mu)$ (or a relative estimator) is very cheap; hence the sample set S_{train} can be chosen much larger than when using a true RB or projection error as indicator. By this, the training set S_{train} can be expected to be much more representative of the complete parameter space in contrast to a smaller training set. No snapshots need to be precomputed. In the complete greedy procedure, only N high-dimensional solves of $(P(\mu))$ are required. Hence, the complete greedy procedure is expected to be rather fast. This version of the greedy algorithm is called *weak greedy*, as will be explained more precisely in the subsequent convergence analysis.

Note that all of these choices for $\Delta(Y,\mu)$ satisfy (2.36): This statement is trivial for (i), the projection error. For the RB error (ii), it is a consequence of the reproduction of solutions, Proposition 2.21, and for the a posteriori error estimators, it is a consequence of the vanishing error bound, Corollary 2.25. Hence, the greedy algorithm is guaranteed to terminate.

Alternatively, one can also use goal-oriented indicators, i.e., $\Delta(Y,\mu) = |s(\mu) - s_N(\mu)|$ or $\Delta_s(\mu)$. One can expect to obtain a rather small basis that approximates $s(\mu)$ very well, but $u(\mu)$ will possibly not be well approximated. In contrast, by using the above indicators (i), (ii), (iii), one can expect to obtain a larger basis, which accurately approximates $u(\mu)$ as well as the output $s(\mu)$.

Note that in general one cannot expect monotonical decay of ε_n for $n = 1,\dots,N$. Only in certain cases can this be proved; see Exercise 2.109.

Remark 2.48 (Overfitting, quality measurement). The error sequence $\{\varepsilon_n\}_{n=0}^N$ generated by the greedy procedure in (2.35) is only a *training error* in statistical learning terminology. The quality of a model, its generalization capabilities, cannot necessarily be concluded from this due to possible *overfitting*. This means that possibly

$$\max_{\mu\in\mathcal{P}} \Delta(X_N,\mu) \gg \varepsilon_N.$$

If $\varepsilon_{\mathrm{tol}}$ is sufficiently small, for example, we obtain $N = |S_{\mathrm{train}}|$ and we will have the training error $\varepsilon_N = 0$. But the model is very likely not exact on the complete parameter set. Hence, it is always recommended to evaluate the quality of a model on an *independent test set* $S_{\mathrm{test}} \subset \mathcal{P}$, which is not related to S_{train}.

We give some hints on the theoretical foundation of the above greedy procedure. Until recently, this algorithm seemed to be a heuristic procedure that worked very well in practice in various cases. Rigorous analysis was not available. But then a useful approximation-theoretic result was formulated, first concerning exponential convergence [9] and then also for algebraic convergence [5]. It states that if $\mathcal{M}$ can be approximated well by some linear subspace, then the greedy algorithm will identify approximation spaces that are only slightly worse than these optimal subspaces. The optimal subspaces are defined via the *Kolmogorov n-width*, defined as the maximum error of the best-approximating linear subspace

$$d_n(\mathcal{M}) := \inf_{\substack{Y\subset X\\ \dim Y=n}} \sup_{u\in\mathcal{M}} \|u - P_Y u\|. \tag{2.37}$$

The convergence statement [5] adapted to our notation and assumptions can then be formulated as follows; we omit the proof.

Proposition 2.49 (Greedy convergence rates). *Let* $S_{\mathrm{train}} = \mathcal{P}$ *be compact and the error indicator* Δ *be chosen such that for suitable* $\gamma \in (0,1]$,

$$\left\|u(\mu^{(n+1)}) - P_{X_n} u(\mu^{(n+1)})\right\| \geq \gamma \sup_{u\in\mathcal{M}} \left\|u - P_{X_n} u\right\|. \tag{2.38}$$

(i) *Algebraic convergence rate: If* $d_n(\mathcal{M}) \leq M n^{-\alpha}$ *for some* $\alpha, M > 0$ *and all* $n \in \mathbb{N}$ *and* $d_0(\mathcal{M}) \leq M$, *then*

$$\varepsilon_n \leq C M n^{-\alpha}, n > 0,$$

with a suitable (explicitly computable) constant $C > 0$.

(ii) *Exponential convergence rate: If* $d_n(\mathcal{M}) \le Me^{-an^\alpha}$ *for* $n \ge 0, M, a, \alpha > 0$*, then*

$$\varepsilon_n \le CMe^{-cn^\beta}, n \ge 0,$$

with $\beta := \alpha/(\alpha+1)$ *and suitable (explicitly computable) constants* $c, C > 0$.

Remark 2.50 (Strong versus weak greedy). If $\gamma = 1$ (e.g., obtained for the choice in Remark 2.47(i), $\Delta(Y,\mu) := \|u(\mu) - P_Y u(\mu)\|$), the algorithm is called *strong* greedy, while for $\gamma < 1$ the algorithm is called *weak* greedy. Note that (2.38) is valid in the case of $\Delta(Y,\mu) := \Delta_u(\mu)$, i.e., Remark 2.47(iii), due to the lemma of Céa (Proposition 2.19), the effectivity (Proposition 2.26), and the error bound property (Proposition 2.24). Using the notation $u_N(\mu), \Delta_u(\mu)$ for the RB solution and estimator and the corresponding intermediate spaces X_n for $1 \le n \le N$, we derive

$$\left\| u(\mu^{(n+1)}) - P_{X_n} u(\mu^{(n+1)}) \right\| = \inf_{v \in X_n} \left\| u(\mu^{(n+1)}) - v \right\|$$

$$\begin{aligned}
&\ge \frac{\alpha(\mu)}{\gamma(\mu)} \left\| u(\mu^{(n+1)}) - u_N(\mu^{(n+1)}) \right\| \ge \frac{\alpha(\mu)}{\gamma(\mu)\eta_u(\mu)} \Delta_u(\mu^{(n+1)}) \\
&= \frac{\alpha(\mu)}{\gamma(\mu)\eta_u(\mu)} \sup_{\mu \in \mathcal{P}} \Delta_u(\mu) \ge \frac{\alpha(\mu)}{\gamma(\mu)\eta_u(\mu)} \sup_{\mu \in \mathcal{P}} \|u(\mu) - u_N(\mu)\| \\
&\ge \frac{\alpha(\mu)}{\gamma(\mu)\eta_u(\mu)} \sup_{\mu \in \mathcal{P}} \left\| u(\mu) - P_{X_n} u(\mu) \right\| \ge \frac{\bar{\alpha}^2}{\bar{\gamma}^2} \sup_{\mu \in \mathcal{P}} \left\| u(\mu) - P_{X_n} u(\mu) \right\|.
\end{aligned}$$

Hence, a weak greedy algorithm with parameter $\gamma := \frac{\bar{\alpha}^2}{\bar{\gamma}^2} \in (0,1]$ is obtained.

As a result, the greedy algorithm is theoretically well founded. Now the question arises as to when "good approximability" is to be expected for a given problem. A positive answer is given in [52, 57]. The assumptions in the statement are, for example, satisfied for the thermal block with $B_1 = 2, B_2 = 1$, fixing $\mu_1 = 1$ and choosing a single scalar parameter $\mu := \mu_2$.

Proposition 2.51 (Global exponential convergence for $p = 1$). *Let* $\mathcal{P} = [\mu_{\min}, \mu_{\max}] \subset \mathbb{R}^+$ *with* $0 < \mu_{\min} < 1, \mu_{\max} = 1/\mu_{\min}$*. Further, assume that* $a(u,v;\mu) := \mu a_1(u,v) + a_2(u,v)$ *is symmetric,* f *is not parameter dependent,* $a := \ln \frac{\mu_{\max}}{\mu_{\min}} > \frac{1}{2e}$*, and* $N_0 := 1 + \lfloor 2ea + 1 \rfloor$*. For* $N \in \mathbb{N}$ *define* S_N *via* $\mu_{\min} = \mu^{(1)} < \cdots < \mu^{(N)} = \mu_{\max}$ *with logarithmically equidistant samples and* X_N *the corresponding Lagrangian RB space. Then,*

$$\frac{\|u(\mu) - u_N(\mu)\|_\mu}{\|u(\mu)\|_\mu} \le e^{-\frac{N-1}{N_0-1}}, \mu \in \mathcal{P}, N \ge N_0.$$

With uniform boundedness of the solution and norm equivalence, we directly obtain the same rate (just with an additional constant factor) for the error $\|u(\mu) - u_N(\mu)\|$.

We proceed with further aspects concerning computational procedures.

Remark 2.52 (Training set treatment). There are several ways to treat the training set slightly differently, leading to improvements.

(i) Multistage greedy: The first approach aims at a runtime acceleration. Instead of working with a fixed large training set S_{train}, which gives rise to $\mathcal{O}(|S_{\text{train}}|)$ runtime complexity in the greedy algorithm, one generates coarser subsets of this large training set:

$$S^{(0)}_{\text{train}} \subset S^{(1)}_{\text{train}} \subset \cdots \subset S^{(m)}_{\text{train}} := S_{\text{train}}.$$

Then, the greedy algorithm is started on $S^{(0)}_{\text{train}}$, resulting in a basis $\Phi_{N^{(0)}}$. This basis is used as the starting basis for the greedy algorithm on the next larger training set. This procedure is repeated until the greedy algorithm is run on the complete training set, but with a large starting basis $\Phi_{N^{(m-1)}}$. The rationale behind this procedure is that many iterations will be performed on small training sets, while the few final iterations still guarantee precision on the complete large training set. Overall, a remarkable runtime improvement can be obtained, while the quality of the basis is not expected to degenerate too much. Such an approach is introduced as the *multistage greedy* procedure in [65].

(ii) Training set adaptation: The next procedure aims at adapting the training set to realize uniform error distribution over the parameter space. For a given problem, it is not clear a priori how the training set should best be chosen. If the training set is chosen too large, the offline runtime may be too high. If the training set is too small, overfitting may easily be obtained; see Remark 2.48. The idea of the adaptive training set refinement [29, 30] is to start the greedy algorithm with a coarse set of training parameters, which are vertices of a mesh on the parameter domain. Then, in the FE spirit, a posteriori error estimators for subdomains are evaluated, grid cells with large error are marked for refinement, the marked cells are refined, and the new vertices are added to the training set. By this procedure, the training set is adapted to the problem at hand. The procedure may adaptively identify "difficult" parameter regions, for example, small diffusion constant values, and refine more in such regions.

(iii) Randomization: A simple idea allows us to implicitly work with a large training set. When working with randomly drawn parameter samples, one can draw a new set S_{train} in each greedy loop iteration. Hereby, the effective parameter set that is involved in the training is virtually enlarged by a factor N. This idea and refinements are presented in [37].

(iv) Full optimization: In special cases, a true (local) optimization over the parameter space in (2.35), i.e., $S_{\text{train}} = \mathcal{P}$, can also be realized [68]. The choice of a large training set is then reduced to the choice of a small set of multiple starting points for the highly nonlinear optimization procedure.

Remark 2.53 (Parameter domain partitioning). The greedy procedure allows us to prescribe accuracy via ε_{tol} and obtain a basis of size N that is a priori unpredictable and hence the final online runtime is unclear. It would be desirable to control both the accuracy (by prescribing ε_{tol}) and the online runtime (by demanding $N \leq N_{\max}$). The main idea to obtain this is via parameter domain partitioning.

(i) hp-RB-approach [19, 20]: Based on adaptive bisection of the parameter domain into subdomains, a partitioning of the parameter domain is generated (h-adaptivity). Then, small local bases can be generated for the different subdomains. If the accuracy and basis size criterion are not both satisfied for a subdomain, this subdomain is again refined and bases on the subdomains are generated. Finally, one has a collection of problems of type $(P_N(\mu))$, where $\mathcal{P}$ is now reduced to each of the subdomains of the partitioning. For a newly given μ in the online phase, only the correct subdomain and model need to be identified by a search in the grid hierarchy. This method balances offline cost, both in terms of computational and storage requirements, against online accuracy and runtime.

(ii) P-partition: A variant of parameter domain partitioning using hexahedral partitioning of the parameter space guarantees shape regularity of the subdomains [29]. The method prevents partitioning into long and thin areas, as can happen in the hp-RB-approach. Instead of two stages of partitioning and then piecewise basis generation, this approach has a single stage. For a given subdomain a basis generation is started. As soon as it can be predicted that the desired accuracy cannot be met with the currently prescribed maximum basis size, the basis generation is stopped (early stopping greedy), the subdomain is uniformly refined into subdomains, and the basis generation is restarted on all child elements. The prediction and early stopping of the greedy procedure is crucial; otherwise $N_{\max}$ basis vectors would have been generated on the coarse element before one detected that the basis must be discarded and the element must be refined. This prediction is therefore based on an extrapolation procedure, estimating the error decay by the decrease of the error for only a few iterations. For more details we refer to [29].

We want to draw attention to the conditioning issue. If $\mu^{(i)} \approx \mu^{(j)}$, it can be expected due to continuity that the two snapshots $u(\mu^{(i)}), u(\mu^{(j)})$ will be almost linearly dependent. Hence, the corresponding rows/columns of the reduced system matrix $\mathbf{A}_N$ will be almost linearly dependent and hence $\mathbf{A}_N$ may be badly conditioned. As seen in Proposition 2.17, orthonormalization of a basis may improve the conditioning of the reduced system. Interestingly, this can be realized via the Gramian matrix, i.e., the matrix $\mathbf{K}_N$ of inner products of the snapshots, and does not involve further expensive high-dimensional operations. For some interesting properties of Gramian matrices, we refer to Exercise 2.110. Using the Gramian matrix, the Gram–Schmidt orthonormalization can then be performed by a Cholesky factorization.

Proposition 2.54 (Orthonormalization of RB). *Assume $\Phi_N = \{\varphi_1, \ldots, \varphi_N\}$ to be an RB with Gramian matrix denoted $\mathbf{K}_N$. Choose $\mathbf{C} := (\mathbf{L}^T)^{-1} \in \mathbb{R}^{N \times N}$, with $\mathbf{L}$ a Cholesky factor of $\mathbf{K}_N = \mathbf{L}\mathbf{L}^T$. We define the transformed basis $\tilde{\Phi}_N := \{\tilde{\varphi}_1, \ldots, \tilde{\varphi}_N\}$ by $\tilde{\varphi}_j := \sum_{i=1}^N C_{ij}\varphi_i$. Then, $\tilde{\Phi}_N$ is the Gram–Schmidt orthonormalized basis.*

Again the proof is skipped and left as Exercise 2.111. If we are working with a discrete $(P(\mu))$, the initial basis is given via the coefficient matrix $\mathbf{\Phi}_N \in \mathbb{R}^{\mathcal{N} \times N}$, and $\mathbf{K} \in \mathbb{R}^{\mathcal{N} \times \mathcal{N}}$ denotes the full inner product matrix, then the Gramian matrix is obtained by (2.27), the matrix $\mathbf{C}$ can be computed as stated in the proposition, and the coefficient matrix of the transformed basis is simply obtained by the matrix product $\tilde{\mathbf{\Phi}}_N = \mathbf{\Phi}\mathbf{C}$. Actually, instead of Gram–Schmidt, other transformations are also viable; see Exercise 2.112.

We briefly comment on basis generation for the primal-dual RB approach.

Remark 2.55 (Basis generation for primal-dual approach). The a posteriori error bound in Proposition 2.32 suggests choosing the dimensionality of X_N, X_N^{du} such that $\Delta_u(\mu) \leq \varepsilon_{\text{tol}}$ and $\Delta_u^{\text{du}}(\mu) \leq \varepsilon_{\text{tol}}$ to obtain the "squared" effect in the error bound. For construction of X_N, X_N^{du} one could proceed as follows: (i) Run independent greedy procedures for the generation of X_N, X_N^{du} by using $(P(\mu))$ and $(P'(\mu))$ as snapshot suppliers, using the same tolerance ε_{tol} and $\Delta_u(\mu)$ and $\Delta_u^{\text{du}}(\mu)$ as error indicator. (ii) Run a single

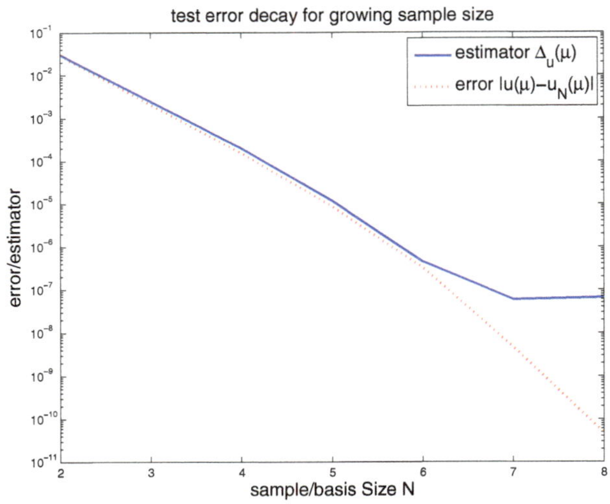

Figure 2.6. *Error convergence for Lagrangian RB with equidistant snapshots.*

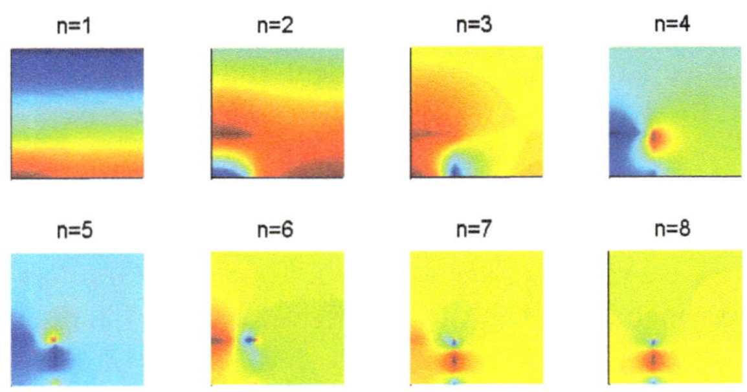

Figure 2.7. *Plot of eight orthogonal basis functions of an equidistant Lagrangian RB.*

greedy procedure based on the output error bound $\Delta_s'(\mu)$ and add snapshots of the solutions of $(P(\mu)), (P'(\mu))$ either in a single space X or in separate spaces X_N, X_N^{du} in each iteration.

We conclude this section with some experiments that are contained in the script `rb_tutorial.m` in our toolbox *RBmatlab*. In particular, we continue the previous experiments on the thermal block model from Section 2.3.1 with a focus on basis generation. In Figure 2.6 we demonstrate the error and error bound convergence when using a Lagrangian RB with equidistantly sampled parameter set. For this example we choose $B_1 = B_2 = 3$ and $\mu = (\mu_1, 1, \ldots, 1)$ with $\mu_1 \in [0.5, 2]$. The error and error bound are measured as a maximum over a random test set of parameters S_{test} with $|S_{\text{test}}| = 100$. We observe nice exponential error decay with respect to the sample size N for both the error and the error bound. A typical effect is the flattening or saturation of the error bound at values of about 10^{-7} when using double values. This is explained by and expected due to numerical accuracy problems, as the error bound is a square

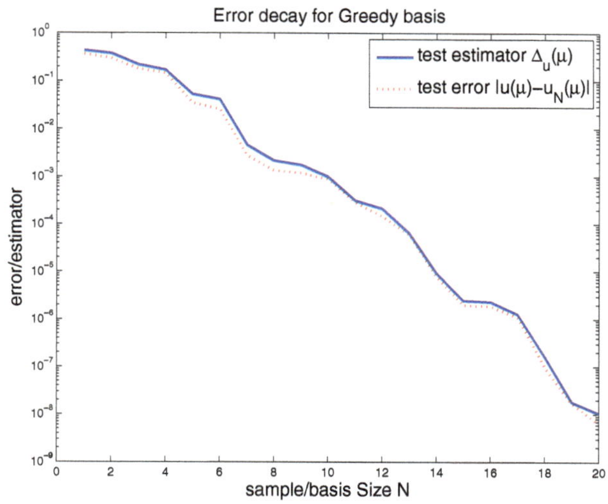

Figure 2.8. *Maximum test error and error bound for greedy RB generation for $B_1 = B_2 = 2$.*

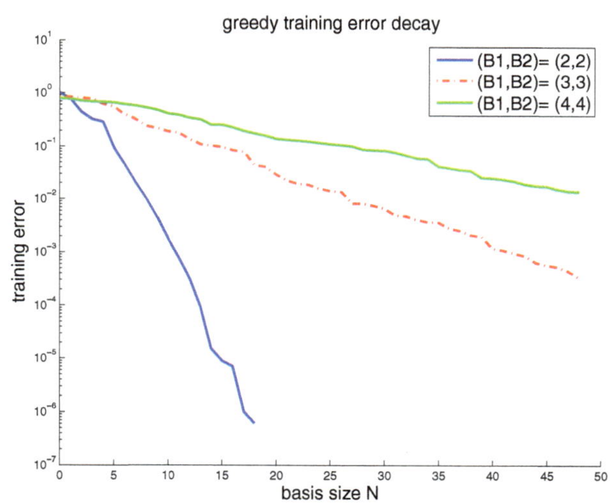

Figure 2.9. *Maximum training error estimator of the greedy procedure for varying block numbers.*

root of a residual norm, the accuracy of which is limited by machine precision. The first eight basis vectors of the corresponding orthonormalized Lagrangian RB are illustrated in Figure 2.7. We clearly see the steep variations of the basis functions around the first subblock Ω_1.

Finally, we investigate the results of the greedy procedure. For $\mu_{\min} = 1/\mu_{\max} = 0.5$ and $B_1 = B_2 = 2$ we choose a training set of 1000 random points, $\varepsilon_{\text{tol}} = 10^{-6}$, and the field error estimator as error indicator, i.e., $\Delta(Y, \mu) := \Delta_u(\mu)$. We again measure the quality by determining the maximum error and error bound over a random test set of size 100. The results are plotted in Figure 2.8 and nicely confirm the exponential convergence of the greedy procedure for smooth problems. With growing block number $B_1 = B_2 = 2, 3, 4$, the problem gets more complicated, as indicated in Figure 2.9, where the slope of the greedy training estimator curves $(\varepsilon_n)_{n\in N}$ is obviously flattening.

2.3.7 ▪ EI method

For the last approach of this section, we comment on an important general interpolation procedure, the empirical interpolation (EI) method [2, 51], which can be used in RB methods for several purposes. In particular, the procedure can be used in the case where the given problem does not allow for a parameter-separable representation. The EI method then generates a parameter-separable approximation, which can approximate the full problem. The notion *empirical* is motivated by the fact that the method is based on function snapshots for parametric function interpolation. The EI method selects a basis that can be used in addition to the RB, which is therefore called the *collateral basis*. The procedure is another instance of a greedy procedure.

Definition 2.56 (EI method, offline phase). *Let $\Omega \subset \mathbb{R}^d$ be an open bounded domain, $G \subset C^0(\bar{\Omega})$ a bounded closed set of functions, and $\varepsilon_{\mathrm{tol,EI}}$ a given error tolerance. Set $Q_0 := \emptyset, X_0 := \{0\}, T_0 := \emptyset$, and $m = 0$ and define iteratively*

$$\begin{aligned}
\textit{while } \varepsilon_m &:= \max_{g\in G} \|g - I_m(g)\|_\infty > \varepsilon_{\mathrm{tol},EI}, \qquad (2.39)\\
g_{m+1} &:= \arg\max_{g\in G} \|g - I_m(g)\|_\infty,\\
r_{m+1} &:= g_{m+1} - I_m(g_{m+1}),\\
x_{m+1} &:= \arg\max_{x\in\bar{\Omega}} |r_{m+1}(x)|,\\
T_{m+1} &:= T_m \cup \{x_{m+1}\},\\
q_{m+1} &:= r_{m+1}/r_{m+1}(x_{m+1}),\\
Q_{m+1} &:= Q_m \cup \{q_{m+1}\},\\
X_{m+1} &:= \mathrm{span}(Q_{m+1}),\\
m &\leftarrow m+1,\\
&\textit{end while.}
\end{aligned}$$

Here, $I_m : C^0(\bar{\Omega}) \to X_m$ denotes the interpolation operator with respect to the interpolation points $T_m \subset \bar{\Omega}$ and the interpolation space X_m, i.e., the unique operator with $I_m(g)(x_{m'}) = g(x_{m'}), m' = 1, \ldots, m, g \in G$.

The algorithm produces an interpolation space X_M and points T_M by setting $M := m+1$ as soon as the loop condition (2.39) *is false.*

We give some remarks on the practical implementation of the procedure, the assumed function regularity, the collateral basis, and the interpolation points.

Remark 2.57 (Practical implementation). First, if $|G| = \infty$, then a finite subset $G_{\mathrm{train}} \subset G$ is used instead of G for practical purposes. Similarly, Ω is usually replaced by a finite set of points $\Omega_{\mathrm{train}} \subset \Omega$ to obtain a computable algorithm. Further, instead of the accuracy termination criterion (2.39), the final dimension M can be specified as input parameter (as long as $M \leq \dim\mathrm{span}(G)$) and the extension loop can be terminated as soon as M is reached.

Remark 2.58 (Function regularity). The formulation assumes continuous functions, which can be directly extended to arbitrary functions, which allow point evaluations. Note that in this respect L^∞ is not sufficient for a well-defined scheme. This sufficiency is frequently wrongly assumed in the literature.

Remark 2.59 (Collateral basis). The sets Q_M are called collateral bases. They are hierarchical in the sense that $Q_M \subset Q_{M+1}$. Furthermore, they have the property that $q_m(x_{m'}) = 0$ for $m' < m$.

Remark 2.60 (Magic points). Surprisingly, when choosing a set of polynomials $G = \{x^i\}_{i=1}^n$ on $\bar{\Omega} = [-1,1]$, the resulting points are such that $\cos^{-1}(T_M)$ is roughly equidistant [27]. This is an interesting fact, as this makes the points have a similar characteristic as the optimal point set for polynomial interpolation, which are the Chebyshev points. This motivates the notion "magic points" [51] for the T_M, as the heuristic procedure "magically" produces point sets that are known to be optimal. Certainly, a fact indicating that this is not a real surprise is that the interpolation points are automatically determined in a greedy fashion to minimize the supremum norm of the resulting interpolation residual. This optimality with respect to the supremum norm is exactly the property of the Chebyshev points. While theoretically optimal interpolation point sets are only known for simple geometries, the EI is straightforwardly applicable to any arbitrarily shaped domain Ω, rendering it a powerful interpolation technique.

Given the offline data Q_M, T_M of the EI, the online phase, i.e., the actual interpolation, can be easily formulated.

Proposition 2.61 (EI, online phase). *Let a collateral basis Q_M and interpolation points T_M be given from the offline phase of the EI. Then, the matrix*

$$\mathbf{Q}_M := (q_j(x_i))_{i,j=1}^M \in \mathbb{R}^{M\times M} \tag{2.40}$$

is a lower triangular matrix with ones on the diagonal, hence regular. For $g \in C^0(\bar{\Omega})$, set the vector $\mathbf{g}_M := (g(x_i))_{i=1}^M \in \mathbb{R}^M$, and let $\boldsymbol{\alpha}_M = (\alpha_i)_{i=1}^M \in \mathbb{R}^M$ be the solution of the linear equation system

$$\mathbf{Q}_M \boldsymbol{\alpha}_M = \mathbf{g}_M.$$

Then the interpolation of g is simply

$$I_M(g) = \sum_{i=1}^M \alpha_i q_i. \tag{2.41}$$

Proof. The fact that the system matrix is lower triangular with ones on the diagonal simply follows from the definition. For the interpolation property we directly see that both sides of (2.41) coincide in the interpolation points $i = 1,\ldots,M$:

$$\sum_{j=1}^M \alpha_j q_j(x_i) = \sum_{j=1}^M (\mathbf{Q}_M)_{ij}\alpha_i = (\mathbf{Q}_M\boldsymbol{\alpha}_M)_i = (\mathbf{g}_M)_i = g(x_i) = I_M(g)(x_i).$$

□

As for other interpolation techniques, the relation to the best approximation is of interest, which is reflected by the Lebesgue constant. For any interpolation operator $I : C^0(\bar{\Omega}) \to X_M$, the Lebesgue constant is defined as

$$\Lambda_M := \max_{x \in \bar{\Omega}} \sum_{i=1}^{M} |\xi_i(x)|,$$

where ξ_i are nodal basis functions with respect to the interpolation points. Then, for any interpolation operator and any $u \in C^0(\bar{\Omega})$,

$$\|u - I(u)\|_\infty \leq (1 + \Lambda_M) \inf_{v \in X_M} \|u - v\|_\infty .$$

Now, for the EI the Lebesgue constant can be upper bounded. The following result holds [51].

Proposition 2.62 (Lebesgue constant bound). *For the EI operator I_M, the Lebesgue constant can be upper bounded by*

$$\Lambda_M \leq 2^M - 1,$$

and the bound is sharp.

This statement is quite pessimistic, as much better Lebesgue constants can be observed in practice. However, the bound on the Lebesgue constant is sharp, as examples can be constructed where this bound is attained.

In contrast, an optimistic result is given in the following. Similar to the RB greedy procedure, approximation rate statements can also be given for the EI offline procedure. For example, a central result of [51] is the following.

Proposition 2.63 (A priori convergence rate). *If there exists a sequence of exponentially approximating subspaces, i.e., $Z_1 \subset Z_2 \subset \cdots \operatorname{span}(G)$ with $\dim Z_M = M$ for $M \in \mathbb{N}$, and there exist $c > 0, \alpha > \log(4)$ such that*

$$\inf_{v \in Z_M} \|u - v\|_\infty \leq c e^{-\alpha M}, \quad \forall u \in G, M \in N,$$

then the EI offline phase yields almost as good spaces in the sense that

$$\|u - I_M(u)\|_\infty \leq c e^{-(\alpha - \log(4))M}.$$

In addition to such a priori convergence statements, a posteriori error control is also possible.

Proposition 2.64 (a posteriori error estimation for EI). *Let $I_M, I_{M'}$ be EI operators for $M' > M$ with corresponding collateral basis and interpolation points. Set the matrix $\mathbf{Q} := (q_j(x_i))_{i,j=M+1}^{M'} \in \mathbb{R}^{(M'-M+1)\times(M'-M+1)}$. For $g \in C^0(\bar{\Omega})$ let $\mathbf{g}' := (g(x_i) - I_M(g)(x_i))_{i=M+1}^{M'}$ and $\boldsymbol{\alpha}' := (\alpha_i')_{i=M+1}^{M'} = \mathbf{Q}^{-1}\mathbf{g}'$. If $g \in \operatorname{span}(Q_{M'})$, then the following a posteriori error bounds hold:*

$$\|g - I_M(g)\|_\infty \leq \Delta_{\mathrm{EI},\infty}(g) := \|\alpha'\|_1 = \sum_{i=M+1}^{M'} |\alpha_i'|, \tag{2.42}$$

$$\|g - I_M(g)\|_{L^2} \leq \Delta_{\mathrm{EI},2}(g) := \sqrt{(\alpha')^T \mathbf{K}_Q \alpha'}, \tag{2.43}$$

where

$$\mathbf{K}_Q := \left(\int_\Omega q_i q_j\right)_{i,j=M+1}^{M'}.$$

The proof is left as a simple exercise using the nestedness of the interpolation matrices Q_M and $Q_{M'}$ and the definitions.

In this bound, a certain exactness for a higher EI index $M' > M$ is assumed for exact certification. This assumption can be widely found, e.g., [24, 73]. An alternative, which requires a certain smoothness knowledge instead of this exactness assumption, is presented in [18].

A useful property is the following, which states conservation of linear operations.

Proposition 2.65 (Conservation property). *Let $f \in \left(C^0(\bar{\Omega})\right)'$ be a linear functional on the continuous functions and $f(g) = 0$ for all $g \in G$. Then, we also have*

$$f(I_M(g)) = 0 \quad \forall g \in G.$$

Proof. By linearity we immediately have $f(g)$ for all $g \in \mathrm{span}(G)$. As $Q_M \subset \mathrm{span}(G)$ by construction, the claim follows by linearity of the interpolation operator. □

An intuitive example of such functionals is conservation of zeros/roots: if f is a point evaluation in $\bar{x}$ and $g(\bar{x}) = 0$ for all $g \in G$, then also $I_M(g)(\bar{x}) = 0$. This can for instance be used in the conservation of zero entries in the interpolation of sparse vectors or (vectorized) sparse matrices [73].

Another example could be the interpolation of zero-mean functions: if $\int_\Omega g = 0$ for all $g \in G$, then also $\int_\Omega I_M(g) = 0$. This can be beneficial in the interpolation of conservative operators [17]. Similarly, one could imagine interpolation of divergence-free velocity fields in fluid dynamics.

Now we will establish the connection to the previous sections, i.e., apply the EI in a parametric context and formulate an EI/RB scheme. In particular, we provide EI approximations for problems with nonseparable data functions.

Definition 2.66 (EI for parametric functionals). *Let $X \subset L^2(\Omega)$ and $f \in X'$ be a parametric continuous linear form of the type*

$$f(v;\mu) = \int_\Omega g(x;\mu)v(x)dx \tag{2.44}$$

for $g(\cdot;\mu) \in C^0(\bar{\Omega})$. In general, g and f may be non-parameter separable. Then, set $G := \{g(\cdot;\mu) | \mu \in \mathcal{P}\}$ and compute the EI offline data $Q_M = \{q_i\}_{i=1}^M$ and $T_M = \{x_i\}_{i=1}^M$ according to Definition 2.56. Then, we obtain parameter-separable approximations $\tilde{g}, \tilde{f}$ for g and f by

$$\tilde{g}(\cdot;\mu) := I_M(g(\cdot;\mu)) = \sum_{m=1}^M \theta_m(\mu)\tilde{g}_m(\cdot), \tag{2.45}$$

$$\tilde{f}(v;\mu) := \int_\Omega \tilde{g}(x;\mu)v(x)dx = \sum_{m=1}^M \theta_m(\mu)\tilde{f}_m, \tag{2.46}$$

with components $\tilde{g}_m := q_m$, $\tilde{f}_m := \int_\Omega \tilde{g}_m v$ *and coefficient function vector*

$$(\theta_1(\mu), \ldots, \theta_M(\mu))^T := \mathbf{Q}_M^{-1}\mathbf{g}(\mu),$$

using the local evaluation vector $\mathbf{g}(\mu) := (g(x_1;\mu), \ldots, g(x_M;\mu))^T$ *and interpolation matrix* $\mathbf{Q}_M$ *as in* (2.40).

Similar parameter-separable approximations can be obtained for non-parameter separable bilinear forms, e.g., $a(u,v;\mu) := \int_\Omega \kappa(x;\mu)\nabla u(x) \cdot \nabla v(x) dx$ by EI for κ.

Hereby, a non-parameter separable problem of type (P) with solution u can be approximated by EI of the data functions resulting in an approximate variational form $(\tilde{P})$ with solution $\tilde{u}$.

For simplicity, in the following we ignore the output functional.

Proposition 2.67 (EI-approximated full problem ($\tilde{P}(\mu)$)). *Assume that* $a(\cdot,\cdot;\mu)$ *is a continuous bilinear form, uniformly coercive in* μ, *and* $f(\cdot;\mu)$ *is uniformly bounded with respect to* μ. *Assume that the EI approximations are sufficiently accurate in the sense that we have* $\varepsilon_a, \varepsilon_f \in \mathbb{R}$ *with* $\varepsilon_a < \bar{\alpha}$ *such that for all* $u, v \in X, \mu \in \mathcal{P}$,

$$|a(u,v;\mu) - \tilde{a}(u,v;\mu)| \leq \varepsilon_a \|u\|\|v\|, \quad |f(v;\mu) - \tilde{f}(v;\mu)| \leq \varepsilon_f \|v\|. \quad (2.47)$$

Then, the forms $\tilde{a}, \tilde{f}$ *are continuous and* $\tilde{a}$ *is coercive with coercivity lower bound* $\tilde{\alpha} := \alpha - \varepsilon_a > 0$; *hence the following problem has a unique solution* $\tilde{u}(\mu) \in X$:

$$\tilde{a}(\tilde{u}, v;\mu) = \tilde{f}(v;\mu), \quad v \in X, \quad (2.48)$$

which satisfies $\|\tilde{u}\| \leq \|\tilde{f}\|_{X'}/\tilde{\alpha}$.

Proof. Continuity of the approximated forms follows simply by

$$|\tilde{f}(v)| \leq |\tilde{f}(v) - f(v)| + |f(v)| \leq \varepsilon_f \|v\| + \|f\|_{X'}\|v\| = (\varepsilon_f + \|f\|_{X'})\|v\|,$$

and similarly for $\tilde{a}$. For the coercivity we obtain

$$\begin{aligned}\frac{\tilde{a}(u,u)}{\|u\|^2} &= \frac{a(u,u) - (a(u,u) + \tilde{a}(u,u))}{\|u\|^2} \geq \frac{a(u,u)}{\|u\|^2} - \frac{|a(u,u) - \tilde{a}(u,u)|}{\|u\|^2}\\ &\geq \alpha - \varepsilon_a = \tilde{\alpha} > 0.\end{aligned}$$

The remaining statements follow by the Lax–Milgram theorem. □

This statement is valid for any approximation procedure. Specifically for EI and the previous example for f in (2.44), it is easy to see that the ε_f can be related to the error bounds (assuming their validity):

$$|f(v) - \tilde{f}(v)| = \left|\int_\Omega (g - \tilde{g})v\right| \leq \|g - \tilde{g}\|_{L^2}\|v\|_{L^2} \leq \Delta_{\mathrm{EI},2}(g(\cdot;\mu))\|v\|_{H^1}.$$

Hence, choosing $\varepsilon_f \geq \sup_{\mu\in\mathcal{P}} \Delta_{\mathrm{EI},2}(g(\cdot;\mu))$ will guarantee the validity of the f approximation assumption in (2.47). Similarly, for the bilinear form $a(u,v) = \int_\Omega \kappa(\cdot;\mu)\nabla u \cdot \nabla v$ we can satisfy the assumption on a in (2.47) by choosing $\varepsilon_a \geq \sup_{\mu\in\mathcal{P}} \Delta_{\mathrm{EI},\infty}(\kappa(\cdot;\mu))$.

So, ε_a can be made small by ensuring that the $\Delta_{\mathrm{EI},\infty}(\kappa(\cdot;\mu))$ are small, i.e., the EI is sufficiently accurate. Now, the RB machinery of the previous sections can be applied to generate an approximate well-posed reduced problem $(\tilde{P}_N)$ with solution $\tilde{u}_N$, and the error $\tilde{u}-\tilde{u}_N$ can be quantified by the presented estimators. However, to control the complete error $u-\tilde{u}_N$, the interpolation error also needs to be estimated. This can be obtained by a disturbance argument.

Proposition 2.68 (A posteriori error estimator for EI/RB approximation). *Let $u(\mu)\in X$ be the solution of the non-parameter separable problem (P) and $\tilde{u}_N(\mu)$ be the RB approximation of the EI-approximated system $(\tilde{P}(\mu))$. Then, we have the error bound*

$$\|u-\tilde{u}_N\| \leq \Delta_{\mathrm{EI}}(\mu)+\Delta_{\tilde{u}}(\mu), \tag{2.49}$$

where $\Delta_{\tilde{u}}(\mu)$ denotes the standard RB error bound for the error $\tilde{u}-\tilde{u}_N$ analogous to (2.8) *and Δ_{EI} is an appropriate EI error contribution*

$$\Delta_{\mathrm{EI}}(\mu) := \frac{1}{\alpha}\varepsilon_f + \frac{\|\tilde{f}\|_{X'}}{\alpha(\alpha-\varepsilon_a)}\varepsilon_a. \tag{2.50}$$

Proof. We first note by the definitions of (P) and $(\tilde{P})$ that

$$a(u-\tilde{u},v)=a(u,v)-a(\tilde{u},v)=f(v)-\tilde{f}(v)+\tilde{a}(\tilde{u},v)-a(\tilde{u},v)=:\tilde{r}(v),$$

where the second equality follows from adding $0=\tilde{f}(v)-\tilde{a}(\tilde{u},v)$. Using the approximation property yields

$$\|\tilde{r}\|_{X'} \leq \varepsilon_f+\varepsilon_a\|\tilde{u}\|.$$

Then, Lax–Milgram together with the bound for $\tilde{u}$ from Proposition 2.67 allows us to conclude that

$$\|u-\tilde{u}\| \leq \frac{1}{\alpha}\|\tilde{r}\| \leq \frac{1}{\alpha}\varepsilon_f + \frac{\|\tilde{f}\|_{X'}}{\alpha(\alpha-\varepsilon_a)}\varepsilon_a = \Delta_{\mathrm{EI}}(\mu). \tag{2.51}$$

The overall bound (2.49) then follows by the triangle inequality. □

2.4 ▪ Instationary problems

In this section, we aim at extending the methodology to time-dependent problems. Historically, time-dependent problems have been the motivation for RB modeling [1, 4, 41, 58], in particular in the context of fluid flow and for understanding complexity in turbulence. However, these techniques were not certified by error estimation and not subject to offline/online decomposition, etc., as presented for the stationary case. The first publication known to us dealing with a *certified* RB method for instationary problems can be found in [24, 26], where parabolic problems were considered. A central result of that work, a space-time energy norm error estimator, will be given in a reformulated fashion in this section. Otherwise, the formulations and techniques of this section are based on our previous work [31] and also comprise new results by following the pattern of the previous section. In particular, we will proceed in parallel to the stationary case and sequentially prove similar results with the same notions.

Instead of giving a variational formulation, we consider an alternative operator-based formulation in the current section. This will in particular allow an RB approach for finite difference or FV discretizations of hyperbolic equations, which are usually not motivated by a variational formulation. However, note that all of the following could as well be formulated in a variational fashion.

We admit that by the focus and choice of the presentation in this section we are clearly biased to our own previous work, as most other articles on RB methods for instationary problems with variational discretizations use corresponding weak forms of the PDE. For such formulations and RB schemes, we refer to [19, 20, 24–26, 44, 47, 67].

2.4.1 ▪ Model problem

As a model problem we consider a linear advection-diffusion problem on a rectangular domain $\Omega = (0,2) \times (0,1)$ with end time $T = 1$, i.e., for given μ find $u(x,t;\mu)$ as a solution of

$$\partial_t u(\mu) + \nabla \cdot (\mathbf{v}(\mu) u(\mu) - d(\mu) \nabla u(\mu)) = 0 \quad \text{in } \Omega \times (0,T)$$

with suitable initial condition $u(x,0;\mu) = u_0(x;\mu)$ and Dirichlet boundary conditions. The initial and time-variant inhomogeneous Dirichlet boundary values $g_D(x,t)$ are based on a nonnegative radial basis function linearly decaying over time with center on the top edge at $x_1 = 1/2$. The velocity field is chosen as a superposition of two divergence-free parabolic velocity fields in the x_1- and x_2-direction with weighting factors one and μ_1, i.e.,

$$\mathbf{v}(x;\mu) = \left(\mu_1 \frac{5}{2}(1 - x_2^2), -\frac{1}{2}(4 - x_1^2)\right)^T .$$

The diffusivity is chosen as $d(\mu) := 0.03 \cdot \mu_2$, resulting in a parametrization of the velocity and diffusivity by $(\mu_1, \mu_2) \in [0,1]^2$. Note that the above problem changes type with the parameter, i.e., for $\mu_2 = 0$ we obtain a hyperbolic problem, while for $\mu_2 > 0$ it is parabolic. Some solution snapshots over time (using an FV-discretization; details will be reported in the experiments at the end of Section 2.4.5) are presented in Figure 2.10, with each column representing the time evolution of a different parameter.

2.4.2 ▪ Full problem

The above problem is an example of a general linear evolution problem of the type

$$\begin{aligned} \partial_t u - \mathcal{L}(u;\mu) &= q(\mu) \quad \text{in } \Omega \times (0,T), \\ u(0) &= u_0(\mu) \quad \text{in } \Omega, \end{aligned}$$

with $\Omega \subset \mathbb{R}^d$ the spatial domain, $[0,T]$ the time interval with final time $T > 0$, $\mathcal{L}$ a linear spatial differential operator, q an inhomogeneity, and u_0 the initial values.

The full problem will be based on a time-discrete formulation based on $K \in \mathbb{N}$ steps in time, step size $\Delta t := T/K$, and time instants $t^k := k\Delta t, k = 0,\ldots,K$. For notational convenience, we assume constant Δt, though varying time-step widths can easily be incorporated in the following.

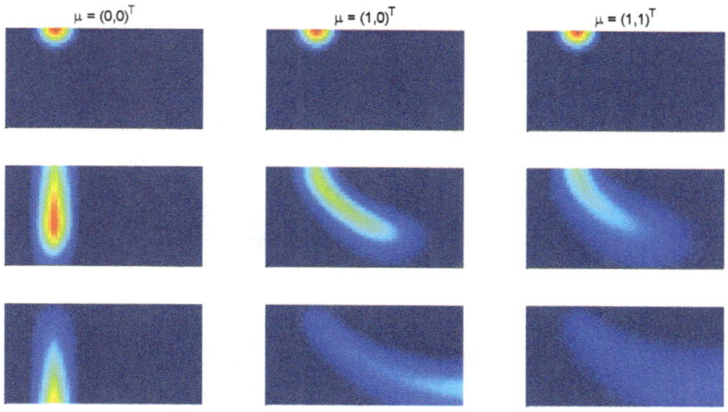

Figure 2.10. *Illustration of snapshots of the advection-diffusion example: initial data (top) and time evolution at time $t = 0.5$ (middle row) and $t = 1$ (bottom) for parameter vectors $\mu = (0,0)^T, (1,0)^T, (1,1)^T$ from left to right.*

We again assume that X is a Hilbert space with inner product $\langle \cdot, \cdot \rangle$ and norm $||\cdot||$, and we seek the solution variable $u^k(\mu) \in X$ with $u^k(x;\mu) \approx u(x, t^k; \mu)$ for $k = 0, \ldots, K$. We assume a general explicit/implicit time discretization with $\mathscr{L}_I^k(\mu), \mathscr{L}_E^k(\mu) : X \to X$ linear continuous operators and $b^k(\mu) \in X$. Note that $\mathscr{L}_I^k, \mathscr{L}_E^k, b^k$ will typically depend on Δt; hence these operators reflect both the time discretization and the space discretization. For simplicity we assume $u^0 \in X$; otherwise an initial data projection needs to be included. Then, the problem for the discrete solution can be formulated as a time-marching evolution scheme.

Definition 2.69 (Full evolution problem ($E(\mu)$)). *For $\mu \in \mathscr{P}$ find a sequence of solutions $\{u^k(\mu)\}_{k=0}^K \subset X$ by starting with $u^0(\mu) \in X$ and iteratively solving the following operator equations for $u^{k+1}(\mu)$:*

$$\mathscr{L}_I^k(\mu) u^{k+1}(\mu) = \mathscr{L}_E^k(\mu) u^k(\mu) + b^k(\mu), \quad k = 0, \ldots, K-1.$$

We omit output estimation here and give comments on this in Remark 2.84. We again formulate some requirements for well-posedness.

Definition 2.70 (Uniform continuity and coercivity). *The parametric operators are assumed to be continuous with continuity constants $\gamma_I^k(\mu) := ||\mathscr{L}_I^k(\mu)||$, $\gamma_E^k(\mu) := ||\mathscr{L}_E^k(\mu)||$. The continuity is assumed to be* uniform with respect to *μ and t in the sense that for some $\bar{\gamma}_I, \bar{\gamma}_E < \infty$, $\gamma_I^k(\mu) \leq \bar{\gamma}_I$, $\gamma_E^k(\mu) \leq \bar{\gamma}_E$ for all μ and k. Furthermore, $\mathscr{L}_I^k$ is assumed to be coercive, i.e., there exists a constant*

$$\alpha_I^k(\mu) := \inf_{v \in X \setminus \{0\}} \frac{\langle \mathscr{L}_I^k(\mu) v, v \rangle}{||v||^2} > 0,$$

and the coercivity is uniform with respect to μ and t in the sense that for some $\bar{\alpha}_I > 0$, $\alpha_I^k(\mu) \geq \bar{\alpha}_I$ for all μ and k. Similarly, for $b^k(\mu)$ we assume uniform boundedness by $||b^k(\mu)|| \leq \bar{\gamma}_b$ for suitable $\bar{\gamma}_b$.

Under these assumptions, one obtains the well-posedness and stability of the problem $(E(\mu))$.

Proposition 2.71 (Well-posedness and stability of $(E(\mu))$). *The solution trajectory $\{u^k(\mu)\}_{k=0}^K$ of $(E(\mu))$ is well defined and bounded by*

$$\left\|u^k(\mu)\right\| \leq \left\|u^0\right\| \left(\frac{\bar{\gamma}_E}{\bar{\alpha}_I}\right)^k + \frac{\bar{\gamma}_b}{\bar{\alpha}_I}\left(\sum_{i=0}^{k-1}\left(\frac{\bar{\gamma}_E}{\bar{\alpha}_I}\right)^i\right). \tag{2.52}$$

Proof. Well-definedness of the solution in iteration k follows by Lax–Milgram and uniform continuity/coercivity and gives the bound

$$\left\|u^{k+1}\right\| \leq \frac{1}{\bar{\alpha}_I}\left(\bar{\gamma}_E\left\|u^k\right\| + \bar{\gamma}_b\right).$$

The bound (2.52) then easily follows by induction. □

The constants $\bar{\gamma}_E, \bar{\gamma}_b$, and $\bar{\alpha}_I$, and hence the constant on the right-hand side of (2.52), depend on Δt. Hence, the behavior for $\Delta t \to 0$ is of interest. One can show that the solution does not diverge with decreasing Δt under some conditions on the continuity and coercivity constant. We leave the proof of the following as Exercise 2.113.

Proposition 2.72 (Uniform boundedness with respect to Δt). *Let $\bar{\gamma}_E \leq 1$, $\bar{\alpha}_I = 1 + \alpha\Delta t$, and $\bar{\gamma}_b = C\Delta t$, with α, C independent of Δt. Then,*

$$\lim_{K\to\infty}\left\|u^K\right\| \leq e^{-\alpha T}\left\|u^0\right\| + \tilde{C}T,$$

with explicitly computable Δt-independent constant $\tilde{C}$.

Note that this is a very coarse qualitative statement. Certainly stronger results such as convergence of $u^K(\mu)$ (and for any other t) are usually expected (and provided) by reasonable discretizations, i.e., instantiations of $(E(\mu))$.

Again, we assume parameter separability analogous to Definition 2.6 for later efficient offline/online decomposition. In contrast to the stationary case, we assume that the time dependency is also encoded in the coefficient functions such that the components are parameter and time independent.

Definition 2.73 (Parameter separability). *We assume the operators $\mathcal{L}_I^k, \mathcal{L}_E^k, b^k$ to be* parameter separable, *i.e., there exist coefficient functions $\theta_{I,q}^k : \mathcal{P} \to \mathbb{R}$ and parameter-independent continuous linear operators $\mathcal{L}_{I,q} : X \to X$ for $q = 1, \ldots, Q_I$ such that*

$$\mathcal{L}_I^k(\mu) = \sum_{q=1}^{Q_I} \theta_{I,q}^k(\mu)\mathcal{L}_{I,q},$$

and similar definitions for $\mathcal{L}_E^k$, b^k, and u^0 with corresponding coefficient functions $\theta_{E,q}^k(\mu)$, $\theta_{b,q}^k(\mu)$, $\theta_{u^0,q}(\mu)$; components $\mathcal{L}_{E,q}, b_q, u_q^0$; and number of components Q_E, Q_b, Q_{u^0}.

Assuming Lipschitz continuity of the coefficient functions with respect to μ, one can derive Lipschitz continuity of the solution, similar to Proposition 2.12. Also, sensitivity equations for instationary problems can be obtained similar to Proposition 2.13, which is made use of in the context of parameter optimization; see [15].

Several examples fit into the framework of the above problem ($E(\mu)$).

Example 2.74 (FE formulation for a diffusion problem). For an implicit FE discretization of a diffusion problem with homogeneous Dirichlet boundary data, we can choose $X := \mathrm{span}\{\phi_i\}_{i=1}^{\mathcal{N}} \subset H_0^1(\Omega)$ as the space of piecewise-linear functions on the grid $\mathcal{T}$ assigned the $H_0^1(\Omega)$ inner product. The variational time-marching form of parabolic problems

$$m(u^{k+1}, v) + \Delta t a(u^{k+1}, v) = m(u^k, v), \quad v \in X,$$

with $m(u,v) := \int_\Omega uv, a(u,v) := \int_\Omega d\nabla u \cdot \nabla v$ and the $L^2(\Omega)$-orthogonal projection of the initial data, $u^0 := P_X u_0$ can then be directly transferred to the operator formulation by defining $b^k := 0$ and the operators implicitly via

$$\left\langle \mathcal{L}_E^k u, v \right\rangle := m(u,v), \quad \left\langle \mathcal{L}_I^k u, v \right\rangle := m(u,v) + \Delta t a(u,v), \quad u, v \in X.$$

Parameter separability of the data functions will result in parameter-separable operators. However, note that this Galerkin formulation does not allow (large) advection terms or vanishing diffusion unless we accept possible instability, as the implicit spatial discretization operator may become noncoercive. ∎

Example 2.75 (FV discretization for advection problem). Given a triangulation $\mathcal{T} = \{T_i\}_{i=1}^{\mathcal{N}}$ of $\Omega \subset \mathbb{R}^d$, one can choose the discrete basis functions $\psi_i := \chi_{T_i}, i = 1, \ldots, \mathcal{N}$, of characteristic functions of the grid elements. Then, we define $X := \mathrm{span}\{\psi_i\}_{i=1}^{\mathcal{N}} \subset L^2(\Omega)$ as an FV space of piecewise-constant functions with the $L^2(\Omega)$ inner product. An explicit Euler forward time discretization of an advection problem with Dirichlet boundary data b_{dir} and velocity field $\mathbf{v}$ can then easily be formulated. For this we first define $u^0 = P_X u_0 \in X$ as the L^2-projection of given initial data u_0 to piecewise-constant functions and define $\mathcal{L}_I^k := Id$ as the identity on X. Using for example a Lax–Friedrichs numerical flux with parameter $\lambda > 0$ [49], we obtain $b^k = \sum_i b_i^k \psi_i \in X$ with

$$b_i^k := -\frac{\Delta t}{|T_i|} \sum_{j \in N_{\mathrm{dir}}(i)} \frac{|e_{ij}|}{2} [\mathbf{v}(\mathbf{c}_{ij}) \cdot \mathbf{n}_{ij} - \lambda^{-1}] b_{\mathrm{dir}}(\mathbf{c}_{ij}),$$

where $N_{\mathrm{dir}}(i)$ is the set of indices enumerating the Dirichlet boundary edges of T_i, $|e_{ij}|$ is the length and $\mathbf{c}_{ij}$ the centroid of the corresponding edge, and $|T_i|$ indicates the volume of element T_i. The operator $\mathcal{L}_E^k$ is specified via its operation on a vector of unknowns, i.e., for all $w = \sum_i w_i \psi_i$ and $w' := \mathcal{L}_E^k w = \sum_i w_i' \psi_i$ with vectors of unknowns $\mathbf{w}, \mathbf{w}' \in \mathbb{R}^{\mathcal{N}}$, we assume $\mathbf{w}' = \hat{\mathbf{L}}_E^k \mathbf{w}$ with matrix $\hat{\mathbf{L}}_E^k \in \mathbb{R}^{\mathcal{N} \times \mathcal{N}}$ defined by entries

$$(\hat{\mathbf{L}}_E^k)_{i,l} := \begin{cases} 1 - \frac{\Delta t}{|T_i|} \sum_{j \in N_{\mathrm{int}}(i) \cup N_{\mathrm{dir}}(i)} \frac{|e_{ij}|}{2} [\mathbf{v}(\mathbf{c}_{ij}) \cdot \mathbf{n}_{ij} + \lambda^{-1}] & \text{for } l = i, \\ -\frac{\Delta t}{|T_i|} \frac{|e_{il}|}{2} [\mathbf{v}(\mathbf{c}_{il}) \cdot \mathbf{n}_{il} + \lambda^{-1}] & \text{for } l \in N_{\mathrm{int}}(i), \\ 0 & \text{otherwise,} \end{cases}$$

where $N_{\mathrm{int}}(i)$ are the neighboring element indices of T_i. Note that this discretization scheme requires a CFL condition, i.e., sufficiently small Δt to guarantee a stable scheme. Similarly, diffusion terms or additional Neumann boundary values can be discretized as explicitly specified in [31]. Our additional requirements are satisfied under suitable assumptions on the data functions: again, parameter separability and uniform continuity of the data functions result in parameter separability and uniform continuity of the operators. The implicit operator (the identity) is clearly uniformly coercive independent of the data functions. In particular, this example gives a discretization and hence results in an RB method for a hyperbolic problem. Also, it can be shown that the assumptions of Proposition 2.72 are satisfied. ∎

Example 2.76 (Finite differences). Similar to the FV example, finite difference (FD) discretizations can also be treated with the current evolution and RB formulation. Given $\Omega_h = \{x_i\}_{i=1}^{\mathcal{N}} \subset \bar{\Omega}$ as a discrete set of grid points used for the FD discretization, we define $X := \operatorname{span}\{\delta_{x_i}\}_{i=1}^{\mathcal{N}}$ as the set of indicator functions $\delta_{x_i} : \Omega_h \to \mathbb{R}$ of the discrete point set and assign it a discrete inner product, e.g., an approximation of an L^2 inner product $\langle u, v\rangle := \sum_{i=1}^{\mathcal{N}} w_i u(x_i) v(x_i)$ for $u, v \in X$ with weights $w_i \in \mathbb{R}$. Again, parameter separability of the data functions will result in parameter-separable FD operators. Uniform boundedness of the data in combination with continuous evaluation functionals results in uniformly bounded operators. ∎

To conclude this section, we summarize that $(E(\mu))$ captures quite general PDEs (hyperbolic, parabolic), which have a first-order time derivative and an arbitrary spatial differential operator. Further, different spatial discretization techniques are allowed: FV, FD, FE, or DG schemes, etc. Different time discretizations are allowed: Euler forward/backward or Crank–Nicolson. Operator splitting is also allowed: different parts of the continuous differential operator $\mathcal{L}$ may be discretized implicitly; while others may be discretized explicitly.

2.4.3 ▪ RB approach

We again assume the availability of an RB space X_N with RB $\Phi_N = \{\varphi_1, \ldots, \varphi_N\}$, $N \in \mathbb{N}$. We give a procedure for basis generation in Section 2.4.5.

Definition 2.77 (RB problem $(E_N(\mu))$). *For $\mu \in \mathcal{P}$ find a sequence of solutions $\{u_N^k(\mu)\} \subset X_N$ by starting with $u_N^0(\mu) := P_{X_N} u^0(\mu)$ and iteratively solving*

$$\mathcal{L}_{I,N}^k(\mu) u_N^{k+1}(\mu) = \mathcal{L}_{E,N}^k(\mu) u_N^k(\mu) + b_N^k(\mu), \quad k = 0, \ldots, K-1,$$

with reduced operators and reduced inhomogeneity

$$\mathcal{L}_{N,I}^k(\mu) = P_{X_N} \circ \mathcal{L}_I^k(\mu), \quad \mathcal{L}_{N,E}^k(\mu) = P_{X_N} \circ \mathcal{L}_E^k(\mu), \quad b_N^k(\mu) = P_{X_N} b^k(\mu),$$

where $P_{X_N} : X \to X_N$ denotes the orthogonal projection with respect to $\langle \cdot, \cdot \rangle$.

The well-posedness and stability of $\{u_N^k\}$ follow identically to Proposition 2.71. We can again state a simple consistency property, which is very useful for validating program code, as mentioned in Remark 2.22.

Proposition 2.78 (Reproduction of solutions). *If for some $\mu \in \mathcal{P}$ we have $\{u^k(\mu)\}_{k=0}^K \subset X_N$, then $u_N^k(\mu) = u^k(\mu)$ for $k = 0, \ldots, K$.*

Proof. The statement follows by induction. For $k=0$ we have $u^0 \in X_N$ and $P_{X_N}|_{X_N} = Id$; therefore, $u_N^0 = P_{X_N} u^0 = u^0$. For the induction step, assume that $u^k = u_N^k$. With $(E_N(\mu))$, we obtain

$$0 = \mathcal{L}_{I,N}^k u_N^{k+1} - \mathcal{L}_{E,N}^k u_N^k - b_N^k = P_{X_N}\left(\mathcal{L}_I^k u_N^{k+1} - \mathcal{L}_E^k u_N^k - b^k\right). \tag{2.53}$$

Using $(E(\mu))$ and $u^k = u_N^k$, we verify $-\mathcal{L}_E^k u_N^k - b^k = -\mathcal{L}_I^k u^{k+1}$, and (2.53) reduces to $P_{X_N}(\mathcal{L}_I^k(u_N^{k+1} - u^{k+1})) = 0$. This means $\mathcal{L}_I^k e^{k+1} \perp X_N$, using the abbreviation $e^{k+1} := u^{k+1} - u_N^{k+1}$. But by the assumption $u^{k+1} \in X_N$, we also have $e^{k+1} \in X_N$. Hence, uniform coercivity implies

$$0 = \left\langle \mathcal{L}_I^k e^{k+1}, e^{k+1}\right\rangle \geq \bar{\alpha}_I \left\|e^{k+1}\right\|^2,$$

proving $e^{k+1} = 0$ and, hence, $u^{k+1} = u_N^{k+1}$. □

Proposition 2.79 (Error residual relation). *For $\mu \in \mathcal{P}$ we define the residual $\mathcal{R}^k(\mu) \in X$ via*

$$\mathcal{R}^k(\mu) := \frac{1}{\Delta t}\left(\mathcal{L}_E^k(\mu) u_N^k(\mu) - \mathcal{L}_I^k(\mu) u_N^{k+1}(\mu) + b^k(\mu)\right), \quad k = 0, \ldots, K-1. \tag{2.54}$$

Then, the error $e^k(\mu) := u^k(\mu) - u_N^k(\mu) \in X$ satisfies the evolution problem

$$\mathcal{L}_I^k(\mu) e^{k+1}(\mu) = \mathcal{L}_E^k(\mu) e^k(\mu) + \mathcal{R}^k(\mu)\Delta t, \quad k = 0, \ldots, K-1. \tag{2.55}$$

Proof. Using $(E(\mu))$ and $(E_N(\mu))$ yields

$$\begin{aligned}\mathcal{L}_I^k e^{k+1} &= \mathcal{L}_I^k u^{k+1} - \mathcal{L}_I^k u_N^{k+1}\\ &= \mathcal{L}_E^k u^k + b^k - \mathcal{L}_E^k u_N^k + \mathcal{L}_E^k u_N^k - \mathcal{L}_I^k u_N^{k+1} = \mathcal{L}_E e^k + \Delta t \mathcal{R}^k.\end{aligned}$$

□

The following a posteriori error bound simply follows by applying the a priori bounding technique of Proposition 2.71 to the error evolution of Proposition 2.79.

Proposition 2.80 (A posteriori error bound X-norm). *Let $\gamma_{\mathrm{UB}}(\mu)$ and $\alpha_{\mathrm{LB}}(\mu)$ be computable upper/lower bounds satisfying*

$$\gamma_E^k(\mu) \leq \gamma_{\mathrm{UB}}(\mu) \leq \bar{\gamma}_E, \quad \alpha_I^k(\mu) \geq \alpha_{\mathrm{LB}}(\mu) \geq \bar{\alpha}_I, \quad \mu \in \mathcal{P}, k = 0, \ldots, K.$$

Then, the RB error can be bounded by

$$\begin{aligned}&\left\|u^k(\mu) - u_N^k(\mu)\right\| \leq \Delta_u^k(\mu) \text{ with}\\ &\qquad \Delta_u^k(\mu) := \left\|e^0\right\| \left(\frac{\gamma_{\mathrm{UB}}(\mu)}{\alpha_{\mathrm{LB}}(\mu)}\right)^k + \sum_{i=0}^{k-1} \left(\frac{\gamma_{\mathrm{UB}}(\mu)}{\alpha_{\mathrm{LB}}(\mu)}\right)^{k-i-1} \frac{\Delta t}{\alpha_{\mathrm{LB}}(\mu)} \left\|\mathcal{R}^i\right\|.\end{aligned}$$

Proof. The error residual relation (2.55) and the Lax–Milgram theorem imply the recursion

$$\left\|e^{k+1}\right\| \leq \frac{\gamma_{\mathrm{UB}}(\mu)}{\alpha_{\mathrm{LB}}(\mu)}\left\|e^k\right\| + \frac{\Delta t}{\alpha_{\mathrm{LB}}(\mu)}\left\|\mathscr{R}^k\right\|.$$

Then, the bound follows by induction. □

The bound can be simplified by ensuring that $u_q^0 \in X_N$. Then, by linear combination $u^0(\mu) \in X_N$ and with reproduction of solutions—see Proposition 2.78—we get $\|e^0\| = 0$; consequently the corresponding term in the error bound vanishes. We will return to this in Remark 2.95.

Note that there is no simple way to obtain effectivity bounds for this error estimator. Numerically, the effectivities of this error bound may be large. Nevertheless, the error bound is rigorous; thus, if the error bound is small, the true error is also ensured to be small, usually even some orders of magnitude better. Additionally, the error bound predicts zero error a posteriori.

Proposition 2.81 (Vanishing error bound). *If $u^k(\mu) = u_N^k(\mu)$ for all $k = 0, \ldots, K$, then $\Delta_u^k(\mu) = 0, k = 0, \ldots, K$.*

Proof. If $u^k = u_N^k, k = 0, \ldots, K$, we conclude with $(E(\mu))$ and the residual definition (2.54) that $\mathscr{R}^k = 0$ and hence $\Delta_u^k(\mu) = 0$. □

The above general estimators can be improved if more structure or knowledge about the problem is available. The following is a result from [24, 26] and applies to implicit discretizations of symmetric parabolic problems; see Example 2.74. For this, we additionally assume

$$\mathscr{L}_E^k(\mu) = \mathscr{L}_m(\mu), \quad \mathscr{L}_I^k(\mu) = \mathscr{L}_m(\mu) + \Delta t \mathscr{L}_a(\mu) \tag{2.56}$$

for $k = 0, \ldots, K$ and $\mathscr{L}_a, \mathscr{L}_m : X \to X$ being continuous linear operators independent of t and Δt. Here $\mathscr{L}_m$ can correspond to a general mass term and $\mathscr{L}_a$ can represent a stiffness term. Correspondingly, we additionally assume symmetry

$$\langle \mathscr{L}_m(\mu)u, v\rangle = \langle u, \mathscr{L}_m(\mu)v\rangle, \; \langle \mathscr{L}_a(\mu)u, v\rangle = \langle u, \mathscr{L}_a(\mu)v\rangle, \quad u, v \in X, \mu \in \mathscr{P}; \tag{2.57}$$

positive definiteness of $\mathscr{L}_m$; and coercivity of $\mathscr{L}_a$:

$$\langle \mathscr{L}_m(\mu)u, u\rangle > 0, u \neq 0, \quad \alpha(\mu) := \inf_{u \neq 0} \frac{\langle \mathscr{L}_a(\mu)u, u\rangle}{\|u\|^2} > 0, \mu \in \mathscr{P}. \tag{2.58}$$

Then, $\mathscr{L}_m, \mathscr{L}_a$ induce scalar products and we can define the following μ-dependent *space-time energy norm* for $u = (u^k)_{k=1}^K \in X^K$:

$$\|u\|_\mu := \left(\left\langle \mathscr{L}_m(\mu)u^K, u^K\right\rangle + \Delta t \sum_{k=1}^K \left\langle \mathscr{L}_a(\mu)u^k, u^k\right\rangle\right)^{1/2}.$$

Then, the following error bound can be proved.

Proposition 2.82 (A posteriori error bound, space-time energy norm). *Under the assumptions (2.56)–(2.58) we have for the solutions $u(\mu) := (u^k(\mu))_{k=1}^K$, $u_N(\mu) := (u_N^k(\mu))_{k=1}^K$ of $(E(\mu))$ and $(E_N(\mu))$*

$$\|u(\mu) - u_N(\mu)\|_\mu \leq \Delta_u^{en}(\mu) \text{ with}$$

$$\Delta_u^{en}(\mu) := \left(\left\langle \mathscr{L}_m(\mu) e^0, e^0 \right\rangle + \frac{\Delta t}{\alpha_{\mathrm{LB}}(\mu)} \sum_{i=0}^{K-1} \left\| \mathscr{R}^i(\mu) \right\|^2 \right)^{1/2},$$

where $\alpha_{\mathrm{LB}}(\mu)$ is a computable lower bound of the coercivity constant of $\mathscr{L}_a$, i.e., $0 < \alpha_{\mathrm{LB}}(\mu) \leq \alpha(\mu)$.

Proof. The proof in [24] makes repeated use of Young's inequality, i.e., for all $\varepsilon, a, b \in \mathbb{R}$, $ab \leq \frac{1}{2\varepsilon^2} a^2 + \frac{1}{2}\varepsilon^2 b^2$. Starting with the error evolution equation (2.55), making use of the additive decomposition (2.56), and taking the scalar product with e^{k+1} yields

$$\left\langle \mathscr{L}_m e^{k+1}, e^{k+1} \right\rangle + \Delta t \left\langle \mathscr{L}_a e^{k+1}, e^{k+1} \right\rangle = \left\langle \mathscr{L}_m e^k, e^{k+1} \right\rangle + \Delta t \left\langle \mathscr{R}^k, e^{k+1} \right\rangle. \tag{2.59}$$

The first term on the right-hand side can be bounded by Cauchy–Schwarz and Young's inequality with $\varepsilon = 1$:

$$\begin{aligned}
\left\langle \mathscr{L}_m e^k, e^{k+1} \right\rangle &\leq \left\langle \mathscr{L}_m e^k, e^k \right\rangle^{1/2} \left\langle \mathscr{L}_m e^{k+1}, e^{k+1} \right\rangle^{1/2} \\
&\leq \frac{1}{2} \left\langle \mathscr{L}_m e^k, e^k \right\rangle + \frac{1}{2} \left\langle \mathscr{L}_m e^{k+1}, e^{k+1} \right\rangle.
\end{aligned}$$

The second term on the right-hand side of (2.59) can be bounded by Young's inequality with $\varepsilon^2 = \alpha$ and coercivity:

$$\begin{aligned}
\Delta t \left\langle \mathscr{R}^k, e^{k+1} \right\rangle &\leq \Delta t \left\| \mathscr{R}^k \right\| \left\| e^{k+1} \right\| \\
&\leq \Delta t \left(\frac{1}{2\alpha} \left\| \mathscr{R}^k \right\|^2 + \frac{1}{2} \alpha \left\| e^{k+1} \right\|^2 \right) \\
&\leq \Delta t \left(\frac{1}{2\alpha} \left\| \mathscr{R}^k \right\|^2 + \frac{1}{2} \left\langle \mathscr{L}_a e^{k+1}, e^{k+1} \right\rangle \right).
\end{aligned}$$

Then, (2.59) implies

$$\frac{1}{2} \left\langle \mathscr{L}_m e^{k+1}, e^{k+1} \right\rangle - \frac{1}{2} \left\langle \mathscr{L}_m e^k, e^k \right\rangle + \frac{1}{2} \Delta t \left\langle \mathscr{L}_a e^{k+1}, e^{k+1} \right\rangle \leq \Delta t \frac{1}{2\alpha(\mu)} \left\| \mathscr{R}^k \right\|^2.$$

Summation over $k = 0, \ldots, K$ yields a telescope sum and simplifies to

$$\frac{1}{2} \left\langle \mathscr{L}_m e^K, e^K \right\rangle - \frac{1}{2} \left\langle \mathscr{L}_m e^0, e^0 \right\rangle + \frac{1}{2} \Delta t \sum_{k=1}^{K} \left\langle \mathscr{L}_a e^k, e^k \right\rangle \leq \sum_{k=0}^{K-1} \frac{\Delta t}{2\alpha} \left\| \mathscr{R}^k \right\|^2.$$

Multiplication by two and adding the e^0-term gives the statement. □

Remark 2.83 (Extensions). Extensions of the error estimator $\Delta_u^{en}(\mu)$ exist. For example, $\mathscr{L}_I$ may be noncoercive [48], or $\mathscr{L}_E$ may be Δt-dependent [31], i.e., one can also allow explicit discretization contributions as obtained in Euler forward or Crank–Nicolson time discretization. More error estimators can be derived by a space-time (Petrov–)Galerkin viewpoint; see [67, 74].

Remark 2.84 (Output estimation). We have not yet addressed output estimation for the instationary case. This can be realized in simple or in more advanced ways, similar to the approaches of Section 2.3. Possible outputs in time-dependent scenarios can be time-dependent outputs $s^k(\mu)$ at each time step or a single scalar quantity $s(\mu)$ for the complete time trajectory. For this, the problem $(E(\mu))$ can be extended by $l^k \in X', k = 0,\ldots,K$, and

$$s^k(\mu) = l^k(u^k;\mu), \quad k = 0,\ldots,K, \quad s(\mu) = \sum_{k=0}^{K} s^k(\mu).$$

Then, one possibility for output estimation is the direct extension of the procedure of Definition 2.14: the reduced problem can be extended by

$$s_N^k(\mu) := l^k(u_N^k;\mu), \quad k = 0,\ldots,K, \quad s_N(\mu) := \sum_{k=0}^{K} s_N^k(\mu).$$

Then, output error bounds are obtained using continuity of the output functionals:

$$\begin{aligned} |s^k(\mu) - s_N^k(\mu)| &\leq \Delta_s^k(\mu) := \left\| l^k(\cdot;\mu) \right\|_{X'} \Delta_u^k(\mu), \quad k = 0,\ldots,K, \\ |s(\mu) - s_N(\mu)| &\leq \Delta_s(\mu) := \sum_{k=0}^{K} \Delta_s^k(\mu). \end{aligned}$$

This procedure again is admittedly very coarse: first, as only a "linear" dependence on the state error bound is obtained, and second, as the possible bad effectivity of the state bounds is passed on to the output bounds. Using a primal-dual technique, better estimates of single outputs can be obtained; see [24]. The dual problem to a scalar output of an instationary problem is a backward-in-time problem, where the inhomogeneities are given by the output functionals. Then, similar to Definition 2.31, an output correction can be performed by the primal residual applied to the dual solution, and output error bounds can be obtained that have the "squared" effect, as in (2.20).

2.4.4 ▪ Offline/online decomposition

The offline/online decomposition is analogous to the stationary case. The main insight is that with time-independent operator components, the offline storage does not grow with K but is independent of the time-step number. Hence, we again obtain an online phase that is independent of the dimension H of the spatial discretization.

We again assume that $X = \operatorname{span}(\psi_i)_{i=1}^{\mathcal{N}}$ is a discrete high-dimensional space of dimension $\mathcal{N}$; we are given the inner product matrix $\mathbf{K}$ and system matrix and vector components

$$\begin{aligned} \mathbf{K} &:= \left(\left\langle \psi_i, \psi_j \right\rangle\right)_{i,j=1}^{\mathcal{N}} \in \mathbb{R}^{\mathcal{N}\times\mathcal{N}}, & \mathbf{b}_q &:= \left(\left\langle b_q, \psi_i \right\rangle\right)_{i=1}^{\mathcal{N}} \in \mathbb{R}^{\mathcal{N}}, \\ \mathbf{L}_{I,q} &:= \left(\left\langle \mathcal{L}_{I,q}\psi_j, \psi_i \right\rangle\right)_{i,j=1}^{\mathcal{N}} \in \mathbb{R}^{\mathcal{N}\times\mathcal{N}}, & \mathbf{L}_{E,q} &:= \left(\left\langle \mathcal{L}_{E,q}\psi_j, \psi_i \right\rangle\right)_{i,j=1}^{\mathcal{N}} \in \mathbb{R}^{\mathcal{N}\times\mathcal{N}} \end{aligned} \tag{2.60}$$

for $q = 1,\dots,Q_b$ and Q_I, Q_E, respectively, and $\mathbf{u}_q^0 = (u_{q,i}^0)_{i=1}^{\mathcal{N}} \in \mathbb{R}^{\mathcal{N}}, q = 1,\dots,Q_{u^0}$, the coefficient vector of $u_q^0 = \sum_{i=1}^{\mathcal{N}} u_{q,i}^0 \psi_i \in X$. For a solution of $(E(\mu))$, one can assemble the system components by evaluating the coefficient functions and linear combinations

$$\mathbf{u}^0(\mu) = \sum_{q=1}^{Q_{u^0}} \theta_{u^0,q}(\mu)\mathbf{u}_q^0, \quad \mathbf{b}^k(\mu) = \sum_{q=1}^{Q_b} \theta_{b,q}^k(\mu)\mathbf{b}_q,$$

$$\mathbf{L}_I^k(\mu) = \sum_{q=1}^{Q_I} \theta_{I,q}^k(\mu)\mathbf{L}_{I,q}, \quad \mathbf{L}_E^k(\mu) = \sum_{q=1}^{Q_E} \theta_{E,q}^k(\mu)\mathbf{L}_{E,q}$$

and iteratively solving

$$\mathbf{L}_I^k(\mu)\mathbf{u}^{k+1} = \mathbf{L}_E^k(\mu)\mathbf{u}^k + \mathbf{b}^k(\mu), \quad k = 0,\dots,K-1, \tag{2.61}$$

to obtain the vector of unknowns $\mathbf{u}^k = (u_i^k)_{i=1}^{\mathcal{N}} \in \mathbb{R}^{\mathcal{N}}$ of $u^k = \sum_{i=1}^{\mathcal{N}} u_i^k \psi_i \in X$.

Remark 2.85 (Time evolution for nonvariational discretizations). For variational discretizations, e.g., FEs, the matrices $\mathbf{L}_{I,q}, \mathbf{L}_{E,q}$ are exactly the components of the FE method mass or stiffness matrices; hence assembly is clear. For nonvariational discretizations such as finite differences or FVs one can apply the current technique by assuming a quadrature approximation of the $L^2(\Omega)$ scalar product based on the current grid points; see Examples 2.75 and 2.76. In both cases, $\mathbf{K}$ will be a diagonal matrix. The discretization for finite difference or FV schemes is frequently not given in terms of the above variational matrices, but matrices and vectors of unknowns are given by point evaluations of the operator results, i.e., if $v = \mathcal{L}u$ for $u, v \in X$, then a linear operator $\mathcal{L} : X \to X$ is realized by a matrix operation $\mathbf{v} = \hat{\mathbf{L}}\mathbf{u}$. The relation to a matrix $\mathbf{L} = (\langle \mathcal{L}\psi_j, \psi_i \rangle)_{i,j=1}^{\mathcal{N}}$ of type (2.60) is simple: set $v = \mathcal{L}\psi_j$; then $\mathbf{v} = \hat{\mathbf{L}}\mathbf{e}_j$ and

$$(\mathbf{L})_{ij} = \mathbf{e}_i^T \mathbf{L}\mathbf{e}_j = \left\langle \mathcal{L}\psi_j, \psi_i \right\rangle = \langle v, \psi_i \rangle = \langle \psi_i, v \rangle = \mathbf{e}_i^T \mathbf{K}\mathbf{v} = \mathbf{e}_i^T \mathbf{K}\hat{\mathbf{L}}\mathbf{e}_j.$$

Hence, $\mathbf{L} = \mathbf{K}\hat{\mathbf{L}}$. Therefore, the evolution step (2.61) reads

$$\mathbf{K}\hat{\mathbf{L}}_I^k(\mu)\mathbf{u}^{k+1} = \mathbf{K}\hat{\mathbf{L}}_E^k(\mu)\mathbf{u}^k + \mathbf{K}\hat{\mathbf{b}}^k(\mu), \quad k = 0,\dots,K-1;$$

hence, one can also omit $\mathbf{K}$ in the evolution of the solution of $(E(\mu))$. In particular, this procedure is implemented in *RBmatlab* and used in the experiments of this section.

Now, the offline/online decomposition of $(E_N(\mu))$ is straightforward.

Proposition 2.86 (Offline/online decomposition of $(E_N(\mu))$).
Offline phase: After computing an RB $\Phi_N = \{\varphi_1,\dots,\varphi_N\}$, compute the parameter- and time-independent matrices and vectors

$$\mathbf{b}_{N,q} := (\left\langle b_q, \varphi_i \right\rangle)_{i=1}^N \in \mathbb{R}^N, \quad \mathbf{L}_{N,I,q} := (\left\langle \mathcal{L}_{I,q}\varphi_j, \varphi_i \right\rangle)_{i,j=1}^N \in \mathbb{R}^{N\times N},$$

$$\mathbf{u}_{N,q}^0 := (\left\langle u_q^0, \varphi_i \right\rangle)_{i=1}^N \in \mathbb{R}^N, \quad \mathbf{L}_{N,E,q} := (\left\langle \mathcal{L}_{E,q}\varphi_j, \varphi_i \right\rangle)_{i,j=1}^N \in \mathbb{R}^{N\times N}.$$

Online phase: For a given $\mu \in \mathcal{P}$, evaluate the coefficient functions $\theta_{I,q}^k(\mu)$, $\theta_{E,q}^k(\mu)$,

$\theta_{u^0}(\mu)$, and $\theta_b^k(\mu)$; assemble the reduced system matrices and vectors

$$\mathbf{L}_{N,I}^k(\mu) := \sum_{q=1}^{Q_I} \theta_I^k(\mu)\mathbf{L}_{N,I,q}, \quad \mathbf{L}_{N,E}^k(\mu) := \sum_{q=1}^{Q_E} \theta_E^k(\mu)\mathbf{L}_{N,E,q},$$

$$\mathbf{b}_N^k(\mu) := \sum_{q=1}^{Q_b} \theta_b^k(\mu)\mathbf{b}_{N,q}, \quad k = 0,\dots,K-1;$$

and solve the discrete reduced evolution system by $\mathbf{u}_N^0 := \sum_{q=1}^{Q_{u^0}} \theta_{u^0,q}(\mu)\mathbf{u}_{N,q}^0$ and

$$\mathbf{L}_{N,I}^k(\mu)\mathbf{u}_N^{k+1} = \mathbf{L}_{N,E}^k(\mu)\mathbf{u}_N^k + \mathbf{b}_N^k(\mu), \quad k = 0,\dots,K-1.$$

Again, the computational procedure for obtaining the components is very simple: if we again assume the RB to be given as coefficient matrix $\mathbf{\Phi}_N \in \mathbb{R}^{\mathcal{N}\times N}$, the components can be computed by

$$\mathbf{L}_{N,E,q} := \mathbf{\Phi}^T\mathbf{L}_{E,q}\mathbf{\Phi}, \quad \mathbf{L}_{N,I,q} := \mathbf{\Phi}^T\mathbf{L}_{I,q}\mathbf{\Phi}, \quad \mathbf{b}_{N,q} := \mathbf{\Phi}^T\mathbf{b}_q, \quad \mathbf{u}_{N,q}^0 := \mathbf{\Phi}^T\mathbf{u}_q^0.$$

The offline/online decomposition of the error estimators can also be realized. Computational procedures for obtaining upper continuity and lower coercivity bounds were addressed in Section 2.3.5. The remaining ingredient for $\Delta_u^k(\mu)$ or Δ_u^{en} is again an efficient computational procedure for the residual norm. Analogous to Proposition 2.36, we obtain parameter separability of the residual.

Proposition 2.87 (Parameter separability of the residual). *Set $Q_R := N(Q_E+Q_I)+Q_b$ and define $\mathcal{R}_q \in X, q = 1,\dots,Q_R$ by*

$$(\mathcal{R}_1,\dots,\mathcal{R}_{Q_R}) := \big(\mathcal{L}_{E,1}\varphi_1,\dots,\mathcal{L}_{E,Q_E}\varphi_1,\dots,\mathcal{L}_{E,1}\varphi_N,\dots,\mathcal{L}_{E,Q_E}\varphi_N,$$
$$\mathcal{L}_{I,1}\varphi_1,\dots,\mathcal{L}_{I,Q_I}\varphi_1,\dots,\mathcal{L}_{I,1}\varphi_N,\dots,\mathcal{L}_{I,Q_I}\varphi_N, b_1,\dots,b_{Q_b}\big).$$

Let $u_N^k(\mu) = \sum_{i=1}^N u_{N,i}^k \varphi_i, k = 0,\dots,K$ be the solution of $(E_N(\mu))$, and define $\theta_{\mathcal{R}}^k(\mu) := (\theta_{\mathcal{R},q}^k(\mu))_{k=0}^{K-1}$ by

$$(\theta_{\mathcal{R},1}^k(\mu),\dots,\theta_{\mathcal{R},Q_R}^k(\mu)) := \frac{1}{\Delta t}\big(\theta_{E,1}^k(\mu)u_{N,1}^k(\mu),\dots,\theta_{E,Q_E}^k(\mu)u_{N,1}^k(\mu),\dots,$$
$$\theta_{E,1}^k(\mu)u_{N,N}^k(\mu),\dots,\theta_{E,Q_E}^k(\mu)u_{N,N}^k(\mu),$$
$$-\theta_{I,1}^k(\mu)u_{N,1}^{k+1}(\mu),\dots,-\theta_{I,Q_I}^k(\mu)u_{N,1}^{k+1}(\mu),\dots,$$
$$-\theta_{I,1}^k(\mu)u_{N,N}^{k+1}(\mu),\dots,-\theta_{I,Q_I}^k(\mu)u_{N,N}^{k+1}(\mu),$$
$$\theta_{b,1}^k(\mu),\dots,\theta_{b,Q_b}^k(\mu)\big).$$

Then the residual $\mathcal{R}^k(\mu)$ defined in (2.54) is parameter separable with

$$\mathcal{R}^k(\mu) = \sum_{q=1}^{Q_R} \theta_{\mathcal{R},q}^k(\mu)\mathcal{R}_q.$$

Hence, the norm $||\mathscr{R}^k(\mu)||$ can be computed efficiently with the offline/online procedure as stated in Proposition 2.38, now using the Gramian matrix $\mathbf{G}_R := (\langle \mathscr{R}_q, \mathscr{R}_{q'} \rangle)_{q,q'=1}^{Q_R}$ and the coefficient vector $\theta_{\mathscr{R}}^k(\mu)$ and computing

$$\left\| \mathscr{R}^k(\mu) \right\| = \sqrt{\theta_{\mathscr{R}}^k(\mu)^T \mathbf{G}_R \theta_{\mathscr{R}}^k(\mu)}.$$

This completes the offline/online computational procedure for the general RB approach for instationary problems.

We close this section on offline/online decomposition with a remark concerning possible coupling of time-step and spatial discretization.

Remark 2.88 (Coupling of spatial and time discretization). Note that a slight dependence of the online phase on the spatial discretization exists via possible time-step constraints: the reduced problem has the identical number of time steps as the full problem. If Δt is constrained to $\mathcal{O}(\Delta x)$ for explicit discretizations of advection terms or $\mathcal{O}(\Delta x^2)$ for explicit discretization of diffusion operators, the full problem cannot be chosen highly accurately without affecting the time discretization and thus the reduced simulation. So in RB approaches to time evolution problems, only the complexity due to spatial but not time discretization is reduced.

2.4.5 ▪ Basis generation

Again, we address basis generation in a separate section, although it is naturally part of the offline phase.

The simplest basis type for the instationary case is obtained by considering time as an additional "parameter" and then using a Lagrangian RB according to Definition 2.45, i.e., $\Phi_N = \{u^{k^{(i)}}(\mu^{(i)})\}_{i=1}^N$.

While this procedure is good for validation purposes (see Proposition 2.78) or for testing an RB scheme, there are serious difficulties with this approach for obtaining a *good and small* basis. First, the time parameter manifold may be more complex and hence many snapshots and thus a larger basis may be required. Further, it is unclear how to choose the time indices $k^{(i)}$, and, technically, to obtain a single $u^{k^{(i)}}(\mu^{(i)})$ one needs to compute the complete trajectory $u^k(\mu^{(i)})$ for $k = 0, \ldots, k^{(i)}$ and discard the unused information of the initial time steps.

The first procedure we will present addresses the first difficulty, the treatment of large snapshot sizes. The so-called proper orthogonal decomposition *(POD)* allows us to compress large snapshot sets to the most important *POD modes*, i.e., a few vectors or functions containing the most important information of the data. Starting with a large number of functions $\{u_i\}_{i=1}^n \subset X$, POD generates a small orthonormal set of basis functions Φ_N with $N \ll n$ by means of the so-called empirical correlation operator. Technically, POD corresponds to principal component analysis [42], the Karhunen–Loève transformation [45, 50], or the Hotelling transformation [38]. We restrict ourselves to the definition and some elementary properties and refer to [42, 72] for details.

Proposition 2.89 (POD). *Let $\{u_i\}_{i=1}^n \subset X$ be a given set of snapshots. Then, define the* empirical correlation operator $R : X \to X$ *by*

$$Ru := \frac{1}{n} \sum_{i=1}^{n} \langle u_i, u \rangle \, u_i, \quad u \in X.$$

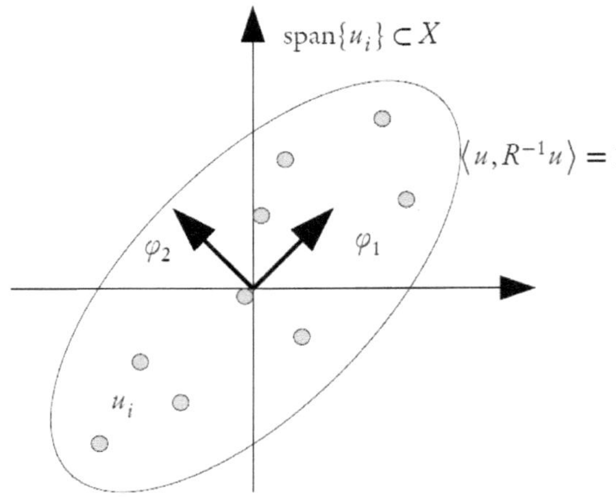

Figure 2.11. *Illustration of POD.*

Then, R is a compact self-adjoint linear operator and there exists an orthonormal set $\{\varphi_i\}_{i=1}^{n'}$ of $n' \leq n$ eigenvectors with real eigenvalues $\lambda_1 \geq \lambda_2 \geq \cdots \geq \lambda_{n'} > 0$ with

$$Ru = \sum_{i=1}^{n'} \lambda_i \langle \varphi_i, u \rangle \varphi_i. \tag{2.62}$$

We denote $\Phi_N := POD_N(\{u_i\}) := \{\varphi_i\}_{i=1}^N$ as the POD basis *of size $N \leq n'$.*

Proof. R is linear and bounded by $||R|| \leq \frac{1}{n}\sum_{i=1}^n ||u_i||^2$ and has finite-dimensional range, so R is compact. Further, R is self-adjoint, as

$$\langle Ru, v \rangle = \frac{1}{n}\sum_i \langle u_i, u \rangle \langle u_i, v \rangle = \langle u, Rv \rangle;$$

hence, by the spectral theorem there exists an orthonormal system satisfying the spectral decomposition (2.62). It must be finite, $n' < \infty$, as the range of R is finite. □

Some interesting properties are the following; see also Figure 2.11 for an illustration:

- $\{\varphi_i\}$ as orthonormal basis is not unique due to a sign change in each vector, or possible rotations if an eigenspace has dimension larger than one.
- The bases obtained by POD are hierarchical, i.e., $\Phi_{N'} \subseteq \Phi_N$ for $N' \leq N$.
- POD does not depend on the order of the data $\{u_i\}_{i=1}^n$ (in contrast to Gram–Schmidt orthonormalization).
- Let $X_{\text{POD},N} := \text{span}(POD_N(\{u_i\}_{i=1}^n))$. Then, φ_1 is the direction of highest variance of $\{u_i\}_{i=1}^n$, φ_2 is the direction of highest variance of the projected data $\{P_{X_{POD,1}^{\perp}} u_i\}_{i=1}^n$, etc.

- The coordinates of the data with respect to the POD basis are uncorrelated; see Exercise 2.114.
- $\{\varphi_i\}$ and $\{\sqrt{\lambda_i}\}$ are the principal axis and axis intercepts of the ellipsoid $\langle u, R^{-1}u\rangle = 1$.
- POD has a best-approximation property with respect to the squared error, and the error can be exactly computed by the truncated eigenvalues; see Exercise 2.115:

$$\inf_{\substack{Y\subset X\\ \dim(Y)=N}} \frac{1}{n}\sum_{i=1}^{n} \|u_i - P_Y u_i\|^2 = \frac{1}{n}\sum_{i=1}^{n} \left\|u_i - P_{X_{\mathrm{POD},N}} u_i\right\|^2 = \sum_{i=N+1}^{n'} \lambda_i. \quad (2.63)$$

The eigenvalue problem for the correlation operator is either very high dimensional ($\dim X = \mathcal{N}$) or even infinite dimensional ($\dim(X) = \infty$). This poses challenges for computational procedures. If the number of snapshots $n < \mathcal{N}$, the $\mathcal{N}$-dimensional eigenvalue problem can be reformulated by an n-dimensional eigenvalue problem for the Gramian matrix and hence provide a more efficient computational procedure. Such a reformulation is sometimes referred to as the *method of snapshots* [66] or the *kernel trick* in machine learning [64]. We leave the proof of the following proposition as Exercise 2.116.

Proposition 2.90 (Computation by Gramian matrix). *Let* $\mathbf{K}_u := (\langle u_i, u_j\rangle)_{i,j=1}^n \in \mathbb{R}^{n\times n}$ *be the Gramian matrix of the snapshot set* $\{u_i\}_{i=1}^n \subset X$. *Then the following are equivalent:*

(i) $\varphi \in X$ *is an eigenvector of* R *for eigenvalue* $\lambda > 0$ *with norm one and a representation* $\varphi = \sum_i a_i u_i$ *with* $\mathbf{a} = (a_i)_{i=1}^n \in \ker(\mathbf{K}_u)^\perp$.

(ii) $\mathbf{a} = (a_i)_{i=1}^n \in \mathbb{R}^n$ *is an eigenvector of* $\frac{1}{n}\mathbf{K}_u$ *with eigenvalue* $\lambda > 0$ *and norm* $\frac{1}{\sqrt{n\lambda}}$.

Remark 2.91 (Difference between greedy and POD). We briefly comment on some differences between the POD of Proposition 2.89 and the strong greedy procedure of Definition 2.46, i.e., using the true projection error as indicator, $\Delta(Y,\mu) := \|u(\mu) - P_Y u(\mu)\|$. Both methods require the set of snapshots to be available; hence we need many full simulations. The main difference is the error measure that guides both procedures. POD aims at minimizing the *mean squared* projection error, while the greedy procedure aims at minimizing the *maximum* projection error. So, "outliers" with single large error are allowed in POD, while the greedy algorithm will prevent such large deviations. Computationally, the greedy procedure produces an RB space spanned by snapshots, i.e., a Lagrangian RB space; POD produces a space that is a *subset* of a span of snapshots, but it is not a Lagrangian RB space.

Now, the greedy and POD procedures can be suitably combined to produce incrementally good bases for time-dependent problems. The resulting algorithm is called the POD-greedy procedure (initially denoted as the PCA-fixspace in [31]) and is standard in RB approaches for time-dependent problems; see [17, 19]. The idea is to "be greedy" with respect to the parameter and use POD with respect to time: we search the currently worst resolved parameter using an error bound or indicator $\Delta(Y,\mu)$ and then compute the complete trajectory of the corresponding solution, orthogonalize this trajectory to the current RB space, perform a POD with respect to time to compress the error trajectory to its most important new information, and add the new POD mode to the current basis.

The use of POD in the POD-greedy procedure eliminates the two remaining problems stated above: we do not need to worry how to select time instants for basis extension, and we do not discard valuable information in the computed trajectory but try to extract maximal new information from the selected trajectory.

Definition 2.92 (POD-greedy procedure). *Let $S_{\text{train}} \subset \mathcal{P}$ be a given training set of parameters and $\varepsilon_{\text{tol}} > 0$ a given error tolerance. Set $X_0 := \{0\}, \Phi_0 := \emptyset, n := 0$ and define iteratively*

$$\begin{aligned}
&\textit{while } \varepsilon_n := \max_{\mu \in S_{\text{train}}} \Delta(X_n, \mu) > \varepsilon_{\text{tol}}, \qquad (2.64)\\
&\quad \mu^{(n+1)} := \arg\max_{\mu \in S_{\text{train}}} \Delta(X_n, \mu),\\
&\quad u_{n+1}^k := u^k(\mu^{(n+1)}), k = 0, \ldots, K, \textit{ solution of } (E(\mu^{(n+1)})),\\
&\quad e_{n+1}^k := u_{n+1}^k - P_{X_n} u_{n+1}^k, k = 0, \ldots, K,\\
&\quad \varphi_{n+1} := POD_1(\{e_{n+1}^k\}_{k=0}^K),\\
&\quad \Phi_{n+1} := \Phi_n \cup \{\varphi_{n+1}\},\\
&\quad X_{n+1} := X_n + \text{span}(\varphi_{n+1}),\\
&\quad n \leftarrow n + 1,\\
&\textit{end while.}
\end{aligned}$$

Thus, the output of the algorithm is the desired RB space X_N and basis Φ_N by setting $N := n + 1$ as soon as (2.64) is false. The algorithm can be used with different error measures, e.g., the true squared projection error $\Delta(Y, \mu) := \sum_{k=0}^K \|u^k(\mu) - P_Y u^k(\mu)\|^2$ or one of the error estimators $\Delta(Y, \mu) := \Delta_u^K(\mu)$ and $\Delta(Y, \mu) := \Delta_u^{en}(\mu)$. In the first case, we denote the algorithm the *strong POD-greedy* procedure, while in the latter cases it is called the *weak POD-greedy* procedure.

Remark 2.93 (Consistency of POD-greedy to POD and greedy). In the case of a single time step $K = 1$, the evolution scheme corresponds to a single system solve and, hence, can be interpreted as solving a stationary problem. The POD-greedy procedure then is just the standard greedy algorithm with included orthogonalization. On the other hand, it can easily be seen that for the case of a single parameter $\mathcal{P} = \{\mu\} \subset \mathbb{R}^p$ but arbitrary $K > 1$ POD and POD-greedy coincide: POD-greedy incrementally finds the next orthogonal basis vector, which is just the next POD mode. Therefore, the POD-greedy basis of a single trajectory is optimal in the least-squares sense. Thus, POD-greedy is a generalization of POD.

Again, not only is the procedure heuristic, but also convergence rates can be derived, and a result analogous to Proposition 2.49 holds after some extensions [28]. First, a space-time norm for trajectories $v := (v^k(\mu))_{k=0}^K \subset X^{K+1}$ is introduced by suitable weights $w_k > 0, \sum_k w_k = T$, and

$$\|v\|_T := \left(\sum_{k=0}^K w_k \left\| v^k \right\|^2 \right)^{1/2}.$$

Then, the manifold of parametric trajectories is introduced as

$$\mathcal{M}_T := \{u(\mu) = (u^k(\mu))_{k=0}^K | \mu \in \mathcal{P}\} \subset X^{K+1},$$

while the *flat* manifold is

$$\mathcal{M} := \{u^k(\mu) | k = 0, \ldots, K, \mu \in \mathcal{P}\} \subset X.$$

The componentwise projection on a subspace $Y \subset X$ is defined by $P_{T,Y} : X^{K+1} \to Y^{K+1}$ via

$$P_{T,Y}u := (P_Y u^k)_{k=0}^K, \quad u \in X^{K+1}.$$

Now the convergence rate statement can be formulated, where again d_n denotes the Kolmogorov n-width as defined in (2.37).

Proposition 2.94 (Convergence rates of POD-greedy procedure). *Let $S_{\text{train}} = \mathcal{P}$ be compact and the error indicator Δ be chosen such that for suitable $\gamma \in (0,1]$,*

$$\left\| u(\mu^{(n+1)}) - P_{T,X_n} u(\mu^{(n+1)}) \right\|_T \geq \gamma \sup_{u \in \mathcal{M}_T} \left\| u - P_{T,X_n} u \right\|_T. \tag{2.65}$$

(i) *Algebraic convergence rate: If $d_n(\mathcal{M}) \leq M n^{-\alpha}$ for some $\alpha, M > 0$ and all $n \in \mathbb{N}$ and $d_0(\mathcal{M}) \leq M$, then*

$$\varepsilon_n \leq C M n^{-\alpha}, n > 0,$$

with suitable (explicitly computable) constant $C > 0$.
(ii) *Exponential convergence rate: If $d_n(\mathcal{M}) \leq M e^{-an^\alpha}$ for $n \geq 0$, $M, a, \alpha > 0$, then*

$$\varepsilon_n \leq C M e^{-cn^\beta}, n \geq 0,$$

with $\beta := \alpha/(\alpha+1)$ and suitable (explicitly computable) constants $c, C > 0$.

Remark 2.95 (Choice of initial basis). The POD-greedy procedure can also be used in a slightly different fashion, namely as an extension of an existing basis $\Phi \neq \emptyset$. For this we simply set $N_0 := |\Phi|$, the initial basis $\Phi_{N_0} := \Phi$, and RB space $X_{N_0} := \text{span}(\Phi)$ and start POD-greedy with N_0 instead of zero. One possible scenario for this could be the improvement of an existing basis: if insufficient accuracy is detected in the online phase, one can return to the offline phase for a basis extension. Another useful application of this variant is choosing the initial basis such that $u_q^0 \in X_{N_0}$. Then the RB error at time $t = 0$ is zero and the a posteriori error bounds are more tight, as the initial error contribution is zero.

Remark 2.96 (Adaptivity, PT partition). As the solution complexity of time-dependent simulations is much higher than for stationary problems, the size of S_{train} is more critical in the instationary case: Although the a posteriori error estimators have complexity polynomial in N, they are still not very cheap due to the linear scaling with K. The size of S_{train} being limited, the choice and adaptation of the training set of parameters [29, 37] becomes an even more important issue for instationary problems.

Especially for time-dependent problems, the solution variability with varying parameter *and* time can be very dramatic (e.g., a transport process with varying directions; see the model problem of Section 2.4.1). Then, the parameter domain partitioning approaches mentioned in Section 2.3.6 can also be applied. In particular, the *hp*-RB-approach is extended to parabolic problems [19] and the P-partition approach is used for hyperbolic problems [29].

If the time interval is large such that solution variations are too high, parameter domain partitioning may be not sufficient. In these cases, time-interval partitioning can also be applied [14, 16]. This means that single bases are constructed for subintervals of the time axis. For the reduced simulation, a suitable switching between the spaces must be realized over time. This can be done either during the Galerkin projection step of the RB scheme or as a separate orthogonal projection step. In both cases, the error estimation procedure can be kept fully rigorous by suitably incorporating the additional projection errors [14]. The division of the time interval can be obtained adaptively: Starting with a large interval, the basis generation with POD-greedy is initiated. As soon as a too large basis is obtained (or anticipated early by extrapolation), POD-greedy is terminated, the time interval is split, and separate bases are constructed on the subintervals.

We also want to conclude this section with some experiments, which again can be reproduced with the script `rb_tutorial.m`. We choose the model problem of Section 2.4.1 and apply an FV discretization. For this we assume a uniform hexahedral grid consisting of 64×32 squares and set $\Delta t = 1/256$. The advection is discretized explicitly with an upwind flux, and the diffusion is discretized implicitly; see [31]. The time-step width is sufficiently small to meet the CFL time-step restriction. The snapshot plots in Figure 2.10 were based on this discretization.

We generate a POD-greedy basis using the initial data field u^0 as starting basis, setting $\varepsilon_{\text{tol}} = 1\cdot 10^{-2}$ and choosing the X-norm error indicator from Proposition 2.80 as selection criterion $\Delta(Y,\mu) := \Delta_u^K(\mu)$ with the (coarse) bound constants $\gamma_{\text{UB}}(\mu) = 1$ and $\alpha_{\text{LB}}(\mu) = 1$, because the explicit and inverted implicit spatial discretization operators are L^2-stable due to our choice of time-step width. As training parameter set we use the vertices of a uniform 10×10 grid of points on $\mathcal{P} = [0,1]^2$.

The resulting POD-greedy training estimator development is illustrated in Figure 2.12(a) by plotting for each basis size N the maximal estimator over the training set of parameters. It nicely shows an exponentially decaying behavior; however, the convergence is much slower than in the stationary case due to the complex parameter and time dependence. In fact the low-diffusion region is very hard to approximate, which causes the relatively large basis sizes. This difficult region is also reflected in Figure 2.12(b), where parameter selection frequencies are plotted over the training set. Larger circles indicate training parameters that are chosen more frequently during basis generation. Note that this is a difference from the greedy algorithm in the stationary case: parameters can be selected multiple times during the POD-greedy procedure, as addition of a single mode to the basis does not necessarily reduce the error to zero.

The first 16 generated basis vectors are plotted in Figure 2.13. One can observe increasing oscillations in the basis functions. One can also observe that the initial condition (see Figure 2.10) was chosen as initial basis.

In Figure 2.14(a) we illustrate the behavior of the a posteriori error bound Δ_u^k and the error

$$e^k(\mu) := \max_{k'=0\dots,k} \left\| u^{k'}(\mu) - u_N^{k'}(\mu) \right\|$$

at final time $k = K$ for a parameter sweep along the diagonal of $\mathcal{P}$ by $\mu = s\cdot(1,1)^T, s \in [0,1]$. It can be verified that indeed the error estimator is below ε_{tol} for the training points obtained by $s = i/10, i = 0,\dots,10$. This bound cannot be guaranteed for test points in the low-diffusivity region. The results indicate that it would be beneficial

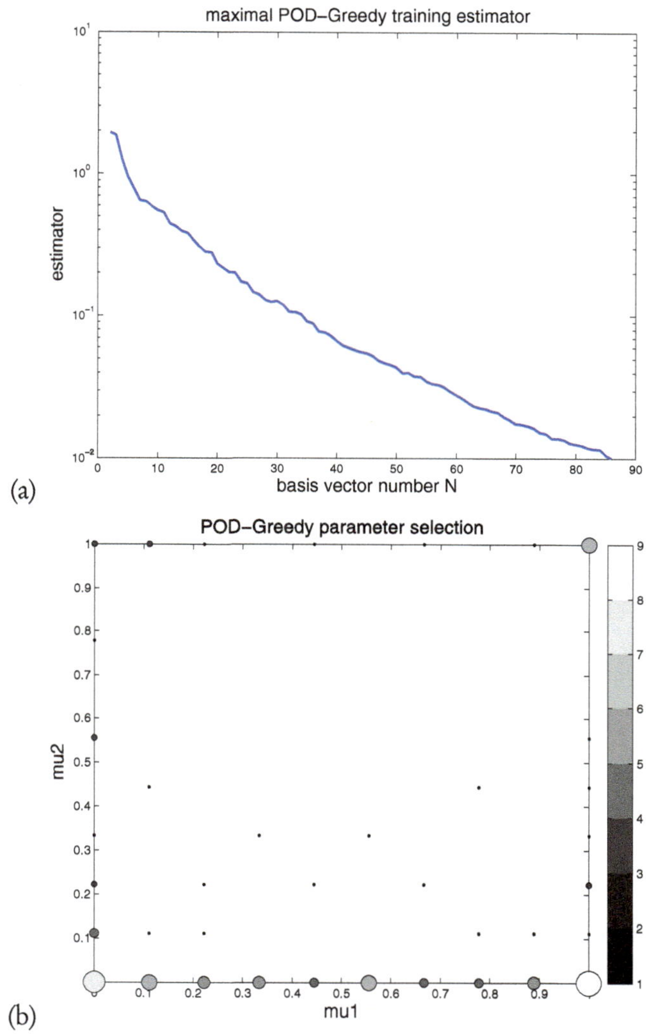

Figure 2.12. *Illustration of POD-greedy results for the advection-diffusion model problem.* (a) *Plot of maximal training estimator decay and* (b) *plot of parameter selection frequency.*

to include more training points in this difficult parameter region. For example, one could choose the diffusivity values log-equidistant in accordance with Proposition 2.51 or apply an adaptive training set extension algorithm [29, 30]. The ratio of the error bound and the true error is about one order of magnitude so can be considered to be quite good. This is made more explicit in Figure 2.14(b), where for a test set of 200 random parameters we plot the effectivities $\eta^k(\mu) := \Delta_u^k(\mu)/e^k(\mu)$ at final time $k = K$ where the test points are sorted according to μ_2. We nicely see that the effectivities are lower bounded by one, i.e., the error estimators are reliable, while the factor of overestimation is not too large. Note, however, that the error estimators are mostly incremental with growing k; hence the effectivities are expected to get worse for larger times T. This can be improved by space-time Galerkin approaches and estimators [67, 74].

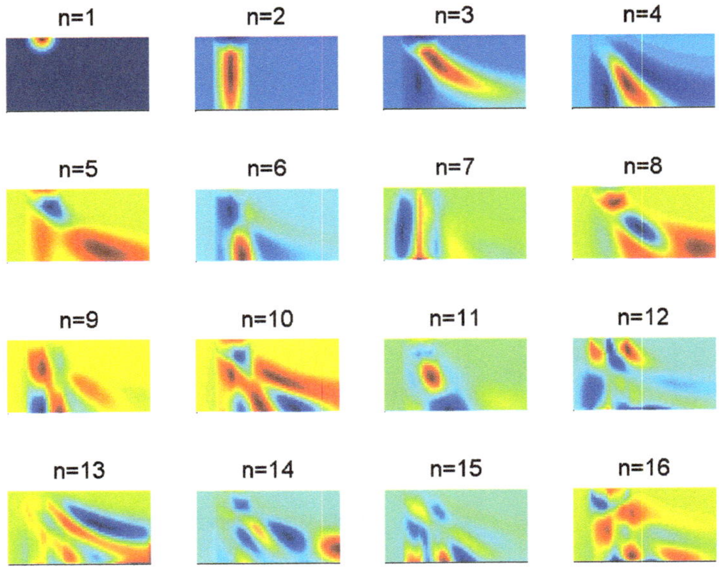

Figure 2.13. *Illustration of the first* 16 *basis vectors produced by the POD-greedy procedure.*

2.5 ▪ Extensions and outlook

We give some comments and references to literature on further aspects and some current developments that could not be covered by this introductory chapter.

Some extensions are well established. The first question is the treatment of nonlinear problems. In the case of simple "polynomial" nonlinearities, which can be written as a multilinear form in the variational form of the PDE, this multilinearity can be effectively used for suitable offline/online decomposition of the Galerkin-reduced system [69] and the Newton-type iteration for solving the fixed-point equation. Also, a posteriori error analysis is possible for these RB approaches, making use of the Brezzi–Rappaz–Raviart theory. Problems that can be treated this way are nonlinear diffusion or nonlinear advection problems, e.g., the Burgers equation [70]. For more general nonlinearities, the EI method can be applied to approximate stationary and instationary nonlinear problems [25]. A specialization of this procedure for discretization operators is the empirical operator interpolation initially used in [33] and then extended to nonlinear problems in [17, 32], which requires local reconstruction of the reduced solution and local evaluation of the differential operator. This procedure was later denoted the discrete empirical interpolation method [10] in the context of nonlinear state-space dynamical systems. Note, however, that the stability of the approximated RB systems involving an EI approximation step is a nontrivial aspect.

The second obvious possibility for extensions of the presented methodology is the treatment of more general linear problems. As harmonic Maxwell's or Helmholtz equations are noncoercive, a more general notion of stability is considered in RB approaches, the *inf-sup* stability. This notion generalizes the coercivity (the inf-sup constant being always at least as large as the coercivity constant), while the RB error bounds have frequently identical structure and the inf-sup constant "replaces" the coercivity constant [61, 71]. A main problem is that inf-sup stability is not inherited

(a)

(b)

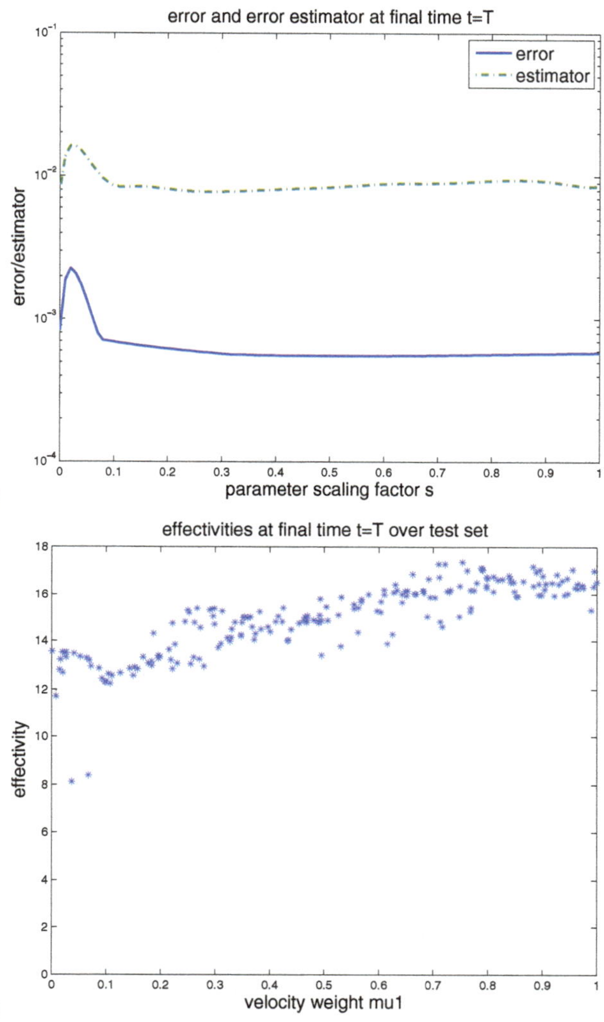

Figure 2.14. *Behavior of error estimator $\Delta_u^k(\mu)$ and error $e^k(\mu)$ at end time $k = K$.* (a) *Error and estimator over parameter sweep along the diagonal of the parameter domain,* $\mu = s \cdot (1,1)^T, s \in [0,1]$, (b) *effectivities $\eta^K(\mu)$ for* 200 *random parameter vectors.*

by subspaces; hence separate test and trial spaces must be constructed to guarantee stability of the reduced systems. By suitable definition of a norm on the test space, optimal stability factors can be obtained by double greedy procedures [11]. In the context of time-dependent problems, an interesting possibility is the formulation as a time-space variational form [67], which represents an inf-sup stable formulation. The resulting RB error estimators are typically very sharp in contrast to the incremental estimators of Section 2.4. In particular, they have provable effectivity bounds. However, the discretization must then be adjusted to be able to cope with the additional dimension by the time variable. The notion of inf-sup stability is also required in the treatment of systems of PDEs, most notably the Stokes system [62] for viscous flow. Parameter dependence can also be obtained by geometry parametrization [63]. As a general solution strategy, the parametrized PDE is mapped to a reference domain that incorporates the geometry parameters in the coefficient functions of the transformed PDE.

Various recent developments can be found that represent active research directions in RB methods. First, the variational problems can be additionally constrained with inequalities. The resulting variational inequalities can also be treated successfully with RB methods, both in stationary and instationary cases [23, 34, 35]. The Stokes system reveals a saddle point structure that is typical for other types of problems. Such general saddle point problems are considered in [22]. Extensions to more complex coupled systems, e.g., Navier–Stokes [13, 69] or the Boussinesq approximation [47], can be found. An important field of application for RB methods is multiquery scenarios in optimization or optimal control. Both parameter optimization [15, 56] and optimal control for elliptic and parabolic equations have been investigated [12, 43, 44]. Parameters can also be considered as a random influence in PDEs; corresponding Monte Carlo approaches using reduced-order models (ROMs) can be realized [7, 36]. Multiscale problems that require multiple evaluation of micromodels for a macroscale simulation can make use of RB micromodels [6, 48], or domain-partitioning approaches can be used to capture global parametric information in local bases [46]. An important branch of past and current application is domain decomposition approaches based on RB models for simple geometries, which are then coupled to more complex geometrical shapes involving a huge number of local parameters. The original RB element method [53] has been extended to an extremely flexible static condensation approach [39] that allows us to construct various online geometries by using RB models as Lego building blocks. Iterative domain decomposition schemes [54] can be used to handle distributed coupled problems.

2.6 ▪ Exercises

Exercise 2.97 (Finite-dimensional X_N for thermal block). *Show that the thermal block model for $B_1 = 1$ has a solution manifold contained in an $N := B_2$-dimensional linear subspace $X_N \subset H^1_{\Gamma_D}$. This can be obtained by deriving an explicit solution representation. In particular, find N snapshot parameters $\mu^{(i)}, i = 1, \ldots, N$, such that $X_N = \text{span}\{u(\mu^{(1)}), \ldots, u(\mu^{(N)})\}$.*

Exercise 2.98 (Many parameters, simple solution). *Devise an instantiation of $(P(\mu))$ with arbitrary number of parameters $p \in \mathbb{N}$ but with the solution contained in a one-dimensional linear subspace.*

Exercise 2.99 (Conditions for uniform coercivity). *Assume a parameter-separable bilinear form $a(\cdot, \cdot; \mu)$. Under which conditions on the coefficient functions θ^a_q and the components a_q can uniform coercivity of a be concluded?*

Exercise 2.100 (Thermal block as instantiation of $(P(\mu))$). *Verify that the thermal block satisfies the assumptions of $(P(\mu))$, i.e., uniform continuity, coercivity, and parameter separability of the bilinear and linear forms. In particular, specify the constants, the coefficient functions, and the components. Argue that it is an example for a* compliant *problem, i.e., symmetric and $f = l$.*

Exercise 2.101 (Finite-dimensional exact approximation for $Q_a = 1$). *Assume a general problem of type $(P(\mu))$ with parameter-separable forms. Show that in the case of $Q_a = 1$ the solution manifold $\mathcal{M}$ is contained in an RB space X_N of dimension at most Q_f, and show that there exist $N \leq Q_f$ parameters $\mu^{(i)}, i = 1, \ldots, N$, such that*

$$X_N = \operatorname{span}(u(\mu^1), \ldots, u(\mu^{(N)})).$$

Exercise 2.102 (Lipschitz continuity of $(P(\mu))$). *Let $\theta_q^a, \theta_q^f, \theta_q^l$ be Lipschitz continuous with respect to μ with Lipschitz constants L_q^a, L_q^f, L_q^l.*

(i) *Show that a, f, l are Lipschitz continuous by computing suitable constants L_a, L_f, L_l such that*

$$\begin{aligned}
|a(u,v;\mu) - a(u,v;\mu')| &\leq L_a \|u\| \|v\| \|\mu - \mu'\|, \\
|f(v;\mu) - f(v;\mu')| &\leq L_f \|v\| \|\mu - \mu'\|, \\
|l(v;\mu) - l(v;\mu')| &\leq L_l \|v\| \|\mu - \mu'\|, \quad u, v \in X, \mu, \mu' \in \mathcal{P}.
\end{aligned}$$

(ii) *Derive suitable constants L_u, L_s such that*

$$\|u(\mu) - u(\mu')\| \leq L_u \|\mu - \mu'\|, \quad |s(\mu) - s(\mu'| \leq L_s \|\mu - \mu'\|, \quad \mu, \mu' \in \mathcal{P}.$$

Hint: $L_u = L_f / \bar{\alpha} + \bar{\gamma}_f L_a / \bar{\alpha}^2$, $L_s = L_l \bar{\gamma}_f / \bar{\alpha} + \bar{\gamma}_l L_u$.

Remark: An identical statement holds for the reduced solutions $u_N(\mu), s_N(\mu)$.

Exercise 2.103 (Differentiability of $u(\mu)$). *Prove Proposition* 2.13.

Exercise 2.104 (Best-approximation bound for symmetric case). *Show that if $a(\cdot,\cdot;\mu)$ is symmetric, for all $\mu \in \mathcal{P}$,*

$$\|u(\mu) - u_N(\mu)\| \leq \sqrt{\frac{\gamma(\mu)}{\alpha(\mu)}} \inf_{v \in X_N} \|u(\mu) - v\|.$$

(This is a sharpening of (2.4), *and hence the lemma of Céa by a square root.)*

Exercise 2.105 (Relative error and effectivity bounds). *Provide proofs for the relative a posteriori error estimate and effectivity of Proposition* 2.27.

Exercise 2.106 (Energy norm error and effectivity bounds). *Provide proofs for the energy norm a posteriori error estimate and effectivity of Proposition* 2.28.

Exercise 2.107 ($\alpha(\mu)$ for thermal block). *Show for the thermal block that $\alpha(\mu) = \min_i \mu_i$. (Therefore, $\alpha_{\mathrm{LB}}(\mu)$ can be chosen as that value in the error bounds.) Similarly, show that $\gamma(\mu) = \max_i \mu_i$. (Hence, the effectivities are always bounded by $\mu_{\max}/\mu_{\min}$ if $\mu \in [\mu_{\min}, \mu_{\max}]^p$.)*

Exercise 2.108 (Max-theta approach for continuity upper bound). *Let a be symmetric and all a_q positive semidefinite and $\theta_q^a(\mu) > 0, q = 1, \ldots, Q_a, \mu \in \mathcal{P}$. Let $\bar{\mu} \in \mathcal{P}$ and $\gamma(\bar{\mu})$ be known. Show that for all $\mu \in \mathcal{P}$,*

$$\gamma(\mu) \leq \gamma_{\mathrm{UB}}(\mu) < \infty,$$

with continuity upper bound

$$\gamma_{\mathrm{UB}}(\mu) := \gamma(\bar{\mu}) \max_{q=1,\ldots,Q_a} \frac{\theta_q^a(\mu)}{\theta_q^a(\bar{\mu})}.$$

Exercise 2.109 (Monotonicity of greedy error). *Prove that the greedy algorithm produces monotonically decreasing error sequences* $(\varepsilon_n)_{n\geq 1}$ *if*

(i) $\Delta(Y,\mu) := \|u(\mu) - P_Y u(\mu)\|$, *i.e., the orthogonal projection error is chosen as error indicator, and*

(ii) *we have the compliant case (*$a(\cdot,\cdot;\mu)$ *symmetric and* $f(\cdot;\mu) = l(\cdot;\mu)$*) and* $\Delta(Y,\mu) := \Delta_u^{en}(\mu)$, *i.e., the energy error estimator from Proposition* 2.28 *is chosen as error indicator.*

Exercise 2.110 (Gramian matrix and properties). *Let* $\{u_1,\ldots,u_n\} \subset X$ *be a finite subset. Define the* Gramian matrix *through*

$$\mathbf{G} := (\langle u_i, u_j \rangle)_{i,j=1}^n \in \mathbb{R}^{n\times n}.$$

Show that the following hold:

(i) $\mathbf{G}$ *is symmetric and positive semidefinite,*

(ii) $\mathrm{rank}(\mathbf{G}) = \dim(\mathrm{span}(\{u_i\}_{i=1}^n))$, *and*

(iii) $\{u_i\}_{i=1}^n$ *are linearly independent* $\Leftrightarrow$ $\mathbf{G}$ *is positive definite.*

(Recall that such Gramian matrices appeared as $\mathbf{G}_l, \mathbf{G}_r, \mathbf{K}_N$ *in the offline/online decomposition in Propositions* 2.38, 2.39, *and* 2.40.*)*

Exercise 2.111 (Gram–Schmidt orthonormalization). *Prove that the procedure given in Proposition* 2.54 *indeed produces the Gram–Schmidt orthonormalized sequence.*

Exercise 2.112 (Orthonormalization of RB). *Prove that the procedure given in Proposition* 2.54 *also produces an orthonormal basis if* $\mathbf{C}$ *is chosen differently, as long as it satisfies* $\mathbf{C}\mathbf{C}^T = \mathbf{G}^{-1}$. *(Hence, more than just Cholesky factorization is possible.)*

Exercise 2.113 (Uniform boundedness with respect to Δt). *Prove the statement of Proposition* 2.72. *Hint:*

$$\left(\frac{1}{1+\alpha\frac{T}{K}}\right)^K = \left(\left(\frac{1}{1+\frac{\alpha T}{K}}\right)^{\frac{K}{\alpha T}}\right)^{\alpha T} \to e^{-\alpha T}$$

as $K \to \infty$.

Exercise 2.114 (Data uncorrelated in POD coordinates). *Verify that the coordinates of the data* $\{u_i\}_{i=1}^n$ *with respect to the POD basis* $\{\varphi_j\}_{j=1}^{n'}$ *are uncorrelated and the mean* *squared coordinates are just the eigenvalues of the correlation operator, i.e., for all* $j,k = 1,\ldots,n'$, $j \neq k$,

$$\sum_{i=1}^{n} \left\langle u_i, \varphi_j \right\rangle \langle u_i, \varphi_k \rangle = 0, \quad \frac{1}{n} \sum_{i=1}^{n} \left\langle u_i, \varphi_j \right\rangle^2 = \lambda_j.$$

Exercise 2.115 (POD mean squared error). *Prove the second equality in* (2.63)*: Let* $\{u_i\}_{i=1}^n$ *be given. Show that the mean squared error of the POD projection can be explicitly obtained by the sum of the truncated eigenvalues, i.e.,*

$$\frac{1}{n} \sum_{i=1}^{n} \left\| u_i - P_{X_{\mathrm{POD},N}} u_i \right\|^2 = \sum_{i=N+1}^{n'} \lambda_i.$$

Exercise 2.116 (POD via Gramian matrix). *Prove Proposition* 2.90*, i.e., the equivalence of computation of POD via the correlation operator or the Gramian matrix.*

Bibliography

[1] B. ALMROTH, P. STERN, AND F. BROGAN, *Automatic choice of global shape functions in structural analysis*, AIAA J., 16 (1978), pp. 525–528.

[2] M. BARRAULT, Y. MADAY, N. NGUYEN, AND A. PATERA, *An "empirical interpolation" method: Application to efficient reduced-basis discretization of partial differential equations*, C. R. Math. Acad. Sci. Paris Series I, 339 (2004), pp. 667–672.

[3] R. BECKER AND R. RANNACHER, *Weighted a-posteriori error estimates in FE methods*, in Proc. ENUMATH-97, 1998, pp. 621–637.

[4] G. BERKOOZ, P. HOLMES, AND J. L. LUMLEY, *The proper orthogonal decomposition in the analysis of turbulent flows*, Ann. Rev. Fluid Mech., 25 (1993), pp. 539–575.

[5] P. BINEV, A. COHEN, W. DAHMEN, R. DEVORE, G. PETROVA, AND P. WOJTASZCZYK, *Convergence rates for greedy algorithms in reduced basis methods*, SIAM J. Math. Anal., 43 (2011), pp. 1457–1472.

[6] S. BOYAVAL, *Reduced-basis approach for homogenization beyond the periodic setting*, Multiscale Modeling Simulation, 7 (2008), pp. 466–494.

[7] S. BOYAVAL, C. L. BRIS, Y. MADAY, N. NGUYEN, AND A. PATERA, *A reduced basis approach for variational problems with stochastic parameters: Application to heat conduction with variable Robin coefficient*, Comput. Methods Appl. Mech. and Eng., 1 (2009), pp. 3187–3206.

[8] D. BRAESS, *Finite Elemente—Theorie, schnelle Löser und Anwendungen in der Elasitzitätstheorie, 4. Aufl.*, Springer, 2007.

[9] A. BUFFA, Y. MADAY, A. PATERA, C. PRUD'HOMME, AND G. TURINICI, *A priori convergence of the greedy algorithm for the parametrized reduced basis*, ESAIM M2AN, Math. Model. Numer. Anal., 46 (2012), pp. 595–603.

[10] S. CHATURANTABUT AND D. SORENSEN, *Nonlinear model reduction via discrete empirical interpolation*, SIAM J. Sci. Comput., 32 (2010), pp. 2737–2764.

[11] W. DAHMEN, C. PLESKEN, AND G. WELPER, *Double greedy algorithms: Reduced basis methods for transport dominated problems*, ESAIM Math. Model. Numer. Anal., 48 (2014), pp. 623–663.

[12] L. DEDE, *Adaptive and Reduced Basis Methods for Optimal Control Problems in Environmental Applications*, PhD thesis, Doctoral School in Mathematical Engineering, Politecnico di Milano, 2008.

[13] S. DEPARIS, *Reduced basis error bound computation of parameter-dependent Navier–Stokes equations by the natural norm approach*, SIAM J. Numer. Anal., 46 (2008), pp. 2039–2067.

[14] M. DIHLMANN, M. DROHMANN, AND B. HAASDONK, *Model reduction of parametrized evolution problems using the reduced basis method with adaptive time-partitioning*, in Proc. of ADMOS 2011.

[15] M. A. DIHLMANN AND B. HAASDONK, *Certified PDE-constrained parameter optimization using reduced basis surrogate models for evolution problems*, Comput. Optim. Appl., 60 (2015), pp. 753–787.

[16] M. DROHMANN, B. HAASDONK, AND M. OHLBERGER, *Adaptive reduced basis methods for nonlinear convection-diffusion equations*, in Proc. FVCA6, 2011.

[17] ——, *Reduced basis approximation for nonlinear parametrized evolution equations based on empirical operator interpolation*, SIAM J. Sci. Comput., 34 (2012), pp. A937–A969.

[18] J. EFTANG, M. GREPL, AND A. PATERA, *A posteriori error bounds for the empirical interpolation method*, C. R. Acad. Sci. Paris Series I 348, (2010), pp. 575–579.

[19] J. EFTANG, D. KNEZEVIC, AND A. PATERA, *An hp certified reduced basis method for parametrized parabolic partial differential equations*, Math. Comput. Model Dyn., 17 (2011), pp. 395–422.

[20] J. L. EFTANG, A. T. PATERA, AND E. M. RØNQUIST, *An "hp" certified reduced basis method for parametrized elliptic partial differential equations*, SIAM J. Sci. Comput., 32 (2010), pp. 3170–3200.

[21] J. FINK AND W. RHEINBOLDT, *On the error behaviour of the reduced basis technique for nonlinear finite element approximations*, ZAMM, 63 (1983), pp. 21–28.

[22] A.-L. GERNER AND K. VEROY, *Certified reduced basis methods for parametrized saddle point problems*, SIAM J. Sci. Comput., 34 (2012), pp. A2812–A2836.

[23] S. GLAS AND K. URBAN, *On noncoercive variational inequalities*, SIAM J. Numer. Anal., 52 (2014), pp. 2250–2271.

[24] M. GREPL, *Reduced-Basis Approximations and a Posteriori Error Estimation for Parabolic Partial Differential Equations*, PhD thesis, Massachusetts Institute of Technology, 2005.

[25] M. GREPL, Y. MADAY, N. NGUYEN, AND A. PATERA, *Efficient reduced-basis treatment of nonaffine and nonlinear partial differential equations*, ESAIM M2AN, Math. Model. Numer. Anal., 41 (2007), pp. 575–605.

[26] M. GREPL AND A. PATERA, *A posteriori error bounds for reduced-basis approximations of parametrized parabolic partial differential equations*, M2AN, Math. Model. Numer. Anal., 39 (2005), pp. 157–181.

[27] B. HAASDONK, *Reduzierte-Basis-Methoden*, Lecture Notes, IANS-Report 4/11, University of Stuttgart, Germany, 2011.

[28] B. HAASDONK, *Convergence rates of the POD-Greedy method*, ESAIM M2AN, Math. Model. Numer. Anal., 47 (2013), pp. 859–873.

[29] B. HAASDONK, M. DIHLMANN, AND M. OHLBERGER, *A training set and multiple basis generation approach for parametrized model reduction based on adaptive grids in parameter space*, Math. Comput. Model. Dyn., 17 (2012), pp. 423–442.

[30] B. HAASDONK AND M. OHLBERGER, *Basis construction for reduced basis methods by adaptive parameter grids*, in Proc. International Conference on Adaptive Modeling and Simulation, ADMOS 2007, P. Díez and K. Runesson, eds., CIMNE, Barcelona, 2007.

[31] ———, *Reduced basis method for finite volume approximations of parametrized linear evolution equations*, ESAIM M2AN, Math. Model. Numer. Anal., 42 (2008), pp. 277–302.

[32] B. HAASDONK AND M. OHLBERGER, *Reduced basis method for explicit finite volume approximations of nonlinear conservation laws*, in Proc. HYP 2008, International Conference on Hyperbolic Problems: Theory, Numerics and Applications, vol. 67 of Proc. Sympos. Appl. Math., AMS, Providence, RI, 2009, pp. 605–614.

[33] B. HAASDONK, M. OHLBERGER, AND G. ROZZA, *A reduced basis method for evolution schemes with parameter-dependent explicit operators*, Electron. Trans. Numer. Anal., 32 (2008), pp. 145–161.

[34] B. HAASDONK, J. SALOMON, AND B. WOHLMUTH, *A reduced basis method for parametrized variational inequalities*, SIAM J. Numer. Anal., 50 (2012), pp. 2656–2676.

[35] B. HAASDONK, J. SALOMON, AND B. WOHLMUTH, *A reduced basis method for the simulation of American options*, in Proc. ENUMATH 2011, 2012.

[36] B. HAASDONK, K. URBAN, AND B. WIELAND, *Reduced basis methods for parameterized partial differential equations with stochastic influences using the Karhunen–Loève expansion*, SIAM/ASA J. Unc. Quant., 1 (2013), pp. 79–105.

[37] J. S. HESTHAVEN, B. STAMM, AND S. ZHANG, *Efficient greedy algorithms for high-dimensional parameter spaces with applications to empirical interpolation and reduced basis methods*, ESAIM: M2AN, Math. Model. Numer. Anal., 48 (2014), pp. 259–283.

[38] H. HOTELLING, *Analysis of a complex of statistical variables into principal components.*, J. Educ. Psychol., 24 (1933), pp. 417–441.

[39] D. HUYNH, D. KNEZEVIC, AND A. PATERA, *A static condensation reduced basis element method: Approximation and a posteriori error estimation*, ESAIM: M2AN, Math. Model. Numer. Anal., 47 (2013), pp. 213–251.

[40] D. HUYNH, G. ROZZA, S. SEN, AND A. PATERA, *A successive constraint linear optimization method for lower bounds of parametric coercivity and inf-sup stability constants*, C. R. Math. Acad. Sci. Paris Series I, 345 (2007), pp. 473–478.

[41] K. ITO AND S. RAVINDRAN, *A reduced order method for simulation and control of fluid flows*, J. Comput. Phys., 143 (1998), pp. 403–425.

[42] I. JOLLIFFE, *Principal Component Analysis*, Springer-Verlag, 2002.

[43] M. KÄRCHER AND M. GREPL, *A certified reduced basis method for parametrized elliptic optimal control problems*, ESAIM: Control Optim. Calc. Var., 20 (2014), pp. 416–441.

[44] ——, *A posteriori error estimation for reduced order solutions of parametrized parabolic optimal control problems*, ESAIM: M2AN, Math. Model. Numer. Anal., 48 (2014), pp. 1615–1638.

[45] K. KARHUNEN, *Über lineare Methoden in der Wahrscheinlichkeitsrechnung*, Ann. Acad. Sri. Fennicae, ser. Al. Math. Phys., 37 (1946).

[46] S. KAULMANN, M. OHLBERGER, AND B. HAASDONK, *A new local reduced basis discontinuous Galerkin approach for heterogeneous multiscale problems*, C. R. Math. Acad. Sci. Paris, 349 (2011), pp. 1233–1238.

[47] D. KNEZEVIC, N. NGUYEN, AND A. PATERA, *Reduced basis approximation and a posteriori error estimation for the parametrized unsteady Boussinesq equations*, Math. Mod. Methods Appl. Sci., 21 (2011), pp. 1415–1442.

[48] D. KNEZEVIC AND A. PATERA, *A certified reduced basis method for the Fokker–Planck equation of dilute polymeric fluids: FENE dumbbells in extensional flow*, SIAM J. Sci. Comput., 32 (2010), pp. 793–817.

[49] D. KRÖNER, *Numerical Schemes for Conservation Laws*, John Wiley & Sons and Teubner, 1997.

[50] M. M. LOEVE, *Probability Theory*, Van Nostrand, Princeton, NJ, 1955.

[51] Y. MADAY, N. NGUYEN, A. PATERA, AND G. PAU, *A General, Multi-Purpose Interpolation Procedure: The Magic Points*, Tech. Rep. RO7037, Laboratoire Jacques-Louis-Lions, Université Pierre et Marie Curie, Paris, 2007.

[52] Y. MADAY, A. PATERA, AND G. TURINICI, *A priori convergence theory for reduced-basis approximations of single-parameter symmetric coercive elliptic partial differential equations*, C. R. Acad Sci. Paris Series I, 335 (2002), pp. 289–294.

[53] Y. MADAY AND E. RØNQUIST, *The reduced basis element method: Application to a thermal fin problem*, SIAM J. Sci. Comput., 26 (2004), pp. 240–258.

[54] I. MAIER AND B. HAASDONK, *A Dirichlet-Neumann reduced basis method for homogeneous domain decomposition*, Appl. Numer. Math., 78 (2014), pp. 31–48.

[55] A. NOOR AND J. PETERS, *Reduced basis technique for nonlinear analysis of structures*, AIAA J., 18 (1980), pp. 455–462.

[56] I. OLIVEIRA AND A. PATERA, *Reduced-basis techniques for rapid reliable optimization of systems described by affinely parametrized coercive elliptic partial differential equations*, Optim. Eng., 8 (2007), pp. 43–65.

[57] A. PATERA AND G. ROZZA, *Reduced Basis Approximation and a Posteriori Error Estimation for Parametrized Partial Differential Equations*, MIT, 2007. Version 1.0, Copyright MIT 2006–2007, to appear in (tentative rubric) MIT Pappalardo Graduate Monographs in Mechanical Engineering.

[58] T. PORSCHING AND M. LEE, *The reduced basis method for initial value problems*, SIAM J. Numer. Anal., 24 (1987), pp. 1277–1287.

[59] C. PRUD'HOMME, D. ROVAS, K. VEROY, L. MACHIELS, Y. MADAY, A. PATERA, AND G. TURINICI, *Reliable real-time solution of parametrized partial differential equations: Reduced-basis output bound methods*, J. Fluids Eng., 124 (2002), pp. 70–80.

[60] W. RHEINBOLDT, *On the theory and error estimation of the reduced-basis method for multi-parameter problems*, Nonlinear Anal., Theory, Methods & Appl., 21 (1993), pp. 849–858.

[61] D. ROVAS, *Reduced-Basis Output Bound Methods for Parametrized Partial Differential Equations*, PhD Thesis, MIT, Cambridge, MA, 2003.

[62] G. ROZZA, *Shape Design by Optimal Flow Control and Reduced Basis Techniques: Applications to Bypass Configurations in Haemodynamics*, PhD Thesis, École Polytechnique Fédérale de Lausanne, 2005.

[63] G. ROZZA, D. HUYNH, AND A. PATERA, *Reduced basis approximation and a posteriori error estimation for affinely parametrized elliptic coercive partial differential equations: Application to transport and continuum mechanics*, Arch. Comput. Methods Eng., 15 (2008), pp. 229–275.

[64] B. SCHÖLKOPF AND A. J. SMOLA, *Learning with Kernels: Support Vector Machines, Regularization, Optimization and Beyond*, MIT Press, 2002.

[65] S. SEN, *Reduced basis approximations and a posteriori error estimation for many-parameter heat conduction problems*, Numer. Heat Transfer, Part B: Fundamentals, 54 (2008), pp. 369–389.

[66] L. SIROVICH, *Turbulence and the dynamics of coherent structures. I. Coherent structures*, Quart. Appl. Math., 45 (1987), pp. 561–571.

[67] K. URBAN AND A. PATERA, *An improved error bound for reduced basis approximation of linear parabolic problems*, Math. Comput., 83 (2014), pp. 1599–1615.

[68] K. URBAN, S. VOLKWEIN, AND O. ZEEB, *Greedy sampling using nonlinear optimization*, in Reduced Order Methods for Modeling and Computational Reductions, A. Quarteroni and G. Rozza, eds., Springer, 2014, pp. 139–157.

[69] K. VEROY AND A. PATERA, *Certified real-time solution of the parametrized steady incompressible Navier-Stokes equations: Rigorous reduced-basis a posteriori error bounds*, Internat. J. Numer. Methods Fluids, 47 (2005), pp. 773–788.

[70] K. VEROY, C. PRUD'HOMME, AND A. PATERA, *Reduced-basis approximation of the viscous Burgers equation: Rigorous a posteriori error bounds*, C. R. Math. Acad. Sci. Paris Series I, 337 (2003), pp. 619–624.

[71] K. VEROY, C. PRUD'HOMME, D. V. ROVAS, AND A. T. PATERA, *A posteriori error bounds for reduced-basis approximation of parametrized noncoercive and nonlinear elliptic partial differential equations*, in Proceedings of 16th AIAA Computational Fluid Dynamics Conference, 2003. Paper 2003-3847.

[72] S. VOLKWEIN, *Model Reduction Using Proper Orthogonal Decomposition*, lecture notes, University of Konstanz, 2013.

[73] D. WIRTZ, D. SORENSEN, AND B. HAASDONK, *A posteriori error estimation for DEIM reduced nonlinear dynamical systems*, SIAM J. Sci. Comput., 36 (2014), pp. A311–A338.

[74] M. YANO AND A. PATERA, *A space-time variational approach to hydrodynamic stability theory*, Proc. Roy. Soc. A, 469 (2013). Article Number 20130036.

Chapter 3
The Theoretical Foundation of Reduced Basis Methods

Ronald A. DeVore[4]

3.1 ▪ Introduction

The main theme of this volume is the efficient solution of families of stochastic or parametric partial differential equations (PDEs). This chapter focuses on the theoretical underpinnings of such methods. It shows how concepts from approximation theory, such as entropy or widths, can help to quantify, a priori, how well such methods can perform. This theory is then used to analyze one of the primary numerical vehicles, reduced basis (RB) methods, for parametric equations. A particular emphasis is placed on understanding the performance of greedy algorithms for selecting bases in RB methods.

RB methods have met with much computational success that is amply described in other contributions of this volume. The present chapter sits at the other end of the spectrum since it is exclusively devoted to the theoretical aspects of this subject. The development of a theory for RB methods is of great interest since it addresses one of the most challenging problems in modern numerical computation, namely the computational recovery of high-dimensional functions. The theory we present here is far from complete, and, indeed, one of the goals of the present exposition is to organize our thinking and illuminate some obvious questions whose solution may advance both the theoretical and the computational aspects of reduction methods.

This chapter will deal exclusively with linear elliptic problems. This restriction was imposed because, quite frankly, not enough is known theoretically in other settings to warrant much discussion. However, let us be clear that theoretical developments for other problems will be extremely interesting and could help advance other application domains. While this chapter will only treat parametric problems, the results put forward have relevance for stochastic problems via chaos expansions.

[4]This research was supported by the Office of Naval Research Contracts ONR N00014-11-1-0712, ONR N00014-12-1-0561, ONR N00014-15-1-2181, and ONR N00014-16-2706, the DARPA Grant HR0011619523 through Oak Ridge National Laboratory, and NSF Grants DMS1222715 and DMS 15-21067.

What is written here is a very personalized view and understanding of this subject. The form of this chapter has been strongly influenced by discussions with many individuals. I mention them here because any of them would justifiably be coauthors of this presentation.

My first exposure to RBs occurred many years ago when visiting Albert Cohen at Paris VI. I had the fortune to be there when Yvon Maday and his collaborators were proving their first results on greedy algorithms and magic points. It was clear this subject had a large intersection with approximation theory and yet seemed to be completely missed by the approximation community. Albert and I began reflecting on reduced modeling and naturally involved our collaborators Wolfgang Dahmen, Peter Binev, Guergana Petrova, and Przemek Wojtaszcsyk. I organized a small seminar on this subject at TAMU, which includes (in addition to Guergana) Andrea Bonito, Bojan Popov, and Gerrit Welper. I thank all of these people for helping my understanding of the subject.

Subsequent to the writing of this chapter, the survey article [8] was written and is already in print. Some topics considered in this chapter appear in expanded form in [8]. So if the reader finds the current exposition too terse on a certain topic, chances are it is dealt with in more detail in [8].

3.2 ▪ Elliptic PDEs

The focal point of this chapter is the study of numerical algorithms for solving a family of elliptic equations. Each of these elliptic equations is of the form

$$-\nabla\cdot(a\nabla u)=f \quad \text{in} \quad D, \qquad u|_{\partial D}=0, \tag{3.1}$$

where $D\subset\mathbb{R}^d$ is a bounded Lipschitz domain, and the right side f is in $H^{-1}(D)$.[5] Here, $a=a(x)$ is a scalar function assumed to be in $L_\infty(D)$ that satisfies the ellipticity assumption: there exist $0<r<R$ such that

$$r\leq a(x)\leq R, \quad x\in D. \tag{3.2}$$

We could just as well consider the case where a is replaced by a positive definite matrix function $A(x)$ with a similar theory and results, only at the expense of more cumbersome notation. In this section, we begin by recalling what is known about the solution to (3.1) when a and f are fixed. The later sections of this chapter will then turn to the question of efficiently solving a family of such problems.

There is a rich theory for existence and uniqueness for equation (3.1), which we briefly recall. A much expanded discussion of this topic can be found in [8]. Central to this theory is the Sobolev space $H_0^1(D,a)$ (called the energy space), which is a Hilbert space equipped with the energy norm

$$||v||_{H_0^1(D,a)}:=||a|\nabla v|||_{L^2(D)}. \tag{3.3}$$

That this is a norm follows from a theorem of Poincaré that says that

$$||v||_{L_2(D)}\leq C_D||v||_{H_0^1(D,a)} \tag{3.4}$$

for every Lipschitz domain D and in particular for every polyhedral domain D.

[5]We use standard notation for Sobolev spaces throughout this chapter. The space $W^s(L_p(D))$ is the Sobolev space with smoothness index s in $L_p(D)$. For the special case $p=2$, this space is typically denoted by H^s in the numerical and PDE communities.

If $a, \tilde{a}$ both satisfy the ellipticity assumption, then the norms for $H_0^1(a)$ and $H_0^1(\tilde{a})$ are equivalent. If we take $a = 1$, we obtain the classical space $H_0^1(D,1)$, which in what follows is simply denoted by $H_0^1 = H_0^1(D)$. The dual of $H_0^1(D)$ consists of all linear functionals defined on this space. It is usually denoted by $H^{-1}(D)$, and its norm is defined by duality. Namely, if $\lambda \in H^{-1}(D)$, then

$$||\lambda||_{H^{-1}(D)} := \sup_{||v||_{H_0^1(D)} \leq 1} |\langle \lambda, v \rangle|. \tag{3.5}$$

The solution u_a of (3.1) is defined in weak form as a function $u \in H_0^1(D)$ that satisfies

$$\int_D a(x) \nabla u_a(x) \cdot \nabla v(x) dx = \int_D f(x) v(x) dx \text{ for all } v \in H_0^1(D). \tag{3.6}$$

This formulation shows that the Lax–Milgram theory applies. In particular, the ellipticity assumption is a sufficient condition for the existence and uniqueness of the solution u_a. Under this assumption, the solution satisfies the estimate

$$||u_a||_{H_0^1(D)} \leq C_0 \frac{||f||_{H^{-1}(D)}}{r}. \tag{3.7}$$

The same theory applies even if a is complex valued. Now the lower ellipticity condition replaces a by $Re(a)$ in (3.2), and the upper condition is that $|a|$ is uniformly bounded.

There is also a general principle of perturbation for elliptic equations that shows to some extent the smooth dependence of the solution on the diffusion coefficient a. If a and $\tilde{a}$ are two such coefficients with the same ellipticity constants r and R, then the solutions u and $\tilde{u}$ with identical right side f will satisfy

$$||u_a - u_{\tilde{a}}||_{H_0^1(D)} \leq C_0 \frac{||a - \tilde{a}||_{L_\infty(D)}}{r}. \tag{3.8}$$

The bound (3.8) shows that the mapping $a \to u_a$ is Lipschitz continuous. Actually, this mapping is in a certain sense analytic, as will be explained in Section 3.3.1. This smooth dependence is at the heart of reduced modeling, so it will be of great concern to us as we proceed.

3.2.1 ▪ Other perturbation results

In some applications, the coefficients a, while bounded, are not continuous. In such applications, they may have discontinuities along curves or higher-dimensional manifolds in $\mathbb{R}^d$. This makes (3.8), more or less, useless since it requires exact matching of the discontinuities of a and $\tilde{a}$. A related issue is that in numerical methods, the diffusion coefficient a is approximated by an $\tilde{a}$, and one will not have that $||a - \tilde{a}||_{L_\infty}$ is small since the discontinuity cannot be matched exactly. Thus, we need other perturbation results that are more amenable to such applications. Results of this type were given in [5], in which L_∞ perturbation is replaced by L_q perturbation for certain q with $q < \infty$, in the form of the following result.

For any $p \geq 2$, the functions u_a and $u_{\tilde{a}}$ satisfy

$$||u_a - u_{\tilde{a}}||_{H_0^1(D)} \leq r^{-1} ||\nabla u_a||_{L_p(D)} ||a - \tilde{a}||_{L_q(D)}, \quad q := \frac{2p}{p-2} \in [2, \infty], \tag{3.9}$$

provided $\nabla u_a \in L_p(D)$. Notice that the case $p = 2$ is (3.8). For (3.9) to be relevant for discontinuous $a,\tilde{a}$, we need ∇u_a to be in L_p for some $p > 2$. It is known that for every Lipschitz domain D, there is $P > 2$ such that for $2 \le p \le P$ one has the following condition.

COND-p. *For each $f \in W^{-1}(L_p(D))$, the solution $u = u_1$ to (3.1) with $a = 1$ and right side f satisfies*

$$|u|_{W^1(L_p(D))} := \|\nabla u\|_{L_p(D)} \le C_p \|f\|_{W^{-1}(L_p(D))}, \tag{3.10}$$

with the constant C_p independent of f.

The assumption $f \in W^{-1}(L_p(\Omega))$ is a rather mild assumption on the right side f and leads to an L_q perturbation with $q := \frac{2p}{p-2}$.

In the special case $a = 1$ (the case of Laplace's equation), the validity of **COND-p** is a well-studied problem in harmonic analysis (see for example Jerison and Kenig [19]). In fact, in this setting, one can take $P > 4$ when $d = 2$ and $P > 3$ when $d = 3$. The case of general a in (3.9) is proved by a perturbation argument (see [5] for details). It is also known that if D is convex, then we can take $P = \infty$.

One can extend the above results from $a = 1$ to general a by using a perturbation result (see Proposition 1 in [5]).

Perturbation property. *If the diffusion coefficients $a,\tilde{a}$ satisfy the strong ellipticity condition for some r,R, then there is a P^* depending on this r,R and on the domain D such that whenever $p \in [2,P^*]$ and $f \in W^{-1}(L_p(D))$, then*

$$\|u_a - u_{\tilde{a}}\|_{H_0^1(D)} \le C_1 \|f\|_{W^{-1}(L_p(D))} \|a - \tilde{a}\|_{L_q(D)}, \tag{3.11}$$

where $q := 2p/(p-1)$.

The strongest perturbation result occurs when we can take $P^* = \infty$. In this case, the L_2-norm appears on the right side of (3.11).

Let us emphasize that assumptions on f other than $f \in W^{-1}(L_p(D))$ may lead to $\nabla u \in L_p$ for a wider range of p. This would, in turn, give the perturbation for a wider range of q.

3.3 ▪ Parametric elliptic equations

We turn now to the principal topic of this article, namely, the solution of a family of elliptic equations. We are not interested in solving (3.1) for just one diffusion coefficient a but rather a family $\mathcal{A}$ of such coefficients. We always assume that the family $\mathcal{A}$ satisfies the following assumption.

Uniform ellipticity assumption. *There exist r,R such that for each $a \in \mathcal{A}$, we have*

$$r \le a(x) \le R, \quad x \in D. \tag{3.12}$$

This family is assumed to be a compact subset of either $L_\infty(D)$ or an appropriate $L_q(D)$ space for which the following property holds.

$L_q(D)$ stability of $\mathscr{A}$. *We say this property holds for the given f if there is a constant C_0 such that for all $a, \tilde{a} \in L_q(D)$ we have*

$$\|u_a - u_{\tilde{a}}\|_{H_0^1(D))} \le C_0 \|a - \tilde{a}\|_{L_q(D)}. \tag{3.13}$$

Of course this property always holds for $q = \infty$. We already discussed in the previous section the fact that this will also hold for a range of $Q \le q \le \infty$, with $2 \le Q < \infty$, under very mild restrictions on f. Notice that the most favorable range is when $Q = 2$.

Our goal is to build a black-box solver such that when presented with any $a \in \mathscr{A}$, the solver will very quickly provide an online computation of $u_a = u_{a,f}$, $a \in \mathscr{A}$. To begin the discussion, let us recall that there are already in existence adaptive finite element solvers that when given $a \in \mathscr{A}$ will rather efficiently approximate the solution u_a. What governs the accuracy of these adaptive solvers is the smoothness of u_a, which in turn is determined by properties of the physical domain D and the regularity of the right side f. Indeed, the performance of such adaptive methods is governed by the regularity of the solution u_a in a certain scale of Besov spaces corresponding to nonlinear approximation [2, 3], and this Besov regularity can be derived from the smoothness of f (see [12]). We do not wish to get into details here but only mention that the typical performance of this approach is to obtain convergence of order $O(n^{-\beta})$ for n computations, where β is typically small. For example, if $f \in L_2(D)$ and the domain is Lipschitz, then $\beta \le 1$ if $d = 3$.

The motivation behind RB methods is the smoothness of the u_a with varying a. We can view the set

$$\mathscr{U} := \mathscr{U}_{\mathscr{A}} := \mathscr{U}_{\mathscr{A},f} := \{u_a : a \in \mathscr{A}\} \tag{3.14}$$

as a manifold in $H_0^1(D)$, and in light of our earlier discussion of perturbation for elliptic equations, this manifold inherits a certain smoothness from that of $\mathscr{A}$. If this smoothness is high enough, then it is reasonable to expect that we can build solvers that perform better than the adaptive PDE solvers since the latter never take advantage of the smoothness of this manifold. Our main interest is to quantify when this is indeed true.

We give two examples of sets $\mathscr{A}$ of diffusion coefficients, which will guide our discussion.

3.3.1 ▪ Affine dependence

In this setting, we are given a family of functions $\psi_j(x)$, $j = 1, 2, \dots$, defined on the physical domain D. We let U be the unit cube in $\ell_\infty := \ell_\infty(\mathbb{N})$. Hence, $y \in U$ means that $y = (y_1, y_2, \dots)$ with $|y_j| \le 1$. For any such $y \in U$, we define

$$a(x,y) = \bar{a}(x) + \sum_{j \ge 1} y_j \psi_j(x) \tag{3.15}$$

and take $\mathscr{A} = \{a(x,y) : y \in U\}$ as our family of diffusion coefficients. Of course, we shall also need additional assumptions to guarantee that the series in (3.15) converges. A typical assumption is that the sequence $(\|\psi_j\|_{L_\infty(D)})$ is in ℓ_p for some $p \le 1$. We assume in going further that the indices have been rearranged so that this sequence $(\|\psi_j\|_{L_\infty(D)})$ is monotonically decreasing.

One may wonder why we consider an infinite number of parameters y_j in (3.15). The answer is twofold. First of all, a standard way of treating stochastic equations is

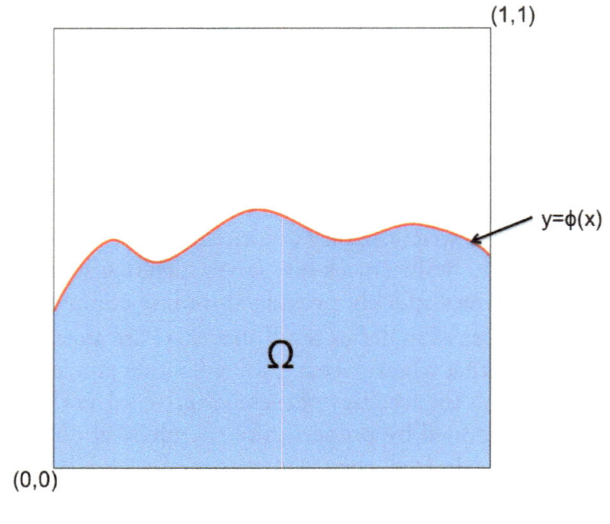

Figure 3.1. *The region marked Ω corresponds to D_-.*

to consider chaos expansions that can be converted to parametric equations but where the number of parameters will be infinite. A second reason is that even when treating a parametric problem with a finite number of parameters m, we want to avoid convergence estimates that blow up with m. By treating the case of an infinite number of parameters, one can sometimes obtain constants independent of m.

3.3.2 ▪ A geometrical setting

Let $D = [0,1]^2$ for simplicity and let $\phi(x)$, $x \in [0,1]$, be a $\mathrm{Lip}_M 1$ function[6] taking values in $[0,1]$. Then, the graph of φ separates D into two domains $D_\pm$ corresponding to the portion D_- of D below the graph and the portion D_+ above the graph (see Figure 3.1). We consider diffusion coefficients

$$a_\phi(x) := \chi_{D_-} + 2\chi_{D_+}. \tag{3.16}$$

These coefficients have a jump across the curve. The purpose of this toy example will be to see how to handle discontinuities in a.

These two examples sit at opposite extremes. The affine model is the most favorable for RB methods since, as we shall see, the manifold $\mathcal{U}$ of solutions is analytic. The geometric model, on the other hand, gives a manifold that is not very smooth because of the discontinuities in a. So, it provides a real challenge for RB methods. This model may, however, prove useful in certain application domains, such as shape optimization.

3.4 ▪ Evaluating numerical methods

Numerical methods for solving PDEs are based on some form of approximation. Understanding the core results of approximation theory not only suggests possible numerical procedures but also determines the optimal performance a numerical method

[6]The space $\mathrm{Lip}_M 1$ consists of all continuous functions satisfying $|f(x)-f(x')| \le M|x-x'|$ for all points $x \in [0,1]$.

can have. This optimal performance is described by the concepts of entropy and widths, which are the subject of this section.

3.4.1 ▪ Linear methods

Methods of approximation are broadly divided into two classes: linear and nonlinear methods. In linear approximation, the approximation process takes place from a sequence of finite-dimensional *linear* spaces X_n, $n = 1, 2, \dots$, with increasing dimensions. By using the space $X_0 := \{0\}$ and, if necessary, repeating the spaces in this sequence, we can assume $\dim(X_n) \le n$. Increasing n results in improved accuracy of the best approximations from X_n.

In our case of parametric equations, we want to choose the linear space X_n so that it approximates well all of the elements $u_a \in \mathcal{U}_{\mathcal{A}}$ in the norm of $H_0^1(D)$ (or perhaps $H_0^1(D,a)$). Once such a space X_n is found, then we can build a numerical method for solving problems. For example, we could use the Galerkin solver corresponding to X_n. This would require the online assembly of the stiffness matrix and the solution of the corresponding matrix problem.

For each $a \in \mathcal{A}$, we have the error

$$E(u_a, X_n) := E(u_a, X_n)_{H_0^1(D)} := \inf_{g \in X_n} \|u_a - g\|_{H_0^1(D)}. \tag{3.17}$$

Notice that because of the *UEA*, the norm $\|\cdot\|_{H_0^1(D)}$ is equivalent to $\|\cdot\|_{H_0^1(D,a)}$, with constants depending only on r and R. Hence, $E(u_a, X_n)$ can also be used to measure the approximation error in $H_0^1(D,a)$. The effectiveness of the space X_n for our parametric problem is given by

$$E(\mathcal{U}_{\mathcal{A}}, X_n) := \sup_{a \in \mathcal{A}} E(u_a, X_n), \tag{3.18}$$

which is the *error on the class* $\mathcal{U}_{\mathcal{A}}$.

The best choice of a linear space X_n is the one that gives the smallest class error. This smallest error for the compact set $\mathcal{U}_{\mathcal{A}}$ is called the *Kolmogorov n-width* of $\mathcal{U}_{\mathcal{A}}$. We can define this width for any compact set K in any Banach space X by

$$d_n(K)_X := \inf_{\dim(Y)=n} \sup_{u \in K} \inf_{g \in Y} \|u - g\|_X, \quad n = 0, 1, \dots. \tag{3.19}$$

So the n-width $d_n(\mathcal{U}_{\mathcal{A}})_{H_0^1(D)}$ gives the optimal performance we can achieve when using linear methods to solve the family of parametric equations (3.1) for all $a \in \mathcal{A}$. Determining d_n and finding an optimal or near-optimal subspace is a difficult problem, and we will return to it later.

3.4.2 ▪ Nonlinear methods

It is now well understood that nonlinear methods of approximation and the numerical methods derived from them often produce superior performance when compared to linear methods. Classical nonlinear methods include approximation by rational functions, free knot splines, n-term approximation, and adaptive partitioning. The basic idea in nonlinear approximation is to replace the linear space X_n by a nonlinear space Σ_n depending on n parameters. Loosely speaking, one can view Σ_n as an n-dimensional manifold.

We discuss, in some detail, the case of n-term approximation in a Banach space X since it has promise in designing numerical algorithms for parametric equations. The starting point for this form of approximation is a collection $\mathscr{D} \subset X$ of functions called a *dictionary*. Given a dictionary, we define the set

$$\Sigma_n(\mathscr{D}) := \left\{ \sum_{g\in\Lambda} c_g g : \Lambda \subset \mathscr{D}, \#(\Lambda) \le n \right\} \tag{3.20}$$

of all n-term linear combinations of elements from $\mathscr{D}$. The elements in Σ_n are said to be *sparse* of order n. Notice that the space Σ_n is not a linear space. If we add two elements from Σ_n, we will generally need $2n$ terms to represent the sum. An important case is the dictionary $\mathscr{D} = \{\varphi_j\}_{j=1}^{\infty}$, where the functions φ_j, $j = 1, 2, \ldots$, form a basis for X. In this case, any function in Σ_n is described by $2n$ parameters, namely the n indices $j \in \Lambda$ and the n coefficients c_{φ_j}.

Suppose now that $X = \mathscr{H}$ is a Hilbert space and $\mathscr{D} = \{\varphi_j\}_{j=1}^{\infty}$ is an orthonormal basis for $\mathscr{H}$. It is very easy to describe the best approximation to a given function $v \in \mathscr{H}$ from Σ_n and the resulting error of approximation. We expand v in its unique series representation

$$v = \sum_{j=1}^{\infty} c_j(v)\varphi_j. \tag{3.21}$$

Given any sequence $(a_j)_{j\ge1}$ of real numbers that tends to zero as $j \to \infty$, we denote by $(a_k^*)_{k\ge1}$ the decreasing rearrangement of the $|a_j|$. Thus, a_k^* is the kth largest of these numbers. For each k, we can find a λ_k such that $a_k^* = |a_{\lambda_k}|$, but the mapping $k \mapsto \lambda_k$ is not unique because of possible ties in the size of the entries. The following discussion is impervious to such differences. If we apply rearrangements to the coordinates $\{c_j(v)\}_{j\ge1}$ and denote by $\Lambda_n := \Lambda_n(v) := \{j_1, \ldots, j_n\}$ the indices of a set of n-largest coefficients, then the best approximation to $u \in \mathscr{H}$ from Σ_n is given by the function

$$G_n v := \sum_{k=1}^{n} c_{j_k}(v)\varphi_{j_k} = \sum_{j\in\Lambda_n} c_j(v)\varphi_j, \tag{3.22}$$

and the resulting error of approximation is

$$\sigma_n(v)_{\mathscr{H}}^2 := \|v - G_n v\|_{\mathscr{H}}^2 = \sum_{j\notin\Lambda_n} |c_j(v)|^2 = \sum_{k>n} (c_k^*(v))^2. \tag{3.23}$$

In particular, $\sigma_0(v) = \|v\|_{\mathscr{H}}$. While the best approximation from Σ_n is not unique, the approximation error $\sigma_n(v)_{\mathscr{H}}$ is uniquely defined. Also, note that $\sigma_n(v)_{\mathscr{H}} = \sigma_n(\tilde{v})_{\mathscr{H}}$ if v and $\tilde{v}$ have the same coefficients up to a permutation of the indices.

We can use (3.23) to characterize the functions $v \in \mathscr{H}$ that can be approximated with order $O(n^{-r})$, $r > 0$, in terms of the coefficients $c_j(v)$. Let us denote by $\mathscr{A}^r = \mathscr{A}_r((\Sigma_n)_{n=1}^{\infty}, \mathscr{H})$ this set of functions ($\mathscr{A}^r$ is called an *approximation class*) and equip it with the norm

$$\|v\|_{\mathscr{A}^r} := \sup_{n\ge0} (n+1)^r \sigma_n(v)_{\mathscr{H}}. \tag{3.24}$$

Given an $r > 0$, we define p by the formula

$$\frac{1}{p} = r + \frac{1}{2}. \tag{3.25}$$

Notice that $p < 2$. The space $w\ell_p$ (*weak* ℓ_p) is defined as the set of all $\mathbf{a} = (a_j)_{j\geq 1}$ whose decreasing rearrangement $(a_k^*)_{k\geq 1}$satisfies

$$k^{1/p} a_k^* \leq M, \quad k \geq 1, \tag{3.26}$$

and the smallest $M = M(\mathbf{a})$ for which (3.26) is valid is the quasi norm $\|\mathbf{a}\|_{w\ell_p}$ of $\mathbf{a}$ in this space. Notice that $w\ell_p$ contains ℓ_p and is slightly larger since it contains sequences whose rearrangement behaves like $k^{-1/p}$ that barely miss being in ℓ_p.

We claim that, with $\mathbf{c} := \mathbf{c}(v) := \{c_j(v)\}_{j\geq 1}$,

$$\mathcal{A}^r := \mathcal{A}^r(\mathcal{H}, (\Sigma_n)_{n\geq 1}) = \{v : \mathbf{c}(v) \in w\ell_p\}, \tag{3.27}$$

and $\|\mathbf{c}(u)\|_{w\ell_p}$ is equivalent to $\|u\|_{\mathcal{A}^r}$. Indeed, if $\mathbf{c}(v) \in w\ell_p$, then for any $n \geq 1$, we have

$$\sigma_n(v) = \sum_{k>n} (c_k^*(v))^2 \leq \|\mathbf{c}(v)\|_{w\ell_p}^2 \sum_{k>n} k^{-2r-1} \leq \frac{1}{2r} \|\mathbf{c}(v)\|_{w\ell_p}^2 n^{-2r}. \tag{3.28}$$

In addition,

$$\|v\|_{\mathcal{H}}^2 = \|\mathbf{c}(v)\|_{\ell^2}^2 \leq \|\mathbf{c}(v)\|_{w\ell^p}^2 \sum_{k\geq 1} k^{-2r-1} \leq (1 + \frac{1}{2r}) \|\mathbf{c}(v)\|_{w\ell_p}^2. \tag{3.29}$$

This shows that $\|v\|_{\mathcal{A}^r} \leq (1 + \frac{1}{2r})^{1/2} \|\mathbf{c}(v)\|_{w\ell_p}$.

To reverse this inequality, we note that for any $k \geq 1$, the monotonicity of $\mathbf{c}^*(v)$ gives

$$2^j (c_{2^{j+1}}^*(v))^2 \leq \sum_{k=2^j+1}^{2^{j+1}} (c_k(v)^*)^2 \leq \sigma_{2^j}(u)^2 \leq |v|_{\mathcal{A}^r}^2 2^{-2jr}. \tag{3.30}$$

For any n, we choose j so that $2^j \leq n < 2^{j+1}$. If $j > 0$, we obtain from the monotonicity of $\mathbf{c}^*(v)$ that

$$c_n^*(v) \leq c_{2^j}^*(v) \leq 2^{r+1/2} |v|_{\mathcal{A}^r} 2^{-(r+1/2)j} = 2^{1/p} |v|_{\mathcal{A}^r} 2^{-j/p} \leq 2^{2/p} |v|_{\mathcal{A}^r} n^{-1/p}. \tag{3.31}$$

On the other hand, we clearly have

$$c_1^*(v) \leq \|v\|_{\mathcal{H}} \leq \|v\|_{\mathcal{A}^r}. \tag{3.32}$$

This gives $\|\mathbf{c}(v)\|_{w\ell_p} \leq 2^{2/p} \|v\|_{\mathcal{A}^r}$ and completes the proof of the equivalence.

In numerical settings, one cannot implement n-term approximation in the form we have just presented since it would require the computation of all coefficients of v and a rearrangement of them. What is done in practice is to choose a value N dependent on n and select the best n-term approximation from the dictionary $\mathcal{D}_N := \{\varphi_j\}_{j=1}^N$. A typical choice of N is $N = n^A$, where A is a fixed integer.

There is another useful view of n-term approximation in this last setting. We can form from $\{\varphi_1, \ldots, \varphi_N\}$ all subsets $\{\varphi_i\}_{i\in\Lambda}$, $\#(\Lambda) = n$, that are linearly independent. Then, each $X_\Lambda := \text{span}\{x_i : i \in \Lambda\}$ is a linear space of dimension n. There are at most $\binom{N}{n} \leq [en^{B-1}]^n$ such subspaces. Then, n-term approximation can be viewed as taking one of these linear spaces and using it to approximate v. The space chosen can depend on v.

3.4.3 ▪ Nonlinear widths

Several definitions of nonlinear widths have been proposed to measure optimal performance of nonlinear methods. We mention the two that seem most relevant for the analysis of RB methods. The first of these is the manifold width [13], which is a good match for numerical algorithms based on nonlinear approximation. Let X be a Banach space and K one of its compact subsets. To define this width, we consider two continuous functions. The first function, b, maps each element $x \in K$ into $\mathbb{R}^n$ and therefore picks out the parameters to be used in approximating x. The second function, M, maps $\mathbb{R}^n$ into the set $\mathcal{M}$ (which we view as an n-dimensional manifold although we do not assume anything about the smoothness of the image $\mathcal{M}$). The manifold width of the compact set K is then defined by

$$\delta_n(K)_X := \inf_{M,b} \sup_{x \in K} \|x - M(b(x))\|_X. \tag{3.33}$$

For typical compact sets K of functions, the manifold widths behave like the entropy numbers defined below. For example, if K is the unit ball of any Besov or Sobolev space of smoothness s that compactly embeds into $L_p(\Omega)$ with $\Omega \subset \mathbb{R}^d$ a Lipschitz domain, then (see [14])

$$C_0 n^{-s/d} \leq \delta_n(K)_{L_p(\Omega)} \leq C_1 n^{-s/d}, \tag{3.34}$$

with C_0, C_1 independent of n. We see in (3.34) the curse of dimensionality in the appearance of d in the exponent of n. To obtain just moderate rates of convergence with $n \to \infty$ we need s to be comparable with d.

A second width, introduced by V. Temlyakov [27], fits the definition of n-term approximation. It considers any collection (called a *library*) $\mathcal{X} := \{X_j\}_{j=1}^N$ of n-dimensional subspaces of X. The approximation error defined by

$$E(v, \mathcal{X})_X := \inf_{1 \leq j \leq N} \operatorname{dist}(v, X_j)_X \tag{3.35}$$

is another type of n-term approximation. This leads us to define the *library widths*

$$d^L_{n,N}(K)_X := \inf_{\mathcal{X}} \sup_{v \in K} E(v, \mathcal{X})_X, \tag{3.36}$$

with the infimum taken over all such collections $\mathcal{X}$. This is another type of nonlinear width. Typically, we would like to eliminate N from the above definition. Similar to the restrictions on dictionaries, one usually assumes that the number N of bases is of the form $N = n^A$ for some fixed integer A. With this assumption, $d_{n,N}$ now only depends on n.

Let us note that the definition of library widths includes approximation from a finite dictionary. Namely, if $\mathcal{D}$ is a dictionary with $\#(\mathcal{D}) = m$, then there are $\binom{m}{n}$ subspaces X_j of dimension $\leq n$ that can be formed using n elements from the dictionary as a spanning set. Thus, with $N = \binom{m}{n}$, the library width allows in the competition all n-term approximations from $\mathcal{D}$. However, the library width allows more general sequences of subspaces X_j in its definition since they do not have to be organized as coming from a fixed dictionary. When the subspaces X_j all come from a fixed dictionary of size m, as described above, then the corresponding width

$$d_{n,N}(K)_X := \inf_{\mathcal{D}} \sup_{v \in K} \sigma_n(K, \mathcal{D})_X \tag{3.37}$$

is called the *dictionary width* of K.

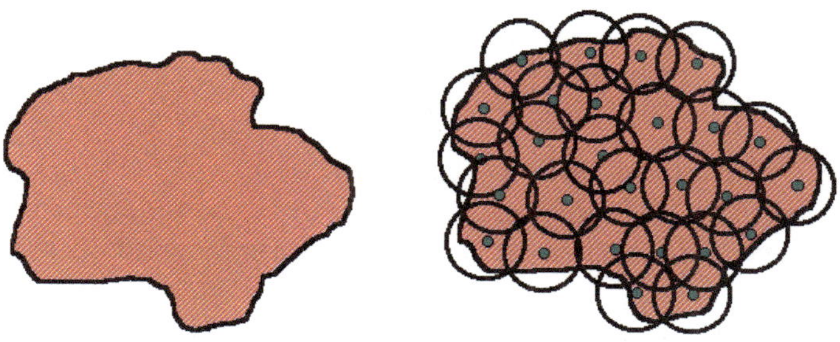

Figure 3.2. *A compact set K and its ϵ cover.*

3.4.4 ▪ Entropy numbers

Another useful concept in our analysis of RB methods is the entropy numbers of a compact set $K \subset X$, where again X is a Banach space. If $\epsilon > 0$, we consider all possible coverings of $K \subset \bigcup_{i=1}^{m} B(x_i, \epsilon)$ using balls $B(x_i, \epsilon)$ of radius ϵ with centers $x_i \in X$. The smallest number $m = N_\epsilon(K)_X$ for which such a covering exists is called the covering number of K. The Kolmogorov entropy of K is then defined as

$$H_\epsilon(K)_X := \log_2(N_\epsilon(K))_X. \tag{3.38}$$

The Kolmogorov entropy measures the size or massivity of K. It has another important property of determining optimal encoding of the elements of K. Namely, if $x \in K$, then we can assign to x the binary bits of an index i for which $x \in B(x_i, \epsilon)$. Each x is then encoded to accuracy ϵ with $\leq \lceil H_\epsilon(K)_X \rceil$ bits, and no other encoder can do better for K.

It is frequently more convenient to consider the entropy numbers

$$\epsilon_n(K)_X := \inf\{\epsilon : H_\epsilon(K)_X \leq n\}. \tag{3.39}$$

Typically, $\epsilon_n(K)_X$ decays like n^{-r} for standard compact sets. Not only does $\epsilon_n(K)_X$ tell us the minimal distortion we can achieve with n-bit encoding, it also says that any numerical algorithm that computes an approximation to each of the elements of K to accuracy $\epsilon_n(K)_X$ will require at least n operations.

An important issue in optimal performance of the standard numerical algorithms for PDEs is the entropy numbers of the classical smoothness spaces. If K is the unit ball $U(W^s(L_p(\Omega)))$ of a Sobolev space, or a unit ball $U(B_q^s(L_p(\Omega)))$ of a Besov space, then for any Lebesgue space $X = L_\mu(\Omega)$,

$$\epsilon_n(K)_X \geq C n^{-s/d}, \quad n = 0, 1, \ldots. \tag{3.40}$$

This result manifests the massivity of these compact sets as the dimension d increases and exhibits fully the curse of dimensionality.

3.4.5 ▪ Comparison of widths

Concepts like n-widths are used to give bounds for the best possible performance of numerical algorithms. Some general comparisons between the different widths are useful in making such evaluations. Let us mention those that will prove most useful for us. For any compact set K in a Banach space X, we always have (see [13])

$$\delta_n(K)_X \leq d_n(K)_X, \quad n \geq 1. \tag{3.41}$$

It is also known that whenever $d_n(K)_X \le Cn^{-r}$, $n \ge 1$ with $r > 0$, then there is a constant C' such that

$$\epsilon_n(K)_X \le C'n^{-r}, \quad n \ge 1. \tag{3.42}$$

This follows from Carl's inequality (see [21]).

In general, it is not possible to compare the nonlinear manifold widths to entropy. However, for the widths of (3.36), we have the same result as (3.42) (see [27]).

3.5 ▪ Comparing widths and entropies of $\mathcal{U}_{\mathcal{A}}$ with those of $\mathcal{A}$

Let us return our discussion to numerical methods for $\mathcal{U}_{\mathcal{A}}$. In trying to understand how well numerical methods can perform in resolving $\mathcal{U}_{\mathcal{A}}$, we would like to know the entropies and widths of $\mathcal{U}_{\mathcal{A}}$. Since we only know this set through the parameter set $\mathcal{A}$, the first question we would like to answer is whether we can bound the widths of $\mathcal{U}_{\mathcal{A}}$ by those of $\mathcal{A}$. The widths and entropies of $\mathcal{A}$ are usually more transparent, and so such bounds give a good indication of how well RB methods might perform. In this section, we shall discuss what is known about such comparisons.

3.5.1 ▪ Comparing entropies

One can utilize the perturbation results (3.9) and (3.13) to compare the entropies of the two classes $\mathcal{A}$ and $\mathcal{U} = \mathcal{U}_{\mathcal{A}}$. We place ourselves in the following situation. We assume that $2 \le q \le \infty$ is a value for which $L_q(D)$ stability is known to hold for $\mathcal{A}$. It follows that any ϵ cover of $\mathcal{A}$ in the $L_q(D)$-norm given by balls $B(a_i,\epsilon)$ will induce a $C_0\epsilon$ cover of $\mathcal{U}$ by the balls $B(u_{a_i}, C_0\epsilon)$ in the $H_0^1(D)$ topology. Therefore, we have

$$\epsilon_n(\mathcal{U}_{\mathcal{A}})_{H_0^1(D)} \le C_0\epsilon_n(\mathcal{A})_{L_q(D)}, \quad n \ge 1. \tag{3.43}$$

It now becomes an interesting question of whether the entropy numbers of $\mathcal{U}_{\mathcal{A}}$ could actually be much better than those of $\mathcal{A}$. We will now show why, in general, we cannot expect this to be the case. We consider the case $q = 2$ and $d = 1$. Our goal is to find an f and many classes $\mathcal{A}$ for which we can reverse (3.43) and thereby see that the entropy of $\mathcal{U}_{\mathcal{A}}$ is not noticeably better than that of $\mathcal{A}$, at least in general. We consider the one-dimensional case $D = [0,1]$, where the PDE is simply

$$-[au']' = f, \quad u(0) = u(1) = 0. \tag{3.44}$$

We will specify a right side f as we proceed. Let $F(x) := -\int_0^x f(s)\,ds$. Then given a,

$$u_a = \int_0^x a^{-1}(s)[F(s) - c_a]\,ds, \quad c_a := \int_0^1 a^{-1}(s)F(s)\,ds. \tag{3.45}$$

Since we are allowed to choose f, we are allowed to choose F as follows. We take F to be a smooth function that is odd with respect to $x = 1/2$ and satisfies $F(0) = F(1/2) = 0$, F is increasing on $[0,1/6]$, $F(x) = +1$ on $J := [1/6,1/3]$, and F is decreasing on $[1/3,1/2]$. We fix this F and assume the following about $\mathcal{A}$.

Assumptions on $\mathcal{A}$. *Each $a \in \mathcal{A}$ is even with respect to $1/2$, and the class $\mathcal{A}_0$ of all $a \in \mathcal{A}$ restricted to J satisfies $\epsilon_n(\mathcal{A}_0)_{L_2(J)} \ge c_0\epsilon_n(\mathcal{A})_{L_2(D)}$ for an absolute constant c_0.*

Returning to (3.45), we see that $u_a' \geq 1/a$ on the interval $J := [1/6, 1/3]$. On J we can now write for any two such $a, \tilde{a}$ that

$$a - \tilde{a} = F/u_a' - F/u_{\tilde{a}}' = \frac{F}{u_a' u_{\tilde{a}}'}[u_{\tilde{a}}' - u_a']. \tag{3.46}$$

This gives the bound on J,

$$\|a - \tilde{a}\|_{L_2(J)} \leq C\|u_a - u_{\tilde{a}}\|_{H_0^1(D)}, \tag{3.47}$$

and therefore

$$\epsilon_n(\mathscr{A})_{L_2(J)} \leq C\epsilon_n(K)_{H_0^1(D)}, \quad n \geq 1, \tag{3.48}$$

and therefore we have reversed (3.43) in the case $q = 2$.

3.5.2 ▪ Comparing Kolmogorov widths

There is a general method (see [7]) for bounding the Kolmogorov n-width of $\mathscr{U}_{\mathscr{A}}$ in terms of the corresponding width of $\mathscr{A}$. Among other results, it gives the following theorem.

Theorem 3.1. *If $\mathscr{A}$ is any set contained in $L_\infty(D)$ whose Kolmgorov width satisfies*

$$d_n(\mathscr{A})_{L_\infty(D)} \leq C_0 n^{-\alpha}, \quad n \geq 1, \tag{3.49}$$

for some $\alpha > 1$ and constant C_0, then for each $\beta < \alpha - 1$,

$$d_n(\mathscr{U}_{\mathscr{A}})_{H_0^1(D)} \leq C_1 n^{-\beta}, \quad n \geq 1, \tag{3.50}$$

for a constant C_1 depending only on C_0 and α.

We can give some ideas behind the proof, which rests on the following result from [10].

Theorem 3.2. *If $\mathscr{A}$ is defined by the affine model* (3.15), *where the ψ_j, $j = 1, 2, \ldots$, satisfy $(\|\psi_j\|_{L_\infty})_{j=1}^\infty \in \ell_p$ with $p < 1$, then $d_n(U_{\mathscr{A}}) \leq M n^{-1+1/p}$.*

We can now sketch out how one proves Theorem 3.1. From our assumption on $\mathscr{A}$, we can find, for each $k \geq 0$, linear spaces Z_k of dimension 2^k that satisfy

$$\text{dist}(\mathscr{A}, Z_k)_{L_\infty(D)} \leq C_0 2^{-k\alpha}, \quad k = 0, 1, 2, \ldots. \tag{3.51}$$

Hence, for each $a \in \mathscr{A}$ there are functions $g_k \in Z_k$ such that

$$\|a - g_k\|_{L_\infty(D)} \leq C_0 2^{-k\alpha}, \quad k = 0, 1, \ldots. \tag{3.52}$$

It follows that

$$a = g_0 + \sum_{k\geq 1}[g_k - g_{k-1}] = \sum_{k=0}^{\infty} h_k \quad \text{and } \|h_k\|_{L_\infty(D)} \leq 2C_0 2^{-k\alpha},\ k = 0, 1, \ldots, \tag{3.53}$$

where $h_k := g_k - g_{k-1}$ and $g_{-1} := 0$. Let us note that each h_k is in the linear space $Y_k := Z_{k-1} + Z_k$, which has dimension $m_k \leq 2^k + 2^{k-1}$.

We now use the following consequence of Auerbach's basis theorem. There is a basis $\psi_{1,k},\dots,\psi_{m_k,k}$ for Y_k, normalized so that $\|\psi_{j,k}\|_{L_\infty(D)}=1$ for all j and such that its dual basis also has norm one. So each h_k can be written $h_k=\sum_{j=1}^{m_k} c_{j,k}\psi_{j,k}$, and the coefficients satisfy

$$\max_{1\le j\le m_k}|c_{j,k}|\le \|h_k\|_{L_\infty(D)}\le 2C_0 2^{-k\alpha},\quad k=1,2,\dots. \tag{3.54}$$

This gives us the decomposition

$$a=\sum_{j=1}^{\infty} y_j\psi_j,\quad ,j=1,2,\dots,\quad |y_j|\le 1, \tag{3.55}$$

where each ψ_j is a renormalization of one of the $\psi_{i,k}$ and satisfies

$$\|\psi_j\|_{L_\infty(D)}\le M_0 C_0 j^{-\alpha},\quad j=1,2,\dots. \tag{3.56}$$

Notice that a function ψ_j is possibly repeated in the sum (3.55).

We now choose K so that

$$M_0\sum_{k>K}\|\psi_j\|_{L_\infty(D)}\le r/2. \tag{3.57}$$

This means that $\sum_{j<K}\|\psi_j\|_{L_\infty(D)}\ge r/2$. It can be shown (details not given) that we can find a finite number, say N, of a_i, each of the form $a_i=\sum_{1\le j\le K} b_j\psi_j$, such that for any $a\in\mathscr{A}$, there is an a_i for which

$$a=a_i+\sum_{j=1}^{\infty} y_j\psi_j, \tag{3.58}$$

where $|y_j|\le 1$ and, moreover, each of the $a_i(x)\ge r/4$ for $x\in D$.

In other words, $\mathscr{U}_{\mathscr{A}}$ is contained in a union of a finite number N of $\mathscr{U}_{\mathscr{A}_i}$, where each set $\mathscr{A}_i$ is of the form (3.58). Because of (3.56), $(\|\psi_j\|_{L_\infty(D)})\in\ell_p$ for each $p>1/\alpha$. This allows us to apply Theorem 3.2 and derive Theorem 3.1.

3.5.3 ▪ Comparing nonlinear widths

We shall see in the sections that follow that nonlinear numerical methods for reduced modeling are not as well developed as their linear counterparts. Nevertheless, it is very desirable to have bounds on the nonlinear widths of $\mathscr{U}_{\mathscr{A}}$ derived from the corresponding widths of $\mathscr{A}$. This would help us understand what nonlinear methods can possibly bring to the table and also perhaps help in the development of nonlinear methods for reduced modeling. In this section, we look at what we know about such comparisons.

Let us begin with the library width $d_{n,N}$ defined in (3.36). We assume that for each n, the library has $N=n^A$ bases with A a fixed integer. Then, building on Theorem 3.1, one can prove (see [7]) that whenever $\mathscr{A}$ satisfies

$$d_{n,N}(\mathscr{A})_{L_\infty(D)}\le Cn^{-\alpha},\quad n\ge 1, \tag{3.59}$$

for some $\alpha>1$, then for any $\beta<\alpha-1$, we have

$$d_{n,N}(\mathscr{U}_{\mathscr{A}})_{H_0^1(D)}\le Cn^{-\beta},\quad n\ge 1. \tag{3.60}$$

For nonlinear manifold widths, general comparisons are not known. However, there is a setting that is sometimes applicable in which we can derive such comparisons. This setting rests on the following assumption.

Uniqueness assumption. *We assume that the right side f and the class $\mathcal{A}$ of diffusion coefficients have the property that whenever $u_a = u_{\tilde{a}}$ with $a, \tilde{a}$ in $\mathcal{A}$, then $a = \tilde{a}$ almost everywhere.*

As we shall discuss in Section 3.6.2, this assumption is satisfied, for example, for the geometric model. Let us note that in the recent paper [4], it is shown that when f is strictly positive then the uniqueness assumption holds under very mild regularity assumptions on the $a \in \mathcal{A}$. Assume now that the uniqueness assumption is valid and that for the given f and a value of q, the stability bound (3.11) holds. Recall that this stability bound always holds for $q = \frac{2p}{p-2}$ and a certain range of $p \in [2, P]$ provided $f \in W^{-1}(L_p(D))$. Consider the mapping Φ from $\mathcal{U}_{\mathcal{A}}$ into $\mathcal{A}$ that takes $u = u_a$ into a. Since $\mathcal{A}$ is assumed to be compact in $L_q(D)$ and the mapping $a \to u_a$ is known to be continuous (see (3.11)), we know from elementary principles that the mapping Φ is also continuous as a mapping from $H_0^1(D)$ to $L_q(D)$.

We now wish to prove that under these assumptions, we have

$$\delta_n(\mathcal{U}_{\mathcal{A}})_{H_0^1(D)} \leq C \delta_n(\mathcal{A})_{L_q(D)}. \tag{3.61}$$

Given n and $\epsilon = \delta_n(\mathcal{A})_{L_q(D)}$, we can choose continuous mappings M, b as in the definition of nonlinear widths so that $b : \mathcal{A} \to \mathbb{R}^n$, $M : \mathbb{R}^n \to L_q(D)$, and

$$\sup_{a \in \mathcal{A}} \|a - M(b(a))\|_{L_q(D)} \leq 2\epsilon. \tag{3.62}$$

We make the additional assumption that M maps $\mathbb{R}^n$ into $\mathcal{A}$ and proceed to construct appropriate mappings for $\mathcal{U}_{\mathcal{A}}$. We can take, for $\mathbf{z} \in \mathbb{R}^n$,

$$\tilde{M}(\mathbf{z}) := u_{M(\mathbf{z})}. \tag{3.63}$$

Since M is continuous as a mapping into $L_q(D)$, the L_q stability (3.11) gives that $\tilde{M}$ is also continuous as a mapping into $H_0^1(D)$. Finally, we define $\tilde{b} : \mathcal{U}_{\mathcal{A}} \to \mathbb{R}_n$ by

$$\tilde{b}(u) = b(\Phi(u)), \quad u \in \mathcal{A}. \tag{3.64}$$

Since Φ is continuous from $\mathcal{U}_{\mathcal{A}}$ in the $H_0^1(D)$ topology to $\mathcal{A}$ in the L_q topology and b is continuous from $\mathcal{A}$ to $\mathbb{R}^n$, we have that $\tilde{b}$ is also continuous.

Given $u \in \mathcal{U}_{\mathcal{A}}$, we have

$$\|u - \tilde{M}(\tilde{b}(u))\|_{H_0^1(D)} = \|u_{\Phi(u)} - u_{M(\Phi(u))}\|_{H_0^1(D)} \leq C\|\Phi(u) - M(\Phi(u))\|_{L_q(D)} \leq 2C\epsilon. \tag{3.65}$$

Since $\epsilon = \delta_n(\mathcal{A})_{L_q(D)}$, we have proved (3.61).

3.6 ▪ Widths of our two model classes

The results of the preceding section were directed at giving general a priori guarantees about the performance of linear and nonlinear methods for reduced modeling. Since the guarantees do not assume any particular structure of the set $\mathscr{A}$, it may be that they can be improved in settings where we assume a specific structure for $\mathscr{A}$. We now discuss what is known in this regard for our two model classes of elliptic equations.

3.6.1 ▪ The affine model

We recall that for the affine model, we assume that

$$a(x,y)=\bar{a}(x)+\sum_{j\geq 1} y_j \psi_j(x), \tag{3.66}$$

where the y_j, $j \geq 1$, are parameters in $[-1,1]$. We can always rearrange the indices so that the sequence $b_j := \|\psi_j\|_{L_\infty(D)}$, $j = 1,2\ldots$, is decreasing. For canonical representation systems $\{\psi_j\}$, such as wavelets or Fourier, the rate of decrease of (b_j) to zero is related to the smoothness of $a(x,y)$ as a function of x. Indeed, smoothness conditions on a translate into decay conditions on the (b_j).

Let us note that if $(b_j)\in \ell_p$, $p<1$, then

$$\sup_{y\in U}\left\| a(\cdot,y)-\sum_{j=1}^{n} y_j\psi_j\right\|_{L_\infty(D)} \leq \sum_{j=n+1}^{\infty} b_j \leq b_{n+1}^{1-p}\sum_{j=n+1}^{\infty} b_j^p \leq Cn^{1-1/p}. \tag{3.67}$$

Here we have used the fact that since (b_j) is decreasing and in ℓ_p, we must have $b_n^p \leq Cn^{-1}$, $n\geq 1$.

The result (3.67) shows that the Kolmogorov width of $\mathscr{A}$ decays like $O(n^{1-1/p})$:

$$(b_j)\in \ell_p \quad \Longrightarrow \quad d_n(\mathscr{A})\leq Cn^{1-1/p}, \quad n\geq 1. \tag{3.68}$$

We could now use Theorem 3.1 to conclude that $d_n(\mathscr{U}_{\mathscr{A}})_{H_0^1(D)} \leq Cn^{2-1/p-\varepsilon}$ for all $\varepsilon > 0$. However, we have already pointed out in Theorem 3.2 that this result is not optimal for the affine case. In fact, we used this stronger result in the derivation of Theorem 3.1. We give a little more detail to illuminate how the stronger result of Theorem 3.2 is proved.

Let $\mathscr{F}$ be the set of all sequences $\nu=(\nu_1,\nu_2,\ldots)$ such that ν has finite support and each entry in ν is a nonnegative integer. So $|\nu|=\sum_{j\geq 1}|\nu_j|$ is always finite when $\nu\in\mathscr{F}$. If $\alpha=(\alpha_j)_{j\geq 1}$ is a sequence of positive numbers, we define, for all $\nu\in\mathscr{F}$,

$$\alpha^\nu := \prod_{j\geq 1}\alpha_j^{\nu_j}.$$

In [10], we showed the following theorem.

Theorem 3.3. *If $(b_j)\in\ell_p$ for some $p<1$, then*

$$u(x,y)=\sum_{\nu\in\mathscr{F}} c_\nu(x)y^\nu, \tag{3.69}$$

where the functions $c_\nu(x)$ are in $H_0^1(D)$ and $(\|c_\nu\|_{H_0^1(D)})\in\ell_p$ for the same value of p.

The line of reasoning for proving this theorem is as follows. The mapping $F: y\mapsto u(\cdot,y)$ takes U into $H_0^1(D)$. One shows that this map is analytic and has a Taylor expansion as a function of y. The complications arise because y consists of an infinite number of variables and the mapping is Banach-space valued. The proof of analyticity is not difficult. For a fixed $y \in U$, we know that for all $v \in H_0^1(D)$,

$$\int_D a(x,y)\nabla u(x,y)\nabla v(x)dx = \int_D f(x)v(x)dx.$$

Differentiating this identity with respect to the variable y_j gives

$$\int_D a(x,y)\nabla \partial_{y_j} u(x,y)\nabla v(x)dx + \int_D \psi_j(x)\nabla u(x,y)\nabla v(x)dx = 0. \tag{3.70}$$

One then shows that more generally,

$$\int_D a(x,y)\nabla \partial_y^\nu u(x,y)\nabla v(x)dx + \sum_{\{j:\nu_j\neq 0\}} \nu_j \int_D \psi_j(x)\nabla \partial_y^{\nu-e_j} u(x,y)\nabla v(x)dx = 0, \tag{3.71}$$

where e_j is the Kronecker sequence with value one at position j and zero elsewhere. The identity (3.71) is proved by induction on $|\nu|$ using the same idea as used in deriving (3.70). From (3.71) it is not difficult to prove

$$\begin{aligned}||\partial_y^\nu u(\cdot,y)||_V &\le C_0 \sum_{\{j:\nu_j\neq 0\}} \nu_j b_j(|\nu|-1)!b^{\nu-e_j} = C_0\left(\sum_{\{j:\nu_j\neq 0\}} \nu_j\right)(|\nu|-1)!b^\nu \\ &= C_0|\nu|!b^\nu,\ \nu\in\mathcal{F}.\end{aligned}$$

One now proves the representation (3.69) with $c_\nu(x) := \frac{D^\nu u(x,0)}{\nu!}$ (see [9, 10] for details). The proof that $(||c_\nu||_{H_0^1(D)}) \in \ell_p$ whenever $(||\psi_j||_{L_\infty(D)}) \in \ell_p$ is far more difficult.

Now let us see how the above theorem gives an estimate for the Kolmogorov n-width of the class $\mathcal{A}$.

Corollary 3.4. *For the affine model $\mathcal{A}$, whenever $(||\psi_j||_{L_\infty}) \in \ell_p$, $p < 1$, the set $\mathcal{U}_\mathcal{A}$ has n-widths*

$$d_n(\mathcal{A})_{H_0^1(D)} \le Cn^{1-1/p}, \quad n \ge 1, \tag{3.72}$$

with C depending only on p and the ellipticity constants r,R.

Indeed, from the fact that $(||c_\nu||_{H_0^1(D)}) \in \ell_p$, one can use similar arguments to that in (3.67) to prove that there is a set $\Lambda \subset \mathcal{F}$ with $\#(\Lambda) = n$ so that

$$\sup_{y\in U} ||u(\cdot,y) - \sum_{\nu\in\Lambda} c_\nu(x)y^\nu||_{H_0^1(D)} \le Cn^{1-1/p}. \tag{3.73}$$

This shows that the n-dimensional space $V := \mathrm{span}\{c_\nu : \nu \in \Lambda\}$ approximates $\mathcal{A}$ with accuracy $Cn^{1-1/p}$ and therefore $d_n(\mathcal{A})_{H_0^1(D)} \le Cn^{1-1/p}$.

One important observation about this bound for widths is that we have broken the curse of dimensionality. Indeed, the parameters $y_1, y_2, \dots$ are infinite. In typical applications, the parameters are finite in number, say d, but then this result shows that the exponent of n in this bound does not depend on d.

3.6.2 ▪ The geometric model

Although, as we shall see, the results about numerical performance for this example are not definitive, it is still instructive to discuss what is known and which questions are still unresolved in the case of the geometric model. Following our usual paradigm, let us first consider $\mathscr{A}$ and try to understand its complexity. It makes no sense to consider the approximation of the functions $a \in \mathscr{A}$ in the $L_\infty(D)$-norm since each of these functions is discontinuous and therefore any approximation would have to match these discontinuities exactly. On the other hand, we can approximate a in an $L_q(\Omega)$-norm and use the perturbation result (3.9). For the convex domain $D = [0,1]^2$, the best possible range of q for the perturbation theorem is $Q \le q \le \infty$, where the smallest value of Q depends on the constants r, R in the uniform ellipticity assumption (see [5]).

We know from our general theory that if we measure the complexity of $\mathscr{A}$ and $\mathscr{U}_{\mathscr{A}}$ in the sense of their entropy then (3.43) always holds. One can rather easily compute the entropy numbers of $\mathscr{A}$ in $L_q(D)$ for any $q \ge 2$:

$$\epsilon_n(\mathscr{A})_{L_q(D)} \sim n^{-1/q}, \quad n \ge 1. \tag{3.74}$$

From our general comparison (3.43), this gives

$$\epsilon_n(\mathscr{U}_{\mathscr{A}})_{H_0^1(D)} \le C n^{-1/q}, \quad n \ge 1, \tag{3.75}$$

with the best bound holding for $q = Q$.

Let us next discuss what is known about linear widths for $\mathscr{A}$ and $\mathscr{U}_{\mathscr{A}}$. The following bounds for n-widths can be shown with appropriate constants $C_1, C_2 > 0$:

$$C_2 n^{-\frac{1}{2q}} \le d_n(\mathscr{A})_{L_q(D)} \le C_1 n^{-\frac{1}{2q}}, \quad n \ge 1, \tag{3.76}$$

with C_1, C_2 absolute constants. To prove the upper estimate, we consider the dictionary $\mathscr{D}_n$ that consists of the functions χ_R, where $R = [(i-1)/n, i/n) \times [0, j/n]$, $1 \le i, j \le n$. Given a function $a \in \mathscr{A}$ corresponding to the Lipschitz function φ, and a value $1 \le i \le n$, we let j_i be the largest integer such that

$$\frac{j_i}{n} \le \varphi(x), \quad x \in \left[\frac{i-1}{n}, \frac{i}{n}\right]. \tag{3.77}$$

We consider the function $g_n := 1 + \sum_{i=1^n} \chi_{R_i}$ with $R_i := [(i-1)/n, i/n) \times [0, j_i/n]$, $1 \le i \le n$. The function g agrees with a except on a set of measure $\le 1/n$. Hence,

$$\|a - g\|_{L_q(D)} \le n^{-1/q}. \tag{3.78}$$

Since the space spanned by $\mathscr{D}_n$ has dimension n^2, we obtain the upper estimate.

The lower estimate is a little more intricate. We first consider the case $q = 2$. Let $V \subset L_2(D)$ be any fixed linear space of dimension $N \le n^2/2$ with n a two power, and let $\varphi_1, \dots, \varphi_N$ be an orthonormal system for V. We assume $\text{dist}(\mathscr{A}, V)_{L_2(D)} \le \epsilon$ and derive a bound from below for ϵ.

We will first construct some functions that can be approximated well by V. Let ψ_k be the piecewise-linear function that is zero outside $I_k := [k/n, (k+1)/n]$ and is the hat function with height $1/(2n)$ on I_k. Then, for any $j > 0$ and any set $\Lambda \subset \{0, 1, \dots, n-1\}$, the function $g_{j,\Lambda} := j/n + \sum_{k \in \Lambda} \psi_k$ is in $\text{Lip}_1 1$. The function

$$f_{j,\Lambda} := a_{g_{j,\Lambda}} - a_{g_{j,\Lambda^c}} \tag{3.79}$$

can be approximated to accuracy 2ϵ by the space V. Each of these functions has support in the strip $j/n \le y \le (j+1)/n$ and has norm $\|f_{j,\Lambda}\|^2_{L^2(D)} = 1/(3n)$. Obviously, those functions with different values of j are orthogonal. Moreover, for a fixed j, we can choose n different sets Λ such that these functions are also orthogonal. Indeed, we take $\Lambda = \{0, 1, \dots, n-1\}$ and then the other $n-1$ choices corresponding to where the Walsh functions of order n are positive. In this way, we get n^2 orthogonal functions. We define the functions $h_1, \dots, h_{n^2}$, where each h_j is one of the functions $\sqrt{3n} f_{i,\Lambda}$ with the n^2 different choices of these function in (3.79). Hence, these functions are an orthonormal system and each one can be approximated to accuracy $2\epsilon\sqrt{3n}$.

We consider the $n^2 \times N$ matrix B whose i,j entry is $b_{i,j} := |\langle h_i, \varphi_j \rangle|^2$. Then, each of the N columns has sum at most one. Hence, one of the rows, i^*, has sum at most $Nn^{-2} \le 1/2$. This means that in approximating h_{i^*} by the elements of V in the $L_2(D)$-norm, we incur an error of at least $1/\sqrt{2}$. It follows that $2\epsilon\sqrt{3n} \ge 1/\sqrt{2}$. In other words, $\epsilon \ge [2\sqrt{6}]^{-1} n^{-1/2}$. Since the only restriction on N is that $N \le n^2/2$, we obtain

$$d_{n^2}(\mathscr{A})_{L_2(D)} \ge C n^{-1/2}, \quad n \ge 1, \tag{3.80}$$

with C an absolute constant. The lower bound in (3.76) for $q = 2$ follows.

Let us mention, without proof, that the lower bound $d_n(\mathscr{A})_{L_q(D)} \ge n^{-\frac{1}{2q}}$, for $2 \le q \le \infty$, can be proved from the L_2 result by using an interpolation argument.

The above results describe how well we can approximate $\mathscr{A}$ but say nothing about approximating $\mathscr{U}_{\mathscr{A}}$. Indeed, our only direct estimate for $d_n(\mathscr{U}_{\mathscr{A}})_{H_0^1(D)}$ in terms of $d_n(\mathscr{A})_{L_2(D)}$ is that given by Theorem 3.1. But multiplying $d_n(\mathscr{A})_{L_2(D)}$ by n does not give a decay to zero, so Theorem 3.1 gives a useless estimate. So, it remains an open problem to determine the n-width of $\mathscr{U}_{\mathscr{A}}$ for the geometric model.

3.7 ▪ Numerical methods for parametric equations

Let us now turn our attention to numerical methods for solving a family of parametric equations with diffusion coefficients coming from the set $\mathscr{A}$. We wish to construct a numerical solver such that given a query $a \in \mathscr{A}$, it produces a function $\hat{u}_a$ that is a good approximation to u_a in the $H_0^1(D)$ norm.

3.7.1 ▪ Online and offline costs

This construction of a solver is decoupled into two tasks.

Offline cost

This is the cost of developing the numerical method that is tailor made for $\mathscr{A}$. For example, in linear methods, it is the computational cost needed to find a good n-dimensional subspace V_n contained in $H_0^1(D)$, which will be used to build an approximation to u_a when given a query $a \in \mathscr{A}$. For a nonlinear method based on n-term approximation from a dictionary of size N, it would be the cost of finding a good dictionary. Notice that the offline cost is a one-time investment for the class $\mathscr{A}$ and does not include the task of actually finding an approximation to u_a given a query $a \in \mathscr{A}$.

We should mention that in some applications, we are not so much interested in finding an approximation $\hat{u}_a$ to u_a as we are in the evaluation of a linear functional ℓ on $H_0^1(D)$ to the solution u_a. This is then the problem of developing a numerical method that approximates the real-valued function $L(a) = \ell(u_a)$ given a query $a \in \mathscr{A}$. In this case, the offline cost would include building an approximation $\hat{L}$ to L. We will not say much more about this important second problem.

Online cost

This is the cost of implementing the solver that was built offline for finding an approximation $\hat{u}_a$ to u_a given a query a. For linear methods, this approximation is usually taken as the Galerkin projection onto V_n, although other projectors may also be reasonable in some scenarios. The Galerkin projector gives the best approximation to u_a from V_n in the $H_0^1(D,a)$-norm. Given the linear space V_n, the Galerkin projection constructs $\hat{u}_a \in V_n$ as the solution to the discrete system of equations

$$\langle \hat{u}_a, v \rangle_a = \langle f, v \rangle \quad \forall v \in V_n, \tag{3.81}$$

where $\langle \cdot, \cdot \rangle_a$ is the $H_0^1(D,a)$ inner product. If we choose a basis $\varphi_1, \dots, \varphi_n$ for V_n, then $\hat{u}_a = \sum_{j=1}^n c_j \varphi_j$, where the coefficients $\mathbf{c} = (c_j)_{j=1}^n$ satisfy

$$A\mathbf{c} = \mathbf{f}, \tag{3.82}$$

where $A = (a_{ij})_{i,j=1}^n$, $a_{ij} := \langle \varphi_i, \varphi_j \rangle_a$, is the so-called stiffness matrix and $\mathbf{f} := (f_i)_{i=1}^n$, with $f_i := \langle f, \varphi_i \rangle$, $i = 1, \dots, n$, is the discretization of the right side f. From the ellipticity assumption, the matrix A is positive definite and so the system is efficiently solved using standard numerical solvers for linear systems. The performance of this numerical method is usually measured by the error in the $H_0^1(D,a)$-norm:

$$\|u_a - \hat{u}_a\|_{H_0^1(D,a)} = \operatorname{dist}(u_a, V_n)_{H_0^1(D,a)} \approx \operatorname{dist}(u_a, V_n)_{H_0^1(D)}. \tag{3.83}$$

When the Galerkin projection is used, then given the query $a \in \mathscr{A}$, one must assemble the stiffness matrix and then solve the corresponding matrix problem. While the assembly of the matrix is a serious problem, we will largely ignore it here since we have nothing to add over what is already known.

Relevance of entropy in online/offline comparisons

Entropy suggests the following extreme setting for online/offline comparisons. Let $\epsilon := C\epsilon_n(\mathscr{A})_{L_q(D)}$ and numerically find a cover $\{B(a_i, \epsilon)\}_{i=1}^N$, $N = 2^n$, for $\mathscr{A}$ and then compute the solutions u_{a_i}, $i = 1, \dots, N$, offline. Since $\mathscr{A}$ is known to us, we can usually find the cover of $\mathscr{A}$ by some form of piecewise-polynomial approximation followed by quantization of the coefficients of the approximation. The offline cost would be proportional to Nm, where m is the computational cost to employ an off-the-shelf solver to find an approximation $\hat{u}_{a_i}$ to u_{a_i} accurate to error ϵ.

For the online computation, given a query a, we find an approximation a_i to a using piecewise-polynomial approximation and quantization. This is then followed by a look-up table to find the approximation $\hat{u}_{a_i}$ of u_a. The accuracy of this method is $C\epsilon_n(\mathscr{A})_{L_q(D)}$. The offline cost is exceedingly high, but the online cost is very low since it does not involve a PDE solve.

The point of this example is to show that it is not only the online cost of the solver that matters. One has to take into consideration the offline investment cost which is extreme in the above example. It is not clear exactly how this balancing should be done, but it would be beneficial to quantify it in some way to advance the theory of RB methods.

3.7.2 ▪ Finding a good linear subspace

The central question in developing linear methods is how to find a good choice for the finite-dimensional space V_n. Since we want V_n to be used for all $a \in \mathcal{A}$, it should be efficient at approximating all of the elements in $\mathcal{U}_{\mathcal{A}}$. Recall that all of the norms $\|\cdot\|_{H_0^1(D,a)}$ are equivalent to $\|\cdot\|_{H_0^1(D)}$. So, essentially, the best choice for V_n is a subspace of $H_0^1(D)$ that achieves the Kolmogorov width $d_n(\mathcal{U}_{\mathcal{A}})_{H_0^1(D)}$. Of course, finding such a subspace may be difficult, but it serves as a benchmark for the optimal performance we can expect.

One of the most prominent and well studied methods for finding a good subspace is the RB method [6, 22, 23, 25, 26, 28]. The general philosophy of such methods is that one is willing to incur high computational costs to determine a good subspace V_n offline. Typically, the space V_n is spanned by n functions $u_{a_i} \in \mathcal{U}_{\mathcal{A}}$, $i = 1, \dots, n$. These functions are called *snapshots* of $\mathcal{U}_{\mathcal{A}}$. The most popular method for finding these snapshots is the following intriguing *greedy algorithm*, introduced first in [28]. While we are primarily interested in this algorithm in the case of a compact set K of a Hilbert space (in our case $K = \mathcal{U}_{\mathcal{A}}$ and the Hilbert space is $H_0^1(D)$), we will formulate the algorithm for any Banach space X.

Let X be a Banach space with norm $\|\cdot\| := \|\cdot\|_X$, and let K be one of its compact subsets. For notational convenience only, we shall assume that the elements f of K satisfy $\|f\|_X \le 1$. We consider the following greedy algorithm for generating approximation spaces for K.

Pure greedy algorithm

We first choose a function f_0 such that

$$\|f_0\| = \max_{f\in K} \|f\|. \tag{3.84}$$

Assuming $\{f_0, \dots, f_{n-1}\}$ and $V_n := \operatorname{span}\{f_0, \dots, f_{n-1}\}$ have been selected, we then take $f_n \in K$ such that

$$\operatorname{dist}(f_n, V_n)_X = \max_{f\in K} \operatorname{dist}(f, V_n)_X \tag{3.85}$$

and define

$$\sigma_n := \sigma_n(K)_X := \operatorname{dist}(f_n, V_n)_X := \sup_{f\in K} \inf_{g\in V_n} \|f - g\|. \tag{3.86}$$

This greedy algorithm was introduced for the case where X is a Hilbert space in [22, 23]. In numerical settings, one cannot find the f_j exactly, and estimates for the error needed in this algorithm are not known precisely. This leads one to consider weaker forms of this algorithm that match better the application.

Weak greedy algorithm. We fix a constant $0 < \gamma \le 1$. At the first step of the algorithm, we choose a function $f_0 \in K$ such that

$$\|f_0\| \ge \gamma \sigma_0(K)_X := \max_{f\in K} \|f\|.$$

At the general step, if $f_0, \dots, f_{n-1}$ have been chosen, we set $V_n := \operatorname{span}\{f_0, \dots, f_{n-1}\}$ and

$$\sigma_n(f)_X := \operatorname{dist}(f, V_n)_X.$$

We now choose $f_n \in \mathcal{F}$ to be the next element in the greedy selection such that

$$\sigma_n(f_n)_X \geq \gamma \max_{f \in K} \sigma_n(f)_X. \tag{3.87}$$

Note that if $\gamma = 1$, then the weak greedy algorithm reduces to the greedy algorithm that we introduced above.

Notice that similar to the greedy algorithm, $(\sigma_n(K)_X)_{n \geq 0}$ is also monotone decreasing. It is also important to note that neither the pure greedy algorithm nor the weak greedy algorithm gives a unique sequence $(f_n)_{n \geq 0}$, nor is the sequence $(\sigma_n(K)_X)_{n \geq 0}$ unique. In all that follows, the notation reflects any sequences that can arise in the implementation of the weak greedy selection for the fixed value of γ.

3.7.3 ▪ Performance of the weak greedy algorithm

We are interested in how well the space V_n generated by the weak greedy algorithm approximates the elements of K. For this purpose we would like to compare its performance with the best possible performance, given by the Kolmogorov width $d_n(K)_X$ of K. Of course, if $(\sigma_n)_{n \geq 0}$ decays at a rate comparable to $(d_n)_{n \geq 0}$, this would mean that the greedy selection provides essentially the best possible accuracy attainable by n-dimensional subspaces. Various comparisons have been given between σ_n and d_n. An early result in this direction, in the case that X is a Hilbert space $\mathcal{H}$, is given in [6], where it is proved that

$$\sigma_n(K)_{\mathcal{H}} \leq C n 2^n d_n(K)_{\mathcal{H}}, \tag{3.88}$$

with C an absolute constant. While this is an interesting comparison, it is only useful if $d_n(K)_{\mathcal{H}}$ decays to zero faster than $n^{-1}2^{-n}$.

Various improvements on (3.88) are given in [1], again in the Hilbert space setting. We mention two of these. It is shown that if $d_n(K)_{\mathcal{H}} \leq C n^{-\alpha}$, $n = 1, 2, \ldots$, then

$$\sigma_n(K)_{\mathcal{H}} \leq C'_\alpha n^{-\alpha}. \tag{3.89}$$

This shows that in the scale of polynomial decay the greedy algorithm performs with the same rates as n-widths. A related result is proved for subexponential decay. If for some $0 < \alpha \leq 1$, we have $d_n(K)_{\mathcal{H}} \leq C e^{-cn^\alpha}$, $n = 1, 2, \ldots$, then

$$\sigma_n(K)_{\mathcal{H}} \leq C'_\alpha e^{-c'_\alpha n^\beta}, \quad \beta = \frac{\alpha}{\alpha + 1}, \quad n = 1, 2, \ldots. \tag{3.90}$$

These results are improved in [15] and extended to the case of a general Banach space X, as we are now discussing. We will outline what is known in this direction and sketch how these results are proved in the following section.

3.7.4 ▪ Results for a Banach space

The analysis of the greedy algorithm is quite simple and executed with elementary results from linear algebra. We provide some details since this may help develop the intuition of the reader. A core result for the analysis of greedy algorithms is the following lemma from [15].

Lemma 3.5. *Let $G=(g_{i,j})$ be a $K\times K$ lower triangular matrix with rows $\mathbf{g}_1,\dots,\mathbf{g}_K$, W be any m-dimensional subspace of $\mathbb{R}^K$, and P be the orthogonal projection of $\mathbb{R}^K$ onto W. Then,*

$$\prod_{i=1}^K g_{i,i}^2 \le \left\{\frac{1}{m}\sum_{i=1}^K \|P\mathbf{g}_i\|_{\ell_2}^2\right\}^m \left\{\frac{1}{K-m}\sum_{i=1}^K \|\mathbf{g}_i - P\mathbf{g}_i\|_{\ell_2}^2\right\}^{K-m}, \tag{3.91}$$

where $\|\cdot\|_{\ell_2}$ is the Euclidean norm of a vector in $\mathbb{R}^K$.

Proof. We choose an orthonormal basis $\varphi_1,\dots,\varphi_m$ for the space W and complete it into an orthonormal basis $\varphi_1,\dots,\varphi_K$ for $\mathbb{R}^K$. If we denote by Φ the $K\times K$ orthogonal matrix whose jth column is φ_j, then the matrix $C:=G\Phi$ has entries $c_{i,j}=\langle \mathbf{g}_i,\varphi_j\rangle$. We denote by $\mathbf{c}_j$ the jth column of C. It follows from the arithmetic geometric mean inequality for the numbers $\{\|\mathbf{c}_j\|_{\ell_2}^2\}_{j=1}^m$ that

$$\prod_{j=1}^m \|\mathbf{c}_j\|_{\ell_2}^2 \le \left\{\frac{1}{m}\sum_{j=1}^m \|\mathbf{c}_j\|_{\ell_2}^2\right\}^m = \left\{\frac{1}{m}\sum_{j=1}^m\sum_{i=1}^K \langle \mathbf{g}_i,\varphi_j\rangle^2\right\}^m = \left\{\frac{1}{m}\sum_{i=1}^K \|P\mathbf{g}_i\|_{\ell_2}^2\right\}^m. \tag{3.92}$$

Similarly,

$$\prod_{j=m+1}^K \|\mathbf{c}_j\|_{\ell_2}^2 \le \left\{\frac{1}{K-m}\sum_{j=m+1}^K \|\mathbf{c}_j\|_{\ell_2}^2\right\}^{K-m} = \left\{\frac{1}{K-m}\sum_{i=1}^K \|\mathbf{g}_i - P\mathbf{g}_i\|_{\ell_2}^2\right\}^{K-m}. \tag{3.93}$$

Now, Hadamard's inequality for the matrix C and relations (3.92) and (3.93) result in

$$(\det C)^2 \le \prod_{j=1}^K \|\mathbf{c}_j\|_{\ell_2}^2 \le \left\{\frac{1}{m}\sum_{i=1}^K \|P\mathbf{g}_i\|_{\ell_2}^2\right\}^m \left\{\frac{1}{K-m}\sum_{i=1}^K \|\mathbf{g}_i - P\mathbf{g}_i\|_{\ell_2}^2\right\}^{K-m}. \tag{3.94}$$

The latter inequality and the fact that $\det G=\prod_{i=1}^K g_{i,i}$ and $|\det C|=|\det G|$ give (3.91). □

Let us now see how this lemma is utilized to derive convergence results for the greedy algorithm. We will for the moment restrict ourselves to the case of a Hilbert space and the weak greedy algorithm with constant γ. Later, we shall say what changes are made when X is a general Banach space.

Note that in general, the weak greedy algorithm does not terminate and we obtain an infinite sequence $f_0,f_1,f_2,\dots$. For consistent notation in what follows, we shall define $f_m:=0$, $m>N$, if the algorithm terminates at N, i.e., if $\sigma_N(K)_{\mathcal{H}}=0$. By $(f_n^*)_{n\ge 0}$ we denote the orthonormal system obtained from $(f_n)_{n\ge 0}$ by Gram–Schmidt orthogonalization. It follows that the orthogonal projector P_n from $\mathcal{H}$ onto V_n is given by

$$P_n f = \sum_{i=0}^{n-1}\langle f,f_i^*\rangle f_i^*,$$

and, in particular,

$$f_n = P_{n+1}f_n = \sum_{j=0}^{n} a_{n,j}f_j^*, \quad a_{n,j}=\langle f_n,f_j^*\rangle,\ j\le n.$$

There is no loss of generality in assuming that the infinite-dimensional Hilbert space $\mathscr{H}$ is $\ell_2(\mathbb{N}\cup\{0\})$ and that $f_j^* = e_j$, where e_j is the vector with a one in the coordinate indexed by j and zeros in all other coordinates, i.e., $(e_j)_i = \delta_{j,i}$.

We consider the lower triangular matrix

$$A := (a_{i,j})_{i,j=0}^{\infty}, \quad a_{i,j} := 0, j > i.$$

This matrix incorporates all the information about the weak greedy algorithm on K. The following two properties characterize any lower triangular matrix A generated by the weak greedy algorithm with constant γ. We use the notation $\sigma_n := \sigma_n(K)_{\mathscr{H}}$.

P1. *The diagonal elements of A satisfy $\gamma\sigma_n \le |a_{n,n}| \le \sigma_n$.*

P2. *For every $m \ge n$, one has $\sum_{j=n}^{m} a_{m,j}^2 \le \sigma_n^2$.*

Indeed, **P1** follows from

$$a_{n,n}^2 = \|f_n\|^2 - \|P_n f_n\|^2 = \|f_n - P_n f_n\|^2,$$

combined with the weak greedy selection property (3.87). To see **P2**, we note that for $m \ge n$,

$$\sum_{j=n}^{m} a_{m,j}^2 = \|f_m - P_n f_m\|^2 \le \max_{f\in K} \|f - P_n f\|^2 = \sigma_n^2.$$

Remark 3.6. If A is any matrix satisfying **P1** and **P2** with $(\sigma_n)_{n\ge 0}$ a decreasing sequence that converges to zero, then the rows of A form a compact subset of $\ell_2(\mathbb{N}\cup\{0\})$. If K is the set consisting of these rows, then one of the possible realizations of the weak greedy algorithm with constant γ will choose the rows in that order and A will be the resulting matrix.

Theorem 3.7. *For the weak greedy algorithm with constant γ in a Hilbert space $\mathscr{H}$ and for any compact set K, we have the following inequalities between $\sigma_n := \sigma_n(K)_{\mathscr{H}}$ and $d_n := d_n(K)_{\mathscr{H}}$ for any $N \ge 0$, $J \ge 1$, and $1 \le m < J$:*

$$\prod_{i=1}^{J} \sigma_{N+i}^2 \le \gamma^{-2J} \left\{\frac{J}{m}\right\}^m \left\{\frac{J}{J-m}\right\}^{J-m} \sigma_{N+1}^{2m} d_m^{2J-2m}. \tag{3.95}$$

Proof. We consider the $J\times J$ matrix $G = (g_{i,j})$ formed by the rows and columns of A with indices from $\{N+1,\dots,N+J\}$. Each row $\mathbf{g}_i$ is the restriction of f_{N+i} to the coordinates $N+1,\dots,N+J$. Let $\mathscr{H}_m$ be the m-dimensional Kolmogorov subspace of $\mathscr{H}$ for which $\mathrm{dist}(K,\mathscr{H}_m) = d_m$. Then, $\mathrm{dist}(f_{N+i},\mathscr{H}_m) \le d_m$, $i = 1,\dots,J$. Let $\widetilde{W}$ be the linear space that is the restriction of $\mathscr{H}_m$ to the coordinates $N+1,\dots,N+J$. In general, $\dim(\widetilde{W}) \le m$. Let W be an m-dimensional space, $W \subset \mathrm{span}\{e_{N+1},\dots,e_{N+J}\}$, such that $\widetilde{W} \subset W$ and P and $\widetilde{P}$ are the projections in $\mathbb{R}^K$ onto W and $\widetilde{W}$, respectively. Clearly,

$$\|P\mathbf{g}_i\|_{\ell_2} \le \|\mathbf{g}_i\|_{\ell_2} \le \sigma_{N+1}, \quad i = 1,\dots,J, \tag{3.96}$$

where we have used property **P2** in the last inequality. Note that

$$\|\mathbf{g}_i - P\mathbf{g}_i\|_{\ell_2} \le \|\mathbf{g}_i - \widetilde{P}\mathbf{g}_i\|_{\ell_2} = \mathrm{dist}(\mathbf{g}_i, \widetilde{W}) \le \mathrm{dist}(f_{N+i},\mathscr{H}_m) \le d_m, \quad i = 1,\dots,J. \tag{3.97}$$

It follows from property **P1** that

$$\prod_{i=1}^{J}|a_{N+i,N+i}| \geq \gamma^J \prod_{i=1}^{J}\sigma_{N+i}. \tag{3.98}$$

We now apply Lemma 3.5 for this G and W and use estimates (3.96), (3.97), and (3.98) to derive (3.95). □

Let us now indicate how one derives some of the performance results for the greedy algorithm from this theorem.

Corollary 3.8. *For the weak greedy algorithm with constant γ in a Hilbert space $\mathscr{H}$, we have the following:*

(i) *For any compact set K and $n \geq 1$, we have*

$$\sigma_n(K) \leq \sqrt{2}\gamma^{-1} \min_{1\leq m<n} d_m^{\frac{n-m}{n}}(K). \tag{3.99}$$

In particular, $\sigma_{2n}(K) \leq \sqrt{2}\gamma^{-1}\sqrt{d_n(K)}$, $n = 1,2,\ldots$.

(ii) *If $d_n(K) \leq C_0 n^{-\alpha}$, $n = 1,2,\ldots$, then $\sigma_n(K) \leq C_1 n^{-\alpha}$, $n = 1,2,\ldots$, with $C_1 := 2^{5\alpha+1}\gamma^{-2}C_0$.*

(iii) *If $d_n(K) \leq C_0 e^{-c_0 n^\alpha}$, $n = 1,2,\ldots$, then $\sigma_n(K) \leq \sqrt{2C_0}\gamma^{-1}e^{-c_1 n^\alpha}$, $n = 1,2,\ldots$, where $c_1 = 2^{-1-2\alpha}c_0$,*

Proof. (i) We take $N = 0$, $J = n$, and any $1 \leq m < n$ in Theorem 3.7 and use the monotonicity of $(\sigma_n)_{n\geq 0}$ and the fact that $\sigma_0 \leq 1$ to obtain

$$\sigma_n^{2n} \leq \prod_{j=1}^{n}\sigma_j^2 \leq \gamma^{-2n}\left\{\frac{n}{m}\right\}^m \left\{\frac{n}{n-m}\right\}^{n-m} d_m^{2n-2m}. \tag{3.100}$$

Since $x^{-x}(1-x)^{x-1} \leq 2$ for $0 < x < 1$, we derive (3.99).

(ii) It follows from the monotonicity of $(\sigma_n)_{n\geq 0}$ and (3.95) for $N = J = n$ and any $1 \leq m < n$ that

$$\sigma_{2n}^{2n} \leq \prod_{j=n+1}^{2n}\sigma_j^2 \leq \gamma^{-2n}\left\{\frac{n}{m}\right\}^m \left\{\frac{n}{n-m}\right\}^{n-m} \sigma_n^{2m} d_m^{2n-2m}.$$

In the case $n = 2s$ and $m = s$ we have

$$\sigma_{4s} \leq \sqrt{2}\gamma^{-1}\sqrt{\sigma_{2s}d_s}. \tag{3.101}$$

Now we prove our claim by contradiction. Suppose it is not true and M is the first value where $\sigma_M(\mathscr{F}) > C_1 M^{-\alpha}$. Let us first assume $M = 4s$. From (3.101), we have

$$\sigma_{4s} \leq \sqrt{2}\gamma^{-1}\sqrt{C_1(2s)^{-\alpha}}\sqrt{C_0 s^{-\alpha}} = \sqrt{2^{1-\alpha}C_0 C_1}\gamma^{-1}s^{-\alpha}, \tag{3.102}$$

where we have used the fact that $\sigma_{2s} \leq C_1(2s)^{-\alpha}$ and $d_s \leq C_0 s^{-\alpha}$. It follows that

$$C_1(4s)^{-\alpha} < \sigma_{4s} \leq \sqrt{2^{1-\alpha}C_0 C_1}\gamma^{-1}s^{-\alpha},$$

and therefore

$$C_1 < 2^{3\alpha+1}\gamma^{-2}C_0 < 2^{5\alpha+1}\gamma^{-2}C_0,$$

which is the desired contradiction. If $M = 4s + q$, $q \in \{1,2,3\}$, then it follows from (3.102) and the monotonicity of $(\sigma_n)_{n\geq 0}$ that

$$C_1 2^{-3\alpha} s^{-\alpha} = C_1 2^{-\alpha}(4s)^{-\alpha} < C_1(4s+q)^{-\alpha} < \sigma_{4s+q} \leq \sigma_{4s} \leq \sqrt{2^{1-\alpha} C_0 C_1}\gamma^{-1} s^{-\alpha}.$$

From this we obtain

$$C_1 < 2^{5\alpha+1}\gamma^{-2} C_0,$$

which is the desired contradiction in this case. This completes the proof of (ii).

(iii) From (i), we have

$$\sigma_{2n+1} \leq \sigma_{2n} \leq \sqrt{2}\gamma^{-1}\sqrt{d_n} \leq \sqrt{2C_0}\gamma^{-1} e^{-\frac{c_0}{2}n^\alpha} = \sqrt{2C_0}\gamma^{-1} e^{-c_0 2^{-1-\alpha}(2n)^\alpha}, \tag{3.103}$$

from which (iii) easily follows. □

Let us now comment on what happens when X is a general Banach space. The analysis is quite similar to that above (see [15]), however, there is some loss in the approximation rate. The precise results are as follows:

(i) For any $n \geq 1$ we have $\sigma_{2n} \leq 2\gamma^{-1}\sqrt{n d_n}$.

(ii) If for $\alpha > 0$, we have $d_n \leq C_0 n^{-\alpha}$, $n = 1,2,\ldots$, then for any $0 < \beta < \min\{\alpha, 1/2\}$, we have $\sigma_n \leq C_1 n^{-\alpha+1/2+\beta}$, $n = 1,2,\ldots$, with

$$C_1 := \max\left\{C_0 4^{4\alpha+1}\gamma^{-4}\left(\frac{2\beta+1}{2\beta}\right)^\alpha, \max_{n=1,\ldots,7}\{n^{\alpha-\beta-1/2}\}\right\}.$$

(iii) If for $\alpha > 0$, we have $d_n \leq C_0 e^{-c_0 n^\alpha}$, $n = 1,2,\ldots$, then $\sigma_n < \sqrt{2C_0}\gamma^{-1}\sqrt{n} e^{-c_1 n^\alpha}$, $n = 1,2,\ldots$, where $c_1 = 2^{-1-2\alpha} c_0$. The factor $\sqrt{n}$ can be deleted by reducing the constant c_1.

In particular, we see that in the estimates (i) and (ii), we lose a factor $\sqrt{n}$ in the approximation rate when compared with the Hilbert space case. It can be shown that in general, this loss cannot be avoided [15].

3.7.5 ▪ Practical considerations in the offline implementation of greedy algorithms

Let us now return to the application of the above greedy algorithms to our parametric PDE problem. On first glance, it appears that the offline implementation of this algorithm is computationally not feasible, since it requires an accurate estimate of $\|u_a - P_{V_n} u_a\|_{H_0^1(D)}$ for all $a \in \mathcal{A}$. On the surface, this would require solving (3.1) for each a, which is of course what we are trying to avoid. Fortunately, as is well known, this norm is equivalent to

$$S(a) := \|f - P_n u_a\|_{H^{-1}(D)}, \tag{3.104}$$

which can be computed (since both f and $P_n u_a$ are available) without computing u_a. (We do not discuss the role of the constants in this equivalence, even though they are an important issue.) We are still left with the problem of having to calculate this surrogate quantity for all a. What we do in practice is the following.

We know that whatever accuracy we have for the discretization of $\mathcal{A}$ in $L_\infty(D)$ (or $L_q(D)$) will be inherited by $\mathcal{U}_{\mathcal{A}}$ because of (3.8) (or (3.11)). Suppose a discretization $\tilde{\mathcal{A}}$ of $\mathcal{A}$ has accuracy ϵ and we find an $a^* \in \tilde{\mathcal{A}}$ such that

$$S(a^*) \geq C_0 \epsilon, \tag{3.105}$$

with C_0 an appropriately large fixed constant (determined by the equivalency constants for the surrogate). Then, we are guaranteed that this discretization is accurate enough for the implementation of the weak greedy algorithm. Hence, we start with a coarse discretization of $\mathscr{A}$ and then decrease the resolution ϵ of the discretization until (3.105) is satisfied.

3.7.6 ▪ Summary of performance of RBs

Let us summarize what we know about the performance of the weak greedy algorithm for our two sample model classes.

Affine model class (see Section 3.3.1)

Assume that $(\|\psi_j\|_{L_\infty(D)}) \in \ell_p$ for some $p < 1$. Then, we know the following:

(i) $d_n(\mathscr{A})_{L_\infty(D)} \leq C n^{1-1/p}$, $n \geq 1$.
(ii) $d_n(\mathscr{U}_{\mathscr{A}})_{H_0^1(D)} \leq C n^{1-1/p}$, $n \geq 1$.
(iii) The weak greedy algorithm generates spaces V_n such that $\sigma_n(\mathscr{U}_{\mathscr{A}})_{H_0^1(D)} \leq C n^{1-1/p}$, $n \geq 1$.

Let us mention that an alternative to the construction of a good basis using the weak greedy algorithm is to utilize a selection based on monotone sets (also known as lower sets), as discussed in [11].

Geometric model class (see Section 3.3.2)

Assume that the stability inequality (3.11) holds for a value of $q \in [2, \infty)$. Then, we know the following:

(i) $d_n(\mathscr{A})_{L_q(D)} \leq C n^{-\frac{1}{2q}}$, $n \geq 1$.
(ii) We do not know any estimate of the form

$$d_n(\mathscr{U}_{\mathscr{A}})_{H_0^1(D)} \leq C n^{-\alpha}, \quad n \geq 1, \tag{3.106}$$

for a value of $\alpha > 0$.
(iii) If we could prove an estimate (3.106), then the weak greedy algorithm would generate spaces V_n such that $\sigma_n(\mathscr{U}_{\mathscr{A}})_{H_0^1(D)} \leq C n^{-\alpha}$, $n \geq 1$.

3.8 ▪ Nonlinear methods in RBs

Generally speaking, there is a large benefit to using nonlinear approximation in the construction of numerical methods for PDEs. Several interesting nonlinear approaches are emerging: hp-RB for elliptic problems [16], [17]; adaptive parameter partitioning [18]; RB selection from dictionaries [20]; and local greedy by parameter distance [24]. We are not aware of any definitive a priori analysis of such algorithms that would substantiate the use of nonlinear methods. In this section, we make some heuristic comments about the possible utilization of nonlinear methods in reduced modeling.

For the affine model, at first glance, there seems to be no advantage in using nonlinear methods since the manifold $\mathscr{U}_{\mathscr{A}}$ is provably smooth. However, it may be that the smoothness is stronger in certain parts of the parameter domain. This means that there may be an advantage to partitioning the parameter domain and using a different linear space on each set of the partition. Concerning the geometric model, it seems ripe for the exploitation of nonlinear methods. We consider only this geometric example in what follows in this section. We have seen that the linear Kolmogorov widths of $\mathscr{A}$ satisfy

$$d_n(\mathscr{A})_{L_2(D)} \geq C n^{-1/4}, \quad n \geq 1. \tag{3.107}$$

This is a good indication that the same lower bound holds for the widths of $\mathscr{U}_{\mathscr{A}}$ in $H_0^1(D)$, although, as we pointed out in the last section, no such results have actually been proved.

3.8.1 ▪ Entropy numbers for the geometric model

As we already noted in (3.74), the entropy numbers $\epsilon_n(\mathscr{A})_{L_q(D)}$ behave like $n^{-1/q}$. It follows from our comparison (3.43) that the entropy numbers $\epsilon_n(\mathscr{U}_{\mathscr{A}})_{H_0^1(D)} \leq C n^{-1/q}$. When $q = 2$, this shows that a nonlinear performance of order $O(n^{-1/2})$ is expected for approximating $\mathscr{U}_{\mathscr{A}}$.

3.8.2 ▪ Nonlinear n-widths for the geometric model

Let us begin with our usual strategy of first trying to understand the nonlinear widths of $\mathscr{A}$ in $L_2(D)$. We have already discussed the dictionary $\mathscr{D}$, which consists of the n^2 functions χ_R, where $R = [(i-1)/n, i/n) \times [0, j/n]$, $1 \leq i, j \leq n$. We have pointed out that each function $a \in \mathscr{A}$ can be approximated to accuracy $C n^{-1/2}$ in $L_2(D)$ by using n elements of $\mathscr{D}$. Namely, any function $a \in \mathscr{A}$ can be approximated by a sum $1+\sum_{R\in\Lambda} \chi_R$ with $\#(\Lambda) = n$ to accuracy $C n^{-1/2}$ in $L_2(D)$. It follows that the dictionary widths defined in (3.36) satisfy

$$d_{n,n^2}(\mathscr{A})_{L_2(D)} \leq C n^{-1/2}, \quad n \geq 1. \tag{3.108}$$

One disadvantage in using the dictionary $\mathscr{D}$ when approximating the elements of $\mathscr{A}$ is that the dictionary elements themselves are not in $\mathscr{A}$. However, it is possible to introduce another dictionary, $\mathscr{D}_0$, with n^2 functions that actually come from $\mathscr{A}$, and when using n-term approximation from $\mathscr{D}_0$ to approximate the elements of $\mathscr{A}$, it still achieves the bound $C n^{-1/2}$ for error measured in $L_2(D)$. Namely, for each point $(i/n, j/n) \in D$, we associate the function $\phi_{i,j}$, which is the characteristic of the region depicted in Figure 3.3. We let $\mathscr{D}_0 := \{\phi_{i,j},\ 1 \leq i, j \leq n\}$. It is easy to see that any $a \in \mathscr{A}$ can be approximated to accuracy $C n^{-1/2}$ by $1+\chi_S$, where S is the region under a piecewise-linear function, which always has slopes ± 1. Such a piecewise function can be written as a linear combination of n-terms from $\mathscr{D}_0$ (see Figure 3.4).

Given the above results for the dictionary n-width of $\mathscr{A}$, one expects correspondingly improved results for $d_{n,n^2}(\mathscr{U}_{\mathscr{A}})_{H_0^1(D)}$. Unfortunately, they do not follow from anything we know. The comparison (3.60) is too debilitating in this case since the factor n kills the decay rate $(n^{-1/2})$ in (3.108). We expect, however, that this is just a defect of our current state of knowledge and that the following problem will have a positive solution.

Open problem. Find n^2 snapshots of $\mathscr{U}_{\mathscr{A}}$ such that any u_a can be approximated to accuracy $C n^{-1/2}$ in $H_0^1(D)$ by using only n of these snapshots. ∎

Let us now turn our discussion to manifold widths. The above dictionary widths are not of the form of nonlinear approximation considered in the definition of the nonlinear manifold width $\delta_n(K)_{L_2(D)}$. This can be remedied, as described in [14], by using

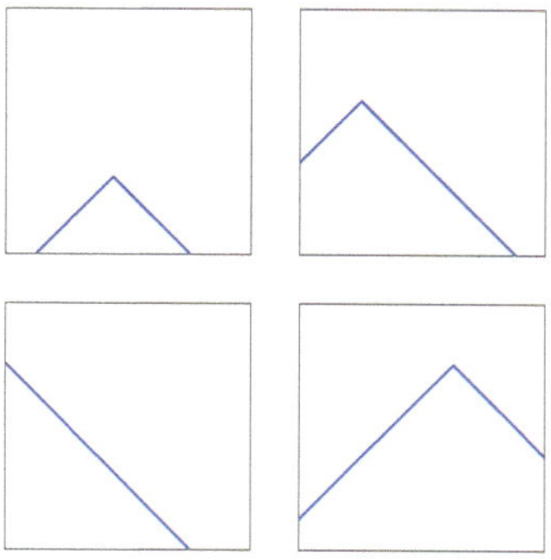

Figure 3.3. *The basis functions $\phi_{i,j}$ with vertex $(i/n, j/n)$. The line segments have slope ± 1.*

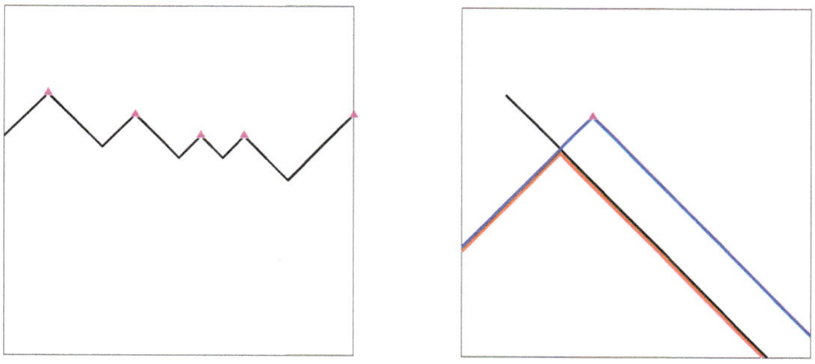

Figure 3.4. *On the left is a typical piecewise-linear function with slopes ± 1 and on the right is a sample decomposition (the region below the red curve).*

the famous Pontryagin–Nöbling lemma on topological embeddings of complexes. We do not go into this in detail here but remark that it allows us to construct mappings b and M of the form described above that achieve the same rate $O(n^{-1/2})$. In other words, we have

$$\delta_n(\mathscr{A})_{L_2(D)} \leq C n^{-1/2}. \tag{3.109}$$

With this result in hand, we can use our general theory to see that at least for certain right sides f, we have

$$\delta_n(\mathcal{U}_{\mathscr{A}}) \leq \delta_n(\mathscr{A})_{L_2(D)} \leq C n^{-1/2}. \tag{3.110}$$

Indeed, this follows from (3.65) provided we show that $\mathscr{A}$ satisfies the uniqueness assumption.

The following simple argument (provided to me by Andrea Bonito) shows that the uniqueness assumption holds for our geometric class $\mathscr{A}$ whenever the right side f is nonzero almost everywhere. By a disjoint open finite covering (DOFC) of D, we mean a collection of open sets D_j, $j = 1,\ldots,K$, such that $D = \cup_{j=1}^K \overline{D_j}$, D_j open, and $D_i \cap D_j = \emptyset$ for $i \neq j$. Define

$$\mathscr{A}_0 := \left\{ a = \sum_{j=1}^{K} c_j \chi_{D_j} \, : \, c_j \geq r > 0, \; j = 1,\ldots,K \right\}.$$

Then, clearly $\mathscr{A} \subset \mathscr{A}_0$.

Now, for any $a \in \mathscr{A}_0$, define $u_a \in H_0^1(\Omega)$ as satisfying

$$\int_\Omega a \nabla u_a \cdot \nabla v = \int_\Omega f v \qquad \forall v \in H_0^1(D).$$

Let $a \in \mathscr{A}_0$ and notice that for each set Ω_i of the DOFC and for each $x \in D_i$, there exists a ball $B(x,\delta)$ of radius $\delta > 0$ centered at x such that $B(x,\delta) \subset\subset D_i$. In this ball, a is constant and the interior regularity property for the elliptic problem implies that $u_a \in C^\infty(B(x,\delta))$. In particular, a satisfies $-a\Delta u_a = f$ in $B(x,\delta)$, which in turn implies $\Delta u_a \neq 0$ almost everywhere (a.e.) in D and

$$a = -f / \Delta u_a \qquad \text{a.e. in } D.$$

Now, given a and $\hat{a}$ in $\mathscr{A}$ with $u_a = u_{\hat{a}} \in H_0^1(D)$, we realize using the above representation that $a = \hat{a}$ a.e.

While the above result is a nice theoretical consequence, it does not provide a reasonable numerical recipe for using nonlinear methods to solve the system of parametric PDEs for the geometric model, even if used in conjunction with the n-term approximation from the dictionary of n^2 elements used to approximate $\mathscr{A}$. Indeed, given a query a, it would identify an $\hat{a}$ that is a good n-term approximation to a but ask to solve for $u_{\hat{a}}$. This would not allow the offline preparation of n^2 snapshots from which an n-term approximation would be constructed for u_a. This returns us to the open problem stated above (3.109).

Bibliography

[1] P. Binev, A. Cohen, W. Dahmen, R. DeVore, G. Petrova, and P. Wojtaszczyk, *Convergence rates for greedy algorithms in reduced basis methods*, SIAM J. Math. Anal., 43 (2011), pp. 1457–1472.

[2] P. Binev, W. Dahmen, and R. DeVore, *Adaptive finite element methods with convergence rates*, Numer. Math., 97 (2004), pp. 219–268.

[3] P. Binev, W. Dahmen, R. DeVore, and P. Petrushev, *Approximation classes for adaptive methods*, Serdica Math. J., 28 (2002), pp. 391–416.

[4] A. Bonito, A. Cohen, R. DeVore, G. Petrova, and G. Welper, *Diffusion coefficients estimation for elliptic partial differential equations*, SIAM J. Math. Anal., 49 (2017), pp. 1570–1592.

[5] A. Bonito, R. DeVore, and R. Nochetto, *Adaptive finite element methods for elliptic problems with discontinuous coefficients*, SIAM J. Numer. Anal., 51 (2013), pp. 3106–3134.

[6] A. BUFFA, Y. MADAY, A.T. PATERA, C. PRUD'HOMME, AND G. TURINICI, *A priori convergence of the greedy algorithm for the parameterized reduced basis*, M2AN Math. Model. Numer. Anal., 46 (2012), pp. 595–603.

[7] A. COHEN AND R. DEVORE, *Kolmogorov widths under holomorphic mappings*, IMA J. Numer. Anal., 36 (2016), pp. 1–12.

[8] A. COHEN AND R. DEVORE, *Approximation of high dimensional parametric PDEs*, Acta Numer., 24 (2015), pp. 1–159.

[9] A. COHEN, R. DEVORE AND C. SCHWAB, *Convergence rates of best N-term Galerkin approximations for a class of elliptic sPDEs*, Found. Comput. Math., 10 (2010), pp. 615–646.

[10] ——, *Analytic regularity and polynomial approximation of parametric stochastic elliptic PDEs*, Anal. Appl., 9 (2011), pp. 11–47.

[11] A. CHKIFA, A. COHEN, R. DEVORE, AND C. SCHWAB, *Sparse adaptive Taylor approximation algorithms for parametric and stochastic elliptic PDEs*, M2AN Math. Model. Numer. Anal., 47 (2013), pp. 253–280.

[12] S. DAHLKE AND R. DEVORE, *Besov regularity for 2-D elliptic boundary value problems with variable coefficients*, Commun. PDEs, 22 (1997) pp. 1–16.

[13] R. DEVORE, R. HOWARD, AND C. MICCHELLI, *Optimal non-linear approximation*, Manuscripta Math., 63 (1989), pp. 469–478.

[14] R. DEVORE, G. KYRIAZIS, D. LEVIATAN, AND V.M. TIKHOMIROV, *Wavelet compression and non linear n-widths*, Adv. Comput. Math., 1 (1993), pp. 197–214.

[15] R. DEVORE, G. PETROVA AND P. WOJTASZCZYK, *Greedy algorithms for reduced bases in Banach spaces*, Constructive Approx., 37 (2013), pp. 455–466.

[16] J.L. EFTANG, A.T. PATERA, AND E.M. RØNQUIST, *An "hp" certified reduced basis method for parametrized elliptic partial differential equations*, SIAM J. Sci. Comput., 32 (2010), pp. 3170–3200.

[17] J.L. EFTANG, D.J. KNEZEVIC, AND A.T. PATERA, *An hp certified reduced basis method for parametrized parabolic partial differential equations*, Math. Comput. Modelling Dynam. Syst., 17 (2011), pp. 395–422.

[18] B. HAASDONK, M. DIHLMANN, AND M. OHLBERGER, *A training set and multiple basis generation approach for parametrized model reduction based on adaptive grids in parameter space*, Math. Comput. Modelling Dynam. Syst., 17 (2011), pp. 423–442.

[19] D. JERISON AND C.E. KENIG, *The inhomogeneous Dirichlet problem in Lipschitz domains*, J. Funct. Anal., 130 (1995), pp. 161–219.

[20] S. KAULMANN, B. HAASDONK, B. MOITINHO DE ALMEIDA, AND J.P.B. DIEZ, *Online greedy reduced basis construction using dictionaries*, VI International Conference on Adaptive Modeling and Simulation (ADMOS 2013), pp. 365–376.

[21] G.G. LORENTZ, M. VON GOLITSCHEK, AND Y. MAKOVOZ, *Constructive Approximation: Advanced Problems*, Springer-Verlag, New York, 1996.

[22] Y. MADAY, A.T. PATERA, AND G. TURINICI, *A priori convergence theory for reduced-basis approximations of single-parametric elliptic partial differential equations*, J. Sci. Comput., 17 (2002), pp. 437–446.

[23] ——, *Global a priori convergence theory for reduced-basis approximations of single-parameter symmetric coercive elliptic partial differential equations*, C. R. Acad. Sci. Paris Series I, Math., 335 (2002), pp. 289–294.

[24] Y. MADAY AND B. STAMM, *Locally adaptive greedy approximations for anisotropic parameter reduced basis spaces*, SIAM J. Sci. Comput., 35 (2013), pp. 2417–2441.

[25] G. ROZZA, D.B.P. HUYNH, AND A.T. PATERA, *Reduced basis approximation and a posteriori error estimation for affinely parametrized elliptic coercive partial differential equations–Application to transport and continuum mechanics*, Arch. Comput. Methods E, 15 (2008), pp. 229–275.

[26] S. SEN, *Reduced-basis approximation and a posteriori error estimation for many-parameter heat conduction problems*, Numer. Heat Tr. B-Fund, 54 (2008), pp. 369–389.

[27] V. TEMLYAKOV, *Nonlinear Kolmogorov widths*, Math. Notes, 63 (1998), pp. 785–795.

[28] K. VEROY, C. PRUD'HOMME, D.V. ROVAS, AND A.T. PATERA, *A posteriori error bounds for reduced-basis approximation of parametrized noncoercive and nonlinear elliptic partial differential equations*, in Proceedings of the 16th AIAA Computational Fluid Dynamics Conference, 2003, 2003–3847.

Part II

Tensor-Based Methods

Part II of this book is concerned with low-rank tensor techniques, in particular for addressing problems of very high dimension without having to face the curse of dimensionality. Typical problems include differential and eigenvalue equations whose solutions are functions of a large number of variables or parameters. Such problems arise in many applications, e.g., quantum physics, computational chemistry, stochastic calculus, uncertainty quantification, and parametric partial differential equations (PDEs).

Tensor techniques generalize low-rank matrix approximation to tensors of higher order. Unfortunately, there does not exist a general decomposition of higher-order tensors with all the nice properties of a matrix singular value decomposition (SVD), and hence different classes of decompositions with partial properties of an SVD have to be considered. The decompositions are based on low-parametric representations of higher-order tensors that are called *formats*.

The key idea of low-rank tensor approximation relies on separation of variables. The sought solution is approximated by some structured linear combination of products of functions with fewer variables. The methods thus exploit the tensor structure of function spaces that is present in many solution spaces of high-dimensional problems.

In the first chapter of Part II, Anthony Nouy introduces complexity reduction based on low-rank approximation methods. Tensor spaces and tensor norms are introduced, and best approximation in subsets of low-rank tensors is discussed. Finally, algorithms for computing approximations of functions in low-rank formats are presented, with a particular focus on applications to solutions of stochastic or parameter-dependent problems.

The second chapter, by Ivan Oseledets, provides a multilinear algebra perspective on tensor techniques, where a tensor is simply considered as a d-dimensional array. Several classical and novel tensor formats are introduced. The chapter particularly focuses on the tensor-train (TT) format, which is based on the SVD. Algorithmic aspects are intensively discussed, such as how to compute approximations in the TT format and how the format can be used to compute approximate solutions of interpolation problems, large linear systems, and eigenvalue problems or even nonstationary problems.

Chapter 4

Low-Rank Methods for High-Dimensional Approximation and Model Order Reduction

Anthony Nouy[7]

Tensor methods are among the most prominent tools for the numerical solution of high-dimensional problems where functions of multiple variables have to be approximated. These methods exploit the tensor structure of function spaces and apply to many problems in computational science that are formulated in tensor spaces, such as problems arising in stochastic calculus, uncertainty quantification, or parametric analyses. Here, we present complexity reduction methods based on low-rank approximation methods. We analyze the problem of best approximation in subsets of low-rank tensors and discuss its connection with the problem of optimal model reduction in low-dimensional reduced spaces. We present different algorithms for computing approximations of a function in low-rank formats. In particular, we present constructive algorithms based either on a greedy construction of an approximation (with successive corrections in subsets of low-rank tensors) or on the greedy construction of tensor subspaces (for subspace-based low-rank formats). These algorithms can be applied for tensor compression, tensor completion, or numerical solution of equations in low-rank tensor formats. A special emphasis is given to the solution of stochastic or parameter-dependent models. Different approaches are presented to approximate vector-valued or multivariate functions (identified with tensors), based on samples of the functions (black-box approaches) or on the model equations that satisfied by the functions.

4.1 ▪ Introduction

Low-rank approximation methods are among the most prominent complexity reduction methods for the solution of high-dimensional problems in computational science and engineering (see the surveys [22, 43, 50, 52] and monograph [44]). Typical problems include the solution of high-dimensional partial differential equations (PDEs) arising in physics or stochastic calculus, or the solution of parameter-dependent or

[7]This research was supported by the French National Research Agency (grants ANR CHORUS MONU-0005 and ANR ICARE MONU-0002).

stochastic equations using a functional approach, where functions of multiple (random) parameters have to be approximated. The construction of reduced-order representations of the solution of complex parameter-dependent equations is of particular importance in parametric analyses (e.g., optimization, control, inverse problems) and uncertainty quantification (uncertainty propagation, sensitivity analyses, statistical inverse problems).

In practical applications, vector-valued or multivariate functions (as elements of tensor spaces) often present low-rank structures that can be efficiently exploited to reduce the complexity of their representation. In this chapter, we introduce the basic concepts of low-rank approximation, first for order-two tensors and then for higher-order tensors. We present different methods of approximating a tensor, based either on complete or incomplete information on the tensor, or on the knowledge of the equations satisfied by the tensor. Particular emphasis is given to the solution of stochastic and parameter-dependent equations.

In Section 4.2, we recall the definition and some useful properties of tensor Banach spaces.

In Section 4.3, we introduce the problem of the best rank-r approximation of order-two tensors, with the singular value decomposition (SVD) as a particular case (the case corresponding to tensor Hilbert spaces equipped with canonical norms). Emphasis is given to the case of Bochner spaces, which are of particular interest for the analysis of parameter-dependent and stochastic equations.

In Section 4.4, we consider the case of higher-order tensors. We first present different notions of rank and the associated low-rank approximation formats, with a special emphasis on subspace-based (or Tucker) formats. Then, we discuss the problem of best approximation in subsets of low-rank tensors and its connection with the problem of finding optimal reduced spaces for the projection of a tensor (for subspace-based tensor formats). Then, we present higher-order versions of the SVD that allow us to obtain quasi-best (and controlled) approximations in subspace-based tensor formats for the particular case of the approximation (compression) of a given tensor in a tensor Hilbert space equipped with a canonical norm.

In Section 4.5, we present constructive algorithms for approximation in low-rank tensor formats. These algorithms rely either on the greedy construction of the approximation, by defining successive corrections in a given subset of low-rank tensors (typically the set of rank-one tensors), or on the greedy construction of subspaces (for approximation in subspace-based tensor formats). The latter approaches yield adaptive algorithms for projection-based model order reduction (MOR). For the case of parameter-dependent equations, these algorithms include the empirical interpolation method (EIM) (at the basis of reduced basis (RB) methods) and some variants of proper generalized decomposition (PGD) methods.

In Section 4.6, we present different approaches of approximating a function (identified with a tensor) in low-rank tensor formats, based on samples of the function. We present least-squares methods and interpolation methods, with the latter related to the problem of tensor completion.

In Section 4.7, we introduce a class of parameter-dependent (or stochastic) models, and we show how these models can be formulated as tensor-structured equations, first by exploiting the order-two tensor structure of Bochner spaces, and then by exploiting higher-order tensor structures of Lebesgue spaces with product measures (e.g., induced by independent random parameters).

Finally, in Section 4.8, we present low-rank methods for the solution of tensor-structured equations, relying either on the use of iterative solvers and standard low-rank compression methods or on the minimization of a certain residual-based distance to the solution (using optimization algorithms in low-rank tensor manifolds or constructive algorithms). Particular emphasis is given to the case of parameter-dependent (or stochastic) equations. In this particular context, greedy algorithms provide adaptive methods for the construction of reduced-order models (ROMs).

4.2 ▪ Tensor spaces

In this section, we introduce basic definitions for tensor Banach spaces and recall some useful properties. For a detailed introduction to tensor analysis, we refer the reader to the monographs [30, 44, 56].

4.2.1 ▪ Tensor Banach spaces

Let us consider vector spaces X_ν, $\nu \in \{1,\ldots,d\}$, equipped with norms $\|\cdot\|_\nu$. For $(v^{(1)},\ldots,v^{(d)}) \in X_1 \times \cdots \times X_d$, we denote by $\bigotimes_{\nu=1}^d v^{(\nu)}$ an elementary tensor. The *algebraic tensor space* $X = \bigotimes_{\nu=1}^d X_\nu$ is defined as the linear span of elementary tensors:

$$X = \bigotimes_{\nu=1}^{d} X_\nu = \operatorname{span}\left\{ \bigotimes_{\nu=1}^{d} v^{(\nu)} : v^{(\nu)} \in X_\nu, 1 \le \nu \le d \right\},$$

so that elements $v \in X$ can be written as finite linear combinations of elementary tensors, i.e.,

$$v = \sum_{i=1}^{m} v_i^{(1)} \otimes \cdots \otimes v_i^{(d)} \tag{4.1}$$

for some $m \in \mathbb{N}$ and some vectors $v_i^{(\nu)} \in X_\nu$, $1 \le i \le m$, $1 \le \nu \le d$. A *tensor Banach space* $X_{\|\cdot\|}$ equipped with a norm $\|\cdot\|$ is defined as the completion of an algebraic tensor space X with respect to the norm $\|\cdot\|$, and we denote $X_{\|\cdot\|} = \overline{X}^{\|\cdot\|} = {}_{\|\cdot\|}\bigotimes_{\nu=1}^d X_\nu$. If the norm $\|\cdot\|$ is associated with an inner product, the resulting space $X_{\|\cdot\|}$ is a *tensor Hilbert space*. In the case of finite-dimensional spaces X_ν, $X_{\|\cdot\|}$ does not depend on the choice of norm and it coincides with the normed algebraic tensor space X.

4.2.2 ▪ Tensor spaces of operators

Let $X = \bigotimes_{\nu=1}^d X_\nu$ and $Y = \bigotimes_{\nu=1}^d Y_\nu$ be two normed algebraic tensor spaces. Let $L(X_\nu, Y_\nu)$ (resp. $\mathcal{L}(X_\nu, Y_\nu)$) denote the set of linear operators (resp. continuous linear operators) from X_ν to Y_ν. For $Y_\nu = \mathbb{R}$, $L(X_\nu, \mathbb{R}) = X_\nu^*$ is the algebraic dual space of X_ν, while $\mathcal{L}(X_\nu, \mathbb{R}) = X_\nu'$ is the continuous dual space of X_ν. For $A^{(\nu)} \in L(X_\nu, Y_\nu)$, $1 \le \nu \le d$, the elementary tensor $A = \bigotimes_{\nu=1}^d A^{(\nu)}$ is defined for elementary tensors $\bigotimes_{\nu=1}^d v^{(\nu)} \in X$ by $A(\bigotimes_{\nu=1}^d v^{(\nu)}) = \bigotimes_{\nu=1}^d A^{\nu}(v^{(\nu)})$ and extended by linearity to the whole space X. The algebraic tensor space $\bigotimes_{\nu=1}^d \mathcal{L}(X_\nu, Y_\nu)$ is defined in the same way. In the particular case where $Y = \mathbb{R}$, with $Y_\nu = \mathbb{R}$ for all ν, an elementary tensor $\bigotimes_{\nu=1}^d \varphi^{(\nu)} \in \bigotimes_{\nu=1}^d X_\nu^*$ is such that for $v = \bigotimes_{\nu=1}^d v^{(\nu)} \in X$, $(\bigotimes_{\nu=1}^d \varphi^{(\nu)})(v) = \prod_{\nu=1}^d \varphi^{(\nu)}(v^{(\nu)})$, and we have $\bigotimes_{\nu=1}^d X_\nu^* \subset X^*$.

4.2.3 ▪ Minimal subspaces

The *minimal subspaces* of a tensor $v \in \bigotimes_{\nu=1}^d X_\nu$, denoted $U_\nu^{min}(v)$ for $1 \le \nu \le d$, are defined by the property that $v \in \bigotimes_{\nu=1}^d U_\nu$ implies $U_\nu^{min}(v) \subset U_\nu$, while $v \in \bigotimes_{\nu=1}^d U_\nu^{min}(v)$. The minimal subspace $U_\nu^{min}(v) \subset X_\nu$ can be equivalently characterized by

$$U_\nu^{min}(v) = \left\{ (id_\nu \otimes \varphi_{\nu^c})(v) : \varphi_{\nu^c} \in \bigotimes_{\beta \neq \nu} X_\beta^* \right\},$$

where $id_\nu \in L(X_\nu, X_\nu)$ is the identity operator on X_ν and where we use the convention $id_\nu \otimes (\bigotimes_{\beta\neq\nu} \varphi_\beta) = \varphi_1 \otimes \cdots \otimes \varphi_{\nu-1} \otimes id_\nu \otimes \varphi_{\nu+1} \otimes \cdots \otimes \varphi_d$. For v having the representation (4.1), $U_\nu^{min}(v) \subset \operatorname{span}\{v_i^{(\nu)}\}_{i=1}^m$, with an equality if the m vectors $\{\otimes_{\beta\neq\nu} v_i^{(\beta)}\}_{i=1}^m$ are linearly independent. A minimal subspace can also be defined for any subset of dimensions $\alpha \subset \{1,\dots,d\}$ such that $1 \le \#\alpha < d$. Letting $X_\alpha = \bigotimes_{\nu\in\alpha} X_\nu$, the minimal subspace $U_\alpha^{min}(v) \subset X_\alpha$ of v is defined by

$$U_\alpha^{min}(v) = \left\{ (id_\alpha \otimes \varphi_{\alpha^c})(v) : \varphi_{\alpha^c} \in \bigotimes_{\beta \notin \alpha} X_\beta^* \right\}.$$

For v having the representation (4.1), $U_\alpha^{min}(v) \subset \operatorname{span}\{v_i^{(\alpha)}\}_{i=1}^m$, with $v_i^{(\alpha)} = \bigotimes_{\nu\in\alpha} v_i^{(\nu)}$, and $U_\alpha^{min}(v) = \operatorname{span}\{v_i^{(\alpha)}\}_{i=1}^m$ if the vectors $\{\otimes_{\nu\notin\alpha} v_i^{(\nu)}\}_{i=1}^m$ are linearly independent. For $\alpha = \dot{\cup}_{k=1}^K \alpha_k$ being the disjoint union of nonempty sets $\alpha_k \subset \{1,\dots,d\}$,

$$U_\alpha^{min}(v) \subset \bigotimes_{k=1}^K U_{\alpha_k}^{min}(v).$$

For a detailed introduction to minimal subspaces and their properties, see [36].

4.2.4 ▪ Tensor norms

A norm $\|\cdot\|$ on X is called a *cross-norm* if $\|\bigotimes_{\nu=1}^d v^{(\nu)}\| = \prod_{\nu=1}^d \|v^{(\nu)}\|_\nu$ for all $(v^{(1)},\dots,v^{(d)}) \in X_1 \times \cdots \times X_d$. For $\nu \in \{1,\dots,d\}$, let $X_\nu' = \mathcal{L}(X_\nu, \mathbb{R})$ denote the continuous dual of X_ν equipped with the dual norm $\|\cdot\|_\nu'$ of $\|\cdot\|_\nu$. If $\|\cdot\|$ is a cross-norm and also the dual norm $\|\cdot\|'$ of $\|\cdot\|$ is a cross-norm on $\bigotimes_{\nu=1}^d X_\nu'$, i.e., $\|\bigotimes_{\nu=1}^d \varphi^{(\nu)}\|' = \prod_{\nu=1}^d \|\varphi^{(\nu)}\|_\nu'$ for all $\varphi^{(\nu)} \in X_\nu'$, then $\|\cdot\|$ is called a *reasonable cross-norm*. For a reasonable cross-norm, the elementary tensor $\bigotimes_{\nu=1}^d \varphi^{(\nu)}$ is in the space $X' = \mathcal{L}(X,\mathbb{R})$ equipped with the dual norm $\|\cdot\|'$, and it can be extended to an element in $(X_{\|\cdot\|})' = \mathcal{L}(X_{\|\cdot\|}, \mathbb{R})$. A norm $\|\cdot\|$ on X is said to be a *uniform cross-norm* if it is a reasonable cross-norm and if for any elementary operator $\bigotimes_{\nu=1}^d A^{(\nu)} \in \bigotimes_{\nu=1}^d \mathcal{L}(X_\nu, X_\nu)$ and for any tensor $v \in X$ it satisfies $\|(\bigotimes_{\nu=1}^d A^{(\nu)})(v)\| \le (\prod_{\nu=1}^d \|A^{(\nu)}\|_{X_\nu \leftarrow X_\nu})\|v\|$, where $\|A^{(\nu)}\|_{X_\nu \leftarrow X_\nu}$ denotes the operator norm of $A^{(\nu)}$. Therefore, when X is equipped with a uniform cross-norm, $A = \bigotimes_{\nu=1}^d A^{(\nu)}$ belongs to the space $\mathcal{L}(X,X)$ of continuous operators from X to X, and the operator norm of A is $\|A\|_{X\leftarrow X} = \prod_{\nu=1}^d \|A^{(\nu)}\|_{X_\nu\leftarrow X_\nu}$. The operator A can then be uniquely extended to a continuous operator $\overline{A} \in \mathcal{L}(X_{\|\cdot\|}, X_{\|\cdot\|})$.

Some norms can be directly defined from the norms $\|\cdot\|_\nu$ on X_ν, $1 \le \nu \le d$. The *injective norm* $\|\cdot\|_\vee$ is a particular uniform cross-norm defined for an algebraic tensor v as

$$\|v\|_\vee = \sup\{(\varphi^{(1)} \otimes \cdots \otimes \varphi^{(d)})(v) : \varphi^{(\nu)} \in X_\nu', \|\varphi^{(\nu)}\|_\nu' = 1, 1 \le \nu \le d\}.$$

The *projective norm* $\|\cdot\|_\wedge$ is another particular uniform cross-norm defined for an algebraic tensor v as

$$\|v\|_\wedge = \inf\left\{ \sum_{i=1}^{m} \prod_{\nu=1}^{d} \|v_i^{(\nu)}\|_\nu : v = \sum_{i=1}^{m} \bigotimes_{\nu=1}^{d} v_i^{(\nu)} \right\},$$

where the infimum is taken over all possible representations of v. The injective and projective norms are respectively the weakest and strongest reasonable cross-norms in the sense that for any reasonable cross-norm $\|\cdot\|$, we have $\|\cdot\|_\vee \lesssim \|\cdot\| \lesssim \|\cdot\|_\wedge$, yielding the following inclusions between the corresponding tensor Banach spaces: $X_{\|\cdot\|_\wedge} \subset X_{\|\cdot\|} \subset X_{\|\cdot\|_\vee}$.

In the case where spaces X_ν, $1 \le \nu \le d$, are Hilbert spaces associated with inner products $\langle \cdot,\cdot \rangle_\nu$, a natural inner product, called the *induced* or *canonical inner product*, can be defined for elementary tensors as

$$\left\langle \bigotimes_{\nu=1}^{d} v^{(\nu)}, \bigotimes_{\nu=1}^{d} w^{(\nu)} \right\rangle = \prod_{\nu=1}^{d} \langle v^{(\nu)}, w^{(\nu)} \rangle_\nu$$

and extended by linearity to the whole algebraic tensor space X. This yields the definition of a natural tensor Hilbert space $X_{\|\cdot\|}$. The associated norm, called the *canonical norm*, is in fact the unique cross-norm associated with an inner product, and it is a uniform cross-norm.

4.2.5 ▪ Examples of tensor Banach spaces

Here, we introduce examples of tensor Banach spaces that are of particular importance in parametric and stochastic analyses.

L^p spaces with product measure

Let (Ξ, Σ, μ) be a measure space with $\Xi \subset \mathbb{R}^s$ and μ a finite measure supported on Ξ (e.g., a probability measure). For $1 \le p \le \infty$, the Lebesgue space $L^p_\mu(\Xi)$ is defined as the Banach space of (equivalence classes of) measurable functions $v : \Xi \to \mathbb{R}$ with finite norm

$$\|v\|_p = \left(\int_\Xi |v(y)|^p \mu(dy) \right)^{1/p} \quad \text{for } 1 \le p < \infty \quad \text{and}$$

$$\|v\|_\infty = \operatorname*{ess\,sup}_{y \in \Xi} |v(y)| \quad \text{for } p = \infty.$$

Now, let us assume that (Ξ, Σ, μ) is the product of measure spaces $(\Xi_\nu, \Sigma_\nu, \mu_\nu)$ where $\Xi_\nu \subset \mathbb{R}^{s_\nu}$ and μ_ν is a finite measure, $1 \le \nu \le d$ ($s = \sum_{\nu=1}^{d} s_\nu$). That means $\Xi = \Xi_1 \times \cdots \times \Xi_d$, $\Sigma = \Sigma_1 \otimes \cdots \otimes \Sigma_d$, and $\mu = \mu_1 \otimes \cdots \otimes \mu_d$. We can define the algebraic tensor space $X = L^p_{\mu_1}(\Xi_1) \otimes \cdots \otimes L^p_{\mu_d}(\Xi_d)$. The natural injection from X to $L^p_\mu(\Xi)$ is such that $(v^{(1)} \otimes \cdots \otimes v^{(d)})(y_1, \dots, y_d) = v^{(1)}(y_1) \dots v^{(d)}(y_d)$ for $(y_1, \dots, y_d) \in \Xi$. X is then the set of functions v in $L^p_\mu(\Xi)$ that can be written

$$v(y_1, \dots, y_d) = \sum_{i=1}^{m} v_i^{(1)}(y_1) \dots v_i^{(d)}(y_d)$$

for some $m \in \mathbb{N}$ and some functions $v_i^{(\nu)} \in L^p_{\mu_\nu}(\Xi_\nu)$. We have the property that the resulting tensor Banach space $X_{\|\cdot\|_p} = {}_{\|\cdot\|_p}\bigotimes_{\nu=1}^d L^p_{\mu_\nu}(\Xi_\nu)$ is such that

$$X_{\|\cdot\|_p} = L^p_\mu(\Xi) \quad \text{for } 1 \leq p < \infty \quad \text{and}$$
$$X_{\|\cdot\|_\infty} \subset L^\infty_\mu(\Xi) \quad \text{for } p = \infty,$$

with equality $X_{\|\cdot\|_\infty} = L^\infty_\mu(\Xi)$ if Ξ is a finite set (see [30]). For any p, the norm $\|\cdot\|_p$ is a reasonable cross-norm. In the case $p = 2$, $L^2_\mu(\Xi)$ is a Hilbert space that can be identified with the tensor Hilbert space $X_{\|\cdot\|_2} = {}_{\|\cdot\|_2}\bigotimes_{\nu=1}^d L^2_{\mu_\nu}(\Xi_\nu)$. The norm $\|\cdot\|_2$ is the canonical inner product norm, which is a uniform cross-norm. For $1 < p < \infty$, $X_{\|\cdot\|_p}$ is reflexive and separable.

Bochner spaces

Bochner spaces are of particular importance in the analysis of parameter-dependent and stochastic equations. Let V denote a Banach space equipped with a norm $\|\cdot\|_V$ and let (Ξ, Σ, μ) denote a measure space, where $\Xi \subset \mathbb{R}^s$ and μ is a finite measure (e.g., a probability measure). For $1 \leq p \leq \infty$, the Bochner space $L^p_\mu(\Xi; V)$ is the Banach space of all (equivalence classes of) Bochner measurable functions[8] $v : \Xi \to V$ with bounded norm

$$\|v\|_p = \left(\int_\Xi \|v(y)\|_V^p \mu(dy)\right)^{1/p} \quad \text{for } 1 \leq p < \infty \quad \text{and}$$
$$\|v\|_\infty = \operatorname*{ess\,sup}_{y \in \Xi} \|v(y)\|_V \quad \text{for } p = \infty.$$

Let us note that $L^p_\mu(\Xi) = L^p_\mu(\Xi; \mathbb{R})$. We can define the algebraic tensor space $X = L^p_\mu(\Xi) \otimes V$ and the natural injection from X to $L^p_\mu(\Xi; V)$ by $\lambda \otimes v \mapsto \lambda(\cdot)v$, such that $(\lambda \otimes v)(y) = \lambda(y)v$ for $y \in \Xi$. The space X is composed by functions that can be written

$$v(y) = \sum_{i=1}^m s_i(y) v_i$$

for some $m \in \mathbb{N}$ and some vectors $v_i \in V$ and functions $s_i \in L^p_\mu(\Xi)$, $1 \leq i \leq m$. We have the property that the resulting tensor Banach space $X_{\|\cdot\|_p} = L^p_\mu(\Xi) \otimes_{\|\cdot\|_p} V$ is such that

$$X_{\|\cdot\|_p} = L^p_\mu(\Xi; V) \quad \text{for } 1 \leq p < \infty \quad \text{and}$$
$$X_{\|\cdot\|_\infty} \subset L^\infty_\mu(\Xi; V) \quad \text{for } p = \infty,$$

with equality $X_{\|\cdot\|_\infty} = L^\infty_\mu(\Xi; V)$ if V is a finite-dimensional space or if Ξ is a finite set.[9] For any $1 \leq p \leq \infty$, the norm $\|\cdot\|_p$ is a reasonable cross-norm. For $p = 2$ and if V is a Hilbert space, then $\|\cdot\|_2$ is an inner product norm, which makes $L^2_\mu(\Xi; V)$ a Hilbert space. Then, $X_{\|\cdot\|_2} = L^2_\mu(\Xi) \otimes_{\|\cdot\|_2} V$ is a tensor Hilbert space and $\|\cdot\|_2$ is the canonical norm, which is a uniform cross-norm. For $1 < p < \infty$, if V is reflexive and separable, then the Bochner tensor space $L^p_\mu(\Xi) \otimes_{\|\cdot\|_p} V$ is reflexive (see [74, Proposition 1.38]).

[8] See, e.g., [74, Section 1.5] for the definition of Bochner measurability and Bochner integrability.

[9] Note that if Ξ is a finite set, then for any $1 \leq p, q \leq \infty$, the norms $\|\cdot\|_p$ and $\|\cdot\|_q$ are equivalent and, therefore, the topological tensor spaces $X_{\|\cdot\|_p}$ and $X_{\|\cdot\|_q}$ coincide.

4.2.6 ▪ Approximation in finite-dimensional tensor spaces

Let $X_{\|\cdot\|} = {}_{\|\cdot\|}\bigotimes_{\nu=1}^{d} X_\nu$ be a tensor Banach space. Approximations of elements of $X_{\|\cdot\|}$ are typically searched in finite-dimensional subspaces of X that can be constructed as follows. Let $\{\psi_{k_\nu}^{(\nu)}\}_{k_\nu \in I_\nu}$ be a set of linearly independent elements in X_ν, with I_ν such that $\# I_\nu = n_\nu$. Let $X_{\nu,I_\nu} = \text{span}\{\psi_{k_\nu}^{(\nu)}\}_{k_\nu \in I_\nu} \subset X_\nu$. Let $I = I_1 \times \cdots \times I_d$. Then,

$$X_I = X_{1,I_1} \otimes \cdots \otimes X_{d,I_d}$$

is a finite-dimensional subspace of X with dimension $\# I = \prod_{\nu=1}^{d} n_\nu$ and with a basis $\{\psi_k\}_{k\in I}$ defined by $\psi_k = \psi_{k_1}^{(1)} \otimes \cdots \otimes \psi_{k_d}^{(d)}$, $k = (k_1, \ldots, k_d) \in I$. An element $u \in X_I$ can be written

$$u = \sum_{k\in I} a_k \psi_k = \sum_{k_1 \in I_1} \cdots \sum_{k_d \in I_d} a_{k_1,\ldots,k_d} \psi_{k_1}^{(1)} \otimes \cdots \otimes \psi_{k_d}^{(d)}, \tag{4.2}$$

where the set of coefficients $a = (a_k)_{k\in I} \in \mathbb{R}^I$ can be identified with a tensor $a \in \mathbb{R}^{n_1} \otimes \cdots \otimes \mathbb{R}^{n_d}$. If X is a Hilbert space with inner product $\langle \cdot, \cdot \rangle$ and if the basis $\{\psi_k\}_{k\in I}$ is orthonormal, then the coefficients in (4.2) are given by $a_{k_1,\ldots,k_d} = \langle \bigotimes_{\nu=1}^{d} \psi_{k_\nu}^{(\nu)}, u \rangle$.

Complexity reduction using sparse and low-rank tensor methods

The approximation space X_I has a dimension that grows exponentially with the dimension d, which makes standard linear approximation methods in X_I for a large d impractical. We can distinguish two main families of complexity reduction methods in tensor spaces: low-rank approximation methods and sparse approximation methods. Sparse approximation methods aim at defining suitable index sets $J \subset I$ with small cardinality to approximate a tensor in the corresponding low-dimensional space $X_J = \text{span}\{\psi_k\}_{k\in J}$. The construction of index sets J can be based on a priori analyses [6, 61] or on adaptive algorithms [8, 23–25, 27, 28, 76]. Sparse and low-rank methods exploit different low-dimensional structures of tensors, and these two complexity reduction methods can also be combined [2, 19]. In this chapter, we only focus on low-rank approximation methods. Note that in practical applications, complexity reduction methods are most often used to approximate tensors in a fixed finite-dimensional space X_I, possibly adapted afterward using a posteriori error estimates (see, e.g., [2]). Thus, low-rank and sparse tensor methods aim at finding a representation of the form (4.2) with a low-dimensional representation of the tensor of coefficients a.

4.2.7 ▪ About best-approximation problems

Here we recall definitions and classical results for the problem of best approximation of an element $u \in X_{\|\cdot\|}$ from a subset M in $X_{\|\cdot\|}$,

$$\min_{v\in M} \|u - v\|. \tag{4.3}$$

A subset M is *proximinal* if for any u there exists an element of best approximation in M. Any finite-dimensional linear subspace of $X_{\|\cdot\|}$ is proximinal. When $X_{\|\cdot\|}$ is reflexive, a sufficient condition for a subset M to be proximinal is that M is weakly closed. In particular, any closed convex set of a normed space is weakly closed. When $X_{\|\cdot\|}$ is finite dimensional or when M is a subset of a finite-dimensional subspace in $X_{\|\cdot\|}$, then a sufficient condition for M to be proximinal is that M is closed.

A subset M is a *unicity set* if for any u there exists at most one element of best approximation of u in M. A subset M is a *Chebyshev set* if it is a proximinal unicity set, i.e., if for any u, there exists a unique element of best approximation of u from M. Any convex subset M of a strictly convex normed space is a unicity set.

4.3 ▪ Low-rank approximation of order-two tensors

In this section, we consider the problem of the low-rank approximation of order-two tensors. We denote by $S \otimes V$ an algebraic tensor space, where S and V are Banach spaces, and by $S \otimes_{\|\cdot\|} V$ the corresponding tensor Banach space equipped with a norm $\|\cdot\|$.

4.3.1 ▪ Best rank-r approximation

The *rank* of $u \in S \otimes V$, denoted rank(u), is the minimal $r \in \mathbb{N}$ such that

$$u = \sum_{i=1}^{r} s_i \otimes v_i \tag{4.4}$$

for some vectors $\{v_i\}_{i=1}^r \in V^r$ and $\{s_i\}_{i=1}^r \in S^r$. We denote by $\mathcal{R}_r$ the set of tensors in $S \otimes V$ with a rank bounded by r,

$$\mathcal{R}_r = \left\{ \sum_{i=1}^{r} s_i \otimes v_i : \{s_i\}_{i=1}^r \in S^r, \{v_i\}_{i=1}^r \in V^r \right\},$$

or, equivalently,

$$\mathcal{R}_r = \left\{ \sum_{i=1}^{r}\sum_{j=1}^{r} a_{ij} s_i \otimes v_j : a = (a_{ij}) \in \mathbb{R}^{r\times r}, \{s_i\}_{i=1}^r \in S^r, \{v_i\}_{i=1}^r \in V^r \right\}.$$

Let $u \in S \otimes_{\|\cdot\|} V$. An element u_r of best approximation of u in $\mathcal{R}_r$ with respect to the norm $\|\cdot\|$ is such that

$$\|u - u_r\| = \min_{v \in \mathcal{R}_r} \|u - v\|. \tag{4.5}$$

If the norm $\|\cdot\|$ is not weaker than the injective norm, then $\mathcal{R}_r$ is weakly closed in $S \otimes_{\|\cdot\|} V$ (see Lemma 8.6 in [44]) and therefore proximinal if $S \otimes_{\|\cdot\|} V$ is reflexive. However, $\mathcal{R}_r$ is not a convex set and there is no guarantee of uniqueness of an element of best approximation.

Example 4.1. As an example, for $1 < p < \infty$ and V a reflexive and separable Banach space, the Bochner tensor space $L^p_\mu(\Xi) \otimes_{\|\cdot\|_p} V$ is reflexive and $\|\cdot\|_p$ is not weaker than the injective norm (see Section 4.2.5). Therefore, $\mathcal{R}_r$ is proximinal in $L^p_\mu(\Xi) \otimes_{\|\cdot\|_p} V$ if $1 < p < \infty$ and V is a reflexive and separable Banach space. ■

4.3.2 ▪ Optimal subspaces

Now, we introduce equivalent reformulations of the best rank-r approximation problem (4.5) by using subspace-based parametrizations of $\mathcal{R}_r$. We first note that $\mathcal{R}_r$ has a simple characterization using minimal subspaces. Indeed,

$$\mathcal{R}_r = \left\{ u \in S \otimes V : \dim(U_1^{min}(u)) = \dim(U_2^{min}(u)) \leq r \right\},$$

where the left and right minimal subspaces are, respectively, $U_1^{min}(u) = \{(id_S \otimes \varphi)(u) : \varphi \in V^*\}$, $U_2^{min}(u) = \{(\psi \otimes id_V)(u) : \psi \in S^*\}$. Let $\mathbb{G}_r(E)$ denote the *Grassmann manifold* of r-dimensional subspaces in the vector space E. First, we have

$$\mathcal{R}_r = \{ u \in S_r \otimes V_r : S_r \in \mathbb{G}_r(S), V_r \in \mathbb{G}_r(V) \}, \tag{4.6}$$

and the best rank-r approximation problem (4.5) can be equivalently written

$$\min_{S_r \in \mathbb{G}_r(S)} \min_{V_r \in \mathbb{G}_r(V)} \min_{v \in S_r \otimes V_r} \|u - v\|. \tag{4.7}$$

The solution of (4.7) yields optimal r-dimensional spaces V_r and S_r to approximate u in the "reduced" tensor space $S_r \otimes V_r$. Also, we have the following parametrization, which involves only subspaces in V:

$$\mathcal{R}_r = \{ u \in S \otimes V_r : V_r \in \mathbb{G}_r(V) \}, \tag{4.8}$$

which yields the following reformulation of the best rank-r approximation problem (4.5):

$$\min_{V_r \in \mathbb{G}_r(V)} \min_{v \in S \otimes V_r} \|u - v\|. \tag{4.9}$$

The solution of (4.9) yields an optimal r-dimensional subspace V_r to approximate u in the "reduced" tensor space $S \otimes V_r$.

Hilbert case

Suppose that S and V are Hilbert spaces and that $S \otimes_{\|\cdot\|} V$ is a Hilbert space with a norm $\|\cdot\|$ associated with an inner product $\langle \cdot, \cdot \rangle$. For a finite-dimensional linear subspace $V_r \subset V$, let $P_{S \otimes V_r}$ denote the orthogonal projection from $S \otimes_{\|\cdot\|} V$ onto $S \otimes V_r$ such that

$$\min_{v \in S \otimes V_r} \|u - v\|^2 = \|u - P_{S \otimes V_r} u\|^2 = \|u\|^2 - \|P_{S \otimes V_r} u\|^2. \tag{4.10}$$

The optimal subspace V_r is the solution of

$$\max_{V_r \in \mathbb{G}_r(V)} \mathcal{R}_u(V_r) \quad \text{with} \quad \mathcal{R}_u(V_r) = \|P_{S \otimes V_r} u\|^2, \tag{4.11}$$

which is an optimization problem on the Grassmann manifold $\mathbb{G}_r(V)$. The application

$$\|\cdot\|_r : u \mapsto \|u\|_r = \max_{V_r \in \mathbb{G}_r(V)} \|P_{S \otimes V_r} u\| = \max_{V_r \in \mathbb{G}_r(V)} \max_{\substack{w \in S \otimes V_r \\ \|w\|=1}} \langle w, u \rangle$$

defines a norm on $S \otimes_{\|\cdot\|} V$, and the best rank-r approximation u_r satisfies

$$\|u - u_r\|^2 = \|u\|^2 - \|u_r\|^2 = \|u\|^2 - \|u\|_r^2.$$

Remark 4.2. In the case where $\langle\cdot,\cdot\rangle$ is the canonical inner product, $P_{S\otimes V_r} = id_S \otimes P_{V_r}$, where P_{V_r} is the orthogonal projection from V to V_r. Then, finding the optimal subspace V_r is equivalent to finding the dominant eigenspace of an operator (see Section 4.3.4).

4.3.3 ▪ Tensors as operators

The following results are taken from [44, Section 4.2.13]. We restrict the presentation to the case where V is a Hilbert space with inner product $\langle\cdot,\cdot\rangle_V$. An element $u = \sum_{i=1}^r s_i \otimes v_i \in S \otimes V$ with rank r can be identified with a rank-r linear operator from V to S such that for $v \in V$,

$$u(v) = \sum_{i=1}^{r} s_i \langle v_i, v\rangle_V.$$

Then, the algebraic tensor space $S \otimes V$ can be identified with the set $\mathscr{F}(V,S)$ of finite rank operators from V to S. The injective norm $||\cdot||_\vee$ coincides with the operator norm, so that the tensor Banach space $S \otimes_{||\cdot||_\vee} V$ can be identified with the closure $\overline{\mathscr{F}(V,S)}$ of $\mathscr{F}(V,S)$ with respect to the operator norm, which coincides with the set of compact operators[10] $\mathscr{K}(V,S)$ from V to S. Therefore, for any norm $||\cdot||$ not weaker than the injective norm, we have

$$S \otimes_{||\cdot||} V \subset S \otimes_{||\cdot||_\vee} V = \mathscr{K}(V,S). \tag{4.12}$$

Also, the tensor Banach space $S \otimes_{||\cdot||_\wedge} V$ equipped with the projective norm $||\cdot||_\wedge$ can be identified with the space of nuclear operators $\mathscr{N}(V,S)$ from V to S. Therefore, for any norm $||\cdot||$ not stronger than the projective norm $||\cdot||_\wedge$, we have

$$\mathscr{N}(V,S) = S \otimes_{||\cdot||_\wedge} V \subset S \otimes_{||\cdot||} V. \tag{4.13}$$

4.3.4 ▪ Singular value decomposition

In this section, we consider the case where V and S are Hilbert spaces. The spaces V and S are identified with their dual spaces V' and S', respectively. Let $||\cdot||$ denote the canonical inner product norm. Let $n = \min\{\dim(V), \dim(S)\}$, with $n < \infty$ or $n = \infty$. Let u be in $S \otimes_{||\cdot||_\vee} V = \mathscr{K}(V,S)$, the set of compact operators.[11] Then, there exists a decreasing sequence of nonnegative numbers $\sigma = \{\sigma_i\}_{i=1}^n$ and two orthonormal systems $\{v_i\}_{i=1}^n \subset V$ and $\{s_i\}_{i=1}^n \subset S$ such that

$$u = \sum_{i=1}^{n} \sigma_i s_i \otimes v_i, \tag{4.14}$$

where in the case $n = \infty$, the only accumulation point of the sequence σ is zero and the series converges with respect to the injective norm $||\cdot||_\vee$, which coincides with the operator norm (see Theorem 4.114 in [44]). The expression (4.14) is the SVD of u, where $(s_i, v_i) \in S \times V$ is a couple of left and right singular vectors of u associated with a singular value σ_i, verifying

$$u(v_i) = \sigma_i s_i \quad \text{and} \quad u^*(s_i) = \sigma_i v_i,$$

[10] $\overline{\mathscr{F}(V,S)}$ coincides with $\mathscr{K}(V,S)$ if the Banach space V has the approximation property, which is the case for V a Hilbert space.

[11] Note that for $n < \infty$, $S \otimes_{||\cdot||_\vee} V = S \otimes V = \mathscr{F}(V,S) = \mathscr{K}(V,S)$.

where $u^* \in \mathscr{K}(S,V)$ is the adjoint operator of u defined by $\langle s, u(v)\rangle_S = \langle u^*(s), v\rangle_V$ for all $(v,s) \in V \times S$. Let u_r be the rank-r truncated SVD defined by

$$u_r = \sum_{i=1}^{r} \sigma_i s_i \otimes v_i.$$

We have

$$\|u\|_\vee = \|\sigma\|_{\ell_\infty} = \sigma_1 \quad \text{and} \quad \|u - u_r\|_\vee = \sigma_{r+1}.$$

If we assume that $\sigma \in \ell_2$, then $u \in S \otimes_{\|\cdot\|} V$ and

$$\|u\| = \|\sigma\|_{\ell_2} = \Big(\sum_{i=1}^{n} \sigma_i^2\Big)^{1/2}, \quad \|u - u_r\| = \Big(\sum_{i=r+1}^{n} \sigma_i^2\Big)^{1/2}.$$

The canonical norm $\|\cdot\|$ coincides with the Hilbert–Schmidt norm of operators. We have the important property that

$$\|u - u_r\| = \min_{w \in \mathscr{R}_r} \|u - w\|,$$

which means that an optimal rank-r approximation of u in the norm $\|\cdot\|$ can be obtained by retaining the first r terms of the SVD. Moreover,

$$\|u - u_r\| = \min_{w \in \mathscr{R}_1} \|u - u_{r-1} - w\| \tag{4.15}$$

and

$$\|u - u_r\|^2 = \|u - u_{r-1}\|^2 - \sigma_r^2 = \|u\|^2 - \sum_{i=1}^{r} \sigma_i^2 = \|u\|^2 - \|u_r\|^2.$$

The r-dimensional subspaces

$$S_r = U_1^{min}(u_r) = \text{span}\{s_i\}_{i=1}^r \in \mathbb{G}_r(S) \text{ and}$$
$$V_r = U_2^{min}(u_r) = \text{span}\{v_i\}_{i=1}^r \in \mathbb{G}_r(V)$$

are respectively left and right dominant singular spaces of u. These subspaces are solutions of problems (4.7) and (4.9), which means that they are optimal r-dimensional subspaces with respect to the canonical norm. Therefore, the SVD defines increasing sequences of optimal subspaces $\{V_r\}_{r\geq 1}$ and $\{S_r\}_{r\geq 1}$ such that

$$V_r \subset V_{r+1} \quad \text{and} \quad S_r \subset S_{r+1}.$$

Note that the optimal subspaces V_r and S_r are uniquely defined if $\sigma_r > \sigma_{r+1}$. Denoting $C_u : V \to V$ as the compact operator defined by $C_u = u^* \circ u$, we have that $(v_i, \sigma_i^2) \in V \times \mathbb{R}^+$ is an eigenpair of C_u, i.e., $C_u v_i = \sigma_i^2 v_i$. An optimal subspace V_r is a dominant r-dimensional eigenspace of C_u. It is a solution of (4.11). Here, the orthogonal projection $P_{S\otimes V_r}$ from $S\otimes_{\|\cdot\|} V$ to $S \otimes V_r$ is such that $P_{S\otimes V_r} = \overline{id_S \otimes P_{V_r}}$, and we have that $\mathscr{R}_u(V_r) = R_u(\mathbf{V}) = \text{Trace}(\{C_u \mathbf{V}, \mathbf{V}\}_V \{\mathbf{V}, \mathbf{V}\}_V^{-1})$, where $\mathbf{V} = \{v_i\}_{i=1}^r \in (V)^r$ is any basis of V_r, $C_u \mathbf{V} = \{C_u v_i\}_{i=1}^r$, and where $\{\{w_i\}_{i=1}^r, \{v_i\}_{i=1}^r\}_V = (\langle w_i, v_j\rangle_V)_{1\leq i,j\leq r} \in \mathbb{R}^{r\times r}$. $R_u(\mathbf{V})$ is the Rayleigh quotient of C_u.

4.3.5 ▪ Low-rank approximations in Bochner spaces

Here, we consider the particular case of low-rank approximations in Bochner spaces $L^p_\mu(\Xi; V)$, $1 \leq p \leq \infty$, where μ is a finite measure. This case is of particular interest for subspace-based MOR of parameter-dependent (or stochastic) problems. Here we consider V as a Hilbert space with norm $\|\cdot\|_V$. The considered algebraic tensor space is $L^p_\mu(\Xi)\otimes V$, and the set $\mathcal{R}_r$ of elements in $L^p_\mu(\Xi)\otimes V$ with rank at most r is identified with the set of functions $u_r : \Xi \to V$ of the form

$$u_r(y) = \sum_{i=1}^{r} s_i(y)v_i, \quad y \in \Xi.$$

For a given $u \in L^p_\mu(\Xi; V)$, let $\rho_r^{(p)}(u)$ denote the error of best rank-r approximation in $L^p_\mu(\Xi) \otimes V$, defined by

$$\rho_r^{(p)}(u) = \inf_{w \in \mathcal{R}_r} \|u - w\|_p$$

or, equivalently, by

$$\rho_r^{(p)}(u) = \inf_{V_r \in \mathbb{G}_r(V)} \inf_{w \in L^p_\mu(\Xi)\otimes V_r} \|u - w\|_p = \inf_{V_r \in \mathbb{G}_r(V)} \|u - P_{V_r} u\|_p,$$

where P_{V_r} is the orthogonal projection from V to V_r and $(P_{V_r} u)(y) = P_{V_r} u(y)$. For $1 \leq p < \infty$,

$$\rho_r^{(p)}(u) = \inf_{V_r \in \mathbb{G}_r(V)} \left(\int_\Xi \|u(y) - P_{V_r} u(y)\|_V^p \mu(dy) \right)^{1/p},$$

and for $p = \infty$,

$$\rho_r^{(\infty)}(u) = \inf_{V_r \in \mathbb{G}_r(V)} \operatorname{ess\,sup}_{y \in \Xi} \|u(y) - P_{V_r} u(y)\|_V.$$

If we assume that μ is a probability measure, we have, for all $1 \leq p \leq q \leq \infty$,

$$\rho_r^{(1)}(u) \leq \rho_r^{(p)}(u) \leq \rho_r^{(q)}(u) \leq \rho_r^{(\infty)}(u).$$

There are two cases of practical importance. The first case is $p = 2$, where $L^2_\mu(\Xi; V) = L^2_\mu(\Xi) \otimes_{\|\cdot\|_2} V$ is a Hilbert space and $\|\cdot\|_2$ is the canonical norm, so that we are in the situation where the best rank-r approximation is the r-term truncated SVD of u (see Section 4.3.4), called in this context Karhunen–Loève decomposition.[12] Then, $\rho_r^{(2)}(u) = (\sum_{i \geq r+1} \sigma_i^2)^{1/2}$, where $\{\sigma_i\}_{i \geq 1}$ is the sequence of decreasing singular values of u. The other important case is $p = \infty$. If we assume that $\Xi = \text{support}(\mu)$ is compact and that u is continuous from Ξ to V, then the set of solutions $u(\Xi) = \{u(y) : y \in \Xi\}$ is a compact subset of V and $\rho_r^{(\infty)}(u)$ coincides with the *Kolmogorov r-width* $d_r(u(\Xi))_V$ of $u(\Xi) \subset V$,

$$\begin{aligned}\rho_r^{(\infty)}(u) &= \inf_{V_r \in \mathbb{G}_r(V)} \sup_{y \in \Xi} \|u(y) - P_{V_r} u(y)\|_V \\ &= \inf_{V_r \in \mathbb{G}_r(V)} \sup_{v \in u(\Xi)} \|v - P_{V_r} v\|_V := d_r(u(\Xi))_V.\end{aligned}$$

[12]Karhunen–Loève decomposition usually corresponds to the SVD of a centered second-order stochastic process u, i.e., of $u - \mathbb{E}_\mu(u) = u - \int_\Xi u(y)\mu(dy)$.

Remark 4.3. In the case $p = 2$, there exists a sequence of nested optimal spaces V_r associated with $\rho_r^{(2)}(u)$. In the case $p \neq 2$, up to the knowledge of the author, it remains an open question to prove whether or not there exists a sequence of nested optimal spaces.

4.4 ▪ Low-rank approximation of higher-order tensors

In this section, we consider the problem of the low-rank approximation of higher-order tensors and we will see how to extend the principles of Section 4.3. Although several concepts apply to general tensor Banach spaces (see [36–38]), we restrict the presentation to the case of tensor Hilbert spaces.

Let X_ν, $\nu \in D := \{1,\dots,d\}$, denote Hilbert spaces equipped with norms $\|\cdot\|_\nu$ and associated inner products $\langle\cdot,\cdot\rangle_\nu$. We denote by $X = \bigotimes_{\nu\in D} X_\nu$ the algebraic tensor space equipped with a norm $\|\cdot\|$ associated with an inner product $\langle\cdot,\cdot\rangle$ and by $X_{\|\cdot\|}$ the corresponding tensor Hilbert space.

4.4.1 ▪ Low-rank tensor formats

A subset $\mathscr{S}_r$ of low-rank tensors in X can be formally defined as a set $\mathscr{S}_r = \{v \in X : \mathrm{rank}(v) \leq r\}$. There is no ambiguity in the case of order-two tensors, for which there is a unique notion of rank and $\mathscr{S}_r = \mathscr{R}_r$, with $r \in \mathbb{N}$. However, there are several notions of rank for higher-order tensors, leading to different subsets $\mathscr{S}_r$. For a detailed introduction to higher-order low-rank tensor formats, see [44, 52]. Here, we briefly recall the main tensor formats, namely the canonical format and the subspace-based (or Tucker) formats. The approximation in the latter formats is closely related to subspace-based MOR.

Canonical rank and canonical format

The *canonical rank* of a tensor $v \in X$ is the minimal integer $r \in \mathbb{N}$ such that

$$v = \sum_{i=1}^{r} v_i^{(1)} \otimes \cdots \otimes v_i^{(d)} \tag{4.16}$$

for some vectors $v_i^{(\nu)}$, $1 \leq i \leq r$, $1 \leq \nu \leq d$. The set of tensors with a canonical rank bounded by r is denoted by $\mathscr{R}_r$.

Remark 4.4. The elements of $\mathscr{R}_r$ can be written $v = F_{\mathscr{R}_r}(\{v_i^{(\nu)} : 1 \leq i \leq r, 1 \leq \nu \leq d\})$, where $F_{\mathscr{R}_r}$ is a multilinear map that parametrizes the subset $\mathscr{R}_r$ with $M(\mathscr{R}_r) = r(\sum_{\nu=1}^{d} \dim(X_\nu))$ real parameters. We have $M(\mathscr{R}_r) \leq dNr$, with $N = \max_\nu \dim(X_\nu)$.

α-rank

A natural notion of rank can be defined for a subset of dimensions based on the notion of minimal subspaces. Let $\alpha \subset D$ be a subset of dimensions and $\alpha^c = D \setminus \alpha$, with α and α^c nonempty. The *α-rank* of v, denoted $\mathrm{rank}_\alpha(v)$, is defined by

$$\mathrm{rank}_\alpha(v) = \dim(U_\alpha^{min}(v)). \tag{4.17}$$

The α-rank coincides with the classical notion of rank for order-two tensors. A tensor $v \in X$ can be identified with a tensor $\mathscr{M}_\alpha(v) \in X_\alpha \otimes X_{\alpha^c}$, where $X_\alpha = \bigotimes_{\nu\in\alpha} X_\nu$ and

$X_{\alpha^c} = \bigotimes_{\nu\in\alpha^c} X_\nu$, such that for v of the form (4.16), $\mathcal{M}_\alpha(v) = \sum_{i=1}^r v_i^{(\alpha)} \otimes v_i^{(\alpha^c)}$, with $v_i^{(\alpha)} = \bigotimes_{\nu\in\alpha} v_i^{(\nu)}$ and $v_i^{(\alpha^c)} = \bigotimes_{\nu\in\alpha^c} v_i^{(\nu)}$. $\mathcal{M}_\alpha : \bigotimes_{\nu\in D} X_\nu \to X_\alpha \otimes X_{\alpha^c}$ is a so-called matricization (or unfolding) operator. The α-rank of v then coincides with the classical rank of the order-two tensor $\mathcal{M}_\alpha(v)$, i.e., $\text{rank}_\alpha(v) = \text{rank}(\mathcal{M}_\alpha(v))$. Subsets of low-rank tensors can now be defined by imposing the α-rank for a collection of subsets $\alpha \in 2^D$.

Remark 4.5. The definition (4.17) of the α-rank also holds for elements $v \in X_{\|\cdot\|}$. In this case, the interpretation as the rank of an order-two tensor requires the extension of the matricization operator to the topological tensor space $X_{\|\cdot\|}$.

Tucker rank and Tucker format

The *Tucker rank* (*multilinear rank*) of a tensor $v \in X$ is defined as the tuple $(\text{rank}_\nu(v))_{\nu\in D} \in \mathbb{N}^d$. The set of tensors with a Tucker rank bounded by $r = (r_\nu)_{\nu\in D}$ is the set of *Tucker tensors*

$$\mathcal{T}_r = \left\{ v \in X : \text{rank}_\nu(v) = \dim(U_\nu^{min}(v)) \le r_\nu, \nu \in D \right\},$$

which can be equivalently characterized by

$$\mathcal{T}_r = \left\{ v \in U_1 \otimes \cdots \otimes U_d : U_\nu \in \mathbb{G}_{r_\nu}(X_\nu), \nu \in D \right\}. \tag{4.18}$$

An element $v \in \mathcal{T}_r$ can be written

$$v = \sum_{i_1=1}^{r_1} \cdots \sum_{i_d=1}^{r_d} C_{i_1,\ldots,i_d} v_{i_1}^{(1)} \otimes \cdots \otimes v_{i_d}^{(d)}$$

for some $C \in \mathbb{R}^{r_1\times\cdots\times r_d}$ (the *core tensor*) and some $v_{i_\nu}^{(\nu)} \in X_\nu$, $1 \le i_\nu \le r_\nu$, $\nu \in D$.

Remark 4.6. The elements of $\mathcal{T}_r$ can be written $v = F_{\mathcal{T}_r}(C, \{v_{i_\nu}^{(\nu)} : 1 \le i_\nu \le r_\nu, 1 \le \nu \le d\})$, where $F_{\mathcal{T}_r}$ is a multilinear map that parametrizes the subset $\mathcal{T}_r$ with $M(\mathcal{T}_r) = \prod_{\nu=1}^d r_\nu + \sum_{\nu=1}^d r_\nu \dim(X_\nu)$ real parameters. We have $M(\mathcal{T}_r) \le R^d + dNR$ with $R = \max_\nu r_\nu$ and $N = \max_\nu \dim(X_\nu)$.

Tree-based rank and tree-based Tucker format

A more general notion of rank can be associated with a tree of dimensions. Let T_D denote a *dimension partition tree* of D, which is a subset of 2^D such that all vertices $\alpha \in T_D$ are nonempty subsets of D, D is the root of T_D, every vertex $\alpha \in T_D$ with $\#\alpha \ge 2$ has at least two children, and the children of a vertex $\alpha \in T_D$ form a partition of α. The set of children of $\alpha \in T_D$ is denoted $S(\alpha)$. A vertex α with $\#\alpha = 1$ is called a leaf of the tree and is such that $S(\alpha) = \emptyset$. The set of leaves of T_D is denoted $\mathcal{L}(T_D)$. The *tree-based Tucker rank* of a tensor u associated with a dimension tree T_D, denoted T_D-rank(u), is a tuple $(\text{rank}_\alpha(u))_{\alpha\in T_D} \in \mathbb{N}^{\#T_D}$. Letting $r = (r_\alpha)_{\alpha\in T_D} \in \mathbb{N}^{\#T_D}$ be a tuple of integers, the subset of *tree-based Tucker tensors* with tree-based Tucker rank bounded by r is defined by

$$\mathcal{BT}_r = \left\{ v \in X : \text{rank}_\alpha(v) = \dim(U_\alpha^{min}(v)) \le r_\alpha, \alpha \in T_D \right\}. \tag{4.19}$$

A tuple $r=(r_\alpha)_{\alpha\in T_D}$ is said to be admissible for T_D if there exists an element $v\in X\setminus\{0\}$ such that $\dim(U_\alpha^{min}(v))=r_\alpha$ for all $\alpha\in T_D$. Here we use the convention $U_D^{min}(v)=\text{span}\{v\}$, so that $r_D=1$ for r admissible. The set $\mathcal{BT}_r$ can be equivalently defined by

$$\mathcal{BT}_r=\left\{v\in\bigotimes_{\alpha\in S(D)}U_\alpha:\begin{array}{c}U_\alpha\subset\bigotimes_{\beta\in S(\alpha)}U_\beta \text{ for all } \alpha\in T_D\setminus\{\mathcal{L}(T_D)\cup D\}\\ \text{and } \dim(U_\alpha)=r_\alpha \text{ for all } \alpha\in T_D\setminus D\end{array}\right\}. \tag{4.20}$$

For an element $v\in\mathcal{BT}_r$ with an admissible tuple r, if $\{v_{i_\alpha}^{(\alpha)}\}_{i_\alpha=1}^{r_\alpha}$ denotes a basis of $U_\alpha^{min}(v)$ for $\alpha\in T_D$, with $v_1^{(D)}=v$, then for all $\alpha\in T_D\setminus\mathcal{L}(T_D)$,

$$v_{i_\alpha}^{(\alpha)}=\sum_{\substack{1\le i_\beta\le r_\beta\\ \beta\in S(\alpha)}}C^{(\alpha)}_{i_\alpha,(i_\beta)_{\beta\in S(\alpha)}}\bigotimes_{\beta\in S(\alpha)}v_{i_\beta}^{(\beta)}$$

for $1\le i_\alpha\le r_\alpha$, where the $C^{(\alpha)}\in\mathbb{R}^{r_\alpha\times(\times_{\beta\in S(\alpha)}r_\beta)}$ are the so-called transfer tensors. Then, proceeding recursively, we obtain the following representation of v:

$$v=\sum_{\substack{1\le i_\nu\le r_\nu\\ \nu\in D}}\left(\sum_{\substack{1\le i_\alpha\le r_\alpha\\ \alpha\in T_D\setminus\mathcal{L}(T_D)}}\prod_{\mu\in T_D\setminus\mathcal{L}(T_D)}C^{(\mu)}_{i_\mu,(i_\beta)_{\beta\in S(\mu)}}\right)\bigotimes_{\nu\in D}v_{i_\nu}^{(\nu)}.$$

Remark 4.7. The elements of $\mathcal{BT}_r$ can be written $v=F_{\mathcal{BT}_r}(\{v_{i_\nu}^{(\nu)}:1\le i_\nu\le r_\nu,1\le\nu\le d\},\{C^{(\alpha)}:\alpha\in T_D\setminus\mathcal{L}(T_D)\})$, where $F_{\mathcal{BT}_r}$ is a multilinear map that parametrizes the subset $\mathcal{BT}_r$ with $M(\mathcal{BT}_r)=\sum_{\nu=1}^d r_\nu\dim(X_\nu)+\sum_{\alpha\in T_D\setminus\mathcal{L}(T_D)}r_\alpha\prod_{\beta\in S(\alpha)}r_\beta$ real parameters. We have $M(\mathcal{BT}_r)\le dNR+R^{\#S(D)}+\sum_{T_D\setminus\{\mathcal{L}(T_D)\cup D\}}R^{\#S(\alpha)+1}\le dNR+R^S+(d-2)R^{S+1}$, with $R=\max_\alpha r_\alpha$, $S=\max_{\alpha\notin\mathcal{L}(T_D)}\#S(\alpha)$, and $N=\max_\nu\dim(X_\nu)$.

Remark 4.8. For a tree T_D such that $S(D)=\mathcal{L}(T_D)=\{\{1\},\dots,\{d\}\}$, the set $\mathcal{BT}_{(1,r_1,\dots,r_d)}$ coincides with the set of Tucker tensors $\mathcal{T}_{(r_1,\dots,r_d)}$. For a binary tree T_D, i.e., such that $\#S(\alpha)=2$ for all $\alpha\notin\mathcal{L}(T_D)$, the set $\mathcal{BT}_r$ coincides with the set of hierarchical Tucker (HT) tensors introduced in [47].

The reader is referred to [37, 44, 47] for a detailed presentation of tree-based Tucker formats and their properties.

Tensor-train rank and tensor-train format

The tensor-train (TT) format (see [69]) is a particular (degenerate) case of tree-based Tucker format associated with a particular binary dimension tree

$$T_D=\{\{k\}:1\le k\le d\}\cup\{\{k,\dots,d\}:1\le k\le d-1\}$$

such that $S(\{k,\dots,d\})=\{\{k\},\{k+1,\dots,d\}\}$ for $1\le k\le d-1$. The TT rank of a tensor u, denoted $\text{rank}_{TT}(u)$, is the tuple $(\text{rank}_{\{k+1,\dots,d\}}(u))_{k=1}^{d-1}$. For a tuple $r=(r_1,\dots,r_d)\in\mathbb{N}^{d-1}$, the set of tensors with TT rank bounded by r is defined by

$$\mathcal{TT}_r=\left\{v\in X:\text{rank}_{\{k+1,\dots,d\}}(v)\le r_k\right\}, \tag{4.21}$$

which corresponds to the definition of a subset of tree-based Tucker tensors with inactive constraints on the ranks $\text{rank}_{\{k\}}(v)$ for $2 \le k \le d-1$.

Remark 4.9. More precisely, $\mathcal{T}\mathcal{T}_r$ coincides with the subset $\mathcal{B}\mathcal{T}_m$ of tree-based Tucker tensors with a tree-based Tucker rank bounded by $m = (m_\alpha)_{\alpha \in T_D}$ if m is such that $m_{\{k+1,\dots,d\}} = r_k$ for $1 \le k \le d-1$ and $m_{\{k\}} \ge r_k r_{k+1}$ for $2 \le k \le d-1$, the latter conditions implying that the constraints $\text{rank}_{\{k\}}(v) \le m_{\{k\}}$ are inactive for $2 \le k \le d-1$.

An element $v \in \mathcal{T}\mathcal{T}_r$ admits the following representation:

$$v = \sum_{i_1=1}^{r_1} \sum_{i_2=1}^{r_2} \cdots \sum_{i_{d-1}}^{r_{d-1}} v^{(1)}_{1,i_1} \otimes v^{(2)}_{i_1,i_2} \otimes \cdots \otimes v^{(d)}_{i_{d-1},1},$$

where $v^{(\nu)}_{i_{\nu-1},i_\nu} \in X_\nu$.

Remark 4.10. The elements of $\mathcal{T}\mathcal{T}_r$ can be written $v = F_{\mathcal{T}\mathcal{T}_r}(\{v^{(\nu)} : 1 \le \nu \le d\})$, with $v^{(\nu)} \in (X_\nu)^{r_{\nu-1} \times r_\nu}$ (using the convention $r_0 = r_d = 1$), where $F_{\mathcal{T}\mathcal{T}_r}$ is a multilinear map that parametrizes the subset $\mathcal{T}\mathcal{T}_r$ with $M(\mathcal{T}\mathcal{T}_r) = \sum_{\nu=1}^d r_{\nu-1} r_\nu \dim(X_\nu)$ real parameters. We have $M(\mathcal{T}\mathcal{T}_r) \le dNR^2$, with $R = \max_k r_k$ and $N = \max_\nu \dim(X_\nu)$.

4.4.2 ▪ Best approximations in subspace-based low-rank tensor formats

Tucker format

Let us first consider the best-approximation problem in Tucker format. A best approximation of $u \in X_{\|\cdot\|}$ in the subset of Tucker tensors $\mathcal{T}_r$ with a rank bounded by $r = (r_1, \dots, r_d)$ is defined by

$$\|u - u_r\| = \min_{v \in \mathcal{T}_r} \|u - v\|. \tag{4.22}$$

Based on the definition (4.18) of $\mathcal{T}_r$, problem (4.22) can be equivalently written

$$\|u - u_r\| = \min_{U_1 \in \mathbb{G}_{r_1}(X_1)} \cdots \min_{U_d \in \mathbb{G}_{r_d}(X_d)} \min_{v \in U_1 \otimes \cdots \otimes U_d} \|u - v\|. \tag{4.23}$$

A solution u_r to problem (4.23) yields optimal subspaces $U_\nu = U_\nu^{min}(u_r)$ with dimension less than r_ν for $1 \le \nu \le d$.

Different conditions ensure that the set $\mathcal{T}_r$ is proximinal, which means that there exists a solution to the best-approximation problem (4.22) for any u (see Section 4.2.7). If the norm $\|\cdot\|$ is not weaker than the injective norm, then $\mathcal{T}_r$ is weakly closed (see [36]) and therefore proximinal if $X_{\|\cdot\|}$ is reflexive (e.g., for $X = \bigotimes_{\nu \in D} L^p_{\mu_\nu}(\Xi_\nu)$ for any $1 < p < \infty$; see Section 4.2.5). In particular, if X is finite dimensional, $\mathcal{T}_r$ is closed and therefore proximinal.

Tree-based Tucker format

Let us now consider the best-approximation problem in the more general tree-based Tucker format. The best approximation of $u \in X_{\|\cdot\|}$ in the subset of tree-based Tucker tensors $\mathcal{BT}_r$ with T_D rank bounded by $r=(r_\alpha)_{\alpha\in T_D}$ is defined by

$$\|u-u_r\| = \min_{v\in\mathcal{BT}_r} \|u-v\|. \tag{4.24}$$

Based on the definition (4.20) of $\mathcal{BT}_r$, problem (4.24) can be equivalently written

$$\|u-u_r\| = \min_{(U_\alpha)_{\alpha\in T_D\setminus D}\in\mathcal{G}_r(T_D)} \min_{v\in\bigotimes_{\alpha\in S(D)} U_\alpha} \|u-v\|, \tag{4.25}$$

where $\mathcal{G}_r(T_D)$ is a set of subspaces defined by

$$\mathcal{G}_r(T_D) = \Bigg\{ (U_\alpha)_{\alpha\in T_D\setminus D} : U_\alpha \in \mathbb{G}_{r_\alpha}(X_\alpha) \text{ for all } \alpha\in T_D\setminus D,$$

$$\text{and } U_\alpha \subset \bigotimes_{\beta\in S(\alpha)} U_\beta \text{ for all } \alpha\in T_D\setminus\{D\cup\mathcal{L}(T_D)\}\Bigg\}.$$

Therefore, a best approximation $u_r\in\mathcal{BT}_r$ yields a collection of optimal subspaces U_α with dimension r_α, $\alpha\in T_D\setminus D$, with a hierarchical structure.

The proof of the existence of a best approximation in $\mathcal{BT}_r$ requires some technical conditions involving norms defined for all the vertices of the tree (see [37]). In particular, these conditions are satisfied in the case of tensor Hilbert spaces equipped with a canonical norm, and also for L^p-spaces.

4.4.3 ▪ Optimization problems in subsets of low-rank tensors

Standard subsets of low-rank tensors $\mathcal{S}_r$ (such as $\mathcal{R}_r$, $\mathcal{T}_r$, $\mathcal{BT}_r$, or $\mathcal{TT}_r$) are neither vector spaces nor convex sets. Therefore, the solution of a best-approximation problem in $\mathcal{S}_r$, or more generally of an optimization problem

$$\min_{v\in\mathcal{S}_r} J(v), \tag{4.26}$$

with $J : X_{\|\cdot\|}\to\mathbb{R}$, requires ad hoc minimization algorithms. Standard subsets of low-rank tensors admit a parametrization of the form

$$\mathcal{S}_r = \{v = F_{\mathcal{S}_r}(p_1,\dots,p_M) : p_i\in P_i, 1\le i\le M\}, \tag{4.27}$$

where $F_{\mathcal{S}_r} : P_1\times\cdots\times P_M\to X$ is a multilinear map and the P_i are vector spaces or standard submanifolds of vector spaces (e.g., Stiefel manifolds) (see Remarks 4.4, 4.6, 4.7, and 4.10, respectively, for $\mathcal{R}_r$, $\mathcal{T}_r$, $\mathcal{BT}_r$ and $\mathcal{TT}_r$). The optimization problem (4.26) is then rewritten as an optimization problem on the parameters

$$\min_{p_1\in P_1,\dots,p_M\in P_M} J(F_{\mathcal{S}_r}(p_1,\dots,p_M)),$$

which allows the use of more or less standard optimization algorithms (e.g., Newton, steepest descent, block coordinate descent), possibly exploiting the manifold structure of $P_1\times\cdots\times P_M$ (see, e.g., [34, 83, 85]). Alternating-minimization algorithms (or block coordinate descent algorithms) transform the initial optimization problem into a succession of simpler optimization problems. They consist of solving successively the minimization problems

$$\min_{p_i\in P_i} J(F_{\mathcal{S}_r}(p_1,\dots,p_M)),$$

each problem being a minimization problem in a linear space (or standard manifold) P_i of a functional $p_i \mapsto J(F_{\mathcal{S}_r}(p_1,\ldots,p_M))$, which inherits some properties of the initial functional J (due to the linearity of the partial map $p_i \mapsto F_{\mathcal{S}_r}(p_1,\ldots,p_M)$ from P_i to X). The available convergence results for these optimization algorithms in a general setting only ensure local convergence or global convergence to critical points (see, e.g., [35, 73]).

4.4.4 ▪ Higher-order singular value decomposition

Higher-order singular value decomposition (HOSVD), introduced in [29] for the Tucker format, in [42] for the HT format, and in [69] for the TT format, constitutes a possible generalization of the SVD for tensors of order $d \geq 3$ that allows us to obtain quasi-best approximations (but not necessarily best approximations) in subsets of low-rank tensors (for tree-based Tucker formats). It relies on the use of the SVD for order-two tensors applied to matricizations of a tensor. Here, we consider a tensor Hilbert space X equipped with the canonical norm $||\cdot||$. For each nonempty subset $\alpha \subset D$, $X_\alpha = \bigotimes_{\nu\in\alpha} X_\nu$ is also equipped with the canonical norm, denoted $||\cdot||_\alpha$.

Let us consider an element u in the algebraic tensor space[13] X. For $\alpha \subset D$, let $u_{\alpha,r_\alpha} \in X$ denote the best approximation of u with α-rank bounded by r_α, i.e.,

$$||u - u_{\alpha,r_\alpha}|| = \min_{\mathrm{rank}_\alpha(v)\leq r_\alpha} ||u - v||.$$

u_{α,r_α} is such that $\mathcal{M}_\alpha(u_{\alpha,r_\alpha})$ is the rank-r_α truncated SVD of $\mathcal{M}_\alpha(u) \in X_\alpha \otimes X_{\alpha^c}$, which can be written

$$u_{\alpha,r_\alpha} = \sum_{i=1}^{r_\alpha} \sigma_i^{(\alpha)} u_i^{(\alpha)} \otimes u_i^{(\alpha^c)},$$

where $\sigma_i^{(\alpha)}$ are the dominant singular values and $u_i^{(\alpha)}$ and $u_i^{(\alpha^c)}$ the corresponding left and right singular vectors of $\mathcal{M}_\alpha(u)$. Let $U_{r_\alpha}^{(\alpha)} = U_\alpha^{min}(u_{\alpha,r_\alpha}) = \mathrm{span}\{u_i^{(\alpha)}\}_{i=1}^{r_\alpha}$ denote the resulting optimal subspace in X_α and $P_{U_{r_\alpha}^{(\alpha)}}$ the corresponding orthogonal projection from X_α to $U_{r_\alpha}^{(\alpha)}$ (associated with the canonical inner product in X_α). The projection is such that $u_{\alpha,r_\alpha} = (P_{U_{r_\alpha}^{(\alpha)}} \otimes id_{\alpha^c})(u)$. We note that $\{U_{r_\alpha}^{(\alpha)}\}_{r_\alpha\geq 1}$ is an increasing sequence of subspaces. We have the orthogonal decomposition $U_{r_\alpha}^{(\alpha)} = \bigoplus_{i_\alpha=1}^{r_\alpha} W_{i_\alpha}^{(\alpha)}$ with $W_{i_\alpha}^{(\alpha)} = \mathrm{span}\{u_{i_\alpha}^{(\alpha)}\}$, and $P_{U_{r_\alpha}^{(\alpha)}} = \sum_{i_\alpha=1}^{r_\alpha} P_{W_{i_\alpha}^{(\alpha)}}$.

HOSVD in Tucker format

Let $r = (r_1,\ldots,r_d) \in \mathbb{N}^d$ such that $r_\nu \leq \mathrm{rank}_\nu(u)$ for $1 \leq \nu \leq d$. For each dimension $\nu \in D$, we define the optimal r_ν-dimensional space $U_{r_\nu}^{(\nu)}$ and the corresponding orthogonal projection $P_{U_{r_\nu}^{(\nu)}}$. Then, we define the space $U_r = \bigotimes_{\nu=1}^d U_{r_\nu}^{(\nu)}$ and the associated orthogonal projection

$$P_{U_r} = P_{U_{r_1}^{(1)}} \otimes \cdots \otimes P_{U_{r_d}^{(d)}}.$$

Then, the truncated HOSVD of u with multilinear rank r is defined by

$$u_r = P_{U_r}(u) \in \mathcal{T}_r.$$

[13]The case where $u \in X_{||\cdot||} \setminus X$ introduces some technical difficulties related to the definition of tree-based topological tensor spaces (see [37]).

We note that subspaces $\{U_r\}_{r\in\mathbb{N}^d}$ are nested: for $s, r \in \mathbb{N}^d$ such that $s \geq r$, we have $U_r \subset U_s$. The approximation u_r can be obtained by truncating a decomposition of u. Indeed, noting that $U_r = \bigoplus_{i\leq r} W_i$, with $W_i = \bigotimes_{\nu\in D} W^{(\nu)}_{i_\nu}$, we have

$$u_r = \sum_{i\leq r} w_i, \quad w_i = P_{W_i}(u),$$

which converges to u when $r_\nu \to \text{rank}_\nu(u)$ for all ν. We have that u_r is a quasi-optimal approximation of u in $\mathcal{T}_r$ (see [44, Theorem 10.3]) such that

$$\|u - u_r\| \leq \sqrt{d} \min_{v\in\mathcal{T}_r} \|u - v\|.$$

Remark 4.11. Another version of the HOSVD can be found in [44, Section 10.1.2], where the spaces $U^{(\nu)}_{r_\nu}$, $1 \leq \nu \leq d$, are computed successively. The space $U^{(\nu)}_{r_\nu}$ is defined as the dominant singular space of $\mathcal{M}_{\{\nu\}}(u^{(\nu-1)})$, with $u^{(\nu-1)} = P_{U^{(1)}_{r_1}} \otimes \cdots \otimes P_{U^{(\nu-1)}_{r_{\nu-1}}} u$.

HOSVD in tree-based Tucker format

Let T_D be a dimension tree and $r = (r_\alpha)_{\alpha\in T_D}$ be an admissible set of ranks, with $r_\alpha \leq \text{rank}_\alpha(u)$ for all $\alpha \in T_D$. For each vertex $\alpha \in T_D$, we define the optimal r_α-dimensional subspace $U^{(\alpha)}_{r_\alpha} \subset X_\alpha$ and the associated projection $P_{U^{(\alpha)}_{r_\alpha}}$. Let $P^{(\alpha)}_{r_\alpha} = P_{U^{(\alpha)}_{r_\alpha}} \otimes id_{\alpha^c}$. Then the truncated HOSVD of u with T_D-rank r is defined by

$$u_r = P^{T_D}_r(u) \in \mathcal{BT}_r,$$

with

$$P^{T_D}_r(u) = P^{T_D,(L)}_r P^{T_D,(L-1)}_r \dots P^{T_D,(1)}_r, \quad P^{T_D,(\ell)}_r = \prod_{\substack{\alpha\in T_D \\ level(\alpha)=\ell}} P^{(\alpha)}_{r_\alpha},$$

where $level(\alpha)$ is the level of a vertex in the tree, with $level(D) = 0$, and where $L = \max_{\alpha\in T_D} level(\alpha)$. We have that u_r is a quasi-optimal approximation of u in $\mathcal{BT}_r$ (see [44, Theorem 11.58]) such that

$$\|u - u_r\| \leq \sqrt{2d - 2 - s} \min_{v\in\mathcal{BT}_r} \|u - v\|,$$

with $s = 1$ if $\#S(D) = 2$ and $s = 0$ if $\#S(D) > 2$.

Remark 4.12. For the TT format, the truncated HOSVD of u with TT rank $r = (r_1, \dots, r_{d-1})$ is defined by $u_r = P^{(\{d\})}_{r_{d-1}} \dots P^{(\{2,\dots,d\})}_{r_1}(u)$, where $P^{(\{k+1,\dots,d\})}_{r_k}$ is the orthogonal projection associated with the optimal r_k-dimensional subspace $U^{\{k+1,\dots,d\}}_{r_k}$ in $X_{\{k+1,\dots,d\}}$, $1 \leq k \leq d-1$ (no projection associated with vertices $\{k\}$, $1 \leq k \leq d-1$). We have that $\|u - u_r\| \leq \sqrt{d-1} \min_{v\in\mathcal{TT}_r} \|u - v\|$.

Remark 4.13. Other versions of HOSVD for tree-based formats can be found in [44, Sections 11.4.2.2 and 11.4.2.3], where the spaces $U^{(\alpha)}_{r_\alpha}$, $\alpha \in T_D$, are computed successively.

4.5 ▪ Greedy algorithms for low-rank approximation

It can be observed in many practical applications that best approximations in low-rank tensor formats present good convergence properties (with respect to the rank). However, the computational complexity for computing best approximations drastically increases with the rank. Also, in general, the sequence of best approximations of a tensor is not associated with a decomposition of the tensor, which means that best approximations cannot be obtained by truncating a decomposition of the tensor. In Sections 4.3.4 and 4.4.4, we saw that the SVD or one of its extensions for higher-order tensors allows one to recover this a notion of decomposition. However, it is restricted to the approximation of a tensor in a tensor Hilbert space equipped with the canonical norm, and it requires an explicit representation of the tensor.

Greedy algorithms (sometimes called PGD methods) aim at recovering the notion of decomposition by relying either on greedy constructions of the approximation (by computing successive corrections in subsets of low-rank tensors) or on greedy constructions of subspaces (for subspace-based low-rank formats). These algorithms are applied in a more general setting where one is interested in constructing low-rank approximations w that minimize some distance $\mathscr{E}(u,w)$ to a tensor u. These constructions, although they are suboptimal, allow one to reduce the computational complexity of high-rank approximations, and they sometimes achieve quasi-optimal convergence (with the rank). These quasi-optimality properties are observed in some practical applications, but they still require theoretical justification.

Remark 4.14. Note that in the particular case where $X = S \otimes V$, with V and S Hilbert spaces and $\mathscr{E}(u,w) = \|u-w\|$ with $\|\cdot\|$ the canonical norm, all the algorithms presented in this section yield the SVD of u (provided that successive minimization problems are solved exactly). In general, when deviating from this particular case, the presented algorithms yield different decompositions.

4.5.1 ▪ Greedy construction of the approximation

A natural way to recover the notion of tensor decomposition is to define a sequence of approximations with increasing canonical rank obtained by successive rank-one corrections. This algorithm constitutes the most prominent version of PGD, and it has been used in many applications (see the review [22] and the monograph [21]). Starting from $u_0 = 0$, a rank-r approximation $u_r \in \mathscr{R}_r$ is defined by

$$u_r = u_{r-1} + w_r,$$

where $w_r = \otimes_{\nu=1}^d w_r^{(\nu)} \in \mathscr{R}_1$ is the optimal rank-one correction of u_{r-1} such that

$$\mathscr{E}(u, u_{r-1} + w_r) = \min_{w \in \mathscr{R}_1} \mathscr{E}(u, u_{r-1} + w). \tag{4.28}$$

This can be interpreted as a greedy algorithm in the dictionary of rank-one tensors $\mathscr{R}_1$ in $X_{\|\cdot\|}$, and it allows one to recover the notion of decomposition, even for higher-order tensors. Indeed, assuming that the sequence $\{u_r\}_{r\geq 1}$ strongly converges to u, then u admits the decomposition

$$u = \sum_{i \geq 1} w_i^{(1)} \otimes \cdots \otimes w_i^{(d)}, \tag{4.29}$$

and the approximation u_r with canonical rank r can be obtained by truncating this series after r terms, therefore justifying the notion of decomposition. When $\mathscr{E}(u,w) = ||u-w||$, conditions for the convergence of greedy algorithms in a general setting can be found in [80]. In the case of the minimization of convex functionals, convergence results can be found in [11, 13, 38, 81]. Note that this greedy construction is not specific to the particular setting of tensor approximation. The available convergence results do not take into account any particular structure of the tensor u and are usually pessimistic. However, except for very particular cases (see Remark 4.14), this algorithm only provides a suboptimal sequence of rank-r approximations. Depending on the properties of $\mathscr{E}(u,w)$, the convergence with the rank r may be strongly deteriorated by this greedy construction, compared with the best-approximation error in canonical format, that is, $\sigma(u;\mathscr{R}_r) = \inf_{v\in\mathscr{R}_r} \mathscr{E}(u,v)$ (which corresponds to the error of best r-term approximation in the dictionary $\mathscr{R}_1$).

A classical improvement of the above construction (known as the orthogonal greedy algorithm) consists of first computing a rank-one correction $w_r = \bigotimes_{\nu=1}^{d} w_r^{(\nu)}$ by solving (4.28) and then (after a possible normalization of w_r) defining

$$u_r = \sum_{i=1}^{r} \sigma_i^{(r)} w_i,$$

where the set of coefficients $(\sigma_i^{(r)})_{i=1}^{r}$ is a solution of

$$\mathscr{E}(u,u_r) = \min_{(\sigma_i^{(r)})_{i=1}^{r}\in\mathbb{R}^r} \mathscr{E}\left(u, \sum_{i=1}^{r} \sigma_i^{(r)} w_i\right).$$

In many applications, it is observed that this additional step does not significantly improve the convergence of the sequence u_r.

Remark 4.15. In the orthogonal greedy construction, the approximation u_r cannot be obtained by truncating a decomposition of the form (4.29), and, therefore, the sequence u_r has to be interpreted as a decomposition in a general sense.

The orthogonal greedy algorithm is analyzed in [38] as a particular case of a family of algorithms using more general dictionaries of low-rank tensors and using improvement strategies that are specific to the context of low-rank tensor approximation. In fact, improvements that seem to be efficient in practice do not rely any more on greedy approximations but rather adopt a subspace point of view in which low-rank corrections are only used for the greedy construction of subspaces. This requires us to move to other tensor formats, as presented in the next section.

Remark 4.16. Note that the above algorithms define sequences of spaces $U_r^{(\nu)} = \text{span}\{w_1^{(\nu)},\dots,w_r^{(\nu)}\}$ in X_ν verifying the nestedness property $U_r^{(\nu)} \subset U_{r+1}^{(\nu)}$ and such that $u_r \in U_r^{(1)} \otimes \cdots \otimes U_r^{(d)}$. However, the algorithms do not exploit this subspace point of view.

4.5.2 ▪ Greedy construction of subspaces for order-two tensors

When using subspace-based tensor formats, other notions of decomposition can be obtained by defining a sequence of approximations in increasing tensor spaces. Here, we present algorithms to approximate an order-two tensor u in $X_{\|\cdot\|}$ with $X = S \otimes V$. Their extensions to the case of higher-order tensors are presented in Section 4.5.3.

Fully greedy construction of subspaces

For an order-two tensor u, the best rank-r approximation problems (4.7), $r \geq 1$, yield a sequence of rank-r approximations

$$u_r = \sum_{i=1}^{r} s_i^{(r)} \otimes v_i^{(r)}.$$

The associated sequences of reduced approximation spaces $S_r = U_1^{min}(u_r) = \mathrm{span}\{s_i^{(r)}\}_{i=1}^r$ and $V_r = U_2^{min}(u_r) = \mathrm{span}\{v_i^{(r)}\}_{i=1}^r$, such that

$$u_r \in S_r \otimes V_r, \tag{4.30}$$

do not necessarily satisfy

$$S_r \subset S_{r+1} \quad \text{and} \quad V_r \subset V_{r+1}. \tag{4.31}$$

The notion of decomposition can be obtained by defining a sequence of rank-r approximations u_r in an increasing sequence of subspaces $S_r \otimes V_r$, which means that minimal subspaces $S_r = U_1^{min}(u_r)$ and $V_r = U_2^{min}(u_r)$ verify the nestedness property (4.31). The resulting approximation u_r is defined as the best approximation in $S_r \otimes V_r$, i.e.,

$$\mathcal{E}(u, u_r) = \min_{w \in S_r \otimes V_r} \mathcal{E}(u, w), \tag{4.32}$$

and can be written in the form

$$u_r = \sum_{i=1}^{r} \sum_{j=1}^{r} \sigma_{ij}^{(r)} s_i \otimes v_j,$$

where $\{s_i\}_{i=1}^r$ and $\{v_i\}_{i=1}^r$ are bases of S_r and V_r, respectively, and where $\sigma^{(r)} \in \mathbb{R}^{r \times r}$ is the solution of

$$\min_{\sigma^{(r)} \in \mathbb{R}^{r \times r}} \mathcal{E}\left(u, \sum_{i=1}^{r} \sum_{j=1}^{r} \sigma_{ij}^{(r)} s_i \otimes v_j \right).$$

Different constructions of nested subspaces can be proposed.

Optimal construction with nested minimal subspaces. A first and natural definition of u_r is such that

$$\mathcal{E}(u, u_r) = \min_{\substack{S_r \in \mathbb{G}_r(S) \\ S_r \supset S_{r-1}}} \min_{\substack{V_r \in \mathbb{G}_r(V) \\ V_r \supset V_{r-1}}} \min_{w \in S_r \otimes V_r} \mathcal{E}(u, w),$$

which corresponds to the definition (4.7) of optimal rank-r approximations with the only additional constraint that minimal subspaces of successive approximations are nested. This definition can be equivalently written in terms of the new elements $s_r \in S$ and $v_r \in V$ and of the matrix of coefficients $\sigma^{(r)} \in \mathbb{R}^{r \times r}$:

$$\mathcal{E}(u, u_r) = \min_{s_r \in S} \min_{v_r \in V} \min_{\sigma^{(r)} \in \mathbb{R}^{r \times r}} \mathcal{E}\left(u, \sum_{i=1}^{r} \sum_{j=1}^{r} \sigma_{ij}^{(r)} s_i \otimes v_j \right). \tag{4.33}$$

Suboptimal construction. A simpler but suboptimal construction (compared to (4.33)) consists of defining the new elements $s_r \in S$ and $v_r \in V$ by computing an optimal rank-one correction of the previous approximation u_{r-1}. More precisely, given $u_{r-1} = \sum_{i=1}^{r-1} \sum_{j=1}^{r-1} \sigma_{ij}^{(r-1)} s_i \otimes v_j$, $s_r \in S$ and $v_r \in V$ are defined by

$$\min_{s_r \in S} \min_{v_r \in V} \mathcal{E}(u, u_{r-1} + s_r \otimes v_r),$$

and then the approximation u_r is obtained by solving (4.32) with spaces $S_r = S_{r-1} + \operatorname{span}\{s_r\}$ and $V_r = V_{r-1} + \operatorname{span}\{v_r\}$.

Partially greedy construction of subspaces

Another notion of decomposition can be obtained by imposing the nestedness property for only one of the minimal subspaces, say $V_r \subset V$, which results in a sequence u_r of the form

$$u_r = \sum_{i=1}^{r} s_i^{(r)} \otimes v_i.$$

This is a nonsymmetric point of view that focuses on the construction of reduced spaces in V. This point of view is of particular interest in the case of Bochner spaces (see Section 4.3.5), for the model reduction of parameter-dependent or stochastic equations (see Section 4.8.5 and references [20, 62, 63, 67, 78]), and also for the model reduction of time-dependent evolution equations (see [54, 55, 65]).

Optimal construction with nested minimal subspaces. The sequence of rank-r approximations u_r can be defined by

$$\mathcal{E}(u, u_r) = \min_{\substack{V_r \in \mathbb{G}_r(V) \\ V_r \supset V_{r-1}}} \min_{w \in S \otimes V_r} \mathcal{E}(u, w). \tag{4.34}$$

This definition corresponds to the definition (4.7) of optimal rank-r approximations with the only additional constraint that the minimal subspaces $V_r = U_2^{min}(u_r)$ are nested. It is equivalent to the minimization problem

$$\min_{v_r \in V} \min_{\{s_i^{(r)}\}_{i=1}^{r} \in S^r} \mathcal{E}\left(u, \sum_{i=1}^{r} s_i^{(r)} \otimes v_i\right), \tag{4.35}$$

which can be solved by an alternating-minimization algorithm (see Section 4.8.5 for the application to parameter-dependent equations).

Suboptimal construction. Suboptimal constructions can also be introduced to reduce the computational complexity, e.g., by computing a rank-one correction of u_{r-1} defined by $\min_{v_r \in V} \min_{s_r \in S} \mathcal{E}(u, u_{r-1} + s_r \otimes v_r)$ and then by solving $\mathcal{E}(u, u_r) = \min_{w \in S \otimes V_r} \mathcal{E}(u, w)$ with $V_r = V_{r-1} + \operatorname{span}\{v_r\}$.

Partially greedy construction of subspaces in Bochner spaces

Let $X = L_\mu^p(\Xi) \otimes V$, and let $\|\cdot\|_p$ denote the Bochner norm. For $1 < p < \infty$ (and in particular for $p = 2$), we can consider the algorithm presented in Section 4.5.2, with

$\mathcal{E}(u,v) = \|u-v\|_p$. It defines the rank-$r$ approximation u_r by

$$\|u-u_r\|_p = \min_{\substack{V_r\in\mathbb{G}_r(V)\\ V_r\supset V_{r-1}}} \min_{w\in S\otimes V_r} \|u-w\|_p = \min_{\substack{V_r\in\mathbb{G}_r(V)\\ V_r\supset V_{r-1}}} \|u-P_{V_r}u\|_p,$$

which is a well-posed optimization problem (as a best-approximation problem in a weakly closed subset of the reflexive Banach space $L^p_\mu(\Xi;V)$; see Sections 4.2.7 and 4.2.5). This algorithm generates an increasing sequence of reduced approximation spaces V_r that are optimal in an "L^p sense." For $p=\infty$, an ideal greedy construction would define $u_r = P_{V_r}u$ with V_r-solution of

$$\inf_{\substack{V_r\in\mathbb{G}_r(V)\\ V_r\supset V_{r-1}}} \|u-P_{V_r}u\|_\infty = \inf_{\substack{V_r\in\mathbb{G}_r(V)\\ V_r\supset V_{r-1}}} \operatorname*{ess\,sup}_{y\in\Xi} \|u(y)-P_{V_r}u(y)\|_V.$$

Suboptimal constructions can be proposed to avoid computational issues related to optimization with respect to the L^∞-norm. Suppose that $u:\Xi\to V$ is continuous and $\Xi = \text{support}(\mu)$ is compact. Then, starting from $V_0 = 0$, one defines $V_r = V_{r-1} + \text{span}\{v_r\}$ with $v_r\in V$ such that

$$\sup_{y\in\Xi}\|u(y)-P_{V_{r-1}}u(y)\|_V = \|v_r - P_{V_{r-1}}v_r\|_V. \tag{4.36}$$

This is the greedy construction used in the EIM [57]. Convergence results for this algorithm can be found in [10, 12, 31], where the error $\|u-u_r\|_\infty = \sup_{y\in\Xi}\|u(y)-P_{V_r}u(y)\|_V$ is compared with the best rank-r approximation error $\rho_r^{(\infty)}(u) = d_r(u(\Xi))_V$.

4.5.3 ▪ Greedy construction of subspaces for higher-order tensors

Here we extend the constructive algorithms presented in Section 4.5.2 to the case of higher-order subspace-based tensor formats.

Greedy construction of subspaces for the Tucker format

The algorithms presented in Section 4.5.2 can be naturally generalized to provide constructive algorithms to approximate tensors in Tucker format. These algorithms construct a sequence of approximations u_m in nested tensor spaces $U_m = U_m^{(1)}\otimes\cdots\otimes U_m^{(d)}$, with $U_m^{(\nu)}\subset U_{m+1}^{(\nu)}$, therefore allowing the notion of decomposition to be recovered.

Construction of subspaces based on rank-one corrections. A first strategy, introduced in [41], consists of progressively enriching the spaces by the factors of rank-one corrections. More precisely, we start with $u_0 = 0$. Then, for $m\geq 1$, we compute a rank-one correction $w_m = \bigotimes_{\nu=1}^d w_m^{(\nu)} \in \mathcal{R}_1$ of u_{m-1}, which is a solution of

$$\mathcal{E}(u, u_{m-1}+w_m) = \min_{w\in\mathcal{R}_1}\mathcal{E}(u,u_{m-1}+w),$$

and then define $U_m^{(\nu)} = \text{span}\{w_i^{(\nu)}\}_{i=1}^m$ for all $\nu\in D$. Then, $u_m\in U_m = \bigotimes_{\nu=1}^d U_m^{(\nu)}$ is defined by

$$\mathcal{E}(u,u_m) = \min_{v\in U_m}\mathcal{E}(u,v)$$

and can be written

$$u_m = \sum_{i_1=1}^{m} \cdots \sum_{i_d=1}^{m} \sigma^{(m)}_{i_1,\ldots,i_d} \bigotimes_{\nu=1}^{d} w^{(\nu)}_{i_\nu}.$$

This construction is also applied to construct an approximate inverse of an operator in low-rank format [40]. For some applications (see [40, 41]), when $\mathcal{E}(u,v) \sim \|u-v\|$, we observe an error $\mathcal{E}(u,u_m)$ that behaves as the best-approximation error in Tucker format $\sigma(u;\mathcal{T}_{r^{(m)}}) = \inf_{v\in\mathcal{T}_{r^{(m)}}} \|u-v\|$ with $r^{(m)} = (m,\ldots,m)$. The theoretical justification of these observations remains an open problem. The above construction is isotropic in the sense that subspaces are enriched in all directions $\nu \in D$ simultaneously. This does not allow us to take advantage of possible anisotropic structures of the tensor u.

Remark 4.17. Of course, computing an approximation in the tensor product space U_m is not tractable in high dimension d without additional complexity reduction techniques. In [40], it is proposed to approximate u_m in a low-rank hierarchical (tree-based) tensor format in the tensor space U_m.

Optimal greedy construction of subspaces. Another natural algorithm consists of simply adding the nestedness property of subspaces in the definition (4.23) of best approximations. Starting from $u_0 = 0$, we let u_{m-1} denote the approximation at step $m-1$ of the construction and $U^{(\nu)}_{m-1} = U^{min}_\nu(u_{m-1})$ for $\nu \in D$. At step m, we select a set of dimensions $D_m \subset D$ to be enriched, we let $U^{(\nu)}_m = U^{(\nu)}_{m-1}$ for $\nu \notin D_m$, and we define u_m by

$$\mathcal{E}(u,u_m) = \min_{\substack{(U^{(\nu)}_m)_{\nu\in D_m} \\ \dim(U^{(\nu)}_m)=\dim(U^{(\nu)}_{m-1})+\Delta r^{(m)}_\nu \\ U^{(\nu)}_m \supset U^{(\nu)}_{m-1}}} \min_{v\in U^{(1)}_m\otimes\cdots\otimes U^{(d)}_m} \mathcal{E}(u,v). \tag{4.37}$$

Choosing $D_m = D$ and $\Delta r^{(m)}_\nu = 1$ for all $\nu \in D$ at each step corresponds to an isotropic enrichment (similar to the previous construction based on rank-one corrections). However, this isotropic construction does not allow any particular structure of the tensor u to be exploited. Choosing $D_m \neq D$ or different values for the $\Delta r^{(m)}_\nu$, $\nu \in D$, yields anisotropic constructions, but the selection of D_m and $\Delta r^{(m)}_\nu$, $\nu \in D$, requires the introduction of some error indicators. This type of construction seems to provide good convergence properties with respect to the rank $r^{(m)} = (r^{(m)}_\nu)_{\nu\in D}$, with $r^{(m)}_\nu = \dim(U^{(\nu)}_m)$. However, it remains an open and challenging question to prove that this type of construction can achieve quasi optimality compared to the best rank-$r^{(m)}$ approximation for certain classes of functions (e.g., associated with a certain decay of the best rank-$r^{(m)}$ approximation error).

Greedy construction of subspaces for the tree-based tensor format

The construction presented above can be extended to more general tree-based Tucker formats, these formats being related to the notion of subspaces. The idea is again to start from the subspace-based formulation of the best-approximation problem in tree-based Tucker format (4.25) and to propose a suboptimal greedy construction of subspaces that consists of adding a nestedness property for the successive minimal subspaces. We start from $u_0 = 0$. Then we let u_{m-1} denote the approximation at step $m-1$ of the construction and we let $U^{(\alpha)}_{m-1} = U^{min}_{\alpha}(u_{m-1})$ denote the current minimal subspaces of dimensions $r^{(m-1)}_{\alpha} = \dim(U^{min}_{\alpha}(u_{m-1}))$, $\alpha \in T_D$. Then, at step m, we select a set of vertices $T_m \subset T_D$ and we define $r^{(m)} = (r^{(m)}_{\alpha})_{\alpha\in T_D}$ with $r^{(m)}_{\alpha} = r^{(m-1)}_{\alpha} + \Delta r^{(m)}_{\alpha}$ for $\alpha \in T_m$ and $r^{(m)}_{\alpha} = r^{(m-1)}_{\alpha}$ for $\alpha \in T_D \setminus T_m$. Then, we define u_m as the solution of

$$\mathcal{E}(u, u_m) = \min_{\substack{(U^{(\alpha)}_m)_{\alpha\in T_D\setminus D}\in \mathcal{G}_{r^{(m)}}(T_D)\\ U^{(\alpha)}_m \supset U^{(\alpha)}_{m-1}}} \min_{v \in \bigotimes_{\alpha\in S(D)} U^{(\alpha)}_m} \mathcal{E}(u, v). \tag{4.38}$$

The selection of vertices T_m and of the $\Delta r^{(m)}_{\alpha}$, $\alpha \in T_m$, requires the introduction of error indicators and strategies of enrichment (preserving admissibility of T_D rank). This type of construction seems to be a good candidate for really exploiting specific tensor structures, but the analysis and the implementation of this type of strategy remain open and challenging issues.

4.5.4 ▪ Remarks on the solution of minimization problems

Constructive algorithms presented in this section require the solution of successive minimization problems in subsets that are neither vector spaces nor convex sets. In practice, one can rely on standard optimization algorithms by exploiting a multilinear parametrization of these approximation subsets (see Section 4.4.3 for optimization in standard low-rank manifolds, e.g., the set of rank-one tensors $\mathcal{R}_1$). As an illustration for a nonstandard subset introduced in the present section, let us consider the solution of (4.33), which is written

$$\min_{s_r\in S} \min_{v_r \in V} \min_{\sigma^{(r)}\in\mathbb{R}^{r\times r}} J(s_r, v_r, \sigma^{(r)}),$$

with $J(s_r, v_r, \sigma^{(r)}) = \mathcal{E}(u, \sum_{i=1}^{r}\sum_{j=1}^{r} \sigma^{(r)}_{ij} s_i \otimes v_j)$. A natural alternating-minimization algorithm then consists of successively solving minimization problems

$$\min_{s_r\in S} J(s_r, v_r, \sigma^{(r)}), \quad \min_{v_r\in V} J(s_r, v_r, \sigma^{(r)}), \quad \text{and} \quad \min_{\sigma^{(r)}\in\mathbb{R}^{r\times r}} J(s_r, v_r, \sigma^{(r)}).$$

Note that in practice, algorithms do not yield exact solutions of optimization problems. The analysis of constructive algorithms presented in this section should therefore take into account these approximations and quantify their impact. Several convergence results are available for weak greedy algorithms [80, 81], which are perturbations of the ideal greedy algorithms presented in Section 4.5.1.

4.6 ▪ Low-rank approximation using samples

In this section, we present methods for the practical construction of low-rank approximations of a vector-valued or multivariate function (identified with a tensor) from sample evaluations of the function.

4.6.1 ▪ Low-rank approximation of vector-valued functions

Let $u : \Xi \to V$ be a vector-valued function, with V a Banach space and Ξ a set equipped with a measure μ, and let us assume that $u \in L^p_\mu(\Xi; V)$. A low-rank approximation of

u can be defined from sample evaluations of u. Let $\Xi_K = \{y^k\}_{k=1}^K$ be a set of sample points in Ξ (e.g., samples drawn according to a probability measure μ on Ξ). Then, for $w \in L^p_\mu(\Xi; V)$, we define

$$\|w\|_{\infty,K} = \sup_{1\leq k\leq K} \|w(y^k)\|_V \text{ and} \tag{4.39}$$

$$\|w\|_{p,K} = \Big(\sum_{k=1}^{K} \omega^k \|w(y^k)\|_V^p\Big)^{1/p} \quad \text{for } p < \infty, \tag{4.40}$$

where $\{\omega^k\}_{k=1}^K$ is a set of positive weights. For $p < \infty$, if the y^k are independent and identically distributed (i.i.d.) samples drawn according the probability measure μ and if $\omega^k = K^{-1}$ for all k, then $\|w\|_{p,K}$ is a statistical estimate of the Bochner norm $\|w\|_p$. For $1 \leq p \leq \infty$, the application $v \mapsto \|v\|_{p,K}$ defines a seminorm on $L^p_\mu(\Xi; V)$. An optimal rank-r approximation u_r of u with respect to the seminorm $\|\cdot\|_{p,K}$ is defined by

$$\|u - u_r\|_{p,K} = \min_{w\in\mathcal{R}_r} \|u - w\|_{p,K} := \rho_r^{(p,K)}(u)$$

or, equivalently, by

$$\|u - u_r\|_{p,K} = \min_{V_r\in\mathbb{G}_r(V)} \|u - P_{V_r} u\|_{p,K}, \tag{4.41}$$

where $(P_{V_r} u)(y^k) = P_{V_r} u(y^k)$. The restriction of a function $w \in L^p_\mu(\Xi; V)$ to the subset Ξ_K, i.e., the tuple $\{w(y^k)\}_{k=1}^K \in V^K$, can be identified with a tensor $\mathbf{w}$ in the tensor space $\mathbb{R}^K \otimes V$ equipped with a norm $\|\cdot\|_p$ such that $\|\mathbf{w}\|_p = \|w\|_{p,K}$. The restriction to Ξ_K of the best rank-r approximation u_r of u is then identified with the best rank-r approximation $\mathbf{u}_r$ of $\mathbf{u}$ in $\mathbb{R}^K \otimes V$ and can be written

$$\mathbf{u}_r = \sum_{i=1}^{r} \mathbf{s}_i \otimes v_i \in \mathbb{R}^K \otimes V,$$

where $\mathbf{s}_i \in \mathbb{R}^K$ can be identified with sample evaluations $\{s_i(y^k)\}_{k=1}^K$ of a certain function $s_i \in L^p_\mu(\Xi)$ such that

$$u_r(y^k) = \sum_{i=1}^{r} s_i(y^k) v_i. \tag{4.42}$$

Any rank-r function u_r whose restriction to Ξ_K is identified with $\mathbf{u}_r$ is a solution of the best-approximation problem (4.41). The selection of a particular solution u_r requires an additional approximation step. Such a particular solution can be obtained by interpolation of functions s_i on the set of points Ξ_K (e.g., using polynomial interpolation on structured grids Ξ_K, or nearest neighbor, Shepard, or radial basis interpolations for unstructured samples).

Remark 4.18. Other approximation methods (e.g., least squares) can be used to approximate functions s_i from their evaluations at sample points Ξ_K. However, if the interpolation property is not satisfied, then the resulting function u_r is not necessarily a solution of (4.41).

Case $p = 2$

For $p = 2$ and V a Hilbert space, the norm $\|\cdot\|_2$ on $\mathbb{R}^K \otimes V$ such that $\|\mathbf{w}\|_2 = \|w\|_{2,K}$ coincides with the canonical inner product norm (when $\mathbb{R}^K$ is equipped with the weighted 2-norm $\|a\|_2 = (\sum_{k=1}^K \omega^k |a_k|^2)^{1/2}$). Therefore, $\mathbf{u}_r$ coincides with the truncated rank-r SVD of $\mathbf{u}$, where vectors $\{v_i\}_{i=1}^r$ are the r dominant eigenvectors of the operator $C_u^K : v \mapsto \sum_{k=1}^K \omega^k u(y^k)\langle u(y^k), v\rangle_V$. The best rank-$r$ approximation error is such that $\rho_r^{(2,K)}(u) = (\sum_{i=r+1}^K \sigma_i)^{1/2}$, where $\{\sigma_i\}_{i=1}^K$ is the set of singular values of $\mathbf{u}$ (eigenvalues of C_u^K) sorted in decreasing order. In a probabilistic setting, when $\{y^k\}_{k=1}^K$ are i.i.d. samples (drawn according to probability measure μ) and $\omega^k = K^{-1}$ for all k, C_u^K is the so-called empirical correlation operator of the V-valued random variable u. Its r-dimensional dominant eigenspace $V_r = \text{span}\{v_i\}_{i=1}^r$ is a statistical estimate of the optimal r-dimensional subspace associated with the best rank-r approximation of u in $L_\mu^2(\Xi; V)$. This corresponds to the standard *principal component analysis*. The obtained reduced approximation space V_r can then be used to compute an approximation of $u(\xi)$ in V_r for all $\xi \in \Xi$. This approach is at the basis of Galerkin *proper orthogonal decomposition* (POD) methods for parameter-dependent equations (see, e.g., [48]).

Case $p = \infty$

For $p = \infty$, the best rank-r approximation is well defined and the corresponding error is

$$\rho_r^{(\infty,K)}(u) = \min_{V_r \in \mathbb{G}_r(V)} \sup_{1 \le k \le K} \|u(y^k) - P_{V_r} u(y^k)\|_V = d_r(u(\Xi_K))_V,$$

where $d_r(u(\Xi_K))_V$ is the Kolmogorov r-width of the finite subset $u(\Xi_K) = \{u(y^k)\}_{k=1}^K$ of V. Suboptimal constructions of low-rank approximations can be proposed. In particular, one can rely on the greedy algorithm (4.36) with Ξ replaced by Ξ_K, which results in a sequence of nested spaces V_r. This algorithm coincides with the EIM for finite parameter sets (see [18, 57]), sometimes called the discrete empirical interpolation method (DEIM). Here also, the reduced approximation space V_r can then be used to compute an approximation of $u(\xi)$ in V_r for all $\xi \in \Xi$.

4.6.2 ▪ Higher-order low-rank approximation of multivariate functions

Here, we consider the approximation of a real-valued multivariate function $g : \Xi \to \mathbb{R}$ from a set of evaluations $\{g(y^k)\}_{k=1}^K$ of g on a set of points $\Xi_K = \{y^k\}_{k=1}^K$ in Ξ. The function g can be a variable of interest that is a function of a solution $u : \Xi \to V$ of a parameter-dependent equation (i.e., $g(\xi) = Q(u(\xi); \xi)$ with $Q(\cdot; \xi) : V \to \mathbb{R}$). It can also be the coefficient of the approximation of a function $u : \Xi \to V$ on a certain basis of a subspace of V (e.g., one of the functions s_i in the representation (4.42)).

Let us assume that $\mu = \mu_1 \otimes \cdots \otimes \mu_d$ is a product measure on $\Xi = \Xi_1 \times \cdots \times \Xi_d$, with μ_ν being a measure on $\Xi_\nu \subset \mathbb{R}$, $1 \le \nu \le d$. Using the notation of Section 4.2.6, we consider the approximation of g in a finite-dimensional subspace $X_I = X_{1,I_1} \otimes \cdots \otimes X_{d,I_d}$ in $X = L_{\mu_1}^2(\Xi_1) \otimes \cdots \otimes L_{\mu_d}^2(\Xi_d)$, where X_{ν,I_ν} is a K_ν-dimensional subspace of $L_{\mu_\nu}^2(\Xi_\nu)$ with basis $\Psi^{(\nu)} = \{\psi_{k_\nu}^{(\nu)}\}_{k_\nu \in I_\nu}$.

Least squares

The standard discrete least-squares method for approximating g in a subset $\mathcal{S}_r$ of low-rank tensors in X_I (see, e.g., [7, 19, 32]) consists of solving

$$\min_{h \in \mathcal{S}_r} \|g - h\|_{2,K}^2, \quad \text{with} \quad \|g - h\|_{2,K}^2 = \frac{1}{K}\sum_{k=1}^{K}(g(y^k) - h(y^k))^2,$$

which is a quadratic convex optimization problem on a nonlinear set. Algorithms presented in Section 4.4.3 can be used to solve this optimization problem. Assuming that $\mathcal{S}_r$ admits a simple parametrization of the form $\mathcal{S}_r = \{v = F_{\mathcal{S}_r}(p_1, \dots, p_M) : p_i \in \mathbb{R}^{N_i}, 1 \leq i \leq M\}$, where $F_{\mathcal{S}_r} : \times_{i=1}^{M} \mathbb{R}^{N_i} \to X_I$ is a multilinear map, the discrete least-squares minimization problem then takes the form

$$\min_{p_1 \in \mathbb{R}^{N_1}, \dots, p_M \in \mathbb{R}^{N_M}} \|g - F_{\mathcal{S}_r}(p_1, \dots, p_M)\|_{2,K}^2,$$

where the function to minimize is quadratic and convex with respect to each argument p_i, $1 \leq i \leq M$. Also, greedy algorithms presented in Section 4.5 can be used to construct approximations in low-rank formats (see [19] for the construction in canonical format).

When the number of available samples is not sufficient to get a stable estimate of the $\sum_{i=1}^{M} N_i$ real parameters, regularization techniques can be used in a quite straightforward way (see, e.g., [32] for the use of ℓ_2 regularization, or [19] for the use of sparsity-inducing regularizations). However, these approaches are still heuristic, and for a given low-rank format, some challenging questions remain open: How many samples are required to get a stable approximation in this format? Are there sampling strategies (not random) that are optimal with respect to this format?

Interpolation

Here we present interpolation methods to approximate g in X_I.

If $\Psi^{(\nu)}$ is a set of interpolation functions associated with a set of points $\Xi_{\nu,K_\nu} = \{y_\nu^{k_\nu}\}_{k_\nu \in I_\nu}$ in Ξ_ν, then $\{\psi_k(y) = \psi_{k_1}^1(y_1) \dots \psi_{k_d}^d(y_d)\}_{k \in I}$ is a set of interpolation functions associated with the tensorized grid $\Xi_K = \Xi_{1,K_1} \times \dots \times \Xi_{d,K_d}$ composed of $K = \prod_{\nu=1}^{d} K_\nu$ points. An interpolation $\mathcal{I}_K(u)$ of u is then given by

$$\mathcal{I}_K(u)(y) = \sum_{k \in I} u(y^k)\psi_k(y),$$

so that $\mathcal{I}_K(u)$ is completely characterized by the order-d tensor $a \in \mathbb{R}^{K_1} \otimes \dots \otimes \mathbb{R}^{K_d}$ whose components $a_{k_1,\dots,k_d} = u(y_1^{k_1}, \dots, y_d^{k_d})$ are the evaluations of u on the interpolation grid Ξ_K.

Then, low-rank approximation methods can be used to approximate the tensor $a \in \mathbb{R}^{K_1} \otimes \dots \otimes \mathbb{R}^{K_d}$ using only a few entries of the tensor (i.e., a few evaluations of the function u). This is related to the problem of tensor completion. A possible approach consists of evaluating some entries of the tensor taken at random and then reconstructing the tensor by minimizing a least-squares functional (which is an algebraic version of the least-squares method described in the previous section) or dual approaches using regularizations of rank minimization problems (see [72]). An algorithm is introduced in [33] for the approximation in canonical format, using least-squares minimization with a structured set of entries selected adaptively. Algorithms have also been proposed for an adaptive construction of low-rank approximations of a in TT format [71] or HT format [4]. These algorithms are extensions of the adaptive cross approximation (ACA) algorithm to high-order tensors and provide approximations that interpolate the tensor a at some adaptively chosen entries.

4.7 ▪ Tensor-structured parameter-dependent or stochastic equations

In this section, we consider a general class of linear parameter-dependent or stochastic equations and we formulate these equations as tensor-structured equations.

4.7.1 ▪ A class of linear parameter-dependent equations

Let ξ denote some parameters taking values in a set $\Xi \subset \mathbb{R}^s$. Ξ is equipped with a finite measure μ (when ξ are random parameters, μ is the probability measure induced by ξ). For an integrable function $g : \Xi \to \mathbb{R}$, we denote by $\int_\Xi g(y)\mu(dy)$ the integral with respect to the measure μ, which is the mathematical expectation $\mathbb{E}_\mu(g(\xi))$ for μ being a probability measure. Let V and W be Hilbert spaces, and let V' and W' be their respective continuous dual spaces. We denote by $\langle \cdot, \cdot \rangle$ the duality pairing. We consider the problem of finding $u : \Xi \to V$ such that

$$b(u(\xi), w; \xi) = \langle f(\xi), w \rangle \quad \forall w \in W \tag{4.43}$$

for almost all $\xi \in \Xi$, where $b(\cdot, \cdot; \xi) : V \times W \to \mathbb{R}$ is a parameter-dependent continuous bilinear form and $f(\xi) \in W'$ is a parameter-dependent continuous linear form. We suppose that $\xi \mapsto b(\cdot, \cdot; \xi)$ is Bochner measurable and we suppose that $b(\cdot, \cdot; \xi)$ is uniformly continuous and uniformly weakly coercive, i.e., there exist constants α and β independent of ξ such that it (for almost all $\xi \in \Xi$),

$$\sup_{v \in V} \sup_{w \in W} \frac{b(v, w; \xi)}{\|v\|_V \|w\|_W} \leq \beta < \infty, \tag{4.44}$$

$$\inf_{v \in V} \sup_{w \in W} \frac{b(v, w; \xi)}{\|v\|_V \|w\|_W} \geq \alpha > 0. \tag{4.45}$$

Also, we assume that for all $w \neq 0 \in W$, we have

$$\sup_{v \in V} b(v, w; \xi) > 0. \tag{4.46}$$

Note that condition (4.46) is deduced from (4.45) when $\dim(V) = \dim(W) < \infty$. When $V = W$, a parametrized family of bilinear forms $b(\cdot, \cdot; \xi) : V \times V \to \mathbb{R}$, $\xi \in \Xi$, is uniformly coercive if there exists a constant independent of ξ such that

$$\inf_{v \in V} \frac{b(v, v; \xi)}{\|v\|_V^2} \geq \alpha > 0, \tag{4.47}$$

which implies both conditions (4.45) and (4.46).

Let $B(\xi) : V \to W'$ denote the parameter-dependent linear operator such that $\langle B(\xi)v, w \rangle = b(v, w; \xi)$ for all $(v, w) \in V \times W$. Problem (4.43) is therefore equivalent to the operator equation

$$B(\xi)u(\xi) = f(\xi), \tag{4.48}$$

where assumptions (4.44), (4.45), and (4.46) are necessary and sufficient conditions for $B(\xi)$ to be an isomorphism from V to W' that satisfies

$$\alpha\|v\|_V \le \|B(\xi)v\|_{W'} \le \beta\|v\|_V \quad \forall v \in V \tag{4.49}$$

for almost all $\xi \in \Xi$. Let $B(\xi)^* : W \to V'$ denote the adjoint of $B(\xi)$, defined by $\langle B(\xi)v, w\rangle = \langle v, B(\xi)^* w\rangle$. Property (4.46) is equivalent to $\|B(\xi)^* w\|_{V'} > 0$ for all $w \neq 0$. Problem (4.43) admits a unique solution $u(\xi)$ satisfying

$$\|u(\xi)\|_V \le \frac{1}{\alpha}\|f(\xi)\|_{W'}. \tag{4.50}$$

From (4.50), it can be deduced that if $f \in L^p_\mu(\Xi; W')$ for a certain $p > 0$, then the solution $u \in L^{p'}_\mu(\Xi; V)$ for any $p' \le p$.

Remark 4.19. Note that the above presentation includes the case of parameter-dependent algebraic equations, for which $V = W = \mathbb{R}^N$, $B(\xi)$ is a matrix in $\mathbb{R}^{N\times N}$, and $f(\xi)$ is a vector in $\mathbb{R}^N$.

Example 1: Elliptic diffusion equation with random coefficients

Let D be an open bounded domain of $\mathbb{R}^m$ with Lipschitz boundary $\partial\Omega$. Let κ be a random field indexed by $x \in D$ defined on a probability space $(\Omega, \mathscr{B}, \mathbb{P})$ and such that it can be expressed as a function of random variables $\xi : \Omega \to \Xi \subset \mathbb{R}^s$, i.e., $\kappa = \kappa(x, \xi)$. We consider the following boundary value problem:

$$-\nabla \cdot (\kappa(\cdot, \xi)\nabla u(\xi)) = g(\cdot, \xi) \quad \text{on} \quad D, \quad u = 0 \quad \text{on} \quad \partial D,$$

with $g(\cdot, \xi) \in L^2(D)$. Let V be an approximation space in $H^1_0(D)$, $W = V$, and $\|v\|_V = \|v\|_W = \left(\int_D |\nabla v|^2\right)^{1/2}$. A Galerkin approximation of the solution, still denoted $u(\xi) \in V$, is the solution of (4.43) where $b(\cdot, \cdot; \xi) : V \times V \to \mathbb{R}$ and $f(\xi) \in V'$ are bilinear and linear forms defined by

$$b(v, w; \xi) = \int_D \kappa(\cdot, \xi)\nabla v \cdot \nabla w, \quad \text{and} \quad \langle f(\xi), w\rangle = \int_D g(\cdot, \xi) w.$$

If κ satisfies almost surely and almost everywhere

$$\alpha \le \kappa(x, \xi) \le \beta, \tag{4.51}$$

then properties (4.44) and (4.47) are satisfied. Let us consider a classical situation where κ admits the representation

$$\kappa(x, \xi) = \kappa_0(x) + \sum_{i=1}^{N} \kappa_i(x)\lambda_i(\xi), \tag{4.52}$$

yielding the following decomposition of the parameter-dependent bilinear form b:

$$b(v, w; \xi) = \int_D \kappa_0 \nabla v \cdot \nabla w + \sum_{i=1}^{N}\left(\int_D \kappa_i \nabla v \cdot \nabla w\right)\lambda_i(\xi).$$

For a spatially correlated second-order random field $\varkappa$, the representation (4.52) can be obtained by using truncated Karhunen–Loève decomposition and truncated polynomial chaos expansions (see, e.g., [68]). This representation also holds in the case where $\varkappa_0$ is a mean diffusion field and the λ_i represent random fluctuations of the diffusion coefficient in subdomains $D_i \subset D$ characterized by their indicator functions $\varkappa_i(x) = I_{D_i}(x)$. This problem has been extensively analyzed; see, e.g., [1, 39, 58].

Remark 4.20. For some problems of interest, the random field $\varkappa$ may not be uniformly bounded (e.g., when considering log-normal random fields) and may only satisfy $0 < \alpha(\xi) \leq \varkappa(x,\xi) \leq \beta(\xi) < +\infty$, where α and β possibly depend on ξ. For the mathematical analysis of such stochastic problems, we refer the reader to [15–17, 60, 68].

Example 2: Evolution equation

Let D denote a bounded domain of $\mathbb{R}^m$ with Lipschitz boundary $\partial\Omega$, and let $I = (0, T)$ denote a time interval. We consider the evolution equation

$$\frac{\partial u}{\partial t} - \nabla \cdot (\varkappa(\cdot,\xi)\nabla u) = g(\cdot,\cdot,\xi) \quad \text{on } D \times I,$$

with initial and boundary conditions

$$u = u_0(\cdot,\xi) \quad \text{on } D \times \{0\} \quad \text{and} \quad u = 0 \quad \text{on } \partial D \times I.$$

We assume that $\varkappa$ satisfies the same properties as in Example 1: $g(\cdot,\cdot,\xi) \in L^2(D \times I)$, and $u_0(\cdot,\xi) \in L^2(D)$. A space-time Galerkin approximation of the solution, still denoted $u(\xi)$, can be defined by introducing an approximation space

$$V \subset L^2(I; H_0^1(D)) \cap H^1(I; L^2(D)) := \mathcal{V}$$

equipped with the norm $\|\cdot\|_V$ such that $\|v\|_V^2 = \|v\|^2_{L^2(I;H_0^1(D))} + \|v\|^2_{H^1(I;L^2(\Omega))}$ and a test space

$$W = W_1 \times W_2 \subset L^2(I; H_0^1(D)) \times L^2(D) := \mathcal{W}$$

equipped with the norm $\|\cdot\|_W$ such that for $w = (w_1, w_2) \in W$, $\|w\|_W^2 = \|w_1\|^2_{L^2(I;H_0^1(D))} + \|w_2\|^2_{L^2(D)}$. Then, the Galerkin approximation $u(\xi) \in V$ is defined by equation (4.43), where the parameter-dependent bilinear form $b(\cdot,\cdot;\xi) : V \times W \to \mathbb{R}$ and the parameter-dependent linear form $f(\xi) : W \to \mathbb{R}$ are defined for $v \in V$ and $w = (w_1, w_2) \in W$ by

$$b(v,w;\xi) = \int_{D\times I} \frac{\partial v}{\partial t} w_1 + \int_{D\times I} \varkappa(\cdot,\xi)\nabla v \cdot \nabla w_1 + \int_D v(\cdot,0) w_2 \quad \text{and}$$

$$\langle f(\xi), w\rangle = \int_{D\times I} g(\cdot,\cdot,\xi) w_1 + \int_D u_0(\cdot,\xi) w_2.$$

For the analysis of this formulation, see [77].

Remark 4.21. $L^2(I; H_0^1(D))$ and $H^1(I; L^2(D))$ are identified with tensor Hilbert spaces $\overline{L^2(I) \otimes H_0^1(D)}^{\|\cdot\|_{L^2(I;H_0^1(D))}}$ and $\overline{H^1(I) \otimes L^2(D)}^{\|\cdot\|_{H^1(I;L^2(D))}}$, respectively, so that the space $\mathcal{V} = L^2(I; H_0^1(D)) \cap H^1(I; L^2(D))$ is an intersection tensor Hilbert space that coincides with $\overline{H^1(I)\otimes H^1_0(D)}^{\|\cdot\|_V}$ (see [44, Section 4.3.6]). Approximation spaces V in $\mathcal{V}$ can be chosen of the form $V = V(I)\otimes V(D)$ in the algebraic tensor space $H^1(I)\otimes H^1_0(D)$, with approximation spaces (e.g., finite element spaces) $V(I) \subset H^1(I)$ and $V(D) \subset H^1_0(D)$. Low-rank methods can also exploit this tensor structure and provide approximations of elements $v \in V$ in the form $v(x,t) = \sum_{i=1}^r a_i(x)b_i(t)$, with $a_i \in V(D)$ and $b_i \in V(I)$. This is the basis of POD methods for evolution problems and of the first versions of PGD methods, which are introduced for solving evolution equations with variational formulations in time in [54, 55, 64, 66].

4.7.2 ▪ Tensor-structured equations

Let us assume that $f \in L^2_\mu(\Xi; W')$, so that the solution of (4.43) is in $L^2_\mu(\Xi; V)$. In this section, we use the notation $\mathbf{V} = L^2_\mu(\Xi; V)$ and $\mathbf{W} = L^2_\mu(\Xi; W)$. The solution $u \in \mathbf{V}$ satisfies

$$a(u,w) = F(w) \quad \forall w \in \mathbf{W}, \tag{4.53}$$

where $a : \mathbf{V} \times \mathbf{W} \to \mathbb{R}$ is the bilinear form defined by

$$a(v,w) = \int_\Xi b(v(y), w(y); y)\mu(dy) \tag{4.54}$$

and $F : \mathbf{W} \to \mathbb{R}$ is the continuous linear form defined by

$$F(w) = \int_\Xi \langle f(y), w(y)\rangle \mu(dy).$$

Under assumptions (4.44), (4.45), and (4.46), it can be proved that a satisfies

$$\sup_{v\in\mathbf{V}} \sup_{w\in\mathbf{W}} \frac{a(v,w)}{\|v\|_{\mathbf{V}}\|w\|_{\mathbf{W}}} \le \beta < \infty, \tag{4.55}$$

$$\inf_{v\in\mathbf{V}} \sup_{w\in\mathbf{W}} \frac{a(v,w)}{\|v\|_{\mathbf{V}}\|w\|_{\mathbf{W}}} \ge \alpha > 0, \tag{4.56}$$

and for all $0 \ne w \in \mathbf{W}$,

$$\sup_{v\in\mathbf{V}} a(v,w) > 0. \tag{4.57}$$

Equation (4.53) can be equivalently rewritten as an operator equation,

$$Au = F, \tag{4.58}$$

where $A : \mathbf{V} \to \mathbf{W}'$ is the continuous linear operator associated with a such that

$$\langle Av, w\rangle = a(v,w) \quad \text{for all } (v,w) \in \mathbf{V}\times\mathbf{W}. \tag{4.59}$$

Properties (4.55), (4.56), and (4.57) imply that A is an isomorphism from $\mathbf{V}$ to $\mathbf{W}'$ such that for all $v \in \mathbf{V}$,

$$\alpha\|v\|_{\mathbf{V}} \le \|Av\|_{\mathbf{W}'} \le \beta\|v\|_{\mathbf{V}}. \tag{4.60}$$

Problem (4.53) therefore admits a unique solution such that $\|u\|_{\mathbf{V}} \le \frac{1}{\alpha}\|F\|_{\mathbf{W}'}$.

Order-two tensor structure

u (resp. f), as an element of the Bochner space $L^2_\mu(\Xi; V)$ (resp. $L^2_\mu(\Xi; W')$), can be identified with a tensor in $L^2_\mu(\Xi) \otimes_{\|\cdot\|_2} V$ (resp. $L^2_\mu(\Xi) \otimes_{\|\cdot\|_2} W'$). Let us further assume that $f \in L^2_\mu(\Xi) \otimes W'$ admits the representation

$$f(\xi) = \sum_{i=1}^{L} \gamma_i(\xi) f_i, \tag{4.61}$$

with $f_i \in W'$ and $\gamma_i \in L^2_\mu(\Xi)$. Then F is identified with the finite rank tensor

$$F = \sum_{i=1}^{L} \gamma_i \otimes f_i. \tag{4.62}$$

Let us now assume that the parameter-dependent operator $B(\xi) : V \to W'$ associated with the parameter-dependent bilinear form $b(\cdot,\cdot;\xi)$ admits the following representation (called *affine representation* in the context of RB methods):

$$B(\xi) = \sum_{i=1}^{R} \lambda_i(\xi) B_i, \tag{4.63}$$

where the $B_i : V \to W'$ are parameter-independent operators associated with parameter-independent bilinear forms b_i, and where the λ_i are real-valued functions defined on Ξ.

Remark 4.22. Let us assume that $\lambda_i \in L^\infty_\mu(\Xi)$, $1 \le i \le R$, and $\lambda_1 \ge 1$. Let us denote by α_i and β_i the constants such that

$$\alpha_i \|v\|_V \le \|B_i v\|_{W'} \le \beta_i \|v\|_V.$$

Property (4.49) is satisfied with $\beta = \sum_{i=1}^R \beta_i \|\lambda_i\|_\infty$ and with $\alpha = \alpha_1 - \sum_{i=2}^R \beta_i \|\lambda_i\|_\infty$ if $\alpha_1 > \sum_{i=2}^R \beta_i \|\lambda_i\|_\infty$. In the case where $V = W$, if all the B_i satisfy $\inf_{v \in V} \frac{\langle B_i v, v\rangle}{\|v\|_V^2} \ge \alpha_i > -\infty$, then property (4.49) is satisfied with either $\alpha = \alpha_1 - \sum_{i=2}^R \alpha_i \|\lambda_i\|_\infty$ if $\alpha_1 > \sum_{i=2}^R \alpha_i \|\lambda_i\|_\infty$ or $\alpha = \alpha_1$ if $\alpha_1 > 0$ and $\alpha_i \ge 0$ and $\lambda_i \ge 0$ for all $i \ge 2$.

Remark 4.23. If the parameter-dependent operator $B(\xi)$ (resp. right-hand side $f(\xi)$) does not admit an affine representation of the form (4.63) (resp. (4.61)), or if the initial affine representation contains a high number of terms, low-rank approximation methods can be used to obtain an affine representation with a small number of terms. For that purpose, one can rely on SVD or on the EIM, the latter approach being commonly used in the context of RB methods.

Assuming that $\lambda_i \in L^\infty_\mu(\Xi)$, the operator $A : V \to W'$ admits the representation[14]

$$A = \sum_{i=1}^{R} \Lambda_i \otimes B_i, \tag{4.64}$$

[14] A is a finite rank tensor in $\mathcal{L}(L^2_\mu(\Xi), L^2_\mu(\Xi)) \otimes \mathcal{L}(V, W')$.

where $\Lambda_i : L^2_\mu(\Xi) \to L^2_\mu(\Xi)$ is a continuous linear operator associated with λ_i such that for $\psi \in L^2_\mu(\Xi)$, $\Lambda_i \psi$ is defined by

$$\langle \Lambda_i \psi, \phi \rangle = \int_\Xi \lambda_i(y)\psi(y)\phi(y)\mu(dy) \quad \text{for all } \phi \in L^2_\mu(\Xi).$$

Therefore, equation (4.58) can be written as the tensor-structured equation

$$\Big(\sum_{i=1}^{R} \Lambda_i \otimes B_i\Big)u = \sum_{i=1}^{L} \gamma_i \otimes f_i. \tag{4.65}$$

Higher-order tensor structure

Let us assume that $\mu = \mu_1 \otimes \cdots \otimes \mu_d$ is a product measure on $\Xi = \Xi_1 \times \cdots \times \Xi_d$, with μ_ν being a measure on $\Xi_\nu \subset \mathbb{R}^{s_\nu}$, $1 \le \nu \le d$, with $s = \sum_{\nu=1}^d s_\nu$. Then $L^2_\mu(\Xi) = {}_{\|\cdot\|_2}\bigotimes_{\nu=1}^d L^2_{\mu_\nu}(\Xi_\nu)$ (see Section 4.2.5). In a probabilistic context, μ would be the measure induced by $\xi = (\xi_1, \dots, \xi_d)$, where the ξ_ν are independent random variables with values in Ξ_ν and probability law μ_ν.

Let us assume that the functions γ_i in (4.61), $1 \le i \le L$, are such that

$$\gamma_i(\xi) = \gamma_i^{(1)}(\xi_1) \dots \gamma_i^{(d)}(\xi_d), \tag{4.66}$$

with $\gamma_i^{(\nu)} \in L^2_{\mu_\nu}(\Xi_\nu)$. Then f is an element of $L^2_{\mu_1}(\Xi_1) \otimes \cdots \otimes L^2_{\mu_d}(\Xi_d) \otimes W'$ and F admits the representation

$$F = \sum_{i=1}^{L} \gamma_i^{(1)} \otimes \cdots \otimes \gamma_i^{(d)} \otimes f_i. \tag{4.67}$$

Let us assume that in the representation (4.63) of $B(\xi)$, the functions λ_i, $1 \le i \le R$, are such that

$$\lambda_i(\xi) = \lambda_i^{(1)}(\xi_1) \dots \lambda_i^{(d)}(\xi_d). \tag{4.68}$$

Assuming that $\lambda_i^{(\nu)} \in L^\infty_{\mu_\nu}(\Xi_\nu)$, $\lambda_i^{(\nu)}$ can be identified with an operator $\Lambda_i^{(\nu)} : S_\nu \to \tilde{S}'_\nu$, where for $\psi \in S_\nu$, $\Lambda_i^{(\nu)} \psi$ is defined by

$$\langle \Lambda_i^{(\nu)} \psi, \phi \rangle = \int_{\Xi_\nu} \lambda_i^{(\nu)}(y_\nu)\psi(y_\nu)\phi(y_\nu)\mu_\nu(dy_\nu) \quad \text{for all } \phi \in \tilde{S}_\nu.$$

Then, λ_i also defines an operator $\Lambda_i : S \to \tilde{S}'$ such that

$$\Lambda_i = \Lambda_i^{(1)} \otimes \cdots \otimes \Lambda_i^{(d)}.$$

Then, the operator A, as an operator from $L^2_{\mu_1}(\Xi_1) \otimes \cdots \otimes L^2_{\mu_d}(\Xi_d) \otimes V$ to $(L^2_{\mu_1}(\Xi_1) \otimes \cdots \otimes L^2_{\mu_d}(\Xi_d) \otimes W)'$, admits the decomposition[15]

$$A = \sum_{i=1}^{R} \Lambda_i^{(1)} \otimes \cdots \otimes \Lambda_i^{(d)} \otimes B_i. \tag{4.69}$$

Therefore, equation (4.58) can be written as the tensor-structured equation

$$\Big(\sum_{i=1}^{R} \Lambda_i^{(1)} \otimes \cdots \otimes \Lambda_i^{(d)} \otimes B_i\Big)u = \sum_{i=1}^{L} \gamma_i^{(1)} \otimes \cdots \otimes \gamma_i^{(d)} \otimes f_i. \tag{4.70}$$

[15] A is a finite rank tensor in $\mathcal{L}(L^2_{\mu_1}(\Xi_1), L^2_{\mu_1}(\Xi_1)) \otimes \cdots \otimes \mathcal{L}(L^2_{\mu_d}(\Xi_d), L^2_{\mu_d}(\Xi_d)) \otimes \mathcal{L}(V, W')$.

4.7.3 ▪ Galerkin approximations

Here, we present Galerkin methods to approximate the solution of (4.43) in a subspace $S \otimes V$ of $L^2_\mu(\Xi; V)$, where S is a finite-dimensional subspace in $L^2_\mu(\Xi)$. In this section, $\mathbf{V} = S \otimes V$ denotes the approximation space in $L^2_\mu(\Xi; V)$, which is equipped with the natural norm in $L^2_\mu(\Xi; V)$, denoted $\|\cdot\|_{\mathbf{V}}$.

Petrov–Galerkin approximation

Let us introduce a finite-dimensional subspace $\tilde{S}$ in $L^2_\mu(\Xi)$, with $\dim(S) = \dim(\tilde{S})$, and let us introduce the tensor space $\mathbf{W} = \tilde{S} \otimes W \subset L^2_\mu(\Xi; W)$, equipped with the natural norm in $L^2_\mu(\Xi; W)$, denoted $\|\cdot\|_{\mathbf{W}}$. A Petrov–Galerkin approximation in $\mathbf{V} = S \otimes V$ of the solution of problem (4.43), denoted u_G, is defined by the equation

$$a(u_G, w) = F(w) \quad \forall w \in \mathbf{W}, \tag{4.71}$$

which can be equivalently rewritten as the operator equation

$$Au_G = F, \tag{4.72}$$

where $A : \mathbf{V} \to \mathbf{W}'$ is associated (through equation (4.59)) with the bilinear form a.

Remark 4.24. Assuming that the approximation space $\mathbf{V}$ and the test space $\mathbf{W}$ are such that properties (4.55), (4.56), and (4.57) are satisfied, then A satisfies (4.60) and equation (4.72) admits a unique solution u_G that is a quasi-optimal approximation of u, with $\|u_G - u\|_{\mathbf{V}} \leq (1 + \frac{\beta}{\alpha}) \min_{v \in \mathbf{V}} \|u - v\|_{\mathbf{V}}$.

Remark 4.25. Letting $\{\psi_i\}_{i=1}^K$ and $\{\phi_i\}_{i=1}^K$ be bases of S and $\tilde{S}$, respectively, the solution u_G of (4.72) can be written $u_G = \sum_{i=1}^K \psi_i \otimes u_i$, where the tuple $\{u_i\}_{i=1}^K \in V^K$ verifies the coupled system of equations

$$\sum_{j=1}^{P} A_{ij} u_j = F_i, \quad 1 \leq i \leq K, \tag{4.73}$$

with $A_{ij} = \int_\Xi B(y)\psi_j(y)\phi_i(y)\mu(dy)$ and $F_i = \int_\Xi f(y)\phi_i(y)\mu(dy)$ for $1 \leq i,j \leq K$. The tuple $\{u_i\}_{i=1}^K \in V^K$ can be identified with a tensor in $\mathbb{R}^K \otimes V$.

Remark 4.26. In practice, integrals over Ξ with respect to the measure μ can be approximated by using a suitable quadrature rule $\{(y^k, \omega^k)\}_{k=1}^K$, replacing (4.71) by

$$\sum_{k=1}^{K} \omega^k \langle B(y^k) u_G(y^k), w(y^k) \rangle = \sum_{k=1}^{K} \omega^k \langle f(y^k), w(y^k) \rangle.$$

Under the assumptions of Section 4.7.2, equation (4.72) can be written in the form of tensor-structured equation (4.65), where the functions γ_i are now identified with elements of $\tilde{S}'$ such that $\langle \gamma_i, \psi \rangle = \int_\Xi \gamma_i(y)\psi(y)\mu(dy)$ for all $\psi \in \tilde{S}$, and where the Λ_i are now considered as operators from S to $\tilde{S}'$ such that for $\psi \in \tilde{S}$, $\Lambda_i \psi$ is defined by $\langle \Lambda_i \psi, \phi \rangle = \int_\Xi \lambda_i(y)\psi(y)\phi(y)\mu(dy)$ for all $\phi \in \tilde{S}$.

Under the stronger assumptions of Section 4.7.2, (4.72) can be written in the form of the tensor-structured equation (4.70), where the functions $\gamma_i^{(\nu)}$ are now identified with elements of $\tilde{S}'_\nu$ such that $\langle \gamma_i^{(\nu)}, \psi \rangle = \int_{\Xi_\nu} \gamma_i(y_\nu)\psi(y_\nu)\mu_\nu(dy_\nu)$ for all $\psi \in \tilde{S}_\nu$, and where the $\Lambda_i^{(\nu)}$ are now considered as operators from S_ν to $\tilde{S}'_\nu$ such that for $\psi \in \tilde{S}_\nu$, $\Lambda_i^{(\nu)}\psi$ is defined by $\langle \Lambda_i^{(\nu)}\psi, \phi \rangle = \int_{\Xi_\nu} \lambda_i^{(\nu)}(y_\nu)\psi(y_\nu)\phi(y_\nu)\mu_\nu(dy_\nu)$ for all $\phi \in \tilde{S}_\nu$.

Minimal residual Galerkin approximation

Let $C(\xi) : W' \to W$ be a symmetric operator that defines on W' an inner product $\langle \cdot, \cdot \rangle_{C(\xi)}$ defined by $\langle g, h \rangle_{C(\xi)} = \langle g, C(\xi)h \rangle = \langle C(\xi)g, h \rangle$ for $g, h \in W'$. Let $\|\cdot\|_{C(\xi)}$ denote the associated norm on W', and assume that

$$\alpha_C \|\cdot\|_{W'} \leq \|\cdot\|_{C(\xi)} \leq \beta_C \|\cdot\|_{W'} \tag{4.74}$$

for some constants $0 < \alpha_C \leq \beta_C < \infty$.

Remark 4.27. A natural choice for C is to take the inverse of the (parameter-independent) Riesz map $R_W : W \to W'$, so that $\|h\|_C = \|h\|_{W'} = \|R_W^{-1}h\|_W$. When $V = W$ and $B(\xi)$ is coercive, another possible choice for $C(\xi)$ is to take the inverse of the symmetric part of $B(\xi)$.

A minimal residual Galerkin approximation in $\boldsymbol{V} = S \otimes V$ of the solution of problem (4.43), denoted u_R, can be defined by

$$u_R = \arg\min_{v \in \boldsymbol{V}} \mathcal{E}(u, v), \tag{4.75}$$

with

$$\mathcal{E}(u, v)^2 = \int_\Xi \|B(y)v(y) - f(y)\|_{C(y)}^2 \mu(dy). \tag{4.76}$$

Let $B(\xi)^* : W \to V'$ denote the adjoint of $B(\xi)$. Then, we define the symmetric bilinear form $\tilde{a} : \boldsymbol{V} \times \boldsymbol{V} \to \mathbb{R}$ such that

$$\tilde{a}(v, w) = \int_\Xi \langle B(y)v(y), B(y)w(y) \rangle_{C(y)} \mu(dy) = \int_\Xi \langle \tilde{B}(y)v(y), w(y) \rangle \mu(dy),$$

with $\tilde{B}(\xi) = B(\xi)^* C(\xi) B(\xi)$, and the linear form $\tilde{F} : \boldsymbol{V} \to \mathbb{R}$ such that

$$\tilde{F}(w) = \int_\Xi \langle f(y), B(y)w(y) \rangle_{C(y)} \mu(dy) = \int_\Xi \langle \tilde{f}(y), w(y) \rangle \mu(dy),$$

with $\tilde{f}(\xi) = B(\xi)^* C(\xi) f(\xi)$. The approximation $u_R \in \boldsymbol{V} = S \otimes V$ defined by (4.75) is equivalently defined by

$$\tilde{a}(u, v) = \tilde{F}(v) \quad \forall v \in \boldsymbol{V}, \tag{4.77}$$

which can be rewritten as the operator equation

$$\tilde{A}u_R = \tilde{F}, \tag{4.78}$$

where $\tilde{A} : V \to V'$ is the operator associated with the bilinear form $\tilde{a}$. The approximation u_R is the standard Galerkin approximation of the solution of the parameter-dependent equation

$$\tilde{B}(\xi)u(\xi) = \tilde{f}(\xi). \tag{4.79}$$

Remark 4.28. Under assumptions (4.44), (4.45), (4.46), and (4.74), we have that

$$\sup_{v\in V}\sup_{w\in V} \frac{\tilde{a}(v,w)}{\|v\|_V\|w\|_V} \leq \tilde{\beta} < \infty, \quad \inf_{v\in V} \frac{\tilde{a}(v,v)}{\|v\|_V^2} \geq \tilde{\alpha} > 0, \tag{4.80}$$

with $\tilde{\alpha} = \alpha_C^2\alpha^2$ and $\tilde{\beta} = \beta_C^2\beta^2$, and u_R is a quasi-optimal approximation of u in V, with $\|u-u_R\|_V \leq \sqrt{\frac{\tilde{\beta}}{\tilde{\alpha}}}\min_{v\in V}\|u-v\|_V$.

Remark 4.29. Letting $\{\psi_i\}_{i=1}^K$ be a basis of S, the solution u_R of (4.78) can be written $u_R = \sum_{i=1}^K \psi_i \otimes u_i$, where the set of vectors $\{u_i\}_{i=1}^K \in V^K$ verifies the coupled system of equations (4.73) with $A_{ij} = \int_\Xi \tilde{B}(y)\psi_j(y)\psi_i(y)\mu(dy)$ and $F_i = \int_\Xi \tilde{f}(y)\psi_i(y)\mu(dy)$ for $1 \leq i,j \leq K$. The tuple $\{u_i\}_{i=1}^K \in V^K$ can be identified with a tensor in $\mathbb{R}^K \otimes V$.

Remark 4.30. In practice, (4.76) can be replaced by

$$\mathcal{E}(u,v) = \sum_{k=1}^K \omega^k \|B(y^k)v(y^k) - f(y^k)\|_{C(y^k)}^2 = \sum_{k=1}^K \omega^k \|v(y^k) - u(y^k)\|_{\tilde{B}(y^k)}^2, \tag{4.81}$$

where $\{(y^k,\omega^k)\}_{k=1}^K$ is a suitable quadrature rule for the integration over Ξ with respect to the measure μ.

Under the assumptions of Section 4.7.2 and if we assume that C admits an affine representation $C(\xi) = \sum_{i=1}^{R_C} C_i\eta_i(\xi)$, then $\tilde{B}(\xi)$ and $\tilde{f}(\xi)$ admit affine representations $\tilde{B}(\xi) = \sum_{i=1}^{\tilde{R}} \tilde{B}_i\tilde{\lambda}_i(\xi)$ and $\tilde{f}(\xi) = \sum_{i=1}^{\tilde{L}} \tilde{f}_i\tilde{\gamma}_i(\xi)$, respectively. Therefore, equation (4.78) can be written in the form of the tensor-structured equation (4.65), where all the quantities are replaced by their tilded versions. Under the stronger assumptions of Section 4.7.2, if we assume that C admits a representation of the form $C(\xi) = \sum_{i=1}^{R_C} C_i\eta_i^{(1)}(\xi_1)\ldots\eta_i^{(d)}(\xi_d)$, then $\tilde{B}(\xi)$ and $\tilde{f}(\xi)$ admit representations of the form $\tilde{B}(\xi) = \sum_{i=1}^{\tilde{R}} \tilde{B}_i\tilde{\lambda}_i^{(1)}(\xi_1)\ldots\tilde{\lambda}_i^{(d)}(\xi_d)$ and $\tilde{f}(\xi) = \sum_{i=1}^{\tilde{L}} \tilde{f}_i\tilde{\gamma}_i^{(1)}(\xi_1)\ldots\tilde{\gamma}_i^{(d)}(\xi_d)$. Therefore, equation (4.78) can be written in the form of the tensor-structured equation (4.70), where all the quantities are replaced by their tilded versions.

4.7.4 ▪ Interpolation (or collocation) method

Let $\Xi_K = \{y_k\}_{k\in I}$ be a set of $K = \#I$ interpolation points in Ξ, and let $\{\phi_k\}_{k\in I}$ be an associated set of interpolation functions. An interpolation $\mathcal{I}_K(u)$ of the solution of (4.43) can then be written

$$\mathcal{I}_K(u)(y) = \sum_{k\in I} u(y^k)\phi_k(y),$$

where $u(y^k) \in V$ is the solution of

$$B(y^k)u(y^k) = f(y^k), \quad k \in I, \tag{4.82}$$

which can be written as the operator equation

$$Au_I = F, \tag{4.83}$$

where $u_I = \{u(y^k)\}_{k\in I} \in V^K$, $F = \{f(y^k)\}_{k\in I} \in (W')^K$, and $A : V^K \to (W')^K$.

Order-two tensor structure

The tuples u_I and F can be identified with tensors in $\mathbb{R}^K \otimes V$ and $\mathbb{R}^K \otimes W'$, respectively. Also, the operator A can be identified with a tensor in $\mathbb{R}^{K\times K} \otimes \mathcal{L}(V, W')$. Under the assumption (4.63) on $B(\xi)$, A can be written in the form (4.64), where $\Lambda_i \in \mathbb{R}^{K\times K}$ is the diagonal matrix $\mathrm{diag}(\lambda_i(y^1),\ldots,\lambda_i(y^K))$. Also, under the assumption (4.61) on f, F can be written in the form (4.62), where $\gamma_i = (\gamma_i(y^1),\ldots,\gamma_i(y^K)) \in \mathbb{R}^K$.

Higher-order tensor structure

Now, using the notation of Section 4.6.2, we consider a tensorized interpolation grid $\Xi_K = \Xi_{1,K_1} \times \cdots \times \Xi_{d,K_d}$ and a corresponding set of interpolation functions $\{\phi_k(y) = \phi^1_{k_1}(y_1)\ldots\phi^d_{k_d}(y_d)\}_{k\in I}$, with $I = \times_{k=1}^d\{1,\ldots,K_k\}$. The tuples u_I and F can now be identified with tensors in $\mathbb{R}^{K_1} \otimes\cdots\otimes \mathbb{R}^{K_d} \otimes V$ and $\mathbb{R}^{K_1} \otimes\cdots\otimes \mathbb{R}^{K_d} \otimes W'$, respectively, and the operator A can be identified with a tensor in $\mathbb{R}^{K_1\times K_1}\otimes\cdots\otimes\mathbb{R}^{K_d\times K_d}\otimes\mathcal{L}(V,W')$. Under the assumptions of Section 4.7.2, A can be written in the form (4.69), with $\Lambda_i^{(\nu)} = \mathrm{diag}(\lambda_i^{(\nu)}(y_\nu^1),\ldots,\lambda_i^{(\nu)}(y_\nu^{K_\nu})) \in \mathbb{R}^{K_\nu\times K_\nu}$, and F can be written in the form (4.67), with $\gamma_i^{(\nu)} = (\gamma_i^{(\nu)}(y_\nu^1),\ldots,\gamma_i^{(\nu)}(y_\nu^{K_\nu})) \in \mathbb{R}^{K_\nu}$.

Remark 4.31. Note that the interpolation (or collocation) method provides an approximation u_I in $S \otimes V$, with $S = \mathrm{span}\{\phi_k\}_{k\in I}$, that coincides with the approximation obtained by a "pseudospectral" Galerkin method where the integrals over Ξ are approximated using a numerical quadrature with $\{y_k\}_{k\in I}$ as the set of integration points (see Remarks 4.26 and 4.30).

4.7.5 ▪ Low-rank structures of the solution of parameter-dependent or stochastic equations

When solving parameter-dependent or stochastic equations with low-rank tensor methods, the first question that should be asked is: does the solution u present a low-rank structure or admit an accurate approximation with low rank? Unfortunately, there are only a few quantitative answers to this question.

When $u \in L^p_\mu(\Xi; V)$ is seen as an order-two tensor in $\overline{L^p_\mu(\Xi)\otimes V}^{\|\cdot\|_p}$ (see Section 4.3.5), there exist some results about the convergence of best rank-r approximations for some classes of functions. For $p = 2$, these results are related to the decay of singular values of u (or equivalently of the compact operator associated with u). For $p = \infty$, these results are related to the convergence of the Kolmogorov widths of the set of solutions $u(\Xi)$. Exploiting these results requires a fine analysis of parameter-dependent (or stochastic) equations to make precise the class of their solutions. The question is more difficult when looking at u as a higher-order tensor in $\overline{L^p_{\mu_1}(\Xi_1)\otimes\cdots\otimes L^p_{\mu_d}(\Xi_d)\otimes V}^{\|\cdot\|_p}$, in particular because of the combinatorial nature of the definition of rank. Some results are available for the convergence of best rank-r approximations of some general classes of functions for canonical or tree-based Tucker formats [45, 75, 79]. However, these results usually exploit some global regularity and do not exploit specific structures of the solution (such as anisotropy), which would again require a detailed analysis of the parameter-dependent equations.

Example 4.32. Low-rank structures can be induced by particular parametrizations of operators and right-hand sides. As an example, consider the equation $B(\xi_1)u(\xi_1,\xi_2) = f(\xi_2)$, where u and f are considered as tensors in $L^2_{\mu_1}(\Xi_1)\otimes L^2_{\mu_2}(\Xi_2)\otimes V$ and $L^2_{\mu_1}(\Xi_1)\otimes L^2_{\mu_2}(\Xi_2)\otimes W'$, respectively. Here, $\text{rank}_1(f) = 1$ and $\text{rank}_2(f) = \text{rank}_3(f)$. Then, $\text{rank}_2(u) = \text{rank}_2(f)$. ∎

Example 4.33. Some structures are particular cases of low-rank structures and can therefore be captured by low-rank methods, such as low effective dimensionality or low-order interactions. For example, a function $u(\xi_1,\ldots,\xi_d)$ that can be well approximated by a low-dimensional function $\tilde{u}(\xi_{\beta_1},\ldots,\xi_{\beta_k})$, with $\{\beta_1,\ldots,\beta_k\} := \beta \subset \{1,\ldots,d\}$, can therefore be approximated with a tensor $\tilde{u}$ with $\text{rank}_\beta(\tilde{u}) = 1$ and $\text{rank}_\alpha(\tilde{u}) = 1$ for any $\alpha \subset \{1,\ldots,d\} \setminus \beta$. When using tree-based tensor formats, the tree should be adapted to reveal these low-rank structures. ∎

Although only a few a priori results are available, it is observed in many applications that the solutions of parameter-dependent (or stochastic) equations can be well approximated using low-rank tensor formats. However, there is a need for a rigorous classification of problems in terms of the expected accuracy of the low-rank approximations of their solutions. In the following section, we set aside the discussion about the good approximability of functions by low-rank tensors and focus on numerical methods for computing low-rank approximations.

Remark 4.34. Note that results on the convergence of best r-term approximations on a polynomial basis for particular classes of parameter-dependent equations [25, 28, 51], which exploit the anisotropy of the solution map $u : \Xi \to V$, provide as a by-product upper bounds for the convergence of low-rank approximations.

4.8 ▪ Low-rank approximation for equations in tensor format

In this section, we present algorithms to approximate in low-rank formats the solution of the operator equation

$$Au = F, \tag{4.84}$$

where u is an element of a finite-dimensional tensor space $V = V_1 \otimes \cdots \otimes V_D$ and A is an operator from V to W', with $W = W_1 \otimes \cdots \otimes W_D$. We can distinguish two main families of algorithms. The first relies on the use of classical iterative methods with efficient low-rank truncations of iterates. The second directly computes a low-rank approximation of the solution based on the minimization of a certain distance between the solution u and its low-rank approximation, using either a direct minimization in subsets of low-rank tensors or suboptimal but constructive (greedy) algorithms.

The algorithms are presented in a general setting and will be detailed for the particular case of the parameter-dependent equations presented in Section 4.7. In this particular case, the space V can be considered as a space $S \otimes V$ of order-two tensors ($D = 2$), where S is a K-dimensional approximation space in either $L^2_\mu(\Xi)$ (for Galerkin methods) or $\mathbb{R}^K$ (for interpolation or collocation methods). The space V can also be considered as a space $S_1 \otimes \cdots \otimes S_d \otimes V$ of higher-order tensors ($D = d + 1$), where

S_ν is a K_ν-dimensional subspace of either $L^2_{\mu_\nu}(\Xi_\nu)$ (for Galerkin methods) or $\mathbb{R}^{K_\nu}$ (for interpolation or collocation methods).

In practice, bases of finite-dimensional tensor spaces V and W are introduced so that (4.84) can be equivalently rewritten

$$\mathbf{A}\mathbf{u} = \mathbf{F}, \tag{4.85}$$

where $\mathbf{u} \in \mathbf{X} = \mathbb{R}^{N_1} \otimes \cdots \otimes \mathbb{R}^{N_D}$ is the set of coefficients of u in the chosen basis of V, and where $\mathbf{A}\mathbf{u}$ and $\mathbf{F}$ are respectively the coefficients of Au and F in the dual basis of the chosen basis in W. Here $\mathbf{A}$ is an operator from $\mathbf{X}$ to $\mathbf{X}$.

4.8.1 ▪ Classical iterative methods using low-rank truncations

Simple iterative algorithms (e.g., Richardson iterations, gradient algorithm) take the form $u^{i+1} = M(u^i)$, where M is an iteration map that involves simple algebraic operations (additions, multiplications) between tensors. Low-rank truncation methods can systematically be used to reduce the storage and computational complexities of these algebraic operations. This results in approximate iterations

$$u^{i+1} \approx M(u^i),$$

where the iterates $\{u^i\}_{i\geq 1}$ are in low-rank format. The resulting algorithm can be analyzed as a perturbed version of the initial algorithm (see, e.g., [46]). The reader is referred to [3, 51, 53] for a detailed introduction to these techniques in a general algebraic setting and to [59] for an application to parameter-dependent equations. Note that these iterative methods usually require constructing good preconditioners in low-rank tensor formats (see [40, 49, 70, 82]).

As an example, let us consider simple Richardson iterations for solving (4.85), where $M(\mathbf{u}) = \mathbf{u} - \alpha(\mathbf{A}\mathbf{u} - \mathbf{F})$. Approximate iterations can take the form

$$\mathbf{u}^{i+1} = \Pi_\epsilon(\mathbf{u}^i - \alpha(\mathbf{A}\mathbf{u}^i - \mathbf{F})),$$

where Π_ϵ is an operator that associates with a tensor $\mathbf{u}$ an approximation $\Pi_\epsilon(\mathbf{u})$ in low-rank format with a certain precision ϵ.

Low-rank truncation controlled in 2-norm

For a control of the error in the 2-norm, the operator Π_ϵ provides an approximation $\Pi_\epsilon(\mathbf{w})$ of a tensor $\mathbf{w}$ such that $||\mathbf{w} - \Pi_\epsilon(\mathbf{w})||_2 \leq \epsilon ||\mathbf{w}||_2$, where $||\mathbf{w}||_2 = (\sum_{i_1,\dots,i_D} |\mathbf{w}_{i_1,\dots,i_D}|^2)^{1/2}$. For order-two tensors, $\Pi_\epsilon(\mathbf{w})$ is obtained by truncating the SVD of $\mathbf{w}$, which can be computed with standard and efficient algorithms. For higher-order tensors, low-rank truncations (in tree-based Tucker format) can be obtained using efficient higher-order SVD algorithms (also allowing a control of the error; see Section 4.4.4) or other optimization algorithms in subsets of low-rank tensors.

Low-rank truncation controlled in ∞-norm

For a control of the error in the ∞-norm, the operator Π_ϵ should provide an approximation $\Pi_\epsilon(\mathbf{w})$ of a tensor $\mathbf{w}$ such that $||\mathbf{w} - \Pi_\epsilon(\mathbf{w})||_\infty \leq \epsilon ||\mathbf{w}||_\infty$, where $||\mathbf{w}||_\infty = \max_{i_1,\dots,i_D} |\mathbf{w}_{i_1,\dots,i_D}|$. A practical implementation of the truncation operator Π_ϵ can be based on the EIM [5, 57] or higher-order extensions of ACA algorithms [71].

Remark 4.35. For the solution of parameter-dependent equations with interpolation methods, simple iterative algorithms take the form

$$u^{i+1}(\xi)=M(\xi)(u^i(\xi)), \quad \xi\in\Xi_K,$$

where $M(\xi)$ is a parameter-dependent iteration map and Ξ_K is a discrete parameter set. For the example of Richardson iterations, exact iterations are $u^{i+1}(\xi) = u^i(\xi) - \alpha(B(\xi)u^i(\xi) - f(\xi))$, $\xi \in \Xi_K$. Low-rank truncations of the iterates $\{u^i(\xi)\}_{\xi\in\Xi_K}$ should therefore be controlled in the 2-norm (resp. ∞-norm) if one is interested in a mean square (resp. uniform) control of the error over the discrete parameter set Ξ_K. However, note that under some assumptions on the regularity of a function, controlling the 2-norm may be sufficient for controlling the ∞-norm.

Remark 4.36. Note that one could be interested in controlling the error with respect to other norms, such as the norm $||\mathbf{w}||_{(\infty,2)} = \max_{i_1}(\sum_{i_2}|\mathbf{w}_{i_1,i_2}|^2)^{1/2}$ for an order-two tensor $\mathbf{w}$. For the solution of parameter-dependent equations with interpolation methods, where $\mathbf{w}\in\mathbb{R}^K\otimes\mathbb{R}^{\dim(V)}$ represents the components on an orthonormal basis of V of samples $\{w(\xi)\}_{\xi\in\Xi_K}\in V^K$ of a function w on a discrete parameter set Ξ_K, we have $||\mathbf{w}||_{(\infty,2)} = \max_{\xi\in\Xi_K}||w(\xi)||_V$. This allows a uniform control over Ξ_K of the error measured in the V-norm. A practical implementation of the truncation operator with a control in norm $||\cdot||_{(\infty,2)}$ can be based on the GEIM [57].

4.8.2 ▪ Minimizing a residual-based distance to the solution

A distance $\mathscr{E}(u,w)$ from w to the solution u of (4.84) can be defined by using a residual norm,

$$\mathscr{E}(u,w)=||Aw-F||_D, \tag{4.86}$$

where $D:W'\to W$ is an operator that defines on W' an inner product norm $||\cdot||_D=\sqrt{\langle D\cdot,\cdot\rangle}$. The operator D plays the role of a preconditioner. It can be chosen such that $||\cdot||_D=||\cdot||_{W'}$, but it can also be defined in a different way. For a linear operator A, $w\mapsto\mathscr{E}(u,w)$ is a quadratic functional.

Remark 4.37. Note that the distance $\mathscr{E}(u,w)$ between the tensors u and w in V corresponds to a distance $\boldsymbol{\mathscr{E}}(\mathbf{u},\mathbf{w})$ between the associated tensors $\mathbf{u}$ and $\mathbf{w}$ in $\mathbb{R}^{N_1}\otimes\cdots\otimes\mathbb{R}^{N_D}$, with $\mathscr{E}(\mathbf{u},\mathbf{w})=||\mathbf{Aw}-\mathbf{F}||_{\mathbf{D}}$, for some operator $\mathbf{D}:\mathbf{X}\to\mathbf{X}$.

Let $\mathscr{S}_r$ denote a subset of tensors in V with bounded rank r. Let u_r denote the minimizer of $w\mapsto\mathscr{E}(u,w)$ over $\mathscr{S}_r$, i.e.,

$$\mathscr{E}(u,u_r)=\min_{w\in\mathscr{S}_r}\mathscr{E}(u,w). \tag{4.87}$$

If A is a linear operator such that $\alpha||v||_V\leq||Av||_{W'}\leq\beta||v||_V$ and if the operator D is such that $\alpha_D||\cdot||_{W'}\leq||\cdot||_D\leq\beta_D||\cdot||_{W'}$, then

$$\tilde{\alpha}||u-w||_V\leq\mathscr{E}(u,v)\leq\tilde{\beta}||u-w||_V, \tag{4.88}$$

with $\tilde{\beta}=\beta_D\beta$ and $\tilde{\alpha}=\alpha_D\alpha$, and the solution u_r of (4.87) is a quasi-best approximation in $\mathscr{S}_r$ with respect to the norm $||\cdot||_V$, with

$$||u-u_r||_V\leq\frac{\tilde{\beta}}{\tilde{\alpha}}\inf_{w\in\mathscr{S}_r}||u-w||_V. \tag{4.89}$$

In practice, an approximation u_r in a certain low-rank format can be obtained by directly solving the optimization problem (4.87) over a subset $\mathcal{S}_r$ of low-rank tensors. Constructive algorithms presented in Section 4.5 (which provide only suboptimal approximations) can also be applied and should be preferred when dealing with complex numerical models.

Remark 4.38. Equation (4.89) highlights the utility of working with well-chosen norms, such that $\tilde{\beta}/\tilde{\alpha} \approx 1$ if one is interested in minimizing the error in the norm $\|\cdot\|_V$. In [9, 26], the authors introduce a norm on W such that the residual norm $\|Aw - F\|_{W'}$ coincides with the error $\|w-u\|_V$, where $\|\cdot\|_V$ is a norm of interest. Quasi-best approximations are then computed using an iterative algorithm.

Remark 4.39. For linear problems, a necessary condition of optimality for problem (4.87) is[16]

$$\langle Au_r - F, DA\delta w\rangle = 0 \quad \text{for all } \delta w \in T_{u_r}\mathcal{S}_r, \tag{4.90}$$

where $T_{u_r}\mathcal{S}_r$ is the tangent space to the manifold $\mathcal{S}_r$ at u_r. Since $\mathcal{S}_r$ is neither a linear space nor a convex set, the condition (4.90) is not a sufficient condition for u_r to be a solution of (4.87).

Remark 4.40. For linear symmetric coercive problems, where $V = W$, A^{-1} defines a norm $\|\cdot\|_{A^{-1}}$ on W' such that $\|F\|_{A^{-1}} = \langle F, A^{-1}F\rangle$. Then, letting $D = A^{-1}$, the distance to the solution can be chosen as

$$\mathcal{E}(u,w) = \|Aw - F\|_{A^{-1}} = \|w-u\|_A, \tag{4.91}$$

where $\|w\|_A^2 = \langle Aw, w\rangle$. In this case, minimizing $w \mapsto \mathcal{E}(u,w)$ on a subset $\mathcal{S}_r$ provides a best approximation of u in $\mathcal{S}_r$ with respect to the operator norm $\|\cdot\|_A$. Denoting $J(w) = \langle Aw, w\rangle - 2\langle F, w\rangle$, we have $\mathcal{E}(u,w)^2 = J(w) - J(u)$, so that minimizing $\mathcal{E}(u,w)$ is equivalent to minimizing the functional $J(w)$, which is a strongly convex quadratic functional.

Parameter-dependent equations

We now consider the particular case of solving parameter-dependent equations using Galerkin or interpolation methods. For Petrov–Galerkin methods (see Section 4.7.3), the distance $\mathcal{E}(u,w)$ can be chosen as in (4.86) with an operator D such that $\|Au\|_D = \|Au\|_{W'}$ or simply $\|Au\|_D = \|\mathbf{Au}\|_2$. For minimal residual Galerkin methods (see Section 4.7.3), the distance $\mathcal{E}(u,w)$ can be chosen such that

$$\mathcal{E}(u,w)^2 = \int_\Xi \|B(y)w(y) - f(y)\|_{C(y)}^2 \mu(dy), \tag{4.92}$$

which corresponds to $\mathcal{E}(u,w) = \|w-u\|_{\tilde{A}}^2 = \|\tilde{A}w - \tilde{F}\|_{\tilde{A}^{-1}}^2$. In the case of interpolation (or collocation) methods (see Section 4.7.4), the distance $\mathcal{E}(u,w)$ can be chosen such that

$$\mathcal{E}(u,w)^2 = \sum_{k=1}^{K} \omega^k \|B(y^k)w(y^k) - f(y^k)\|_{C(y^k)}^2, \tag{4.93}$$

[16]This stationarity condition reveals the importance of the analysis of the manifold structure of subsets of tensors with bounded rank (see, e.g., [37, 84]).

with suitable weights $\{\omega^k\}_{k=1}^K$.[17] In both case (4.92) and case (4.93), with a linear operator $B(\xi)$ satisfying the assumptions of Section 4.7.3, property (4.88) is satisfied, where in the case of interpolation methods, $\|\cdot\|_V$ coincides with the norm $\|\cdot\|_{2,K}$ defined by (4.40).

Remark 4.41. The distance could also be chosen as

$$\mathscr{E}(u,w) = \sup_{1\leq k\leq K} \|B(y^k)w(y^k) - f(y^k)\|_{C(y^k)},$$

thereby moving from a Hilbert setting to a Banach setting. This is the classical framework for the RB methods. With a linear operator $B(\xi)$ satisfying the assumptions of Section 4.7.3, Property (4.88) is satisfied, where $\|\cdot\|_V$ coincides with the norm $\|\cdot\|_{\infty,K}$ defined by (4.39). The optimal rank-r approximation u_r in $\mathbb{R}^K \otimes V$, such that $\mathscr{E}(u,u_r) = \min_{w\in\mathscr{R}_r} \mathscr{E}(u,w)$, satisfies

$$\|u - u_r\|_{\infty,K} \leq \frac{\tilde{\beta}}{\tilde{\alpha}} \min_{v\in\mathscr{R}_r} \|u - w\|_{\infty,K} = \frac{\tilde{\beta}}{\tilde{\alpha}} d_r(u(\Xi_K))_V,$$

where $d_r(u(\Xi_K))_V$ is the Kolmogorov r-width of the discrete set of solutions $u(\Xi_K)$ in V. In practice, one can rely on an algorithm based on a greedy construction of subspaces $V_r \subset V$, such as presented in Section 4.6.1 (replacing $\|w(y) - u(y)\|_V$ by $\|B(y)w(y) - f(y)\|_{C(y)}$). This is the so-called offline phase of RB methods, and convergence results for this algorithm can be found in [10, 12, 31], where the error $\|u - u_r\|_{\infty,K} = \sup_{y\in\Xi_K} \|u(y) - P_{V_r} u(y)\|_V$ is compared with the best rank-r approximation error $\rho_r^{(\infty,K)}(u) = d_r(u(\Xi_K))_V$.

4.8.3 ▪ Coupling iterative methods and residual norm minimizations

Methods presented in Sections 4.8.1 and 4.8.2 can be combined. This allows the use of a larger class of iterative solvers for which one iteration takes the form $u^{i+1} = M(u^i)$, with $M(u^i) = C_i^{-1}G(u^i)$, where C_i is an operator given in low-rank format whose inverse C_i^{-1} is not computable explicitly. At iteration i, a low-rank approximation u^{i+1} of $C_i^{-1}G(u^i)$ can be obtained by minimizing the functional $w \mapsto \mathscr{E}(C_i^{-1}G(u^i), w) = \|C_i w - G(u^i)\|_\star$, where $\|\cdot\|_\star$ is some computable residual norm, either by a direct optimization in subsets of low-rank tensors or by using greedy algorithms.

The above iterations can be associated with an advanced iterative method for solving the linear system $Au = F$ (C_i could be the inverse of a known preconditioner of the operator A or a piece of the operator A in a method based on operator splitting) or with a nonlinear iterative solver for solving the nonlinear equation $A(u) = F$, with A being a nonlinear map. For example, for a Newton solver, C_i would be the differential of A (tangent operator) at u^i.

4.8.4 ▪ Galerkin approaches for low-rank approximations

The minimal residual-based approaches presented in Section 4.8.2 are robust approaches that guarantee the convergence of low-rank approximations. However, when an ap-

[17] By identifying an element $F \in (\mathbb{R}^K \otimes W)'$ with an element $\{f_k\}_{k=1}^K \in (W')^K$, (4.93) corresponds to (4.86) with an operator D such that $\|F\|_D^2 = \sum_{k=1}^K \omega^k \|f_k\|_{C(y^k)}^2$, i.e., $D = \sum_{k=1}^K \Omega_k \otimes C(y^k)$, with $\Omega_k \in \mathbb{R}^{K\times K}$ such that $(\Omega^k)_{ij} = \delta_{ik}\delta_{jk}\omega^k$. For $C(y) = C$ independent of y, $D = \Omega \otimes C$, with $\Omega = \mathrm{diag}(\omega^1, \dots, \omega^k) \in \mathbb{R}^{K\times K}$.

proximation u_r is defined as the minimizer of the residual norm $||Au_r - F||_D = \langle u_r - u, A^*DA(u_r - u)\rangle^{1/2}$ with a certain operator D, these approaches may suffer from ill-conditioning and they may induce high computational costs since they require operations between objects with a possibly high rank (operator A^*DA and right-hand side $A^*DAu = A^*DF$). Moreover, in the context of parameter-dependent (or stochastic) equations, they require solving problems that do not in general have the structure of standard parameter-independent problems, and, therefore, they cannot rely on standard parameter-independent (or deterministic) solvers (see Section 4.8.5). To address these issues, low-rank approximations can also be defined using other Galerkin orthogonality criteria (see [14, 64]).

Galerkin orthogonality

Let us assume that $V = W$. Let $\mathscr{S}_r$ denote a subset of tensors in V with a rank bounded by r. An approximation u_r in $\mathscr{S}_r$ can be searched such that the residual $Au_r - F$ is orthogonal to the tangent space $T_{u_r}\mathscr{S}_r$ to the manifold $\mathscr{S}_r$ at u_r, i.e., such that

$$\langle Au_r - F, \delta w\rangle = 0 \quad \text{for all } \delta w \in T_{u_r}\mathscr{S}_r. \tag{4.94}$$

Remark 4.42. If $\mathscr{S}_r$ admits a multilinear parametrization of the form (4.27), then $u_r \in \mathscr{S}_r$ can be written $u_r = F_{\mathscr{S}_r}(p_1, \ldots, p_M)$. Assuming that the parameters are in vector spaces P_i, (4.94) is equivalent to a set of M coupled equations on the parameters $(p_1, \ldots, p_M) \in P_1 \times \cdots \times P_M$:

$$\langle AF_{\mathscr{S}_r}(p_1, \ldots, p_M) - F, F_{\mathscr{S}_r}(p_1, \ldots, \delta p_i, \ldots, p_M)\rangle = 0 \quad \forall \delta p_i \in P_i, \tag{4.95}$$

$1 \leq i \leq M$.

To simplify the presentation of this formulation, let us consider the particular case where an approximation u_{r-1} of rank $r-1$ in $S \otimes V$ is given and let us define $\mathscr{S}_r = u_{r-1} + \mathscr{R}_1$. Then, $u_r \in \mathscr{R}_r$ is searched under the form $u_r = u_{r-1} + w_r$, where $w_r = s_r \otimes v_r \in \mathscr{R}_1$ is a rank-one correction of u_{r-1} that must satisfy the Galerkin orthogonality condition (4.94). Since $T_{u_r}\mathscr{S}_r = T_{w_r}\mathscr{R}_1 = \{\delta w = s \otimes v_r + s_r \otimes v : s \in S, v \in V\}$, the condition (4.94) becomes

$$\langle Aw_r - (F - Au_{r-1}), \delta w\rangle = 0 \quad \text{for all } \delta w \in T_{w_r}\mathscr{R}_1 \tag{4.96}$$

or, equivalently,

$$\langle As_r \otimes v_r - (F - Au_{r-1}), s \otimes v_r\rangle = 0 \quad \text{for all } s \in S, \tag{4.97a}$$

$$\langle As_r \otimes v_r - (F - Au_{r-1}), s_r \otimes v\rangle = 0 \quad \text{for all } v \in V. \tag{4.97b}$$

Equation (4.96) may not have any solution w_r or may have many solutions (possibly infinitely many for problems formulated in infinite-dimensional spaces), with particular solutions that are not relevant for approximating u. In practice, heuristic algorithms are used to solve equation (4.96), such as alternating-direction algorithms. This consists of successively solving equation (4.97a) with a fixed v_r and equation (4.97b) with a fixed s_r. It has to be noted that when (4.96) admits solutions, this heuristic algorithm selects particular solutions u_r that are usually relevant. It can be understood in the case where A is symmetric and defines a norm, since the alternating-direction algorithm coincides with an alternating-minimization algorithm. This explains why solutions that minimize a certain residual norm are selected.

Remark 4.43. To illustrate, let us consider the case where A is such that $\langle Av, w\rangle = \langle v, w\rangle$, with $\langle\cdot,\cdot\rangle$ the canonical norm on $S \otimes V$. For $r = 1$, equation (4.94) becomes $\langle s_1 \otimes v_1 - u, s \otimes v_1 + s_1 \otimes v\rangle = 0$ for all $(s, v) \in S \times V$. This implies that $u(v_1) = \|v_1\|_V^2 s_1$ and $u^*(s_1) = \|s_1\|_S^2 v_1$, i.e., the solutions of equation (4.94) are the tensors $s_1 \otimes v_1$, where v_1 and s_1 are right and left singular vectors of u associated with a certain nonzero singular value $\|v_1\|_V^2 = \|s_1\|_S^2$. In this case, the alternating-direction algorithm corresponds to a power method for finding the dominant eigenvector of $u^* \circ u$. This means that the algorithm allows us to select the optimal rank-one approximation with respect to the canonical norm $\|\cdot\|$ among possible solutions of equation (4.96). See [63, 64] for further details on algorithms and their interpretation for solving invariant subspace problems associated with generalizations of SVDs.

In many applications, Galerkin orthogonality criteria (in conjunction with suitable algorithms for solving (4.94)) provide rather good low-rank approximations, although they are not associated with minimizing a certain distance to the solution and therefore do not guarantee convergence with the rank. Note that these Galerkin approaches have also been applied successfully to some nonlinear problems (A being a nonlinear map); see [67, 78].

Parameter-dependent equations. For the case of parameter-dependent equations, equation (4.97b) for $v_r \in V$ (with fixed s_r) is a parameter-independent equation of the form

$$\widehat{B}_{r,r} v_r = \widehat{f}_r - \sum_{i=1}^{r-1} \widehat{B}_{r,i} v_i,$$

where $\widehat{B}_{r,i} = \int_\Xi B(y) s_r(y) s_i(y) \mu(dy)$ and $\widehat{f}_r = \int_\Xi f(y) s_r(y) \mu(dy)$ for Galerkin methods, or $\widehat{B}_{r,i} = \sum_{k=1}^K \omega^k B(y^k) s_r(y^k) s_i(y^k)$ and $\widehat{f}_r = \sum_{k=1}^K \omega^k f(y^k) s_r(y^k)$ for interpolation (or collocation) methods. When $B(\xi)$ and $f(\xi)$ admit affine representations of the form (4.63) and (4.61), respectively, then $\widehat{B}_{r,i}$ and $\widehat{f}_r$ can be interpreted as evaluations of B and f for particular values of parameter-dependent functions λ_i and γ_i. Therefore, equation (4.97b) can usually be solved with a standard solver for parameter-independent or deterministic models (see Section 4.8.5 for further discussion and illustration of model example 1 from Section 4.7.1).

Petrov–Galerkin orthogonality

For nonsymmetric problems, possibly with $V \neq W$, an alternative construction based on a Petrov–Galerkin orthogonality criterion is proposed in [64]. At step r, assuming that $u_{r-1} \in \mathcal{R}_{r-1} \subset V$ is known, a rank-one correction $w_r = s_r \otimes v_r \in V$ and an auxiliary rank-one element $\tilde{w}_r = \tilde{s}_r \otimes \tilde{v}_r \in W$ are constructed such that

$$\begin{aligned} &\langle Aw_r - (F - Au_{r-1}), \delta\tilde{w}\rangle = 0 \quad \text{for all } \delta\tilde{w} \in T_{\tilde{w}_r}\mathcal{R}_1 \subset W, \\ &\langle A\delta w, \tilde{w}_r\rangle = \langle \delta w, w_r\rangle_V \quad \text{for all } \delta w \in T_{w_r}\mathcal{R}_1 \subset V \end{aligned}$$

or, equivalently,

$$\langle As_r \otimes v_r - (F - Au_{r-1}), \tilde{s} \otimes \tilde{v}_r\rangle = 0 \quad \text{for all } \tilde{s} \in \tilde{S}, \tag{4.98a}$$

$$\langle As \otimes v_r, \tilde{s}_r \otimes \tilde{v}_r\rangle = \langle s \otimes v_r, s_r \otimes v_r\rangle_V \quad \text{for all } s \in S, \tag{4.98b}$$

$$\langle As_r \otimes v_r - (F - Au_{r-1}), \tilde{s}_r \otimes \tilde{v}\rangle = 0 \quad \text{for all } \tilde{v} \in W, \tag{4.98c}$$

$$\langle As_r \otimes v, \tilde{s}_r \otimes \tilde{v}_r\rangle = \langle s_r \otimes v, s_r \otimes v_r\rangle_V \quad \text{for all } v \in V. \tag{4.98d}$$

A heuristic alternating-direction algorithm can be used that consists of solving successively equations (4.98a) to (4.98d) respectively for s_r, $\tilde{s}_r$, v_r, $\tilde{v}_r$ (see [64] for the application to evolution problems and [14] for the application to parameter-dependent equations).

Parameter-dependent equations. For the case of parameter-dependent equations, we note that problems (4.98c) and (4.98d) are parameter-independent equations, respectively, of the form

$$\widehat{B}_{r,r} v_r = \hat{f}_r - \sum_{i=1}^{r-1} \widehat{B}_{r,i} v_i \quad \text{and} \quad \widehat{B}^*_{r,r} \tilde{v}_r = \hat{s}_r R_V v_r,$$

where $R_V : V \to V'$ is such that $\langle R_V v, \hat{v}\rangle = \langle v, \hat{v}\rangle_V$, and where $\widehat{B}_{r,i} = \int_\Xi B(y) s_i(y) \tilde{s}_r(y) \mu(dy)$, $\widehat{f}_r = \int_\Xi f(y) \tilde{s}_r(y) \mu(dy)$, and $\hat{s}_r = \int_\Xi s_r(y)^2 \mu(dy)$ for the case of Galerkin methods, or $\widehat{B}_{r,i} = \sum_{k=1}^K \omega^k B(y^k) s_i(y^k) \tilde{s}_r(y^k)$, $\widehat{f}_r = \sum_{k=1}^K \omega^k f(y^k) \tilde{s}_r(y^k)$, and $\hat{s}_r = \sum_{k=1}^K \omega^k s_r(y^k)^2$ for the case of interpolation (or collocation) methods. When $B(\xi)$ and $f(\xi)$ admit affine representations of the form (4.63) and (4.61), respectively, then $\widehat{B}_{r,i}$ and $\widehat{f}_r$ can be interpreted as evaluations of B and f for particular values of parameter-dependent functions λ_i and γ_i. Therefore, equations (4.98c) and (4.98d) can usually be solved with a standard solver for parameter-independent or deterministic models.[18]

4.8.5 ▪ Greedy construction of subspaces for parameter-dependent equations

Any of the algorithms presented in Section 4.5 can be applied to construct a low-rank approximation w of the solution u when a suitable measure $\mathcal{E}(u,w)$ of the error has been defined. However, the variants based on the progressive construction of reduced spaces (see Section 4.5.2) are particularly pertinent in the context of parameter-dependent problems since they only involve the solution of a sequence of problems with the complexity of a parameter-independent problem, and of reduced-order parameter-dependent models that are the projections of the initial model on the reduced spaces V_r. Moreover, specific algorithms can take advantage of the particular structure of the parameter-dependent model, so that parameter-independent equations have the structure of standard problems that can be solved with available solution codes for parameter-independent models. The reader is referred to [20, 62, 63, 65] for practical implementations and illustrations of the behavior of these algorithms.

Here, we illustrate how to apply the algorithm presented in Section 4.5.2 for computing a sequence of low-rank approximations in $S \otimes V$. The algorithm relies on the construction of optimal nested subspaces $V_r \subset V$. For simplicity, we only consider the case of a symmetric coercive problem (e.g., model example 1 described in Section 4.7.1), with a residual-based error $\mathcal{E}(u,w)$ defined by (4.91), which corresponds to $\mathcal{E}(u,w)^2 = J(w) - J(u)$ with $J(w) = \langle Aw, w\rangle - 2\langle F, w\rangle$. We have

$$J(w) = \int_\Xi \big(\langle B(y)w(y), w(y)\rangle - 2\langle f(y), w(y)\rangle\big)\mu(dy)$$

[18] For example, when applied to model example 2 in Section 4.7.1, equation (4.98c) is a weak form of the deterministic evolution equation $\hat{\alpha}\frac{\partial v_r}{\partial t} - \nabla\cdot(\hat{\kappa}\nabla v_r) = \hat{g}$, with initial condition $v_r(\cdot,0) = \mathbb{E}_\mu(u_0(\cdot,\xi)\tilde{s}_r)$, and where $\hat{\alpha} = \mathbb{E}_\mu(s_i(\xi)\tilde{s}_r(\xi))$ and $\hat{\kappa}(\cdot) = \mathbb{E}_\mu(\kappa(\cdot,\xi)s_r(\xi)\tilde{s}_r(\xi))$.

for the case of Galerkin methods, or

$$J(w)=\sum_{k=1}^{K}\omega^k\big(\langle B(y^k)w(y^k),w(y^k)\rangle-2\langle f(y^k),w(y^k)\rangle\big)$$

for the case of interpolation (or collocation) methods. To simplify the presentation, $\mathbb{E}_\mu(g(\xi))$ will denote either $\int_\Xi g(y)\mu(dy)$ in the case of Galerkin methods or $\sum_{k=1}^{K}\omega^k g(y^k)$ in the case of interpolation (or collocation) methods. With this notation, we have

$$J(w)=\mathbb{E}_\mu(\langle B(\xi)w(\xi),w(\xi)\rangle-2\langle f(\xi),w(\xi)\rangle).$$

Remark 4.44. For nonsymmetric problems, such as model example 2 described in Section 4.7.1, one can adopt the formulation presented in Section 4.7.3 and use the expression (4.92) (or (4.93)) for $\mathcal{E}(u,w)$. We apply the algorithm by following the same lines, simply replacing operators and right-hand sides $(A,F,B(\xi),f(\xi))$ by their tilded versions $(\tilde{A},\tilde{F},\tilde{B}(\xi)=B(\xi)^*C(\xi)B(\xi),\tilde{f}(\xi)=B(\xi)^*C(\xi)f(\xi))$, and searching the approximation as the minimizer of the functional $J(w)=\langle\tilde{A}w,w\rangle-2\langle\tilde{F},w\rangle$.

The algorithm is defined by (4.34). At iteration r, the $(r-1)$-dimensional RB $\{v_1,\dots,v_{r-1}\}$ of the subspace $V_{r-1}\subset V$ being given, the rank-r approximation $u_r=\sum_{i=1}^{r}s_i^{(r)}\otimes v_i$ is defined by

$$J(u_r)=\min_{\substack{V_r\in\mathbb{G}_r(V)\\ V_r\supset V_{r-1}}}\ \min_{w\in S\otimes V_r}J(w)=\min_{v_r\in V_r}\ \min_{\{s_i\}_{i=1}^r\in S^r}J\left(\sum_{i=1}^{r}s_i\otimes v_i\right).$$

To solve this optimization problem, we can use an alternating-minimization algorithm, solving alternately

$$\min_{v_r\in V}J\left(\sum_{i=1}^{r}s_i\otimes v_i\right)\quad\text{or}\tag{4.99a}$$

$$\min_{\{s_i\}_{i=1}^r\in S^r}J\left(\sum_{i=1}^{r}s_i\otimes v_i\right).\tag{4.99b}$$

Solution of problem (4.99a) (a parameter-independent equation)

Problem (4.99a) is equivalent to solving the equation

$$\widehat{B}_{r,r}v_r=\widehat{f}_r-\sum_{i=1}^{r-1}\widehat{B}_{r,i}v_i,\tag{4.100}$$

where the operators $\widehat{B}_{r,i}:V\to W'$ and the vector $\widehat{f}_r\in W'$ are defined by

$$\widehat{B}_{r,i}=\mathbb{E}_\mu(B(\xi)s_r(\xi)s_i(\xi))\quad\text{and}\quad\widehat{f}_r=\mathbb{E}_\mu(f(\xi)s_r(\xi)).$$

When $B(\xi)$ and $f(\xi)$ admit affine representations of the form (4.63) and (4.61), respectively, then $\widehat{B}_{r,i}$ and $\widehat{f}_r$ take the forms

$$\widehat{B}_{r,i}=\sum_{l=1}^{R}B_l\widehat{\lambda}_{l,r,i}\quad\text{and}\quad\widehat{f}_r=\sum_{l=1}^{L}f_l\widehat{\gamma}_{l,r},$$

where

$$\widehat{\lambda}_{l,r,i} = \mathbb{E}_\mu(\lambda_l(\xi)s_r(\xi)s_i(\xi)) \quad \text{and} \quad \widehat{\gamma}_{l,r} = \mathbb{E}_\mu(\gamma_l(\xi)s_r(\xi)).$$

Let us emphasize that the operator $\widehat{B}_{r,i} = \sum_{l=1}^R B_l \widehat{\lambda}_{l,r,i}$ has the same structure as the parameter-dependent operator $B(\xi) = \sum_{l=1}^R B_l \lambda_l(\xi)$, but $\widehat{\lambda}_{l,r,i}$ does not correspond to an evaluation of the function $\lambda_l(\xi)$ at some particular values of ξ. However, looking at B as a family of operators parametrized by the λ_l, then $\widehat{B}_{r,i}$ corresponds to an evaluation of B at some given values $\widehat{\lambda}_{l,r,i}$ of the parameters λ_l. In practical applications, this means that this problem can be solved with standard solvers (for parameter-independent or deterministic models).

Example 4.45. When this algorithm is applied to model example 1 (see Section 4.7.1), $\widehat{f}_r$ and $\widehat{B}_{r,i}$ are such that

$$\langle \widehat{f}_r, w\rangle = \int_D \widehat{g}_r\, w \quad \text{and} \quad \langle \widehat{B}_{r,i} v, w\rangle = \int_D \nabla w \cdot \widehat{\varkappa}_{r,i} \cdot \nabla v,$$

with $\widehat{g}_r(\cdot) = \mathbb{E}_\mu(g(\cdot,\xi)s_r(\xi)) \in L^2(D)$ and $\varkappa_{r,i}(\cdot) = \mathbb{E}_\mu(\varkappa(\cdot,\xi)s_r(\xi)s_i(\xi))$. Problem (4.99a) therefore corresponds to the solution of the deterministic diffusion equation $-\nabla \cdot (\widehat{\varkappa}_{r,r} \nabla v_r) = \hat{g}_r + \sum_{i=1}^{r-1} \nabla \cdot (\widehat{\varkappa}_{r,i} \nabla v_i)$. ■

Solution of problem (4.99b) (a reduced-order parameter-dependent equation)

Problem (4.99b) is equivalent to approximating the solution in $S \otimes V_r$, where V_r is a reduced space with basis $\{v_1, \ldots, v_r\}$. Let $\mathbf{B}(\xi) \in \mathbb{R}^{r\times r}$ be the parameter-dependent matrix defined by $\mathbf{B}(\xi) = (\langle B(\xi)v_j, v_i\rangle)_{i,j=1}^r = (b(v_j, v_i; \xi))_{i,j=1}^r$, and let $\mathbf{f}(\xi) \in \mathbb{R}^r$ be the parameter-dependent vector defined by $\mathbf{f}(\xi) = (\langle f(\xi), v_i\rangle)_{i=1}^r$. If $B(\xi)$ and $f(\xi)$ admit affine representations of the forms (4.63) and (4.61), respectively, then $\mathbf{B}(\xi) = \sum_{l=1}^R \mathbf{B}_l \lambda_l(\xi)$ and $\mathbf{f}(\xi) = \sum_{l=1}^L \mathbf{f}_l \gamma_l(\xi)$, where the matrices $\mathbf{B}_l \in \mathbb{R}^{r\times r}$ and vectors $\mathbf{f}_l \in \mathbb{R}^r$ are associated with projections on the reduced spaces V_r of operators B_l and vectors f_l, respectively. Then, denoting $\mathbf{s} = (s_i)_{i=1}^r \in S^r = S \otimes \mathbb{R}^r$, problem (4.99b) is equivalent to

$$\mathbb{E}_\mu(\mathbf{t}(\xi)^T \mathbf{B}(\xi)\mathbf{s}(\xi)) = \mathbb{E}_\mu(\mathbf{t}(\xi)^T \mathbf{f}(\xi)) \quad \forall\, \mathbf{t} \in S \otimes \mathbb{R}^r, \tag{4.101}$$

which requires solving a system of $\dim(S) \times r$ equations. When $\dim(S)$ is large, order reduction methods can also be used at this step to obtain a reduced-order approximation of the solution $\mathbf{s}$. For example, for the case of high-dimensional parameter-dependent models with a projection on a tensor-structured approximation space S or with an interpolation on a tensor-structured grid, we can rely on sparse approximation methods or the higher-order low-rank methods presented in Section 4.4. Note that in the case of Galerkin methods, equation (4.101) defines the Galerkin approximation of the reduced-order parameter-dependent equation

$$\mathbf{B}(\xi)\mathbf{s}(\xi) = \mathbf{f}(\xi), \tag{4.102}$$

so that an approximation of $\mathbf{s}$ can also be obtained by sampling-based approaches, based on many sample evaluations $\mathbf{s}(y^k) = \mathbf{B}(y^k)^{-1}\mathbf{f}(y^k)$ (only requiring the solution of reduced systems of equations).

Remark 4.46. As mentioned in Remark 4.44, this algorithm can be applied to nonsymmetric problems such as model example 2 (described in Section 4.7.1) by using minimal residual formulations. However, when applied to this evolution problem, the algorithm requires solving parameter-independent problems of the form (4.100) that are global over the space-time domain (time-stepping methods cannot be used) and may be computationally intractable. An additional order reduction can be introduced by also exploiting the tensor structure of the space $V = V(D) \otimes V(I)$ of space-time functions (see Remark 4.21). Low-rank methods that exploit this structure allow the complexity of the representations of space-time functions to be reduced.

Bibliography

[1] I. BABUSKA, R. TEMPONE, AND G. E. ZOURARIS, *Solving elliptic boundary value problems with uncertain coefficients by the finite element method: The stochastic formulation*, Computer Methods in Applied Mechanics and Engineering, 194 (2005), pp. 1251–1294.

[2] M. BACHMAYR AND W. DAHMEN, *Adaptive near-optimal rank tensor approximation for high-dimensional operator equations*, Foundations of Computational Mathematics, 15 (2015), pp. 839–898.

[3] J. BALLANI AND L. GRASEDYCK, *A projection method to solve linear systems in tensor format*, Numerical Linear Algebra with Applications, 20 (2013), pp. 27–43.

[4] J. BALLANI, L. GRASEDYCK, AND M. KLUGE, *Black box approximation of tensors in hierarchical Tucker format*, Linear Algebra and its Applications, 438 (2013), pp. 639–657.

[5] M. BARRAULT, Y. MADAY, N. C. NGUYEN, AND A. T. PATERA, *An empirical interpolation method: Application to efficient reduced-basis discretization of partial differential equations*, Comptes Rendus Mathematique, 339 (2002), pp. 667–672.

[6] J. BECK, F. NOBILE, L. TAMELLINI, AND R. TEMPONE, *Convergence of quasi-optimal stochastic Galerkin methods for a class of PDES with random coefficients*, Computers and Mathematics with Applications, 67 (2014), pp. 732–751.

[7] G. BEYLKIN, B. GARCKE, AND M. J. MOHLENKAMP, *Multivariate regression and machine learning with sums of separable functions*, Journal of Computational Physics, 230 (2011), pp. 2345–2367.

[8] M. BIERI, R. ANDREEV, AND C. SCHWAB, *Sparse tensor discretization of elliptic SPDEs*, SIAM Journal on Scientific Computing, 31 (2009), pp. 4281–4304.

[9] M. BILLAUD-FRIESS, A. NOUY, AND O. ZAHM, *A tensor approximation method based on ideal minimal residual formulations for the solution of high-dimensional problems*, ESAIM: Mathematical Modelling and Numerical Analysis, 48 (2014), pp. 1777–1806.

[10] P. BINEV, A. COHEN, W. DAHMEN, R. DEVORE, G. PETROVA, AND P. WOJTASZCZYK, *Convergence rates for greedy algorithms in reduced basis methods*, SIAM Journal on Mathematical Analysis, 43 (2011), pp. 1457–1472.

[11] C. LE BRIS, T. LELIEVRE, AND Y. MADAY, *Results and questions on a nonlinear approximation approach for solving high-dimensional partial differential equations*, Constructive Approximation, 30 (2009), pp. 621–651.

[12] A. BUFFA, Y. MADAY, A. T. PATERA, C. PRUD'HOMME, AND G. TURINICI, *A priori convergence of the greedy algorithm for the parametrized reduced basis method*, ESAIM: Mathematical Modelling and Numerical Analysis, 46 (2012), pp. 595–603. Special volume in honor of Professor David Gottlieb.

[13] E. CANCES, V. EHRLACHER, AND T. LELIEVRE, *Convergence of a greedy algorithm for high-dimensional convex nonlinear problems*, Mathematical Models and Methods in Applied Sciences, 21 (2011), pp. 2433–2467.

[14] E. CANCES, V. EHRLACHER, AND T. LELIEVRE, *Greedy algorithms for high-dimensional non-symmetric linear problems*, In *ESAIM: Proceedings*, volume 41, pages 95–131. EDP Sciences, 2013.

[15] J. CHARRIER, *Strong and weak error estimates for elliptic partial differential equations with random coefficients*, SIAM Journal on Numerical Analysis, 50 (2012), pp. 216–246.

[16] J. CHARRIER AND A. DEBUSSCHE, *Weak truncation error estimates for elliptic PDEs with lognormal coefficients*, *Stochastic Partial Differential Equations: Analysis and Computations*, 1 (2013), pp. 63–93.

[17] J. CHARRIER, R. SCHEICHL, AND A. TECKENTRUP, *Finite element error analysis of elliptic PDEs with random coefficients and its application to multilevel Monte Carlo methods*, SIAM Journal on Numerical Analysis, 51 (2013), pp. 322–352.

[18] S. CHATURANTABUT AND D. C. SORENSEN, *Nonlinear model reduction via discrete empirical interpolation*, SIAM Journal on Scientific Computing, 32 (2010), pp. 2737–2764.

[19] M. CHEVREUIL, R. LEBRUN, A. NOUY, AND P. RAI, *A Least-Squares Method for Sparse Low Rank Approximation of Multivariate Functions*, ArXiv e-prints arXiv:1305.0030, April 2013.

[20] M. CHEVREUIL AND A. NOUY, *Model order reduction based on proper generalized decomposition for the propagation of uncertainties in structural dynamics*, International Journal for Numerical Methods in Engineering, 89 (2012), pp. 241–268.

[21] F. CHINESTA, R. KEUNINGS, AND A. LEYGUE, *The Proper Generalized Decomposition for Advanced Numerical Simulations—A Primer*, SpringerBriefs in Applied Sciences and Technology, Springer, 2014.

[22] F. CHINESTA, P. LADEVEZE, AND E. CUETO, *A short review on model order reduction based on proper generalized decomposition*, Archives of Computational Methods in Engineering, 18 (2011), pp. 395–404.

[23] A. CHKIFA, A. COHEN, R. DEVORE, AND C. SCHWAB, *Sparse adaptive Taylor approximation algorithms for parametric and stochastic elliptic PDEs*, ESAIM: Mathematical Modelling and Numerical Analysis, 47 (2013), pp. 253–280.

[24] A. CHKIFA, A. COHEN, AND C. SCHWAB, *High-dimensional adaptive sparse polynomial interpolation and applications to parametric PDEs*, Foundations of Computational Mathematics, 14 (2014), pp. 601–633.

[25] A. CHKIFA, A. COHEN, AND C. SCHWAB, *Breaking the curse of dimensionality in sparse polynomial approximation of parametric PDEs*, Journal de Mathématiques Pures et Appliquées, 103 (2015), pp. 400–428.

[26] A. COHEN, W. DAHMEN, AND G. WELPER, *Adaptivity and variational stabilization for convection-diffusion equations*, ESAIM: Mathematical Modelling and Numerical Analysis, 46 (2012), pp. 1247–1273.

[27] A. COHEN, R. DEVORE, AND C. SCHWAB, *Convergence rates of best* n-*term Galerkin approximations for a class of elliptic SPDEs*, Foundations of Computational Mathematics, 10 (2010), pp. 615–646.

[28] A. COHEN, R. DEVORE, AND C. SCHWAB, *Analytic regularity and polynomial approximation of parametric and stochastic elliptic PDEs*, Analysis and Applications, 09 (2011), pp. 11–47.

[29] L. DE LATHAUWER, B. DE MOOR, AND J. VANDEWALLE, *A multilinear singular value decomposition*, SIAM Journal on Matrix Analysis and Applications, 21 (2000), pp. 1253–1278.

[30] A. DEFANT AND K. FLORET, *Tensor Norms and Operator Ideals*, North-Holland, Amsterdam, New York, 1993.

[31] R. DEVORE, G. PETROVA, AND P. WOJTASZCZYK, *Greedy algorithms for reduced bases in Banach spaces*, Constructive Approximation, 37 (2013), pp. 455–466.

[32] A. DOOSTAN, A. VALIDI, AND G. IACCARINO, *Non-intrusive low-rank separated approximation of high-dimensional stochastic models*, Computer Methods in Applied Mechanics and Engineering, 263 (2013), pp. 42–55.

[33] M. ESPIG, L. GRASEDYCK, AND W. HACKBUSCH, *Black box low tensor-rank approximation using fiber-crosses*, Constructive Approximation, 30 (2009), pp. 557–597.

[34] M. ESPIG AND W. HACKBUSCH, *A regularized Newton method for the efficient approximation of tensors represented in the canonical tensor format*, Numerische Mathematik, 122 (2012), pp. 489–525.

[35] M. ESPIG, W. HACKBUSCH, AND A. KHACHATRYAN, *On the Convergence of Alternating Least Squares Optimisation in Tensor Format Representations*, ArXiv e-prints, May 2015.

[36] A. FALCÓ AND W. HACKBUSCH, *On minimal subspaces in tensor representations*, Foundations of Computational Mathematics, 12 (2012), pp. 765–803.

[37] A. FALCO, W. HACKBUSCH, AND A. NOUY, *Geometric Structures in Tensor Representations*, arXiv preprint arXiv:1505.03027, 2015.

[38] A. FALCÓ AND A. NOUY, *Proper generalized decomposition for nonlinear convex problems in tensor Banach spaces*, Numerische Mathematik, 121 (2012), pp. 503–530.

[39] P. FRAUENFELDER, C. SCHWAB, AND R. A. TODOR, *Finite elements for elliptic problems with stochastic coefficients*, Computer Methods in Applied Mechanics and Engineering, 194 (2005), pp. 205–228.

[40] L. GIRALDI, A. NOUY, AND G. LEGRAIN, *Low-rank approximate inverse for preconditioning tensor-structured linear systems*, SIAM Journal on Scientific Computing, 36 (2014), pp. A1850–A1870.

[41] L. GIRALDI, A. NOUY, G. LEGRAIN, AND P. CARTRAUD, *Tensor-based methods for numerical homogenization from high-resolution images*, Computer Methods in Applied Mechanics and Engineering, 254 (2013), pp. 154–169.

[42] L. GRASEDYCK, *Hierarchical singular value decomposition of tensors*, SIAM Journal on Matrix Analysis and Applications, 31 (2010), pp. 2029–2054.

[43] L. GRASEDYCK, D. KRESSNER, AND C. TOBLER, *A literature survey of low-rank tensor approximation techniques*, GAMM-Mitteilungen, 36 (2013), pp. 53–78.

[44] W. HACKBUSCH, *Tensor Spaces and Numerical Tensor Calculus*, volume 42 of *Springer Series in Computational Mathematics*, Springer, Heidelberg, 2012.

[45] W. HACKBUSCH, *The use of sparse grid approximation for the r-term tensor representation*, In J. Garcke and M. Griebel, eds., *Sparse Grids and Applications*, volume 88 of *Lecture Notes in Computational Science and Engineering*, pages 151–159. Springer, Berlin, Heidelberg, 2013.

[46] W. HACKBUSCH, B. KHOROMSKIJ, AND E. TYRTYSHNIKOV, *Approximate iterations for structured matrices*, Numerische Mathematik, 109 (2008), pp. 365–383.

[47] W. HACKBUSCH AND S. KUHN, *A new scheme for the tensor representation*, Journal of Fourier Analysis and Applications, 15 (2009), pp. 706–722.

[48] M. KAHLBACHER AND S. VOLKWEIN, *Galerkin proper orthogonal decomposition methods for parameter dependent elliptic systems*, Discussiones Mathematicae: Differential Inclusions, Control and Optimization, 27 (2007), pp. 95–117.

[49] B. KHOROMSKIJ, *Tensor-structured preconditioners and approximate inverse of elliptic operators in $\mathbb{R}^d$*, Constructive Approximation, 30 (2009), pp. 599–620.

[50] B. KHOROMSKIJ, *Tensors-structured numerical methods in scientific computing: Survey on recent advances*, Chemometrics and Intelligent Laboratory Systems, 110 (2012), pp. 1–19.

[51] B. N. KHOROMSKIJ AND C. SCHWAB, *Tensor-structured Galerkin approximation of parametric and stochastic elliptic PDEs*, SIAM Journal on Scientific Computing, 33 (2011), pp. 364–385.

[52] T. G. KOLDA AND B. W. BADER, *Tensor decompositions and applications*, SIAM Review, 51 (2009), pp. 455–500.

[53] D. KRESSNER AND C. TOBLER, *Low-rank tensor Krylov subspace methods for parametrized linear systems*, SIAM Journal on Matrix Analysis and Applications, 32 (2011), pp. 1288–1316.

[54] P. LADEVÈZE, *Nonlinear Computational Structural Mechanics—New Approaches and Non-incremental Methods of Calculation*, Springer-Verlag, 1999.

[55] P. LADEVÈZE, J.C. PASSIEUX, AND D. NÉRON, *The LATIN multiscale computational method and the proper generalized decomposition*, Computer Methods in Applied Mechanics and Engineering, 199 (2010), pp. 1287–1296

[56] W.A. LIGHT AND E.W. CHENEY, *Approximation Theory in Tensor Product Spaces*, volume 1169, Springer-Verlag, Berlin, 1985.

[57] Y. MADAY, N. C. NGUYEN, A. T. PATERA, AND G. S. H. PAU, *A general multipurpose interpolation procedure: The magic points*, Communications on Pure and Applied Analysis, 8 (2009), pp. 383–404.

[58] H. G. MATTHIES AND A. KEESE, *Galerkin methods for linear and nonlinear elliptic stochastic partial differential equations*, Computer Methods in Applied Mechanics and Engineering, 194 (2005), pp. 1295–1331.

[59] H. G. MATTHIES AND E. ZANDER, *Solving stochastic systems with low-rank tensor compression*, Linear Algebra and Its Applications, 436 (2012).

[60] A. MUGLER AND H.-J. STARKLOFF, *On the convergence of the stochastic Galerkin method for random elliptic partial differential equations*, ESAIM: Mathematical Modelling and Numerical Analysis, 47 (2013), pp. 1237–1263.

[61] F. NOBILE, R. TEMPONE, AND C. WEBSTER, *An anisotropic sparse grid stochastic collocation method for partial differential equations with random input data*, SIAM Journal on Numerical Analysis, 46 (2008), pp. 2411–2442.

[62] A. NOUY, *A generalized spectral decomposition technique to solve a class of linear stochastic partial differential equations*, Computer Methods in Applied Mechanics and Engineering, 196 (2007), pp. 4521–4537.

[63] A. NOUY, *Generalized spectral decomposition method for solving stochastic finite element equations: Invariant subspace problem and dedicated algorithms*, Computer Methods in Applied Mechanics and Engineering, 197 (2008), pp. 4718–4736.

[64] A. NOUY, *A priori model reduction through proper generalized decomposition for solving time-dependent partial differential equations*, Computer Methods in Applied Mechanics and Engineering, 199 (2010), pp. 1603–1626.

[65] A. NOUY, *Proper generalized decompositions and separated representations for the numerical solution of high dimensional stochastic problems*, Archives of Computational Methods in Engineering, 17 (2010), pp. 403–434.

[66] A. NOUY AND P. LADEVÈZE, *Multiscale computational strategy with time and space homogenization: A radial-type approximation technique for solving micro problems*, International Journal for Multiscale Computational Engineering, 170 (2004), pp. 557–574.

[67] A. NOUY AND O. P. LE MAÎTRE, *Generalized spectral decomposition method for stochastic nonlinear problems*, Journal of Computational Physics, 228 (2009), pp. 202–235.

[68] A. NOUY AND C. SOIZE, *Random field representations for stochastic elliptic boundary value problems and statistical inverse problems*, European Journal of Applied Mathematics, 25 (2014), pp. 339–373.

[69] I. OSELEDETS, *Tensor-train decomposition*, SIAM Journal on Scientific Computing, 33 (2011), pp. 2295–2317.

[70] I. OSELEDETS AND S. DOLGOV, *Solution of linear systems and matrix inversion in the TT-format*, SIAM Journal on Scientific Computing, 34 (2012), pp. A2718–A2739.

[71] I. OSELEDETS AND E. TYRTYSHNIKOV, *TT-cross approximation for multidimensional arrays*, Linear Algebra and Its Applications, 432 (2010), pp. 70–88, JAN.

[72] H. RAUHUT, R. SCHNEIDER, AND Z. STOJANAC, *Tensor Completion in Hierarchical Tensor Representations*, ArXiv e-prints, April 2014.

[73] T. ROHWEDDER AND A. USCHMAJEW, *On local convergence of alternating schemes for optimization of convex problems in the tensor train format*, SIAM Journal on Numerical Analysis, 51 (2013), pp. 1134–1162.

[74] T. ROUBÍČEK, *Nonlinear Partial Differential Equations with Applications*, volume 153, Springer Science and Business Media, 2013.

[75] R. SCHNEIDER AND A. USCHMAJEW, *Approximation rates for the hierarchical tensor format in periodic Sobolev spaces*, Journal of Complexity, 30 (2014), pp. 56–71. Dagstuhl 2012.

[76] C. SCHWAB AND C. GITTELSON, *Sparse tensor discretizations of high-dimensional parametric and stochastic PDEs*, Acta Numerica, 20 (2011), pp. 291–467.

[77] C. SCHWAB AND R. STEVENSON, *Space-time adaptive wavelet methods for parabolic evolution problems*, Mathematics of Computation, 78 (2009), pp. 1293–1318.

[78] L. TAMELLINI, O. LE MAÎTRE, AND A. NOUY, *Model reduction based on proper generalized decomposition for the stochastic steady incompressible Navier–Stokes equations*, SIAM Journal on Scientific Computing, 36 (2014), pp. A1089–A1117.

[79] V. TEMLYAKOV, *Estimates of best bilinear approximations of periodic functions*, Proceedings of the Steklov Institute of Mathematics, pages 275–293, 1989.

[80] V. TEMLYAKOV, *Greedy Approximation*, Cambridge Monographs on Applied and Computational Mathematics, Cambridge University Press, 2011.

[81] V. TEMLYAKOV, *Greedy approximation in convex optimization*, Constructive Approximation, 41 (2012), pp. 269–296.

[82] A. TOUZENE, *A tensor sum preconditioner for stochastic automata networks*, INFORMS Journal on Computing, 20 (2008), pp. 234–242.

[83] A. USCHMAJEW AND B. VANDEREYCKEN, *The Geometry of Algorithms Using Hierarchical Tensors*, Technical report, ANCHP-MATHICSE, Mathematics Section, Ecole Polytechnique Fédérale de Lausanne, 2012.

[84] A. USCHMAJEW AND B. VANDEREYCKEN, *The geometry of algorithms using hierarchical tensors*, Linear Algebra Appl., 439 (2013), pp. 133–166.

[85] B. VANDEREYCKEN, *Low-rank matrix completion by Riemannian optimization*, SIAM Journal on Optimization, 23 (2013), pp. 1214–1236.

Chapter 5

Model Reduction for High-Dimensional Parametric Problems by Tensor Techniques

Ivan Oseledets[19]

5.1 ▪ Introduction

Approximating multiparametric dependencies often boils down to approximating functions of many variables. The term "curse of dimensionality" was probably first used by Richard E. Bellmann to characterize the exponential complexity growth in the number of variables for a particular problem in dynamic programming. There are several results on the *tractability of multivariate function approximation*—see, for example, the book [41]—and many of them are negative. Nevertheless, despite these negative results, when a high-dimensional problem is encountered in an application, a good algorithm is often found! The most vivid example is the solution of the Schrödinger equation in quantum chemistry, where intricate methods have been proposed to solve it for large systems and with high accuracy. These methods are still in active development.

In this chapter we will discuss the current state of the art of *numerical tensor methods*. We introduce common low-parametric representations (also called *formats*) and outline basic and advanced algorithms for working with these representations.

Besides tensor techniques, one has to mention other approaches that have been successfully used for a wide range of high-dimensional problems:[20] sparse grids [6, 59], radial basis functions [5], best N-term approximations [61], and, of course, Monte Carlo and quasi-Monte Carlo methods. They have their own advantages and disadvantages, which should be taken into account for each particular case.

The chapter is organized as follows. First we briefly describe two classical tensor formats, namely the canonical format and the Tucker format, and discuss their properties. Then, we move on to the novel tensor formats: the tensor-train (TT) format and the hierarchical Tucker (HT) format, which are based on singular value decomposition (SVD) and do not have the intrinsic curse of dimensionality. Then, we describe how

[19]This work was supported by Russian Science Foundation grant 14-11-00659.

[20]The list of references for other methods is not complete. It just gives a few snapshots of possible directions; a comprehensive literature survey is beyond the scope of this chapter.

to solve different basic problems (linear systems, eigenvalue problems, interpolation), where the solution can be well approximated by a tensor in the TT format.

5.2 ▪ The concept of tensor formats

In this chapter, by "tensor" we will mean an array with d indices. This terminology, which may look awkward to experts in such fields as theoretical physics and differential geometry, is standard in the multilinear algebra community; see the review [34]. To do numerical computations we have to discretize the problem first, thus replacing a function of several variables by a tensor. Often this can be straightforwardly done by introducing a tensor product basis set. Then, the tensor is naturally defined as an array of coefficients of the expansion of a function with respect to this basis set. Other possibilities are not that straightforward. For example, the initial array is mapped somehow to a high-dimensional tensor. Even "one-dimensional" objects can be treated this way! This leads to the idea of quantization (also called tensorization), which can lead to tremendous complexity reduction.

So we assume that the discretization step has been done and we have a d-dimensional tensor $\mathbf{A}$. We will denote its elements by $A(i_1,\dots,i_d)$, where each i_k varies from one to n_k. The total storage required for $\mathbf{A}$ is $n_1 \dots n_d$. For simplicity, from now on in the complexity estimates we will assume that all mode sizes n_k are approximately the same, $n_k \approx n$. Even for $d \sim 10$ and $n \sim 32$ it is impossible to store the full tensor, so a certain approximation is required. A typical situation is that we allow for a certain approximation accuracy ε in some norm. The most common choice is the *Frobenius* norm, which is defined as

$$||\mathbf{A}||_F = \sqrt{\sum_{i_1,\dots,i_d} |A(i_1,\dots,i_d)|^2}.$$

An approximation of a tensor $\mathbf{A}$ by a tensor $\mathbf{B}$ is considered sufficiently good if

$$||\mathbf{A}-\mathbf{B}||_F \leq \varepsilon ||\mathbf{B}||_F.$$

The tensor $\mathbf{B}_F$ should be in a low-parametric *tensor format*. The choice of a good tensor format is crucial, and although it depends on the application, it should satisfy several basic requirements. First of all, the number of parameters should be small compared to the total number of elements of a tensor. The low-parametric representation should be computable at low cost, and basic operations in the tensor format should be performed without forming the full tensor.

The most widely used tensor formats are based on the idea of separation of variables. We will describe four of them: the canonical polyadic (CP) format, the Tucker format, and the more recent HT and TT representations, with the main focus on the TT format.

Where can these formats be useful? If the tensor is already in the low-parametric format, we may want to do basic linear algebra operations: add tensors, multiply them elementwise, compute scalar products and norms, find maximal and minimal elements. For all of the basic formats, these operations are easy to implement. What is not an easy task is how to find a low-parametric representation. The answer depends on how the tensor is actually defined. It can be defined either implicitly (as a solution of some problem) or explicitly (by a subroutine that allows us to compute certain elements). Below we summarize the most common cases.

1. Best-approximation problem: Given a tensor Y, find Z such that $||Y - Z||$ is minimal.

2. General optimization problems: Given a functional $F(X)$, find $X \in \mathcal{M}$, where $\mathcal{M}$ is the manifold of tensors in the given format, such that $F(X)$ is minimal.

3. Dynamical approximation: Given a trajectory $A(t)$, find a trajectory $X(t) \in \mathcal{M}$ that approximates $A(t)$ in a certain optimal sense.

4. Black-box approximation: Given a subroutine that computes any prescribed element of a tensor, approximate the tensor using the least samples possible.

In different tensor format the complexity and robustness of solving problems 1 to 4 varies significantly. We will start by introducing two classical tensor formats (the canonical and Tucker formats), describe their properties, present basic algorithms, and discuss their disadvantages. Then, the HT and TT formats will be introduced. While the TT format is a special case of the HT format, many algorithms are much easier to formulate in the TT format, so we will describe the TT format in detail and point out similar results for the HT format when appropriate. For the TT format we will briefly cover all the main computational techniques: the idea of alternating least-squares (ALS) algorithms, the density matrix renormalization group (DMRG), algorithm, and the alternating minimal energy (AMEN) algorithm for the adaptive selection of the number of parameters, a very important concept of cross-approximation both in two and many dimensions and also dynamical low-rank approximation of matrices and tensors.

5.3 ▪ Canonical format

One of the basic ideas of model reduction is the idea of separation of variables. In its most straightforward form it leads to the canonical format [7, 24]. In different fields different names are used for this representation (proper generalized decomposition (PGD), canonical polyadic decomposition, separated representation). The simplest case is the rank-one tensor of the form

$$A(i_1,\dots,i_d) = U_1(i_1)\dots U_d(i_d). \tag{5.1}$$

Note that an analogous representation can be written for functions of many variables by formally replacing discrete indices by continuous ones, but for illustration purposes we will consider only the discrete case.

The rank-R tensor is defined as a sum of rank-one tensors:

$$A(i_1,\dots,i_d) = \sum_{\alpha=1}^{r} U_1(i_1,\alpha)\dots U_d(i_d,\alpha). \tag{5.2}$$

The factors U_k in (5.2) are referred to as *canonical factors*, and the number R is referred to as the *canonical rank* of the representation (often the canonical rank is defined as the minimal number R for which the representation (5.2) exists; however, for data compression purposes it is not required to have R minimal, but just sufficiently small).

The canonical format is often considered as a *multidimensional generalization* of the *singular value decomposition* (SVD). Indeed, for $d = 2$ the matrix of the form (5.2) is just a matrix of rank not higher than R,

$$A = U_1 U_2^\top,$$

where U_1 and U_2 are matrices of sizes $n_1 \times R$ and $n_2 \times R$, respectively, and the canonical rank coincides with the classical matrix rank.

5.3.1 ▪ Approximation by low-rank tensors and the ALS algorithm

Consider the following basic tensor approximation problem: given a tensor A we would like to find a best rank-R approximation to A. This can be written as the minimization problem

$$||A-B||_F^2 \to \min,$$

where B is a tensor of canonical rank R. The minimization is performed over the factors $U_k(i_k, \alpha)$. In total there are $R\sum_{k=1}^{d} n_k$ parameters to be determined. The first-order optimality conditions take the form

$$\begin{aligned} M_k U_k' &= F_k, \\ M_k = \circ_{j\neq k}\Gamma_j, \quad \Gamma_j &= U_j^* U_j, \\ F_k(i_k,\alpha) = \sum_{\substack{1\leq i_s\leq n_s \\ s\neq k}} A(i_1,\ldots,i_d) &\prod_{\substack{1\leq l\leq d \\ l\neq k}} U_l(i_l,\alpha). \end{aligned} \tag{5.3}$$

This is a system of polynomial equations. Unless $d=2$, there are no robust algorithms to solve this system for several reasons that will be discussed later on. Still, one of the most popular ways to find an approximation to the solution is the *ALS algorithm*. It makes use of the polylinear structure of the problem. If all the factors except U_k are fixed, then the problem becomes a quadratic minimization problem and the first-order optimality condition reduces to a linear system of equations, which is cheap to solve. The iteration is then organized as a sweep: first we optimize over the first factor, then over the second, and so on. This is also a block Gauss–Seidel method for the system of equations (5.3). Local convergence of such a method is studied in [54], but it is well known that it can converge to local minima. Block coordinate descent can be replaced by much more sophisticated numerical methods, like Newton or quasi-Newton algorithms, but this does not solve the problem with local minima. TensorLab Toolbox in MATLAB [60] contains efficient implementation of several optimization algorithms for fitting the canonical format to the given data.

5.3.2 ▪ Why the canonical format can be bad

Nonexistence of the best approximation

One of the problems with the canonical format is that the best possible approximation may not exist [11], and the canonical rank can be different over the real and complex fields, so many of the classical matrix results do not generalize to many dimensions. The computation of the canonical rank is NP-complete [25]. As an example, consider the following case, which frequently arises in practice. Let T be a tensor of the form

$$T = a\otimes b\otimes\cdots\otimes b+\cdots+b\otimes\cdots\otimes a,$$

where a and b are two linearly independent vectors of length n. It can be shown[21] that T cannot be exactly represented as a sum of less than d rank-one terms, i.e., its

[21]This is not a trivial result and requires sophisticated techniques from algebraic geometry [35, 36].

canonical rank is equal to d. However, this tensor can be approximated arbitrarily well by a tensor of rank two. Consider an auxiliary tensor

$$P(t) = (b + ta) \otimes \cdots \otimes (b + ta),$$

and notice that

$$P'(0) = T.$$

The dependence on the parameter t is polynomial, so the derivative can be approximated by a finite difference and

$$T = \frac{P(h) - P(0)}{h} + \mathcal{O}(h),$$

which proves the claim. That means that for any $\varepsilon > 0$ there exists a rank-two tensor such that

$$||T - T_\varepsilon||_F \leq \varepsilon.$$

But an exact rank-two representation does not exist.

Dependence of the canonical rank on the field

If a matrix has real entries, its rank over the real field is equal to its rank over the complex field. This is not true for tensors. Consider the tensor generated by the function

$$\sin(x_1 + \cdots + x_d)$$

on a tensor product grid. It can be shown that it can be represented by a sum of d rank-one terms with real entries [3, 4]. This is a nontrivial trigonometric identity. No representations with smaller rank over the real field are known (although no formal proofs are known up to now either). Over the complex field, due to the identity

$$\sin x = \frac{e^{ix} - e^{-ix}}{2i},$$

the canonical rank is equal to two and it is obvious that this is the minimal rank.

To summarize, computing the canonical approximation can be hard. Despite the fact that it may lead to a very compact representation, if the rank r is small, sometimes it is very difficult to find this good representation and there are no guarantees that the approximation will be found even if it exists. Thus, alternatives have been proposed that use a larger number of parameters but allow for more stable representations. The second classical tensor format is the Tucker format.

5.4 ▪ Tucker format

The Tucker format [62] was introduced in statistics to deal with multiway data. It is a factor model for multiway data, and it still remains an important tool in chemometrics and psychometrics. As a general numerical tool to approximate functions, it was first considered in [9, 10] and studied extensively in [30, 48] and many other papers. A tensor is said to be in the Tucker format if

$$A(i_1, \ldots, i_d) = \sum_{\alpha_1, \ldots, \alpha_d} G(\alpha_1, \ldots, \alpha_d) U_1(i_1, \alpha_1) \ldots U_d(i_d, \alpha_d), \tag{5.4}$$

where the summation in α_k goes from 1 to r_k, $k = 1, \dots, n$. If we assume that $r_k \sim r$, then the number of parameters is equal to $\mathcal{O}(dnr + r^d)$; thus, the exponential factor is still there. However, for small dimensions, especially for $d = 3$, the Tucker format can be very efficient [9, 10, 30, 48]. So, the Tucker format can potentially be used for not-so-large d. Now we have instead of a single-rank parameter the *multilinear rank* $\mathbf{r} = (r_1, \dots, r_k)$.

An obvious relation between the canonical rank and the multilinear rank is $r_k \leq R$. Upper bounds for the canonical rank are very tricky to get.

5.4.1 ▪ Higher-order SVD

The quasi-optimal Tucker approximation can be computed using the SVD of auxiliary matrices: this is its main advantage. To show this, consider d unfoldings $\mathbf{A}_k = A(i_k; i_1 \dots i_{k-1} i_{k+1} \dots i_d)$, i.e., the kth index enumerates the rows of A_k, and all other indices enumerate the columns of it. The elements of the unfolding are taken from the tensor in lexicographical order. It can be shown that $\operatorname{rank}\mathbf{A}_k \leq r_k$. Moreover, the opposite is true: if

$$\operatorname{rank}\mathbf{A}_k = r_k, \quad k = 1, \dots, d,$$

then there exists a Tucker decomposition with the multilinear rank $(r_1, \dots, r_d)$. The proof is constructive and gives the algorithm to compute the factors. For each unfolding we compute the truncated SVD of the form

$$\mathbf{A}_k = \mathbf{U}_k \Lambda_k \mathbf{V}_k^\top,$$

where $\mathbf{U}_k$ is an $n_k \times r_k$ orthonormal matrix. Then, the matrices U_k can be chosen as Tucker factors, and the core $\mathbf{G}$ can be computed as a multidimensional contraction

$$G(\alpha_1, \dots, \alpha_d) = \sum_{i_1, \dots, i_d} A(i_1, \dots, i_d) U_1(i_1, \alpha_1) \dots U_d(i_d, \alpha_d).$$

Note that for $d = 2$ the core $\mathbf{G}$ can be made diagonal; thus, the canonical format and the Tucker format are the same low-rank format for matrices. For $d > 2$, the orthogonal reduction of the tensor to the diagonal form is typically not possible except for special cases.

5.5 ▪ SVD-based tensor formats

The Tucker format reduces the tensor approximation problem to a low-rank approximation of d matrices—unfoldings. The manifold of tensors with fixed multilinear ranks is in fact an intersection of low-rank matrix manifolds, and the quasi-optimal approximation can be recovered from the SVD of auxiliary matrices. This is not the only option for tensor-to-matrix reduction. In multilinear algebra, these ideas were first studied in the tree-Tucker [49] and HT [23] formats. Both formats were based on a similar idea of dimensionality reduction based on a dimension tree, but the final representations are different. The tree-Tucker format is shown [50] to be algebraically equivalent to the TT format, which has a remarkably simple structure:

$$A(i_1, \dots, i_d) = G_1(i_1) \dots G_d(i_d), \tag{5.5}$$

where $G_k(i_k)$ is an $r_{k-1} \times r_k$ matrix for each fixed $i_k = 1, \dots, n_k$ and $r_0 = r_d = 1$. It is convenient to denote the elements of the matrix $G_k(i_k)$ by $G_k(\alpha_{k-1}, i_k, \alpha_k)$, leading to an index representation of the form

$$A(i_1,\ldots,i_d)=\sum_{\alpha_0,\ldots,\alpha_d} G_0(\alpha_0,i_1,\alpha_1)G_1(\alpha_1,i_1,\alpha_2)\ldots G_d(\alpha_{d-1},i_d,\alpha_d). \tag{5.6}$$

This representation has been known for many years as the *matrix product state* in condensed matter theory, where it has been used to represent *wavefunctions* of spin systems. The HT format is introduced in [23], and basic algorithms are considered in [20]. The HT format is based on successively applying the Tucker decomposition. First, the indices are merged into tuples (i.e., the dimension of the tensor is decreased), and then the Tucker decomposition is computed and the procedure is applied to the resulting Tucker core again. The only parameters to be stored are the *transfer tensors* from the children of the given node to the node itself. For a binary tree, the transfer tensors are three-dimensional.

The TT format can be considered as a special case of the HT format with linear-dimension reduction tree, but the computational algorithms are typically the same (i.e., if there is a robust and fast algorithm in the TT format, there is a similar algorithm in the HT format). The relation between the TT and HT formats is thoughtfully studied in [22]. In this chapter, we will use the TT format to illustrate the basic concepts since it is much easier for the presentation. However, many results have been obtained for the HT format as well.

Shortly after the introduction of new tensor formats, several researchers (Thomas Huckle probably being the first to present this explicitly in 2010 at a GAMM Seminar in Leipzig) pointed out that these formats are not actually new. Similar approaches were under active development in different seemingly unconnected areas. In solid state physics and quantum information theory, matrix product states (MPSs) [31, 53, 57] are used to approximate many-body quantum systems. MPSs are equivalent to the TT format. This connection has already been very fruitful in developing new algorithms in both areas. The concept of the *density matrix renormalization group* (DMRG) leads to an absolutely nonstandard numerical algorithm for approximately solving optimization problems in tensor formats. However, the linear algebra viewpoint gives new algorithms that do not directly correspond to some physical problems, most notably the idea of cross-approximation.

It is also worth noting that the HT format can also be found in chemistry; for example, the ML-MCTDH method [39, 40] utilizes the same idea. Again, these results lay dormant for the general mathematical public. A general concept beyond tensor decompositions is *tensor networks* (for a discussion, see, for example, [27]), which are the main object of study in quantum information theory. For the TT/MPS format, the network is linear; for the HT format it is a tree. More complicated networks can be considered, such as a two-dimensional grid or a periodic network (tensor chain). However, it is now quite clear that these networks correspond to tensor formats that are not stable, i.e., only networks without cycles can be used if stability is required.

Despite its simplicity, the TT format can be very powerful. As was said above, the algorithms for the TT/MPS format can be generalized to the case of a more general tree; thus, from now on we will focus on describing the algorithms in the TT format in detail.

5.6 ▪ TT format

In this section, we describe the basic properties of the TT format. All of the algorithms have open-source implementations in MATLAB, at `http://github.com/oseledets/TT-Toolbox`, and in Python, at `http://github.com/oseledets/ttpy`.

5.6.1 ▪ Prerequisites

When dealing with tensors, convenient and consistent notation is very important. Several different variants have been proposed, and there is no standard at the moment. In this section, we will follow the notation introduced in the paper [37]. The idea is to use as many well-established matrix operations as possible and to reduce explicit usage of indices. First, we introduce several important definitions.

Definition 5.1 (Unfoldings). *The kth unfolding of a tensor $X \in \mathbb{R}^{n_1 \times \cdots \times n_d}$ is the matrix $\mathbf{X}^{\langle k \rangle} \in \mathbb{R}^{(n_1 \cdots n_k) \times (n_{k+1} \cdots n_d)}$ that aligns all entries $X(i_1, \ldots, i_d)$ with fixed $i_1, \ldots, i_k$ in a row of $\mathbf{X}^{\langle k \rangle}$, and rows and columns are ordered colexicographically. The inverse of unfolding is reconstructing, which we denote as*

$$X = \mathrm{Ten}_k(\mathbf{X}^{\langle k \rangle}),$$

that is, the tensor $X \in \mathbb{R}^{n_1 \times \cdots \times n_d}$ has the kth unfolding $\mathbf{X}^{\langle k \rangle} \in \mathbb{R}^{(n_1 \ldots n_k) \times (n_{k+1} \ldots n_d)}$.

Definition 5.2 (TT/MPS format). *A tensor $X \in \mathbb{R}^{n_1 \times \cdots \times n_d}$ is in TT/MPS format if there exist* core tensors *$C_k \in \mathbb{R}^{r_{k-1} \times n_k \times r_k}$ with $r_0 = r_d = 1$ such that*

$$X(i_1, \ldots, i_d) = \sum_{j_1=1}^{r_1} \cdots \sum_{j_{d-1}}^{r_{d-1}} C_1(1, i_1, j_1) \cdot C_2(j_1, i_2, j_2) \cdots C_d(j_{d-1}, i_d, 1)$$

for $i_k = 1, \ldots, n_i$ and $i = 1, \ldots, d$. Equivalently, we have

$$X(i_1, \ldots, i_d) = \mathbf{C}_1(i_1) \cdots \mathbf{C}_d(i_d),$$

where the $r_{k-1} \times r_k$ matrices $\mathbf{C}_k(i_k)$ are defined as the slices $C_k(:, i_k, :)$.

Observe that X can be parametrized by $\sum_{k=1}^{d} n_k r_{k-1} r_k \leq d n r^2$ degrees of freedom, where $n = \max\{n_k\}$ and $r = \max\{r_k\}$. In high-dimensional applications where TT/MPS tensors are practically relevant, n is constant or only mildly dependent on d. Hence, for large d, one obtains a considerable reduction in the degrees of freedom compared to a general tensor of size n^d.

Definition 5.3 (Left and right unfoldings of the cores). *For any core tensor $C_k \in \mathbb{R}^{r_{k-1} \times n_k \times r_k}$, we denote*

$$\mathbf{C}_k^{<} = \begin{bmatrix} C_k(:, 1, :) \\ \vdots \\ C_k(:, n_k, :) \end{bmatrix} \in \mathbb{R}^{(r_{k-1} n_k) \times r_k}, \qquad \mathbf{C}_k^{>} = \begin{bmatrix} C_k(:, :, 1)^\top \\ \vdots \\ C_k(:, :, r_k)^\top \end{bmatrix} \in \mathbb{R}^{(r_k n_k) \times r_{k-1}}.$$

The matrix $\mathbf{C}_k^{<}$ is called the left unfolding *of C_k and $\mathbf{C}_k^{>}$ is the* right unfolding.

Definition 5.4 (TT/MPS rank). *We call a vector $\mathbf{r} = (1, r_1, \ldots, r_{d-1}, 1)$ the* TT/MPS rank *of a tensor $X \in \mathbb{R}^{n_1 \times \cdots \times n_d}$ if*

$$\mathrm{rank} \mathbf{X}^{\langle k \rangle} = r_k, \qquad k = 1, \ldots, d-1.$$

If $r_k \leq \min\{\prod_{j=1}^{k} n_j, \prod_{j=k+1}^{d} n_j\}$, then X can be represented in TT/MPS format with core tensors $C_k \in \mathbb{R}^{r_{k-1} \times n_k \times r_k}$ of full multilinear rank, that is,

$$\mathrm{rank} \mathbf{C}_k^{<} = r_k \qquad \textit{and} \qquad \mathrm{rank} \mathbf{C}_k^{>} = r_{k-1}, \qquad k = 1, \ldots, d.$$

In addition, it is known (see [26, Lemma 4]) that for fixed $\mathbf{r}$ *such a full-rank condition on the core tensors implies that the set*

$$\mathcal{M} = \{X \in \mathbb{R}^{n_1 \times \cdots \times n_d} : \text{TT/MPS rank of } X \text{ is } \mathbf{r}\} \tag{5.7}$$

is a smooth embedded submanifold in $\mathbb{R}^{n_1 \times \cdots \times n_d}$.

Definition 5.5 (Partial products). *Define the left partial product* $X_{\leq k} \in \mathbb{R}^{n_1 \times \cdots \times n_k \times r_k}$ *as*

$$X_{\leq k}(i_1, \ldots, i_k, :) = \mathbf{C}_1(i_1) \ldots \mathbf{C}_i(i_k)$$

and the right partial product $X_{\geq k+1} \in \mathbb{R}^{n_{k+1} \times \cdots \times n_d \times r_k}$ *as*

$$X_{\geq k+1}(i_{k+1}, \ldots, i_d, :) = \mathbf{C}_{k+1}(i_{k+1}) \ldots \mathbf{C}_d(i_d).$$

Let a particular unfolding of each of these partial products be denoted as

$$\mathbf{X}_{\leq k} \in \mathbb{R}^{(n_1 \cdots n_k) \times r_k}, \quad \mathbf{X}_{\geq k+1} \in \mathbb{R}^{(n_{k+1} \cdots n_d) \times r_k}.$$

The obvious elementwise relation $X(i_1, \ldots, i_d) = X_{\leq k}(i_1, \ldots, i_k) X_{\geq i+1}(i_{k+1}, \ldots, i_d)$ *then translates into*

$$\mathbf{X}^{\langle k \rangle} = \mathbf{X}_{\leq k} \mathbf{X}_{\geq k+1}^{\top}.$$

Definition 5.6 (Recursive construction). *We note the recurrence relations*

$$\mathbf{X}_{\leq k} = (\mathbf{I}_{n_k} \otimes \mathbf{X}_{\leq k-1}) \mathbf{C}_k^{<}, \quad k = 1, \ldots, d, \tag{5.8}$$

starting from $\mathbf{X}_{\leq 0} = 1$, *and*

$$\mathbf{X}_{\geq k} = (\mathbf{X}_{\geq k+1} \otimes \mathbf{I}_{n_k}) \mathbf{C}_k^{>}, \quad k = 1, \ldots, d, \tag{5.9}$$

with $\mathbf{X}_{\geq d+1} = 1$. *Here* $\otimes$ *denotes the standard Kronecker product.*

Combining the above formulas, we note that

$$\mathbf{X}^{\langle k \rangle} = (\mathbf{I}_{n_k} \otimes \mathbf{X}_{\leq k-1}) \mathbf{C}_k^{<} \mathbf{X}_{\geq k+1}^{\top}, \tag{5.10}$$

which will be an important formula later. Using the recurrence relations for $\mathbf{X}_{\geq k}$, *we also obtain*

$$\mathbf{X}^{\langle k-1 \rangle} = \mathbf{X}_{\leq k-1} \mathbf{C}_k^{>\top} (\mathbf{X}_{\geq k+1} \otimes \mathbf{I}_{n_k})^{\top}, \tag{5.11}$$

which together with the previous formula allows us to pass from the $(k-1)$*th to the* k*th unfolding.*

5.6.2 ▪ Left and right orthogonalizations

Thanks to the recursive relations (5.8) and (5.9), it is possible to compute the QR decompositions of the matrices $\mathbf{X}_{\leq k}$ and $\mathbf{X}_{\geq k}$ efficiently.

Let us explain the case for $\mathbf{X}_{\leq k}$ in detail. First, compute a QR factorization (the $^<$ in $\mathbf{Q}_1^{<}$ is just notational for now but will become clear in Section 5.6.3):

$$\mathbf{X}_{\leq 1} = \mathbf{C}_1^{<} = \mathbf{Q}_1^{<} \mathbf{R}_1, \quad \text{with} \quad \mathbf{Q}_1^{<\top} \mathbf{Q}_1^{<} = \mathbf{I}_{r_1},\ \mathbf{Q}_1^{<} \in \mathbb{R}^{n_1 \times r_1},\ \mathbf{R}_1 \in \mathbb{R}^{r_1 \times r_1},$$

and insert it into the recurrence relation (5.8) to obtain

$$\mathbf{X}_{\le 2} = (\mathbf{I}_{n_2} \otimes \mathbf{Q}_1^< \mathbf{R}_1)\mathbf{C}_2^< = (\mathbf{I}_{n_2} \otimes \mathbf{Q}_1^<)(\mathbf{I}_{n_2} \otimes \mathbf{R}_1)\mathbf{C}_2^<.$$

Next, make another QR decomposition

$$(\mathbf{I}_{n_2} \otimes \mathbf{R}_1)\mathbf{C}_2^< = \mathbf{Q}_2^< \mathbf{R}_2, \quad \text{with} \quad \mathbf{Q}_2^{<\top}\mathbf{Q}_2^< = \mathbf{I}_{r_2}, \ \mathbf{Q}_2^< \in \mathbb{R}^{(r_1 n_2)\times r_2}, \ \mathbf{R}_2 \in \mathbb{R}^{r_2 \times r_2},$$

so that we have obtained a QR decomposition of

$$\mathbf{X}_{\le 2} = \mathbf{Q}_{\le 2}\mathbf{R}_2 \qquad \text{with} \quad \mathbf{Q}_{\le 2} = (\mathbf{I}_{n_2} \otimes \mathbf{Q}_1^<)\mathbf{Q}_2^<.$$

These orthogonalizations can be continued in the same way for $k = 2, 3, \ldots$. Putting $\mathbf{Q}_{\le 0} = 1$, we have obtained for each $k = 1, \ldots, d$ the QR decompositions

$$\mathbf{X}_{\le k} = \mathbf{Q}_{\le k}\mathbf{R}_k \qquad \text{with} \qquad \mathbf{Q}_{\le k} = (\mathbf{I}_{n_k} \otimes \mathbf{Q}_{\le k-1})\mathbf{Q}_k^<,$$

where the matrices $\mathbf{Q}_k^< \in \mathbb{R}^{(r_{k-1}n_k)\times r_k}$ and $\mathbf{R}_k \in \mathbb{R}^{r_k \times r_k}$ are obtained recursively from QR decompositions of lower-dimensional matrices $(\mathbf{I}_{n_k} \otimes \mathbf{R}_{k-1})\mathbf{C}_k^< = \mathbf{Q}_k^< \mathbf{R}_k$. We call the left partial product $\mathbf{X}_{\le k}$ in that case *left orthogonalized.*

In a completely analogous way, we can obtain a *right-orthogonalized* $\mathbf{X}_{\ge k}$ as follows. Denote $\mathbf{Q}_{\ge d+1} = 1$. Then, starting with $\mathbf{X}_{\ge d} = \mathbf{C}_d^> = \mathbf{Q}_d^> \mathbf{R}_d$, we can use (5.9) to obtain the QR decompositions

$$\mathbf{X}_{\ge k} = \mathbf{Q}_{\ge k}\mathbf{R}_k \qquad \text{with} \qquad \mathbf{Q}_{\ge k} = (\mathbf{Q}_{\ge k+1} \otimes \mathbf{I}_{n_k})\mathbf{Q}_k^>,$$

where the matrices $\mathbf{Q}_k^> \in \mathbb{R}^{(r_k n_k)\times r_{k-1}}$ and $\mathbf{R}_k \in \mathbb{R}^{r_{k-1}\times r_{k-1}}$ are recursively obtained from $(\mathbf{R}_{k+1} \otimes \mathbf{I}_{n_k})\mathbf{C}_k^> = \mathbf{Q}_k^> \mathbf{R}_k$.

Observe that when $\mathbf{X}_{\le k}$ is left orthogonalized, then so is $\mathbf{X}_{\le s}$ for any $s < k$. Since $\mathbf{X}_{\le d} = \mathbf{X}^{\langle d \rangle}$, we call X *left orthogonal* if $\mathbf{X}_{\le d}$ is left orthogonalized. A left orthogonal X is recursively computed by modifying the cores C_k from left to right during a *forward sweep*. Likewise, we call X *right orthogonal* if $\mathbf{X}_{\ge 1} = \mathbf{X}^{\langle 1 \rangle}$ is right orthogonalized, which is obtained by a *backward sweep* from right to left.

5.6.3 ▪ TT-SVD

A quasi-optimal TT approximation can be compute by an algorithm that is analogous to the higher-order singular value decomposition (HOSVD) [46, 64]. First, we start from the exact case. Suppose that the ranks of the unfoldings are equal to r_k:

$$\operatorname{rank}\mathbf{X}^{\langle k \rangle} = r_k, \qquad k = 1, \ldots, d-1.$$

Then, there exists a TT decomposition of $\mathbf{X}$ with TT ranks r_k. The proof is constructive. First,

$$\mathbf{X}^{\langle 1 \rangle} = \mathbf{X}_1^> \mathbf{X}_{\le 1}, \tag{5.12}$$

where $\mathbf{X}_1^>$ is a unitary matrix of size $n_1 \times r_1$ and $\mathbf{X}_{\le 1}$ is a matrix of size $r_1 \times (n_2 \ldots n_d)$. Using Kronecker product notation, we can rewrite (5.12) for the second unfolding as

$$\mathbf{X}^{\langle 2 \rangle} = (I_{n_1} \otimes \mathbf{X}_1^>)\widehat{\mathbf{X}}_{\le 1},$$

where $\widehat{\mathbf{X}}_{\leq 1}$ is a matrix of size $(r_1 n_2) \times (n_3 \dots n_d)$ obtained from $\mathbf{X}_{\leq 1}$ by reshaping. Thus, the rank of $\widehat{\mathbf{X}}_{\leq 1}$ is equal to the rank of $\mathbf{X}^{\langle 2 \rangle}$, and it can be factorized in a similar fashion:

$$\widehat{\mathbf{X}}_{\leq 1} = \mathbf{X}_2^{>} \mathbf{X}_{\leq 2},$$

where $\mathbf{X}_2^{>}$ is a unitary matrix of size $(r_1 n_2) \times r_2$ and $\mathbf{X}_{\leq 2}$ is a matrix of size $r_2 \times (n_3 \dots n_d)$. Reshaping the last core, we arrive at the TT decomposition of the tensor $\mathbf{X}$.

For the approximate case, the error analysis is again similar to the Tucker case. Since only unitary matrices (left factors of the SVD) are involved, the process is stable and the error does not accumulate. More specifically, it can be shown [51] that if at each step the error ε_k in the Frobenius norm is made, the final error satisfies

$$||\mathbf{X} - \mathbf{X}_{\mathrm{TT}}|| \leq \sqrt{\sum_{k=1}^{d-1} \varepsilon_k^2}.$$

In practice, to get the controllable error bound during the TT-SVD procedure, it is sufficient to truncate the singular values so the Frobenius norm error is less than $\frac{\varepsilon}{\sqrt{d-1}}$.

5.6.4 ▪ Rounding procedure

The TT-SVD procedure described in the previous subsection requires knowledge of the full tensor. However, if the original tensor is already in TT format but with suboptimal ranks, the approximation can be computed in $\mathcal{O}(dnr^3)$ operations via left and right orthogonalization of the cores. This process is similar to rounding of floating point numbers, but instead of digits we have TT cores. Another possible name for this procedure is truncation.

Suppose that $\mathbf{X}_{\leq k} = \mathbf{Q}_{\leq k} \mathbf{R}_k$ and $\mathbf{X}_{\geq k+1} = \mathbf{Q}_{\geq k+1} \mathbf{R}_{k+1}$ are QR decompositions obtained from left and right orthogonalizations; we then have the following SVD-like decomposition:

$$\mathbf{X}^{\langle k \rangle} = \mathbf{Q}_{\leq k} \mathbf{S}_k \mathbf{Q}_{\geq k+1}^{\top}, \qquad \text{with} \quad \mathbf{S}_k = \mathbf{R}_k \mathbf{R}_{k+1}^{\top} \in \mathbb{R}^{r_k \times r_k}. \tag{5.13}$$

We will call (5.13) the *recursive SVD* of the tensor. The recursive nature becomes especially apparent in the SVD of $\mathbf{X}^{\langle k+1 \rangle}$. We interpret the factor $\mathbf{Q}_{k+1}^{>}$ from the QR decomposition as the right unfolding of a tensor $Q_{k+1} \in \mathbb{R}^{r_k \times n_{k+1} \times r_{k+1}}$. By identifying

$$\mathbf{X}^{\langle k \rangle} = \mathbf{Q}_{\leq k} \mathbf{S}_k \mathbf{Q}_{\geq k+1}^{\top} = (\mathbf{Q}_{\leq k} \mathbf{S}_k) \mathbf{Q}_{k+1}^{>\top} (\mathbf{Q}_{\geq k+2} \otimes \mathbf{I}_{n_{k+1}})^{\top} \tag{5.14}$$

with (5.11), we obtain, using (5.10) and the left unfolding of Q_{k+1}, that

$$\mathbf{X}^{\langle k+1 \rangle} = (\mathbf{I}_{n_{k+1}} \otimes \mathbf{Q}_{\leq k} \mathbf{S}_k) \mathbf{Q}_{k+1}^{<} \mathbf{Q}_{\geq k+2}^{\top} = (\mathbf{I}_{n_{k+1}} \otimes \mathbf{Q}_{\leq k})(\mathbf{I}_{n_{k+1}} \otimes \mathbf{S}_k) \mathbf{Q}_{k+1}^{<} \mathbf{Q}_{\geq k+2}^{\top}. \tag{5.15}$$

Hence, after a QR decomposition

$$(\mathbf{I}_{n_{k+1}} \otimes \mathbf{S}_k) \mathbf{Q}_{k+1}^{<} = \overline{\mathbf{Q}}_{k+1}^{<} \overline{\mathbf{S}}_{k+1}, \tag{5.16}$$

we obtain the $(k+1)$th recursive SVD

$$\mathbf{X}^{\langle k+1 \rangle} = \overline{\mathbf{Q}}_{\leq k+1} \overline{\mathbf{S}}_{k+1} \mathbf{Q}_{\geq k+2}^{\top} \qquad \text{with} \quad \overline{\mathbf{Q}}_{\leq k+1} = (\mathbf{I}_{n_{k+1}} \otimes \mathbf{Q}_{\leq k}) \overline{\mathbf{Q}}_{k+1}^{<}.$$

A similar relation holds between $\mathbf{X}^{\langle k\rangle}$ and $\mathbf{X}^{\langle k-1\rangle}$. Let

$$\mathbf{X}^{\langle k\rangle} = \mathbf{Q}_{\le k}\mathbf{S}_k\mathbf{Q}_{\ge k+1}^\top = (\mathbf{I}_{n_k}\otimes\mathbf{Q}_{\le k-1})\mathbf{Q}_k^<(\mathbf{Q}_{\ge k+1}\mathbf{S}_k^\top)^\top. \tag{5.17}$$

Then, using the QR decomposition

$$(\mathbf{S}_k^\top\otimes\mathbf{I}_{n_k})\mathbf{Q}_k^> = \overline{\mathbf{Q}}_k^>\overline{\mathbf{S}}_k,$$

we can write

$$\mathbf{X}^{\langle k-1\rangle} = \mathbf{Q}_{\le k-1}\overline{\mathbf{S}}_k\overline{\mathbf{Q}}_{\ge k}^\top, \qquad \text{where} \qquad \overline{\mathbf{Q}}_{\ge k} = (\mathbf{Q}_{\ge k+1}\otimes\mathbf{I}_{n_k})\overline{\mathbf{Q}}_k^<. \tag{5.18}$$

Thus, the rounding procedure can be computed without going to the full tensor case using only stable QR and SVD subroutines.

5.6.5 ▪ Basic linear algebra operations

The TT format is good for several basic linear algebra operations.

Addition and elementwise product

Several basis operations are very easy to implement in the TT format. For example, two tensors in the TT format,

$$\mathbf{A} = \mathbf{A}_1(i_1)\dots\mathbf{A}_d(i_d), \quad \mathbf{B} = \mathbf{B}_1(i_1)\dots\mathbf{B}_d(i_d),$$

can be added as

$$\mathbf{C}_k(i_k) = \begin{bmatrix}\mathbf{A}_k(i_k) & 0\\ 0 & \mathbf{B}_k(i_k)\end{bmatrix}, \quad k=2,\dots,d-1,$$

and

$$\mathbf{C}_1(i_1) = \begin{bmatrix}\mathbf{A}_1(i_1) & \mathbf{B}_1(i_1)\end{bmatrix}, \quad \mathbf{C}_d(i_d) = \begin{bmatrix}\mathbf{A}_d(i_d)\\ \mathbf{B}_d(i_d)\end{bmatrix}.$$

Another obvious representation is for the elementwise product of two tensors:

$$\mathbf{C} = \mathbf{A}\circ\mathbf{B}, \quad \mathbf{C}_k(i_k) = \mathbf{A}_k(i_k)\otimes\mathbf{B}_k(i_k).$$

Note that the ranks increase after such operations; thus, the rounding procedure described in the previous section should be very handy.

Norms and scalar products

The Frobenius norm of a TT tensor can be computed in $\mathcal{O}(dnr^3)$ by using the orthogonalization procedure: all cores except the last one are orthogonalized from the left, and the Frobenius norm is equal to the norm of the last (nonorthogonal) core. Computing the scalar product is a little bit more involved:

$$\langle\mathbf{A},\mathbf{B}\rangle_F = \sum_{i_1,\dots,i_d} A(i_1,\dots,i_d)B(i_1,\dots,i_d) = \prod_{k=1}^{d}\sum_{i_k=1}^{n_k}\Big(\mathbf{A}_k(i_k)\otimes\mathbf{B}_k(i_k)\Big),$$

where we have used the formula for the cores of the elementwise product. The result can be computed by sequential evaluation of

$$\phi_0 = 1, \quad \phi_k = \phi_{k-1}\sum_{i_k=1}^{n_k}\Big(\mathbf{A}_k(i_k)\otimes\mathbf{B}_k(i_k)\Big), \quad k=1,\dots,d.$$

Conversion from the canonical format

If the tensor is given in the canonical format with rank r and factor matrices U_k, then the TT representation of this tensor has the diagonal cores

$$\Lambda_k(i_k) = \mathrm{diag}(U_k(i_k,1)\dots U_k(i_k,r)), \quad k = 2,\dots,d-1,$$

and

$$\Lambda_1(i_1,\alpha) = U_1(i_1,\alpha), \quad \Lambda_d(\alpha,i_d) = U_d(i_d,\alpha), \quad \alpha = 1,\dots,d-1.$$

After this conversion is computed, we can apply the rounding (recursive SVD) procedure to compute the quasi-optimal TT representation.

5.6.6 ▪ Matrix representation in the TT format

The TT format can be used to represent matrices as well. In the MPS community it is known under the name *matrix product operator*, or MPO. A linear operator acting on a space of d-dimensional tensors can naturally be represented as a $2d$-dimensional (for simplicity we assume only the square case) tensor

$$M(i_1,\dots,i_d;j_1,\dots,j_d),$$

where $(i_1,\dots,i_d)$ enumerates the rows of the matrix M and $(j_1,\dots,j_d)$ enumerates the columns of it. The matrix-vector product in this case is defined as a contraction of the form

$$Y(i_1,\dots,i_d) = \sum_{j_1,\dots,j_d} M(i_1,\dots,i_d;j_1,\dots,j_d)X(j_1,\dots,j_d).$$

The d-level matrix M is said to be in the TT/MPO format if

$$M(i_1,\dots,i_d;j_1,\dots,j_d) = \mathbf{M}_1(i_1,j_1)\mathbf{M}_2(i_2,j_2)\dots\mathbf{M}_d(i_d,j_d), \tag{5.19}$$

where $\mathbf{M}_k(i_k,j_k)$ is an $R_{k-1} \times R_k$ matrix for each fixed pair (i_k,j_k), and $R_0 = R_d = 1$. i.e., a TT format for the tensor obtained from M by permuting the indices. The representation (5.19) is a natural generalization of the TT format to operators, since if $R_k = 1$, $k = 0,\dots,d$, then it reduces to the Kronecker product of small matrices:

$$M = M_1 \otimes M_2 \otimes \dots \otimes M_d.$$

Many important operators can be represented in the TT format. For example, the Laplace operator discretized by second-order differences on a tensor product uniform grid can be represented as a sum of d rank-one operators:

$$\Delta_d = \Delta_1 \otimes I \otimes \dots \otimes I + \dots + I \otimes \dots \otimes \Delta_1,$$

where Δ_1 is an $n \times n$ discretization matrix of a one-dimensional Laplace operator and I is the identity matrix of order n. It is obvious that the TT rank of the matrix Δ_d does not exceed d; however, it can be shown that it is equal to two, and the explicit representation is obtained in [28].

Strong Kronecker product

Let A be a $p \times q$ block matrix with $n \times m$ blocks. Let B be a $q \times r$ block matrix with $k \times l$ blocks. Then, by $C = A \bowtie B$ we denote a $p \times r$ block matrix with blocks

$$C_{\alpha\beta} = \sum_{\gamma=1}^{q} A_{\alpha\gamma} \otimes B_{\gamma\beta},$$

i.e., the block matrices are multiplied as in the usual matrix-vector product, but the inner multiplication is replaced by a Kronecker product of matrices. If we organize the ith core M_i of the TT matrix into an $R_{i-1} \times R_i$ block matrix $\mathbf{M}_i$ with $n_i \times n_i$ blocks defined as

$$((\mathbf{M}_i)_{\alpha\beta})_{kl} = M_i(\alpha, k, l, \beta),$$

then the TT matrix (5.19) can be written as

$$M = \mathbf{M}_1 \bowtie \cdots \bowtie \mathbf{M}_d.$$

An important property of the strong Kronecker product (SKP) is that

$$(A \otimes B)(C \bowtie D)(E \otimes F) = (ACE) \bowtie (BDF),$$

where A, B, C, D, E, F have consistent block sizes and sizes of the blocks.

5.7 ▪ Optimization algorithms in TT format

Basic linear algebra operations (such as matrix-vector product, addition, or elementwise multiplication) in the TT format can be implemented in a finite amount of operations, as was shown above. Together with a robust rounding procedure this gives many opportunities to develop tensor-structured methods. For example, an iterative method that involves matrix-vector products and additions can be written in the same form, where exact operations are replaced with approximate ones with controlled accuracy. The disadvantage of this approach is that the convergence of the iterative methods can be slow, and preconditioners are often required. In many cases, it is better to formulate the initial problem as a minimization problem in a suitable tensor format. This has the effect of "preconditioning," since the number of parameters to be optimized is much smaller. This comes at the cost of a more sophisticated optimization problem to be solved. Now we will discuss approaches that efficiently solve optimization problems in the TT format.

Many practical problems can be formulated as minimization problems. For example, a linear system

$$Ax = f$$

with symmetric positive definite matrix A can be formulated as a minimization of the functional

$$J(x) = \langle Ax, x \rangle - 2\langle f, x \rangle.$$

The general strategy is always the same. We assume that there is a one-to-one correspondence between the elements of x and some d-dimensional tensor $X(i_1, \ldots, i_d)$, and we know in advance that the solution can be well approximated in the TT format. Then, the approximate solution is sought as a minimization of the functional $J(X)$ over the set of all tensors with bounded TT ranks.

In the same way, we can formulate the eigenvalue problem as a minimization of the Rayleigh quotient

$$J(x) = \frac{\langle Ax, x \rangle}{\langle x, x \rangle}.$$

An important case is when $J(X)$ is a quadratic functional, since it includes both linear systems and eigenvalue problems. However, even in this case, the optimization over the TT manifold is nonconvex. However, similar to the canonical format, we can use the polylinear structure of the manifold and get a fast algorithm that is based on alternating minimization. Given a current approximation to the solution,

$$X(i_1, \dots, i_d) = \mathbf{X}_1(i_1) \dots \mathbf{X}_d(i_d),$$

we fix all cores except the kth core. The functional is then quadratic, and the problem can be reduced to a small *local problem* with $r_{k-1} n_k r_k$ degrees of freedom, either a linear system or an eigenvalue problem, depending on the initial functional. The process is then organized in a sweep: first we update the first core, then the second, and so on. The method is typically initialized by a random tensor or by a right-hand side. The local system matrix can be computed quickly if the matrix A has a TT structure. Let

$$A = A_1 \bowtie \cdots \bowtie A_d.$$

During the sweep, we orthogonalize the cores from 1 to $k-1$ from the left, and the cores from $(k+1)$ to d from the right, yielding the representation of the current approximation in the form

$$X^{\langle k \rangle} = (I_{n_k} \otimes \mathbf{Q}_{\leq (k-1)}) \mathbf{C}_k^{<} \mathbf{Q}_{\geq (k+1)}^{\top},$$

and if we vectorize X and $\mathbf{C}_k(i_k)$, we get

$$\operatorname{vec}(X) = \left(\mathbf{Q}_{\geq (k+1)} \otimes I_{n_k} \otimes \mathbf{Q}_{\leq (k-1)} \right) \operatorname{vec}(\mathbf{C}_k),$$

and the matrix $\mathbf{Q} = \mathbf{Q}_{\geq (k+1)} \otimes I_{n_k} \otimes \mathbf{Q}_{\leq (k-1)}$ has orthonormal columns. Minimizing the Rayleigh quotient over $\mathbf{C}_k$ reduces this to the solution of the eigenvalue problem of the form

$$\left(\mathbf{Q}^{\top} A \mathbf{Q} \right) \operatorname{vec}(\mathbf{C}_k) = \lambda \operatorname{vec}(\mathbf{C}_k).$$

The matrix A can be written as

$$A = \mathbf{A}_{\leq k-1} \bowtie \mathbf{A}_k \bowtie \mathbf{A}_{\geq k+1},$$

so we can define

$$B_k = \left(\mathbf{Q}_{\leq k-1}^{\top} A_{\leq k-1} \mathbf{Q}_{\leq k-1} \right) \bowtie \mathbf{A}_k \bowtie \left(\mathbf{Q}_{\geq k+1}^{\top} \mathbf{A}_{\geq k+1} \mathbf{Q}_{\geq k+1} \right).$$

Finally, we introduce matrices

$$\Phi_{\leq k-1} = \mathbf{Q}_{\leq k-1}^{\top} A_{\leq k-1} \mathbf{Q}_{\leq k-1}, \quad \Phi_{\geq k+1} = \mathbf{Q}_{\geq k+1}^{\top} \mathbf{A}_{\geq k+1} \mathbf{Q}_{\geq k+1}.$$

They can be computed cheaply during the sweep. Suppose we are going from left to right and compute everything at the kth core and want to compute the next left matrix $\Phi_{\leq k}$ provided that $\Phi_{\leq k-1}$ is known. The recursion for $\mathbf{Q}_{\leq k}$ gives us

$$\mathbf{Q}_{\leq k} = (I_{n_{k+1}} \otimes \mathbf{Q}_{\leq k-1}) \mathbf{Q}_k^{<}$$

and

$$\mathbf{A}_{\leq k} = \mathbf{A}_{\leq k-1} \bowtie \mathbf{A}_k.$$

Therefore

$$\begin{aligned}\Phi_{\leq k} &= (\mathbf{Q}_k^<)^\top (I_{n_{k+1}} \otimes Q_{\leq k-1}^\top)(\mathbf{A}_{\leq k-1} \bowtie \mathbf{A}_k)(I_{n_{k+1}} \otimes \mathbf{Q}_{\leq k-1}^\top)\mathbf{Q}_k^< \\ &= (\mathbf{Q}_k^<)^\top (\Phi_{\leq k-1} \bowtie \mathbf{A}_k)\mathbf{Q}_k.\end{aligned} \tag{5.20}$$

A similar formula holds for $\Phi_{\geq k}$:

$$\Phi_{\geq k} = (\mathbf{Q}_k^>)^\top (\mathbf{A}_k \bowtie \Phi_{\geq k+1})\mathbf{Q}_k^>,$$

and this update should be used for the right-to-left sweep, when right orthogonalization of cores in ensured. In Algorithm 5.1, the general scheme for the ALS algorithm for the eigenvalue problem is presented. In the same way, any quadratic functional can be minimized, including the solution of linear systems with symmetric positive definite matrices. The difference is that the local problem will be a linear system, not an eigenvalue problem.

***ALGORITHM* 5.1. ALS method to compute the minimal eigenvalue.**

Data: Matrix A in the TT format; initial approximation X in the TT format.

begin

- Set $\Phi_{\geq d+1} = 1, \quad R_{d+1} = 1, \quad R_1 = 1, \quad \Phi_{\leq 1} = 1.$
- **for** $k = d$ **to** 2 **do**
  - $\mathbf{X}_k^> := \mathbf{X}_k^> R_{k+1}^\top$
  - $[\mathbf{X}_k^<, R_k] := \text{QR}(\mathbf{X}_k^<)$ ▷ `Orthogonalization`
  - $\Phi_{\geq k} = (\mathbf{Q}_k^>)^\top (\mathbf{A}_k \bowtie \Phi_{\geq k+1})\mathbf{Q}_k^>$
- **for** $k = 1$ **to** d **do**
  - Solve the local eigenvalue problem and update $\mathbf{X}_k$ $B_k \text{vec}(\mathbf{X}_k) = \lambda \text{vec}(\mathbf{X}_k)$, where $B_k = \Phi_{\leq k-1} \bowtie \mathbf{A}_k \bowtie \Phi_{\geq k+1}$.
  - $[\mathbf{Q}_k^>, R_k] := \text{QR}(\mathbf{X}_k^>)$ ▷ `Orthogonalization`
  - If $k < d$, $\mathbf{X}_{k+1}^< := R_k \mathbf{X}_{k+1}^<$
  - $\Phi_{\leq k} = (\mathbf{Q}_k^<)^\top (\Phi_{\leq k-1} \bowtie \mathbf{A}_k)\mathbf{Q}_k^<$
- Do a right-left sweep.

5.7.1 ▪ Adaptation of the ranks: Density matrix renormalization group approach

One of the major problems of the ALS algorithm is that the TT ranks should be fixed in advance. If we underestimate the ranks, the solution will have poor quality; if we overestimate the ranks, the complexity will be high. One of the most successful approaches is the DMRG method, which was initially proposed for solving eigenvalue problems [53, 65]. It has a very interesting mathematical structure that uses not only the polylinear structure of the TT model but also the fact that only two neighboring indices are connected. We can merge two adjacent cores into one:

$$\mathbf{W}_k = \mathbf{X}_k^> \mathbf{X}_{k+1}^<,$$

and the new core is an $(r_{k-1}n_k)\times(n_{k+1}r_{k+1})$ matrix. The resulting tensor is a $(d-1)$-dimensional tensor, for which we optimize over the superblock using the same formulas as in Algorithm 5.1. Once the new $\mathbf{W}_k$ is found, we split it back by computing the decomposition

$$\mathbf{W}_k \approx \widehat{\mathbf{X}}_k^{>} \widehat{\mathbf{X}}_{k+1}^{<},$$

with either a fixed rank r_k or a required accuracy ε. This can be done by computing the SVD of $\mathbf{W}_k$, and the new rank r_k is determined in an adaptive fashion! Such an approach is uncommon in optimization, where typically the number of parameters is fixed, and only their values are optimized; in the DMRG approach the manifold itself is modified at each iteration step, and the number of parameters can either increase or decrease. The DMRG method is very well suited for problems with small mode sizes (i.e., $n=2$) since it works with squared mode sizes. For n of order 100 it becomes computationally demanding, and more efficient approaches are often required. One of the promising research directions is the AMEN method [14], which we will discuss in the following section.

5.7.2 ▪ Enrichment of the basis

In this section we discuss the AMEN method. Instead of solving the local problem for the enlarged block $\mathbf{W}_k$, we can try to use its low-rank structure. If we expand the core $\mathbf{X}_k$ by some vectors

$$\widehat{\mathbf{X}}_k^{>} = \begin{bmatrix} \mathbf{X}_k^{>} & \mathbf{S}_k^{>} \end{bmatrix},$$

then the enrichment of the next core,

$$\widehat{\mathbf{X}}_{k+1}^{<} = \begin{bmatrix} \mathbf{X}_{k+1}^{<} \\ 0 \end{bmatrix},$$

does not change the product. In [14] it is proposed to use the *projected residual*

$$\mathrm{vec}(\mathbf{R}_k) = \mathrm{vec}(\mathbf{F}_k) - B_k \mathrm{vec}(\mathbf{X}_k)$$

(or its low-rank approximation) for enrichment. The coefficients $\widehat{X}_{k+1}$ are then updated using the standard ALS step. In this way we can adapt the ranks with local problems with $\mathcal{O}(n)$ unknowns; the original paper [14] also has a theoretical proof that the resulting method (without truncation) converges at least with the speed of the steepest descent method. In practice, however, the convergence is much faster and typically eight or nine sweeps are required. The AMEN method is implemented in the TT-Toolbox both in MATLAB and in Python [42, 43], and it has been generalized to other problems as well [12]. However, if nonstationary problems have to be solved, then the variational setting should be different, and we will discuss this in the next section.

5.8 ▪ Dynamical low-rank approximation

The solution of nonstationary problems plays a crucial role in many applications; thus, is natural to consider two closely related approximation problems. First is the *dynamical low-rank approximation* [32, 33]: given a trajectory $\mathbf{A}(t)$ where $\mathbf{A}(t)$ is a d-dimensional tensor, find a low TT rank approximation $\mathbf{X}(t)$ that approximates $\mathbf{A}(t)$ in a certain optimal way. Second is when the time-dependent tensor is defined implicitly as a solution of a certain system of differential equations:

$$\frac{dY}{dt} = F(Y).$$

Such applications appear in high-dimensional partial differential equation (PDE) modeling. The naive approach would be just to take the best possible approximation pointwise, i.e.,

$$Y(t) = \arg\min_{Y \in \mathcal{M}} \|A(t) - Y\|,$$

where $\mathcal{M}$ is our low-parametric manifold. There are several problems with such an approach, best illustrated using the two-dimensional (matrix) case. Consider the parametrization of a rank-r matrix as

$$X(t) = U(t)S(t)V^\top(t).$$

If we use the SVD to compute the best possible approximation, then $U(t), S(t), V^\top(t)$ will depend on time in a nonsmooth way. Thus, if some noise accumulates during the time evolution, the low-rank approximation might be spoiled since it does not take into account any previous values of $X(t)$: for example, an instant change of $A(t)$ at some time point t might be undesirable. The *Dirac–Frenkel principle* provides a viable variational formulation for the low-rank approximation of time-varying tensors. Given time-varying tensor $A(t)$, the dynamical low-rank approximation $X(t)$ should satisfy

$$\left\langle \frac{dX}{dt} - \frac{dA}{dt}, v \right\rangle = 0 \tag{5.21}$$

for all v in the *tangent space* of the low-rank manifold.

The set of all tensors with bounded TT ranks is an embedded submanifold [26]. The Dirac–Frenkel principle leads to a system of nonlinear ordinary differential equations (ODEs) for the TT cores $X_k(i_k)$ of the approximation. For simplicity, consider the case $d = 2$. The generalization to the case of TT tensors with $d > 2$ will be described shortly, but the main properties are the same as in two dimensions. We parametrize a low-rank matrix $X(t)$ in the form

$$\widehat{Y}(t) = U(t)S(t)V^*(t),$$

where $U(t)$ is an $n \times r$ matrix with orthonormal columns, $V(t)$ is an $m \times r$ matrix with orthonormal columns, and $S(t)$ is an $r \times r$ matrix that is not required to be diagonal. The variational principle leads [32] to the system of ODEs of the form

$$\frac{dU}{dt}S = \frac{dY}{dt}V, \quad \frac{dV}{dt}S^* = \frac{dY^*}{dt}U, \quad \frac{dS}{dt} = U^*\frac{dY}{dt}V. \tag{5.22}$$

System (5.22) can be quite difficult to integrate, especially if S is singular or almost singular. However, a very efficient integrator is proposed in [38] that is based on the splitting scheme applied to the projector onto the tangent space. To use it, we have to write the equation in the form

$$\frac{d\widehat{Y}}{dt} = P_Y\left(\frac{dY}{dt}\right),$$

where P_Y is a projector onto the tangent space with the form

$$P_Y(Z) = UU^*Z + ZVV^* - UU^*ZVV^* = P_1(Z) + P_2(Z) - P_3(Z).$$

Then, we apply the splitting scheme. For each projector $P_i, i = 1,2,3$, the equations can be integrated exactly. What is interesting is that the order of the splitting is very important. The step in $P_1(Z) = UU^*Z$ reduces to the equations

$$\frac{dU}{dt} = 0, \quad \frac{d(VS^*)}{dt} = \frac{dY^*}{dt}U,$$

which can be solved as $VS^* = L = (Y(t+\tau) - Y(t))U$. To get new V and S, it is sufficient to compute any orthogonal factorization. The integration with the right-hand side $P_2(Z) = ZVV^*$ is similar. It is interesting that the integration of the ODE with the right-hand side $P_3(Z) = -UU^*ZVV^*$ leaves U and V unchanged, whereas the matrix S is recomputed using the formula $S := S - U^*(Y(t+\tau) - Y(t))V$, which can be considered as an integration of the original equation for S *back in time*. The scheme has first order in time; however, it is very easy to get second order in time by using Strang–Marchuk splitting.

Moreover, as shown in [38], the KSL order of the splitting (i.e., first step in P_2, then in P_3, then in P_1) has a wonderful *exactness property*. If $Y(t)$ and $Y(t+\tau)$ are both of rank r, then the scheme is exact. Experimentally, for $Y(t)$ that can be approximated by rank r with accuracy ε, the error obtained by the splitting scheme is of order $\mathcal{O}(\varepsilon\tau)$.

Since the TT format can be represented as a sequential rank-r approximation, an analogous scheme can be derived for the TT format as well. One of the most important cases is the application of the Dirac–Frenkel principle for the solution of a time-dependent linear system of ODEs of the form

$$\frac{dy}{dt} = Ay, \quad y(0) = y_0,$$

where the matrix A and the initial condition y_0 are both in the TT format. To get the equation of motion in this case, it is sufficient to replace $Y(t)$ by $A(Y)$. The integration process then proceeds as in the usual sweep algorithm, similar to Algorithm 5.1, with several minor changes. First, the local problem is a local linear system of ODEs. Second, we also have to solve the equation for the S matrix back in time, which also reduces to a small linear system. The local problem can be integrated by any suitable Krylov-type method. One of the most widely used packages is the Expokit software package [58]. A detailed description can be found in [37].

5.9 ▪ Black-box approximation of tensors

One of the basic problems in the approximation of multivariate functions and related tensors is the problem of interpolation. Assume the following situation:

1. We are given a subroutine that computes any prescribed element of a given tensor
2. We have a priori knowledge that the tensor can be approximated by a tensor with small TT ranks.

Is it possible to compute a few "samples" of a tensor to recover the full tensor? A positive answer was first given in the paper [51]. If all the TT ranks are bounded by r, it is possible to recover the tensor from $\mathcal{O}(dnr^2)$ entries. The multidimensional formula comes from a generalization of the celebrated *skeleton decomposition* [18] for low-rank matrices. If $A \in \mathbb{R}^{m \times n}$ has rank r, it can be written as

$$A = C\widehat{A}^{-1}R,$$

where $C \in \mathbb{R}^{m\times r}$ are some r columns of A, $R \in \mathbb{R}^{r\times n}$ are some r rows of A, and $\widehat{A} \in \mathbb{R}^{r\times r}$ is the submatrix on the intersection of these rows and columns.

5.9.1 ▪ Skeleton decomposition in two dimensions and maximal volume principle

The skeleton decomposition of the matrix is the key ingredient for the tensor approximation by sampling of its elements. If the matrix A is of rank r, then it can be decomposed as

$$A = C\widehat{A}^{-1}R,$$

where C are some r columns of A, R are some rows of r, and $\widehat{A}$ is a nonsingular submatrix on the intersection of these rows and columns. A convenient notation for the skeleton decomposition is to use submatrices of identity matrices.

Index selection

Let $\mathcal{I}$ be an index set with r indices from 1 to n, and by $P_{\mathcal{I}}$ we denote an $r \times n$ submatrix of an $n \times n$ identity matrix comprising rows with indices from $\mathcal{I}$. The column size of $P_{\mathcal{I}}$ should be clear from the context.

Let $\mathcal{I}$ and $\mathcal{J}$ be row and column numbers selected in the skeleton decomposition. Then,

$$C = AP_{\mathcal{J}}^{\top}, \quad R = P_{\mathcal{I}}A, \quad \widehat{A} = P_{\mathcal{I}}AP_{\mathcal{J}}^{\top}.$$

The guiding principle for selecting basis rows and columns is the *maximum volume principle*. By that volume of an $r \times r$ submatrix we mean the absolute value of its determinant. If $\widehat{A}$ is selected as the submatrix of maximal volume, then by [17] the skeleton decomposition satisfies the error bound

$$||A - C\widehat{A}^{-1}R||_C \leq (r+1)\sigma_{r+1}(A), \tag{5.23}$$

where $\sigma_{r+1}(A)$ is the $(r+1)$th singular value of A, and the Chebyshev norm of a matrix is defined as $||X||_C = \max_{ij} |X_{ij}|$. Finding the maximal volume submatrix is an NP-hard problem [2], but there exist very powerful heuristic algorithms for finding quasi-optimal submatrices. We will use the `maxvol` algorithm, which computes a submatrix of sufficiently large volume in a tall $n \times r$ matrix. This algorithm is very robust and cheap; for details see [16].

5.9.2 ▪ Skeleton decomposition in many dimensions

In many dimensions, the skeleton decomposition is generalized in [51]. Let $\mathbf{A}$ be a given tensor, and let its unfoldings $A_k, k = 1, \dots, d-1$, satisfy

$$\operatorname{rank} A_k = r_k,$$

where r_k are sufficiently small, namely we will need that $r_k \leq r_{k-1}n_k$. Let

$$C_k = AP_{\mathcal{J}_k}^{\top}$$

be some *basis columns* of A_k and let

$$R_k = P_{\mathcal{I}_k}A.$$

be some *basis rows* of it. The matrices A_k and A_{k+1} are unfoldings of the same tensor, and it is not difficult to see that the columns of the matrix $C_k \otimes I_{n_{k+1}}$ span the column space of A_{k+1}. This can be seen, for example, from the index form

$$A_{k+1}(i_1 \dots i_{k+1}; i_{k+1} \dots i_d) = \sum_{\alpha=1}^{r} C(i_1 \dots i_k, \alpha) D(\alpha, i_{k+1} \dots i_d)$$
$$= \sum_{\alpha=1}^{r} \sum_{\gamma=1}^{n_{k+1}} C(i_1 \dots i_k, \alpha) \delta(\alpha, \gamma) D(\gamma, i_{k+1} \dots i_d).$$

Therefore, there exists a matrix M_k of size $r_k n_k \times r_{k+1}$ such that

$$C_{k+1} = (C_k \otimes I_{n_{k+1}}) M_k, \quad k = 1, \dots, d-2. \tag{5.24}$$

The equation (5.24) immediately gives us the TT decomposition of $\mathbf{A}$. Indeed,

$$A_{d-1} = C_{d-1} \widehat{A}_{d-1}^{-1} R_{d-1} = (C_{d-2} \otimes I_{n_{d-1}}) \widehat{A}_{d-1}^{-1} R_{d-1} M_{d-2} = \dots;$$

therefore

$$A(i_1, \dots, i_d) = A_1(i_1) \dots A_d(i_d),$$

where

$$A_1 = C_1, \quad A_k^{<} = M_k, \quad i = 2, \dots, d-1, \quad A_d^{<} = \widehat{A}_{d-1}^{-1} R_{d-1}.$$

Thus, there is no need to store the matrices C_k. Only the transfer matrices M_k need to be stored. The matrices M_k that define the decomposition can be computed without accessing all the entries of $\mathbf{A}$. To show this, consider the equation (5.24). It is an overdetermined system of linear equations for M_k. Let us consider only part of these equations to get a square linear system for M_k. Since R_k are basis rows in A_k, the matrix

$$\widehat{C}_k = P_{\mathcal{I}_k} C_k$$

is square and nonsingular. Therefore,

$$(P_{\mathcal{I}_k} \otimes I_{n_{k+1}}) C_{k+1} = (\widehat{C}_k \otimes I_{n_{k+1}}) M_k \Rightarrow M_k = (\widehat{C}_k^{-1} \otimes I_{n_k}) \Big((P_{\mathcal{I}_k} \otimes I_{n_{k+1}}) C_{k+1} \Big). \tag{5.25}$$

The matrix $\widetilde{C}_{k+1} = (P_{\mathcal{I}_k} \otimes I_{n_{k+1}}) C_{k+1}$ is $r_k n_{k+1} \times r_{k+1}$, and its entries are some elements of the tensor $\mathbf{A}$: we select r_{k+1} columns of the A_{k+1} unfolding and then select $r_k n_k$ rows from the resulting matrix.

5.9.3 ▪ Cross-approximation in many dimensions

To compute the cross-approximation in many dimensions, we may use different approaches. All of them have a certain iterative structure. Let us fix some columns $\mathcal{J}_k$ in each unfolding. Then, the columns C_1 of the first unfolding make up a small matrix of size $n_1 \times r_1$. Suppose we know good rows $\mathcal{I}_k$ in C_k. Then, due to the equality

$$C_{k+1} = (C_k \otimes I_{n_{k+1}}) M_k,$$

basis rows for A_{k+1} can be taken from the rows of the $r_k n_{k+1} \times r_{k+1}$ matrix $\widetilde{C}_{k+1}$. In the exact case, if the selected rows at the previous step span the whole row space, this procedure will guarantee that the selected rows again span the required space. In the approximate case, this is a heuristic procedure. Since $\widetilde{C}_k$ is a small matrix, rows can be selected in a standard way by first computing the SVD and then applying the `maxvol` algorithm. The rows computed in this way will be nested. Let $\widetilde{\mathcal{I}}_k$ be the selected basis rows in the matrix $\widetilde{C}_k$; then the row selection matrix $P_{\mathcal{I}_{k+1}}$ for A_{k+1} is equal to

$$P_{\mathcal{I}_{k+1}} = (P_{\mathcal{I}_k} \otimes I_{n_{k+1}}) P_{\widetilde{\mathcal{I}}_k}. \tag{5.26}$$

Equation (5.26) has a simple interpretation. The basis rows for A_{k+1} are selected not from all possible rows, but from a subset of $r_k n_k$ rows. Thus, in principle in the computations we only need to store the indices $\mathcal{I}_k$ to compute the approximation. In practice, however, to access the required elements of $\mathbf{A}$ we also need $\mathcal{I}_k$ and $\mathcal{J}_k$. Equations (5.26) and (5.25) suggest a simple algorithm for computing $\mathcal{I}_k$ once $\mathcal{J}_k$ are chosen. First, we compute $\mathcal{I}_1$ and M_1, then we compute $\mathcal{I}_2$ and M_2, and so on. At each step, only a small subtensor of $\mathbf{A}$ is computed. This process is usually referred to as a *sweep*. Once $\mathcal{I}_k$ are recomputed, the column indices $\mathcal{J}_k$ can be updated in the same way from right to left. The iteration is absolutely crucial in case the tensor is only approximately of low rank. The method described above is suggested in [51].

The TT cross method is summarized in Algorithm 5.2.

ALGORITHM **5.2. Pseudocode of the TT cross algorithm.**

Data: Subroutine that computes any prescribed element of the tensor $A(i_1,\dots,i_d)$, $1 \le i_k \le n_k$. Initial approximation X in the TT format with ranks r_k.
Result: Approximation $X = X_1(i_1)\dots X_d(i_d)$ to $\mathbf{A}$ in the TT format

begin
  Set $P_{\mathcal{I}_0} = 1, P_{\mathcal{J}_d} = 1, r_0 = r_d = 1.$ ▷ `Initialization`
  for $k = 1$ **to** $d-1$ **do**
    $r_k := \min(r_{k-1} n_k, r_k)$
    Randomly generate $\mathcal{J}_k = (i_{k+1}^{(\alpha)},\dots,i_d^{(\alpha)}), \alpha = 1,\dots,r_k$.
  for swp $= 1$ **to** n_{swp} **do**
    ▷ `Forward`
    for $k = 1$ **to** $d-1$ **do**
      Compute local block $\mathbf{C}_k^{>} = P_{\mathcal{I}_{k-1}} A_k P_{\mathcal{J}_k}^{\top}$.
      Compute QR of $\mathbf{C}_k^{>}$: $[\mathbf{Q}_k^{>}, R_k] = \text{QR}(\mathbf{C}_k)$.
      Compute basis rows: $\widetilde{I}_k = \mathbf{maxvol}(\mathbf{Q}_k^{>})$.
      Update next core: $\mathbf{X}_{k+1}^{<} := R_k \mathbf{X}^{k+1}$.
      Recompute indices: $P_{\mathcal{I}_k} = (P_{\mathcal{I}_{k-1}} \otimes I_{n_k}) P_{\widetilde{I}_k}$.
    ▷ `Backward`
    Do backward sweep to update $\mathcal{J}_k$.

The TT cross method is easy to implement and often works fine, but it has no adaptation in the TT ranks, which have to be chosen in advance. If they are underestimated, this would lead to a bad approximation; if they are overestimated, this would lead to increased complexity. Several other approaches have been proposed, namely the DMRG approach [56] and also a greedy-type technique [55]. A variant of the cross method was also proposed for the HT format [1]. We will not describe them in detail and just note that such algorithms are in active development right now, but the core idea is the interpolation formula. Prototype implementations of the cross methods (with rank adaptation) are available in MATLAB and Python versions of the TT-Toolbox.

Theoretical estimates

In the two-dimensional (matrix) case, the existence of a good sampling points is guaranteed by the maximal volume principle. It is a good question whether the same result holds for the general d-dimensional case. A straightforward application of the matrix result gives an estimate

$$||\mathbf{A}-\mathbf{C}||_C \leq C r^{d-1}\varepsilon,$$

where ε is the best approximation of the tensor $\mathbf{A}$ by a tensor with TT ranks $r_1,\ldots,r_{d-1}$ and $||\cdot||_C$ denotes the Chebyshev (elementwise) norm. Recently, a stronger result was obtained in [55] with the constant

$$C = r^{\log d}$$

and also an asymptotical estimate: there exists a skeleton tensor decomposition such that

$$||\mathbf{A}-\mathbf{C}||_C = dnr\varepsilon + \mathcal{O}(\varepsilon^2).$$

These are existence results, i.e., they show that the required rows and columns exist, but they say nothing about the convergence of corresponding numerical algorithms. Such algorithms are heuristical, and even in the matrix case it is quite easy to come up with a counterexample for which any sampling strategy will fail. In practice, however, cross-approximation methods work extremely well, which means that matrices and tensors from applications belong to a certain "good" subclass in the class of tensors with small TT ranks. One of the promising directions is related to the restricted isometry property (RIP) of the singular vectors. In the two-dimensional case, the result of [8] shows that if a matrix A can be written as

$$A = U\Phi V^\top + E,$$

with $U\in\mathbb{R}^{n\times r}$, $\Phi\in\mathbb{R}^{r\times r}$, $V\in\mathbb{R}^{m\times r}$, where U and V are orthonormal matrices that are μ-incoherent, then it is sufficient to sample $r+\mathcal{O}(\log n)$ columns of the matrix. By μ-incoherence of an $n\times r$ matrix with orthonormal columns we mean that $||X||_C \leq (\mu/n)^{1/2}$. The meaning of μ-coherence is that the matrix has *no gaps* in the elements and there are no zeros.

5.10 ▪ Quantized TT format

A very interesting research direction is the idea of *tensorization*, or introduction of virtual dimensions in arrays of small dimensions. Such ideas have been used for a long time in different applications, but only recently have they become a powerful computational tool for function approximation. The idea of introducing additional dimensions is used by Tyrtyshnikov [63], where matrices, corresponding to one-dimensional operators, are treated as multilevel matrices, leading to better compression. This idea is presented in [44] (a full version is published in [45]), where it is used to represent $2^d \times 2^d$ matrices; Khoromskij in [29] proposes to use it as a tool to represent one-dimensional functions.[22]

The idea of quantized TT (QTT) can be illustrated on a very simple example. Consider a function $f(x)$ (say, $f(x) = \sin x$) defined on an interval $[a, b]$. Introduce a uniform grid on $[a, b]$ with 2^d nodes and a vector of values of this function on this grid:

$$v_k = f(x_k).$$

Any integer $0 \le i \le 2^d$ can be uniquely represented by its binary digits $i_1, \dots, i_d, i_k \in \{0, 1\}$, and

$$i = \sum_{k=1}^{d} i_k 2^{k-1}.$$

This mapping allows us to have a one-to-one correspondence between v and a $2 \times 2 \times 2 \times \cdots \times 2$ d-dimensional tensor $\mathbf{V}$ with elements

$$V(i_1, \dots, i_d) = v(i).$$

In MATLAB and Python, this can be achieved by one call to the `reshape` function. The QTT decomposition is obtained by applying the TT decomposition to the tensor $\mathbf{V}$. If the resulting TT ranks are bounded ($r_k \le r$), the complexity is then $\mathcal{O}(2dr^2) = \log(\mathcal{N})$, where $N = 2^d$ is the total number of elements in the vector, i.e., we get *logarithmic complexity*. First theoretical results for the existence of good QTT approximations are given in [29]. If $f(x) = e^{-\lambda x}$, then all QTT ranks are equal to one, for the function $f(x) = \sin(\alpha x + \beta)$, the QTT ranks are equal to two, so every function that can be approximated by a sum of exponents and trigonometric functions can be well approximated in the QTT format. In [21], the estimates for piecewise-polynomial functions are obtained and in [47] the explicit QTT decompositions for standard functions are obtained and a general characterization theorem is proved.

Theorem 5.7. *Let f be a function such that $f(x + y)$ can be represented as a sum of rank-one functions:*

$$f(x+y) = \sum_{\alpha=1}^{r} u_\alpha(x) v_\alpha(y).$$

Then, the QTT ranks of f on any uniform grids are bounded by r.

Later, the QTT format was interpreted as a wavelet-type transform [52], which allows its use for the compression of large data.

5.11 ▪ Numerical illustrations

We will illustrate how some of the algorithms described above work for several model problems.

[22]The original name was quantics tensor train, but later it was proposed to interpret QTT as quantized tensor train, the convention we follow.

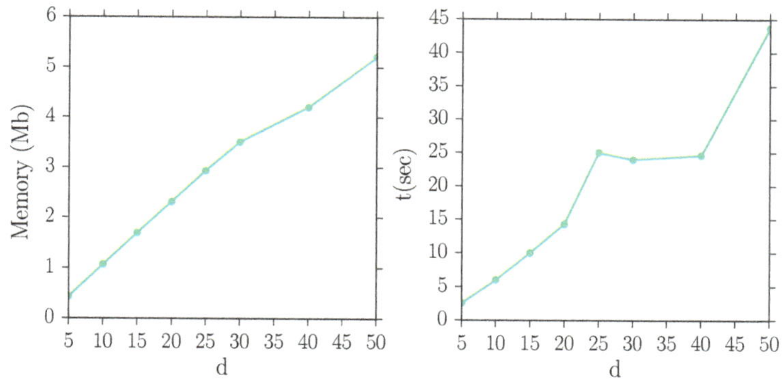

Figure 5.1. *Left: memory to store the approximate solution from the dimension of the problem. Right: time to solve the optimization problem using the AMEN method from the dimension of the problem.*

5.11.1 ▪ High-dimensional Poisson equation

We consider the Poisson equation of the form

$$\Delta_d u = f$$

on $[0,1]^d$ with homogeneous Dirichlet boundary conditions and constant right-hand side one. We discretize the problem on a tensor product uniform mesh using central differences, and the matrix has the form

$$\Delta_d = \Delta_1 \otimes I \otimes \cdots \otimes I + \cdots + I \otimes \cdots \otimes \Delta_1,$$

where

$$\Delta_1 = \frac{1}{h^2}\text{tridiag}[-1,2,-1].$$

For the experiments, we choose $h = \frac{1}{n+1}$ and $n = 2^D + 1$, with $D = 8$, i.e., the one-dimensional mesh size is 256. It can be shown [15, 19] that the solution u can be well approximated in the canonical format with rank

$$r = \mathcal{O}(\log n \log \varepsilon^{-1}).$$

To solve this linear system, we first put the matrix into the QTT format (the QTT ranks are bounded by four) and then run the black-box AMEN solver. The memory and time required for the solution for different d are presented in Figure 5.1.

5.11.2 ▪ High-dimensional eigenvalue problem

We consider computing several minimal eigenvalues of the Laplace operator via the block ALS method proposed in [13]. The same discretization as in the previous section is used. Again the matrix is already in QTT format, and it is sufficient to run the black-box solver. Eigenfunctions of the Laplace operators are separable, and for the one-dimensional Laplace operator they are sine functions that have an exact rank-two representation. The results are presented in Figure 5.2.

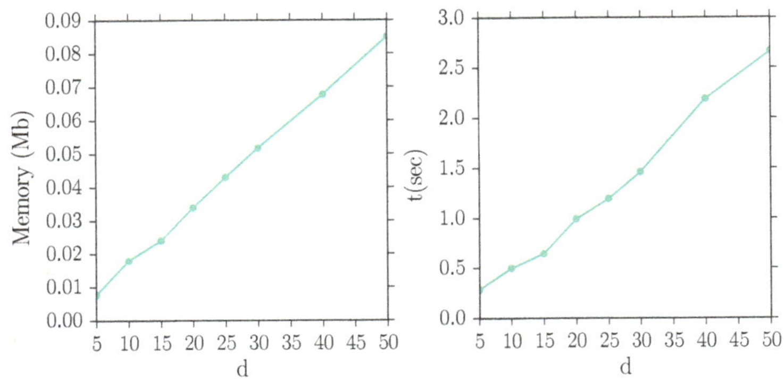

Figure 5.2. *Left: memory to store the approximate solution from the dimension of the problem. Right: time to solve the optimization problem using the ALS method from the dimension of the problem.*

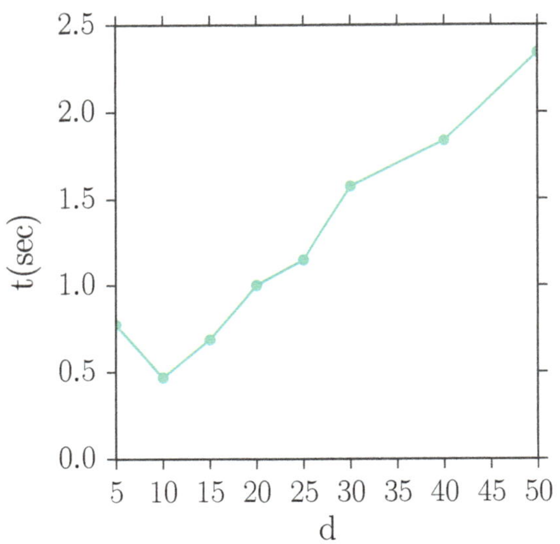

Figure 5.3. *Time for the KSL integration. The rank for the manifold is set to four (although the rank of the solution is one, this does not influence the final result).*

5.11.3 ▪ Time-dependent problem

Here we consider time evolution for the Laplace operator as well (see Figure 5.3):

$$\frac{dy}{dt} = \Delta_d y, \quad y(0) = y_0.$$

The initial condition is a vector of all ones, and we use the time step $\tau = 10^{-2}$ and compute the time evolution up to $T = 1$ (i.e., making 100 time steps). We integrate using the TT-Toolbox in Python and the second-order KSL scheme from [37]. Note that due to the structure of the Laplace operator, the matrix exponent can be represented as a product of one-dimensional exponents. The QTT ranks, however, are not preserved.

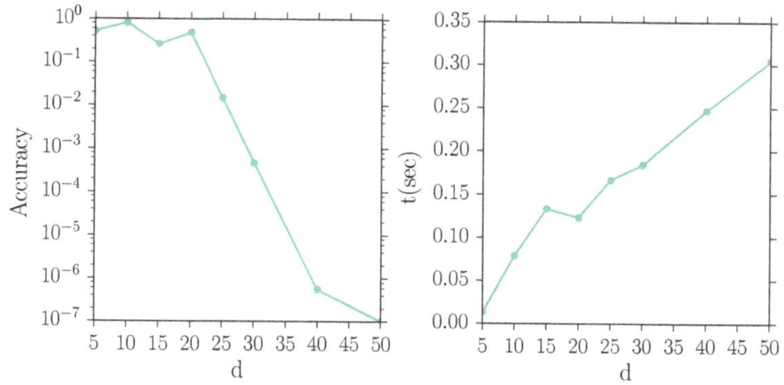

Figure 5.4. *Left: time for the cross-approximation from d. Right: Accuracy of the integral computation.*

5.11.4 ▪ Cross-approximation

Finally, we present a model problem where cross-approximation can be used. We consider the function

$$f(x) = \frac{\sin x}{x}$$

defined on a uniform grid on $[0, 10^6]$ with 2^d points. After tensorization, we get a $2 \times \cdots \times 2$ d-dimensional tensor, which we approximate by the cross method. Then, we use the obtained approximation to compute the approximation of the integral

$$\int_0^\infty \frac{\sin x}{x} dx = \frac{\pi}{2} \approx \sum_{k=1}^{N} w_k f(x_k),$$

where $N = 2^d$. In Figure 5.4 we present both the time for the cross method and the computation of the sum and the accuracy of the integral computation.

Bibliography

[1] J. BALLANI, L. GRASEDYCK, AND M. KLUGE, *Black box approximation of tensors in hierarchical Tucker format*, Linear Alg. Appl., 428 (2013), pp. 639–657.

[2] J. J. BARTHOLDI III, *A good submatrix is hard to find*, Operations Research Lett., 1 (1982), pp. 190–193.

[3] G. BEYLKIN AND M. J. MOHLENKAMP, *Numerical operator calculus in higher dimensions*, Proc. Nat. Acad. Sci. USA, 99 (2002), pp. 10246–10251.

[4] ——, *Algorithms for numerical analysis in high dimensions*, SIAM J. Sci. Comput., 26 (2005), pp. 2133–2159.

[5] M. BUHMANN, *Radial basis functions*, Acta Numer., 9 (2000), pp. 1–38.

[6] H.-J. BUNGARTZ AND M. GRIEBEL, *Sparse grids*, Acta Numer., 13 (2004), pp. 147–269.

[7] R. CATTELL, *"Parallel proportional profiles" and other principles for determining the choice of factors by rotation*, Psychometrika, 9 (1944), pp. 267–283.

[8] J. CHIU AND L. DEMANET, *Sublinear randomized algorithms for skeleton decompositions*, SIAM J. Matrix Anal. Appl., 34 (2013), pp. 1361–1383.

[9] L. DE LATHAUWER, B. DE MOOR, AND J. VANDEWALLE, *A multilinear singular value decomposition*, SIAM J. Matrix Anal. Appl., 21 (2000), pp. 1253–1278.

[10] ——, *On the best rank-*1 *and rank-*$(R_1, R_2, \ldots, R_N)$ *approximation of higher-order tensors*, SIAM J. Matrix Anal. Appl., 21 (2000), pp. 1324–1342.

[11] V. DE SILVA AND L.-H. LIM, *Tensor rank and the ill-posedness of the best low-rank approximation problem*, SIAM J. Matrix Anal. Appl., 30 (2008), pp. 1084–1127.

[12] S. V. DOLGOV, *Alternating Minimal Energy Approach to ODEs and Conservation Laws in Tensor Product Formats*, arXiv preprint 1403.8085, 2014.

[13] S. V. DOLGOV, B. N. KHOROMSKIJ, I. V. OSELEDETS, AND D. V. SAVOSTYANOV, *Computation of extreme eigenvalues in higher dimensions using block tensor train format*, Comput. Phys. Comm., 185 (2014), pp. 1207–1216.

[14] S. V. DOLGOV AND D. V. SAVOSTYANOV, *Alternating minimal energy methods for linear systems in higher dimensions*, SIAM J. Sci. Comput., 36 (2014), pp. A2248–A2271.

[15] I. P. GAVRILYUK, W. HACKBUSCH, AND B. N. KHOROMSKIJ, *Hierarchical tensor-product approximation to the inverse and related operators for high-dimensional elliptic problems*, Computing, 74 (2005), pp. 131–157.

[16] S. A. GOREINOV, I. V. OSELEDETS, D. V. SAVOSTYANOV, E. E. TYRTYSHNIKOV, AND N. L. ZAMARASHKIN, *How to find a good submatrix*, in Matrix Methods: Theory, Algorithms, Applications, V. Olshevsky and E. Tyrtyshnikov, eds., World Scientific, Hackensack, NY, 2010, pp. 247–256.

[17] S. A. GOREINOV AND E. E. TYRTYSHNIKOV, *The maximal-volume concept in approximation by low-rank matrices*, Contemporary Mathematics, 208 (2001), pp. 47–51.

[18] S. A. GOREINOV, E. E. TYRTYSHNIKOV, AND N. L. ZAMARASHKIN, *A theory of pseudo-skeleton approximations*, Linear Algebra Appl., 261 (1997), pp. 1–21.

[19] L. GRASEDYCK, *Existence and computation of low Kronecker-rank approximations for large systems in tensor product structure*, Comput., 72 (2004), pp. 247–265.

[20] ——, *Hierarchical singular value decomposition of tensors*, SIAM J. Matrix Anal. Appl., 31 (2010), pp. 2029–2054.

[21] ——, *Polynomial Approximation in Hierarchical Tucker Format by Vector-Tensorization*, DFG-SPP1324 Preprint 43, Philipps-Univ., Marburg, 2010.

[22] L. GRASEDYCK AND W. HACKBUSCH, *An introduction to hierarchical* ($\mathcal{H}$-) *and TT-rank of tensors with examples*, Comput. Methods Appl. Math., 3 (2011), pp. 291–304.

[23] W. HACKBUSCH AND S. KÜHN, *A new scheme for the tensor representation*, J. Fourier Anal. Appl., 15 (2009), pp. 706–722.

[24] R. A. HARSHMAN, *Foundations of the PARAFAC procedure: Models and conditions for an explanatory multimodal factor analysis*, UCLA Working Papers in Phonetics, 16 (1970), pp. 1–84.

[25] J. HASTAD, *Tensor rank is NP–complete*, J. Algorithms, 11 (1990), pp. 644–654.

[26] S. HOLTZ, T. ROHWEDDER, AND R. SCHNEIDER, *On manifolds of tensors of fixed TT-rank*, Numer. Math., 120 (2012), pp. 701–731.

[27] T. K. HUCKLE, K. WALDHERR, AND T. SCHULTE-HERBRÜGGEN, *Exploiting matrix symmetries and physical symmetries in matrix product states and tensor trains*, Linear Multilinear Algebra, 61 (2013), pp. 91–122.

[28] V. A. KAZEEV AND B. N. KHOROMSKIJ, *Low-rank explicit QTT representation of the Laplace operator and its inverse*, SIAM J. Matrix Anal. Appl., 33 (2012), pp. 742–758.

[29] B. N. KHOROMSKIJ, *$\mathcal{O}(d \log n)$-quantics approximation of N-d tensors in high-dimensional numerical modeling*, Construct. Approx., 34 (2011), pp. 257–280.

[30] B. N. KHOROMSKIJ AND V. KHOROMSKAIA, *Low rank Tucker-type tensor approximation to classical potentials*, Central European J. Math., 5 (2007), pp. 523–550.

[31] A. KLÜMPER, A. SCHADSCHNEIDER, AND J. ZITTARTZ, *Matrix product ground states for one-dimensional spin-1 quantum antiferromagnets*, Europhys. Lett., 24 (1993), pp. 293–297.

[32] O. KOCH AND C. LUBICH, *Dynamical low-rank approximation*, SIAM J. Matrix Anal. Appl., 29 (2007), pp. 434–454.

[33] ———, *Dynamical tensor approximation*, SIAM J. Matrix Anal. Appl., 31 (2010), pp. 2360–2375.

[34] T. G. KOLDA AND B. W. BADER, *Tensor decompositions and applications*, SIAM Rev., 51 (2009), pp. 455–500.

[35] J. M. LANDSBERG, *Tensors: Geometry and Applications*, vol. 128, American Mathematical Society, Providence, RI, 2012.

[36] J. M. LANDSBERG, Y. QI, AND K. YE, *On the geometry of tensor network states*, Quantum Inform. Comput., 12 (2012), pp. 346–354.

[37] C. LUBICH, I. OSELEDETS, AND B. VANDEREYCKEN, *Time Integration of Tensor Trains*, arXiv preprint 1407.2042, 2014.

[38] C. LUBICH AND I.V. OSELEDETS, *A projector-splitting integrator for dynamical low-rank approximation*, BIT, 54 (2014), pp. 171–188.

[39] U. MANTHE, *A multilayer multiconfigurational time-dependent Hartree approach for quantum dynamics on general potential energy surfaces.*, J. Chem. Phys., 128 (2008), p. 164116.

[40] H.-D. MEYER, U. MANTHE, AND L. S. CEDERBAUM, *The multi-configurational time-dependent Hartree approach*, Chem. Phys. Lett., 165 (1990), pp. 73–78.

[41] E. Novak and H. Woźniakowski, *Tractability of Multivariate Problems: Standard Information for Functionals*, vol. 12, European Mathematical Society, 2010.

[42] I. Oseledets, S. Dolgov, D. Savostyanov, and V. Kazeev, *TT-Toolbox v2.2*, 2014.

[43] I. Oseledets, S. Dolgov, D. Savostyanov, and T. Saluev, *ttpy v.0.9*, 2014.

[44] I.V. Oseledets, *Approximation of matrices with logarithmic number of parameters*, Doklady Math., 428 (2009), pp. 23–24.

[45] ——, *Approximation of $2^d \times 2^d$ matrices using tensor decomposition*, SIAM J. Matrix Anal. Appl., 31 (2010), pp. 2130–2145.

[46] ——, *Tensor-train decomposition*, SIAM J. Sci. Comput., 33 (2011), pp. 2295–2317.

[47] ——, *Constructive representation of functions in low-rank tensor formats*, Construct. Approx., 37 (2013), pp. 1–18.

[48] I. V. Oseledets, D. V. Savostianov, and E. E. Tyrtyshnikov, *Tucker dimensionality reduction of three-dimensional arrays in linear time*, SIAM J. Matrix Anal. Appl., 30 (2008), pp. 939–956.

[49] I. V. Oseledets and E. E. Tyrtyshnikov, *Breaking the curse of dimensionality, or how to use SVD in many dimensions*, SIAM J. Sci. Comput., 31 (2009), pp. 3744–3759.

[50] ——, *Tensor Tree Decomposition Does Not Need a Tree*, preprint 2009-04, INM RAS, Moscow, 2009.

[51] ——, *TT-cross approximation for multidimensional arrays*, Linear Algebra Appl., 432 (2010), pp. 70–88.

[52] ——, *Algebraic wavelet transform via quantics tensor train decomposition*, SIAM J. Sci. Comput., 33 (2011), pp. 1315–1328.

[53] S. Östlund and S. Rommer, *Thermodynamic limit of density matrix renormalization*, Phys. Rev. Lett., 75 (1995), pp. 3537–3540.

[54] T. Rohwedder and A. Uschmajew, *On local convergence of alternating schemes for optimization of convex problems in the tensor train format*, SIAM J. Numer. Anal., 51 (2013), pp. 1134–1162.

[55] D. V. Savostyanov, *Quasioptimality of maximum-volume cross interpolation of tensors*, Linear Algebra Appl., 458 (2014), pp. 217–244.

[56] D. V. Savostyanov and I. V. Oseledets, *Fast adaptive interpolation of multi-dimensional arrays in tensor train format*, in Proceedings of the 7th International Workshop on Multidimensional Systems (nDS), IEEE, 2011.

[57] U. Schollwöck, *The density-matrix renormalization group in the age of matrix product states*, Ann. Phys., 326 (2011), pp. 96–192.

[58] R. B. SIDJE, *Expokit: A software package for computing matrix exponentials*, ACM Trans. Math. Software (TOMS), 24 (1998), pp. 130–156.

[59] S. A. SMOLYAK, *Quadrature and interpolation formulas for tensor products of certain classes of functions*, Soviet Math. Doklady, 4 (1963), pp. 240–243.

[60] L. SORBER, M. VAN BAREL, AND L. DE LATHAUWER, *Tensorlab v1.0*, 2013.

[61] V. N. TEMLYAKOV, *Nonlinear methods of approximation*, Found. Comput. Math., 3 (2003), pp. 33–107.

[62] L. R. TUCKER, *Some mathematical notes on three-mode factor analysis*, Psychometrika, 31 (1966), pp. 279–311.

[63] E. E. TYRTYSHNIKOV, *Tensor approximations of matrices generated by asymptotically smooth functions*, Sbornik: Math., 194 (2003), pp. 941–954.

[64] G. VIDAL, *Efficient classical simulation of slightly entangled quantum computations*, Phys. Rev. Lett., 91 (2003), p. 147902.

[65] S. R. WHITE, *Density-matrix algorithms for quantum renormalization groups*, Phys. Rev. B, 48 (1993), pp. 10345–10356.

Part III

System-Theoretic Methods

Part III of this book is concerned with system-theoretic methods. It covers balancing-type methods, interpolatory methods, and the data-driven Loewner model reduction framework. The advantages of balancing reduction methods include preservation of stability and an a priori computable error bound, while interpolatory methods tend to be numerically efficient and have appealing optimality properties. The Loewner framework is an attractive option in the data-driven setting where the high-fidelity model is unavailable.

In the first chapter of Part III, Peter Benner and Tobias Breiten introduce model order reduction (MOR) techniques based on the concept of balanced truncation (BT). They review methods for constructing low-rank approximations to the system Gramians and the associated balanced reduced-order models (ROMs), and discuss connections among the methods. They also discuss generalizations of BT to linear stochastic and bilinear control systems and to frequency-weighted balancing.

In the second chapter, Christopher Beattie and Serkan Gugercin survey interpolatory model reduction methods. The chapter covers basic principles for the novice, including deriving locally optimal models, and then advances to recent developments, including structure-preserving methods based on generalized coprime representations and reduction of systems of differential-algebraic equations (DAEs). The chapter also extends interpolatory methods to reduction of parametrized systems.

In the third chapter of Part III, Athanasios Antoulas and co-authors present a tutorial introduction to the Loewner framework for model reduction. The Loewner framework uses the interpolatory model reduction setting of the previous chapter, but in contrast to the methods presented in earlier chapters, it is entirely data driven. In the case of a linear dynamical system, the Loewner framework uses frequency response measurements and recovers an interpolatory ROM without needing an explicit high-fidelity model. The Loewner framework also provides a trade-off between accuracy of fit and complexity of the model.

In the final chapter of Part III, Ulrike Baur and co-authors compare methods for parametric model reduction (PMOR) of nonstationary problems. The chapter compares many of the methods discussed in this book—POD, POD-greedy, several interpolatory methods, and the empirical cross-Gramian method. The methods are each applied to three benchmarks selected from the MOR Wiki benchmark collection. A discussion of the results lends insight into the advantages and disadvantages of the methods.

Chapter 6

Model Order Reduction Based on System Balancing

Peter Benner and Tobias Breiten

We introduce balancing-based model order reduction (MOR) techniques for large-scale dynamical systems. We emphasize the efficient implementation of state-of-the-art techniques for systems with up to millions of degrees of freedom. In particular, we briefly review recent results on low-rank methods that compute approximate balanced reduced-order models (ROMs). In particular, we discuss a connection between several of these approximate balancing techniques. We further present generalizations of the classical balanced truncation (BT) method for linear systems. This will include balancing of linear stochastic and bilinear control systems. For these systems, generalized Lyapunov equations play an important role. We also survey possible generalizations of low-rank solution methods initially derived for the linear case, such as the ADI iteration or Krylov subspace methods for the tensorized linear system. Moreover, we discuss frequency-weighted balancing of linear systems. Again, the more complicated structure of the underlying Lyapunov equations is studied and problems that might occur are addressed. A brief outlook, with current and future topics of research, is provided.

6.1 ▪ Introduction

In this chapter, we consider balancing-based model reduction methods for large-scale dynamical systems. We start with an introduction of the classical BT approach for linear time-invariant (LTI) control systems of the form

$$\Sigma\colon \begin{cases} \dot{\mathbf{x}}(t) = \mathbf{A}\mathbf{x}(t) + \mathbf{B}\mathbf{u}(t), \quad \mathbf{x}(0) = 0, \\ \mathbf{y}(t) = \mathbf{C}\mathbf{x}(t) + \mathbf{D}\mathbf{u}(t), \end{cases} \tag{6.1}$$

with $\mathbf{A} \in \mathbb{R}^{n\times n}, \mathbf{B} \in \mathbb{R}^{n\times m}, \mathbf{C} \in \mathbb{R}^{p\times n}$, and $\mathbf{D} \in \mathbb{R}^{p\times m}$. Here, $\mathbf{x}(t) \in \mathbb{R}^n, \mathbf{u}(t) \in \mathbb{R}^m$, and $\mathbf{y}(t) \in \mathbb{R}^p$ are called the *state, input,* and *output* of the system. These kinds of systems typically arise from a spatial semidiscretization of a more general partial differential equation (PDE). As a consequence, the number of states, n, can become very large, preventing us from efficiently simulating the system. The idea of *MOR* is to replace Σ by a structurally similar system $\hat{\Sigma}$ of the form

$$\hat{\Sigma}: \begin{cases} \dot{\hat{\mathbf{x}}}(t) = \hat{\mathbf{A}}\hat{\mathbf{x}}(t) + \hat{\mathbf{B}}\mathbf{u}(t), & \hat{\mathbf{x}}(0) = \hat{\mathbf{x}}_0, \\ \hat{\mathbf{y}}(t) = \hat{\mathbf{C}}\hat{\mathbf{x}}(t) + \hat{\mathbf{D}}\mathbf{u}(t), \end{cases} \tag{6.2}$$

where $\hat{\mathbf{A}} \in \mathbb{R}^{r\times r}, \hat{\mathbf{B}} \in \mathbb{R}^{r\times m}, \hat{\mathbf{C}} \in \mathbb{R}^{p\times r}$, and $\hat{\mathbf{D}} \in \mathbb{R}^{p\times m}$. The main goals we are particularly interested in are $r \ll n$ and a small output error $||\mathbf{y}-\hat{\mathbf{y}}||$ for all admissible input functions $\mathbf{u}(\cdot)$, e.g., $\mathbf{u} \in L_2([0,\infty),\mathbb{R}^m)$. Obviously, the first demand ensures a significant speedup in simulation time. The second demand means that we want to construct a ROM that reproduces the original quantity of interest as closely as possible. Depending on the specific norm we are interested in, different techniques have been developed over the last decades. We refer to [3, 4, 10, 26, 92, 109] and also Chapters 7 and 8 in this volume, where detailed overviews of several general MOR techniques are given.

Remark 6.1. In many applications, mathematical modeling leads to systems with an additional matrix $\mathbf{E} \in \mathbb{R}^{n\times n}$ in front of the state's time derivative, i.e., the dynamics are described by

$$\mathbf{E}\dot{\mathbf{x}}(t) = \mathbf{A}\mathbf{x}(t) + \mathbf{B}\mathbf{u}(t), \quad \mathbf{E}\mathbf{x}(0) = \mathbf{x}_0. \tag{6.3}$$

If the model comes from a finite element (FE) discretization, then $\mathbf{E}$ denotes the mass matrix of the ansatz space and is therefore usually symmetric positive definite and in particular invertible. Then, we can formulate a new realization of Σ by premultiplying (6.3) with $\mathbf{E}^{-1}$. This changes neither inputs or outputs nor the input-output relation, and thus everything discussed in this chapter can be applied in this way. Of course, this should be done for formal derivation of the methods only; in numerical algorithms, one should work with the initial data as much as possible, and it is almost always possible to formulate the algorithms discussed in this chapter without ever inverting $\mathbf{E}$ or even solving linear systems with $\mathbf{E}$. E.g., in some approaches, the repeated solution of linear systems of the form

$$\left(\mathbf{E}^{-1}\mathbf{A} - \mu\mathbf{I}\right)z = w, \quad \mu \in \mathbb{C},$$

will be necessary. This is easy to reformulate into

$$(\mathbf{A} - \mu\mathbf{E})\,z = \mathbf{E}w, \quad \mu \in \mathbb{C},$$

so that only a (sparse) matrix-vector multiplication needs to be performed; for more details on how to implement algorithms with an invertible $\mathbf{E}$ matrix, see [12, 30, 107].

The case that $\mathbf{E}$ is singular often occurs when modeling constrained mechanical systems or circuits (see [85]), leading to a *descriptor system*, i.e., (6.3) is a system of linear differential-algebraic equations (DAEs). Most model reduction methods can also be applied in this situation, but this requires using certain projectors and becomes quite technically involved. Therefore, for tutorial purposes, we restrict ourselves to the easier case of the standard state-space system (6.1) and refer to the recent survey [31] for details of model reduction for descriptor systems.

In what follows, we always assume that the system under consideration is (asymptotically) stable, meaning that the eigenvalues of $\mathbf{A}$ are located in the open left complex plane $\mathbb{C}_-$. For these systems, we can define the $\mathcal{H}_\infty$-norm as:

$$||\mathbf{H}||_\infty := \sup_{\omega\in\mathbb{R}} \sigma_{\max}(\mathbf{H}(\imath\omega)), \tag{6.4}$$

where $\imath$ is the imaginary unit and $\sigma_{\max}$ denotes the maximum singular value of a matrix. For a better understanding of the relevance of this norm, we have to consider the system in the so-called frequency domain. Applying the Laplace transform to the system in (6.1) and assuming that $\mathbf{x}(0) = 0$ (see Remark 6.5) yields a set of algebraic equations that can be solved to get an explicit input-output relation as follows:

$$\mathbf{y}(s) = \big(\underbrace{\mathbf{C}(s\mathbf{I}-\mathbf{A})^{-1}\mathbf{B}+\mathbf{D}}_{\mathbf{H}(s)}\big)\mathbf{u}(s), \tag{6.5}$$

where $\mathbf{u}(s)$ and $\mathbf{y}(s)$ denote the Laplace transforms of the input and output functions $\mathbf{u}(t)$ and $\mathbf{y}(t)$. Using Plancherel's theorem, we can bound the approximation error via

$$||\mathbf{y}-\hat{\mathbf{y}}||_2 \leq ||\mathbf{H}-\hat{\mathbf{H}}||_{\mathcal{H}_\infty}||\mathbf{u}||_2, \tag{6.6}$$

where the 2-norm can be interpreted in time as well as in the frequency domain. In other words, minimizing the error between the original and the reduced transfer function implies minimizing the output error. A very important observation regarding the transfer function is its invariance with respect to *restricted equivalence transformations*

$$(\mathbf{A},\mathbf{B},\mathbf{C},\mathbf{D}) \mapsto (\mathbf{TAZ},\mathbf{TB},\mathbf{CZ},\mathbf{D}), \tag{6.7}$$

where $\mathbf{T},\mathbf{Z} \in \mathbb{R}^{n\times n}$ are arbitrary nonsingular matrices. Indeed, one can easily show that

$$(\mathbf{CZ})(s\mathbf{TZ}-\mathbf{TAZ})^{-1}(\mathbf{TB})+\mathbf{D} = \mathbf{C}(s\mathbf{I}-\mathbf{A})^{-1}\mathbf{B}+\mathbf{D} = \mathbf{H}(s).$$

The intrinsic idea of balancing and balancing-related techniques is to find matrices $\mathbf{T}$ and $\mathbf{Z}$ such that the system is balanced, meaning that certain *Gramians* $\mathbf{P}$ and $\mathbf{Q}$ related to (6.1) are equal and diagonal. While the transfer function remains invariant (as seen above), balanced systems exhibit useful properties from a system-theoretic point of view. For example, if the system is balanced with respect to the controllability Gramian $\mathbf{P}$ and the observability Gramian $\mathbf{Q}$, one can immediately *read off* related control and output energies of the system. In this way, the balancing step itself is worthwhile in its own right. If one is further interested in approximating the original input-output behavior by a reduced system, one can perform a truncation step, which actually means discarding the least relevant states with respect to the balanced basis of the system. While in exact arithmetic the complexity for computing a balanced realization and performing the truncation step is $\mathcal{O}(n^3)$, several current tools from numerical linear algebra allow us to handle systems with millions of degrees of freedom.

The structure of this chapter is as follows. In Section 6.2, we review the standard BT approach for LTI systems of the form (6.1). We introduce controllability and observability concepts and discuss the interpretation of the Gramians as optimal control functions. Based on the idea of balancing, we present the method of singular perturbation approximation as well as frequency-weighted BT in Section 6.3. While the first method typically produces models that are accurate for lower frequencies, the latter allows us to define a frequency range of interest. In Section 6.4 we explain how balancing (related) concepts can be used in view of MOR of more general systems, such as linear stochastic or bilinear control systems. Motivated by all these methods, in Section 6.5 we provide an overview of recent developments regarding the numerical solution of large-scale linear matrix equations. This includes the relatively new concept of Riemannian optimization and its application to linear matrix equations. We provide possible solution techniques for the involved and more general linear matrix equations. In Section 6.6 we compare the methods for some numerical examples. Finally, in Section 6.7 we conclude with a summary of very recent and current research topics of interest.

6.2 ▪ BT for LTI systems

As we already mentioned above, the idea of BT is to find a suitable basis transformation of the system such that we can *read off* the states that are less important than others. Of course, we first have to define what we consider to be important. For this, we need some basic system-theoretic concepts such as controllability and observability; see also any standard textbook on control theory, e.g., [64, 73, 112]. In what follows, we assume for the state space $\mathbb{X}$ of the system that $\mathbb{X} = \mathbb{R}^n$.

Definition 6.2. *The system* (6.1) *is called* controllable *if for all* $\mathbf{x}_0, \mathbf{x}_1 \in \mathbb{X}$ *there exists* T *and an admissible input function* $\mathbf{u}\colon [0, T] \to \mathbb{R}^m$ *such that the solution* $\mathbf{x}(t)$ *of* (6.1) *fulfills* $\mathbf{x}(0) = \mathbf{x}_0$ *and* $\mathbf{x}(T) = \mathbf{x}_1$.

Definition 6.3. *A state* $\mathbf{x}_1 \in \mathbb{X}$ *is* unobservable *if for* $\mathbf{x}(0) = \mathbf{x}_1$ *in* (6.1), *we have* $\mathbf{y}(t) = 0$ *for all* $t \geq 0$, *i.e., if* $\mathbf{x}_1$ *is indistinguishable from the zero state for all* $t \geq 0$. *The* unobservable subspace $\mathbb{X}^{\text{unobs}}$ *is the set of all unobservable states of the system. The system is observable if* $\mathbb{X}^{\text{unobs}} = \{0\}$.

Intuitively speaking, we expect states that are neither controllable nor observable to be irrelevant for the task of controller and observer design. In fact, discarding these states will not change the transfer function of the system at all. Systems without such states, i.e., systems that are controllable and observable, are *minimal* systems. This means there does not exist another realization $(\tilde{\mathbf{A}}, \tilde{\mathbf{B}}, \tilde{\mathbf{C}}, \tilde{\mathbf{D}})$ of dimension $\tilde{n} < n$ that has the same transfer function $\tilde{\mathbf{G}}(s) \equiv \mathbf{G}(s)$. In our considerations, we now always assume the system to be minimal. For the general case, see also Remark 6.6.

There might still exist states that are hard to control or hard to observe. To characterize this mathematically, we introduce the infinite controllability and observability Gramians

$$\begin{aligned} \mathbf{P} &= \int_0^\infty e^{\mathbf{A}t}\mathbf{B}\mathbf{B}^T e^{\mathbf{A}^T t}\,dt, \\ \mathbf{Q} &= \int_0^\infty e^{\mathbf{A}^T t}\mathbf{C}^T\mathbf{C}e^{\mathbf{A}t}\,dt. \end{aligned} \tag{6.8}$$

For asymptotically stable and minimal systems, it is well known [3, 64] that both $\mathbf{P}$ and $\mathbf{Q}$ are symmetric positive definite matrices satisfying

$$\begin{aligned} \mathbf{A}\mathbf{P} + \mathbf{P}\mathbf{A}^T + \mathbf{B}\mathbf{B}^T &= 0, \\ \mathbf{A}^T\mathbf{Q} + \mathbf{Q}\mathbf{A} + \mathbf{C}^T\mathbf{C} &= 0. \end{aligned} \tag{6.9}$$

Moreover, given an initial state $\mathbf{x}_0 \in \mathbb{R}^n$, the solutions of these matrix equations characterize the input and output energy in the following way:

$$\begin{aligned} \mathbf{x}_0^T \mathbf{P}^{-1} \mathbf{x}_0 &= \min_{\substack{\mathbf{u} \in L_2(-\infty,0] \\ \mathbf{x}(-\infty,x_0,\mathbf{u})=0}} \int_{-\infty}^{0} \|\mathbf{u}(t)\|^2 \mathrm{d}t, \\ \mathbf{x}_0^T \mathbf{Q} \mathbf{x}_0 &= \int_0^\infty \|\mathbf{y}(t,\mathbf{x}_0,0)\|^2 \mathrm{d}t. \end{aligned} \tag{6.10}$$

Hence, for systems with $\mathbf{P} = \mathbf{Q} = \operatorname{diag}(\sigma_1, \ldots, \sigma_n)$, $\sigma_1 \geq \sigma_2 \geq \cdots \geq \sigma_n > 0$, a small *Hankel singular value* σ_i characterizes states that require a large amount of energy to be controlled while leading only to a small amount of energy when observed. The Gramians of *balanced systems* now exactly exhibit this special property. The goal is thus to find a state-space transformation

$$\mathscr{T}: \begin{cases} \mathbf{x} \mapsto \mathbf{T}\mathbf{x}, \\ (\mathbf{A},\mathbf{B},\mathbf{C},\mathbf{D}) \mapsto (\mathbf{T}\mathbf{A}\mathbf{T}^{-1}, \mathbf{T}\mathbf{B}, \mathbf{C}\mathbf{T}^{-1}, \mathbf{D}), \end{cases} \tag{6.11}$$

with $\mathbf{T} \in \mathbb{R}^{n \times n}$ nonsingular such that the resulting system is balanced. Noting that a state-space transformation changes the Gramians according to

$$(\mathbf{P},\mathbf{Q}) \mapsto (\mathbf{P}_{\mathscr{T}}, \mathbf{Q}_{\mathscr{T}}) = (\mathbf{T}\mathbf{P}\mathbf{T}^T, \mathbf{T}^{-T}\mathbf{Q}\mathbf{T}^{-1}),$$

a balancing transformation can be shown to be given by $\mathbf{T}_{\mathscr{B}} = \Sigma^{-\frac{1}{2}} \mathbf{V}^T \mathbf{R}$, where $\mathbf{P} = \mathbf{S}^T\mathbf{S}, \mathbf{Q} = \mathbf{R}^T\mathbf{R}$, and $\mathbf{S}\mathbf{R}^T = \mathbf{U}\Sigma\mathbf{V}^T$ is the singular value decomposition (SVD) of $\mathbf{S}\mathbf{R}^T$. Next, we assume that the balanced system is partitioned as

$$\mathbf{A}_{\mathscr{B}} = \begin{bmatrix} \mathbf{A}_{11} & \mathbf{A}_{12} \\ \mathbf{A}_{21} & \mathbf{A}_{22} \end{bmatrix}, \; \mathbf{B}_{\mathscr{B}} = \begin{bmatrix} \mathbf{B}_1 \\ \mathbf{B}_2 \end{bmatrix}, \; \mathbf{C}_{\mathscr{B}} = \begin{bmatrix} \mathbf{C}_1 & \mathbf{C}_2 \end{bmatrix}, \; \mathbf{D}_{\mathscr{B}} = \mathbf{D}. \tag{6.12}$$

In the balanced basis, the truncation step is simply obtained by setting the ROM to $(\mathbf{A}_{11}, \mathbf{B}_1, \mathbf{C}_1, \mathbf{D})$, with corresponding system dimension r. Model reduction by BT was first discussed in [91] and later on considered within a control theory framework in [90]. One of the reasons that BT is such a popular model reduction method for LTI systems is that it preserves several system properties, e.g., stability (see [99]), and yields an $\mathscr{H}_\infty$-error bound (see [47, 50]).

Theorem 6.4 ([3, Theorem 7.9]). *Let $\Sigma = (\mathbf{A},\mathbf{B},\mathbf{C},\mathbf{D})$ be an asymptotically stable and minimal control system. Assume that for the balanced system $(\mathbf{A}_{\mathscr{B}}, \mathbf{B}_{\mathscr{B}}, \mathbf{C}_{\mathscr{B}}, \mathbf{D})$,*

$$\mathbf{P}_{\mathscr{B}} = \mathbf{Q}_{\mathscr{B}} = \operatorname{diag}(\sigma_1 \mathbf{I}_{n_1}, \ldots, \sigma_k \mathbf{I}_{n_k}) \quad \text{with} \quad \sigma_1 > \sigma_2 > \cdots > \sigma_k > 0.$$

Consider a reduced-order system $\hat{\Sigma} = (\mathbf{A}_{11}, \mathbf{B}_1, \mathbf{C}_1, \mathbf{D})$ of dimension $r = n_1 + \cdots + n_r$. Then $\hat{\Sigma}$ is asymptotically stable, minimal, and balanced with $\hat{\mathbf{P}} = \hat{\mathbf{Q}} = \operatorname{diag}(\sigma_1 \mathbf{I}_{n_1}, \ldots, \sigma_r \mathbf{I}_{n_r})$. It moreover holds that

$$\|\Sigma - \hat{\Sigma}\|_{\mathscr{H}_\infty} \leq 2 \sum_{i=r+1}^{k} \sigma_i. \tag{6.13}$$

An important observation for numerical implementation is that we do not have to compute the balanced realization explicitly. Instead, we can use the truncated SVD of the product of the factors of the Gramians. Assume the partitioning

$$\mathbf{S}\mathbf{R}^T = \begin{bmatrix} \mathbf{U}_1 & \mathbf{U}_2 \end{bmatrix} \begin{bmatrix} \Sigma_1 & \\ & \Sigma_2 \end{bmatrix} \begin{bmatrix} \mathbf{V}_1^T \\ \mathbf{V}_2^T \end{bmatrix}, \tag{6.14}$$

where $\mathbf{U}_1, \mathbf{V}_1 \in \mathbb{R}^{n \times r}$ and $\Sigma_1 \in \mathbb{R}^{r \times r}$. Then the reduced model $(\mathbf{A}_{11}, \mathbf{B}_1, \mathbf{C}_1, \mathbf{D})$ from (6.12) is obtained by a Petrov–Galerkin projection $\mathscr{P} = \mathbf{V}\mathbf{W}^T$:

$$\mathbf{A}_{11} = \mathbf{W}^T \mathbf{A} \mathbf{V}, \ \ \mathbf{B}_1 = \mathbf{W}^T \mathbf{B}, \ \ \mathbf{C}_1 = \mathbf{C}\mathbf{V}, \tag{6.15}$$

with $\mathbf{V} = \mathbf{S}^T \mathbf{U}_1 \Sigma_1^{-\frac{1}{2}}$ and $\mathbf{W} = \mathbf{R}^T \mathbf{V}_1 \Sigma_1^{-\frac{1}{2}}$. This approach is known as square root balancing. While there also exist so-called balancing-free approaches, here we refrain from a more detailed discussion and instead refer to [3].

The main ingredient for a computational realization is obviously the solution of the linear matrix equations (6.9). In Section 6.5 we provide an overview of existing and recently proposed algorithms allowing us to perform balanced truncation for very large-scale systems.

Remark 6.5. We assumed a zero initial condition $\mathbf{x}_0 = 0$ in (6.1). If this is not the case, the Laplace transform of the system will result in an additional term of the form $\mathbf{C}(s\mathbf{I} - \mathbf{A})^{-1}\mathbf{x}_0$, and the error bound in Theorem 6.4 is no longer valid. However, there exist suitable adaptations for the general case. The basic idea is to consider a transformed system for $\tilde{\mathbf{x}}(t) = \mathbf{x}(t) - \mathbf{x}_0$. For more detail and different ways of incorporating nonzero initial conditions, we refer to [10, 63].

Remark 6.6. For systems that are not minimal, we can still perform a contragredient transformation leading to a balanced system with diagonal Gramians $\mathbf{P}$ and $\mathbf{Q}$ satisfying

$$\mathbf{P}\mathbf{Q} = \operatorname{diag}(\sigma_1^2, \ldots, \sigma_N^2, 0, \ldots, 0).$$

Note that, as long as $\mathbf{A}$ is Hurwitz, the Gramians $\mathbf{P}$ and $\mathbf{Q}$ are unique even for nonminimal systems. Hence, one can employ the above procedure without any modification to compute a minimal realization by truncating the system in accordance with the nonzero Hankel singular values. For more detail, interested readers are referred to [79, 116], where the original ideas for this are proposed.

6.3 ▪ Balancing-related model reduction

Though the standard BT approach ensures stability of the ROM, there exist several modifications for special types of systems. Since all of them share the idea of balancing two Gramian or Gramian-like matrices by a contragredient transformation, they are sometimes referred to as balancing-related methods. For example, in the case of unstable systems, replacing the Gramians $\mathbf{P}$ and $\mathbf{Q}$ by the solutions of two dual Riccati equations leads to linear-quadratic Gaussian (LQG) balancing; see [70, 120, 121]. For preservation of additional system properties such as passivity, one might use positive real BT; see [13, 14, 43, 58]. Below, we discuss two further related methods: (a) singular perturbation approximation (SPA) and (b) frequency-weighted BT. The first approach is of particular importance since we later on discuss its applicability in the case of bilinear control systems.

6.3.1 ▪ SPA

One of the disadvantages when using the classical BT approach is the possible mismatch between the original and reduced-order transfer functions for smaller frequencies. Hence, for simulations in the time domain, this implies deviations for the steady-state approximation. A remedy for this problem is given by SPA. Starting from a balanced realization, we decompose the system into a *slow* variable $\mathbf{x}_1$ and a *fast* variable $\mathbf{x}_2$:

$$\begin{aligned} \frac{\mathrm{d}}{\mathrm{d}t}\begin{bmatrix}\mathbf{x}_1(t)\\ \mathbf{x}_2(t)\end{bmatrix} &= \begin{bmatrix}\mathbf{A}_{11} & \mathbf{A}_{12}\\ \mathbf{A}_{21} & \mathbf{A}_{22}\end{bmatrix}\begin{bmatrix}\mathbf{x}_1(t)\\ \mathbf{x}_2(t)\end{bmatrix} + \begin{bmatrix}\mathbf{B}_1\\ \mathbf{B}_2\end{bmatrix}\mathbf{u}(t), \\ \mathbf{y}(t) &= \begin{bmatrix}\mathbf{C}_1 & \mathbf{C}_2\end{bmatrix}\begin{bmatrix}\mathbf{x}_1(t)\\ \mathbf{x}_2(t)\end{bmatrix} + \mathbf{D}\mathbf{u}(t). \end{aligned} \tag{6.16}$$

Assuming that the fast variable $\mathbf{x}_2$ obtains a steady state with $\dot{\mathbf{x}}_2 = 0$ allows us to explicitly solve for $\mathbf{x}_2$ in the second block row. As a consequence, we can derive an ROM without $\mathbf{x}_2$ via

$$\begin{aligned} \dot{\mathbf{x}}_1(t) &= (\mathbf{A}_{11} - \mathbf{A}_{12}\mathbf{A}_{22}^{-1}\mathbf{A}_{21})\mathbf{x}_1(t) + (\mathbf{B}_1 - \mathbf{A}_{12}\mathbf{A}_{22}^{-1}\mathbf{B}_2)\mathbf{u}(t), \\ \hat{\mathbf{y}}(t) &= (\mathbf{C}_1 - \mathbf{C}_2\mathbf{A}_{22}^{-1}\mathbf{A}_{21})\mathbf{x}_1(t) + (\mathbf{D} - \mathbf{C}_2\mathbf{A}_{22}^{-1}\mathbf{B}_2)\mathbf{u}(t). \end{aligned} \tag{6.17}$$

Due to the inversion of block matrices, one now can easily show that the transfer function of this system interpolates the original transfer function at $s = 0$. However, note that the reduced feed-through term is nonzero in general even if $\mathbf{D} = 0$. Moreover, to derive the reduced model, we need not only the full balanced realization but also the inversion of $\mathbf{A}_{22} \in \mathbb{R}^{(n-r)\times(n-r)}$. Since this matrix reflects the dynamics of the less important states with respect to the balanced basis, it is typically ill conditioned. Thus, for numerical computations, one first has to compute a numerically minimal realization. Roughly speaking, this means we first truncate all states corresponding to Hankel singular values that are smaller than machine precision ε relative to σ_1. Based on the resulting (numerically) balanced and minimal realization we then construct the ROM according to (6.16). For the reduced model, one can show the same properties as for the classical BT method; see [83].

Theorem 6.7 ([83, Theorem 3.1/3.2]). *Let* $(\mathbf{A},\mathbf{B},\mathbf{C},\mathbf{D})$ *be asymptotically stable and minimal. Assume that* Σ_1 *and* Σ_2 *have no common element. Then the reduced system* (6.17) *is asymptotically stable and minimal. Moreover, we have the error bound*

$$\|\Sigma - \hat{\Sigma}\|_{\mathcal{H}_\infty} \leq 2\mathrm{tr}(\Sigma_2).$$

Since the ROM is derived from a balanced realization, the major computational obstacle is again computing the Gramians $\mathbf{P}$ and $\mathbf{Q}$. Additionally, SPA can be seen as a counterpart of classical BT in the sense that it is exact for $s = 0$ but deviates for $s \to \infty$. This is due to the difference of the feed-through terms of the original and the reduced-order system. On the contrary, classical BT is exact for $s \to \infty$ but may suffer from inaccuracies at $s = 0$.

6.3.2 ▪ Frequency-weighted BT

Both BT and SPA yield an error bound for the whole frequency range. In real-life applications, a specific frequency range of interest is often known, allowing us to appropriately modify model reduction techniques. Here, we introduce frequency-weighted BT based on input and output weighting functions $\mathbf{W}_i(s)$ and $\mathbf{W}_o(s)$. However, we stick to the classical method from Enns [47] and only refer to (recent) developments in this direction. Let us assume that rational matrix-valued weighting functions

$\mathbf{W}_i(s) = \mathbf{C}_i(s\mathbf{I}-\mathbf{A}_i)^{-1}\mathbf{B}_i + \mathbf{D}_i$ and $\mathbf{W}_o(s) = \mathbf{C}_o(s\mathbf{I}-\mathbf{A}_o)^{-1}\mathbf{B}_o + \mathbf{D}_o$ are given. For a given system $\mathbf{H}(s)$ we want to construct $\hat{\mathbf{H}}(s)$ such that the weighted error is minimized, i.e.,

$$\left\|\mathbf{W}_o\left(\mathbf{H}-\hat{\mathbf{H}}\right)\mathbf{W}_i\right\|_{\mathcal{H}_\infty}. \tag{6.18}$$

By choosing $\mathbf{W}_i(s)$ and $\mathbf{W}_o(s)$, we can determine the frequency range of interest. For example, $\mathbf{W}_i(s)$ and $\mathbf{W}_o(s)$ can be low-pass filters, indicating that we are interested in an accurate approximation for lower frequencies. Similarly, combining a low-pass filter $\mathbf{W}_i(s)$ and a high-pass filter $\mathbf{W}_o(s)$ can serve as a band-pass filter $\mathbf{W}_o(s)\mathbf{H}(s)\mathbf{W}_i(s)$. To understand the computation of the Gramians required for frequency-weighted BT, it helps to consider explicit realizations for $\mathbf{H}(s)\mathbf{W}_i(s)$:

$$\mathbf{H}\cdot\mathbf{W}_i: \begin{cases} \dfrac{\mathrm{d}}{\mathrm{d}t}\begin{bmatrix}\mathbf{x}(t)\\ \mathbf{w}_i(t)\end{bmatrix} = \underbrace{\begin{bmatrix}\mathbf{A} & \mathbf{BC}_i\\ 0 & \mathbf{A}_i\end{bmatrix}}_{\mathcal{A}_i}\begin{bmatrix}\mathbf{x}(t)\\ \mathbf{w}_i(t)\end{bmatrix} + \underbrace{\begin{bmatrix}\mathbf{BD}_i\\ \mathbf{B}_i\end{bmatrix}}_{\mathcal{B}_i}\mathbf{u}(t),\\[2ex] \mathbf{y}_i(t) = \underbrace{\begin{bmatrix}\mathbf{C} & 0\end{bmatrix}}_{\mathcal{C}_i}\begin{bmatrix}\mathbf{x}(t)\\ \mathbf{w}_i(t)\end{bmatrix} + \underbrace{\mathbf{DD}_i}_{\mathcal{D}_i}\mathbf{u}(t), \end{cases} \tag{6.19}$$

and for $\mathbf{W}_o(s)\mathbf{H}(s)$:

$$\mathbf{W}_o\cdot\mathbf{H}: \begin{cases} \dfrac{\mathrm{d}}{\mathrm{d}t}\begin{bmatrix}\mathbf{x}(t)\\ \mathbf{w}_o(t)\end{bmatrix} = \underbrace{\begin{bmatrix}\mathbf{A} & 0\\ \mathbf{B}_o\mathbf{C} & \mathbf{A}_o\end{bmatrix}}_{\mathcal{A}_o}\begin{bmatrix}\mathbf{x}(t)\\ \mathbf{w}_o(t)\end{bmatrix} + \underbrace{\begin{bmatrix}\mathbf{B}\\ 0\end{bmatrix}}_{\mathcal{B}_o}\mathbf{u}(t),\\[2ex] \mathbf{y}_o(t) = \underbrace{\begin{bmatrix}\mathbf{D}_o\mathbf{C} & \mathbf{C}_o\end{bmatrix}}_{\mathcal{C}_o}\begin{bmatrix}\mathbf{x}(t)\\ \mathbf{w}_o(t)\end{bmatrix} + \underbrace{\mathbf{D}_o\mathbf{D}}_{\mathcal{D}_o}\mathbf{u}(t). \end{cases} \tag{6.20}$$

One can directly verify that

$$\mathbf{H}(s)\mathbf{W}_i(s) = \mathcal{C}_i(s\mathbf{I}-\mathcal{A}_i)^{-1}\mathcal{B}_i + \mathcal{D}_i$$

and

$$\mathbf{W}_o(s)\mathbf{H}(s) = \mathcal{C}_o(s\mathbf{I}-\mathcal{A}_o)^{-1}\mathcal{B}_o + \mathcal{D}_o.$$

Next, consider the solution of the associated Lyapunov equations

$$\begin{aligned} \mathcal{A}_i\mathcal{P} + \mathcal{P}\mathcal{A}_i^T + \mathcal{B}_i\mathcal{B}_i^T &= 0,\\ \mathcal{A}_o^T\mathcal{Q} + \mathcal{Q}\mathcal{A}_o + \mathcal{C}_o^T\mathcal{C}_o &= 0, \end{aligned} \tag{6.21}$$

where the Gramians $\mathcal{P}$ and $\mathcal{Q}$ are partitioned as

$$\mathcal{P} = \begin{bmatrix}\mathcal{P}_{11} & \mathcal{P}_{12}\\ \mathcal{P}_{12}^T & \mathcal{P}_{22}\end{bmatrix}, \quad \mathcal{Q} = \begin{bmatrix}\mathcal{Q}_{11} & \mathcal{Q}_{12}\\ \mathcal{Q}_{12}^T & \mathcal{Q}_{22}\end{bmatrix}. \tag{6.22}$$

The balancing step finally consists of the simultaneous diagonalization of $\mathcal{P}_{11}$ and $\mathcal{Q}_{11}$. Hence, we have to solve two Lyapunov equations of dimensions $n+n_i$ and $n+n_o$, respectively. Instead of using a direct method for solving (6.21) one should rather use the fact that the coefficient matrices $\mathcal{A}_i$ and $\mathcal{A}_o$ are block triangular. This can be exploited in a numerically efficient implementation, as we briefly discuss in Section 6.5. In contrast to the unweighted case, the standard approach presented above is not always stability preserving. However, this is the case for one-sided weighting, i.e., if $\mathbf{W}_i(s) = \mathbf{I}$ or $\mathbf{W}_o(s) = \mathbf{I}$. If the ROM is asymptotically stable, then one can prove an error bound, as is done in [71]. There exist several modifications of the method by Enns. For example, in [82] the authors replace $\mathcal{P}_{11}$ and $\mathcal{Q}_{11}$ by the Schur complements $\mathcal{P}_{11} - \mathcal{P}_{12}\mathcal{P}_{22}^{-1}\mathcal{P}_{21}$ and $\mathcal{Q}_{11} - \mathcal{Q}_{12}\mathcal{Q}_{22}^{-1}\mathcal{Q}_{21}$. This way, stability of the ROM is preserved as long as the original system is minimal and stable. There also exists a provable error bound; see [82]. In [119], both approaches are combined and $\mathcal{P}_{11} - \alpha\mathcal{P}_{12}\mathcal{P}_{22}^{-1}\mathcal{P}_{21}$ and $\mathcal{Q}_{11} - \alpha\mathcal{Q}_{12}\mathcal{Q}_{22}^{-1}\mathcal{Q}_{21}$ are used. Other modifications of the classical method by Enns can be found in [124, 126]. Due to the limitations of the scope of this survey, we dispense with a more detailed discussion at this point.

6.4 ▪ BT for generalized systems

Though LTI control systems are nowadays understood quite well, their practical use in real-life applications is limited. This is simply because dynamical processes are rarely purely linear but instead exhibit nonlinear or uncertain behavior. There do exist methods that aim at generalizing balancing-based concepts to a nonlinear setting; see, e.g., [108]. While these methods are fairly general and allow for handling arbitrary (smooth) nonlinear systems, applicability is rather limited since they involve the solutions of Hamilton–Jacobi equations. On the other hand, there do exist generalized (but structured) systems that share properties that are very similar to the standard LTI case. In particular, due to numerical tractability by iterative linear solvers, here we are interested in performing the balancing step by solving an algebraic (generalized) linear matrix equation. As it turns out, the class of linear stochastic and bilinear control systems fits particularly well into the framework from the previous sections.

6.4.1 ▪ Linear stochastic control systems

We follow the presentation in [17]. As we already indicated, the modeling of a system will often not result in an exact representation of the underlying dynamics. Instead, noise effects and modeling errors have to be taken into account. A first step in this direction is to consider the linear stochastic differential equation

$$\begin{aligned} d\mathbf{x} &= \mathbf{A}\mathbf{x}\,\mathrm{d}t + \sum_{j=1}^{K} \mathbf{A}_j \mathbf{x}\,\mathrm{d}\omega_j + \mathbf{B}\mathbf{u}\,\mathrm{d}t, \\ \mathbf{y} &= \mathbf{C}\mathbf{x}, \end{aligned} \tag{6.23}$$

where $\mathbf{A}_j \in \mathbb{R}^{n\times n}, j = 1,\ldots,K$, and $\omega_j(t)$ are independent zero-mean real Wiener processes. All other matrices coincide with the setting of (6.1). Note that $\omega_j(t)$ is nowhere differentiable and the above formulation should be understood as x being a solution in the sense of the Itô calculus; see, e.g., [6]. Moreover, for a stochastic process $\mathbf{v}$ with values in $\mathbb{R}^q$, we define $L^2_\omega(\mathbb{R}_+, \mathbb{R}^q)$ as the space of those $\mathbf{v}$ that are finite with respect to

$$\|\mathbf{v}(\cdot)\|^2_{L^2_\omega} := \mathbb{E}\left(\int_0^\infty \|\mathbf{v}(t)\|^2 \mathrm{d}t\right) < \infty, \tag{6.24}$$

where $\mathbb{E}$ denotes the expectation. As shown in [41], for mean square stable systems, i.e., systems that exhibit $\mathbb{E}(\|\mathbf{x}(t)\|^2) \xrightarrow{t\to 0} 0$ for all initial conditions $\mathbf{x}_0$, there exist nonnegative definite solutions $\mathbf{P} \succeq 0$ and $\mathbf{Q} \succeq 0$ to the linear matrix equations

$$\begin{aligned} \mathbf{AP} + \mathbf{PA}^T + \sum_{j=1}^{K} \mathbf{A}_j \mathbf{PA}_j^T + \mathbf{BB}^T &= 0, \\ \mathbf{A}^T\mathbf{Q} + \mathbf{QA} + \sum_{j=1}^{K} \mathbf{A}_j^T \mathbf{QA}_j + \mathbf{C}^T\mathbf{C} &= 0. \end{aligned} \tag{6.25}$$

Moreover, the solutions $\mathbf{P}$ and $\mathbf{Q}$ can be computed via integrals of the form

$$\begin{aligned} \mathbf{P} &= \mathbb{E}\left(\int_0^\infty \Phi(t)\mathbf{BB}^T\Phi^T(t)\mathrm{d}t \right), \\ \mathbf{Q} &= \mathbb{E}\left(\int_0^\infty \Phi(t)^T\mathbf{C}^T\mathbf{C}\Phi(t)\mathrm{d}t \right), \end{aligned} \tag{6.26}$$

where $\Phi(t)$ denotes the fundamental solution matrix of the homogeneous problem corresponding to (6.23). Despite their analogy to the Gramians for linear deterministic systems, computing $\mathbf{P}$ and $\mathbf{Q}$ is much more complicated. In particular, for general $\mathbf{A}_j$, a direct solution method for the underlying tensorized linear system would require $\mathcal{O}(n^6)$ operations. Theoretically, however, the interpretation of the Gramians as certain energy functionals remains valid for stochastic systems as well. For given initial condition $\mathbf{x}_0 \in \mathbb{R}^n$ and final time T, similar to (6.10), it can be shown [17] that

$$\begin{aligned} E_c(\mathbf{x}_0) &:= \inf_{\substack{\mathbf{u}\in L^2_\omega[0,T],\, T>0 \\ \mathbf{x}(T,\mathbf{x}_0,\mathbf{u})=0}} \mathbb{E}\left(\int_0^T \|\mathbf{u}(t)\|^2 \mathrm{d}t \right) = \mathbf{x}_0^T\mathbf{P}^{-1}\mathbf{x}_0, \\ E_o(\mathbf{x}_0) &:= \mathbb{E}\left(\int_0^\infty \|\mathbf{y}(t,\mathbf{x}_0,0)\|^2 \mathrm{d}t \right) = \mathbf{x}_0^T\mathbf{Q}\mathbf{x}_0. \end{aligned} \tag{6.27}$$

Hence, for a balanced system, states that are hard to reach are also difficult to observe and thus may be neglected to obtain a reduced-order system. The balancing step itself is exactly the same as discussed in the previous section. Once $\mathbf{P}$ and $\mathbf{Q}$ are computed, an SVD of the product factors allows for constructing a balanced and truncated model. We refer to [28], where some recent theoretical and numerical results concerning balanced model reduction are given. In particular, the authors prove an error bound for the reduced system. For this, we again assume a partitioning of the balanced system as

$$\mathbf{A}_{\mathscr{B}} = \begin{bmatrix} \mathbf{A}_{11} & \mathbf{A}_{12} \\ \mathbf{A}_{21} & \mathbf{A}_{22} \end{bmatrix}, \mathbf{A}_{j,\mathscr{B}} = \begin{bmatrix} \mathbf{A}_{j,11} & \mathbf{A}_{j,12} \\ \mathbf{A}_{j,21} & \mathbf{A}_{j,22} \end{bmatrix}, \mathbf{B}_{\mathscr{B}} = \begin{bmatrix} \mathbf{B}_1 \\ \mathbf{B}_2 \end{bmatrix}, \mathbf{C}_{\mathscr{B}} = \begin{bmatrix} \mathbf{C}_1 & \mathbf{C}_2 \end{bmatrix}. \tag{6.28}$$

Theorem 6.8 ([28, Theorem 4.5]). *Let* $(\mathbf{A}, \mathbf{A}_j, \mathbf{B}, \mathbf{C})$ *be a realization of* (6.23). *Suppose that the ROM with coefficients* $(\mathbf{A}_{11}, \mathbf{A}_{j,11}, \mathbf{B}_1, \mathbf{C}_1)$ *is asymptotically mean square stable; then, for every* $T > 0$,

$$\sup_{t\in[0,T]} \mathbb{E}(\|\hat{\mathbf{y}}(t) - \mathbf{y}(t)\|_2) \le \left(\mathrm{tr}(\mathbf{CPC}^T) + \mathrm{tr}(\mathbf{C}_1\hat{\mathbf{P}}\mathbf{C}_1^T) - 2\mathrm{tr}(\mathbf{CRC}_1^T)\right)^{\frac{1}{2}} \|\mathbf{u}\|_{L^2[0,T]},$$

where $\hat{\mathbf{P}}$ *and* $\mathbf{R}$ *are the solutions of*

$$\mathbf{A}_{11}\hat{\mathbf{P}}+\hat{\mathbf{P}}\mathbf{A}_{11}^T+\sum_{j=1}^{K}\mathbf{A}_{j,11}\hat{\mathbf{P}}\mathbf{A}_{j,11}^T+\mathbf{B}_1\mathbf{B}_1^T=0,$$
$$\mathbf{A}\mathbf{R}+\mathbf{R}\mathbf{A}_{11}^T+\sum_{j=1}^{K}\mathbf{A}_j\mathbf{R}(\mathbf{A}_{j,11})^T+\mathbf{B}\mathbf{B}_1^T=0.$$

We remark that the previous result generalizes the $\mathcal{H}_2$-norm error bound for balanced truncation of linear systems as specified in [3, Section 7.2.2]. Unfortunately, the generalized Gramians in (6.25) do not allow us to show an $\mathcal{H}_\infty$-norm error bound of the form

$$||\mathbf{y}-\hat{\mathbf{y}}||_{L^2_\omega(\mathbb{R}_+,\mathbb{R}^p)}\leq \alpha\,\mathrm{tr}(\Sigma_2)||\mathbf{u}||_{L^2_\omega(\mathbb{R}_+,\mathbb{R}^m)}.$$

We refer to [18], where it is in fact shown that no general α exists.

On the other hand, in [18], a slightly modified definition of the Gramians allows us to prove an error bound analogous to the one known for deterministic systems. In particular, the Gramians in (6.25) are replaced by

$$\begin{aligned}\mathbf{A}\bar{\mathbf{P}}+\bar{\mathbf{P}}\mathbf{A}^T+\sum_{j=1}^{K}\bar{\mathbf{P}}\mathbf{A}_j^T\bar{\mathbf{P}}^{-1}\mathbf{A}_j\bar{\mathbf{P}}+\mathbf{B}\mathbf{B}^T&=0,\\ \mathbf{A}^T\mathbf{Q}+\mathbf{Q}\mathbf{A}+\sum_{j=1}^{K}\mathbf{A}_j^T\mathbf{Q}\mathbf{A}_j+\mathbf{C}^T\mathbf{C}&=0.\end{aligned}\tag{6.29}$$

Note that the second equation remains invariant. Using a balancing and truncation step with respect to $\bar{\mathbf{P}}$ and $\mathbf{Q}$ rather than $\mathbf{P}$ and $\mathbf{Q}$ yields a reduced system satisfying the following estimate.

Theorem 6.9 ([18, Theorem III.3]). *If* $\mathbf{x}(0)=0$ *and* $\hat{\mathbf{x}}(0)=0$, *then for all* $T>0$, *we have*

$$||\mathbf{y}-\hat{\mathbf{y}}||_{L^2_\omega[0,T]}\leq 2(\sigma_{r+1}+\cdots+\sigma_\nu)||\mathbf{u}||_{L^2_\omega[0,T]}.$$

The major bottleneck of this approach, however, is that the required matrix equations are in fact nonlinear, and, up to now, no existence results exist for that case.

6.4.2 ▪ Bilinear control systems

While linear systems provide only marginal approximations, considering the full nonlinear dynamics on the other hand is often not feasible at all. An important subclass between linear and nonlinear systems consists of so-called bilinear control systems. A common state-space representation is

$$\Sigma:\begin{cases}\dot{\mathbf{x}}(t)=\mathbf{A}\mathbf{x}(t)+\sum_{j=1}^{m}\mathbf{N}_j\mathbf{u}_j(t)\mathbf{x}(t)+\mathbf{B}\mathbf{u}(t),\\ \mathbf{y}(t)=\mathbf{C}\mathbf{x}(t),\quad \mathbf{x}(0)=\mathbf{x}_0,\end{cases}\tag{6.30}$$

where $\mathbf{N}_j \in \mathbb{R}^{n\times n}$ and $\mathbf{u}_j(t)$ denotes the jth component of the input $\mathbf{u}(t)$. Obviously, the terminology *bilinear* is motivated by the fact that the system is linear in the state $\mathbf{x}$ and in the control $\mathbf{u}$, but not jointly. Some of the first discussions on bilinear control systems can be found in, e.g., [34, 35, 40, 87]. For a detailed overview, see also [88, 89]. Note the structural similarity to the linear stochastic control system (6.23). While this allows for some of the previously discussed concepts, theoretical interpretations are far less obvious than in the other cases. The relevance of bilinear control systems now stems from their use in approximating more general nonlinear control systems. By means of the *Carleman linearization*, it is theoretically possible to generate bilinear approximations of arbitrary accuracy; see [74, 105]. However, the practical feasibility is rather limited due to the exponential growth of the approximations. Still, bilinear control systems have been shown to appear for boundary-controlled parabolic PDEs [42] as well as for the Fokker–Planck equation and open quantum systems [62].

The first ideas for balancing and balancing-based model reduction of bilinear control systems are given in [1, 2, 67]. Again, we consider generalized Lyapunov equations of the form

$$\begin{aligned} \mathbf{A}\mathbf{P}+\mathbf{P}\mathbf{A}^T+\sum_{j=1}^{m}\mathbf{N}_j\mathbf{P}\mathbf{N}_j^T+\mathbf{B}\mathbf{B}^T&=0,\\ \mathbf{A}^T\mathbf{Q}+\mathbf{Q}\mathbf{A}+\sum_{j=1}^{m}\mathbf{N}_j^T\mathbf{Q}\mathbf{N}_j+\mathbf{C}^T\mathbf{C}&=0. \end{aligned} \tag{6.31}$$

At this point, we can already see the difficulty with the interpretation of $\mathbf{P}$ and $\mathbf{Q}$. In contrast to systems (6.1) and (6.23), there is no one-to-one correspondence between definiteness of the Gramians and (asymptotic) stability of $\mathbf{H}$. In particular, system (6.30) is locally asymptotically stable if $\mathbf{A}$ is Hurwitz. This, however, does not imply that $\mathbf{P}\succ 0$ and $\mathbf{Q}\succ 0$. Roughly speaking, the influence of the bilinear coupling matrices $\mathbf{N}_j$ should be *small*, i.e., we require that $||\mathbf{N}_j||$ be negligible compared to the eigenvalues of $\mathbf{A}$. For more details, see also [17, 42]. There exist several discussions on a relation between $\mathbf{P}$ and $\mathbf{Q}$ and certain energy functionals associated with the bilinear system (6.31); see, e.g., [38, 56]. Once more, we follow the presentation in [17] and only state the most important facts. Given a state $\mathbf{x}_0\in\mathbb{R}^n$, we consider the control energy functional

$$E_c(\mathbf{x}_0)=\min_{\substack{\mathbf{u}\in L_2(-\infty,0]\\ \mathbf{x}(-\infty,\mathbf{x}_0,\mathbf{u})=0}}\int_{-\infty}^{0}||\mathbf{u}(t)||^2_{L_2((-\infty,0])}dt. \tag{6.32}$$

For an adequate output energy functional, according to [17], we consider the homogeneous equation

$$\begin{aligned} \dot{\mathbf{x}}(t)&=\mathbf{A}\mathbf{x}(t)+\sum_{j=1}^{m}\mathbf{N}_j\mathbf{u}_j(t)\mathbf{x}(t),\\ \mathbf{y}(t)&=\mathbf{C}\mathbf{x}(t), \end{aligned} \tag{6.33}$$

as well as the dual antistable bilinear system

$$\dot{\mathbf{z}}(t)=-\mathbf{A}^T\mathbf{z}(t)-\sum_{j=1}^{m}\mathbf{N}_j^T\mathbf{u}_j(t)\mathbf{z}(t)+\mathbf{C}^T\mathbf{u}(t). \tag{6.34}$$

In case the numbers of inputs and outputs are not equal, i.e., $m\neq p$, we can simply add zero columns or rows to $\mathbf{B}$ and $\mathbf{C}$, respectively. For given $\mathbf{x}_0$, let $\mathbf{u}_{\mathbf{x}_0}$ denote the control of minimal L_2-norm such that $\lim_{t\to\infty} \mathbf{z}(t,\mathbf{x}_0,\mathbf{u}) = 0$ and define

$$E_o(\mathbf{x}_0) = ||\mathbf{y}(\cdot,\mathbf{x}_0,\mathbf{u}_{\mathbf{x}_0})||^2_{L_2([0,\infty))} \tag{6.35}$$

associated with the output of the homogeneous bilinear system (6.33). With these definitions, for a balanced bilinear system, i.e., a system with $\mathbf{P} = \mathbf{Q} = \operatorname{diag}(\sigma_1,\ldots,\sigma_n)$, in [17], the authors show the following result.

Proposition 6.10 ([17]). *Let* $\mathbf{H}$ *be a balanced bilinear control system with positive definite Gramians* $\mathbf{P} = \mathbf{Q} = \operatorname{diag}(\sigma_1,\ldots,\sigma_n)$. *Then, there exists* $\varepsilon > 0$ *such that for all unit vectors* $\mathbf{e}_i$,

$$E_c(\varepsilon \mathbf{e}_i) > \frac{\varepsilon^2}{\sigma_i}, \quad E_o(\varepsilon \mathbf{e}_i) < \varepsilon^2 \sigma_i. \tag{6.36}$$

Once $\mathbf{P}$ and $\mathbf{Q}$ are computed, we can construct a reduced bilinear control system

$$\hat{\Sigma}: \begin{cases} \dot{\hat{\mathbf{x}}}(t) = \hat{\mathbf{A}}\hat{\mathbf{x}}(t) + \sum_{j=1}^{m} \hat{\mathbf{N}}_j \mathbf{u}_j(t)\hat{\mathbf{x}}(t) + \hat{\mathbf{B}}\mathbf{u}(t), \\ \hat{\mathbf{y}}(t) = \hat{\mathbf{C}}\hat{\mathbf{x}}(t), \end{cases} \tag{6.37}$$

with $\hat{\mathbf{A}} = \mathbf{W}^T\mathbf{A}\mathbf{V}, \hat{\mathbf{N}}_j = \mathbf{W}^T\mathbf{N}_j\mathbf{V}, \hat{\mathbf{B}} = \mathbf{W}^T\mathbf{B}$, and $\hat{\mathbf{C}} = \mathbf{C}\mathbf{V}$. For the projection matrices we have $\mathbf{V} = \mathbf{S}^T\mathbf{U}_1\Sigma_1^{-\frac{1}{2}}$ and $\mathbf{W} = \mathbf{R}^T\mathbf{V}_1\Sigma_1^{-\frac{1}{2}}$, with the same partitioning as in (6.14) or (6.28). Unfortunately, proceeding this way, the reduced model is in general not balanced. Moreover, even for minimal (in the linear sense) systems, the preservation of controllability and observability is not guaranteed, as we show in the following example.

Example 6.11. Consider the bilinear control system defined by

$$\mathbf{A} = \begin{bmatrix} -\frac{1}{4} & 1 \\ 1 & -9 \end{bmatrix}, \quad \mathbf{N} = \begin{bmatrix} 0 & 1 \\ 1 & -3 \end{bmatrix}, \quad \mathbf{B} = \begin{bmatrix} 0 \\ \sqrt{7} \end{bmatrix} = C^T.$$

A simple calculation shows that the linear system $(\mathbf{A},\mathbf{B},\mathbf{C})$ is both controllable and observable, hence minimal. The eigenvalues of $\mathbf{A}$ are located in the open left complex half-plane and the system is already balanced, with $\mathbf{P} = \begin{bmatrix} 2 & 0 \\ 0 & 1 \end{bmatrix} = \mathbf{Q}$. For the reduced model we thus have $\hat{\mathbf{A}} = -\frac{1}{4}$ and $\hat{\mathbf{N}} = \hat{\mathbf{B}} = \hat{\mathbf{C}} = 0$. ∎

The previous example shows that controllability and observability properties might get lost during the reduction process. On the other hand, the reduced system is still stable. Depending on how we precisely define stability for a bilinear control system, we have different results. In addition to requiring $\mathbf{A}$ to be asymptotically stable, it makes sense to demand that the spectrum of the operator

$$\underbrace{\mathbf{A}\mathbf{X} + \mathbf{X}\mathbf{A}^T}_{:=\mathscr{L}_{\mathbf{A}}(\mathbf{X})} + \underbrace{\sum_{j=1}^{m} \mathbf{N}_j\mathbf{X}\mathbf{N}_j^T}_{:=\Pi_{\mathbf{N}}(\mathbf{X})} \tag{6.38}$$

be located in the open left complex plane, i.e., $\sigma(\mathscr{L}_{\mathbf{A}} + \Pi_{\mathbf{N}})$. Based on this characterization, we have a result concerning the stability preservation of a balancing-based reduction process.

Theorem 6.12 ([19, Theorem 2.3]). *Let*

$$\mathbf{A} = \begin{bmatrix} \mathbf{A}_{11} & \mathbf{A}_{12} \\ \mathbf{A}_{21} & \mathbf{A}_{22} \end{bmatrix}, \quad \mathbf{N} = \begin{bmatrix} \mathbf{N}_{11} & \mathbf{N}_{12} \\ \mathbf{N}_{21} & \mathbf{N}_{22} \end{bmatrix}$$

be given and assume there exists a block diagonal matrix $\Sigma = \operatorname{diag}(\Sigma_1, \Sigma_2) \succ 0$ *with* $\Sigma_1 \in \mathbb{R}^{r \times r}$ *and* $\sigma(\Sigma_1) \cap \sigma(\Sigma_2) = \emptyset$ *so that*

$$\mathscr{L}_{\mathbf{A}}(\Sigma) + \Pi_{\mathbf{N}}(\Sigma) + \mathbf{B}\mathbf{B}^T = 0, \quad \mathscr{L}_{\mathbf{A}^*}(\Sigma) + \Pi_{\mathbf{N}^*}(\Sigma) + \mathbf{C}^T\mathbf{C} = 0.$$

Then, $\sigma(\mathscr{L}_{\mathbf{A}_{11}} + \Pi_{\mathbf{N}_{11}}) \subset \mathbb{C}_-$ *if* $\sigma(\mathscr{L}_{\mathbf{A}} + \Pi_{\mathbf{N}}) \subset \mathbb{C}_-$.

In summary, we conclude that balancing concepts for linear stochastic and bilinear control systems are very similar. However, the interpretation of the Gramians for stochastic systems is less ambiguous. We can still choose between two different Gramians, influencing the properties of the reduced system. While the Gramians in (6.25) preserve stability and yield an error bound with respect to a generalized $\mathscr{H}_2$-error measure, the recently defined Gramians (6.29) can be used to prove a generalization of the classical $\mathscr{H}_\infty$-error bound, with the disadvantage of the latter being the more complicated (or impossible) computation.

Balancing-related methods

In analogy to the closely related linear case, the question arises whether known balancing-related concepts have some interpretation in the bilinear setting as well. As a first step in this direction, in [62], the authors propose an averaging principle where the dynamics are split into fast and slow subspaces. Despite some differences, the method can be seen as a generalization of the SPA method to bilinear systems. In particular, for a SISO system, the reduced system matrices are determined as

$$\hat{\Sigma}: \begin{cases} \dot{\hat{\mathbf{x}}}(t) = \hat{\mathbf{A}}\hat{\mathbf{x}}(t) + \hat{\mathbf{N}}\mathbf{u}(t)\hat{\mathbf{x}}(t) + \hat{\mathbf{B}}\mathbf{u}(t), \\ \hat{\mathbf{y}}(t) = \hat{\mathbf{C}}\hat{\mathbf{x}}(t), \end{cases} \tag{6.39}$$

with

$$\hat{\mathbf{A}} = \mathbf{A}_{11} - \mathbf{A}_{12}\mathbf{A}_{22}^{-1}\mathbf{A}_{21}, \quad \hat{\mathbf{N}} = \mathbf{N}_{11} - \mathbf{N}_{12}\mathbf{A}_{22}^{-1}\mathbf{A}_{21},$$
$$\hat{\mathbf{B}} = \mathbf{B}_1, \quad \text{and} \quad \hat{\mathbf{C}} = \mathbf{C}_1 - \mathbf{C}_2\mathbf{A}_{22}^{-1}\mathbf{A}_{21}.$$

Here, we assume a partitioning of the balanced system similar to the one in (6.12). Note the different construction of $\hat{\mathbf{B}}$ compared to system (6.17). For a detailed description of the method and a proof of the averaging principle, we refer to [62, Theorem 3.1].

To the authors' knowledge, other extensions of balancing-related methods do not exist and might be a topic of future research.

6.5 ▪ Numerical solution of linear matrix equations

We have seen that the key ingredient for balancing (related) model reduction is the solution $\mathbf{X}$ of a linear matrix equation of the form

$$\sum_{i=1}^{d} \mathbf{A}_i \mathbf{X} \mathbf{B}_i = \mathbf{C}, \tag{6.40}$$

where $\mathbf{A}_i, \mathbf{B}_i, \mathbf{C} \in \mathbb{R}^{n \times n}$ are given matrices. Using the Kronecker product $\otimes$ [52, 66] and the vectorization operator vec, instead of (6.40) we can also solve the system of linear equations

$$\underbrace{\left(\sum_{i=1}^{d} \mathbf{B}_i^T \otimes \mathbf{A}_i\right)}_{\mathcal{A}} \underbrace{\operatorname{vec}(\mathbf{X})}_{\mathcal{X}} = \underbrace{\operatorname{vec}(\mathbf{C})}_{\mathcal{C}}. \tag{6.41}$$

This way, by simply using Gaussian elimination, we can directly compute $\mathbf{X}$. Of course, the fact that we have to deal with a system of dimension n^2 and, thus, a complexity of $\mathcal{O}(n^6)$ makes this approach infeasible for all but very small problems. Note also that proceeding this way, the symmetry of the underlying problem is not taken into account. For the special case $d = 2$, it is well known that there exist methods to obtain $\mathbf{X}$ in only $\mathcal{O}(n^3)$ operations. For example, we refer to the Bartels–Stewart algorithm (see [7]) and the sign function method (see [104]). Since the involved matrices often come from spatially discretized PDEs, cubic complexity is often still not acceptable. Also, the storage complexity of $\mathcal{O}(n^2)$ for the general dense solution matrix $\mathbf{X}$ can easily cause problems in a large-scale setting. A common misjudgment is that this makes balancing-related methods impossible for realistic examples. In the following, we summarize several methods that aim to replace $\mathbf{X}$ by a *low-rank* approximation $\mathbf{Z}_k \mathbf{Z}_k^T$, where $\mathbf{Z}_k \in \mathbb{R}^{n \times k}$ and $k \ll n$, allowing us to (numerically) balance very large-scale systems.

The basic idea to solve for a (Cholesky) factor of $\mathbf{X}$ rather than $\mathbf{X}$ itself has already been pursued in an extension of the Bartels–Stewart method known as Hammarling's method; see [61]. Here, an $n \times n$ matrix $\mathbf{Z}_k$ is still computed, i.e., the full Cholesky factor. The same idea in connection with the sign function method is suggested in [78]. The idea of computing a low-rank factor $\mathbf{Z}_k$ in the above-mentioned sense directly with the sign function iteration was first proposed in [27] for Lyapunov equations, and in [11] for Sylvester equations. None of these methods overcome the $\mathcal{O}(n^3)$ barrier in the overall computational complexity. This was achieved for special cases of coefficient matrices resulting from finite and boundary element methods using the sign function matrix approach and a data-sparse arithmetic for so-called hierarchical matrices in [9]; see also the related work for Riccati and Sylvester equations [8, 54, 55]. We omit the details of these specialized methods here and in the following only focus on methods that are generally applicable for sparse coefficient matrices and low-rank right-hand sides.

6.5.1 ▪ The Lyapunov equation

Let us start with the common Lyapunov equation

$$\mathbf{A}\mathbf{X} + \mathbf{X}\mathbf{A}^T + \mathbf{B}\mathbf{B}^T = 0. \tag{6.42}$$

As we already mentioned, the key tool in numerically efficient solution techniques for (6.42) is to construct rank-k approximations $\mathbf{X}_k = \mathbf{Z}_k\mathbf{Z}_k^T$, with $\mathbf{Z}_k$ as above. Of course, one might think about the question of whether it is actually reasonable to expect there to exist approximations with $\mathbf{X}_k \approx \mathbf{X}$. Considering

$$\mathbf{A} = \begin{bmatrix} -\frac{1}{2} & 0 & \dots & 0 \\ -1 & -\frac{1}{2} & \ddots & \vdots \\ \vdots & \ddots & \ddots & 0 \\ -1 & \dots & -1 & -\frac{1}{2} \end{bmatrix}, \quad \mathbf{B} = \begin{bmatrix} 1 \\ 1 \\ \vdots \\ 1 \end{bmatrix},$$

we immediately see that $\mathbf{X} = \mathbf{I}$ is the unique solution of the Lyapunov equation (6.42). Hence, for $k < n$, no approximations $\mathbf{X}_k$ will yield sufficient results. However, this phenomenon is not generic, and it has been shown that the singular values of $\mathbf{X}$ often decay reasonably fast such that the numerical rank of $\mathbf{X}$ is much smaller than n. We refer to, e.g., [5, 53, 97, 106, 114]. More recently, this has even been studied from an operator point of view; see [57, 93].

Instead of computing $\mathbf{X}$ and constructing an approximation by a truncated SVD, the idea of successful low-rank methods is to perform all operations on the factor $\mathbf{Z}_k$ itself. Proceeding this way, for sparse matrices (with few entries) it is often possible to compute very accurate approximations in $\mathcal{O}(n)$.

ADI-based methods

The first method we want to consider is the alternating directions implicit (ADI) iteration, which has its origin in the numerical solution of elliptic and parabolic differential equations; see [95]. Similar to the idea of splitting the 2D Laplacian into its 1D counterparts, in [122], the author proposed exploiting the implicit Kronecker structure of (6.42). As a result, we obtain the classical ADI iteration for (6.42) as

$$\begin{aligned} (\mathbf{A} + p_j\mathbf{I})\mathbf{X}_{j+\frac{1}{2}} &= -\mathbf{B}\mathbf{B}^T - \mathbf{X}_j(\mathbf{F} - p_j\mathbf{I})^T, \\ (\mathbf{A} + \overline{p}_j\mathbf{I})\mathbf{X}_{j+1}^T &= -\mathbf{B}\mathbf{B}^T - \mathbf{X}_{j+\frac{1}{2}}^T(\mathbf{A} - \overline{p}_j\mathbf{I})^T, \end{aligned} \tag{6.43}$$

where $\mathbf{X}_0 = 0$ and $\mathbf{p} = \{p_1, \dots, p_k\} \subset \mathbb{C}_-$ is a given set of *shift parameters*. Solving the first equation for $\mathbf{X}_{j+\frac{1}{2}}$ and inserting the result into the second allows for a one-step formulation of the method. Moreover, using commutation properties of the coefficient matrices $(\mathbf{A} + p_j\mathbf{I})^{-1}$ and $(\mathbf{A} - p_j\mathbf{I})$, we get the *factored ADI iteration*

$$\begin{aligned} \mathbf{V}_1 &= \sqrt{-2\mathrm{Re}(p_1)}(\mathbf{A} + p_1\mathbf{I})^{-1}\mathbf{B}, \quad \mathbf{Z}_1 = \mathbf{V}_1, \\ \mathbf{V}_j &= \frac{\sqrt{\mathrm{Re}(p_j)}}{\sqrt{\mathrm{Re}(p_{j-1})}}\left(\mathbf{V}_{j-1} - (p_j + \overline{p}_{j-1})(\mathbf{A} + p_j\mathbf{I})^{-1}\mathbf{V}_{j-1}\right), \\ \mathbf{Z}_j &= [\mathbf{Z}_{j-1} \ \mathbf{V}_j]. \end{aligned} \tag{6.44}$$

For more details on the classical low-rank version, see [24, 80, 81, 96, 98] and the two recent survey articles [30, 111]. Since the factored ADI iteration only requires linear solves with matrices of the form $(\mathbf{A} + p_j\mathbf{I})$, for sparse matrices it often scales linearly with n. The speed of convergence to the true solution $\mathbf{X}$, however, heavily depends on the choice of the shift parameters p_j. Though the derivation of *optimal* shifts is discussed in [123], these so-called Wachspress shifts can only be efficiently computed for matrices with real spectra. Several other methods for computing (sub)optimal shifts have been proposed over the last few years. For a complete overview and some new developments in this area of research, see [23]. Later on, we briefly come back to the shift selection problem and discuss the special role taken by $\mathcal{H}_2$ (pseudo)optimal parameters.

Besides extensions of the original ADI method to Sylvester equations (see, e.g., [25, 36, 113]), more recent ideas aim to avoid complex arithmetic and efficiently compute the associated Lyapunov residual

$$\mathbf{R}_k = \mathbf{A}(\mathbf{Z}_k\mathbf{Z}_k^T) + (\mathbf{Z}_k\mathbf{Z}_k^T)\mathbf{A}^T + \mathbf{B}\mathbf{B}^T;$$

see [20, 21]. Although it is well known that the ADI subspaces span a rational Krylov subspace [80], below we see that under certain assumptions, the ADI method can also be interpreted as a projection-based method, meaning that it leads to the exact same approximation.

Projection-based methods

Another very important class of low-rank solution methods for the Lyapunov equation (6.42) consists of projection-based methods. Initially proposed even before the factored ADI method in [106], the main idea is to project onto some *smaller* subspace $\mathcal{V}$ where one can easily find the exact solution by an explicit method such as the Bartels–Stewart algorithm. An approximation is then obtained by injection of the reduced solution to the original space. More precisely, let us assume that we have chosen a subspace $\mathcal{V}$ of dimension k that is characterized by its orthonormal basis vectors $\mathbf{V} = [\mathbf{v}_1, \dots, \mathbf{v}_k] \in \mathbb{R}^{n\times k}$. Consider next the reduced Lyapunov equation

$$\hat{\mathbf{A}}\hat{\mathbf{X}} + \hat{\mathbf{X}}\hat{\mathbf{A}}^T + \hat{\mathbf{B}}\hat{\mathbf{B}}^T = 0, \tag{6.45}$$

where $\hat{\mathbf{A}} = \mathbf{V}^T\mathbf{A}\mathbf{V}$ and $\hat{\mathbf{B}} = \mathbf{V}^T\mathbf{B}$. After the solution $\hat{\mathbf{X}}$ is computed, we set the rank-k approximation as $\mathbf{X}_k = \mathbf{V}\hat{\mathbf{X}}\mathbf{V}^T$. Following [106], for the associated residual $\mathbf{R}_k = \mathbf{A}\mathbf{X}_k + \mathbf{X}_k\mathbf{A}^T + \mathbf{B}\mathbf{B}^T$, we get

$$\begin{aligned}\mathbf{V}^T\mathbf{R}_k\mathbf{V} &= \mathbf{V}^T\left(\mathbf{A}(\mathbf{V}\hat{\mathbf{X}}\mathbf{V}^T) + (\mathbf{V}\hat{\mathbf{X}}\mathbf{V}^T)\mathbf{A}^T + \mathbf{B}\mathbf{B}^T\right)\mathbf{V}\\ &= \hat{\mathbf{A}}\hat{\mathbf{X}} + \hat{\mathbf{X}}\hat{\mathbf{A}}^T + \hat{\mathbf{B}}\hat{\mathbf{B}}^T = 0.\end{aligned}$$

In other words, the residual satisfies a classical *Galerkin* condition. We naturally expect the quality of the method to depend on the projection subspace $\mathcal{V}$ and its basis $\mathbf{V}$, respectively. Over the past few years, several choices have been proposed in the literature. Due to space limitations, we only point to the most prominent approaches in that direction and references therein. In [106], the author suggests using the (block) Krylov space associated with the system matrix $\mathbf{A}$ and the input matrix $\mathbf{B}$, i.e.,

$$\mathcal{V} = \mathcal{K}_q(\mathbf{A},\mathbf{B}) := \operatorname{span}\{\mathbf{B}, \mathbf{A}\mathbf{B}, \dots, \mathbf{A}^{q-1}\mathbf{B}\}.$$

While this method only requires matrix-vector multiplications, it often yields only moderate approximations. Additionally, one might consider an extended subspace that also uses the inverse of $\mathbf{A}$, as is discussed in [110]. There, the author uses

$$\mathcal{V} = \mathcal{K}_q(\mathbf{A},\mathbf{B}) \cup \mathcal{K}_q(\mathbf{A}^{-1},\mathbf{A}^{-1}\mathbf{B}).$$

The resulting method is sometimes called the Krylov-plus-inverted-Krylov (K-PIK) method or extended Krylov subspace method (EKSM). In fact, the latter name is motivated by the fact that the combination of the above Krylov subspaces has already been discussed in a slightly different context in [44]. For a rigorous convergence analysis of this method, we also refer to [72]. Finally, both of the above-mentioned techniques can be seen as special cases of the rational Krylov subspace method (RKSM); see, e.g., [45]. Here, for a given set of shifts $S=\{s_1,\ldots,s_q\}$ (interpolation points), the subspace $\mathcal{V}$ is determined via

$$\mathcal{V}=\mathcal{K}_q(\mathbf{A},\mathbf{B},S):=\operatorname{span}\{(s_1\mathbf{I}-\mathbf{A})^{-1}\mathbf{B},\ldots,(s_q\mathbf{I}-\mathbf{A})^{-1}\mathbf{B}\}.$$

In terms of interpolation techniques, the common Krylov subspace $\mathcal{K}_q(\mathbf{A},\mathbf{B})$ now corresponds to the special case of $s_i\equiv\infty$, while the inverse Krylov subspace corresponds to $s_i\equiv 0$. Obviously, since computing the projection subspaces requires q solutions of linear systems with different coefficient matrices, this method is the most expensive one. On the other hand, the flexibility of choosing q different shifts also allows for very accurate low-rank approximations. Again, we refer to [45, 46] for more details. As we indicated previously, for the ADI method a special role is taken by $\mathcal{H}_2$ (pseudo)optimal shifts or interpolation points [59, 84], respectively. In fact, by choosing these shifts, it is shown in [45, 49] that the projection-based approximation coincides with the one from the ADI iteration. Moreover, for symmetric system matrices $\mathbf{A}=\mathbf{A}^T\prec 0$, these shifts are even optimal [16, Theorem 3.1] with respect to the energy norm (see [117]) induced by the negative Lyapunov operator $\mathcal{L}(\mathbf{X})=-\mathbf{AX}-\mathbf{XA}$. The major drawback of these shifts is that their computation is rather expensive. For more details on the corresponding iterative algorithm, see Chapter 7 in this volume and the original reference in [59]. For other projection-based methods, see also [68, 69].

Preconditioned iterative linear solvers

We have already pointed out that explicitly computing the vectorized solution vec($\mathbf{X}$) of the tensorized linear system (6.41) is infeasible for realistic problem sizes. On the other hand, implicitly working with this formulation allows for the incorporation of truncation techniques within an iterative Krylov-based solver such as CG, BiCG, or MinRes. One of the first references for this approach is [65], where the authors test different preconditioners (e.g., the ADI iteration) within the QMR method applied to the Lyapunov equation. However, in that paper, no low-rank implementation is incorporated and the applicability is thus limited to medium-size problems. In [32, 48], the authors discuss a low-rank GMRES version with ADI preconditioning. In a similar way, the discussions in [75, 76] show how to treat more general linear systems of tensor product structure. A slightly different use of the Kronecker system is proposed in [33]. In [54], the authors suggest using a multigrid method to solve the Lyapunov equation.

Riemannian optimization

Rather recently, a conceptually different idea was proposed in [117, 118]. Instead of iteratively increasing the rank of the approximation $\mathbf{X}_k$, the authors aim to find a (locally) optimal approximation for a given and prescribed rank k. The method in [118] addresses the special case of a symmetric matrix $\mathbf{A}=\mathbf{A}^T\prec 0$. As pointed out by the authors, in this setting, for a matrix $\mathbf{X} \in \mathbb{R}^{n\times n}$ a natural norm is defined by

$$\|\mathbf{X}\|_{\mathcal{L}}^2 := \mathrm{vec}(\mathbf{X})^T \underbrace{(-\mathbf{I}\otimes\mathbf{A}-\mathbf{A}\otimes\mathbf{I})}_{\mathcal{L}} \mathrm{vec}(\mathbf{X}). \tag{6.46}$$

Given a rank k, the authors suggest using the Riemannian geometry of the manifold of symmetric positive semidefinite matrices to find approximations that are locally optimal with respect to that energy norm. Using a Riemannian trust region method with suitable preconditioning, the results in [118] indicate the potential of this rather new approach. In particular, the authors discuss the superiority of their method when the rank of the right-hand side is large. While this specific method is restricted to the symmetric case, there do exist modifications for the unsymmetric case by using the residual instead of the energy norm. For more details, we refer to [117]. Interestingly enough, in [16, Theorem 3.1], it is shown that the projection-based approximations resulting from $\mathcal{H}_2$ optimal shifts minimize the energy norm as well. Hence, under certain assumptions, rather different frameworks yield the exact same approximation. We remark that there also exist generalizations to the unsymmetric case where the goal is to locally minimize the residual. We again refer to [16, 117].

6.5.2 ▪ Frequency-weighted Lyapunov equations

Recall from Section 6.3 that in the case of frequency-weighted BT according to Enns, we have to solve two Lyapunov equations of size $(n+n_i)$ and $(n+n_o)$, respectively, independent of the specific choice of the input/output weights. Let us, for example, focus on the corresponding controllability Lyapunov equation

$$\mathcal{A}_i\mathcal{P}+\mathcal{P}\mathcal{A}_i^T+\mathcal{B}_i\mathcal{B}_i^T=0, \tag{6.47}$$

where we use the notation introduced in Section 6.3. A straightforward extension of the first two low-rank methods from above would thus require the solutions of shifted linear systems of the form

$$(\mathcal{A}_i+p\mathbf{I})\mathbf{x}=\mathcal{B}_i. \tag{6.48}$$

Obviously, in addition to the fact that we have an enlarged system size, in general the sparsity pattern of $\mathcal{A}_i$ will get lost due to the term $\mathbf{BC}_i$. An easy remedy is to exploit the block triangular structure of $\mathcal{A}_i$, allowing us to obtain $\mathbf{x}$ by two sparse linear system solves of dimensions n and n_i, respectively. In fact, the same trick could be applied to the Lyapunov equation itself. Using the partitioning (6.22), we find that $\mathcal{P}$ is determined via

$$\begin{aligned}\mathbf{A}_i\mathbf{P}_{22}+\mathbf{P}_{22}\mathbf{A}_i^T+\mathbf{B}_i\mathbf{B}_i^T&=0,\\ \mathbf{A}\mathbf{P}_{12}+\mathbf{P}_{12}\mathbf{A}_i^T+\mathbf{B}\mathbf{D}_i\mathbf{B}_i^T+\mathbf{B}\mathbf{C}_i\mathbf{P}_{22}&=0,\\ \mathbf{A}\mathbf{P}_{11}+\mathbf{P}_{11}\mathbf{A}^T+\mathbf{B}\mathbf{C}_i\mathbf{P}_{12}^T+\mathbf{P}_{12}\mathbf{C}_i^T\mathbf{B}^T+(\mathbf{B}\mathbf{D}_i)(\mathbf{B}\mathbf{D}_i)^T&=0.\end{aligned} \tag{6.49}$$

Hence, we first solve a Lyapunov equation of dimension n_i to obtain $\mathbf{P}_{22}$. The result can be used to compute $\mathbf{P}_{12}$ by solving a Sylvester equation. Finally, we combine $\mathbf{P}_{22}$ and $\mathbf{P}_{12}$ for deriving $\mathbf{P}_{11}$ in the last equation. Proceeding this way, we can work with the original matrices and, more important, with their sparsity patterns. Note, however, that the last Lyapunov equation contains a possibly indefinite constant term, which should be paid attention to within an implementation.

6.5.3 • Generalized Lyapunov equations

Finally, let us turn to the more general setting (6.40). Due to the relevance for balancing of linear stochastic and bilinear control systems, let us again assume the splitting into a common Lyapunov operator $\mathscr{L}_{\mathbf{A}}(\mathbf{X}) = \mathbf{AX}+\mathbf{XA}^T$ and a positive operator $\Pi_{\mathbf{N}}(\mathbf{X}) = \sum_{j=1}^{m} \mathbf{N}_j\mathbf{X}\mathbf{N}_j^T$. For more details on positive operators, see also, e.g., [41]. Consider now the generalized Lyapunov equation

$$\mathscr{L}_{\mathbf{A}}(\mathbf{X}) + \Pi_{\mathbf{N}}(\mathbf{X}) + \mathbf{BB}^T = 0. \tag{6.50}$$

Despite the similarity to the common Lyapunov equation, computing $\mathbf{X}$ for (6.50) is far more complicated. Roughly speaking, the noncommutativity of the matrices $\mathbf{A}$ and $\mathbf{N}_j$ does not allow us to diagonalize the equation by pre- and postmultiplication. Hence, there do not exist straightforward extensions of the Bartels–Stewart algorithm or Hammarling's method. To the authors' knowledge, besides solving the tensorized linear system, the only existing direct method to compute $\mathbf{X}$ is discussed in [42]. The method is based on using the Sherman–Morrison–Woodbury formula and is only applicable in situations where the sum of the ranks of $\mathbf{N}_i$ is very small compared to n. In the same paper, Damm also proposes a fixed-point iteration

$$\begin{aligned} \mathbf{AX}_1 + \mathbf{X}_1\mathbf{A}^T + \mathbf{BB}^T &= 0, \\ \mathbf{AX}_i + \mathbf{X}_i\mathbf{A}^T + \Pi_{\mathbf{N}}(\mathbf{X}_{i-1}) + \mathbf{BB}^T &= 0, \quad i = 2,3,\ldots. \end{aligned} \tag{6.51}$$

Under certain assumptions on the matrices $\mathbf{N}_i$, this method can be shown to converge to the true solution $\mathbf{X}$. However, the speed of convergence is often very slow, so that many standard Lyapunov equations have to be solved. A third way of solving (6.50) is given by an ADI-like splitting, as in the linear case. Again, we follow the presentation in [42]. Observe that for any parameter $p > 0$, we can shift the operator $\mathscr{L}(\mathbf{X})$ according to

$$\mathbf{AX} + \mathbf{XA}^T = \frac{1}{2p}\left((\mathbf{A}+p\mathbf{I})\mathbf{X}(\mathbf{A}+p\mathbf{I})^T - (\mathbf{A}-p\mathbf{I})\mathbf{X}(\mathbf{A}-p\mathbf{I})^T\right). \tag{6.52}$$

In particular, we can thus transform equation (6.50) into

$$\begin{aligned} \mathbf{X} = {} & (\mathbf{A}-p\mathbf{I})^{-1}(\mathbf{A}+p\mathbf{I})\mathbf{X}(\mathbf{A}+p\mathbf{I})^T(\mathbf{A}-p\mathbf{I})^{-T} \\ & + 2p(\mathbf{A}-p\mathbf{I})^{-1}\left(\sum_{j=1}^{m}\mathbf{N}_j\mathbf{X}\mathbf{N}_j^T + \mathbf{BB}^T\right)(\mathbf{A}-p\mathbf{I})^{-T}. \end{aligned} \tag{6.53}$$

Given $p_\ell > 0, \ell = 1,\ldots,L$, and $\mathbf{X}_k$, in [42], the author defines an ADI iteration procedure via

$$\begin{aligned} \mathbf{X}_{k+\frac{\ell}{L}} = {} & (\mathbf{A}-p_\ell\mathbf{I})^{-1}(\mathbf{A}+p_\ell\mathbf{I})\mathbf{X}_{k+\frac{\ell-1}{L}}(\mathbf{A}+p_\ell\mathbf{I})^T(\mathbf{A}-p_\ell\mathbf{I})^{-T} \\ & + 2p_\ell(\mathbf{A}-p_\ell\mathbf{I})^{-1}\left(\sum_{j=1}^{m}\mathbf{N}_j\mathbf{X}_k\mathbf{N}_j^T + \mathbf{BB}^T\right)(\mathbf{A}-p_\ell\mathbf{I})^{-T}. \end{aligned} \tag{6.54}$$

Of course, as in the linear case, the choice of the shift parameters is a nontrivial task itself and has to be taken into account for the performance of this method. While the latter method certainly allows us to compute approximations for medium-size systems, we still have to store the full matrix $\mathbf{X}_{k+\frac{\ell}{L}}$ rather than appropriate low-rank factors. Under similar assumptions on the rank of the $\mathbf{N}_j$ as above, in [15], we have shown the existence of low-rank approximations. The main idea is based on the Sherman–Morrison–Woodbury formula as in [42]. Besides a low-rank implementation of (6.54), we also proposed appropriate generalizations of the projection-based methods known for the linear case. Similarly, we tested a low-rank BiCG version in combination with ADI preconditioning. Note that this approach is also investigated in a slightly different context in [77]. It should be mentioned that, at least in the authors' opinion, the theoretical and numerical understanding of this field of research seems to be insufficient. In particular, though several examples suggest (see [15, 17, 62]) that even for full-rank matrices $\mathbf{N}_j$, the singular values of $\mathbf{X}$ decay rapidly, no theoretical explanation has been given so far.

6.6 ▪ Numerical examples

We provide two numerical examples illustrating the advantages and disadvantages of balancing and balancing-related reduction techniques. The benchmarks we study are standard within the MOR community and belong to the SLICOT benchmark collection.[23] We emphasize that neither of the examples is very large scale and should only be understood for illustrative purposes. Moreover, we only use SISO examples.

6.6.1 ▪ CD player

The first model describes the dynamics of a CD player and consists of $n = 120$ states. The original model from the SLICOT collection has two inputs and two outputs. Here, we choose $\mathbf{B}$ to be only the first column of the input matrix and $\mathbf{C}$ to be the second row of the output matrix. We then obtain a SISO system for which we compute the transfer function (FOM) evaluated on the imaginary axis in Figure 6.1. The performance of three ROMs resulting from BT and balancing-type methods is compared in Figure 6.1 and also in Figure 6.2. For the frequency-weighted BT (FWBT) method, we use a weight given by the combination of a low-pass filter (cutoff frequency 10^4 rad/sec) and a high-pass filter (cutoff frequency $2.5 \cdot 10^4$ rad/sec). Together, this acts as a bandpass filter and emphasizes the frequency interval $[10^4, 2.5 \cdot 10^4]$. As shown in Figure 6.3 and Figure 6.4, the frequency-weighted balanced reduced model is very accurate in this range but strongly deviates otherwise; see Figure 6.1. On the other hand, the example also confirms the fact that SPA interpolates the original transfer function at $s = 0$; see Figure 6.2.

6.6.2 ▪ ISS 1R module

We perform a similar test for the ISS 1R module with $n = 1412$ states. The bandpass is again obtained by a combination of a low-pass filter and a high-pass filter and puts a weighting on the frequency interval $[4, 6]$. The transfer functions of the original model and three ROMs of dimension $r = 40$ are shown in Figure 6.5. The pointwise relative error is visualized in Figure 6.6. Once more, we see that the approximation from the weighted model is more accurate than those resulting from classical BT and SPA; see Figure 6.7 and Figure 6.8. Moreover, we see that SPA yields good approximations for low frequencies and BT yields good approximations for higher frequencies.

[23] `http://www.slicot.org/index.php?site=benchmodred`.

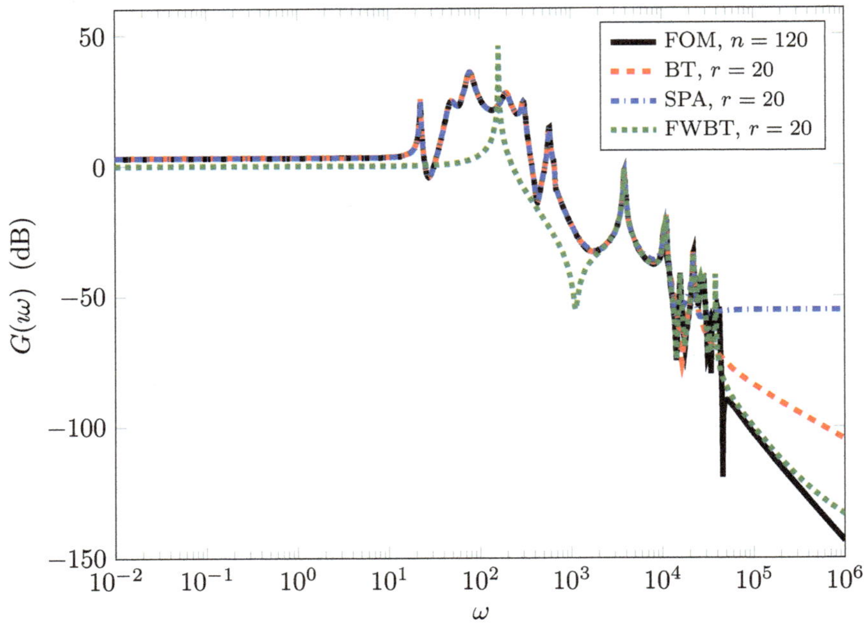

Figure 6.1. *Bode plot for the CD player.*

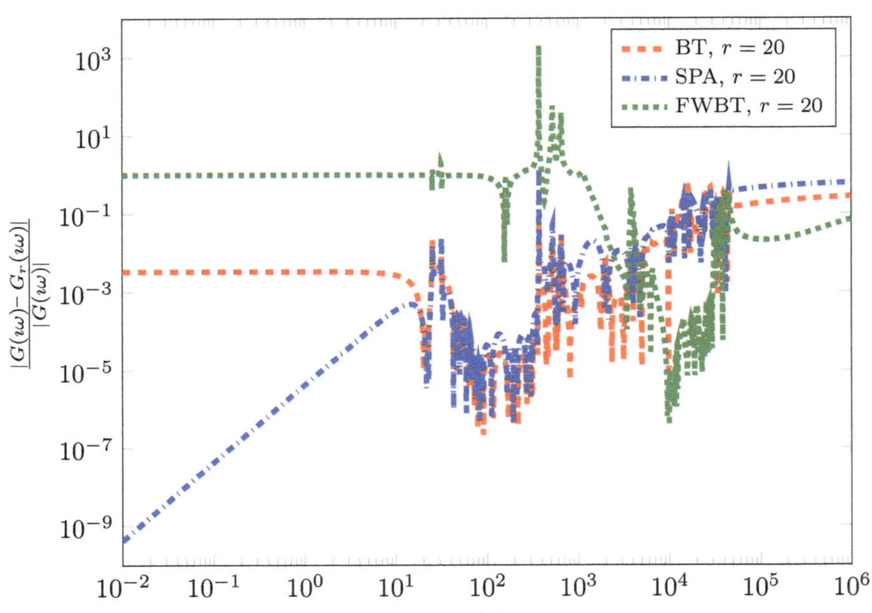

Figure 6.2. *Comparison of relative errors for the CD player.*

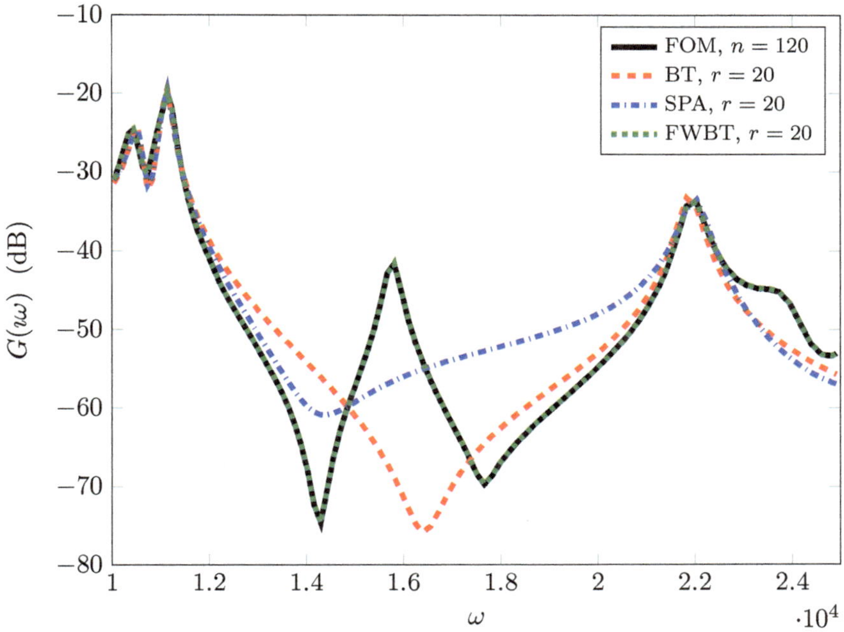

Figure 6.3. *Bode plot for the CD player (zoom).*

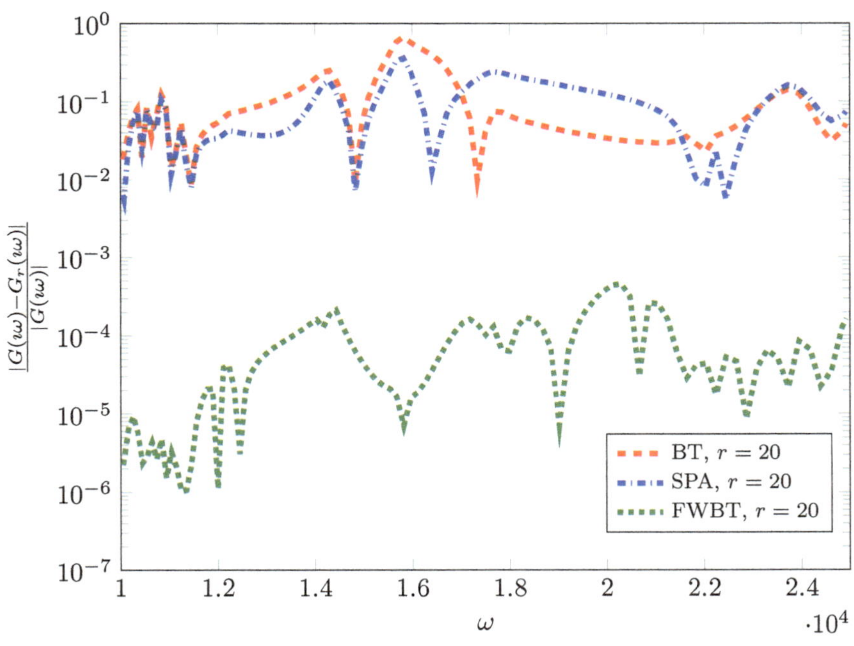

Figure 6.4. *Comparison of relative errors for the CD player (zoom).*

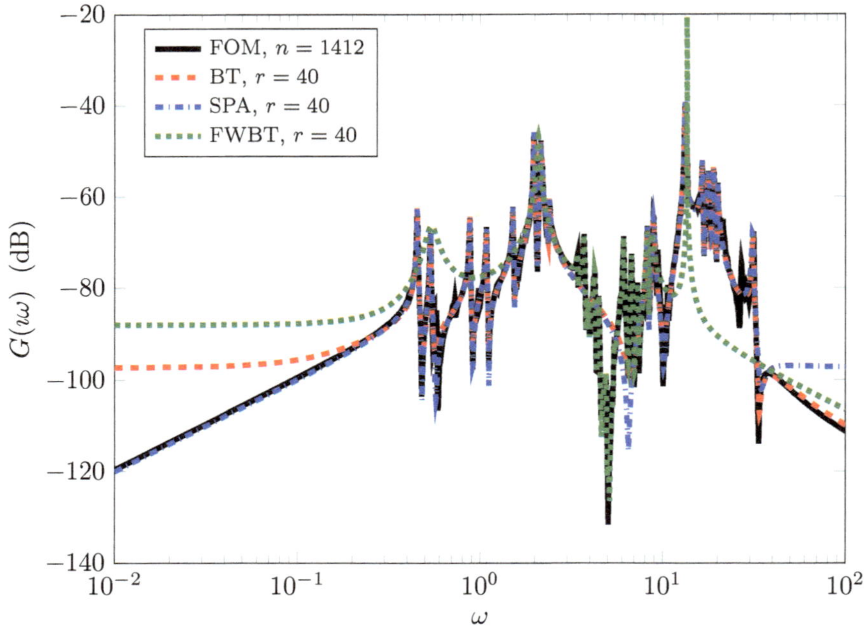

Figure 6.5. *Bode plot for the ISS.*

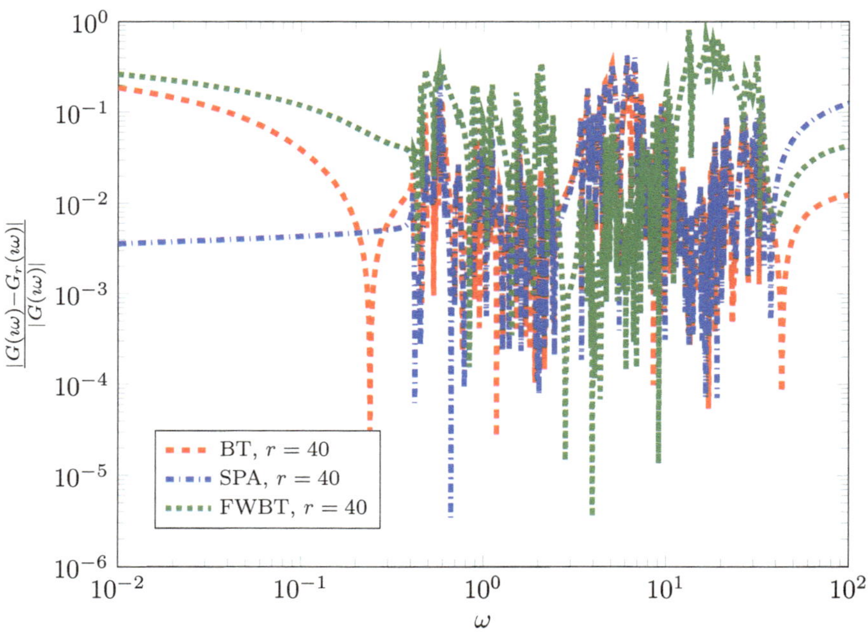

Figure 6.6. *Comparison of relative errors for the ISS.*

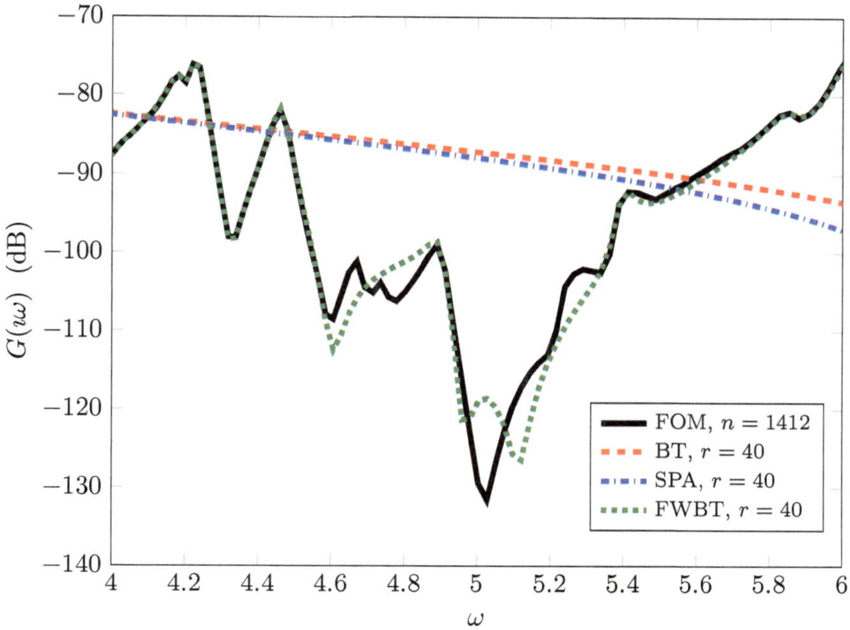

Figure 6.7. *Bode plot for the ISS (zoom).*

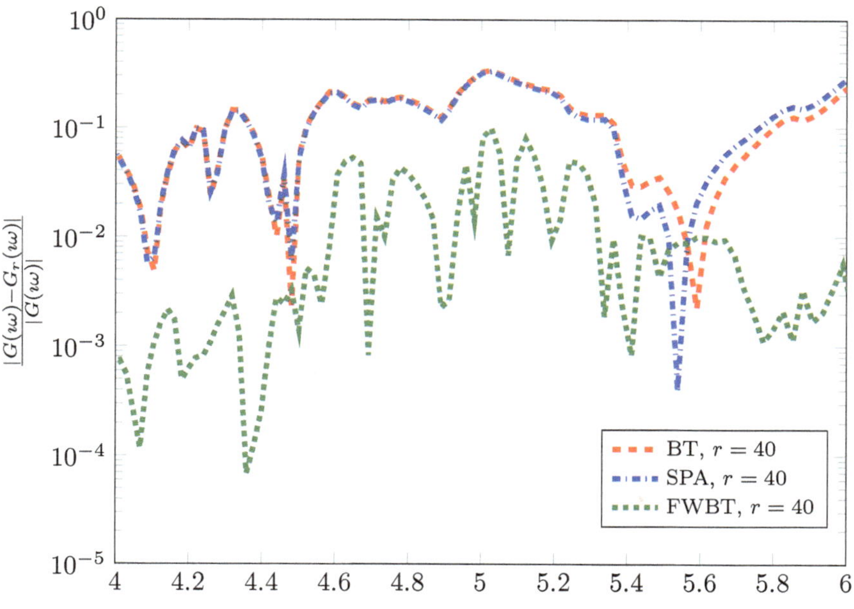

Figure 6.8. *Comparison of relative errors for the ISS (zoom).*

6.7 ▪ Conclusions and outlook

We have presented a state-of-the-art survey of balancing-based and balancing-related model reduction techniques. Since the introduction of the standard BT method in [91], several improvements and extensions have been suggested. Besides balancing methods for systems that require accuracy in known frequency ranges, we have reviewed recent developments toward balancing-based model reduction for linear stochastic and bilinear control systems. While in the first case a reasonable interpretation of the Gramians as energy functionals is possible, concepts for the latter case are ambiguous since they only hold locally. In the linear as well as in the bilinear case, the Gramians have to be computed as the solutions of (generalized) linear matrix equations whose dimensions scale with the number of system states. Despite this computational bottleneck, we have seen that low-rank methods can still approximate solutions up to dimensions $n \sim 10^6$ because, essentially, the methods scale with the cost for solving a linear system of equations with multiple right-hand sides with (a shifted) $\mathbf{A}$ as coefficient matrix and the number of right-hand sides being equal to the number of inputs/outputs. Hence, whenever sparse systems can be solved efficiently, these model reduction techniques are applicable. In this context, recent discussions have focused on new low-rank methods based on Riemannian optimization. Moreover, several papers in the literature have presented relations among most of the existing methods. In particular, the importance of $\mathcal{H}_2$ optimal interpolation points for approximate balancing techniques has been pointed out. Future challenges for bilinear deterministic and linear stochastic control systems might include the (possible) fast decay of the singular values of the generalized Gramians as well as their numerically efficient computation.

Here, we only focused on parts of balancing (related) techniques. In particular, we did not present methods for DAEs

$$\Sigma: \begin{cases} \mathbf{E}\dot{\mathbf{x}}(t) = \mathbf{A}\mathbf{x}(t) + \mathbf{B}\mathbf{u}(t), \quad \mathbf{x}(0) = \mathbf{x}_0, \\ \mathbf{y}(t) = \mathbf{C}\mathbf{x}(t) + \mathbf{D}\mathbf{u}(t), \end{cases} \tag{6.55}$$

where the matrix $\mathbf{E} \in \mathbb{R}^{n \times n}$ might be singular, leading to additional algebraic constraints on the solution $\mathbf{x}(t)$. Most of the above concepts have been studied for this more general setting, and a complete overview is beyond the scope of this chapter. However, we refer to, e.g., [31, 86, 102, 103, 115] and references therein.

While we restricted ourselves to the continuous-time case, balancing concepts have also been discussed for discrete-time systems of the form

$$\begin{aligned} \mathbf{x}(k+1) &= \mathbf{A}\mathbf{x}(k) + \mathbf{B}\mathbf{u}(k), \\ \mathbf{y}(k) &= \mathbf{C}\mathbf{x}(k) + \mathbf{D}\mathbf{u}(k). \end{aligned} \tag{6.56}$$

For more details on this topic, we refer to, e.g., [3].

Another interesting area of research is second-order systems

$$\begin{aligned} \mathbf{M}\ddot{\mathbf{x}}(t) + \mathbf{D}\dot{\mathbf{x}}(t) + \mathbf{K}\mathbf{x}(t) &= \mathbf{B}\mathbf{u}(t), \\ \mathbf{y}(t) &= \mathbf{C}_1\dot{\mathbf{x}}(t) + \mathbf{C}_2\mathbf{x}(t), \end{aligned} \tag{6.57}$$

where $\mathbf{M}, \mathbf{D}, \mathbf{K} \in \mathbb{R}^{n \times n}$ are mass, damping, and stiffness matrices. These systems typically arise in, e.g., multibody systems or circuit simulation. Balancing-based model reduction concepts tailored to this structure are studied in, e.g., [22, 29, 37, 101, 125].

Recently, more and more attention has been paid to directly reducing the PDE instead of using a spatial discretization to obtain a large-scale system of ODEs. In this setting, the situation is typically that we have an abstract Cauchy problem of the form

$$\begin{aligned} \dot{z} &= \mathcal{A}z + \mathcal{B}u, \\ y &= \mathcal{C}z, \end{aligned} \tag{6.58}$$

where $\mathcal{X}, \mathcal{U}, \mathcal{Y}$ are given Hilbert spaces; $\mathcal{A}\colon \mathrm{dom}(\mathcal{A}) \subset \mathcal{X} \to \mathcal{X}$ is an infinitesimal generator of a strongly continuous semigroup on $\mathcal{X}$; and $\mathcal{B}\colon \mathcal{U} \to \mathcal{X}$ and $\mathcal{C}\colon \mathcal{X} \to \mathcal{Y}$ are (possibly unbounded) linear control/output operators. For an introduction to the concepts of balancing and truncation and also the existence of finite rank approximations for computing the solutions of the infinite-dimensional Lyapunov equation, we refer to [39, 51, 60, 94, 100].

Bibliography

[1] S. AL-BAIYAT AND M. BETTAYEB, *A new model reduction scheme for k-power bilinear systems*, in Proceedings of the 32nd IEEE Conference on Decision and Control, 1993, IEEE, pp. 22–27.

[2] S. AL-BAIYAT, M. BETTAYEB, AND U. AL-SAGGAF, *New model reduction scheme for bilinear systems*, International Journal of Systems Science, 25 (1994), pp. 1631–1642.

[3] A. ANTOULAS, *Approximation of Large-Scale Dynamical Systems*, SIAM, Philadelphia, 2005.

[4] A. ANTOULAS, D. SORENSEN, AND S. GUGERCIN, *A survey of model reduction methods for large-scale systems*, Contemporary Mathematics, 280 (2001), pp. 193–220.

[5] A. ANTOULAS, D. SORENSEN, AND Y. ZHOU, *On the decay rate of Hankel singular values and related issues*, Systems and Control Letters, 46 (2002), pp. 323–342.

[6] L. ARNOLD, *Stochastic Differential Equations: Theory and Applications*, John Wiley and Sons, New York, 1974.

[7] R. BARTELS AND G. STEWART, *Solution of the matrix equation $AX + XB = C$: Algorithm 432*, Communications of the ACM, 15 (1972), pp. 820–826.

[8] U. BAUR, *Low rank solution of data-sparse Sylvester equations*, Numerical Linear Algebra with Applications, 15 (2008), pp. 837–851.

[9] U. BAUR AND P. BENNER, *Factorized solution of Lyapunov equations based on hierarchical matrix arithmetic*, Computing, 78 (2006), pp. 211–234.

[10] U. BAUR, P. BENNER, AND L. FENG, *Model order reduction for linear and nonlinear systems: A system-theoretic perspective*, Archives of Computational Methods in Engineering, 21 (2014), pp. 331–358.

[11] P. BENNER, *Factorized solution of Sylvester equations with applications in control*, in Proceedings of the International Symposium on Mathematical Theory of Networks and Systems, MTNS 2004.

[12] ——, *Solving large-scale control problems*, IEEE Control Systems Magazine, 24 (2004), pp. 44–59.

[13] ——, *Advances in balancing-related model reduction for circuit simulation*, in Scientific Computing in Electrical Engineering SCEE 2008, Mathematics in Industry, J. Roos and L. R. J. Costa, eds., vol. 14, Springer-Verlag, Berlin, Heidelberg, 2010, pp. 469–482.

[14] ——, *System-theoretic methods for model reduction of large-scale systems: Simulation, control, and inverse problems*, in Proceedings of MathMod, Vienna, I. Troch and F. Breitenecker, eds., vol. 35 of ARGESIM Report, February 11–13, 2009, pp. 126–145.

[15] P. BENNER AND T. BREITEN, *Low rank methods for a class of generalized Lyapunov equations and related issues*, Numerische Mathematik, 124 (2013), pp. 441–470.

[16] ——, *On optimality of approximate low rank solutions of large-scale matrix equations*, Systems and Control Letters, 67 (2014), pp. 55–64.

[17] P. BENNER AND T. DAMM, *Lyapunov equations, energy functionals, and model order reduction of bilinear and stochastic systems*, SIAM Journal on Control and Optimization, 49 (2011), pp. 686–711.

[18] ——, *Balanced truncation for stochastic linear systems with guaranteed error bound*, in Proceedings of the 21st International Symposium on Mathematical Theory of Networks and Systems, 2014, pp. 1492–1497.

[19] P. BENNER, T. DAMM, M. REDMANN, AND Y. CRUZ, *Positive operators and stable truncation*, Linear Algebra and Its Applications, 498 (2014), pp. 74–87.

[20] P. BENNER AND P. KÜRSCHNER, *Computing real low-rank solutions of Sylvester equations by the factored ADI method*, Computers and Mathematics with Applications, 67 (2014), pp. 1656–1672.

[21] P. BENNER, P. KÜRSCHNER, AND J. SAAK, *Efficient handling of complex shift parameters in the low-rank Cholesky factor ADI method*, Numerical Algorithms, 62 (2013), pp. 225–251.

[22] ——, *An improved numerical method for balanced truncation for symmetric second-order systems*, Mathematical and Computer Modelling of Dynamical Systems, 19 (2013), pp. 593–615.

[23] ——, *Self-generating and efficient shift parameters in ADI methods for large Lyapunov and Sylvester equations*, Electronic Transactions on Numerical Analysis, 43 (2014), pp. 142–162.

[24] P. BENNER, J.-R. LI, AND T. PENZL, *Numerical solution of large Lyapunov equations, Riccati equations, and linear-quadratic control problems*, Numerical Linear Algebra with Applications, 15 (2008), pp. 755–777.

[25] P. BENNER, R. LI, AND N. TRUHAR, *On the ADI method for Sylvester equations*, Journal of Computational and Applied Mathematics, 233 (2009), pp. 1035–1045.

[26] P. BENNER, V. MEHRMANN, AND D. SORENSEN, *Dimension Reduction of Large-Scale Systems*, vol. 45 of Lecture Notes in Computational Science and Engineering, Springer-Verlag, Berlin, Heidelberg, 2005.

[27] P. BENNER AND E. QUINTANA-ORTI, *Solving stable generalized Lyapunov equations with the matrix sign function*, Numerical Algorithms, 20 (1999), pp. 75–100.

[28] P. BENNER AND M. REDMANN, *Model reduction for stochastic systems*, Stochastic Partial Differential Equations: Analysis and Computation, 3 (2015), pp. 291–338.

[29] P. BENNER AND J. SAAK, *Efficient balancing-based MOR for large-scale second-order systems*, Mathematical and Computer Modeling of Dynamical Systems, 17 (2011), pp. 123–143.

[30] ——, *Numerical solution of large and sparse continuous time algebraic matrix Riccati and Lyapunov equations: A state of the art survey*, GAMM Mitteilungen, 36 (2013), pp. 32–52.

[31] P. BENNER AND T. STYKEL, *Model order reduction for differential-algebraic equations: A survey*, in Surveys in Differential-Algebraic Equations IV, Differential-Algebraic Equations Forum, A. Ilchmann and T. Reis, eds., Springer International, Berlin, 2017, pp. 107–160.

[32] M. BOLLHÖFER AND A. EPPLER, *A structure preserving FGMRES method for solving large Lyapunov equations*, in Progress in Industrial Mathematics at ECMI 2010, M. Günther, A. Bartel, M. Brunk, S. Schöps, and M. Striebel, eds., Mathematics in Industry, Springer-Verlag, Berlin, Heidelberg, 2012, pp. 131–136.

[33] A. BOUHAMIDI, K. JBILOU, L. REICHEL, AND H. SADOK, *A generalized global Arnoldi method for ill-posed matrix equations*, Journal of Computational and Applied Mathematics, 236 (2012), pp. 2078–2089.

[34] C. BRUNI, G. DIPILLO, AND G. KOCH, *On the mathematical models of bilinear systems*, Automatica, 2 (1971), pp. 11–26.

[35] ——, *Bilinear systems: An appealing class of nearly linear systems in theory and applications*, IEEE Transactions on Automatic Control, 19 (1974), pp. 334–348.

[36] D. CALVETTI AND L. REICHEL, *Application of ADI iterative methods to the restoration of noisy images*, SIAM Journal on Matrix Analysis and Applications, 17 (1996), pp. 165–186.

[37] Y. CHAHLAOUI, D. LEMONNIER, A. VANDENDORPE, AND P. V. DOOREN, *Second-order balanced truncation*, Linear Algebra and Its Applications, 415 (2006), pp. 373 – 384.

[38] M. CONDON AND R. IVANOV, *Nonlinear systems—Algebraic Gramians and model reduction*, COMPEL: The International Journal for Computation and Mathematics in Electrical and Electronic Engineering, 24 (2005), pp. 202–219.

[39] R. CURTAIN AND K. GLOVER, *Balanced realisations for infinite-dimensional systems*, Operator Theory: Advances and Applications, 19 (1986), pp. 87–104.

[40] P. D'ALESSANDRO, A. ISIDORI, AND A. RUBERTI, *Realization and structure theory of bilinear dynamical systems*, SIAM Journal on Control, 12 (1974), pp. 517–535.

[41] T. DAMM, *Rational Matrix Equations in Stochastic Control*, vol. 297, Springer, 2004.

[42] T. DAMM, *Direct methods and ADI-preconditioned Krylov subspace methods for generalized Lyapunov equations*, Numerical Linear Algebra with Applications, 15 (2008), pp. 853–871.

[43] U. DESAI AND D. PAL, *A transformation approach to stochastic model reduction*, IEEE Transactions on Automatic Control, 29 (1984), pp. 1097–1100.

[44] V. DRUSKIN AND L. KNIZHNERMAN, *Extended Krylov subspaces: Approximation of the matrix square root and related functions*, SIAM Journal on Matrix Analysis and Applications, 19 (1998), pp. 755–771.

[45] V. DRUSKIN, L. KNIZHNERMAN, AND V. SIMONCINI, *Analysis of the rational Krylov subspace and ADI methods for solving the Lyapunov equation*, SIAM Journal on Numerical Analysis, 49 (2011), pp. 1875–1898.

[46] V. DRUSKIN AND V. SIMONCINI, *Adaptive rational Krylov subspaces for large-scale dynamical systems*, Systems and Control Letters, 60 (2011), pp. 546–560.

[47] D. ENNS, *Model reduction with balanced realizations: An error bound and a frequency weighted generalization*, in 23rd IEEE Conference on Decision and Control, vol. 23, IEEE, 1984, pp. 127–132.

[48] A. EPPLER AND M. BOLLHÖFER, *An alternative way of solving large Lyapunov equations*, Proceedings in Applied Mathematics and Mechanics, 10 (2010), pp. 547–548.

[49] G. FLAGG AND S. GUGERCIN, *On the ADI method for the Sylvester equation and the optimal-$\mathcal{H}_2$ points*, Applied Numerical Mathematics, 64 (2013), pp. 50–58.

[50] K. GLOVER, *All optimal Hankel-norm approximations of linear multivariable systems and their $\mathcal{L}_\infty$-error bounds*, International Journal of Control, 39 (1984), pp. 1115–1193.

[51] K. GLOVER, R. CURTAIN, AND J. PARTINGTON, *Realisation and approximation of linear infinite-dimensional systems with error bounds*, SIAM Journal on Control and Optimization, 26 (1988), pp. 863–898.

[52] G. GOLUB AND C. V. LOAN, *Matrix Computations*, vol. 3, Johns Hopkins University Press, 1996.

[53] L. GRASEDYCK, *Existence of a low rank or $\mathcal{H}$-matrix approximant to the solution of a Sylvester equation*, Numerical Linear Algebra with Applications, 11 (2004), pp. 371–389.

[54] L. GRASEDYCK AND W. HACKBUSCH, *A multigrid method to solve large scale Sylvester equations*, SIAM Journal on Matrix Analysis and Applications, 29 (2007), pp. 870–894.

[55] L. GRASEDYCK, W. HACKBUSCH, AND B. N. KHOROMSKIJ, *Solution of large scale algebraic matrix Riccati equations by use of hierarchical matrices*, Computing, 70 (2003), pp. 121–165.

[56] W. GRAY AND J. MESKO, *Energy functions and algebraic Gramians for bilinear systems*, in Preprints of the 4th IFAC Nonlinear Control Systems Design Symposium, 1998, pp. 103–108.

[57] L. GRUBIŠIĆ AND D. KRESSNER, *On the eigenvalue decay of solutions to operator Lyapunov equations*, Systems and Control Letters, 73 (2014), pp. 42–47.

[58] S. GUGERCIN, A. ANTOULAS, AND S. BEATTIE, *A survey of model reduction by balanced truncation and some new results*, International Journal of Control, 77 (2004), pp. 748–766.

[59] ——, *$\mathcal{H}_2$ model reduction for large-scale linear dynamical systems*, SIAM Journal on Matrix Analysis and Applications, 30 (2008), pp. 609–638.

[60] C. GUIVER AND M. OPMEER, *Model reduction by balanced truncation for systems with nuclear Hankel operators*, SIAM Journal on Control and Optimization, 52 (2014), pp. 1366–1401.

[61] S. HAMMARLING, *Numerical solution of the stable, non-negative definite Lyapunov equation*, IMA Journal of Numerical Analysis, 2 (1982), pp. 303–323.

[62] C. HARTMANN, B. SCHÄFER-BUNG, AND A. THÖNS-ZUEVA, *Balanced averaging of bilinear systems with applications to stochastic control*, SIAM Journal on Control and Optimization, 51 (2013), pp. 2356–2378.

[63] M. HEINKENSCHLOSS, T. REIS, AND A. ANTOULAS, *Balanced truncation model reduction for systems with inhomogeneous initial conditions*, Automatica, 47 (2011), pp. 559–564.

[64] D. HINRICHSEN AND A. PRITCHARD, *Mathematical Systems Theory I: Modelling, State Space Analysis, Stability and Robustness*, vol. 1, Springer-Verlag, 2005.

[65] M. HOCHBRUCK AND G. STARKE, *Preconditioned Krylov subspace methods for Lyapunov matrix equations*, SIAM Journal on Matrix Analysis and Applications, 16 (1995), pp. 156–171.

[66] R. HORN AND C. JOHNSON, *Matrix Analysis*, Cambridge University Press, 1990.

[67] C. HSU, U. DESAI, AND C. CRAWLEY, *Realization algorithms and approximation methods of bilinear systems*, in The 22nd IEEE Conference on Decision and Control, 1983, vol. 22, pp. 783–788.

[68] I. JAIMOUKHA AND E. KASENALLY, *Krylov subspace methods for solving large Lyapunov equations*, SIAM Journal on Numerical Analysis, 31 (1994), pp. 227–251.

[69] K. JBILOU AND A. J. RIQUET, *Projection methods for large Lyapunov matrix equations*, Linear Algebra and Its Applications, 415 (2006), pp. 344–358.

[70] E. JONCKHEERE AND L. SILVERMAN, *A new set of invariants for linear systems—Application to reduced order compensator design*, IEEE Transactions on Automatic Control, 28 (1983), pp. 953–964.

[71] S. KIM, B. ANDERSON, AND A. MADIEVSKI, *Error bound for transfer function order reduction using frequency weighted balanced truncation*, Systems and Control Letters, 24 (1995), pp. 183–192.

[72] L. KNIZHNERMAN AND V. SIMONCINI, *Convergence analysis of the extended Krylov subspace method for the Lyapunov equation*, Numerische Mathematik, 118 (2011), pp. 567–586.

[73] H. KNOBLOCH AND H. KWAKERNAAK, *Lineare Kontrolltheorie*, Springer-Verlag, Berlin, 1985. In German.

[74] A. KRENER, *Linearization and bilinearization of control systems*, in Proceedings of the 1974 Allerton Conference on Circuit and System Theory, vol. 834, Monticello, 1974.

[75] D. KRESSNER AND C. TOBLER, *Krylov subspace methods for linear systems with tensor product structure*, SIAM Journal on Matrix Analysis and Applications, 31 (2010), pp. 1688–1714.

[76] ——, *Low-rank tensor Krylov subspace methods for parametrized linear systems*, SIAM Journal on Matrix Analysis and Applications, 32 (2011), pp. 1288–1316.

[77] ——, *Preconditioned low-rank methods for high-dimensional elliptic PDE eigenvalue problems*, Computational Methods in Applied Mathematics, 11 (2011), pp. 363–381.

[78] V. B. LARIN AND F. A. ALIEV, *Construction of square root factor for solution of the Lyapunov matrix equation*, Systems Control Letters, 20 (1993), pp. 109–112.

[79] A. LAUB, M. HEATH, C. PAIGE, AND R. WARD, *Computation of system balancing transformations and other applications of simultaneous diagonalization algorithms*, IEEE Transactions on Automatic Control, 32 (1987), pp. 115–122.

[80] J.-R. LI, *Model Reduction of Large Linear Systems via Low Rank System Gramians*, PhD thesis, Massachusettes Institute of Technology, September 2000.

[81] J.-R. LI AND J. WHITE, *Low rank solution of Lyapunov equations*, SIAM Journal on Matrix Analysis and Applications, 24 (2002), pp. 260–280.

[82] C.-A. LIN AND T.-Y. CHIU, *Model reduction via frequency weighted balanced realization*, in Proceedings of the American Control Conference, 1990, pp. 2069–2070.

[83] Y. LIU AND B. ANDERSON, *Singular perturbation approximation of balanced systems*, in Proceedings of the 28th Conference on Decision and Control, 1989, pp. 1355–1360.

[84] L. MEIER AND D. LUENBERGER, *Approximation of linear constant systems*, IEEE Transactions on Automatic Control, 12 (1967), pp. 585–588.

[85] V. MEHRMANN AND T. STYKEL, *Descriptor systems: A general mathematical framework for modelling, simulation and control*, at-Automatisierungstechnik, 54 (2006), pp. 405–415.

[86] J. MÖCKEL, T. REIS, AND T. STYKEL, *Linear-quadratic Gaussian balancing for model reduction of differential-algebraic systems*, International Journal of Control, 84 (2011), pp. 1627–1643.

[87] R. MOHLER, *Bilinear Control Processes*, Academic Press, New York, 1973.

[88] ——, *Nonlinear Systems (vol. 2): Applications to Bilinear Control*, Prentice-Hall, Upper Saddle River, NJ, 1991.

[89] ——, *Natural bilinear control processes*, IEEE Transactions on Systems Science and Cybernetics, 6 (2007), pp. 192–197.

[90] B. C. MOORE, *Principal component analysis in linear systems: Controllability, observability, and model reduction*, IEEE Transactions on Automatic Control, AC-26 (1981), pp. 17–32.

[91] C. MULLIS AND R. ROBERTS, *Synthesis of minimum roundoff noise fixed point digital filters*, IEEE Transactions on Circuits and Systems, 23 (1976), pp. 551–562.

[92] G. OBINATA AND B. ANDERSON, *Model Reduction for Control System Design*, Springer-Verlag, 2000.

[93] M. OPMEER, *Decay of Hankel singular values of analytic control systems*, Systems and Control Letters, 59 (2010), pp. 635–638.

[94] M. OPMEER, T. REIS, AND W. WOLLNER, *Finite-rank ADI iteration for operator Lyapunov equations*, SIAM Journal on Control and Optimization, 51 (2013), pp. 4084–4117.

[95] D. PEACEMAN AND H. RACHFORD, *The numerical solution of parabolic and elliptic differential equations*, Journal of the Society for Industrial and Applied Mathematics, 3 (1955), pp. pp. 28–41.

[96] T. PENZL, *A cyclic low-rank Smith method for large sparse Lyapunov equations*, SIAM Journal on Scientific Computing, 21 (2000), pp. 1401–1418.

[97] ——, *Eigenvalue decay bounds for solutions of Lyapunov equations: The symmetric case*, Systems and Control Letters, 40 (2000), pp. 139–144.

[98] ——, *Algorithms for model reduction of large dynamical systems*, Linear Algebra and Its Applications, 415 (2006), pp. 322 – 343.

[99] L. PERNEBO AND L. SILVERMAN, *Model reduction via balanced state space representations*, IEEE Transactions on Automatic Control, 27 (1982), pp. 382–387.

[100] T. REIS AND T. SELIG, *Balancing transformations for infinite-dimensional systems with nuclear Hankel operator*, Integral Equations and Operator Theory, 79 (2014), pp. 67–105.

[101] T. REIS AND T. STYKEL, *Balanced truncation model reduction of second-order systems*, Mathematical and Computer Modelling of Dynamical Systems, 14 (2008), pp. 391–406.

[102] ——, *Positive real and bounded real balancing for model reduction of descriptor systems*, International Journal of Control, 83 (2010), pp. 74–88.

[103] T. REIS AND E. VIRNIK, *Positivity preserving balanced truncation for descriptor systems*, SIAM Journal on Control and Optimization, 48 (2009), pp. 2600–2619.

[104] J. D. ROBERTS, *Linear model reduction and solution of the algebraic Riccati equation by use of the sign function*, International Journal of Control, 32 (1980), pp. 677–687. (Reprint of Technical Report No. TR-13, CUED/B-Control, Cambridge University, Engineering Department, 1971).

[105] W. RUGH, *Nonlinear System Theory*, The Johns Hopkins University Press, 1982.

[106] Y. SAAD, *Numerical solution of large Lyapunov equation*, in Signal Processing, Scattering, Operator Theory and Numerical Methods, M. A. Kaashoek, J. H. van Schuppen, and A. C. M. Ran, eds., Birkhäuser, 1990, pp. 503–511.

[107] J. SAAK, *Efficient Numerical Solution of Large Scale Algebraic Matrix Equations in PDE Control and Model Order Reduction*, PhD thesis, TU Chemnitz, July 2009. Available from `http://nbn-resolving.de/urn:nbn:de:bsz:ch1-200901642`.

[108] J. SCHERPEN, *Balancing for Nonlinear Systems*, PhD thesis, Faculty of Applied Mathematics, University of Twente, 1994.

[109] W. SCHILDERS, H. V. D. VORST, AND J. ROMMES, *Model Order Reduction: Theory, Research Aspects and Applications*, vol. 13, Springer-Verlag, 2008.

[110] V. SIMONCINI, *A new iterative method for solving large-scale Lyapunov matrix equations*, SIAM Journal on Scientific Computing, 29 (2007), pp. 1268–1288.

[111] ——, *Computational methods for linear matrix equations*, SIAM Review, 58 (2016), pp. 377–441.

[112] E. SONTAG, *Mathematical Control Theory: Deterministic Finite Dimensional Systems*, vol. 6, Springer-Verlag, 1998.

[113] D. SORENSEN AND A. ANTOULAS, *The Sylvester equation and approximate balanced reduction*, Linear Algebra and Its Applications, 351–352 (2002), pp. 671–700.

[114] D. SORENSEN AND Y. ZHOU, *Bounds on Eigenvalue Decay Rates and Sensitivity of Solutions of Lyapunov Equations*, Technical Report 7, Rice University, 2002.

[115] T. STYKEL, *Gramian-based model reduction for descriptor systems*, Mathematics of Control, Signals and Systems, 16 (2004), pp. 297–319.

[116] M. TOMBS AND I. POSTLEWAITHE, *Truncated balanced realization of a stable non-minimal state-space system*, International Journal of Control, 46 (1987), pp. 1319–1330.

[117] B. VANDEREYCKEN, *Riemannian and Multilevel Optimization for Rank-Constrained Matrix Problems*, PhD thesis, Department of Computer Science, Katholieke Universiteit Leuven, 2010.

[118] B. VANDEREYCKEN AND S. VANDEWALLE, *A Riemannian optimization approach for computing low-rank solutions of Lyapunov equations*, SIAM Journal on Matrix Analysis and Applications, 31 (2010), pp. 2553–2579.

[119] A. VARGA AND B. ANDERSON, *Accuracy-enhancing methods for balancing-related frequency-weighted model and controller reduction*, Automatica, 39 (2003), pp. 919 - 927.

[120] E. VERRIEST, *On Generalized Balanced Realizations*, PhD thesis, Stanford University, 1980.

[121] ——, *Low sensitivity design and optimal order reduction for the LQG-problem*, in Proceedings of the 1981 Midwest Symposium on Stochastic Systems, 1981, pp. NM 365–369.

[122] E. WACHSPRESS, *Iterative solution of the Lyapunov matrix equation*, Applied Mathematics Letters, 107 (1988), pp. 87–90.

[123] ——, *The ADI Model Problem*, Springer-Verlag, Berlin, 2013.

[124] G. WANG, V. SREERAM, AND W.-Q. LIU, *A new frequency-weighted balanced truncation method and an error bound*, IEEE Transactions on Automatic Control, 44 (1999), pp. 1734–1737.

[125] B. YAN, S. TAN, AND B. MCGAUGHY, *Second-order balanced truncation for passive-order reduction of RLCK circuits*, IEEE Transactions on Circuits and Systems II: Express Briefs, 55 (2008), pp. 942–946.

[126] K. ZHOU, *Frequency-weighted $\mathscr{L}_\infty$ norm and optimal Hankel norm model reduction*, IEEE Transactions on Automatic Control, 40 (1995), pp. 1687–1699.

Chapter 7

Model Reduction by Rational Interpolation

Christopher Beattie and Serkan Gugercin[24]

The last two decades have seen major developments in interpolatory methods for model reduction of large-scale linear dynamical systems. Notable advances include greater ability to produce (locally) optimal reduced models at modest cost, refined methods for deriving interpolatory reduced models directly from input/output measurements, and extensions for the reduction of parametrized systems. This chapter offers a survey of interpolatory model reduction methods starting from basic principles and ranging up through recent developments that include weighted model reduction and structure-preserving methods based on generalized coprime representations. Our discussion is supported by an assortment of numerical examples.

7.1 ▪ Introduction

Numerous techniques exist for model reduction of large-scale dynamical systems, among which are proper orthogonal decomposition (POD; see Chapter 1), balanced truncation (BT; see Chapter 6), and interpolatory methods, to be discussed both here and in Chapter 8. Interpolatory model reduction methods include methods referred to as rational Krylov methods but should be viewed as distinct for reasons we describe later. Over the past two decades, major progress has been made in interpolation-based model reduction approaches, and, as a result, these methods have emerged as one of the leading choices for reducing large-scale dynamical systems.

This chapter surveys projection-based interpolatory methods for model reduction. Section 7.2 introduces the model reduction problem setting that we consider. In Section 7.3, we give general projection results for interpolatory model reduction, followed by a discussion of $\mathcal{H}_2$ optimal model reduction by interpolation in Section 7.4. Up until Section 7.4, we assume that the original system to be reduced is in a standard first-order state-space form. Beginning in Section 7.5, we discuss how interpolatory methods can be extended with ease to much more general settings, including systems

[24]This work was supported in part by the National Science Foundation under DMS-1217156.

with delays and systems with polynomial structure. For the most part, we assume that the internal dynamics of the full system are specified, accessible, and manifested either in a known state-space or generalized coprime representation. This assumption will be relaxed in Section 7.6 (and also in Chapter 8), where a data-driven framework for interpolatory methods is introduced, useful for situations with no direct access to internal dynamics and where only input/output measurements are available. Finally, in Section 7.7, we show how to use interpolatory methods to reduce parametric dynamical systems.

The methods we discuss in this chapter have been applied with great success to very large scale dynamical systems. Motivated by brevity, we do not present such examples here, preferring instead to illustrate the ideas with simple and approachable (albeit more academic) examples that may better reveal details of the process. For those with a hunger for more, we refer to the original papers in which large-scale case studies are presented.

7.2 ▪ Model reduction via projection

In this section, we introduce the basic concepts of projection-based model reduction. We also discuss the main error measures with which the approximation error will be quantified.

7.2.1 ▪ The problem setting

We consider linear dynamical systems represented in state-space form as

$$\begin{aligned} \mathbf{E}\dot{\mathbf{x}}(t) &= \mathbf{A}\mathbf{x}(t)+\mathbf{B}\mathbf{u}(t), \\ \mathbf{y}(t) &= \mathbf{C}\mathbf{x}(t)+\mathbf{D}\mathbf{u}(t), \end{aligned} \qquad \text{with } \mathbf{x}(0)=0, \tag{7.1}$$

where $\mathbf{A},\mathbf{E} \in \mathbb{R}^{n\times n}$, $\mathbf{B} \in \mathbb{R}^{n\times m}$, $\mathbf{C} \in \mathbb{R}^{q\times n}$, and $\mathbf{D} \in \mathbb{R}^{q\times m}$ are constant matrices. In (7.1), $\mathbf{x}(t) \in \mathbb{R}^n$ is the *internal variable*, or the *state variable* if $\mathbf{E}$ is nonsingular. The length n of $\mathbf{x}$ is called the *dimension* of the underlying dynamical system. $\mathbf{u}(t) \in \mathbb{R}^m$ and $\mathbf{y}(t) \in \mathbb{R}^q$ are, respectively, the *inputs* and *outputs* of the system. Dynamical systems with $m = p = 1$ will be called SISO systems (*single-input single-output*) while all other cases will be grouped together and referred to as MIMO systems (*multi-input multi-output*).

For cases where the dimension n is large, e.g., $n \geq 10^5$ or 10^6, the simulation and control of the system can lead to a huge computational burden, especially when the system must be resimulated over and over again, say, using different input selections $\mathbf{u}(t)$. Our goal is to replace (7.1) with a simpler reduced model of the form

$$\begin{aligned} \widehat{\mathbf{E}}\dot{\hat{\mathbf{x}}}(t) &= \widehat{\mathbf{A}}\hat{\mathbf{x}}(t)+\widehat{\mathbf{B}}\mathbf{u}(t), \\ \hat{\mathbf{y}}(t) &= \widehat{\mathbf{C}}\hat{\mathbf{x}}(t)+\widehat{\mathbf{D}}\mathbf{u}(t), \end{aligned} \qquad \text{with } \hat{\mathbf{x}}(0)=0, \tag{7.2}$$

where $\widehat{\mathbf{A}},\widehat{\mathbf{E}} \in \mathbb{R}^{r\times r}$, $\widehat{\mathbf{B}} \in \mathbb{R}^{r\times m}$, $\widehat{\mathbf{C}} \in \mathbb{R}^{q\times r}$, and $\widehat{\mathbf{D}} \in \mathbb{R}^{q\times m}$ collectively represent the corresponding reduced system realization with $r \ll n$, designed such that over a wide range of system inputs $\mathbf{u}(t)$, the corresponding output of the reduced system, $\hat{\mathbf{y}}(t)$, will be a good approximation to the (original) true output, $\mathbf{y}(t)$. In all that follows, we endeavor to denote quantities associated with *reduced models* with a "hat": $\widehat{\mathbf{A}}$, $\widehat{\mathbf{B}}$, $\hat{\mathbf{x}}(t)$, $\hat{\mathbf{y}}(t)$, etc. Quantities associated with the original (large-order) model generally remain unadorned.

7.2.2 ▪ Transfer functions and the frequency domain error

For the linear dynamical systems considered here, the frequency domain representation is a powerful tool for quantifying the model reduction error. In this setting, error measures in the frequency domain may be related directly to error measures in the time domain.

Let $\mathbf{Y}(s)$, $\widehat{\mathbf{Y}}(s)$, and $\mathbf{U}(s)$ denote the Laplace transforms of $\mathbf{y}(t)$, $\hat{\mathbf{y}}(t)$, and $\mathbf{u}(t)$, respectively. Taking the Laplace transforms of (7.1) and (7.2) yields

$$\mathbf{Y}(s) = \left(\mathbf{C}(s\mathbf{E}-\mathbf{A})^{-1}\mathbf{B}+\mathbf{D}\right)\mathbf{U}(s), \tag{7.3}$$

$$\widehat{\mathbf{Y}}(s) = \left(\widehat{\mathbf{C}}\left(s\widehat{\mathbf{E}}-\widehat{\mathbf{A}}\right)^{-1}\widehat{\mathbf{B}}+\widehat{\mathbf{D}}\right)\mathbf{U}(s). \tag{7.4}$$

The mapping from $\mathbf{U}(s)$ to $\mathbf{Y}(s)$ is called the *transfer function*. Likewise, the mapping from $\mathbf{U}(s)$ to $\widehat{\mathbf{Y}}(s)$ is the *transfer function of the reduced model*. They will be denoted as $\mathbf{H}(s)$ and $\widehat{\mathbf{H}}(s)$, respectively:

$$\mathbf{H}(s) = \mathbf{C}(s\mathbf{E}-\mathbf{A})^{-1}\mathbf{B}+\mathbf{D} \text{ and} \tag{7.5}$$

$$\widehat{\mathbf{H}}(s) = \widehat{\mathbf{C}}\left(s\widehat{\mathbf{E}}-\widehat{\mathbf{A}}\right)^{-1}\widehat{\mathbf{B}}+\widehat{\mathbf{D}}. \tag{7.6}$$

Note that $\mathbf{H}(s)$ is a $q\times m$ matrix-valued degree-n rational function in s and $\widehat{\mathbf{H}}(s)$ is a $q\times m$ matrix-valued degree-r rational function in s.

It follows immediately that

$$\mathbf{Y}(s)-\widehat{\mathbf{Y}}(s) = \left[\mathbf{H}(s)-\widehat{\mathbf{H}}(s)\right]\mathbf{U}(s).$$

The frequency domain error in the outputs, $\mathbf{Y}-\widehat{\mathbf{Y}}$, will be seen to directly relate to the time domain error between original and reduced systems, $\mathbf{y}(t)-\hat{\mathbf{y}}(t)$. To make this error small, we should seek $\widehat{\mathbf{H}}$ so that $\widehat{\mathbf{H}}\approx\mathbf{H}$ in an appropriate sense, and model reduction could be viewed as a rational approximation problem in the complex domain. This perspective is emphasized in Section 7.3.

7.2.3 ▪ Error Measures

As in any approximation problem, error measures are important in understanding and quantifying the quality of the approximation. We have seen that the closeness of $\widehat{\mathbf{Y}}(s)$ to $\mathbf{Y}(s)$ is related to the closeness of the reduced transfer function, $\widehat{\mathbf{H}}(s)$, to the original, $\mathbf{H}(s)$. How does this relate to the time domain error of the system outputs, $\mathbf{y}(t)-\hat{\mathbf{y}}(t)$? The $\mathcal{H}_2$- and $\mathcal{H}_\infty$-norms are the most common measures of closeness in this context.

The $\mathcal{H}_\infty$-Norm

Let $\mathbf{H}(s)$ be the transfer function of a stable dynamical system. The $\mathcal{H}_\infty$-norm is defined as

$$\|\mathbf{H}\|_{\mathcal{H}_\infty} = \sup_{\omega\in\mathbb{R}}\|\mathbf{H}(\imath\omega)\|_2, \tag{7.7}$$

where $\|\mathbf{M}\|_2$ denotes the spectral (Euclidean-induced) norm of the complex matrix $\mathbf{M}$. When $\mathbf{E}$ is nonsingular, all eigenvalues of the matrix pencil $\lambda\mathbf{E}-\mathbf{A}$ must lie in the left half-plane. When $\mathbf{E}$ is singular, we assume additionally that zero is not a defective eigenvalue of $\mathbf{E}$. This guarantees that $\mathbf{H}(s)$ remains bounded as $s\to\infty$.

The importance of the $\mathcal{H}_\infty$-norm stems from it being the (L_2-L_2)-induced operator norm of an underlying convolution operator mapping the system inputs $\mathbf{u}$ to system outputs $\mathbf{y}$: $||\mathbf{H}||_{\mathcal{H}_\infty} = \sup_{\mathbf{u}\in L_2} \frac{||\mathbf{y}||_{L_2}}{||\mathbf{u}||_{L_2}}$, where $||\mathbf{z}||_{L_2} = \sqrt{\int_0^\infty ||\mathbf{z}(t)||_2^2 dt}$. With respect to the model reduction error, one directly obtains

$$||\mathbf{y}-\hat{\mathbf{y}}||_{L_2} \leq \left\|\mathbf{H}-\widehat{\mathbf{H}}\right\|_{\mathcal{H}_\infty} ||\mathbf{u}||_{L_2}.$$

If one wishes to produce reduced models that generate outputs $\hat{\mathbf{y}}(t)$ that are always close (with respect to the L_2-norm) to the corresponding true outputs $\mathbf{y}(t)$ uniformly over all L_2-bounded inputs $\mathbf{u}(t)$, then one should apply a model reduction technique that produces a small $\mathcal{H}_\infty$-error.

The $\mathcal{H}_2$-Norm

Let $\mathbf{H}(s)$ be the transfer function of a stable dynamical system. Then, the $\mathcal{H}_2$-norm is defined as

$$||\mathbf{H}||_{\mathcal{H}_2} := \left(\frac{1}{2\pi}\int_{-\infty}^{\infty} ||\mathbf{H}(\imath\omega)||_{\mathrm{F}}^2\right)^{1/2}, \tag{7.8}$$

where $||\mathbf{M}||_{\mathrm{F}}$ denotes the Frobenius norm of a complex matrix $\mathbf{M}$.

When $\mathbf{E}$ is nonsingular, we require that all the eigenvalues of the matrix pencil $\lambda\mathbf{E}-\mathbf{A}$ lie in the left half-plane and $\mathbf{D}=0$ for the $\mathcal{H}_2$-norm to be finite. When $\mathbf{E}$ is singular, we assume in addition that zero is not a defective eigenvalue of $\mathbf{E}$ and that $\lim_{s\to\infty} \mathbf{H}(s) = 0$. The $\mathcal{H}_2$-norm bears a direct relationship to the time domain norm of $\mathbf{y}(t)$:

$$||\mathbf{y}||_{L_\infty} = \sup_{t>0} ||\mathbf{y}(t)||_\infty \leq ||\mathbf{H}||_{\mathcal{H}_2} ||\mathbf{u}||_{L_2},$$

and this bound is the best possible for MISO (multi-input single-output) systems ($p=1$), SIMO (single-input multi-output) systems ($m=1$), and SISO systems ($m=p=1$), reflecting the fact that the $\mathcal{H}_2$-norm is simply the (L_2-L_∞)-induced norm of the underlying convolution operator in these cases. With respect to the model reduction error, we have, in general,

$$||\mathbf{y}-\hat{\mathbf{y}}||_{L_\infty} \leq \left\|\mathbf{H}-\widehat{\mathbf{H}}\right\|_{\mathcal{H}_2} ||\mathbf{u}||_{L_2}.$$

So, if one wishes to produce reduced models that generate outputs $\hat{\mathbf{y}}(t)$ that are uniformly and instantaneously close to the corresponding true outputs $\mathbf{y}(t)$ uniformly over all L_2-bounded inputs $\mathbf{u}(t)$, then one should apply a model reduction technique that produces small $\mathcal{H}_2$-error. Note that the $\mathcal{H}_2$-error may be made small even for original systems $\mathbf{H}$ with $\lim_{s\to\infty}\mathbf{H}(s)\neq 0$ with an appropriate choice of $\widehat{\mathbf{D}}$ and $\widehat{\mathbf{E}}$; $||\mathbf{H}-\widehat{\mathbf{H}}||_{\mathcal{H}_2}$ may be small even if $||\mathbf{H}||_{\mathcal{H}_2}$ is unboundedly large. This is discussed in more detail in Section 7.3.2.

7.2.4 ▪ Petrov–Galerkin projections

Most model reduction methods can be formulated with the aid of Petrov–Galerkin or Galerkin projections. Even though the original internal variable $\mathbf{x}(t)$ evolves in a (large) n-dimensional space, it is often the case that it hews rather closely to some (typically unknown) r-dimensional subspace. Let $\mathbf{V} \in \mathbb{R}^{n\times r}$ be a basis for this subspace, which is as yet undetermined. The original state may be approximated as $\mathbf{x}(t) \approx \mathbf{V}\hat{\mathbf{x}}(t)$ for some $\hat{\mathbf{x}}(t) \in \mathbb{R}^r$, and this expression may be used to represent the reduced model dynamics. Plug the approximation $\mathbf{x}(t) \approx \mathbf{V}\hat{\mathbf{x}}(t)$ into (7.1) to obtain a residual

$$\mathsf{R}(\hat{\mathbf{x}}(t)) = \mathbf{E}\mathbf{V}\dot{\hat{\mathbf{x}}}(t) - \mathbf{A}\mathbf{V}\hat{\mathbf{x}}(t) - \mathbf{B}\,\mathbf{u}(t) \tag{7.9}$$

and the approximate output

$$\hat{\mathbf{y}}(t) = \mathbf{C}\mathbf{V}\hat{\mathbf{x}}(t) + \mathbf{D}\mathbf{u}(t). \tag{7.10}$$

The reduced state trajectory $\hat{\mathbf{x}}(t)$ is determined by enforcing a Petrov–Galerkin orthogonality condition on the residual $\mathsf{R}(\hat{\mathbf{x}}(t))$ in (7.9): we pick a second r-dimensional subspace with a basis $\mathbf{W} \in \mathbb{R}^{n\times r}$ and impose the Petrov–Galerkin condition:

$$\mathbf{W}^T \mathsf{R}(\hat{\mathbf{x}}(t)) = \mathbf{W}^T \left(\mathbf{E}\mathbf{V}\dot{\hat{\mathbf{x}}}(t) - \mathbf{A}\mathbf{V}\hat{\mathbf{x}}(t) - \mathbf{B}\,\mathbf{u}(t)\right) = 0.$$

This leads to a reduced model as in (7.2),

$$\widehat{\mathbf{E}}\dot{\hat{\mathbf{x}}}(t) = \widehat{\mathbf{A}}\hat{\mathbf{x}}(t) + \widehat{\mathbf{B}}\mathbf{u}(t), \quad \hat{\mathbf{y}}(t) = \widehat{\mathbf{C}}\hat{\mathbf{x}}(t) + \widehat{\mathbf{D}}\mathbf{u}(t),$$

with reduced model quantities defined as

$$\widehat{\mathbf{E}} = \mathbf{W}^T\mathbf{E}\mathbf{V}, \quad \widehat{\mathbf{A}} = \mathbf{W}^T\mathbf{A}\mathbf{V}, \quad \widehat{\mathbf{B}} = \mathbf{W}^T\mathbf{B}, \quad \widehat{\mathbf{C}} = \mathbf{C}\mathbf{V}, \quad \text{and} \quad \widehat{\mathbf{D}} = \mathbf{D}. \tag{7.11}$$

One critical observation to make here is that the reduced model does not depend on the specific basis selection made for $\mathbf{V}$ and $\mathbf{W}$, but only on the subspaces themselves. To see this, let $\widetilde{\mathbf{V}} = \mathbf{V}\mathbf{T}_1$ and $\widetilde{\mathbf{W}} = \mathbf{W}\mathbf{T}_2$, with $\mathbf{T}_1, \mathbf{T}_2 \in \mathbb{R}^{r\times r}$ being nonsingular matrices corresponding to separate changes of bases for each of $\mathbf{V}$ and $\mathbf{W}$. This leads to a change in reduced model quantities:

$$\widetilde{\mathbf{E}} = \mathbf{T}_2^T\widehat{\mathbf{E}}\mathbf{T}_1, \quad \widetilde{\mathbf{A}} = \mathbf{T}_2^T\widehat{\mathbf{A}}\mathbf{T}_1, \quad \widetilde{\mathbf{B}} = \mathbf{T}_2^T\widehat{\mathbf{B}}, \quad \widetilde{\mathbf{C}} = \widehat{\mathbf{C}}\mathbf{T}_1, \quad \text{and} \quad \widetilde{\mathbf{D}} = \widehat{\mathbf{D}},$$

where $\widehat{\mathbf{E}}$, $\widehat{\mathbf{A}}$, $\widehat{\mathbf{B}}$, $\widehat{\mathbf{C}}$, and $\widehat{\mathbf{D}}$ are as defined in (7.11) and $\widetilde{\mathbf{E}}$, $\widetilde{\mathbf{A}}$, $\widetilde{\mathbf{B}}$, $\widetilde{\mathbf{C}}$, and $\widetilde{\mathbf{D}}$ are the modified reduced model quantities. If $\widetilde{\mathbf{H}}$ is used to denote the associated transfer function, then a straightforward comparison reveals

$$\begin{aligned}
\widetilde{\mathbf{H}}(s) &= \widetilde{\mathbf{C}}\left(s\widetilde{\mathbf{E}} - \widetilde{\mathbf{A}}\right)^{-1}\widetilde{\mathbf{B}} + \widetilde{\mathbf{D}} \\
&= \widehat{\mathbf{C}}\mathbf{T}_1\left(s\mathbf{T}_2^T\widehat{\mathbf{E}}\mathbf{T}_1 - \mathbf{T}_2^T\widehat{\mathbf{A}}\mathbf{T}_1\right)^{-1}\mathbf{T}_2^T\widehat{\mathbf{B}} + \widehat{\mathbf{D}} \\
&= \widehat{\mathbf{C}}\left(s\widehat{\mathbf{E}} - \widehat{\mathbf{A}}\right)^{-1}\widehat{\mathbf{B}} + \widehat{\mathbf{D}} = \widehat{\mathbf{H}}(s).
\end{aligned}$$

7.3 ▪ Model reduction by interpolation

In this section, we present the fundamental projection theorems used in interpolatory model reduction and illustrate their use with some simple examples.

7.3.1 ▪ Tangential interpolation problem

One easily observes that the model reduction problem for linear, time-invariant dynamical systems can be formulated as a rational approximation problem: given a degree-n rational function $\mathbf{H}(s)$ (the full model), find a degree-r rational function $\widehat{\mathbf{H}}(s)$ (the reduced model) that approximates $\mathbf{H}(s)$ accurately with respect to either the $\mathcal{H}_\infty$- or the $\mathcal{H}_2$-norm. Interpolation is a commonly applied tool for function approximation; typically, effective polynomial interpolants are easy to calculate. Here, we will develop straightforward methods for obtaining rational interpolants; however, the precise notion of "interpolation" that will be used must be clarified. Since $\mathbf{H}(s)$ is a $q \times m$ matrix-valued rational function, the immediate extension of pointwise interpolation to matrix-valued functions suggests that we attempt to enforce conditions such as $\mathbf{H}(s_0) = \widehat{\mathbf{H}}(s_0)$ at each interpolation point $s_0 \in \mathbb{C}$. But viewed elementwise, this would require, in effect, $q \times m$ interpolation conditions at every interpolation point. For systems with even a modestly large number of input and output dimensions m and p, this will lead to a large number of interpolation conditions requiring, as a result, quite a large reduced order r. Thus, for MIMO systems, instead of this notion of "full matrix interpolation," we require that the interpolating matrix function match the original only along certain directions, using "tangential interpolation." We will show later that this relaxed notion of interpolation is adequate to characterize necessary conditions for optimal approximation in the $\mathcal{H}_2$-norm.

Tangential interpolation involves choosing interpolation directions in addition to interpolation points. We separate the interpolation points and directions into two categories: left and right. We say that $\widehat{\mathbf{H}}(s)$ is a right tangential interpolant to $\mathbf{H}(s)$ at $s = \sigma_i$ along the right tangent direction $\mathbf{r}_i \in \mathbb{C}^m$ if

$$\mathbf{H}(\sigma_i)\mathbf{r}_i = \widehat{\mathbf{H}}(\sigma_i)\mathbf{r}_i.$$

Similarly, we say that $\widehat{\mathbf{H}}(s)$ is a left tangential interpolant to $\mathbf{H}(s)$ at $s = \mu_i$ along the left tangent direction $\boldsymbol{\ell}_i \in \mathbb{C}^q$ if

$$\boldsymbol{\ell}_i^T \mathbf{H}(\mu_i) = \boldsymbol{\ell}_i^T \widehat{\mathbf{H}}(\mu_i).$$

Our model reduction task can now be formulated as tangential interpolation as follows: given a set of r right interpolation points $\{\sigma_i\}_{i=1}^r \in \mathbb{C}$, r left interpolation points $\{\mu_i\}_{i=1}^r$, r right tangential directions $\{\mathbf{r}_i\}_{i=1}^r \in \mathbb{C}^m$, and r left tangential directions $\{\boldsymbol{\ell}_i\}_{i=1}^r \in \mathbb{C}^p$, find a degree-$r$ reduced transfer function $\widehat{\mathbf{H}}(s)$ so that

$$\begin{aligned} \mathbf{H}(\sigma_i)\mathbf{r}_i &= \widehat{\mathbf{H}}(\sigma_i)\mathbf{r}_i \\ \boldsymbol{\ell}_i^T \mathbf{H}(\mu_i) &= \boldsymbol{\ell}_i^T \widehat{\mathbf{H}}(\mu_i) \end{aligned} \quad \text{for} \quad i = 1, \ldots, r. \tag{7.12}$$

We say that $\widehat{\mathbf{H}}(s)$ is a bitangential Hermite interpolant to $\mathbf{H}(s)$ at $s = \sigma_i$ along the right tangent direction $\mathbf{r}_i \in \mathbb{C}^m$ and the left tangent direction $\boldsymbol{\ell}_i \in \mathbb{C}^q$ if

$$\boldsymbol{\ell}_i^T \mathbf{H}'(\sigma_i)\mathbf{r}_i = \boldsymbol{\ell}_i^T \widehat{\mathbf{H}}'(\sigma_i)\mathbf{r}_i,$$

where $'$ denotes differentiation with respect to s. Therefore, in addition to (7.12), we may also require $\widehat{\mathbf{H}}(s)$ to satisfy

$$\boldsymbol{\ell}_i^T \mathbf{H}'(\sigma_i)\mathbf{r}_i = \boldsymbol{\ell}_i^T \widehat{\mathbf{H}}'(\sigma_i)\mathbf{r}_i \quad \text{for} \quad i = 1, \ldots, r. \tag{7.13}$$

In Section 7.4, we will show how to choose interpolation points and tangent directions to produce optimal approximation with respect to the $\mathcal{H}_2$-norm.

7.3.2 ▪ Petrov–Galerkin projections for tangential interpolation

Our goal here is to appropriately pick the model reduction bases $\mathbf{V}$ and $\mathbf{W}$ so that the reduced model obtained by a Petrov–Galerkin projection as in (7.11) satisfies the tangential interpolation conditions (7.12) and (7.13). We first present the projection theorem, followed by a historical perspective.

Theorem 7.1. *Given the transfer function* $\mathbf{H}(s) = \mathbf{C}(s\mathbf{E}-\mathbf{A})^{-1}\mathbf{B}+\mathbf{D}$, *let* $\widehat{\mathbf{H}}(s)$ *denote a reduced transfer function obtained by projection as in* (7.11) *using the model reduction bases* $\mathbf{V}$ *and* $\mathbf{W}$. *For interpolation points* $\sigma, \mu \in \mathbb{C}$, *suppose that* $\sigma\mathbf{E}-\mathbf{A}$ *and* $\mu\mathbf{E}-\mathbf{A}$ *are invertible. Let* $\mathbf{r} \in \mathbb{C}^m$ *and* $\boldsymbol{\ell} \in \mathbb{C}^q$ *be designated (nontrivial) tangent directions. Then,*

(a) *if*

$$(\sigma\mathbf{E}-\mathbf{A})^{-1}\mathbf{B}\mathbf{r} \in \mathsf{Ran}(\mathbf{V}), \tag{7.14}$$

then

$$\mathbf{H}(\sigma)\mathbf{r} = \widehat{\mathbf{H}}(\sigma)\mathbf{r}; \tag{7.15}$$

(b) *if*

$$\left(\boldsymbol{\ell}^T\mathbf{C}(\mu\mathbf{E}-\mathbf{A})^{-1}\right)^T \in \mathsf{Ran}(\mathbf{W}), \tag{7.16}$$

then

$$\boldsymbol{\ell}^T\mathbf{H}(\mu) = \boldsymbol{\ell}^T\widehat{\mathbf{H}}(\mu); \tag{7.17}$$

(c) *if both* (7.14) *and* (7.16) *hold, and* $\sigma = \mu$, *then*

$$\boldsymbol{\ell}^T\mathbf{H}'(\sigma)\mathbf{r} = \boldsymbol{\ell}^T\widehat{\mathbf{H}}'(\sigma)\mathbf{r} \tag{7.18}$$

as well.

Theorem 7.1 illustrates that imposing either a left or a right tangential interpolation condition requires adding only one vector either to the left or right model reduction basis. For the case of repeated left and right interpolation points, the bitangential Hermite condition is satisfied for free, in the sense that no additional vectors need to be included in the model reduction bases. Notice that the values that are interpolated are never explicitly computed; this is a significant advantage of the Petrov–Galerkin projection framework as used in interpolatory model reduction.

A projection framework for interpolatory model reduction is introduced by Skelton et al. in [38, 92, 93]. This approach was put into a robust numerical framework by Grimme [52], who employed the rational Krylov subspace method of Ruhe [80]. The tangential interpolation framework of Theorem 7.1 was developed by Gallivan et al. [51]. For SISO systems, the model reduction bases $\mathbf{V}$ and $\mathbf{W}$ produced from Theorem 7.1 become rational Krylov subspaces, and so interpolatory model reduction is sometimes referred to as *rational Krylov methods*. However, the connection to rational Krylov subspaces is lost for general MIMO systems unless all tangent directions are the same. So, we prefer the simpler descriptive label *interpolatory methods*. Indeed, for the more general systems that we consider in Section 7.5, the direct extensions we develop for interpolatory methods have no connection to rational Krylov subspaces even in the SISO case. Another term that is used to refer to interpolatory model reduction methods is *moment matching methods*. The "kth moment of $\mathbf{H}(s)$ around σ" refers to the kth derivative of $\mathbf{H}(s)$ evaluated at $s = \sigma$. In the SISO case, the reduced transfer function obtained via rational interpolation will match these moments; this is, in effect, generalized Hermite interpolation. The notion of moment matching for MIMO systems with respect to tangent directions is not so unambiguous, however. See [9, 10, 12, 26, 43, 48, 50] and the references therein for related works on model reduction by interpolation.

Example 7.2. Consider the following linear dynamical system with $n = 3$, $m = q = 2$, $\mathbf{E} = \mathbf{I}_3$, $\mathbf{D} = 0$,

$$\mathbf{A} = \begin{bmatrix} -6 & -11 & -6 \\ 1 & 0 & 0 \\ 0 & 1 & 0 \end{bmatrix}, \quad \mathbf{B} = \begin{bmatrix} -1 & 1 \\ 0 & 1 \\ 1 & 0 \end{bmatrix}, \quad \text{and} \quad \mathbf{C} = \begin{bmatrix} 1 & 0 & 1 \\ 1 & -1 & 0 \end{bmatrix}.$$

The transfer function of this dynamical system can be computed as

$$\mathbf{H}(s) = \frac{1}{s^3 + 6s^2 + 11s + 6} \begin{bmatrix} 10 & s^2 - 10s + 1 \\ -s^2 - 5s + 6 & -18s - 6 \end{bmatrix}.$$

Let $\sigma_1 = \mu_1 = 0$ be the left and right interpolation points together with tangent directions

$$\mathbf{r}_1 = \begin{bmatrix} 1 \\ 2 \end{bmatrix} \quad \text{and} \quad \boldsymbol{\ell}_1 = \begin{bmatrix} 3 \\ 1 \end{bmatrix}.$$

Using Theorem 7.1, we compute the interpolatory model reduction bases:

$$\mathbf{V} = (\sigma_1 \mathbf{E} - \mathbf{A})^{-1} \mathbf{B} \mathbf{r}_1 = \begin{bmatrix} -2 \\ -1 \\ 4 \end{bmatrix} \quad \text{and}$$

$$\mathbf{W} = (\sigma_1 \mathbf{E} - \mathbf{A})^{-T} \mathbf{C}^T \boldsymbol{\ell}_1 = \begin{bmatrix} 0.5 \\ -1 \\ 6.5 \end{bmatrix}.$$

Then, the Petrov–Galerkin projection in (7.11) leads to the reduced model quantities

$$\widehat{\mathbf{E}} = 26, \; \widehat{\mathbf{A}} = -5, \; \widehat{\mathbf{B}} = \begin{bmatrix} 6 & -0.5 \end{bmatrix}, \; \widehat{\mathbf{C}} = \begin{bmatrix} 2 \\ -1 \end{bmatrix}, \; \widehat{\mathbf{D}} = 0,$$

and consequently to the reduced model transfer function

$$\widehat{\mathbf{H}}(s) = \frac{1}{26s + 5} \begin{bmatrix} 12 & -1 \\ -6 & 0.5 \end{bmatrix}.$$

Now that we have $\widehat{\mathbf{H}}(s)$, we can check the interpolation conditions explicitly:

$$\mathbf{H}(\sigma_1)\mathbf{r}_1 = \widehat{\mathbf{H}}(\sigma_1)\mathbf{r}_1 = \begin{bmatrix} 2 \\ -1 \end{bmatrix}, \quad \checkmark$$

$$\boldsymbol{\ell}_1^T \mathbf{H}(\sigma_1) = \boldsymbol{\ell}_1^T \widehat{\mathbf{H}}(\sigma_1) = \begin{bmatrix} 6 & -0.5 \end{bmatrix}, \quad \checkmark$$

$$\boldsymbol{\ell}_1^T \mathbf{H}'(\sigma_1)\mathbf{r}_1 = \boldsymbol{\ell}_1^T \widehat{\mathbf{H}}'(\sigma_1)\mathbf{r}_1 = -26. \quad \checkmark$$

Notice that $\widehat{\mathbf{H}}(s)$ does not fully interpolate $\mathbf{H}(s)$ at $s = \sigma_1 = 0$:

$$\mathbf{H}(\sigma_1) = \begin{bmatrix} 5/3 & 1/6 \\ 1 & -1 \end{bmatrix} \neq \begin{bmatrix} 2.4 & -0.2 \\ -1.2 & 0.1 \end{bmatrix} = \widehat{\mathbf{H}}(\sigma_1).$$

To enforce full matrix interpolation, we need to modify the construction of the model reduction bases and remove the tangential vectors. Denote the new bases for full matrix interpolation by $\mathbf{V}_m$ and $\mathbf{W}_m$:

$$\mathbf{V}_m = (\sigma_1\mathbf{E}-\mathbf{A})^{-1}\mathbf{B} = \begin{bmatrix} 0 & -1 \\ -1 & 0 \\ 5/3 & 7/6 \end{bmatrix},$$

$$\mathbf{W}_m = (\sigma_1\mathbf{E}-\mathbf{A})^{-T}\mathbf{C}^T = \begin{bmatrix} 1/6 & 0 \\ 0 & -1 \\ 11/6 & 1 \end{bmatrix}.$$

Note that even with only a single interpolation point, full matrix interpolation leads to reduction spaces having dimension two, which in turn leads to a degree-two reduced model with transfer function

$$\widehat{\mathbf{H}}_m(s) = \begin{bmatrix} 10 & 1 \\ 6 & -6 \end{bmatrix}\left(s\begin{bmatrix} 110 & 71 \\ 96 & 42 \end{bmatrix} - \begin{bmatrix} -60 & -6 \\ -36 & 36 \end{bmatrix}\right)^{-1}\begin{bmatrix} 10 & 1 \\ 6 & -6 \end{bmatrix}.$$

This new reduced model fully interpolates $\mathbf{H}(s)$ and $\mathbf{H}'(s)$ at $s = \sigma_1 = 0$:

$$\mathbf{H}(\sigma_1) = \widehat{\mathbf{H}}_m(\sigma_1) = \begin{bmatrix} 5/3 & 1/6 \\ 1 & -1 \end{bmatrix},$$

$$\mathbf{H}'(\sigma_1) = \widehat{\mathbf{H}}'_m(\sigma_1) = \begin{bmatrix} -55/18 & -71/36 \\ -8/3 & -7/6 \end{bmatrix}.$$

This simple example illustrates the fundamental difference between tangential and full interpolation. In the case of tangential interpolation, each interpolation condition contributes only one additional order to the reduced dimension. However, in the case of full transfer function interpolation, a single (full matrix) interpolation condition will generically add m or p dimensions to the reduced model. We will see in Section 7.4 that optimality requires consideration only of tangential interpolation conditions. ■

The computations in the previous example can be extended easily to the case of r interpolation points: given $\mathbf{H}(s) = \mathbf{C}(s\mathbf{E}-\mathbf{A})^{-1}\mathbf{B}+\mathbf{D}$, r right interpolation points $\{\sigma_i\}_{i=1}^r$ and right directions $\{\mathbf{r}_k\}_{k=1}^r \in \mathbb{C}^m$, and r left interpolation points $\{\mu_j\}_{j=1}^r$ and left directions $\{\boldsymbol{\ell}_k\}_{k=1}^r \in \mathbb{C}^q$, construct

$$\mathbf{V} = \left[(\sigma_1\mathbf{E}-\mathbf{A})^{-1}\mathbf{B}\mathbf{r}_1,\ \ldots,\ (\sigma_r\mathbf{E}-\mathbf{A})^{-1}\mathbf{B}\mathbf{r}_r\right] \text{ and} \tag{7.19}$$

$$\mathbf{W} = \left[(\mu_1\mathbf{E}-\mathbf{A})^{-T}\mathbf{C}^T\boldsymbol{\ell}_1,\ \ldots,\ (\mu_r\mathbf{E}-\mathbf{A})^{-T}\mathbf{C}^T\boldsymbol{\ell}_r\right]. \tag{7.20}$$

Then, $\widehat{\mathbf{H}}(s) = \widehat{\mathbf{C}}(s\widehat{\mathbf{E}}-\widehat{\mathbf{A}})^{-1}\widehat{\mathbf{B}}$ constructed by a Petrov–Galerkin projection as in (7.11) satisfies the Lagrange tangential interpolation conditions (7.12) and the bitangential Hermite interpolation conditions (7.13) if in addition $\sigma_i = \mu_i$ (provided that $\sigma_i\widehat{\mathbf{E}}-\widehat{\mathbf{A}}$ and $\mu_i\widehat{\mathbf{E}}-\widehat{\mathbf{A}}$ are nonsingular for each $i = 1, \ldots, r$).

Theorem 7.1 can be extended readily to include higher-order Hermite interpolation.

Theorem 7.3. *Given a full model with transfer function*

$$\mathbf{H}(s) = \mathbf{C}(s\mathbf{E}-\mathbf{A})^{-1}\mathbf{B}+\mathbf{D},$$

let $\widehat{\mathbf{H}}(s)$ *denote a reduced transfer function obtained by projection as in* (7.11), *using model reduction bases* $\mathbf{V}$ *and* $\mathbf{W}$. *Let* $\mathbf{H}^{(k)}(\sigma)$ *denote the kth derivative of* $\mathbf{H}(s)$ *with respect to s evaluated at* $s=\sigma$. *For interpolation points* $\sigma, \mu \in \mathbb{C}$, *suppose* $\sigma\,\mathbf{E}-\mathbf{A}$ *and* $\mu\,\mathbf{E}-\mathbf{A}$ *are invertible, and that* $\mathbf{r}\in\mathbb{C}^m$ *and* $\boldsymbol{\ell}\in\mathbb{C}^q$ *are given (nontrivial) tangent directions. Then,*

(a) *if*

$$\left((\sigma\,\mathbf{E}-\mathbf{A})^{-1}\mathbf{E}\right)^{j-1}(\sigma\,\mathbf{E}-\mathbf{A})^{-1}\mathbf{B}\mathbf{r}\in\mathsf{Ran}(\mathbf{V}),\ \textit{for}\ j=1,\ldots,N \tag{7.21}$$

then

$$\mathbf{H}^{(k)}(\sigma)\mathbf{r}=\widehat{\mathbf{H}}^{(k)}(\sigma)\mathbf{r}\ \textit{for}\ k=0,1,\ldots,N-1; \tag{7.22}$$

(b) *if*

$$\left((\mu\,\mathbf{E}-\mathbf{A})^{-T}\mathbf{E}^T\right)^{j-1}(\mu\,\mathbf{E}-\mathbf{A})^{-T}\mathbf{C}^T\boldsymbol{\ell}\in\mathsf{Ran}(\mathbf{W})\ \textit{for}\ j=1,\ldots,M, \tag{7.23}$$

then

$$\boldsymbol{\ell}^T\mathbf{H}^{(k)}(\mu)=\boldsymbol{\ell}^T\widehat{\mathbf{H}}^{(k)}(\mu)\ \textit{for}\ k=0,1,\ldots,M-1; \tag{7.24}$$

(c) *if* $\sigma=\mu$ *and both* (7.21) *and* (7.23) *hold, then*

$$\boldsymbol{\ell}^T\mathbf{H}^{(k)}(\sigma)\mathbf{r}=\boldsymbol{\ell}^T\widehat{\mathbf{H}}^{(k)}(\sigma)\mathbf{r}\ \textit{for}\ k=0,\ldots,M+N-1 \tag{7.25}$$

as well.

The main cost in interpolatory model reduction originates from the need to solve large-scale (typically sparse) shifted linear systems. There is no need to solve large-scale Lyapunov or Riccati equations, giving interpolatory methods a computational advantage over competing methods. The discussion here assumes that these linear systems are solved by direct methods (e.g., Gaussian elimination). However, for systems with millions of degrees of freedom, one would prefer to incorporate iterative solution strategies to construct the model reduction bases $\mathbf{V}$ and $\mathbf{W}$. We refer to [17, 22, 90] for detailed analyses of the effects of iterative solves on interpolatory model reduction and to [2–4, 22, 24] for development of effective iterative solvers in the context of interpolatory model reduction.

Rational interpolants with $\widehat{\mathbf{D}} \neq \mathbf{D}$

So far, we have assumed that $\widehat{\mathbf{D}}=\mathbf{D}$. This is the logical choice if one is interested in minimizing the $\mathcal{H}_2$-norm of the error system. For the case of ordinary differential equations (ODEs) where $\mathbf{E}$ is nonsingular, choosing $\widehat{\mathbf{D}}\neq\mathbf{D}$ will lead to an unbounded $\mathcal{H}_2$-error norm. However, if, instead, one is interested in the $\mathcal{H}_\infty$-error, then flexibility in choosing $\widehat{\mathbf{D}}$ will be necessary as the optimal $\mathcal{H}_\infty$-approximation will have $\widehat{\mathbf{D}}\neq\mathbf{D}$ (see, e.g., [16, 47]).

Another case that may require choosing $\widehat{\mathbf{D}}\neq\mathbf{D}$ is the case of an index-one system of differential algebraic equations (DAEs). In our setting, this means that the $\mathbf{E}$ matrix in (7.1) has a nondefective eigenvalue at zero. (Interpolatory projection methods for DAEs are considered in detail in Section 7.3.3.) In this case, $\lim_{s\to\infty} \mathbf{H}(s) \neq \mathbf{D}$, so for $\widehat{\mathbf{H}}(s)$ to match $\mathbf{H}(s)$ asymptotically well at high frequencies, we require

$$\widehat{\mathbf{D}} = \lim_{s\to\infty} \left(\mathbf{H}(s) - \widehat{\mathbf{C}}(s\widehat{\mathbf{E}} - \widehat{\mathbf{A}})^{-1}\widehat{\mathbf{B}}\right).$$

Since $\widehat{\mathbf{E}}$ will be generically nonsingular (assuming $r < \text{rank}(\mathbf{E})$), setting $\widehat{\mathbf{D}} = \lim_{s\to\infty} \mathbf{H}(s)$ will guarantee $\lim_{s\to\infty} \mathbf{H}(s) = \lim_{s\to\infty} \widehat{\mathbf{H}}(s)$.

The next theorem shows how one may construct reduced models with $\widehat{\mathbf{D}} \neq \mathbf{D}$ without losing interpolation properties. Without loss of generality, we assume $\mathbf{D} = 0$, i.e.,

$$\mathbf{H}(s) = \mathbf{C}(s\mathbf{E} - \mathbf{A})^{-1}\mathbf{B}. \tag{7.26}$$

In the general case with $\mathbf{D} \neq 0$, one simply needs to replace $\widehat{\mathbf{D}}$ with $\widehat{\mathbf{D}} - \mathbf{D}$. The result below was first given in [72] and later generalized in [19].

Theorem 7.4. *Given a full model with transfer function $\mathbf{H}(s)$ as in (7.26), r distinct left interpolation points $\{\mu_i\}_{i=1}^r$ together with r left tangent directions $\{\boldsymbol{\ell}_i\}_{i=1}^r \subset \mathbb{C}^q$, and r distinct right interpolation points $\{\sigma_j\}_{j=1}^r$ together with r right tangent directions $\{\mathbf{r}_j\}_{j=1}^r \subset \mathbb{C}^m$. Let the model reduction bases $\mathbf{V} \in \mathbb{C}^{n\times r}$ and $\mathbf{W} \in \mathbb{C}^{n\times r}$ be constructed as in (7.19) and (7.20), respectively. Define $\widetilde{\mathbf{r}}$ and $\widetilde{\boldsymbol{\ell}}$ as*

$$\widetilde{\mathbf{r}} = [\mathbf{r}_1, \mathbf{r}_2, \ldots, \mathbf{r}_r] \quad \text{and} \quad \widetilde{\boldsymbol{\ell}}^T = [\boldsymbol{\ell}_1, \boldsymbol{\ell}_2, \ldots, \boldsymbol{\ell}_r]^T$$

For any $\widehat{\mathbf{D}} \in \mathbb{C}^{q\times m}$, define

$$\begin{aligned} \widehat{\mathbf{E}} &= \mathbf{W}^T\mathbf{E}\mathbf{V}, & \widehat{\mathbf{A}} &= \mathbf{W}^T\mathbf{A}\mathbf{V} + \widetilde{\boldsymbol{\ell}}^T\widehat{\mathbf{D}}\widetilde{\mathbf{r}}, \\ \widehat{\mathbf{B}} &= \mathbf{W}^T\mathbf{B} - \widetilde{\boldsymbol{\ell}}^T\widehat{\mathbf{D}}, & \text{and } \widehat{\mathbf{C}} &= \mathbf{C}\mathbf{V} - \widehat{\mathbf{D}}\widetilde{\mathbf{r}} \end{aligned} \tag{7.27}$$

Then, the reduced model $\widehat{\mathbf{H}}(s) = \widehat{\mathbf{C}}(s\widehat{\mathbf{E}} - \widehat{\mathbf{A}})^{-1}\widehat{\mathbf{B}} + \widehat{\mathbf{D}}$ satisfies

$$\mathbf{H}(\sigma_i)\mathbf{r}_i = \widehat{\mathbf{H}}(\sigma_i)\mathbf{r}_i \quad \text{and} \quad \boldsymbol{\ell}_i^T\mathbf{H}(\mu_i) = \boldsymbol{\ell}_i^T\widehat{\mathbf{H}}(\mu_i) \quad \text{for } i = 1, \ldots, r.$$

Theorem 7.4 shows how to construct a rational tangential interpolant with an *arbitrary* $\widehat{\mathbf{D}}$ term. Thus, $\widehat{\mathbf{D}}$ could be chosen to satisfy additional design goals.

7.3.3 ▪ Interpolatory projections for differential algebraic systems

The interpolation conditions in Theorems 7.1 and 7.3 are valid regardless of whether or not the matrix $\mathbf{E}$ is singular, as long as $s\mathbf{E} - \mathbf{A}$ and $s\widehat{\mathbf{E}} - \widehat{\mathbf{A}}$ are invertible matrices for $s = \sigma, \mu$. When $\mathbf{E}$ is nonsingular, the underlying model is a system of ODEs; when $\mathbf{E}$ is singular, the underlying model is a system of DAEs. Thus, from a pure interpolation perspective, the distinction does not make a difference. However, from the perspective of error measures, there is a crucial difference.

A crucial difference between a DAE system and an ODE system is that the transfer function of a DAE system could contain a polynomial part that may grow unboundedly as $s \to \infty$. In the case of ODE systems, the polynomial part is simply the constant feed-forward term, $\mathbf{D}$.

Let $\mathbf{H}(s)$ be the transfer function of a DAE system. We decompose $\mathbf{H}(s)$ as

$$\mathbf{H}(s) = \mathbf{C}(s\mathbf{E}-\mathbf{A})^{-1}\mathbf{B}+\mathbf{D} = \mathbf{G}(s)+\mathbf{P}(s), \tag{7.28}$$

where $\mathbf{G}(s)$ is the strictly proper rational part, i.e., $\lim_{s\to\infty}\mathbf{G}(s)=0$, and $\mathbf{P}(s)$ is the polynomial part. Now, assume that the Petrov–Galerkin projection is applied to $\mathbf{H}(s)$ as in (7.11). Then, even though $\mathbf{E}$ is singular, the reduced matrix $\widehat{\mathbf{E}} = \mathbf{W}^T\mathbf{E}\mathbf{V}$ will generically be a nonsingular matrix for $r \le \text{rank}(\mathbf{E})$. This means that unlike $\mathbf{H}(s)$, which contains a polynomial part $\mathbf{P}(s)$, the reduced model will correspond to an ODE and the polynomial part of the reduced transfer function $\widehat{\mathbf{H}}(s)$ will be $\mathbf{D}$. Decompose $\widehat{\mathbf{H}}(s)$ in a similar way as

$$\widehat{\mathbf{H}}(s) = \widehat{\mathbf{G}}(s)+\mathbf{D},$$

where $\widehat{\mathbf{G}} = \widehat{\mathbf{C}}(s\widehat{\mathbf{E}}-\widehat{\mathbf{A}})^{-1}\widehat{\mathbf{B}}$ is strictly proper. Then, the error transfer function

$$\mathbf{H}(s)-\widehat{\mathbf{H}}(s) = (\mathbf{G}(s)-\widehat{\mathbf{G}}(s))+(\mathbf{P}(s)-\mathbf{D})$$

has a polynomial part $\mathbf{P}(s)-\mathbf{D}$. Even when $\mathbf{P}(s)$ is a polynomial of degree one, the difference $\mathbf{P}(s)-\mathbf{D}$ will grow without bound as $s\to\infty$, leading to unbounded $\mathcal{H}_\infty$- and $\mathcal{H}_2$-error norms. Even when $\mathbf{P}(s)$ is a constant polynomial (i.e., degree zero), unless $\mathbf{P}(s)=\mathbf{D}$, this will still lead to unbounded $\mathcal{H}_2$-error. The only way to guarantee bounded error norms is to make sure that the reduced transfer function $\widehat{\mathbf{H}}(s)$ has exactly the same polynomial part as $\mathbf{H}(s)$, i.e., $\widehat{\mathbf{H}}(s) = \widehat{\mathbf{G}}(s)+\mathbf{P}(s)$, so that the error function is simply $\mathbf{H}(s)-\widehat{\mathbf{H}}(s) = \mathbf{G}(s)-\widehat{\mathbf{G}}(s)$, with only a null polynomial component. Based on these observations, [58, 90] discuss how to modify the interpolatory projection bases $\mathbf{V}$ and $\mathbf{W}$ to achieve this goal. As expected, the left and right deflating subspaces of the pencil $\lambda\mathbf{E}-\mathbf{A}$ corresponding to finite and infinite eigenvalues play a crucial role.

Theorem 7.5 ([58]). *Suppose the transfer function* $\mathbf{H}(s) = \mathbf{C}(s\mathbf{E}-\mathbf{A})^{-1}\mathbf{B}+\mathbf{D} = \mathbf{G}(s)+\mathbf{P}(s)$ *is associated with a DAE, where* $\mathbf{G}(s)$ *and* $\mathbf{P}(s)$ *are, respectively, the strictly proper and the polynomial parts of* $\mathbf{H}(s)$. *Let* $\mathbb{P}_l$ *and* $\mathbb{P}_r$ *be the spectral projectors onto the left and right deflating subspaces of the pencil* $\lambda\mathbf{E}-\mathbf{A}$ *corresponding to the finite eigenvalues. Also, let the columns of* $\mathbf{W}_\infty$ *and* $\mathbf{V}_\infty$ *span the left and right deflating subspaces of* $\lambda\mathbf{E}-\mathbf{A}$ *corresponding to the eigenvalue at infinity. For interpolation points* $\sigma, \mu \in \mathbb{C}$, *suppose* $\sigma\mathbf{E}-\mathbf{A}$ *and* $\mu\mathbf{E}-\mathbf{A}$ *are invertible and* $\mathbf{r}\in\mathbb{C}^m$ *and* $\ell\in\mathbb{C}^q$ *are given (nontrivial) tangent directions. Suppose further that* $\widehat{\mathbf{H}}(s)$ *is the reduced transfer function obtained by projection as in* (7.11) *using the model reduction bases* $\mathbf{V}$ *and* $\mathbf{W}$. *Construct* $\mathbf{V}_f$ *and* $\mathbf{W}_f$ *so that*

$$\left((\sigma\mathbf{E}-\mathbf{A})^{-1}\mathbf{E}\right)^{j-1}(\sigma\mathbf{E}-\mathbf{A})^{-1}\mathbb{P}_l\mathbf{B}\mathbf{r} \in \mathsf{Ran}(\mathbf{V}_f) \tag{7.29}$$

for $j=1,\ldots,N$ *and*

$$\left((\mu\mathbf{E}-\mathbf{A})^{-T}\mathbf{E}^T\right)^{j-1}(\mu\mathbf{E}-\mathbf{A})^{-T}\mathbb{P}_r^T\mathbf{C}^T\ell \in \mathsf{Ran}(\mathbf{W}_f) \tag{7.30}$$

for $j=1,\ldots,M$. *Define* $\mathbf{W}$ *and* $\mathbf{V}$ *as* $\mathbf{W} = [\mathbf{W}_f,\ \mathbf{W}_\infty]$ *and* $\mathbf{V} = [\mathbf{V}_f,\ \mathbf{V}_\infty]$. *Then,* $\widehat{\mathbf{H}}(s) = \widehat{\mathbf{G}}(s)+\widehat{\mathbf{P}}(s)$ *satisfies* $\widehat{\mathbf{P}}(s) = \mathbf{P}(s)$ *together with* (7.22) *and* (7.24). *If, in addition,* $\sigma=\mu$, *then* (7.25) *holds as well.*

Theorem 7.5 addresses the question of how to reduce DAEs with projection-based tangential interpolation in the most general case where both the index of the DAE and the selected interpolation points are arbitrary. By appropriately incorporating the deflating projectors $\mathbb{P}_r$ and $\mathbb{P}_l$ in the model reduction bases, the polynomial part of $\mathbf{H}(s)$ is exactly matched as desired while simultaneously enforcing interpolation. For the special case of DAEs with proper transfer functions and interpolation around $s = \infty$, a solution is given in [28]. For descriptor systems of index one, [10] offers a solution that uses an appropriately chosen $\widehat{\mathbf{D}}$ term.

Remark 7.6. A fundamental difficulty in the reduction of DAEs is the need to compute deflating projectors $\mathbb{P}_r$ and $\mathbb{P}_l$. For large-scale DAEs, construction of $\mathbb{P}_r$ and $\mathbb{P}_l$ is at best very costly, if even feasible. However, for the cases of semiexplicit descriptor systems of index one and Stokes-type descriptor systems of index two, it is possible to apply interpolatory projections without forming $\mathbb{P}_r$ and $\mathbb{P}_l$ explicitly [58, 90]. Thus, no greater effort is required to produce reduced models than for the case of ODEs. The Stokes-type descriptor systems of index two are also studied in [63] in a BT setting. Recently, [1] extended the work of [58] to address the reduction of index-three DAEs without forming projectors explicitly. For some structured problems arising in circuit simulation, multibody systems, or computational fluid dynamics, these projectors can be constructed without much computational effort [82]. We choose to omit these details from the present discussion but refer the interested reader to the original sources.

7.4 ▪ Interpolatory projections for $\mathcal{H}_2$ optimal approximation

When interpolation points and tangent directions are specified, Section 7.3 presents an approach that one may follow to construct a reduced model satisfying the desired (tangential) conditions. Notably, this development does not suggest a strategy for choosing interpolation points and tangent directions that lead to high-fidelity reduced models. In this section, we approach this issue by developing interpolatory conditions that are necessary for optimal approximation with respect to the $\mathcal{H}_2$-norm.

7.4.1 ▪ Interpolatory $\mathcal{H}_2$ optimality conditions

Consider the following optimization problem: Given a full system $\mathbf{H}(s)$, find a reduced model $\widehat{\mathbf{H}}(s)$ that minimizes the $\mathcal{H}_2$ error, i.e.,

$$\left\|\mathbf{H}-\widehat{\mathbf{H}}\right\|_{\mathcal{H}_2} = \min_{\dim(\widetilde{\mathbf{H}})=r} \left\|\mathbf{H}-\widetilde{\mathbf{H}}\right\|_{\mathcal{H}_2}. \tag{7.31}$$

As we pointed out in Section 7.2.3, a small $\mathcal{H}_2$-error induces a small time domain error $\|\mathbf{y}-\hat{\mathbf{y}}\|_{L_\infty}$, so attempting to *minimize* the $\mathcal{H}_2$-error is a worthy goal.

The $\mathcal{H}_2$ optimization problem (7.31) is nonconvex; finding a global minimizer is typically infeasible. A common approach used instead involves finding *locally* optimal reduced models that satisfy first-order necessary conditions for optimality. The (simpler) problem of finding locally optimal $\mathcal{H}_2$ reduced models has been studied extensively. Optimality conditions have been formulated either in terms of Lyapunov and Sylvester equations [27, 61, 64, 81, 89, 91, 96] or in terms of rational (tangential) interpolation conditions [18, 20, 21, 35, 54, 56, 57, 66, 67, 73, 76, 83]. [57] shows the equivalence between the Lyapunov/Sylvester equation conditions and the interpolation framework that we describe here.

We will first assume that $\mathbf{E}$ is nonsingular, so that $\mathbf{H}(s) = \mathbf{C}(s\mathbf{E}-\mathbf{A})^{-1}\mathbf{B}+\mathbf{D}$ will correspond to a system of ODEs and $\lim_{s\to\infty}\mathbf{H}(s) = \mathbf{D}$. To have a bounded $\mathcal{H}_2$-error norm, $\|\mathbf{H}-\widehat{\mathbf{H}}\|_{\mathcal{H}_2}$, it is necessary that $\widehat{\mathbf{D}} = \mathbf{D}$. Therefore, without loss of generality, one may take $\widehat{\mathbf{D}} = \mathbf{D} = 0$.

For MIMO systems, interpolatory first-order conditions for $\mathcal{H}_2$ optimality are best understood from the pole-residue expansion for $\widehat{\mathbf{H}}(s)$. We write $\widehat{\mathbf{H}}(s)$ in the following way:

$$\widehat{\mathbf{H}}(s) = \widehat{\mathbf{C}}(s\widehat{\mathbf{E}}-\widehat{\mathbf{A}})^{-1}\widehat{\mathbf{B}} = \sum_{i=1}^{r}\frac{\boldsymbol{\ell}_i\mathbf{r}_i^T}{s-\lambda_i}, \tag{7.32}$$

where we have assumed that the λ_i's are distinct. We refer to $\boldsymbol{\ell}_i \in \mathbb{C}^q$ and $\mathbf{r}_i \in \mathbb{C}^m$ in (7.32), respectively, as left/right residue directions associated with the pole λ_i of $\widehat{\mathbf{H}}(s)$; $\boldsymbol{\ell}_i\mathbf{r}_i^T$ is the (matrix) residue of $\widehat{\mathbf{H}}(s)$ at $s = \lambda_i$. The pole-residue expansion in (7.32) can be computed effectively by computing a generalized eigenvalue decomposition for the matrix pencil $\lambda\widehat{\mathbf{E}}-\widehat{\mathbf{A}}$, which is a trivial computation for the small to modest orders of r typically encountered. Note that finding such a representation for the full model $\mathbf{H}(s)$ will generally be infeasible.

Theorem 7.7. *Let $\widehat{\mathbf{H}}(s)$ in (7.32) be the best rth-order rational approximation of $\mathbf{H}(s)$ with respect to the $\mathcal{H}_2$-norm. Then,*

$$\mathbf{H}(-\lambda_k)\mathbf{r}_k = \widehat{\mathbf{H}}(-\lambda_k)\mathbf{r}_k, \tag{7.33a}$$

$$\boldsymbol{\ell}_k^T\mathbf{H}(-\lambda_k) = \boldsymbol{\ell}_k^T\widehat{\mathbf{H}}(-\lambda_k), \text{ and} \tag{7.33b}$$

$$\boldsymbol{\ell}_k^T\mathbf{H}'(-\lambda_k)\mathbf{r}_k = \boldsymbol{\ell}_k^T\widehat{\mathbf{H}}'(-\lambda_k)\mathbf{r}_k \tag{7.33c}$$

for $k = 1, 2, \ldots, r$.

In particular, any optimal $\mathcal{H}_2$ approximation $\widehat{\mathbf{H}}(s)$ must be a bitangential Hermite interpolant to $\mathbf{H}(s)$, and this theorem directly connects optimal model reduction to interpolation. The optimal interpolation points and tangent directions are derived from the pole-residue representation of $\widehat{\mathbf{H}}(s)$: the optimal interpolation points are the mirror images of the poles of $\widehat{\mathbf{H}}(s)$ reflected across the imaginary axis, and the optimal tangent directions are the residue directions associated with that pole.

Interpolatory conditions for SISO systems were initially introduced by Meier and Luenberger [73]. However, until recently, effective numerical algorithms to find reduced systems that satisfy these conditions were lacking, especially for large-scale settings. Gugercin et al. in [54, 56] introduce such an algorithm, called the *iterative rational Krylov algorithm* (IRKA). In practice, IRKA has significantly expanded the utility of optimal $\mathcal{H}_2$ model reduction. The optimality conditions for MIMO systems as presented in Theorem 7.7 were developed in [35, 57, 83] and led to an analogous algorithm for IRKA in the MIMO case. This is the main focus of Section 7.4.2. Recall that we have assumed that $\widehat{\mathbf{H}}(s)$ has distinct (reduced) poles $\lambda_1, \ldots, \lambda_r$. Optimality conditions for cases when $\widehat{\mathbf{H}}(s)$ has repeated poles are derived in [84].

7.4.2 ▪ IRKA for optimal $\mathcal{H}_2$ approximation

Theorem 7.7 gives optimality conditions that depend on the poles and residues of a reduced system, a locally $\mathcal{H}_2$-optimal reduced system that is yet to be determined. IRKA utilizes the construction of Theorem 7.1 to force interpolation at the mirror images of successive sets of reduced poles, iteratively correcting the reduced model until the optimality conditions of Theorem 7.7 hold. The method proceeds as follows: Given some initial interpolation points $\{\sigma_i\}_{i=1}^r$ and directions $\{\mathbf{r}_i\}_{i=1}^r$ and $\{\boldsymbol{\ell}_i\}_{i=1}^r$, construct $\mathbf{V}$ and $\mathbf{W}$ as in (7.19) and (7.20), respectively, and construct an intermediate reduced model $\widehat{\mathbf{H}}(s)$ using (7.11). Then, compute the pole-residue decomposition of $\widehat{\mathbf{H}}(s)$,

$$\widehat{\mathbf{H}}(s) = \sum_{i=1}^{r} \frac{\widehat{\boldsymbol{\ell}}_i \widehat{\mathbf{r}}_i^T}{s - \lambda_i}$$

(by solving a small $r \times r$ generalized eigenvalue problem). For $\widehat{\mathbf{H}}(s)$ to satisfy the first-order necessary conditions, we need $\sigma_i = -\lambda_i$, $\mathbf{r}_i = \widehat{\mathbf{r}}_i$, and $\boldsymbol{\ell}_i = \widehat{\boldsymbol{\ell}}_i$ for $i = 1, \ldots, r$. Therefore, set

$$\sigma_i \longleftarrow -\lambda_i, \quad \mathbf{r}_i \longleftarrow \widehat{\mathbf{r}}_i, \quad \text{and} \quad \boldsymbol{\ell}_i \longleftarrow \widehat{\boldsymbol{\ell}}_i \quad \text{for} \quad i = 1, \ldots, r$$

as the next interpolation data point, and iterate until convergence is reached. A brief sketch of IRKA is given in Algorithm 7.1.

***ALGORITHM* 7.1. MIMO $\mathcal{H}_2$-optimal tangential interpolation (IRKA).**

1. Make an initial r-fold shift selection $\{\sigma_1, \ldots, \sigma_r\}$ that is closed under conjugation (i.e., $\{\sigma_1, \ldots, \sigma_r\} \equiv \{\overline{\sigma_1}, \ldots, \overline{\sigma_r}\}$ viewed as sets) and initial tangent directions $\mathbf{r}_1, \ldots, \mathbf{r}_r$ and $\boldsymbol{\ell}_1, \ldots, \boldsymbol{\ell}_r$, also closed under conjugation.
2. $\mathbf{V} = \left[(\sigma_1 \mathbf{E} - \mathbf{A})^{-1}\mathbf{B}\mathbf{r}_1 \ \cdots \ (\sigma_r \mathbf{E} - \mathbf{A})^{-1}\mathbf{B}\mathbf{r}_r\right]$.
3. $\mathbf{W} = \left[(\sigma_1 \mathbf{E} - \mathbf{A}^T)^{-1}\mathbf{C}^T\boldsymbol{\ell}_1 \cdots (\sigma_r \mathbf{E} - \mathbf{A}^T)^{-1}\mathbf{C}^T\boldsymbol{\ell}_1\right]$.
4. while (not converged)
   (a) $\widehat{\mathbf{A}} = \mathbf{W}^T\mathbf{A}\mathbf{V}$, $\widehat{\mathbf{E}} = \mathbf{W}^T\mathbf{E}\mathbf{V}$, $\widehat{\mathbf{B}} = \mathbf{W}^T\mathbf{B}$, and $\widehat{\mathbf{C}} = \mathbf{C}\mathbf{V}$.
   (b) Compute a pole-residue expansion of $\widehat{\mathbf{H}}(s)$:
   $$\widehat{\mathbf{H}}(s) = \widehat{\mathbf{C}}(s\widehat{\mathbf{E}} - \widehat{\mathbf{A}})^{-1}\widehat{\mathbf{B}} = \sum_{i=1}^{r} \frac{\widehat{\boldsymbol{\ell}}_i \widehat{\mathbf{r}}_i^T}{s - \lambda_i}.$$
   (c) $\sigma_i \longleftarrow -\lambda_i$, $\mathbf{r}_i \longleftarrow \widehat{\mathbf{r}}_i$, and $\boldsymbol{\ell}_i \longleftarrow \widehat{\boldsymbol{\ell}}_i$ for $i = 1, \ldots, r$.
   (d) $\mathbf{V} = \left[(\sigma_1 \mathbf{E} - \mathbf{A})^{-1}\mathbf{B}\mathbf{r}_1 \ \cdots \ (\sigma_r \mathbf{E} - \mathbf{A})^{-1}\mathbf{B}\mathbf{r}_r\right]$.
   (e) $\mathbf{W} = \left[(\sigma_1 \mathbf{E} - \mathbf{A}^T)^{-1}\mathbf{C}^T\boldsymbol{\ell}_1 \cdots (\sigma_r \mathbf{E} - \mathbf{A}^T)^{-1}\mathbf{C}^T\boldsymbol{\ell}_1\right]$.
5. $\widehat{\mathbf{A}} = \mathbf{W}^T\mathbf{A}\mathbf{V}$, $\widehat{\mathbf{E}} = \mathbf{W}^T\mathbf{E}\mathbf{V}$, $\widehat{\mathbf{B}} = \mathbf{W}^T\mathbf{B}$, $\widehat{\mathbf{C}} = \mathbf{C}\mathbf{V}$.

Upon convergence, the reduced model $\widehat{\mathbf{H}}(s)$ satisfies the interpolatory first-order necessary conditions (7.33) for $\mathcal{H}_2$ optimality by construction. Convergence is generally observed to be rapid, though it slows as input/output orders grow. Convergence may be guaranteed a priori in some circumstances [46], yet there are known cases where convergence may fail [46, 57]. When convergence occurs, the resulting reduced model is guaranteed to be a local $\mathcal{H}_2$-minimizer since the local maxima of the $\mathcal{H}_2$ minimization problem are known to be repellent [66]. Overall in practice, IRKA has seen significant success in computing high-fidelity (locally) optimal reduced models and has been successfully applied in large-scale settings to find $\mathcal{H}_2$-optimal reduced models for systems with hundreds of thousands of state variables; for example, see [65] for application in cellular neurophysiology, [30] for energy-efficient building design to produce accurate compact models for the indoor air environment, and [57] for optimal cooling for steel profiles. Moreover, [23] has extended IRKA to the reduction of *bilinear* dynamical systems, a special class of weakly nonlinear dynamical systems.

Our analysis so far has assumed that $\mathbf{E}$ is a nonsingular matrix. Interpolatory optimal $\mathcal{H}_2$ model reduction for the case of singular $\mathbf{E}$, i.e., for systems of DAEs, is developed in [58], and IRKA has been extended to DAEs. Similar to the ODE case, where we require $\mathbf{D} = \widehat{\mathbf{D}}$, the DAE case requires that the polynomial part of $\widehat{\mathbf{H}}(s)$ match that of $\mathbf{H}(s)$ exactly and the strictly proper part of $\widehat{\mathbf{H}}(s)$ be an optimal tangential interpolant to the strictly proper part of $\mathbf{H}(s)$. For details, we refer the reader to [58].

7.4.3 ▪ Interpolatory weighted $\mathcal{H}_2$ model reduction

The error measures we have considered thus far give the same weight to all frequencies, and they are global in nature to the extent that degradation in fidelity is penalized in the same way throughout the full frequency spectrum. However, some applications require that certain frequencies be weighted more than others. For example, certain dynamical systems, such as mechanical systems or electrical circuits, might operate only in certain frequency bands, so retaining fidelity outside this frequency band carries no value. This leads to the problem of weighted model reduction. We formulate it here in terms of a weighted $\mathcal{H}_2$-norm.

Let $\mathbf{W}(s)$ be an input weighting function, a "shaping filter." We will assume that $\mathbf{W}(s)$ is a rational function itself in the form

$$\mathbf{W}(s) = \mathbf{C}_w (s\mathbf{I} - \mathbf{A}_w)^{-1} \mathbf{B}_w + \mathbf{D}_w = \sum_{k=1}^{n_w} \frac{\mathbf{e}_k\, \mathbf{f}_k^T}{s - \gamma_k} + \mathbf{D}_w, \tag{7.34}$$

where n_w denotes the dimension of $\mathbf{W}(s)$. Then, given the full model $\mathbf{H}(s)$ and the weight $\mathbf{W}(s)$, define the weighted $\mathcal{H}_2$-error norm as

$$\|\mathbf{H} - \widehat{\mathbf{H}}\|_{\mathcal{H}_2(W)} \stackrel{\text{def}}{=} \|\left(\mathbf{H} - \widehat{\mathbf{H}}\right) \cdot \mathbf{W}\|_{\mathcal{H}_2}. \tag{7.35}$$

In addition to the input weighting $\mathbf{W}(s)$, one may also define a filter for output weighting. For simplicity of presentation, we focus here on one-sided weighting only. The goal is to find a reduced model $\widehat{\mathbf{H}}(s)$ that minimizes the weighted error (7.35):

$$\|\mathbf{H} - \widehat{\mathbf{H}}\|_{\mathcal{H}_2(W)} = \min_{\dim(\widetilde{\mathbf{H}})=r} \|\mathbf{H} - \widetilde{\mathbf{H}}\|_{\mathcal{H}_2(W)}. \tag{7.36}$$

Weighted $\mathcal{H}_2$ model reduction is considered in [61] and [81] with a framework that uses Riccati and/or Lyapunov equations. A numerically more efficient, interpolation-based approach is introduced in [6] for the SISO case. This initial interpolatory framework is significantly extended and placed on a more rigorous theoretical footing (which allows straightforward extension to MIMO systems) in [32], where the equivalence of the Riccati and interpolation-based frameworks is also proved. Our presentation below follows [32]. We use the notation $\mathbf{H} \in \mathcal{H}_2$ to indicate that $\mathbf{H}(s)$ is a stable dynamical system with $\|\mathbf{H}\|_{\mathcal{H}_2} < \infty$. We use the notation $\mathbf{H} \in \mathcal{H}_2(W)$ analogously.

The interpolatory framework for the weighted $\mathcal{H}_2$ problem is best understood by defining a new linear transformation [6, 32]

$$\mathfrak{F}[\mathbf{H}](s) = \mathbf{H}(s)\mathbf{W}(s)\mathbf{W}(-s)^T + \sum_{k=1}^{n_w} \mathbf{H}(-\gamma_k)\mathbf{W}(-\gamma_k)\frac{\mathbf{f}_k\mathbf{e}_k^T}{s+\gamma_k}, \tag{7.37}$$

where $\mathbf{H} \in \mathcal{H}_2(W)$ and $\mathbf{f}_k$ and $\mathbf{e}_k$ are as defined in (7.34). $\mathfrak{F}[\mathbf{H}](s)$ is a bounded linear transformation from $\mathcal{H}_2(W)$ to $\mathcal{H}_2$ [32]. A state-space representation for $\mathfrak{F}[\mathbf{H}](s)$ is given by

$$\begin{aligned}\mathfrak{F}[\mathbf{H}](s) &= \mathfrak{C}_{\mathfrak{F}}(s\mathbf{I}-\mathcal{A}_{\mathfrak{F}})^{-1}\mathfrak{B}_{\mathfrak{F}} \qquad (7.38)\\ &= \underbrace{\begin{bmatrix}\mathbf{C} & \mathbf{D}\mathbf{C}_w\end{bmatrix}}_{\mathfrak{C}_{\mathfrak{F}}}\left(s\mathbf{I}-\underbrace{\begin{bmatrix}\mathbf{A} & \mathbf{B}\mathbf{C}_w\\ 0 & \mathbf{A}_w\end{bmatrix}}_{\mathcal{A}_{\mathfrak{F}}}\right)^{-1}\underbrace{\begin{bmatrix}\mathbf{Z}\mathbf{C}_w^T+\mathbf{B}\mathbf{D}_w\mathbf{D}_w^T\\ \mathbf{P}_w\mathbf{C}_w^T+\mathbf{B}_w\mathbf{D}_w^T\end{bmatrix}}_{\mathfrak{B}_{\mathfrak{F}}},\end{aligned}$$

where $\mathbf{P}_w$ and $\mathbf{Z}$ solve, respectively,

$$\mathbf{A}_w\mathbf{P}_w + \mathbf{P}_w\mathbf{A}_w^T + \mathbf{B}_w\mathbf{B}_w^T = 0 \quad \text{and} \tag{7.39}$$

$$\mathbf{A}\mathbf{Z} + \mathbf{Z}\mathbf{A}_w^T + \mathbf{B}(\mathbf{C}_w\mathbf{P}_w + \mathbf{D}_w\mathbf{B}_w^T) = 0. \tag{7.40}$$

For $\mathbf{H}$, $\widehat{\mathbf{H}} \in \mathcal{H}_2(W)$, denote the impulse responses corresponding to $\mathfrak{F}[\mathbf{H}](s)$ and $\mathfrak{F}[\widehat{\mathbf{H}}](s)$, respectively, by $\mathbf{F}(t)$ and $\widehat{\mathbf{F}}(t)$, so that $\mathfrak{F}[\mathbf{H}] = \mathcal{L}\{\mathbf{F}\}$ and $\mathfrak{F}[\widehat{\mathbf{H}}] = \mathcal{L}\{\widehat{\mathbf{F}}\}$, where $\mathcal{L}\{\cdot\}$ denotes the Laplace transform.

Theorem 7.8. *Given the input weighting $\mathbf{W}(s)$, let $\widehat{\mathbf{H}} \in \mathcal{H}_2(W)$ be the best order-r rational approximation to $\mathbf{H}$ in the weighted $\mathcal{H}_2$-norm. Suppose that $\widehat{\mathbf{H}}$ has the form*

$$\widehat{\mathbf{H}}(s) = \widehat{\mathbf{C}}\left(s\mathbf{I}-\widehat{\mathbf{A}}\right)^{-1}\widehat{\mathbf{B}} + \widehat{\mathbf{D}} = \sum_{k=1}^{n_r}\frac{\boldsymbol{\ell}_k\,\mathbf{r}_k^T}{s-\lambda_k} + \widehat{\mathbf{D}}, \tag{7.41}$$

where $\lambda_1, \ldots, \lambda_r$ are assumed to be distinct. Then,

$$\mathfrak{F}[\mathbf{H}](-\lambda_k)\mathbf{r}_k = \mathfrak{F}[\widehat{\mathbf{H}}](-\lambda_k)\mathbf{r}_k, \tag{7.42a}$$

$$\boldsymbol{\ell}_k^T\mathfrak{F}[\mathbf{H}](-\lambda_k) = \boldsymbol{\ell}_k^T\mathfrak{F}[\widehat{\mathbf{H}}](-\lambda_k), \tag{7.42b}$$

$$\boldsymbol{\ell}_k^T\mathfrak{F}'[\mathbf{H}](-\lambda_k)\mathbf{r}_k = \boldsymbol{\ell}_k^T\mathfrak{F}'[\widehat{\mathbf{H}}](-\lambda_k)\mathbf{r}_k, \text{ and} \tag{7.42c}$$

$$\mathbf{F}(0)\mathbf{n} = \widehat{\mathbf{F}}(0)\mathbf{n} \tag{7.42d}$$

for $k = 1, 2, \ldots, r$ and for all $\mathbf{n} \in \mathsf{Ker}(\mathbf{D}_w^T)$, where $\mathfrak{F}'[\,\cdot\,](s) = \frac{d}{ds}\mathfrak{F}[\,\cdot\,](s)$.

Bitangential Hermite interpolation once again appears as a necessary condition for optimality. However, unlike the unweighted $\mathcal{H}_2$ case, the interpolation conditions need to be satisfied by the maps $\mathfrak{F}[\mathbf{H}](s)$ and $\mathfrak{F}[\widehat{\mathbf{H}}](s)$ as opposed to $\mathbf{H}(s)$ and $\widehat{\mathbf{H}}(s)$. For $\mathbf{W}(s) = \mathbf{I}$, (7.42a)–(7.42c) simplify to (7.33a)–(7.33c), and (7.42d) is automatically satisfied since $\mathsf{Ker}(\mathbf{D}_w^T) = \{0\}$ when $\mathbf{W}(s) = \mathbf{I}$.

Guided by how IRKA is employed to satisfy the $\mathcal{H}_2$ optimality conditions (7.33a)–(7.33c), one might consider simply applying IRKA to the state-space representation of $\mathfrak{F}[\mathbf{H}](s)$ given in (7.38) to satisfy the weighted interpolation conditions (7.42a)–(7.42c). Unfortunately, this solves a different problem: if IRKA is applied to $\mathfrak{F}[\mathbf{H}](s)$ directly, then one obtains a reduced model $\widehat{\mathbf{H}}(s)$ that interpolates $\mathfrak{F}[\mathbf{H}](s)$. That is, instead of (7.42a), one obtains instead $\mathfrak{F}[\mathbf{H}](-\lambda_k)\mathbf{r}_k = \widehat{\mathbf{H}}(-\lambda_k)\mathbf{r}_k$, which is clearly not appropriate. The need to preserve the structure in the maps $\mathfrak{F}[\mathbf{H}](s)$ and $\mathfrak{F}[\widehat{\mathbf{H}}](s)$ while satisfying interpolatory conditions makes the development of an IRKA-like algorithm for the weighted $\mathcal{H}_2$ problem quite nontrivial. Breiten et al. in [32] propose an algorithm, called nearly optimal weighted interpolation (NOWI), that *nearly satisfies* the interpolatory optimality conditions (7.42a)–(7.42c) while preserving the structure in $\mathfrak{F}[\mathbf{H}](s)$ and $\mathfrak{F}[\widehat{\mathbf{H}}](s)$. The deviation from exact interpolation is quantified explicitly. Even though NOWI proves itself to be a very effective numerical algorithm in many circumstances, the development of an algorithm that satisfies (7.42a)–(7.42c) exactly remains an important future goal.

7.4.4 ▪ Descent algorithms for $\mathcal{H}_2$ model reduction

At its core, IRKA is a fixed-point iteration. Excepting the special case of symmetric state-space systems,[25] where convergence is guaranteed, convergence of IRKA for general systems is not guaranteed (see [46, 57]), although superior performance is commonly observed. More significantly, IRKA is not a descent algorithm; that is, the $\mathcal{H}_2$-error might fluctuate during intermediate steps, and premature termination of the algorithm could result (at least in principle) in a worse approximation than what was provided for initialization. To address these issues, Beattie and Gugercin [20] developed an $\mathcal{H}_2$ descent algorithm that reduces the $\mathcal{H}_2$-error at each step of the iteration and assures global convergence to a local minimum.

The key to their approach is the following representation of the $\mathcal{H}_2$-error norm for MIMO systems [20].

Theorem 7.9. *Given a full model* $\mathrm{H}(s)$*, let* $\widehat{\mathbf{H}}(s)$ *have the form in* (7.32)*, i.e.,*

$$\widehat{\mathbf{H}}(s) = \sum_{i=1}^{r} \frac{\boldsymbol{\ell}_i \mathbf{r}_i{}^T}{s - \lambda_i}.$$

Then, the $\mathcal{H}_2$*-norm of the error system is given by*

$$\left\|\mathbf{H} - \widehat{\mathbf{H}}\right\|_{\mathcal{H}_2}^2 = \|\mathbf{H}\|_{\mathcal{H}_2}^2 - 2\sum_{k=1}^{r} \boldsymbol{\ell}_k^T \mathbf{H}(-\lambda_k)\mathbf{r}_k + \sum_{k,j=1}^{r} \frac{\boldsymbol{\ell}_k^T \boldsymbol{\ell}_j \mathbf{r}_j^T \mathbf{r}_k}{-\lambda_k - \lambda_j}. \tag{7.43}$$

For SISO systems, Krajewski et al. [66] developed and proved a similar expression, which was later rediscovered in [9, 53, 55].

[25] $\mathbf{E} = \mathbf{E}^T$ is positive definite, $\mathbf{A} = \mathbf{A}^T$ is negative definite, and $\mathbf{B} = \mathbf{C}^T$.

If one considers $\widehat{\mathbf{H}}(s)=\sum_{i=1}^{r}\frac{\boldsymbol{\ell}_i\mathbf{r}_i^T}{s-\lambda_i}$ in the pole-residue form, the variables defining the reduced model are the residue directions $\boldsymbol{\ell}_i$, $\mathbf{r}_i$ and the poles λ_i for $i=1,\ldots,r$. The formula (7.43) expresses the error in terms of these variables. Thus, one can compute the gradient and Hessian of the error with respect to unknowns and construct globally convergent descent (optimization) algorithms. Gradient and Hessian expressions are derived in [20]. For brevity of presentation, we include only the gradient expressions here.

Theorem 7.10. *Given the full model* $\mathbf{H}(s)$ *and the reduced model* $\widehat{\mathbf{H}}(s)$ *as in* (7.32), *define*

$$\mathcal{J} \stackrel{\text{def}}{=} \left\|\mathbf{H}-\widehat{\mathbf{H}}\right\|_{\mathcal{H}_2}^2.$$

Then, for $i=1,\ldots,r$,

$$\frac{\partial \mathcal{J}}{\partial \lambda_i} = -2\boldsymbol{\ell}_i^T\left(\widehat{\mathbf{H}}'(-\lambda_i)-\mathbf{H}'(-\lambda_i)\right)\mathbf{r}_i. \tag{7.44}$$

Moreover, the gradient of $\mathcal{J}$ *with respect to residue directions listed as*

$$\{\mathbf{r},\boldsymbol{\ell}\}=[\mathbf{r}_1^T,\boldsymbol{\ell}_1^T,\mathbf{r}_2^T,\boldsymbol{\ell}_2^T,\ldots,\mathbf{r}_r^T,\boldsymbol{\ell}_r^T]^T$$

is given by $\nabla_{\{\mathbf{r},\boldsymbol{\ell}\}}\mathcal{J}$, *a vector of length* $r(m+q)$, *partitioned into* r *vectors of length* $m+q$ *as*

$$\left(\nabla_{\{\mathbf{r},\boldsymbol{\ell}\}}\mathcal{J}\right)_k = \begin{pmatrix} 2\left(\boldsymbol{\ell}_k^T\widehat{\mathbf{H}}(-\lambda_k)-\boldsymbol{\ell}_k^T\mathbf{H}(-\lambda_k)\right)^T \\ 2\left(\widehat{\mathbf{H}}(-\lambda_k)\mathbf{r}_k-\mathbf{H}(-\lambda_k)\mathbf{r}_k\right) \end{pmatrix} \tag{7.45}$$

for $k=1,2,\ldots,r$.

One may observe that setting the gradient expression in (7.44) and (7.45) to zero leads immediately to the interpolatory optimality conditions (7.33). Having gradient and Hessian expressions at hand, one may then develop a globally convergent descent algorithm for $\mathcal{H}_2$ optimal reduction, as done in [20], where the optimization algorithm is put in a trust-region framework. Unlike in IRKA, the intermediate reduced models are not interpolatory. However, upon convergence, they satisfy the interpolatory optimality conditions.

In a recent paper, for SISO systems $H(s)=\mathbf{c}^T(s\mathbf{E}-\mathbf{A})^{-1}\mathbf{b}$ where $\mathbf{b},\mathbf{c}\in\mathbb{R}^n$ are length-n vectors, Panzer et al. [76] apply a descent-type algorithm successively. Instead of designing a degree-r rational function directly, [76] first constructs a SISO degree-two rational function $\widehat{H}(s)=\hat{\mathbf{c}}^T(s\widehat{\mathbf{E}}-\widehat{\mathbf{A}})^{-1}\hat{\mathbf{b}}$, where $\widehat{\mathbf{A}},\widehat{\mathbf{E}}\in\mathbb{R}^{2\times 2}$ and $\mathbf{b}_r,\mathbf{c}_r^T\in\mathbb{R}^2$, by a descent method where only Lagrange optimality conditions are enforced (without Hermite conditions). Then, the error transfer function is decomposed into the multiplicative form

$$H-\widehat{H}(s)=\underbrace{\left(\mathbf{c}^T(s\mathbf{E}-\mathbf{A})^{-1}\mathbf{b}_\perp\right)}_{H_\perp(s)}\left(\hat{\mathbf{c}}^T(s\widehat{\mathbf{E}}-\widehat{\mathbf{A}})^{-1}\hat{\mathbf{b}}\right),$$

where $\mathbf{b}_\perp=\mathbf{b}-\mathbf{E}\mathbf{V}(\mathbf{W}^T\mathbf{E}\mathbf{V})^{-1}\hat{\mathbf{b}}$, and the method proceeds by constructing another degree-two approximation to $H_\perp(s)$ in a descent framework, once more only enforcing the Lagrange optimality conditions. At the end, all the intermediate degree-two approximants are put together in a special way to form the final reduced model of degree r. For details, we refer the reader to [76]. The final reduced model will not generally satisfy the full set of interpolatory $\mathcal{H}_2$ optimality conditions—only the Lagrange conditions are satisfied. Moreover, the incremental approach means optimization over a smaller set; thus, for a given r, optimization directly over a degree-r rational function using the gradient and Hessian expressions in [20] will lead to a smaller model reduction error than an incremental search. However, since [76] performs the optimization over a very small number of variables in each step, this approach can provide some numerical advantages.

Druskin et al. in [39] and [40] suggest alternative greedy-type algorithms for interpolatory model reduction. Instead of constructing r interpolation points (and directions at every step), as done in IRKA or in the descent framework of [20], [39] and [40] start instead with an interpolation point and corresponding tangent directions. Then, a greedy search on the residual determines the next set of interpolation data. Since the greedy search is not done in a descent setting, this is not a descent method and at the end optimality conditions will not typically be satisfied. Nonetheless, the final reduced model is still an interpolatory method. Even though the resulting reduced models will not generally be as accurate as those obtained by IRKA, the methods of [39] and [40] provide satisfactory approximants at relatively low cost.

Descent-type algorithms have been extended to the weighted $\mathcal{H}_2$-norm minimization as well; for details see [31, 77, 86].

7.5 ▪ Model reduction with generalized coprime realizations

So far, we have assumed that the original transfer function has a generic first-order state-space representation: $\mathbf{H}(s) = \mathbf{C}(s\mathbf{E}-\mathbf{A})^{-1}\mathbf{B}$. This representation is quite general, and a wide range of linear dynamical systems can be converted to this form, at least in principle. However, problem formulations often lead to somewhat different structures that reflect the underlying physics or other important system features. One may wish to retain such structural features, but conversion to standard first-order representations often obscures them and may lead to "unphysical" reduced models. Neutral and delay differential equations present another interesting class of dynamical systems that are generically of infinite order, so they do not accept a standard first-order representation using a finite-dimensional state space. In this section, we follow the discussion of Beattie and Gugercin [19] and show how interpolatory methods can be used to preserve relevant system structure in reduced models, often entirely avoiding the need to convert the system to an equivalent first-order state-space representation.

7.5.1 ▪ Polynomial and delay systems

A common example of a situation where conversion to the standard state-space form is possible but may not be prudent is the case of constant-coefficient ODEs of order two or more, with dynamics given by

$$\begin{aligned} \mathbf{A}_0\frac{d^{\nu}\mathbf{x}}{dt^{\nu}} + \mathbf{A}_1\frac{d^{\nu-1}\mathbf{x}}{dt^{\nu-1}} + \cdots + \mathbf{A}_{\nu}\mathbf{x}(t) &= \mathbf{B}\mathbf{u}(t), \\ \mathbf{y}(t) = \mathbf{C}_1\frac{d^{\nu-1}\mathbf{x}}{dt^{\nu-1}} + \mathbf{C}_2\frac{d^{\nu-2}\mathbf{x}}{dt^{\nu-2}} + \cdots + \mathbf{C}_{\nu}\mathbf{x}(t)&, \end{aligned} \tag{7.46}$$

where $\mathbf{A}_i \in \mathbb{R}^{n\times n}$ for $i = 0,\ldots,\nu$, $\mathbf{B} \in \mathbb{R}^{n\times m}$, and $\mathbf{C}_i \in \mathbb{R}^{q\times n}$ for $i = 1,\ldots,\nu$. By defining a state vector $\mathbf{q} = [\mathbf{x}^T, \dot{\mathbf{x}}^T, \ddot{\mathbf{x}}^T, \ldots, (\mathbf{x}^{(\nu-1)})^T]^T$, one may easily convert (7.46) into the equivalent first-order form (7.1). This has two major disadvantages:

1. By forming the vector $\mathbf{q}(t)$, the physical meaning of the state vector $\mathbf{x}(t)$ is lost in the model reduction state since the model reduction process will mix physical quantities such as displacement and velocity; the reduced state loses its physical significance.

2. The dimension of the aggregate state $\mathbf{q}(t)$ is νn. As a consequence, conversion to first-order form has made the model reduction problem numerically much harder. For example, if reduction is approached via interpolation, we now need to solve linear systems of size $(\nu n) \times (\nu n)$.

Therefore, it is desirable to perform model reduction in the original state space associated with the original representation (7.46); we wish to preserve the structure of (7.46) in the reduced model and produce a reduced model of the form

$$\begin{aligned} \widehat{\mathbf{A}}_0 \frac{d^\nu \hat{\mathbf{x}}}{dt^\nu} + \widehat{\mathbf{A}}_1 \frac{d^{\nu-1}\hat{\mathbf{x}}}{dt^{\nu-1}} + \cdots + \widehat{\mathbf{A}}_\nu \hat{\mathbf{x}}(t) &= \widehat{\mathbf{B}}\mathbf{u}(t), \\ \hat{\mathbf{y}}(t) = \widehat{\mathbf{C}}_1 \frac{d^{\nu-1}\hat{\mathbf{x}}}{dt^{\nu-1}} + \widehat{\mathbf{C}}_2 \frac{d^{\nu-2}\hat{\mathbf{x}}}{dt^{\nu-2}} &+ \cdots + \widehat{\mathbf{C}}_\nu \hat{\mathbf{x}}(t), \end{aligned} \tag{7.47}$$

where $\widehat{\mathbf{A}}_i \in \mathbb{R}^{r\times r}$ for $i = 0,\ldots,\nu$, $\widehat{\mathbf{B}} \in \mathbb{R}^{r\times m}$, and $\widehat{\mathbf{C}}_i \in \mathbb{R}^{q\times r}$ for $i = 1,\ldots,\nu$.

Another example where the structure of a dynamical system presents an obstacle to reduction using methods that depend on availability of a standard first-order form is the class of delay differential equations. Consider a linear dynamical system with an internal delay, given in state-space form as

$$\mathbf{E}\dot{\mathbf{x}}(t) = \mathbf{A}_0\mathbf{x}(t) + \mathbf{A}_1\mathbf{x}(t-\tau) + \mathbf{B}\mathbf{u}(t), \quad \mathbf{y}(t) = \mathbf{C}\mathbf{x}(t), \tag{7.48}$$

with $\tau > 0$ and $\mathbf{E}, \mathbf{A}_0, \mathbf{A}_1 \in \mathbb{R}^{n\times n}$, $\mathbf{B} \in \mathbb{R}^{n\times m}$, and $\mathbf{C} \in \mathbb{R}^{q\times n}$. The system in (7.48) is not associated with a rational transfer function due to the delay term; it is intrinsically of infinite order. Preserving the delay structure in the reduced model is crucial for accurate representation, so we seek a reduced model of the form

$$\widehat{\mathbf{E}}\dot{\hat{\mathbf{x}}}(t) = \widehat{\mathbf{A}}_0\hat{\mathbf{x}}(t) + \widehat{\mathbf{A}}_1\hat{\mathbf{x}}(t-\tau) + \widehat{\mathbf{B}}\mathbf{u}(t), \quad \hat{\mathbf{y}}(t) = \widehat{\mathbf{C}}\hat{\mathbf{x}}(t), \tag{7.49}$$

with $\tau > 0$ and $\widehat{\mathbf{E}}, \widehat{\mathbf{A}}_0, \widehat{\mathbf{A}}_1 \in \mathbb{R}^{r\times r}$, $\widehat{\mathbf{B}} \in \mathbb{R}^{r\times m}$, and $\mathbf{C} \in \mathbb{R}^{q\times r}$. We want to perform this reduction step without needing to approximate the delay term with an additional rational approximation.

7.5.2 ▪ Generalized coprime representations

Examples such as these lead us to consider transfer functions with the following *generalized coprime representation*:

$$\mathcal{H}(s) = \mathcal{C}(s)\mathcal{K}(s)^{-1}\mathcal{B}(s) + \mathcal{D}, \tag{7.50}$$

where $\mathcal{D}$ is a constant $q \times m$ matrix, both $\mathcal{C}(s) \in \mathbb{C}^{q\times n}$ and $\mathcal{B}(s) \in \mathbb{C}^{n\times m}$ are analytic in the right half-plane, and $\mathcal{K}(s) \in \mathbb{C}^{n\times n}$ is analytic and full rank throughout the right half-plane. Note that both (7.46) and (7.48) fit this framework: for the polynomial system (7.47), we obtain

$$\mathcal{H}(s)=\underbrace{\left(\sum_{i=1}^{\nu} s^{\nu-i}\mathbf{C}_i\right)}_{\mathcal{C}(s)}\underbrace{\left(\sum_{i=0}^{\nu} s^{\nu-i}\mathbf{A}_i\right)^{-1}}_{\mathcal{K}(s)^{-1}}\underbrace{\mathbf{B}}_{\mathcal{B}(s)}, \tag{7.51}$$

and for the delay system (7.48), we have

$$\mathcal{H}(s)=\underbrace{\mathbf{C}}_{\mathcal{C}(s)}\underbrace{\left(s\,\mathbf{E}-\mathbf{A}_0-e^{-\tau s}\mathbf{A}_1\right)^{-1}}_{\mathcal{K}(s)^{-1}}\underbrace{\mathbf{B}}_{\mathcal{B}(s)}. \tag{7.52}$$

Our model reduction goals are the same: construct a reduced (generalized coprime) transfer function that tangentially interpolates the original one. To this end, we choose two model reduction bases $\mathbf{V}\in\mathbb{R}^{n\times r}$ and $\mathbf{W}\in\mathbb{R}^{n\times r}$, as before. This leads to a reduced transfer function

$$\widehat{\mathcal{H}}(s)=\widehat{\mathcal{C}}(s)\widehat{\mathcal{K}}(s)^{-1}\widehat{\mathcal{B}}(s)+\widehat{\mathcal{D}}, \tag{7.53}$$

where $\widehat{\mathcal{C}}(s)\in\mathbb{C}^{q\times r}$, $\widehat{\mathcal{B}}(s)\in\mathbb{C}^{r\times m}$, $\widehat{\mathcal{D}}\in\mathbb{C}^{q\times m}$, and $\widehat{\mathcal{K}}(s)\in\mathbb{C}^{r\times r}$ are obtained by Petrov–Galerkin projection:

$$\begin{aligned}\widehat{\mathcal{K}}(s)&=\mathbf{W}^T\mathcal{K}(s)\mathbf{V}, \quad &\widehat{\mathcal{B}}(s)&=\mathbf{W}^T\mathcal{B}(s),\\ \widehat{\mathcal{C}}(s)&=\mathcal{C}(s)\mathbf{V}, &\text{and }\ \widehat{\mathcal{D}}&=\mathcal{D}.\end{aligned} \tag{7.54}$$

7.5.3 ▪ Interpolatory projections for generalized coprime representations

Interpolatory projections for generalized coprime representations of transfer functions are introduced in [19]. We follow the notation in [19] and use $\mathscr{D}_\sigma^k f$ to denote the kth derivative of the univariate function $f(s)$ evaluated at $s=\sigma$, with the usual convention that $\mathscr{D}_\sigma^0 f=f(\sigma)$.

Theorem 7.11. *Given the original model transfer function*

$$\mathcal{H}(s)=\mathcal{C}(s)\mathcal{K}(s)^{-1}\mathcal{B}(s)+\mathcal{D},$$

let $\widehat{\mathcal{H}}(s)$ denote the reduced transfer function in (7.53) *obtained by projection as in* (7.54) *using the model reduction bases $\mathbf{V}$ and $\mathbf{W}$. For the interpolation points $\sigma,\mu\in\mathbb{C}$, suppose that $\mathcal{B}(s)$, $\mathcal{C}(s)$, and $\mathcal{K}(s)$ are analytic at $\sigma\in\mathbb{C}$ and $\mu\in\mathbb{C}$. Also let $\mathcal{K}(\sigma)$ and $\mathcal{K}(\mu)$ have full rank. Also, let $\mathbf{r}\in\mathbb{C}^m$ and $\boldsymbol{\ell}\in\mathbb{C}^q$ be nontrivial tangential direction vectors. The following implications hold:*

(a) *If*

$$\mathscr{D}_\sigma^i[\mathcal{K}(s)^{-1}\mathcal{B}(s)]\mathbf{r}\in\mathsf{Ran}(\mathbf{V}) \quad \textit{for } i=0,\ldots,N, \tag{7.55}$$

then

$$\mathcal{H}^{(k)}(\sigma)\mathbf{r}=\widehat{\mathcal{H}}^{(k)}(\sigma)\mathbf{r} \quad \textit{for } k=0,\ldots,N. \tag{7.56}$$

(b) *If*

$$\left(\ell^T \mathscr{D}_\mu^j[\mathcal{C}(s)\mathcal{K}(s)^{-1}]\right)^T \in \mathsf{Ran}(\mathcal{W}) \quad \textit{for } j=0,\ldots,M, \tag{7.57}$$

then

$$\ell^T \mathcal{H}^{(k)}(\mu) = \ell^T \widehat{\mathcal{H}}^{(k)}(\mu) \quad \textit{for } k=0,\ldots,M. \tag{7.58}$$

(c) *If both* (7.55) *and* (7.57) *hold and if* $\sigma=\mu$, *then*

$$\ell^T \mathcal{H}^{(k)}(\sigma)\mathbf{r} = \ell^T \widehat{\mathcal{H}}^{(k)}(\sigma)\mathbf{r} \quad \textit{for } k=0,\ldots,M+N-1, \tag{7.59}$$

provided $\widehat{\mathcal{K}}(\sigma)=\mathcal{W}^T\mathcal{K}(\sigma)\mathcal{V}$ *and* $\widehat{\mathcal{K}}(\mu)=\mathcal{W}^T\mathcal{K}(\mu)\mathcal{V}$ *both have full rank.*

Theorem 7.11 proves the power and flexibility of the interpolatory framework for model reduction. The earlier interpolation result, Theorem 7.3, directly extends to this much more general class of transfer function, requiring very similar subspace conditions. Moreover, this structure guarantees that the reduced transfer function will have a similar generalized coprime representation. Computational complexity is comparable; one need only solve $n\times n$ (often sparse) linear systems.

Recall the delay example appearing in (7.48). Assume that $r=2$ interpolation points, $\{\sigma_1,\sigma_2\}$, together with right directions $\{\mathbf{r}_1,\mathbf{r}_2\}$ and left directions $\{\ell_1,\ell_2\}$, are given. Based on Theorem 7.11, we construct $\mathcal{V}\in\mathbb{C}^{n\times 2}$ and $\mathcal{W}\in\mathbb{C}^{n\times 2}$ using

$$\mathcal{V}=\left[\left(\sigma_1\mathbf{E}-\mathbf{A}_0-e^{-\tau\sigma_1}\mathbf{A}_1\right)^{-1}\mathbf{B}\mathbf{r}_1 \quad \left(\sigma_2\mathbf{E}-\mathbf{A}_0-e^{-\tau\sigma_2}\mathbf{A}_1\right)^{-1}\mathbf{B}\mathbf{r}_2\right]$$

and

$$\mathcal{W}=\left[\left(\sigma_1\mathbf{E}-\mathbf{A}_0-e^{-\tau\sigma_1}\mathbf{A}_1\right)^{-T}\mathbf{C}^T\ell_1 \quad \left(\sigma_2\mathbf{E}-\mathbf{A}_0-e^{-\tau\sigma_2}\mathbf{A}_1\right)^{-T}\mathbf{C}^T\ell_2\right].$$

Note that structure is preserved; the reduced model of dimension two has the same internal delay structure,

$$\begin{aligned}\mathcal{W}^T\mathbf{E}\mathcal{V}\dot{\hat{\mathbf{x}}}(t) &= \mathcal{W}^T\mathbf{A}_0\mathcal{V}\hat{\mathbf{x}}(t)+\mathcal{W}^T\mathbf{A}_1\mathcal{V}\hat{\mathbf{x}}(t-\tau)+\mathcal{W}^T\mathbf{B}\,\mathbf{u}(t)\\ \mathbf{y}(t) &= \mathbf{C}\mathcal{V}\hat{\mathbf{x}}(t),\end{aligned} \tag{7.60}$$

with a correspondingly structured (reduced) transfer function

$$\widehat{\mathcal{H}}(s)=\mathbf{C}\mathcal{V}\left(s\,\mathcal{W}^T\mathbf{E}\mathcal{V}-\mathcal{W}^T\mathbf{A}_0\mathcal{V}-e^{-\tau s}\,\mathcal{W}^T\mathbf{A}_1\mathcal{V}\right)^{-1}\mathcal{W}^T\mathbf{B}.$$

Moreover, due to the interpolation-based construction of $\mathcal{V}$ and $\mathcal{W}$, the reduced transfer function is a Hermite bitangential interpolant:

$$\mathcal{H}(\sigma_i)\mathbf{r}_i=\widehat{\mathcal{H}}(\sigma_i)\mathbf{r}_i,\ \ell_i^T\mathcal{H}(\sigma_i)=\ell_i^T\widehat{\mathcal{H}}(\sigma_i),\ \text{and}\ \ell_i^T\mathcal{H}'(\sigma_i)\mathbf{r}_i=\ell_i^T\widehat{\mathcal{H}}'(\sigma_i)\mathbf{r}_i$$

for $i=1,2$. The reduced transfer function fully incorporates the delay structure and *exactly* interpolates the original transfer function. This would not be true if the delay term $e^{-\tau s}$ had been approximated via a rational approximation, such as Padé approximation, as is commonly done while responding to the urge to convert to standard first-order form.

Remark 7.12. The reduced system (7.60) is not strictly speaking a "*reduced-order*" system since "order" traditionally refers to the number of poles of a system, and in this sense, both $\widehat{\mathcal{H}}(s)$ and $\mathcal{H}(s)$ have *infinite* order. Nonetheless, (7.60) evolves in a state space of reduced dimension and so still earns the distinction of being a reduced system.

Remark 7.13. The construction of rational interpolants $\widehat{\mathcal{H}}(s) = \widehat{\mathcal{C}}(s)\widehat{\mathcal{K}}(s)^{-1}\widehat{\mathcal{B}}(s) + \widehat{\mathcal{D}}$ with $\widehat{\mathcal{D}} \neq \mathcal{D}$ can be achieved for generalized coprime representations similarly as described in Theorem 7.4 for standard first-order realizations. For details, we refer the reader to the original source [19].

7.6 ▪ Realization-independent optimal $\mathcal{H}_2$ approximation

We described IRKA in Section 7.4.2 and promoted it as an effective tool for constructing locally optimal rational $\mathcal{H}_2$ approximations at modest cost. One may observe, however, that the formulation of IRKA, as it appears in Algorithm 7.1, assumes that a first-order realization for $\mathbf{H}(s)$ is available: $\mathbf{H}(s) = \mathbf{C}(s\mathbf{E}-\mathbf{A})^{-1}\mathbf{B}$. As we found in Section 7.5, there are several important examples where the original transfer function $\mathbf{H}(s)$ is not naturally represented in this way. To address these situations, among others, Beattie and Gugercin in [21] removed the need for *any* particular realization and extended applicability of IRKA to any evaluable $\mathcal{H}_2$ transfer function. We focus on this extension of IRKA in the present section. Since $\mathbf{H}(s)$ is not required to have a first-order realization here, we will follow the notation of Section 7.5 and use $\mathcal{H}(s)$ to denote the original transfer function; however, we do not require knowledge of a generalized coprime representation, only that we can evaluate $\mathcal{H}(s)$ for any value of s. Note that the reduced model will still be a rational function of order r and will be delivered in a first-order form. Hence, we will continue to use $\widehat{\mathbf{H}}(s)$ to denote the final reduced transfer function.

There are two main observations behind the methodology offered in [21]. The first observation is based on the first-order $\mathcal{H}_2$ optimality conditions (7.33) in Theorem 7.7. Recall that Theorem 7.7 does not put any restrictions on $\mathcal{H}(s)$; the only assumption is that the approximant $\widehat{\mathbf{H}}(s)$ is a rational function; thus, the theorem and the bitangential Hermite optimality conditions apply equally if $\mathcal{H}(s)$ has the form, for example, $\mathcal{H}(s) = \mathbf{C}(s^2\mathbf{M} + s\mathbf{G} + \mathbf{K})^{-1}\mathbf{B}$. A second observation is related to how IRKA constructs the solution: for the current set of interpolation points and tangent directions, IRKA constructs a bitangential Hermite interpolant and updates the interpolation data. Thus, the key issue becomes, given a set of interpolation data, how we can construct a rational approximant $\widehat{\mathbf{H}}(s)$ that is a Hermite bitangential interpolant to $\mathcal{H}(s)$ (which may not be presented as a first-order state-space model). The Loewner interpolatory framework introduced by Mayo and Antoulas [72] (discussed in detail in Chapter 8) is the right tool.

For the balance of this section, we assume that both $\mathcal{H}(s)$ and its derivative, $\mathcal{H}'(s)$, are only accessible through the evaluation $s \mapsto (\mathcal{H}(s), \mathcal{H}'(s))$. No particular system realizations are assumed. Suppose we are given interpolation points $\{\sigma_1, \ldots, \sigma_r\}$ with the corresponding tangential directions $\{\mathbf{r}_1, \ldots, \mathbf{r}_r\}$ and $\{\boldsymbol{\ell}_1, \ldots, \boldsymbol{\ell}_r\}$. We want to construct a degree-r rational approximant $\widehat{\mathbf{H}}(s)$ that is a bitangential Hermite interpolant to $\mathcal{H}(s)$:

$$\mathcal{H}(\sigma_k)\mathbf{r}_k = \widehat{\mathbf{H}}(\sigma_k)\mathbf{r}_k, \tag{7.61a}$$

$$\mathbf{r}_k^T\mathcal{H}(\sigma_k) = \mathbf{r}_k^T\widehat{\mathbf{H}}(\sigma_k), \text{ and} \tag{7.61b}$$

$$\mathbf{r}_k^T\mathcal{H}'(\sigma_k)\mathbf{r}_k = \mathbf{r}_k^T\widehat{\mathbf{H}}'(\sigma_k)\mathbf{r}_k \tag{7.61c}$$

for $k = 1, 2, \ldots, r$. As seen in Chapter 8, the framework of [72] allows one to achieve this goal requiring only the evaluation $\mathcal{H}(s)$ and $\mathcal{H}'(s)$ at σ_k without any constraint on the structure of $\mathcal{H}(s)$: simply construct the matrices $\widehat{\mathbf{E}}$, $\widehat{\mathbf{A}}$, $\widehat{\mathbf{B}}$, and $\widehat{\mathbf{C}}$ using

$$\left(\widehat{\mathbf{E}}\right)_{i,j} := \begin{cases} -\dfrac{\boldsymbol{\ell}_i^T\left(\mathcal{H}(\sigma_i)-\mathcal{H}(\sigma_j)\right)\mathbf{r}_j}{\sigma_i-\sigma_j} & \text{if } i\neq j \\ -\boldsymbol{\ell}_i^T\mathcal{H}'(\sigma_i)\mathbf{r}_i & \text{if } i=j \end{cases}, \tag{7.62}$$

$$\left(\widehat{\mathbf{A}}\right)_{i,j} := \begin{cases} -\dfrac{\boldsymbol{\ell}_i^T\left(\sigma_i\mathcal{H}(\sigma_i)-\sigma_j\mathcal{H}(\sigma_j)\right)\mathbf{r}_j}{\sigma_i-\sigma_j} & \text{if } i\neq j \\ -\boldsymbol{\ell}_i^T\,[s\mathcal{H}(s)]'|_{s=\sigma_i}\,\mathbf{r}_i & \text{if } i=j \end{cases}, \tag{7.63}$$

and

$$\widehat{\mathbf{C}} = [\mathcal{H}(\sigma_1)\mathbf{r}_1,\ldots,\mathcal{H}(\sigma_r)\mathbf{r}_r],\ \ \widehat{\mathbf{B}} = \begin{bmatrix} \boldsymbol{\ell}_1^T\mathcal{H}(\sigma_1) \\ \vdots \\ \boldsymbol{\ell}_r^T\mathcal{H}(\sigma_r) \end{bmatrix}. \tag{7.64}$$

Then $\widehat{\mathbf{H}}(s) = \widehat{\mathbf{C}}(s\widehat{\mathbf{E}}-\widehat{\mathbf{A}})^{-1}\widehat{\mathbf{B}}$ satisfies (7.61). The matrices $\widehat{\mathbf{E}}$ and $\widehat{\mathbf{A}}$ are, respectively, a *Loewner matrix* and a *shifted Loewner matrix* (see Chapter 8 and [72]).

To use IRKA for $\mathcal{H}_2$ approximation without any structural constraints on $\mathcal{H}(s)$, one need only replace the projection-based construction of the intermediate Hermite interpolant with a Loewner-based construction. This is exactly what [21] introduces, leading to the realization-independent optimal $\mathcal{H}_2$ approximation methodology of Algorithm 7.2.

***ALGORITHM* 7.2. TF-IRKA: IRKA using transfer function evaluations.**

1. Make an initial r-fold shift selection $\{\sigma_1,\ldots,\sigma_r\}$ that is closed under conjugation (i.e., $\{\sigma_1,\ldots,\sigma_r\} \equiv \{\overline{\sigma_1},\ldots,\overline{\sigma_r}\}$ viewed as sets) and initial tangent directions $\mathbf{r}_1,\ldots,\mathbf{r}_r$ and $\boldsymbol{\ell}_1,\ldots,\boldsymbol{\ell}_r$, also closed under conjugation.
2. while (not converged)
   (a) Construct $\widehat{\mathbf{E}}$, $\widehat{\mathbf{A}}$, $\widehat{\mathbf{C}}$ and $\widehat{\mathbf{B}}$ as in (7.62)–(7.64).
   (b) Compute a pole-residue expansion of $\widehat{\mathbf{H}}(s)$:
   $$\widehat{\mathbf{H}}(s) = \widehat{\mathbf{C}}(s\widehat{\mathbf{E}}-\widehat{\mathbf{A}})^{-1}\widehat{\mathbf{B}} = \sum_{i=1}^{r}\frac{\boldsymbol{\ell}_i\mathbf{r}_i^T}{s-\lambda_i}.$$
   (c) $\sigma_i \longleftarrow -\lambda_i$, $\quad \mathbf{r}_i \longleftarrow \widehat{\mathbf{r}}_i$, $\quad$ and $\quad \boldsymbol{\ell}_i \longleftarrow \widehat{\boldsymbol{\ell}}_i$, $\quad$ for $\quad i = 1,\ldots,r$.
3. Construct $\widehat{\mathbf{E}}$, $\widehat{\mathbf{A}}$, $\widehat{\mathbf{C}}$ and $\widehat{\mathbf{B}}$ as in (7.62)–(7.64).

As for the original formulation of IRKA, upon convergence the rational approximant resulting from Algorithm 7.2 will satisfy the first-order necessary conditions (7.33) for $\mathcal{H}_2$ optimality.

Example 7.14 (An optimal rational approximation for a delay system). Consider the delay system given in (7.52), i.e.,

$$\mathcal{H}(s) = \mathbf{C}\big(s\,\mathbf{E} - \mathbf{A}_0 - e^{-\tau s}\,\mathbf{A}_1\big)^{-1}\mathbf{B}.$$

Following [19], we take $\mathbf{E} = \varkappa\mathbf{I} + \mathbf{T}$, $\mathbf{A}_0 = \frac{3}{\tau}(\mathbf{T} - \varkappa\mathbf{I})$, and $\mathbf{A}_1 = \frac{1}{\tau}(\mathbf{T} - \varkappa\mathbf{I})$ for any $\varkappa > 2$ and delay $\tau > 0$, where $\mathbf{T}$ is an $n \times n$ matrix with ones on the first superdiagonal, on the first subdiagonal, at the $(1,1)$ entry, and at the (n,n) entry. The remaining entries of $\mathbf{T}$ are zero. We take the internal delay as $\tau = 0.1$ and a SISO system with $n = 1000$, i.e., $\mathbf{E}, \mathbf{A}_0, \mathbf{A}_1 \in \mathbb{R}^{1000\times 1000}$, and $\mathbf{B}, \mathbf{C}^T \in \mathbb{R}^{1000\times 1}$. Then, we use TF-IRKA as illustrated in Algorithm 7.2 to construct a degree $r = 20$ (locally) $\mathcal{H}_2$-optimal rational approximation $\widehat{\mathbf{H}}(s)$. TF-IRKA requires evaluating $\mathcal{H}(s)$ and $\mathcal{H}'(s)$. For this delay model, $\mathcal{H}'(s)$ is given by

$$\mathcal{H}'(s) = -\mathbf{C}(s\mathbf{E} - \mathbf{A}_1 - e^{-\tau s}\mathbf{A}_2)^{-1}(\mathbf{E} + \tau e^{-\tau s}\mathbf{A}_2)(s\mathbf{E} - \mathbf{A}_1 - e^{-\tau s}\mathbf{A}_2)^{-1}\mathbf{B}.$$

Another approach to obtaining a rational approximation for such a delay system is to replace the exponent $e^{-\tau s}$ with a rational approximation and then reduce the resulting rational large-scale model with standard techniques. Here we will use the second-order Padé approximation toward this goal where we replace $e^{-\tau s}$ by $\frac{12 - 6\tau s + \tau^2 s^2}{12 + 6\tau s + \tau^2 s^2}$, obtaining the large-scale approximate rational transfer function

$$\mathcal{H}^{[P_2]}(s) = (12\mathbf{C} + s6\tau\mathbf{C} + s^2\tau^2\mathbf{C})(\mathbf{N}s^3 + \widetilde{\mathbf{M}}s^2 + \widetilde{\mathbf{G}}s + \widetilde{\mathbf{K}})^{-1}\mathbf{B}, \tag{7.65}$$

where $\mathbf{N} = \tau^2\mathbf{E}$, $\widetilde{\mathbf{M}} = 6\tau\mathbf{E} - \tau^2(\mathbf{A}_0 + \mathbf{A}_1)$, $\widetilde{\mathbf{G}} = 12\mathbf{E} + 6\tau(-\mathbf{A}_0 + \mathbf{A}_1)$, and $\widetilde{\mathbf{K}} = -12(\mathbf{A}_0 + \mathbf{A}_1)$. We use the notation $\mathcal{H}^{[P_2]}(s)$ to denote the resulting large-scale rational approximation due to the second-order Padé approximation. We note that the resulting approximation has a term s^3 and will result in an $N = 3000$ first-order model. Once an interpolatory model reduction technique is applied to $\mathcal{H}^{[P_2]}(s)$, the reduced model will be an exact interpolant to $\mathcal{H}^{[P_2]}(s)$, but not to $\mathcal{H}(s)$. This contrasts starkly to TF-IRKA, where the resulting rational approximation exactly interpolates the original delay model $\mathcal{H}(s)$. In Figure 7.1 below, we show the amplitude Bode plots of $\mathcal{H}(s)$ (denoted by "Full"), order $r = 20$ TF-IRKA approximant (denoted by "TF-IRKA"), and order $N = 3000$ Padé model (denoted by "Pade"). The figure clearly illustrates that the locally optimal $\mathcal{H}_2$ approximation due to TF-IRKA almost exactly replicates the original model. On the other hand, even the order $N = 3000$ Padé model $\mathcal{H}^{[P_2]}(s)$ is a very poor approximation. ∎

7.7 ▪ Interpolatory model reduction of parametric systems

All dynamical systems considered here are *linear time-invariant* systems; the system properties are presumed to be *constant*, at least with respect to time. Very often system properties will depend on external parameters and system dynamics will vary as parameter values change. Parameters enter naturally into the models in various ways, representing changes in boundary conditions, material properties, system geometry, etc. Producing a new reduced model for every new set of parameter values could be very costly, so a natural goal is to generate parametrized reduced models that provide high-fidelity approximations throughout a wide range of parameter values. This is usually referred to as *parametric model reduction (PMOR)*. It has found immediate applications in inverse problems [37, 41, 49, 69, 87], optimization [7, 8, 11, 94, 95], and

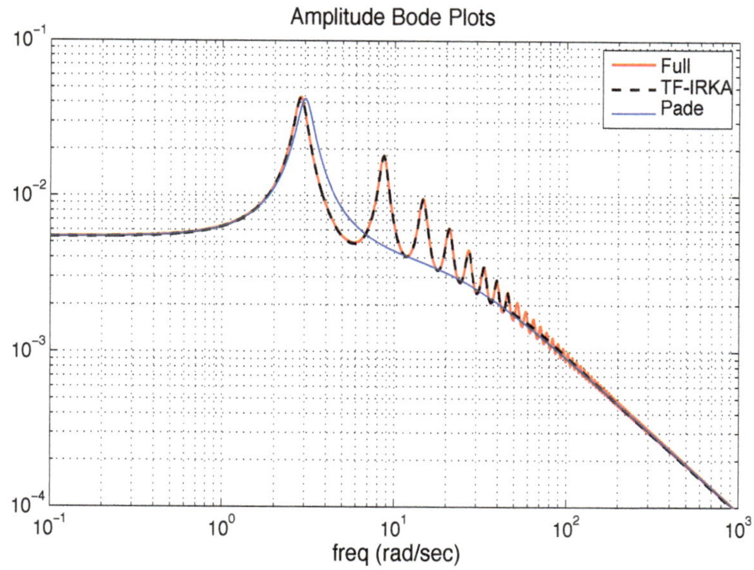

Figure 7.1. *Amplitude Bode plots for $\mathcal{H}(s)$ ("Full"), order $r = 20$ TF-IRKA approximant, and order $N = 3000$ Padé model.*

design and control [5, 15, 36, 45, 62, 70, 71]. There are various approaches to PMOR methods; see, e.g., [13, 33, 34, 60, 62, 75, 78, 79, 85] and the references therein. In this section, we focus on interpolatory methods. For a recent, detailed survey on PMOR, we refer the reader to [25].

7.7.1 ▪ Parametric structure

We consider MIMO transfer functions that are parametrized with p parameters $\mathsf{p} = [\mathsf{p}_1, \ldots, \mathsf{p}_p]$:

$$\mathcal{H}(s,\mathsf{p}) = \mathcal{C}(s,\mathsf{p})\mathcal{K}(s,\mathsf{p})^{-1}\mathcal{B}(s,\mathsf{p}), \tag{7.66}$$

with $\mathcal{K}(s,\mathsf{p}) \in \mathbb{C}^{n\times n}$, $\mathcal{B}(s,\mathsf{p}) \in \mathbb{C}^{n\times m}$, and $\mathcal{C}(s,\mathsf{p}) \in \mathbb{C}^{q\times n}$. The standard case of parametric linear dynamical systems of the form

$$\mathbf{E}(\mathsf{p})\,\dot{\mathbf{x}}(t;\mathsf{p}) = \mathbf{A}(\mathsf{p})\,\mathbf{x}(t;\mathsf{p}) + \mathbf{B}(\mathsf{p})\,\mathbf{u}(t), \quad \mathbf{y}(t;\mathsf{p}) = \mathbf{C}(\mathsf{p})\,\mathbf{x}(t;\mathsf{p}) \tag{7.67}$$

then becomes a special case of the more general form (7.66) we consider here, with $\mathcal{K}(s,\mathsf{p}) = s\mathbf{E}(\mathsf{p}) - \mathbf{A}(\mathsf{p})$, $\mathcal{B}(s,\mathsf{p}) = \mathbf{B}(\mathsf{p})$, and $\mathcal{C}(s,\mathsf{p}) = \mathbf{C}(\mathsf{p})$.

Even though the theoretical discussion applies to general parametric dependency, we assume an affine parametric form:

$$\begin{aligned}
\mathcal{K}(s,\mathsf{p}) &= \mathcal{K}^{[0]}(s) + k_1(\mathsf{p})\mathcal{K}^{[1]}(s) + \cdots + k_\nu(\mathsf{p})\mathcal{K}^{[\nu]}(s),\\
\mathcal{B}(s,\mathsf{p}) &= \mathcal{B}^{[0]}(s) + b_1(\mathsf{p})\mathcal{B}^{[1]}(s) + \cdots + b_\nu(\mathsf{p})\mathcal{B}^{[\nu]}(s),\\
\mathcal{C}(s,\mathsf{p}) &= \mathcal{C}^{[0]}(s) + c_1(\mathsf{p})\mathcal{C}^{[1]}(s) + \cdots + c_\nu(\mathsf{p})\mathcal{C}^{[\nu]}(s),
\end{aligned} \tag{7.68}$$

where $\{k_i(\mathsf{p})\}$,$\{b_i(\mathsf{p})\}$, and $\{c_i(\mathsf{p})\}$ for $i = 1, \ldots, \nu$ are scalar-valued nonlinear (or linear) parameter functions. Even though we have linear dynamics with respect to the state variable, we allow nonlinear parametric dependency in the state-space representation.

Our reduction framework remains the same: use a Petrov–Galerkin projection to construct, in this case, a reduced parametric model. Thus, we will pick two model reduction bases, $\mathcal{V} \in \mathbb{C}^{n \times r}$ and $\mathcal{W} \in \mathbb{C}^{n \times r}$, and obtain the reduced parametric model

$$\widehat{\mathcal{H}}(s,\mathsf{p}) = \widehat{\mathcal{C}}(s,\mathsf{p})\widehat{\mathcal{K}}(s,\mathsf{p})^{-1}\widehat{\mathcal{B}}(s,\mathsf{p}), \tag{7.69}$$

where we use a Petrov–Galerkin projection to obtain the reduced quantities $\widehat{\mathcal{C}}(s,\mathsf{p}) \in \mathbb{C}^{q \times r}$, $\widehat{\mathcal{B}}(s,\mathsf{p}) \in \mathbb{C}^{r \times m}$, and $\widehat{\mathcal{K}}(s) \in \mathbb{C}^{r \times r}$, i.e.,

$$\widehat{\mathcal{C}}(s,\mathsf{p}) = \mathcal{C}(s,\mathsf{p})\mathcal{V}, \quad \widehat{\mathcal{B}}(s,\mathsf{p}) = \mathcal{W}^T\mathcal{B}(s,\mathsf{p}), \quad \widehat{\mathcal{K}}(s,\mathsf{p}) = \mathcal{W}^T\mathcal{K}(s,\mathsf{p})\mathcal{V}. \tag{7.70}$$

Applying (7.70) to the affine parametric structure (7.68) yields

$$\widehat{\mathcal{K}}(s,\mathsf{p}) = \mathcal{W}^T\mathcal{K}^{[0]}(s)\mathcal{V} + \sum_{i=1}^{\nu} k_i(\mathsf{p})\,\mathcal{W}^T\mathcal{K}^{[i]}(s)\mathcal{V}, \tag{7.71}$$

$$\widehat{\mathcal{B}}(s,\mathsf{p}) = \mathcal{W}^T\mathcal{B}^{[0]}(s) + \sum_{i=1}^{\nu} b_i(\mathsf{p})\,\mathcal{W}^T\mathcal{B}^{[i]}(s), \tag{7.72}$$

$$\widehat{\mathcal{C}}(s,\mathsf{p}) = \mathcal{C}^{[0]}(s)\mathcal{V} + \sum_{i=1}^{\nu} c_i(\mathsf{p})\,\mathcal{C}^{[i]}(s)\mathcal{V}. \tag{7.73}$$

The advantages are clear. The affine structure allows fast online evaluation of the reduced model: all the reduced-order coefficient matrices can be precomputed, and for a new parameter value, only the scalar nonlinear parametric coefficients need to be recomputed. No operation in the original dimension n is required.

7.7.2 ▪ Interpolatory projections for PMOR

The question we want to answer in this section is how to choose $\mathcal{W}$ and $\mathcal{V}$ so that the reduced parametric model interpolates the original one. The main difference from the earlier cases is that we now have two variables with which to interpolate, namely the frequency variable $s \in \mathbb{C}$ and the parameter vector $\mathsf{p} \in \mathbb{R}^{\nu}$. Thus, we will require $\widehat{\mathcal{H}}(s,\mathsf{p})$ to (tangentially) interpolate $\mathcal{H}(s,\mathsf{p})$ at selected s and p values. In particular, this will require choosing both frequency interpolation points and parameter interpolation points.

Interpolatory PMOR has been studied in various papers; see, e.g., [29, 36, 42, 44, 45, 59, 68, 74, 88]. These papers focus on matrix interpolation (as opposed to tangential interpolation) and in some cases are restricted to special cases where parametric dependence is allowed only within a subset of state-space matrices. See Chapter 9 for a careful comparison of various parametrized model reduction strategies. Baur et al. [14] provide quite a general projection-based framework for approaching structure-preserving PMOR via tangential interpolation. Our discussion below follows [14] closely. However, we note that instead of the standard first-order framework (7.67) that is considered there, we present results for more general *parametrized* generalized coprime representations as in (7.66). To keep the presentation concise, we only list the zeroth- and first-order interpolation conditions.

Theorem 7.15. *Given* $\mathcal{H}(s,\mathsf{p}) = \mathcal{C}(s,\mathsf{p})\mathcal{K}(s,\mathsf{p})^{-1}\mathcal{B}(s,\mathsf{p})$, *let* $\widehat{\mathcal{H}}(s,\mathsf{p})$ *denote the reduced transfer function in* (7.69) *obtained by projection as in* (7.70) *using the model reduction bases* $\mathcal{V}$ *and* $\mathcal{W}$. *For the frequency interpolation points* σ, $\mu \in \mathbb{C}$ *and the parameter interpolation point* $\pi \in \mathbb{R}^p$, *suppose that* $\mathcal{B}(s,\mathsf{p})$, $\mathcal{C}(s,\mathsf{p})$, *and* $\mathcal{K}(s,\mathsf{p})$ *are analytic with respect* *to s at σ and $\mu \in \mathbb{C}$ and are continuously differentiable with respect to* p *in a neighborhood of π. Also let $\mathcal{K}(\sigma,\pi)$ and $\mathcal{K}(\mu,\pi)$ have full rank and let $\mathbf{r} \in \mathbb{C}^m$ and $\ell \in \mathbb{C}^q$ be the nontrivial tangential directions vectors. Then, we have the following:*

(a) *If*

$$\mathcal{K}(\sigma,\pi)^{-1}\mathcal{B}(\sigma,\pi)\mathbf{r} \in \mathsf{Ran}(\mathcal{V}), \tag{7.74}$$

then

$$\mathcal{H}(\sigma,\pi)\mathbf{r} = \widehat{\mathcal{H}}(\sigma,\pi)\mathbf{r}. \tag{7.75}$$

(b) *If*

$$\left(\ell^T\mathcal{C}(\mu,\pi)\mathcal{K}(\mu,\pi)^{-1}\right)^T \in \mathsf{Ran}(\mathcal{W}), \tag{7.76}$$

then

$$\ell^T\mathcal{H}(\mu,\pi) = \ell^T\widehat{\mathcal{H}}(\mu,\pi). \tag{7.77}$$

(c) *If both* (7.74) *and* (7.76) *hold and if $\sigma = \mu$, then*

$$\ell^T\mathcal{H}'(\sigma,\pi)\mathbf{r} = \ell^T\widehat{\mathcal{H}}'(\sigma,\pi)\mathbf{r} \tag{7.78}$$

and

$$\nabla_{\mathsf{p}}\ell^T\mathcal{H}(\sigma,\pi)\mathbf{r} = \nabla_{\mathsf{p}}\ell^T\widehat{\mathcal{H}}(\sigma,\pi)\mathbf{r}. \tag{7.79}$$

For the above, we assume $\widehat{\mathcal{K}}(\sigma,\pi) = \mathcal{W}^T\mathcal{K}(\sigma,\pi)\mathcal{V}$ and $\widehat{\mathcal{K}}(\mu,\pi) = \mathcal{W}^T\mathcal{K}(\mu,\pi)\mathcal{V}$ each have full rank.

Once again, the basic interpolatory projection theorem extends directly to a more general setting, in this case to the reduction of parametric systems. Possibly the most important property here is that, as (7.79) shows, interpolatory projection matches the parameter sensitivity without ever computing the parameters, i.e., the subspaces $\mathcal{V}$ and $\mathcal{W}$ do not contain any information about the parameter sensitivity. Nonetheless, the two-sided projection forces a match with this quantity. Indeed, the Hessian with respect to the parameter vector can be matched similarly by adding more vectors to the subspace; see [14] for details.

Example 7.16. Consider a mass-spring-damper system where two masses, m_1 and m_2, are connected with a spring-dashpot pair with spring constant k_2 and damping constant p_2. Further assume that the mass m_1 is connected to the ground by another spring-dashpot pair with spring constant k_1 and damping constant p_1. Also, suppose that a point external force $u(t)$ is applied to m_1 and we are interested in the displacement of the mass m_2. Let the state vector $\mathbf{x}(t) = [\mathbf{x}_1(t)\ \mathbf{x}_2(t)]$ consist of the displacements of both masses. Then, the corresponding differential equation is given by

$$\mathbf{M}\ddot{\mathbf{x}} + \mathbf{G}\dot{\mathbf{x}} + \mathbf{K}\mathbf{x} = \mathbf{b}u(t)\ , y(t) = \mathbf{c}\mathbf{x}(t),$$

where $\mathbf{b} = [1\ \ 0]^T$, $\mathbf{c} = [0\ \ 1]$,

$$\mathbf{M} = \begin{bmatrix} m_1 & 0 \\ 0 & m_2 \end{bmatrix}, \mathbf{G} = \begin{bmatrix} \mathsf{p}_1 + \mathsf{p}_2 & -\mathsf{p}_2 \\ -\mathsf{p}_2 & \mathsf{p}_2 \end{bmatrix}, \text{ and } \mathbf{K} = \begin{bmatrix} k_1 + k_2 & -k_2 \\ -k_2 & k_2 \end{bmatrix}.$$

Let $m_1 = m_2 = 1$, $k_1 = 2$, and $k_2 = 2$. Also, let the damping constants be parametric and vary as $\mathsf{p}_1 \in [0.15, 0.25]$ and $\mathsf{p}_2 \in [0.25, 0.35]$. Define the parameter vector $\mathsf{p} = [\mathsf{p}_1\ \mathsf{p}_2]^T$. Then, the damping matrix can be written as

$$\mathbf{G}(\mathsf{p}) = \mathsf{p}_1 \begin{bmatrix} 1 & 0 \\ 0 & 0 \end{bmatrix} + \mathsf{p}_2 \begin{bmatrix} 1 & -1 \\ -1 & 1 \end{bmatrix} = \mathsf{p}_1 \mathcal{K}_1 + \mathsf{p}_2 \mathcal{K}_2.$$

Then, the underlying system becomes a parametric dynamical system with a transfer function of the form (7.66), i.e., $\mathcal{H}(s,\mathsf{p}) = \mathcal{C}(s,\mathsf{p})\mathcal{K}(s,\mathsf{p})^{-1}\mathcal{B}(s,\mathsf{p})$ with

$$\mathcal{K}(s,\mathsf{p}) = \underbrace{s^2\mathbf{M} + \mathbf{K}}_{\mathcal{K}^{[0]}(s)} + \underbrace{\mathsf{p}_1}_{k_1(\mathsf{p})} \underbrace{s\mathbf{K}_1}_{\mathcal{K}^{[1]}(s)} + \underbrace{\mathsf{p}_2}_{k_2(\mathsf{p})} \underbrace{s\mathbf{K}_2}_{\mathcal{K}^{[2]}(s)}, \tag{7.80}$$

$$\mathcal{B}(s,\mathsf{p}) = \mathbf{b} = \mathcal{B}^{[0]}(s), \quad \text{and } \mathcal{C}(s,\mathsf{p}) = \mathbf{c} = \mathcal{C}^{[0]}(s). \tag{7.81}$$

We would like to construct a degree-one parametric reduced model using the frequency interpolation point $\sigma = 1$ and the parameter interpolation vector $\pi = [0.2\ 0.3]^T$. Note that since the system is SISO, no direction vectors are needed. Then,

$$\mathcal{V} = \mathcal{K}(1,\pi)^{-1}\mathcal{B}(1,\pi) = \begin{bmatrix} 2.5661 \times 10^{-1} \\ 1.7885 \times 10^{-1} \end{bmatrix},$$

$$\mathcal{W} = \mathcal{K}(1,\pi)^{-T}\mathcal{C}(1,\pi)^T = \begin{bmatrix} 1.7885 \times 10^{-1} \\ 4.2768 \times 10^{-1} \end{bmatrix}.$$

This leads to a reduced parametric model $\widehat{\mathcal{H}}(s,\mathsf{p}) = \widehat{\mathcal{C}}(s,\mathsf{p})\widehat{\mathcal{K}}(s,\mathsf{p})^{-1}\widehat{\mathcal{B}}(s,\mathsf{p})$ with

$$\widehat{\mathcal{K}}(s,\mathsf{p}) = \left(s^2\mathcal{W}^T\mathbf{M}\mathcal{V} + \mathcal{W}^T\mathbf{K}\mathcal{V}\right) + \mathsf{p}_1\left(s\mathcal{W}^T\mathbf{K}_1\mathcal{V}\right) + \mathsf{p}_2\left(s\mathcal{W}^T\mathbf{K}_2\mathcal{V}\right),$$

$$\widehat{\mathcal{B}}(s,\mathsf{p}) = \mathcal{W}^T\mathbf{b}, \quad \text{and } \mathcal{C}(s,\mathsf{p}) = \mathbf{c}\mathcal{V}.$$

One can directly check that at $\sigma = 1$ and $\pi = [0.2\ 0.3]^T$,

$$\mathcal{H}(\sigma,\pi) = \widehat{\mathcal{H}}(\sigma,\pi) = 1.7885 \times 10^{-1};$$

thus, (7.75) holds. Since the system is SISO, (7.75) and (7.77) are equivalent. Note that

$$\mathcal{H}'(s,\mathsf{p}) = -\mathbf{c}\mathcal{K}(s,\mathsf{p})^{-1}(2s\mathbf{M} + \mathsf{p}_1\mathbf{K}_1 + \mathsf{p}_2\mathbf{K}_2)\mathcal{K}(s,\mathsf{p})^{-1}\mathbf{b}$$

and similarly for $\widehat{\mathcal{H}}'(s,\mathsf{p})$. Then, by substituting $s = \sigma = 1$ and $\mathsf{p} = \pi = [0.2\ 0.3]^T$, we obtain

$$\mathcal{H}'(\sigma,\pi) = \widehat{\mathcal{H}}'(\sigma,\pi) = -2.4814 \times 10^{-1};$$

thus, (7.78) holds. We are left with the parametric sensitivity matching condition (7.79). One can directly compute the parameter gradients as

$$\nabla_{\mathsf{p}}\mathcal{H}(s,\mathsf{p}) = \begin{bmatrix} -\mathbf{c}\mathcal{K}(s,\mathsf{p})^{-1}(s\mathbf{K}_1)\mathcal{K}(s,\mathsf{p})^{-1}\mathbf{b} \\ -\mathbf{c}\mathcal{K}(s,\mathsf{p})^{-1}(s\mathbf{K}_2)\mathcal{K}(s,\mathsf{p})^{-1}\mathbf{b} \end{bmatrix}.$$

A direct computation yields that at $s = \sigma = 1$ and $\mathsf{p} = \pi = [0.2\ 0.3]^T$,

$$\nabla_{\mathsf{p}}\mathcal{H}(\sigma,\pi) = \nabla_{\mathsf{p}}\widehat{\mathcal{H}}(\sigma,\pi) = \begin{bmatrix} -4.5894 \times 10^{-2} \\ 1.9349 \times 10^{-2} \end{bmatrix}.$$

As this simple example illustrates, by adding one vector to each subspace, in addition to matching the transfer function and its s-derivative, we were able to match the parameter gradients for free; once again we emphasize that no parameter gradient information was added to the subspaces. However, we still match them by employing a two-sided Petrov–Galerkin projection. ■

Theorem 7.15 reveals how to proceed in the case of multiple frequency and parameter interpolation points. Given two sets of frequency points, $\{\sigma_i\}_{i=1}^K \in \mathbb{C}$ and $\{\mu_i\}_{i=1}^K \in \mathbb{C}$, and the parameter points $\{\pi^{(j)}\}_{j=1}^L \in \mathbb{C}^p$ together with the right directions $\{\mathbf{r}_{ij}\}_{i=1,j=1}^{K,L} \in \mathbb{C}^m$ and the left directions $\{\boldsymbol{\ell}_{ij}\}_{i=1,j=1}^{K,L} \in \mathbb{C}^q$, compute

$$\mathbf{v}_{ij} = \mathcal{K}(\sigma_i, \pi^{(j)})^{-1}\mathcal{B}(\sigma_i, \pi^{(j)})\mathbf{r}_{i,j} \text{ and } \mathbf{w}_{ij} = \mathcal{K}(\mu_i, \pi^{(j)})^{-T}\mathcal{C}(\mu_i, \pi^{(j)})^T\boldsymbol{\ell}_{i,j}$$

for $i = 1, \ldots, K$ and $j = 1, \ldots, L$ and construct

$$\mathcal{V} = [\mathbf{v}_{11}, \ldots, \mathbf{v}_{1L}, \mathbf{v}_{21}, \ldots, \mathbf{v}_{2L}, \ldots, \mathbf{v}_{K1}, \ldots, \mathbf{v}_{KL}] \in \mathbb{C}^{n\times(KL)}$$

and

$$\mathcal{W} = [\mathbf{w}_{11}, \ldots, \mathbf{w}_{1L}, \mathbf{w}_{21}, \ldots, \mathbf{w}_{2L}, \ldots, \mathbf{w}_{K1}, \ldots, \mathbf{w}_{KL}] \in \mathbb{C}^{n\times(KL)},$$

and apply projection as in (7.70). In practice, $\mathcal{V}$ and $\mathcal{W}$ might have linearly dependent columns. In these cases, applying a rank-revealing QR or a singular value decomposition (SVD) to remove these linearly independent columns will be necessary and will also help decrease the reduced model dimension.

Remark 7.17. We have focused here on a global basis approach to interpolatory PMOR in the sense that we assume that the reduction bases $\mathcal{V}$ and $\mathcal{W}$ are constant with respect to parameter variation and rich enough to carry global information for the entire parameter space. As for other PMOR approaches, interpolatory model reduction can also be formulated with p-dependent model reduction bases $\mathcal{V}(\mathsf{p})$ and $\mathcal{W}(\mathsf{p})$. These parameter-dependent bases can be constructed in several ways, say by interpolating local bases that correspond to parameter samples $\pi^{(i)}$. Such considerations are not specific to interpolatory approaches and occur in other PMOR approaches where the bases might be computed via POD, BT, etc. Similar questions arise as to how best to choose the parameter samples $\pi^{(i)}$. This is also a general consideration for all PMOR methods. Common approaches such as greedy sampling can be applied here as well. For a detailed discussion of these general issues related to PMOR, we refer the reader to [25]. We mention in passing that [14] introduces an *optimal* joint parameter and frequency interpolation point selection strategy for a special case of parametric systems.

7.8 ▪ Conclusions

We have provided here a brief survey of interpolatory methods for model reduction of large-scale dynamical systems. In addition to a detailed discussion of basic principles for generic first-order realizations, we have presented an interpolation framework for more general system classes that include generalized coprime realizations and parametrized systems. Reduction of systems of DAEs was also discussed. An overview of optimal interpolation methods in the $\mathcal{H}_2$-norm, including the weighted case, has also been provided.

Bibliography

[1] M. AHMAD AND P. BENNER, *Interpolatory model reduction techniques for linear second-order descriptor systems*, in Proceedings of the 2014 European Control Conference (ECC), 2014, IEEE Publications, pp. 1075–1079.

[2] M. AHMAD, D. SZYLD, AND M. VAN GIJZEN, *Preconditioned multishift BiCG for $\mathcal{H}_2$-optimal model reduction*, SIAM J. Matrix Anal. Appl., 38 (2017), pp. 401–424.

[3] K. AHUJA, P. BENNER, E. DE STURLER, AND L. FENG, *Recycling BiCGSTAB with an application to parametric model order reduction*, SIAM Journal on Scientific Computing, 37 (2015), pp. S429–S446.

[4] K. AHUJA, E. DE STURLER, S. GUGERCIN, AND E. CHANG, *Recycling BiCG with an application to model reduction*, SIAM Journal on Scientific Computing, 34 (2012), pp. A1925–A1949.

[5] D. AMSALLEM AND C. FARHAT, *Interpolation method for the adaptation of reduced-order models to parameter changes and its application to aeroelasticity*, AIAA Journal, 46 (2008), pp. 1803–1813.

[6] B. ANIĆ, C. BEATTIE, S. GUGERCIN, AND A. ANTOULAS, *Interpolatory weighted-$\mathcal{H}_2$ model reduction*, Automatica, 49 (2013), pp. 1275–1280.

[7] H. ANTIL, M. HEINKENSCHLOSS, AND R. H. W. HOPPE, *Domain decomposition and balanced truncation model reduction for shape optimization of the Stokes system*, Optimization Methods and Software, 26 (2011), pp. 643–669.

[8] H. ANTIL, M. HEINKENSCHLOSS, R. H. W. HOPPE, C. LINSENMANN, AND A. WIXFORTH, *Reduced order modeling based shape optimization of surface acoustic wave driven microfluidic biochips*, Mathematics and Computers in Simulation, 82 (2012), pp. 1986–2003.

[9] A. ANTOULAS, *Approximation of Large-Scale Dynamical Systems*, Advances in Design and Control 6, SIAM, Philadelphia, 2005.

[10] A. ANTOULAS, C. BEATTIE, AND S. GUGERCIN, *Interpolatory model reduction of large-scale dynamical systems*, in Efficient Modeling and Control of Large-Scale Systems, J. Mohammadpour, and K. Grigoriadis, eds., Springer-Verlag, 2010, pp. 2–58.

[11] E. ARIAN, M. FAHL, AND E. SACHS, *Trust-region proper orthogonal decomposition models by optimization methods*, in Proceedings of the 41st IEEE Conference on Decision and Control, Las Vegas, NV, 2002, pp. 3300–3305.

[12] Z. BAI, *Krylov subspace techniques for reduced-order modeling of large-scale dynamical systems*, Applied Numerical Mathematics, 43 (2002), pp. 9–44.

[13] U. BAUR AND P. BENNER, *Model reduction for parametric systems using balanced truncation and interpolation*, at-Automatisierungstechnik, 57 (2009), pp. 411–420.

[14] U. BAUR, C. BEATTIE, P. BENNER, AND S. GUGERCIN, *Interpolatory projection methods for parameterized model reduction*, SIAM Journal on Scientific Computing, 33 (2011), pp. 2489–2518.

[15] U. BAUR, P. BENNER, A. GREINER, J. KORVINK, J. LIENEMANN, AND C. MOOSMANN, *Parameter preserving model order reduction for MEMS applications*, Mathematical and Computer Modelling of Dynamical Systems, 17 (2011), pp. 297–317.

[16] C. BEATTIE, G. FLAGG, AND S. GUGERCIN, *An interpolation-based approach to optimal $\mathcal{H}_\infty$ model reduction*, in SIAM Conference on Computational Science and Engineering, Miami, 2009.

[17] C. BEATTIE AND S. GUGERCIN, *Inexact solves in Krylov-based model reduction*, in 45th IEEE Conference on Decision and Control, IEEE, 2006, pp. 3405–3411.

[18] C. BEATTIE AND S. GUGERCIN, *Krylov-based minimization for optimal $\mathcal{H}_2$ model reduction*, 46th IEEE Conference on Decision and Control, IEEE, 2007, pp. 4385–4390.

[19] C. BEATTIE AND S. GUGERCIN, *Interpolatory projection methods for structure-preserving model reduction*, Systems and Control Letters, 58 (2009), pp. 225–232.

[20] C. BEATTIE AND S. GUGERCIN, *A trust region method for optimal $\mathcal{H}_2$ model reduction*, 48th IEEE Conference on Decision and Control, IEEE, 2009.

[21] C. BEATTIE AND S. GUGERCIN, *Realization-independent $\mathcal{H}_2$ approximation*, in Proceedings of the 51st IEEE Conference on Decision and Control, IEEE, 2012, pp. 4953–4958.

[22] C. BEATTIE, S. GUGERCIN, AND S. WYATT, *Inexact solves in interpolatory model reduction*, Linear Algebra and Its Applications, 436 (2012), pp. 2916–2943.

[23] P. BENNER AND T. BREITEN, *Interpolation-based $\mathcal{H}_2$-model reduction of bilinear control systems*, SIAM Journal on Matrix Analysis and Applications, 33 (2012), pp. 859–885.

[24] P. BENNER AND L. FENG, *Recycling Krylov subspaces for solving linear systems with successively changing right-hand sides arising in model reduction*, in Model Reduction for Circuit Simulation, P. Benner, M. Hinze, and E. ter Maten, eds., Springer-Verlag, 2011, pp. 125–140.

[25] P. BENNER, S. GUGERCIN, AND K. WILLCOX, *A survey of projection-based model reduction methods for parametric dynamical systems*, SIAM Review, 57 (2015), pp. 483–531.

[26] P. BENNER, M. HINZE, AND E. TER MATEN, eds., *Model Reduction for Circuit Simulation*, vol. 74 of Lecture Notes in Electrical Engineering, Springer-Verlag, Dordrecht, Netherlands, 2011.

[27] P. BENNER, M. KÖHLER, AND J. SAAK, *Sparse-Dense Sylvester Equations in $\mathcal{H}_2$-Model Order Reduction*, Technical Report MPIMD/11-11, Max Planck Institute Magdeburg Preprints, December 2011.

[28] P. BENNER AND V. SOKOLOV, *Partial realization of descriptor systems*, Systems and Control Letters, 55 (2006), pp. 929–938.

[29] B. BOND AND L. DANIEL, *Parameterized model order reduction of nonlinear dynamical systems*, in IEEE/ACM International Conference on Computer-Aided Design, 2005, ICCAD-2005, pp. 487–494.

[30] J. BORGGAARD, E. CLIFF, AND S. GUGERCIN, *Model reduction for indoor-air behavior in control design for energy-efficient buildings*, in American Control Conference (ACC), IEEE, 2012, pp. 2283–2288.

[31] T. BREITEN, *A Descent Method for the Frequency Weighted $\mathcal{H}_2$ Model Reduction Problem*, Talk given at the Centre International de Rencontres Mathématiques, Luminy, France 2013.

[32] T. BREITEN, C. BEATTIE, AND S. GUGERCIN, *Near-optimal frequency-weighted interpolatory model reduction*, Systems and Control Letters, 78 (2013), pp. 8–18.

[33] T. BUI-THANH, K. WILLCOX, AND O. GHATTAS, *Model reduction for large-scale systems with high-dimensional parametric input space*, SIAM Journal on Scientific Computing, 30 (2008), pp. 3270–3288.

[34] ———, *Parametric reduced-order models for probabilistic analysis of unsteady aerodynamic applications*, AIAA Journal, 46 (2008), pp. 2520–2529.

[35] A. BUNSE-GERSTNER, D. KUBALINSKA, G. VOSSEN, AND D. WILCZEK, *$\mathcal{H}_2$-norm optimal model reduction for large scale discrete dynamical MIMO systems*, Journal of Computational and Applied Mathematics, 233 (2009), pp. 1202–1216. doi:10.1016/j.cam.2008.12.029.

[36] L. DANIEL, O. SIONG, S. LOW, K. LEE, AND J. WHITE, *A multiparameter moment matching model reduction approach for generating geometrically parameterized interconnect performance models*, IEEE Transactions on Computer-Aided Design of Integrated Circuits and Systems, 23 (2004), pp. 678–693.

[37] E. DE STURLER, S. GUGERCIN, M. KILMER, S. CHATURANTABUT, C. BEATTIE, AND M. O'CONNELL, *Nonlinear parametric inversion using interpolatory model reduction*, SIAM Journal on Scientific Computing, 37 (2015), B495–B517.

[38] C. DE VILLEMAGNE AND R. SKELTON, *Model reductions using a projection formulation*, International Journal of Control, 46 (1987), pp. 2141–2169.

[39] V. DRUSKIN AND V. SIMONCINI, *Adaptive rational Krylov subspaces for large-scale dynamical systems*, Systems and Control Letters, 60 (2011), pp. 546–560.

[40] V. DRUSKIN, V. SIMONCINI, AND M. ZASLAVSKY, *Adaptive tangential interpolation in rational Krylov subspaces for MIMO dynamical systems*, SIAM J. Matrix Anal. Appl., 35 (2014), pp. 476–498.

[41] ———, *Solution of the time-domain inverse resistivity problem in the model reduction framework part I. One-dimensional problem with SISO data*, SIAM Journal on Scientific Computing, 35 (2013), pp. A1621–A1640.

[42] O. FARLE, V. HILL, P. INGELSTRÖM, AND R. DYCZIJ-EDLINGER, *Multiparameter polynomial order reduction of linear finite element models*, Mathematical and Computer Modelling of Dynamical Systems, 14 (2008), pp. 421–434.

[43] P. FELDMANN AND R. FREUND, *Efficient linear circuit analysis by Padé approximation via the Lanczos process*, IEEE Transactions on Computer-Aided Design of Integrated Circuits and Systems, 14 (1995), pp. 639–649.

[44] L. FENG AND P. BENNER, *A robust algorithm for parametric model order reduction based on implicit moment matching*, Proceedings in Applied Mathematics and Mechanics, 7 (2008), pp. 10215.01–10215.02.

[45] L. FENG, E. RUDNYI, AND J. KORVINK, *Preserving the film coefficient as a parameter in the compact thermal model for fast electrothermal simulation*, IEEE Transactions on Computer-Aided Design of Integrated Circuits and Systems, 24 (2005), pp. 1838–1847.

[46] G. FLAGG, C. BEATTIE, AND S. GUGERCIN, *Convergence of the iterative rational Krylov algorithm*, Systems and Control Letters, 61 (2012), pp. 688–691.

[47] G. FLAGG, C. A. BEATTIE, AND S. GUGERCIN, *Interpolatory H_∞ model reduction*, Systems and Control Letters, 62 (2013), pp. 567–574.

[48] R. FREUND, *Model reduction methods based on Krylov subspaces*, Acta Numerica, 12 (2003), pp. 267–319.

[49] D. GALBALLY, K. FIDKOWSKI, K. WILLCOX, AND O. GHATTAS, *Nonlinear model reduction for uncertainty quantification in large-scale inverse problems*, International Journal for Numerical Methods in Engineering, 81 (2010), pp. 1581–1608.

[50] K. GALLIVAN, E. GRIMME, AND P. VAN DOOREN, *Asymptotic waveform evaluation via a Lanczos method*, Applied Mathematics Letters, 7 (1994), pp. 75–80.

[51] K. GALLIVAN, A. VANDENDORPE, AND P. VAN DOOREN, *Model reduction of MIMO systems via tangential interpolation*, SIAM Journal on Matrix Analysis and Applications, 26 (2004), pp. 328–349.

[52] E. GRIMME, *Krylov Projection Methods for Model Reduction*, PhD thesis, Coordinated-Science Laboratory, University of Illinois at Urbana-Champaign, 1997.

[53] S. GUGERCIN, *Projection Methods for Model Reduction of Large-Scale Dynamical Systems*, PhD thesis, ECE Dept., Rice University, December 2002.

[54] ——, *An iterative rational Krylov algorithm (IRKA) for optimal $\mathcal{H}_2$ model reduction*, in Householder Symposium XVI, Seven Springs Mountain Resort, PA, May 2005.

[55] S. GUGERCIN AND A. ANTOULAS, *An $\mathcal{H}_2$ error expression for the Lanczos procedure*, in Proceedings of the 42nd IEEE Conference on Decision and Control, IEEE, 2003.

[56] S. GUGERCIN, A. ANTOULAS, AND C. BEATTIE, *A rational Krylov iteration for optimal $\mathcal{H}_2$ model reduction*, in Proceedings of MTNS, 2006.

[57] S. GUGERCIN, A. ANTOULAS, AND C. BEATTIE, *$\mathcal{H}_2$ model reduction for large-scale linear dynamical systems*, SIAM Journal on Matrix Analysis and Applications, 30 (2008), pp. 609–638.

[58] S. GUGERCIN, T. STYKEL, AND S. WYATT, *Model reduction of descriptor systems by interpolatory projection methods*, SIAM Journal on Scientific Computing, 35 (2013), pp. B1010–B1033.

[59] P. GUNUPUDI, R. KHAZAKA, AND M. NAKHLA, *Analysis of transmission line circuits using multidimensional model reduction techniques*, IEEE Transactions on Advanced Packaging, 25 (2002), pp. 174–180.

[60] B. HAASDONK AND M. OHLBERGER, *Efficient reduced models and a posteriori error estimation for parametrized dynamical systems by offline/online decomposition*, Mathematical and Computer Modelling of Dynamical Systems, 17 (2011), pp. 145–161.

[61] Y. HALEVI, *Frequency weighted model reduction via optimal projection*, IEEE Transactions on Automatic Control, 37 (1992), pp. 1537–1542.

[62] A. HAR, J. BORGGAARD, AND D. PELLETIER, *Local improvements to reduced-order models using sensitivity analysis of the proper orthogonal decomposition*, Journal of Fluid Mechanics, 629 (2009), pp. 41–72.

[63] M. HEINKENSCHLOSS, D. SORENSEN, AND K. SUN, *Balanced truncation model reduction for a class of descriptor systems with application to the Oseen equations*, SIAM Journal on Scientific Computing, 30 (2008), pp. 1038–1063.

[64] D. HYLAND AND D. BERNSTEIN, *The optimal projection equations for model reduction and the relationships among the methods of Wilson, Skelton, and Moore*, IEEE Transactions on Automatic Control, 30 (1985), pp. 1201–1211.

[65] A. KELLEMS, D. ROOS, N. XIAO, AND S. COX, *Low-dimensional, morphologically accurate models of subthreshold membrane potential*, Journal of Computational Neuroscience, 27 (2009), pp. 161–176.

[66] W. KRAJEWSKI, A. LEPSCHY, M. REDIVO-ZAGLIA, AND U. VIARO, *A program for solving the $\mathcal{L}_2$ reduced-order model problem with fixed denominator degree*, Numerical Algorithms, 9 (1995), pp. 355–377.

[67] D. KUBALINSKA, A. BUNSE-GERSTNER, G. VOSSEN, AND D. WILCZEK, *$\mathcal{H}_2$-optimal interpolation based model reduction for large-scale systems*, in Proceedings of the 16th International Conference on System Science, Poland, 2007.

[68] A.-M. LEUNG AND R. KHAZAKA, *Parametric model order reduction technique for design optimization*, in IEEE International Symposium on Circuits and Systems (ISCAS), May 2005, pp. 1290–1293.

[69] C. LIEBERMAN, K. WILLCOX, AND O. GHATTAS, *Parameter and state model reduction for large-scale statistical inverse problems*, SIAM Journal on Scientific Computing, 32 (2010), pp. 2523–2542.

[70] T. LIEU AND C. FARHAT, *Adaptation of aeroelastic reduced-order models and application to an F-16 configuration*, AIAA Journal, 45 (2007), pp. 1244–1257.

[71] T. LIEU, C. FARHAT, AND M. LESOINNE, *Reduced-order fluid/structure modeling of a complete aircraft configuration*, Computer Methods in Applied Mechanics and Engineering, 195 (2006), pp. 5730–5742.

[72] A. MAYO AND A. ANTOULAS, *A framework for the solution of the generalized realization problem*, Linear Algebra and Its Applications, 425 (2007), pp. 634–662.

[73] L. MEIER III AND D. LUENBERGER, *Approximation of linear constant systems*, IEEE Transactions on Automatic Control, 12 (1967), pp. 585–588.

[74] K. MOOSMANN AND J. KORVINK, *Automatic parametric MOR for MEMS design*, in Tagungsband GMA-FA 1.30 "Modellbildung, Identifikation und Simulation in der Automatisierungstechnik," Workshop am Bostalsee, B. Lohmann and A. Kugi, eds., 2006, pp. 89–99.

[75] N. NGUYEN, A. PATERA, AND J. PERAIRE, *A best points interpolation method for efficient approximation of parametrized functions*, International Journal for Numerical Methods in Engineering, 73 (2008), pp. 521–543.

[76] H. PANZER, S. JAENSCH, T. WOLF, AND B. LOHMANN, *A greedy rational Krylov method for $\mathcal{H}_2$-pseudooptimal model order reduction with preservation of stability*, in American Control Conference (ACC), 2013, pp. 5512–5517.

[77] D. PETERSSON, *A Nonlinear Optimization Approach to $\mathcal{H}_2$-Optimal Modeling and Control*, PhD thesis, Linköping University, 2013.

[78] C. PRUD'HOMME, D. ROVAS, K. VEROY, L. MACHIELS, Y. MADAY, A. PATERA, AND G. TURINICI, *Reliable real-time solution of parametrized partial differential equations: Reduced-basis output bound methods*, Journal of Fluids Engineering, 124 (2002), pp. 70–80.

[79] G. ROZZA, D. HUYNH, AND A. PATERA, *Reduced basis approximation and a posteriori error estimation for affinely parametrized elliptic coercive partial differential equations: Application to transport and continuum mechanics*, Archives of Computational Methods in Engineering, 15 (2008), pp. 229–275.

[80] A. RUHE, *Rational Krylov algorithms for nonsymmetric eigenvalue problems. II: Matrix pair*, Linear Algebra and Its Applications, 197–198 (1984), pp. 282–295.

[81] J. SPANOS, M. MILMAN, AND D. MINGORI, *A new algorithm for L^2 optimal model reduction*, Automatica (Journal of IFAC), 28 (1992), pp. 897–909.

[82] T. STYKEL, *Low-rank iterative methods for projected generalized Lyapunov equations*, Electronic Transactions on Numerical Analysis, 30 (2008), pp. 187–202.

[83] P. VAN DOOREN, K. GALLIVAN, AND P. ABSIL, *$\mathcal{H}_2$-optimal model reduction of MIMO systems*, Applied Mathematics Letters, 21 (2008), pp. 1267–1273.

[84] P. VAN DOOREN, K. A. GALLIVAN, AND P.-A. ABSIL, *$\mathcal{H}_2$-optimal model reduction with higher-order poles*, SIAM Journal on Matrix Analysis and Applications, 31 (2010), pp. 2738–2753.

[85] K. VEROY, C. PRUD'HOMME, D. ROVAS, AND A. PATERA, *A posteriori error bounds for reduced-basis approximation of parametrized noncoercive and nonlinear elliptic partial differential equations*, in Proceedings of the 16th AIAA Computational Fluid Dynamics Conference, 2003.

[86] P. VUILLEMIN, C. POUSSOT-VASSAL, AND D. ALAZARD, *A Spectral Expression for the Frequency-Limited $\mathcal{H}_2$-Norm*, arXiv preprint arXiv:1211.1858 (2012).

[87] J. WANG AND N. ZABARAS, *Using Bayesian statistics in the estimation of heat source in radiation*, International Journal of Heat and Mass Transfer, 48 (2005), pp. 15–29.

[88] D. S. WEILE, E. MICHIELSSEN, E. GRIMME, AND K. GALLIVAN, *A method for generating rational interpolant reduced order models of two-parameter linear systems*, Applied Mathematics Letters, 12 (1999), pp. 93–102.

[89] D. WILSON, *Optimum solution of model-reduction problem*, Proceedings of the IEE, 117 (1970), pp. 1161–1165.

[90] S. WYATT, *Issues in Interpolatory Model Reduction: Inexact Solves, Second-Order Systems and DAEs*, PhD thesis, Virginia Polytechnic Institute and State University, 2012.

[91] W. YAN AND J. LAM, *An approximate approach to $\mathcal{H}_2$ optimal model reduction*, IEEE Transactions on Automatic Control, 44 (1999), pp. 1341–1358.

[92] A. YOUSUFF AND R. SKELTON, *Covariance equivalent realizations with applications to model reduction of large-scale systems*, Control and Dynamic Systems, 22 (1985), pp. 273–348.

[93] A. YOUSUFF, D. WAGIE, AND R. SKELTON, *Linear system approximation via covariance equivalent realizations*, Journal of Mathematical Analysis and Applications, 106 (1985), pp. 91–115.

[94] Y. YUE AND K. MEERBERGEN, *Using Krylov-Padé model order reduction for accelerating design optimization of structures and vibrations in the frequency domain*, International Journal for Numerical Methods in Engineering, 90 (2012), pp. 1207–1232.

[95] ———, *Accelerating optimization of parametric linear systems by model order reduction*, SIAM Journal on Optimization, 23 (2013), pp. 1344–1370.

[96] D. ZIGIC, L. WATSON, AND C. BEATTIE, *Contragredient transformations applied to the optimal projection equations*, Linear Algebra and Its Applications, 188 (1993), pp. 665–676.

Chapter 8

A Tutorial Introduction to the Loewner Framework for Model Reduction

Athanasios C. Antoulas, Sanda Lefteriu, and A. Cosmin Ionita[26]

One of the main approaches to model reduction of both linear and nonlinear dynamical systems is by means of interpolation. Data-driven model reduction constitutes a special case. The Loewner matrix, originally developed for rational interpolation, has been recently extended to the Loewner framework and constitutes a versatile approach to data-driven model reduction. Its main attribute is that it provides a trade-off between accuracy of fit and complexity of the model. Furthermore, constructing models from the given data is quite natural. The purpose of this chapter is to present the fundamentals of the Loewner framework for data-driven reduction of linear systems.

8.1 ▪ Introduction

Model reduction seeks to replace a large-scale system described in terms of differential or difference equations by a system of much lower dimension that has nearly the same response characteristics.

Model (order) reduction (MOR) is commonly used to simulate and control complex physical processes. The systems that inevitably arise in such cases are often too complex to meet the expediency requirements of interactive design, optimization, or real-time control. MOR was devised as a means to reduce the dimensionality of these complex systems to a level that is amenable to such requirements. The ensuing methods are an indispensable tool for speeding up the simulations arising in various engineering applications involving large-scale dynamical systems.

Generally, large systems arise due to accuracy requirements on the spatial discretization of partial differential equations (PDEs) for fluids or structures or in the context of lumped-circuit approximations of distributed circuit elements. For some applications, see, e.g., [3].

[26]The authors wish to thank the referees for their careful reading of the manuscript and their useful suggestions.

The work of A. C. Antoulas was supported by NSF Grants CCF-1017401 and CCF-1320866 as well as DFG Grant AN-693/1-1.

Approaches to model reduction

Model reduction methods can be classified in two broad categories, namely, *SVD-based* and *Krylov-based* or *moment matching* methods. The former category derives its name from the fact that the corresponding reduction methods are related to SVD (singular value decomposition) and the associated 2-norm. The most prominent among them is *balanced truncation* (BT). These methods are based on computing controllability and observability *Gramians* (or generalized/empirical versions thereof), which leads to the elimination of states that are difficult to reach and observe. The bottleneck in applying them is due to the high computational cost required to obtain the Gramians by solving Lyapunov or Riccati equations. However, recent developments make it possible to obtain the approximate solution (i.e., approximate balancing) of realistic size problems; see, e.g., [10, 13, 14, 34] and references therein.

The reduction methods in the second category are based on *moment matching*, that is, matching of the coefficients of power series expansions of the transfer function at selected points in the complex plane; these coefficients are the values of the underlying transfer function, together with the values of its derivatives. The main underlying problem is *rational interpolation*. These methods are closely related to the so-called Krylov iteration, encountered in numerical linear algebra, as well as the Arnoldi or the Lanczos procedures, and multipoint (rational) versions thereof.

The advantages of balancing reduction methods include preservation of stability and an a priori computable error bound. Krylov-based methods are numerically efficient and have lower computational cost, but, in general, the preservation of other properties is not automatic and depends on the choice of the expansion points and the form of the projector (e.g., orthogonal or oblique).

For details on the above, as well as other issues in model reduction, we refer to the book [3].

8.1.1 ▪ Main notation

$\mathbb{R}, \mathbb{C}$	real numbers, complex numbers
$\imath$	$= \sqrt{-1}$
$(\cdot)^T$	transposition
$(\cdot)^*$	transposition and complex conjugation
$(\mathbf{E}, \mathbf{A}, \mathbf{B}, \mathbf{C}, \mathbf{D})$	descriptor system realization
$(\mathbf{E}_\mathbb{L}, \mathbf{A}_\mathbb{L}, \mathbf{B}_\mathbb{L}, \mathbf{C}_\mathbb{L})$	descriptor system realization as in (8.5)
$\mathbf{H}(s) \in \mathbb{R}^{p\times m}$	transfer function
$\mathbf{M} = \text{diag}(\mu_j) \in \mathbb{C}^{q\times q}$	matrix of left interpolation frequencies
$\boldsymbol{\ell}_i \in \mathbb{C}^p$	left tangential directions
$\mathbf{L} \in \mathbb{C}^{q\times p}$	matrix of left directions
$\mathbf{v}_i \in \mathbb{C}^m$	left tangential values
$\mathbf{V} \in \mathbb{C}^{q\times m}$	matrix of left values
$\mathbf{\Lambda} = \text{diag}(\lambda_j) \in \mathbb{C}^{k\times k}$	matrix of right interpolation frequencies

(continued on next page)

$\mathbf{r}_j \in \mathbb{C}^m$	right tangential directions
$\mathbf{R} \in \mathbb{C}^{m\times k}$	matrix of right directions
$\mathbf{w}_j \in \mathbb{C}^p$	right tangential values
$\mathbf{W} \in \mathbb{C}^{p\times k}$	matrix of right values
$\mathbb{L} \in \mathbb{C}^{q\times k}$	Loewner matrix
$\mathbb{L}_s \in \mathbb{C}^{q\times k}$	shifted Loewner matrix
$\phi(s)$	scalar rational interpolant
$\mathbf{c} = [\alpha_j] \in \mathbb{C}^k$	element in the null space of $\mathbb{L}$

8.1.2 ▪ Model reduction of linear descriptor systems

A linear time-invariant dynamical system Σ with m inputs, p outputs, and n internal variables in *descriptor-form representation* is given by a set of differential algebraic equations (DAEs):

$$\Sigma: \ \mathbf{E}\frac{d}{dt}\mathbf{x}(t) = \mathbf{A}\mathbf{x}(t) + \mathbf{B}\mathbf{u}(t), \quad \mathbf{y}(t) = \mathbf{C}\mathbf{x}(t) + \mathbf{D}\mathbf{u}(t), \tag{8.1}$$

where $\mathbf{x}(t) \in \mathbb{R}^n$ is the internal variable (the state if $\mathbf{E}$ is invertible); $\mathbf{u}(t) \in \mathbb{R}^m$ and $\mathbf{y}(t) \in \mathbb{R}^p$ are the input and output functions, respectively; and

$$\mathbf{E}, \mathbf{A} \in \mathbb{R}^{n\times n}, \ \ \mathbf{B} \in \mathbb{R}^{n\times m}, \ \ \mathbf{C} \in \mathbb{R}^{p\times n}, \ \ \mathbf{D} \in \mathbb{R}^{p\times m}$$

are constant matrices. The matrix pencil $(\mathbf{A}, \mathbf{E})$ is *regular* if the matrix $\mathbf{A} - \lambda\mathbf{E}$ is nonsingular for some finite $\lambda \in \mathbb{C}$. In this case the *transfer function* of Σ is the $p \times m$ rational matrix function

$$\mathbf{H}(s) = \mathbf{C}(s\mathbf{E} - \mathbf{A})^{-1}\mathbf{B} + \mathbf{D}. \tag{8.2}$$

It is *proper* if its value at infinity is finite and *strictly proper* if that value is zero. The *poles* of the system are the eigenvalues of the matrix pencil $(\mathbf{A}, \mathbf{E})$. Σ is *stable* if all its finite poles are in the left half of the complex plane.

The quintuple $(\mathbf{E}, \mathbf{A}, \mathbf{B}, \mathbf{C}, \mathbf{D})$ is called a *descriptor realization* of $\mathbf{H}(s)$; realizations are not unique, and those with the smallest possible dimension n are called *minimal realizations*; furthermore, rank $\mathbf{E}$ is the McMillan degree of Σ (regardless of the minimality of the realization) [28]. A realization is minimal if it is *completely controllable* and *observable*. A descriptor system with $(\mathbf{A}, \mathbf{E})$ regular is *completely controllable* [37] if $\operatorname{rank}[\mathbf{A} - \lambda\mathbf{E}, \mathbf{B}] = n$ for all finite $\lambda \in \mathbb{C}$ and $\operatorname{rank}[\mathbf{E}, \mathbf{B}] = n$. This is equivalent to the matrix

$$\mathcal{R}_n = \left[\begin{array}{ccc} (\lambda_1\mathbf{E} - \mathbf{A})^{-1}\mathbf{B} & \cdots & (\lambda_n\mathbf{E} - \mathbf{A})^{-1}\mathbf{B} \end{array}\right] \tag{8.3}$$

having full rank n for any set of distinct $\lambda_i \in \mathbb{C}$ that are not eigenvalues of the pencil $(\mathbf{A}, \mathbf{E})$. It is called *completely observable* if $\operatorname{rank}\left[\mathbf{A}^T - \mu\mathbf{E}^T, \mathbf{C}^T\right] = n$ for all finite $\mu \in \mathbb{C}$ and $\operatorname{rank}\left[\mathbf{E}^T, \mathbf{C}^T\right] = n$ (where $(\cdot)^T$ denotes transpose); this is equivalent to the matrix

$$\mathcal{O}_n = \left[\begin{array}{c} \mathbf{C}(\mu_1\mathbf{E} - \mathbf{A})^{-1} \\ \vdots \\ \mathbf{C}(\mu_n\mathbf{E} - \mathbf{A})^{-1} \end{array}\right] \tag{8.4}$$

having full rank n for any set of distinct $\mu_i \in \mathbb{C}$ that are not eigenvalues of the pencil $(\mathbf{A}, \mathbf{E})$. For details on these concepts we refer to [15].

Remark 8.1. **(a) The D-term.** In description (8.1) we can eliminate $\mathbf{D}$ by incorporating it in the remaining matrices. To achieve this, we have to allow the dimension of the realization to increase by rank $\mathbf{D}$. Consider a rank-revealing factorization

$$\mathbf{D} = \mathbf{D}_1\mathbf{D}_2 \quad \text{where} \quad \mathbf{D}_1 \in \mathbb{R}^{p\times\rho},\ \mathbf{D}_2 \in \mathbb{R}^{\rho\times m},$$

and $\rho = \operatorname{rank} \mathbf{D}$. It readily follows that

$$\mathbf{E}_\delta = \begin{bmatrix} \mathbf{E} & \\ & 0_{\rho\times\rho} \end{bmatrix},\ \mathbf{A}_\delta = \begin{bmatrix} \mathbf{A} & \\ & -\mathbf{I}_\rho \end{bmatrix},\ \mathbf{B}_\delta = \begin{bmatrix} \mathbf{B} \\ \mathbf{D}_2 \end{bmatrix},\ \mathbf{C}_\delta = \begin{bmatrix} \mathbf{C} & \mathbf{D}_1 \end{bmatrix} \tag{8.5}$$

is a descriptor realization of the same system with no $\mathbf{D}$-term (i.e., $\mathbf{D}_\delta = 0$). If the original realization is controllable and observable, this one is R-controllable and R-observable (see [15, 37] for a survey of these concepts).

The reason for introducing descriptor realizations where the $\mathbf{D}$*-term is incorporated in the remaining matrices is that the Loewner framework yields precisely such descriptor realizations.*

(b) Poles at infinity. The eigenvalues of the pencil $(\mathbf{A}_\delta, \mathbf{E}_\delta)$ at infinity are also referred to as the poles of the system Σ at infinity (under the assumption that the realization is minimal). A pole at infinity of the same algebraic and geometric multiplicity corresponds to the existence of a constant $\mathbf{D}$-term in $\mathbf{H}$, while a pole of algebraic multiplicity $\nu \geq 1$ and geometric multiplicity one implies the existence in $\mathbf{H}$ of a polynomial term of degree $\nu - 1$.

The *model reduction* problem consists of constructing reduced-order DAE systems of the form

$$\hat{\Sigma}:\ \hat{\mathbf{E}}\frac{d}{dt}\hat{\mathbf{x}}(t) = \hat{\mathbf{A}}\hat{\mathbf{x}}(t) + \hat{\mathbf{B}}\mathbf{u}(t),\ \ \hat{\mathbf{y}}(t) = \hat{\mathbf{C}}\hat{\mathbf{x}}(t) + \hat{\mathbf{D}}\mathbf{u}(t), \tag{8.6}$$

where $\hat{\mathbf{x}}(t) \in \mathbb{R}^r$ is the internal variable (the state if $\hat{\mathbf{E}}$ is invertible), $\hat{\mathbf{y}}(t) \in \mathbb{R}^p$ is the output of $\hat{\Sigma}$ corresponding to the same input $\mathbf{u}(t)$, and

$$\hat{\mathbf{E}}, \hat{\mathbf{A}} \in \mathbb{R}^{r\times r},\quad \hat{\mathbf{B}} \in \mathbb{R}^{r\times m},\quad \hat{\mathbf{C}} \in \mathbb{R}^{p\times r},\quad \hat{\mathbf{D}} \in \mathbb{R}^{p\times m}.$$

Thus, the number of inputs m and outputs p remains the same while $r \ll n$.

Some general *goals* for reduced-order models (ROMs) are as follows: (1) the reduced input-output map should be uniformly "close" to the original: for the same $\mathbf{u}$, $\mathbf{y} - \hat{\mathbf{y}}$ should be "small" in an appropriate sense; (2) critical system features and structure should be preserved, e.g., stability, passivity, Hamiltonian structure, subsystem interconnectivity, or second-order structure; (3) strategies for computing the reduced system should lead to robust, numerically stable algorithms and require minimal application-specific tuning.

8.1.3 ▪ Interpolatory reduction for linear systems

The exposition in this section follows [8]. See also [11]. These papers should be consulted for an overview of interpolatory reduction methods in general.

Consider the system Σ and its transfer function $\mathbf{H}(s)$ defined by (8.2). We are given *left interpolation points* $\{\mu_i\}_{i=1}^q \subset \mathbb{C}$, with *left tangential directions* $\{\boldsymbol{\ell}_i\}_{i=1}^q \subset \mathbb{C}^p$, and *right interpolation points* $\{\lambda_i\}_{i=1}^k \subset \mathbb{C}$, with *right tangential directions* $\{\mathbf{r}_i\}_{i=1}^k \subset \mathbb{C}^m$; for simplicity we assume that the left and right interpolation points are distinct. We seek a reduced-order system $\hat{\Sigma}$ such that the associated transfer function $\hat{\mathbf{H}}(s)$ is a *tangential interpolant* to $\mathbf{H}(s)$:

$$\left.\begin{array}{ccc} \boldsymbol{\ell}_j^T \hat{\mathbf{H}}(\mu_j) = \boldsymbol{\ell}_j^T \mathbf{H}(\mu_j) & \text{and} & \hat{\mathbf{H}}(\lambda_i)\mathbf{r}_i = \mathbf{H}(\lambda_i)\mathbf{r}_i \\ \text{for } j = 1,\dots,q & & \text{for } i = 1,\dots,k \end{array}\right\}. \tag{8.7}$$

Interpolation points and tangential directions are selected to realize the model reduction goals stated.

If, instead of descriptor-form data as in (8.1), we are given *input/output data* (measured or generated by DNS[27]), the resulting problem is modified as follows. Given a set of input-output response measurements specified by *left driving frequencies* $\{\mu_i\}_{i=1}^q \subset \mathbb{C}$, using *left input or tangential directions* $\{\boldsymbol{\ell}_i\}_{i=1}^q \subset \mathbb{C}^p$, producing *left responses* $\{\mathbf{v}_i\}_{i=1}^q \subset \mathbb{C}^m$, and *right driving frequencies* $\{\lambda_i\}_{i=1}^k \subset \mathbb{C}$, using *right input or tangential directions* $\{\mathbf{r}_i\}_{i=1}^k \subset \mathbb{C}^m$, producing *right responses* $\{\mathbf{w}_i\}_{i=1}^k \subset \mathbb{C}^p$, find a (low-order) system $\hat{\Sigma}$ such that the resulting transfer function, $\hat{\mathbf{H}}(s)$, is an *(approximate) tangential interpolant* to the data:

$$\left.\begin{array}{ccc} \boldsymbol{\ell}_j^T \hat{\mathbf{H}}(\mu_j) = \mathbf{v}_j^T & \text{and} & \hat{\mathbf{H}}(\lambda_i)\mathbf{r}_i = \mathbf{w}_i \\ \text{for } j = 1,\dots,q & & \text{for } i = 1,\dots,k \end{array}\right\}. \tag{8.8}$$

As before, interpolation points and tangential directions are determined by the problem. It should be noted that for SISO systems, i.e., systems with a single input and a single output ($m = p = 1$), left and right directions can be taken equal to one ($\boldsymbol{\ell}_j = 1$, $\mathbf{r}_i = 1$), and hence conditions (8.7) become

$$\hat{\mathbf{H}}(\mu_j) = \mathbf{H}(\mu_j),\ j = 1,\dots,q, \quad \hat{\mathbf{H}}(\lambda_i) = \mathbf{H}(\lambda_i), i = 1,\dots,k, \tag{8.9}$$

while conditions (8.8) become

$$\hat{\mathbf{H}}(\mu_j) = \mathbf{v}_j,\ j = 1,\dots,q, \quad \hat{\mathbf{H}}(\lambda_i) = \mathbf{w}_i,\ i = 1,\dots,k. \tag{8.10}$$

In the following we will consider exclusively *interpolatory model reduction methods* for systems described either by (i) descriptor realizations or (ii) data, measured or computed via DNS. Roughly speaking, we will seek reduced models whose transfer function matches that of the original system at selected frequencies.

Remark 8.2. System identification from frequency response measurements is common in many engineering applications, for instance in electronic design, where S-(scattering), Y-(admittance), or Z-(impedance) *parameters* of chips, packages, or boards are considered, or in civil engineering, where frequency response functions (FRFs), of mechanical structures are given. For instance, given a Z-parameter representation $\mathbf{Y}(s) = \mathbf{H}(s)\mathbf{U}(s)$, $m = p$, the associated *S-parameter representation* is

$$\bar{\mathbf{Y}}(s) = \mathbf{S}(s)\bar{\mathbf{U}}(s) = [\mathbf{H}(s) + \mathbf{I}][\mathbf{H}(s) - \mathbf{I}]^{-1}\bar{\mathbf{U}}(s),$$

[27]DNS stands for direct numerical simulation.

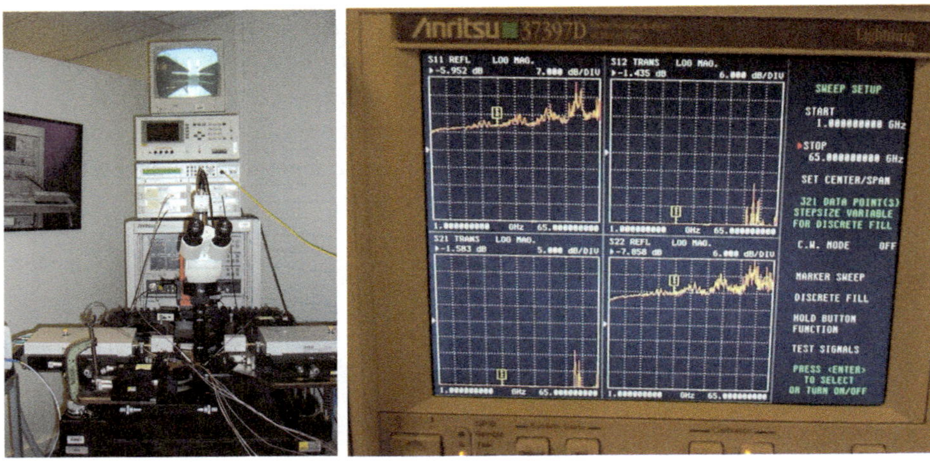

Figure 8.1. *Left pane: VNA. Right pane: screen showing the magnitude of the S-parameters for a two-port (two-input, two-output) device.*

where $\bar{\mathbf{Y}} = \frac{1}{2}(\mathbf{Y}+\mathbf{U})$ and $\bar{\mathbf{U}} = \frac{1}{2}(\mathbf{Y}-\mathbf{U})$ are the *transmitted and reflected waves*, respectively.[28] Important for our purposes is that S-parameters can be measured using *vector network analyzers* (VNAs), as shown in Figure 8.1.

In terms of the problems formulated above, (8.8) applies: given measured S-parameters at given frequencies λ_j, μ_i, we seek to construct a model of low order (complexity) for the underlying (unknown) system.

8.1.4 ▪ Content overview

This chapter starts with the definition of Loewner matrices $\mathbb{L}$, followed by a historical account thereof (Section 8.2.1). Section 8.2.2 discusses the basics in the scalar case. Our starting point is a Lagrange-type interpolation approach that is closely related to the so-called barycentric interpolation formula. Making use of the resulting degrees of freedom leads to the introduction of the Loewner matrix. Next we explore the relationship between rational functions and Loewner matrices, which leads to the fundamental result that given enough data, the rank of any Loewner matrix is equal to the complexity (McMillan degree) of the associated rational function. There follow three ways of constructing rational functions given a Loewner matrix. In Section 8.2.3 the framework is generalized to matrix and tangential interpolation. The new quantity introduced is the shifted Loewner matrix $\mathbb{L}_s$, which, together with $\mathbb{L}$, forms the Loewner pencil. Section 8.2.4 discusses the issues of positive real interpolation in the Loewner framework; the fact that projected systems satisfy interpolation conditions as well as pole and zero placement of interpolants; error expressions; and, last, how to work in real arithmetic. Section 8.2.5 presents several examples illustrating aspects of the Loewner framework for exact data. Section 8.3 introduces the model reduction problem for (approximate or noisy) measured data. The resulting data-driven reduction framework has the unique feature that it provides a trade-off between accuracy of fit and model complexity. Two numerical examples are presented, the second one treating measured data without a known underlying system.

[28]Given a function of time $\mathbf{f}(t)$, its Laplace transform is denoted by $\mathbf{F}(s)$.

What is not treated

Several topics are not covered in this chapter: (a) the multiple-point case (Hermite interpolation) [9, 33]; (b) the recursive framework, in which data are not provided all at once [29, 30]; (c) the generalization of the Loewner framework to linear parametric systems [9, 24, 25]; and (d) generalization to bilinear systems [23]. For the complete Loewner story, we refer the reader to the book [22].

8.2 ▪ The Loewner framework for linear systems

The main ingredient of our approach is the Loewner matrix, which was developed in a series of papers by one of the authors of this chapter (see, e.g., [2, 5, 6]). More recently, important contributions have been made in [33] as well as [9, 24, 25, 29, 30]. In the following, we provide an overview of the Loewner framework.

Definition 8.3 (The Loewner matrix). *Given a row array of pairs of complex numbers* $(\mu_j, \mathbf{v}_j)$, $j = 1,\ldots,q$, *and a column array of pairs of complex numbers* $(\lambda_i, \mathbf{w}_i)$, $i = 1,\ldots,k$, *with* λ_i, μ_j *distinct, the associated Loewner or divided-differences matrix is*

$$\mathbb{L} = \begin{bmatrix} \frac{\mathbf{v}_1-\mathbf{w}_1}{\mu_1-\lambda_1} & \cdots & \frac{\mathbf{v}_1-\mathbf{w}_k}{\mu_1-\lambda_k} \\ \vdots & \ddots & \vdots \\ \frac{\mathbf{v}_q-\mathbf{w}_1}{\mu_q-\lambda_1} & \cdots & \frac{\mathbf{v}_q-\mathbf{w}_k}{\mu_q-\lambda_k} \end{bmatrix} \in \mathbb{C}^{q\times k}. \tag{8.11}$$

If there is a known underlying function $\boldsymbol{\phi}$, *then* $\mathbf{w}_i = \boldsymbol{\phi}(\lambda_i)$ *and* $\mathbf{v}_j = \boldsymbol{\phi}(\mu_j)$.

This matrix was introduced by Karel Löwner (later Charles Loewner) in his seminal paper [32]. For a biography of Loewner, see, e.g., [31].

In the following, we will present the Loewner framework in connection with rational interpolation and consequently in connection with reduced-order modeling of linear dynamical systems given frequency domain data (either measured or computed by DNS). The main property of the Loewner matrix is that its rank encodes information about the minimal admissible complexity of the solutions of the interpolation problem. In the case of measured data, the *numerical rank* of an appropriate Loewner matrix or (as we will see later) of a Loewner pencil needs to be determined.

8.2.1 ▪ Historical remarks

The Loewner matrix was introduced in [32] for the study of *operator convex functions*. The main contribution of [32] is the solution of the operator convexity problem involving the Loewner matrix $\mathbb{L}$ as the main tool. In the process of proving this result, Loewner established the connection of $\mathbb{L}$ with rational interpolation, also known as *Cauchy interpolation*. Since then, the connection of $\mathbb{L}$ with operator monotonicity has been studied extensively; we refer to papers by R. Bhatia and co-authors for details [18, 19]; see also [20]. The connection of $\mathbb{L}$ with rational interpolation (the Cauchy problem) was subsequently pursued by Belevitch [12]. The author rederives the result that, if we attach a large enough Loewner matrix to a given rational function, its rank is equal to the McMillan degree of the given rational function; as remarked by the same author, however, the opposite does not always hold true. This open problem was taken up later by Antoulas and Anderson, and their paper [5] provides the solution. All the above contributions construct interpolants based on computing determinants of submatrices of $\mathbb{L}$. An advance in this respect is constructing interpolants in state-space form in [2].

A breakthrough in the Loewner framework came with the publication of [33]. This paper introduces an additional quantity, the *shifted Loewner matrix*, denoted by $\mathbb{L}_s$. The advantage of this new component comes from the fact that the Loewner pencil $(\mathbb{L}_s, \mathbb{L})$ consisting of the Loewner and the shifted Loewner matrices constitutes a high-order realization of the underlying $(\mathbf{A}, \mathbf{E})$ pencil.

We would also like to refer the reader to the works by Meinguet [35] and Schneider and Werner [36] for a classical view of interpolation. Finally, the work of Amstutz [1] should be mentioned; an approach to rational interpolation based on the Loewner matrix is proposed therein. However, the author was apparently not aware of Loewner's work. More recently, Loewner matrices have also been studied by Fiedler; see, e.g., [21].

8.2.2 ▪ Scalar rational interpolation and the Loewner matrix

Consider the array of pairs of points

$$P = \{(x_i, y_i) : i = 1, \ldots, N,\ x_i \neq x_j,\ i \neq j\}. \tag{8.12}$$

We are looking for rational functions

$$\phi(x) = \frac{\mathbf{n}(x)}{\mathbf{d}(x)}, \tag{8.13}$$

where the numerator and denominator polynomials $\mathbf{n}, \mathbf{d}$ have no common factors, that *interpolate* the points of the array P, i.e.,

$$\phi(x_i) = y_i,\ i = 1, \ldots, N. \tag{8.14}$$

For what follows we will make use of this definition.

Definition 8.4. *The order or complexity of a scalar rational function $\phi(s)$ is*

$$\deg \phi = \max\{\deg \mathbf{n}, \deg \mathbf{d}\}.$$

This is sometimes referred to as the McMillan degree of ϕ.

Minimal degree solutions of the interpolation problem are of particular interest.

Remark 8.5. In array (8.12), the points x_i have been assumed *distinct*. In terms of the interpolation problem, this means that only the value of the underlying rational function is prescribed at each x_i. For the sake of presenting the main ideas as clearly as possible, in this section, only the scalar, distinct-point interpolation problem will be discussed.

A rational Lagrange-type formula

The idea behind the present approach to rational interpolation is to use a formula for rational interpolants that is similar to the one defining the Lagrange polynomial. First we partition the array P into two disjoint subarrays:

$$P_c = \{(\lambda_i, \mathbf{w}_i) : i = 1, \ldots, k\}, \quad P_r = \{(\mu_j, \mathbf{v}_j) : j = 1, \ldots, q\},$$

where, for simplicity of notation, the points have been redefined as

$$\left.\begin{array}{lll} \lambda_i = x_i, & \mathbf{w}_i = y_i, & i = 1,\ldots,k \\ \mu_j = x_{k+j}, & \mathbf{v}_j = y_{k+j}, & j = 1,\ldots,q \end{array}\right\} \quad \text{and} \quad k+q = N.$$

Thus, the first k pairs of points are denoted by λ_i and $\mathbf{w}_i$ and the rest by μ_j and $\mathbf{v}_j$. Next we define a *Lagrange basis* for polynomials of degree at most $k-1$: given $\lambda_i \in \mathbb{C}$, $i = 1,\ldots,q$: $\lambda_i \neq \lambda_j$, $i \neq j$,

$$\mathbf{q}_i(s) = \prod_{\substack{i'=1 \\ i' \neq i}}^{i'=k} (s - \lambda_{i'}),\ i = 1,\ldots,k.$$

For constants α_i, $\mathbf{w}_i$, $i = 1,\ldots,k$, consider $\phi(s)$ defined by

$$\sum_{i=1}^{k} \alpha_i \frac{\phi(s) - \mathbf{w}_i}{s - \lambda_i} = 0, \quad \alpha_i \neq 0. \tag{8.15}$$

Solving for ϕ, we obtain

$$\phi(s) = \frac{\sum_{i=1}^{k} \frac{\alpha_i \mathbf{w}_i}{s-\lambda_i}}{\sum_{i=1}^{k} \frac{\alpha_i}{s-\lambda_i}} = \frac{\sum_{i=1}^{k} \alpha_i \mathbf{w}_i \mathbf{q}_i(s)}{\sum_{i=1}^{k} \alpha_i \mathbf{q}_i(s)}, \quad \alpha_i \neq 0. \tag{8.16}$$

It follows that $\phi(\lambda_i) = \mathbf{w}_i$. This is the *barycentric (rational) Lagrange interpolation* formula. The reference to Lagrange follows from the fact that, if we choose $\alpha_i = \frac{1}{\mathbf{q}_i(\lambda_i)}$, and since $\sum_{i=1}^{k} \frac{\mathbf{q}_i(s)}{\mathbf{q}_i(\lambda_i)} = 1$, we obtain

$$\phi(s) = \phi_{\text{lag}}(s) = \sum_{i=1}^{k} \mathbf{w}_i \frac{\mathbf{q}_i(s)}{\mathbf{q}_i(\lambda_i)};$$

in other words, the rational function $\phi(s)$ becomes the *Lagrange polynomial* $\phi_{\text{lag}}(s)$ interpolating the array P_c. For more information on barycentric rational interpolation, we refer to [16, 17] and references therein.

The free parameters α_i can be determined so that the additional constraints contained in array P_r are satisfied:

$$\phi(\mu_j) = \mathbf{v}_j,\ j = 1,\ldots,q.$$

After inserting these conditions into (8.15), the following equation results:

$$\mathbb{L}\mathbf{c} = 0, \tag{8.17}$$

where $\mathbb{L}$ is the Loewner matrix defined by (8.11) and the vector $\mathbf{c}$ contains the unknowns α_i:

$$\mathbb{L} = \begin{bmatrix} \frac{\mathbf{v}_1 - \mathbf{w}_1}{\mu_1 - \lambda_1} & \cdots & \frac{\mathbf{v}_1 - \mathbf{w}_k}{\mu_1 - \lambda_k} \\ \vdots & \ddots & \vdots \\ \frac{\mathbf{v}_q - \mathbf{w}_1}{\mu_q - \lambda_1} & \cdots & \frac{\mathbf{v}_q - \mathbf{w}_k}{\mu_q - \lambda_k} \end{bmatrix} \in \mathbb{C}^{q \times k}, \ \mathbf{c} = \begin{bmatrix} \alpha_1 \\ \vdots \\ \alpha_k \end{bmatrix} \in \mathbb{C}^k. \tag{8.18}$$

Here, the *Loewner matrix* is constructed by means of the *row array* $(\mu_j, \mathbf{v}_j)$, $j = 1,\ldots,q$, and the *column array* $(\lambda_i, \mathbf{w}_i)$, $i = 1,\ldots,k$.

Remark 8.6 (Hankel and Loewner matrices). If the value of successive derivatives at the same points is prescribed, we are dealing with the so-called *Hermite* or interpolation problem with *multiplicities*.

As shown in [5], the (generalized) Loewner matrix associated with an array P as in (8.12), but consisting of *one point with multiplicity* N, has *Hankel* structure. This matrix is actually the same as the Hankel matrix of the corresponding realization problem, hinting at the fact that the Loewner matrix is the right tool to generalize realization theory to rational interpolation.

The theory, part of which is presented in the following, has been worked out for the multiple-point as well as for the (more general) matrix and tangential interpolation problems; for details, the reader is referred to [2, 5, 30, 33].

From rational functions to Loewner matrices

Consider a rational function together with some samples. The key result in connection with the Loewner matrix is the following.

Lemma 8.7. Main property. *Given the rational function ϕ and an array of points P, where $y_i = \phi(x_i)$ and x_i is not a pole of ϕ, let $\mathbb{L}$ be a $q \times k$ Loewner matrix for some partitioning P_c, P_r of P. Then,*

$$q, k \geq \deg\phi \;\;\Rightarrow\;\; \operatorname{rank}\mathbb{L} = \deg\phi.$$

It follows that every square sub-Loewner matrix of size $\deg\phi$ *is nonsingular.*

This is a pivotal result in our approach. The proof presented next is based on the following known result.

Proposition 8.8. *Let $(\mathbf{E}, \mathbf{A}, \mathbf{B})$ be a controllable descriptor triple and λ_i, $i = 1,\ldots,r$, be distinct scalars that are not eigenvalues of the pencil of $n \times n$ matrices $(\mathbf{A}, \mathbf{E})$. It follows that*

$$\operatorname{rank}\left[(\lambda_1\mathbf{E}-\mathbf{A})^{-1}\mathbf{B} \;\;\cdots\;\; (\lambda_r\mathbf{E}-\mathbf{A})^{-1}\mathbf{B}\right] = n$$

provided that $r \geq n$.

For a proof of a similar version of this result, see [2]. Based on this proposition, we can now provide a state-space proof of Lemma 8.7.

Proof (Lemma 8.7). Let $(\mathbf{E}_\delta, \mathbf{A}_\delta, \mathbf{B}_\delta, \mathbf{C}_\delta)$ be a minimal descriptor realization of ϕ of order n (recall notation (8.5)):

$$\phi(s) = \mathbf{C}_\delta(s\mathbf{E}_\delta - \mathbf{A}_\delta)^{-1}\mathbf{B}_\delta.$$

This implies

$$(\mathbb{L})_{i,j} = \frac{\mathbf{v}_i - \mathbf{w}_j}{\mu_i - \lambda_j} = -\mathbf{C}_\delta(\mu_i\mathbf{E}_\delta - \mathbf{A}_\delta)^{-1}\mathbf{E}_\delta(\lambda_j\mathbf{E}_\delta - \mathbf{A}_\delta)^{-1}\mathbf{B}_\delta$$

for $i = 1,\ldots,q$, $j = 1,\ldots,k$. Consequently, $\mathbb{L}$ can be factorized as

$$\mathbb{L} = -\mathscr{O}_q\,\mathbf{E}_\delta\,\mathscr{R}_k,$$

where matrices $\mathcal{R}_k$, $\mathcal{O}_q$ are defined in (8.3), (8.4), respectively. In analogy with the realization problem (where the Hankel matrix factors in a product of an observability times a controllability matrix), we will call $\mathcal{O}_q$ the *generalized observability matrix* and $\mathcal{R}_k$ the *generalized controllability matrix* associated with the sets of scalars λ_i and μ_j.

Because of Proposition 8.8, the rank of both $\mathcal{O}_q$ and $\mathcal{R}_k$ is n. This implies that the rank of $\mathbb{L}$ is equal to the rank of $\mathbf{E}_\delta$, which, in turn, is equal to the rank of $\mathbf{E}$ (recall (8.5)). Then, according to [28], the rank of $\mathbf{E}$ is equal to the McMillan degree of $\phi(s)$.

Finally, since every square sub-Loewner matrix is a Loewner matrix for some subset of the row and column arrays, it has the same full-rank property. This completes the proof of the main lemma. □

The above result was derived by Löwner [32]; a different proof can be found in [1] and [7].

From Loewner matrices to interpolating functions

Given interpolation data (8.12), we are now ready to tackle the interpolation problem (8.13), (8.14). In this section, we construct interpolating functions. Three construction methods will be detailed: the first is based on the barycentric formula and constructs the numerator and denominator polynomials of interpolants, while the remaining two methods describe descriptor (generalized state-space) approaches to constructing interpolants.

Polynomial construction of interpolants. Given the array of points P defined by (8.12), the developments in this section follow the references cited earlier as well as [12]. The following definition will be needed.

Definition 8.9. *The rank of the array P is*

$$\textit{rank}\, P = \max_{\mathbb{L}} \{\textit{rank}\, \mathbb{L}\} = n,$$

where the maximum is taken over all possible Loewner matrices that can be formed from P.

A consequence of Lemma 8.7 is that the rank of all Loewner matrices with at least n rows and n columns is equal to n. Assume that $2n < N$. For any Loewner matrix with $\operatorname{rank} \mathbb{L} = n$ and n rows, there exists a column vector $\mathbf{c}$ satisfying

$$\mathbb{L}\mathbf{c} = 0, \quad \mathbf{c} \in \mathbb{C}^k, \quad k = N - n. \tag{8.19}$$

In this case, we can attach to $\mathbb{L}$ a rational function denoted by

$$\phi_{\mathbb{L}}(s) = \frac{\mathbf{n}_{\mathbb{L}}(s)}{\mathbf{d}_{\mathbb{L}}(s)}, \tag{8.20}$$

using the barycentric formula (8.16), where $(\mathbf{c})_{j,1} = \alpha_j$, i.e.,

$$\mathbf{n}_{\mathbb{L}}(s) = \sum_{j=1}^{k} \frac{\alpha_j \mathbf{w}_j}{s - \lambda_j}, \quad \mathbf{d}_{\mathbb{L}}(s) = \sum_{j=1}^{k} \frac{\alpha_j}{s - \lambda_j}. \tag{8.21}$$

The rational function $\phi_{\mathbb{L}}$ has the following properties.

Lemma 8.10. (a) $\deg \phi_{\mathbb{L}} \leq n < N$.
(b) *There is a unique* $\phi_{\mathbb{L}}$, *attached to all* $\mathbb{L}$ *and* $\mathbf{c}$ *satisfying* (8.19), *as long as* $\operatorname{rank} \mathbb{L} = n$.
(c) *The numerator and denominator polynomials* $\mathbf{n}_{\mathbb{L}}$, $\mathbf{d}_{\mathbb{L}}$ *have* $n - \deg \phi_{\mathbb{L}}$ *common factors of the form* $s - \lambda_i$.
(d) $\phi_{\mathbb{L}}$ *interpolates exactly* $N - n + \deg \phi_{\mathbb{L}}$ *points of the array* P.

As a consequence of Lemma 8.10 and Lemma 8.7, we obtain the following.

Corollary 8.11. *The rational function* $\phi_{\mathbb{L}}$ *interpolates all given points if and only if* $\deg \phi_{\mathbb{L}} = n$ *if and only if all* $n \times n$ *Loewner matrices that can be formed from the data array* P *are nonsingular.*

We are now ready to state the main result in [5].

Theorem 8.12. *Given the array of* N *pairs of points* P, *let* $\operatorname{rank} P = n$.
(a) *If* $2n < N$ *and all square Loewner matrices of size* n *that can be formed from* P *are nonsingular, there is a unique interpolating function of minimal degree denoted by* $\phi^{\min}(s)$ *and* $\deg \phi^{\min} = n$.
(b) *Otherwise,* $\phi^{\min}(s)$ *is not unique and* $\deg \phi^{\min} = N - n$.

Corollary 8.13. *Under the assumptions of Theorem 8.12*(a), *the admissible degrees of interpolants are* n *and any integer larger than or equal to* $N - n$. *Otherwise, if Theorem 8.12*(b) *holds, i.e.,* $2n > N$, *the admissible degrees are all integers greater than or equal to* $N - n$.

The proof of the four results quoted above can be found in [5].

Remark 8.14. (a) If $2n = N$, the only solution $\mathbf{c}$ of (8.19) is $\mathbf{c} = 0$. Hence, $\phi_{\mathbb{L}}$ defined by (8.20) does not exist, and part (b) of Theorem 8.12 applies.
(b) To distinguish between case (a) and case (b) of Theorem 8.12, we only need to check the nonsingularity of $2n + 1$ Loewner matrices. For details, see [5].
(c) Given the array P, let r be an admissible degree. For the polynomial construction, we need to form *any* Loewner matrix with $r + 1$ columns,

$$\mathbb{L}_r \in \mathbb{R}^{(N-r-1)\times(r+1)},$$

and determine a parametrization of all $\mathbf{c}_r$ such that $\mathbb{L}_r \mathbf{c}_r = 0$. A parametrization of all interpolating functions of degree r is then $\phi_{\mathbb{L}_r}(s) = \frac{\mathbf{n}_{\mathbb{L}_r}(s)}{\mathbf{d}_{\mathbb{L}_r}(s)}$, where the numerator and denominator polynomials are defined by (8.21). If $r \geq N - n$, we have to make sure that there are no common factors between the numerator and denominator of $\phi_{\mathbb{L}_r}$; this is the case for almost all $\mathbf{c}_r$. More precisely, the $2r + 1 - N$ (scalar) parameters that parametrize all $\mathbf{c}_r$ have to avoid the hypersurfaces defined by the equations

$$\mathbf{d}_{\mathbb{L}_r}(\lambda_i) = 0, \ i = 1, \ldots, r + 1.$$

Since we can always make sure that $\mathbf{c}_r$ depends affinely on these parameters, we are actually dealing with hyperplanes. For details and examples, see [5, 33].

Descriptor realization of interpolants based on c. We wish to derive a descriptor realization of the rational interpolants obtained above by means of the solution $\mathbf{c}$ of $\mathbb{L}\mathbf{c} = \mathbf{0}$. For details we refer to [9].

Assume that $k = q+1$, and recall the rational function $\phi(s)$ expressed in barycentric form (8.16):

$$\phi(s) = \frac{\sum_{i=0}^{q} \beta_i \, \mathbf{q}_i(s)}{\sum_{i=0}^{q} \alpha_i \, \mathbf{q}_i(s)}, \quad \beta_i = \alpha_i \mathbf{w}_i.$$

We define

$$\mathbf{J}_{\text{lag}}(\xi;q) = \begin{bmatrix} \xi - x_1 & x_2 - \xi & & & \\ \xi - x_1 & 0 & x_3 - \xi & & \\ \vdots & & \ddots & \ddots & \\ \xi - x_1 & & & 0 & x_{q+1} - \xi \end{bmatrix} \in \mathbb{C}^{q\times(q+1)},$$

where, for simplicity, we assume that $x_i \neq x_j$, $i \neq j$, and

$$\mathbf{a} = [\alpha_0, \ \alpha_1, \ \dots, \ \alpha_q], \ \mathbf{b} = [\beta_0, \ \beta_1, \ \dots, \ \beta_q] \in \mathbb{R}^{1\times(q+1)}.$$

Then, we have the following.

Lemma 8.15. *The following triple constitutes a descriptor realization of* ϕ*:*

$$\mathbf{C} = \mathbf{b}, \ \ \mathbf{\Phi}(s) = \begin{bmatrix} \mathbf{J}_{\text{lag}}(s;q) \\ \mathbf{a} \end{bmatrix}, \ \ \mathbf{B} = \mathbf{e}_{q+1}, \tag{8.22}$$

i.e., $\phi(s) = \mathbf{C}\mathbf{\Phi}(s)^{-1}\mathbf{B}$*, where* $x_i = \lambda_i$*,* $i = 1, \dots, q+1$*. The dimension of this realization is* $q+1$ *and it can represent arbitrary rational functions, including polynomials. Also, it is R-controllable and R-observable, that is,* $[\mathbf{\Phi}(s), \ \mathbf{B}]$ *and* $[\mathbf{\Phi}^T(s), \ \mathbf{C}^T]$ *have full rank for all* $s \in \mathbb{C}$*, provided that the numerator and denominator of* ϕ *have no common factors.*

Descriptor realization of interpolants without the use of c. Given the interpolation data left (row) array $(\mu_i, \mathbf{v}_i)$, $i = 1, \dots, q$, and right (column) array $(\lambda_j, \mathbf{w}_j)$, $j = 1, \dots, q+1$, let $\mathbb{L} \in \mathbb{C}^{q\times(q+1)}$ be the associated Loewner matrix. We will assume that all $q \times q$ Loewner matrices constructed from these data are nonsingular and hence there exists a unique interpolant of degree q. In this section we will show how to construct this interpolant *without* the explicit use of $\mathbf{c}$ satisfying $\mathbb{L}\mathbf{c} = \mathbf{0}$. Toward this goal, we partition the Loewner matrix as

$$\mathbb{L} = [\hat{\mathbb{L}} \quad \ell], \ \text{ where } \mathbb{L} \in \mathbb{C}^{q\times(q+1)}, \ \hat{\mathbb{L}} \in \mathbb{C}^{q\times q}, \ \ell \in \mathbb{C}^{q}. \tag{8.23}$$

Define the matrix containing the factors making up the Lagrange basis as

$$\mathbf{\Pi}(s) = \begin{bmatrix} \mathbf{p}_1(s) & & \\ & \ddots & \\ & & \mathbf{p}_q(s) \\ -\mathbf{p}_{q+1}(s) & \cdots & -\mathbf{p}_{q+1}(s) \end{bmatrix} \in \mathbb{C}^{(q+1)\times q}, \ \ \mathbf{p}_i(s) = s - \lambda_i.$$

It turns out that realization (8.24) below can be interpreted in terms of *one-sided shifted Loewner matrices*. We stress the *one-sided* aspect as, later on, we will encounter *two-sided* shifted Loewner matrices (see (8.30)). Explicitly, given $\mathbb{L} \in \mathbb{C}^{q\times k}$ defined by (8.11), the *one-sided shifted Loewner matrix* is

$$\mathbb{L}_\Lambda = \begin{bmatrix} \lambda_1 \frac{\mathbf{v}_1-\mathbf{w}_1}{\mu_1-\lambda_1} & \cdots & \lambda_k \frac{\mathbf{v}_1-\mathbf{w}_k}{\mu_1-\lambda_k} \\ \vdots & \ddots & \vdots \\ \lambda_1 \frac{\mathbf{v}_q-\mathbf{w}_1}{\mu_q-\lambda_1} & \cdots & \lambda_k \frac{\mathbf{v}_q-\mathbf{w}_k}{\mu_q-\lambda_k} \end{bmatrix} = \mathbb{L}\mathbf{\Lambda} \in \mathbb{C}^{q\times k},$$

where $\mathbf{\Lambda} = \operatorname{diag}[\lambda_1,\ldots,\lambda_k]$. In this framework a realization is as follows.

Lemma 8.16. *For $k=q+1$, define the $(q+1)\times q$ matrix $\mathbb{J} = \begin{bmatrix} \mathbb{I} \\ -\mathbf{1} \end{bmatrix}$, where $\mathbb{I}$ is the identity of size q and $\mathbf{1} = ones(1,q)$. Partition $\mathbb{L}$ as in (8.23). With $\mathbf{W} = [\mathbf{w}_1 \ \cdots \ \mathbf{w}_q \ \ \mathbf{w}_{q+1}]$ and $\mathbf{V} = [\mathbf{v}_1 \cdots \mathbf{v}_q]^T$, the associated descriptor realization is*

$$\mathbf{H}(s) = \mathbf{w}_{q+1} - \mathbf{p}_{q+1}(s)(\mathbf{W}\mathbb{J})[(s\mathbb{L}-\mathbb{L}_\Lambda)\mathbb{J}]^{-1}\boldsymbol{\ell}. \tag{8.24}$$

If the interpolant is strictly proper rational (i.e., the sum of the entries of the vector $\mathbf{c}$ *satisfying* $[\hat{\mathbb{L}} \ \ \boldsymbol{\ell}]\mathbf{c} = \mathbf{0}$ *is not equal to zero), the interpolant can be expressed as*

$$\mathbf{H}(s) = (\mathbf{W}\mathbb{J})[(s\mathbb{L}-\mathbb{L}_\Lambda)\mathbb{J}]^{-1}\mathbf{V}. \tag{8.25}$$

Proof. First, recall that, given $\mathbb{L}$ and $\mathbf{c}$ such that $\mathbb{L}\mathbf{c} = \mathbf{0}$, according to the barycentric formula, the corresponding interpolant can be written as in (8.21), namely

$$\boldsymbol{\phi}(s) = \frac{\mathbf{n}(s)}{\mathbf{d}(s)} = \frac{\sum_{i=1}^{q+1} \frac{\alpha_i \mathbf{w}_i}{s-\lambda_i}}{\sum_{i=1}^{q+1} \frac{\alpha_i}{s-\lambda_i}}.$$

Assuming, without loss of generality, that $\alpha_{q+1} = 1$, we can write $[\hat{\mathbb{L}} \ \ \boldsymbol{\ell}][\alpha_1,\ldots,\alpha_q,1]^T = \mathbf{0}$. Hence, $[\alpha_1,\ldots,\alpha_q]^T = -\hat{\mathbb{L}}^{-1}\boldsymbol{\ell}$. Thus, the numerator polynomial of $\boldsymbol{\phi}(s)$ times $(s-\lambda_{q+1})$ can be written as:

$$(s-\lambda_{q+1})\mathbf{n}(s) = \mathbf{w}_{q+1} - (s-\lambda_{q+1})\big[\mathbf{w}_1 \ \cdots \ \mathbf{w}_q\big]\Big(\operatorname{diag}\big[s-\lambda_1,\ldots,s-\lambda_q\big]\Big)^{-1}\hat{\mathbb{L}}^{-1}\boldsymbol{\ell},$$

while the denominator polynomial times the same factor can be written as

$$(s-\lambda_{q+1})\mathbf{d}(s) = 1 - (s-\lambda_{q+1})[1 \ \cdots \ 1]\Big(\operatorname{diag}\big[s-\lambda_1,\ldots,s-\lambda_q\big]\Big)^{-1}\hat{\mathbb{L}}^{-1}\boldsymbol{\ell}.$$

Consequently, the quotient is

$$\boldsymbol{\phi}(s) = \frac{\mathbf{n}(s)}{\mathbf{d}(s)}$$

$$= \mathbf{w}_{q+1} - \frac{(s-\lambda_{q+1})\big[\mathbf{w}_1 - \mathbf{w}_{q+1} \ \cdots \ \mathbf{w}_q - \mathbf{w}_{q+1}\big]\Big(\operatorname{diag}\big[s-\lambda_1,\ldots,s-\lambda_q\big]\Big)^{-1}\hat{\mathbb{L}}^{-1}\boldsymbol{\ell}}{1-(s-\lambda_{q+1})[1 \ \cdots \ 1]\Big(\operatorname{diag}\big[s-\lambda_1,\ldots,s-\lambda_q\big]\Big)^{-1}\hat{\mathbb{L}}^{-1}\boldsymbol{\ell}}. \tag{$*$}$$

The *Sherman–Morrison–Woodbury formula*[29] implies

$$\frac{\left(\mathrm{diag}\left[(s-\lambda_1),\cdots,(s-\lambda_q)\right]\right)^{-1}\hat{\mathbb{L}}^{-1}\ell}{1-(s-\lambda_{q+1})[1\ \cdots\ 1]\left(\mathrm{diag}\left[(s-\lambda_1),\ldots,(s-\lambda_q)\right]\right)^{-1}\hat{\mathbb{L}}^{-1}\ell}$$

$$=\left[\mathrm{diag}\left[(s-\lambda_1),\ldots,(s-\lambda_q)\right]-(s-\lambda_{q+1})\hat{\mathbb{L}}^{-1}\ell[1\cdots 1]\right]^{-1}\hat{\mathbb{L}}^{-1}\ell=[(s\mathbb{L}-\mathbb{L}_\Lambda)\mathbb{J}]^{-1}\ell.$$

Combining this with expression $(*)$ yields the desired (8.24). The second expression for $\mathbf{H}(s)$ is proved similarly. □

Remark 8.17. The method of constructing interpolants by means of the barycentric formula (8.21) and by means of the descriptor realization in Lemma 8.15 can be used to obtain any interpolant of admissible degree. The method presented in Lemma 8.16, howwever, applies only to minimal interpolants.

8.2.3 ▪ Matrix rational interpolation and the Loewner pencil

In this section, we will formulate the results outlined above for the more general *tangential interpolation* problem defined in (8.8). We are given, respectively, the *right* or *column data* and the *left* or *row data*:

$$(\lambda_i;\mathbf{r}_i,\mathbf{w}_i),\ i=1,\ldots,k,\quad (\mu_j;\ell_j^T,\mathbf{v}_j^T),\ j=1,\ldots,q.$$

It is assumed for simplicity that all points, i.e., λ_i and μ_j, are distinct (for the general case, see [33]). The *right data* are organized as

$$\left.\begin{array}{l}\Lambda=\mathrm{diag}\,[\lambda_1,\ \ldots,\ \lambda_k]\in\mathbb{C}^{k\times k}\\ \mathbf{R}=[\mathbf{r}_1\ \mathbf{r}_2\ \cdots\ \mathbf{r}_k]\in\mathbb{C}^{m\times k}\\ \mathbf{W}=[\mathbf{w}_1\ \mathbf{w}_2\ \cdots\ \mathbf{w}_k]\in\mathbb{C}^{p\times k}\end{array}\right\},\tag{8.26}$$

and the *left data* are organized as

$$\left.\begin{array}{l}\mathbf{M}=\mathrm{diag}\left[\mu_1,\ \ldots,\ \mu_q\right]\in\mathbb{C}^{q\times q}\\ \mathbf{L}^T=[\ \ell_1\ \ \cdots\ \ \ell_q\ ]\in\mathbb{C}^{p\times q}\\ \mathbf{V}^T=[\ \mathbf{v}_1\ \ \cdots\ \ \mathbf{v}_q\ ]\in\mathbb{C}^{m\times q}\end{array}\right\}.\tag{8.27}$$

The associated *Loewner* and *shifted Loewner* matrices, referred to as the *Loewner pencil*, are constructed next. The *Loewner matrix* for tangential data (first introduced in [33]) is

$$\mathbb{L}=\begin{bmatrix}\frac{\mathbf{v}_1^T\mathbf{r}_1-\ell_1^T\mathbf{w}_1}{\mu_1-\lambda_1} & \cdots & \frac{\mathbf{v}_1^T\mathbf{r}_k-\ell_1^T\mathbf{w}_k}{\mu_1-\lambda_k}\\ \vdots & \ddots & \vdots\\ \frac{\mathbf{v}_q^T\mathbf{r}_1-\ell_q^T\mathbf{w}_1}{\mu_q-\lambda_1} & \cdots & \frac{\mathbf{v}_q^T\mathbf{r}_k-\ell_q^T\mathbf{w}_k}{\mu_q-\lambda_k}\end{bmatrix}\in\mathbb{C}^{q\times k}.\tag{8.28}$$

[29]Namely $[\mathbf{A}-\mathbf{u}\mathbf{v}^T]^{-1}=\mathbf{A}^{-1}+\frac{1}{1-\mathbf{v}^T\mathbf{A}^{-1}\mathbf{u}}\mathbf{A}^{-1}\mathbf{u}\mathbf{v}^T\mathbf{A}^{-1}\Rightarrow[\mathbf{A}-\mathbf{u}\mathbf{v}^T]^{-1}\mathbf{u}=[\frac{1}{1-\mathbf{v}^T\mathbf{A}^{-1}\mathbf{u}}]\,\mathbf{A}^{-1}\mathbf{u}.$

Notice that the quantities $\mathbf{v}_i^T \mathbf{r}_j$ and $\ell_i^T \mathbf{w}_j$ are (complex) scalars. It is readily checked that $\mathbb{L}$ satisfies the Sylvester equation

$$\mathbf{M}\mathbb{L} - \mathbb{L}\mathbf{\Lambda} = \mathbf{V}\mathbf{R} - \mathbf{L}\mathbf{W}. \tag{8.29}$$

The *shifted Loewner matrix*, first introduced in [33] for both scalar and tangential (and consequently also matrix) data, is defined as

$$\mathbb{L}_s = \begin{bmatrix} \frac{\mu_1 \mathbf{v}_1^T \mathbf{r}_1 - \ell_1^T \mathbf{w}_1 \lambda_1}{\mu_1 - \lambda_1} & \cdots & \frac{\mu_1 \mathbf{v}_1^T \mathbf{r}_k - \ell_1^T \mathbf{w}_k \lambda_k}{\mu_1 - \lambda_k} \\ \vdots & \ddots & \vdots \\ \frac{\mu_q \mathbf{v}_q^T \mathbf{r}_1 - \ell_q^T \mathbf{w}_1 \lambda_1}{\mu_q - \lambda_1} & \cdots & \frac{\mu_q \mathbf{v}_q^T \mathbf{r}_k - \ell_q^T \mathbf{w}_k \lambda_k}{\mu_q - \lambda_k} \end{bmatrix} \in \mathbb{C}^{q \times k}, \tag{8.30}$$

and it is straightforward to check that it satisfies the Sylvester equation

$$\mathbf{M}\mathbb{L}_s - \mathbb{L}_s\mathbf{\Lambda} = \mathbf{M}\mathbf{V}\mathbf{R} - \mathbf{L}\mathbf{W}\mathbf{\Lambda}. \tag{8.31}$$

If the data (8.26), (8.27) are sampled from a system with transfer function $\mathbf{H}(s) = \mathbf{C}_\delta(s\mathbf{E}_\delta - \mathbf{A}_\delta)^{-1}\mathbf{B}_\delta$[30] of size $p \times m$ with $\mathbf{A}_\delta, \mathbf{E}_\delta \in \mathbb{R}^{n \times n}$, we define

$$\left.\begin{aligned} \mathcal{O}_q &= \begin{bmatrix} \ell_1^T \mathbf{C}_\delta(\mu_1 \mathbf{E}_\delta - \mathbf{A}_\delta)^{-1} \\ \vdots \\ \ell_q^T \mathbf{C}_\delta(\mu_q \mathbf{E}_\delta - \mathbf{A}_\delta)^{-1} \end{bmatrix} \\ \mathcal{R}_k &= \left[(\lambda_1 \mathbf{E}_\delta - \mathbf{A}_\delta)^{-1}\mathbf{B}_\delta \mathbf{r}_1, \cdots, (\lambda_k \mathbf{E}_\delta - \mathbf{A}_\delta)^{-1}\mathbf{B}_\delta \mathbf{r}_k\right] \end{aligned}\right\} \tag{8.32}$$

of size $q \times n$, $n \times k$, respectively. Similarly to their scalar counterparts, these are called the *generalized tangential observability* and *generalized tangential controllability* matrices. It follows that the Loewner pencil constructed from tangential data has a system-theoretic interpretation in terms of the tangential controllability and observability matrices. Given left interpolation data $\mathbf{v}_j^T = \ell_j^T \mathbf{H}(\mu_j)$ and right interpolation data $\mathbf{w}_i = \mathbf{H}(\lambda_i)\mathbf{r}_i$,

$$\begin{aligned} (\mathbb{L})_{j,i} &= \frac{\mathbf{v}_j^T \mathbf{r}_i - \ell_j^T \mathbf{w}_i}{\mu_j - \lambda_i} = \frac{\ell_j^T \mathbf{H}(\mu_j)\mathbf{r}_i - \ell_j^T \mathbf{H}(\lambda_i)\mathbf{r}_i}{\mu_j - \lambda_i} \\ &= -\ell_j^T \mathbf{C}_\delta(\mu_j \mathbf{E}_\delta - \mathbf{A}_\delta)^{-1}\mathbf{E}_\delta(\lambda_i \mathbf{E}_\delta - \mathbf{A}_\delta)^{-1}\mathbf{B}_\delta \mathbf{r}_i \quad \text{and} \\ (\mathbb{L}_s)_{j,i} &= \frac{\mu_j \mathbf{v}_i^T \mathbf{r}_i - \lambda_i \ell_j^T \mathbf{w}_i}{\mu_j - \lambda_i} = \frac{\mu_j \ell_j^T \mathbf{H}(\mu_j)\mathbf{r}_i - \lambda_i \ell_j^T \mathbf{H}(\lambda_i)\mathbf{r}_i}{\mu_j - \lambda_i} \\ &= -\ell_j^T \mathbf{C}_\delta(\mu_j \mathbf{E}_\delta - \mathbf{A}_\delta)^{-1}\mathbf{A}_\delta(\lambda_i \mathbf{E}_\delta - \mathbf{A}_\delta)^{-1}\mathbf{B}_\delta \mathbf{r}_i. \end{aligned}$$

Thus, with notation (8.32), we obtain

$$\mathbb{L} = -\mathcal{O}_q \mathbf{E}_\delta \mathcal{R}_k \quad \text{and} \quad \mathbb{L}_s = -\mathcal{O}_q \mathbf{A}_\delta \mathcal{R}_k.$$

[30]Following Remark 8.1, we assume that a possible $\mathbf{D}$-term is incorporated in the remaining matrices, as in (8.5).

Lemma 8.18. *Given (tangential) samples of a rational function defined in terms of a minimal descriptor realization* $(\mathbf{E}_{\mathcal{S}}, \mathbf{A}_{\mathcal{S}}, \mathbf{B}_{\mathcal{S}}, \mathbf{C}_{\mathcal{S}})$ *as in Remark* 8.1, *construct the associated Loewner and shifted Loewner matrices* $\mathbb{L}$, $\mathbb{L}_s$. *Assuming that we have enough samples, and that the left and right tangential directions* $\boldsymbol{\ell}_j$, $\mathbf{r}_i$ *are chosen so that* $\mathcal{O}_q$ *and* $\mathcal{R}_k$ *have full rank, we have the following:*

(a) $rank\,\mathbb{L} = rank\,\mathbf{E}_{\mathcal{S}} = rank\,\mathbf{E} = \begin{cases} \text{McMillan degree of the} \\ \text{underlying rational function.} \end{cases}$

(b) $rank\,\mathbb{L}_s = rank\,\mathbf{A}_{\mathcal{S}} = rank\,\mathbf{A} + rank\,\mathbf{D}$.

Remark 8.19. (a) In the SISO case, i.e., $m = p = 1$, in other words, the case where the associated transfer function is rational and scalar, the Loewner pencil contains $\mathbb{L}$ defined by (8.11) and

$$\mathbb{L}_s = \begin{bmatrix} \frac{\mu_1 \mathbf{v}_1 - \mathbf{w}_1 \lambda_1}{\mu_1 - \lambda_1} & \cdots & \frac{\mu_1 \mathbf{v}_1 - \mathbf{w}_k \lambda_k}{\mu_1 - \lambda_k} \\ \vdots & \ddots & \vdots \\ \frac{\mu_q \mathbf{v}_q - \mathbf{w}_1 \lambda_1}{\mu_q - \lambda_1} & \cdots & \frac{\mu_q \mathbf{v}_q - \mathbf{w}_k \lambda_k}{\mu_q - \lambda_k} \end{bmatrix}. \tag{8.33}$$

(b) The issue of choice of tangential directions mentioned in Lemma 8.18 is illustrated in Example 8.36 below.
(c) Loewner matrices are sometimes referred to as *generalized Cauchy matrices*. See, e.g., [26, p. 343], and references therein. Such matrices possess what is known as *displacement structure*. In our case, this is a direct consequence of the fact that $\mathbb{L}$ and $\mathbb{L}_s$ satisfy the Sylvester equations (8.29), (8.31), where the right-hand side has low rank ($m + p$ is general).

We are now ready to state the main result concerning the construction of interpolants using the Loewner pencil.

Theorem 8.20 (Minimal amount of data). *Assume that* $k = q$ *and let* $(\mathbb{L}_s, \mathbb{L})$ *be a regular pencil with no* μ_i *or* λ_j *being an eigenvalue.*
(a) *The quadruple*

$$\mathbf{E}_{\mathcal{S}} = -\mathbb{L}, \;\; \mathbf{A}_{\mathcal{S}} = -\mathbb{L}_s, \;\; \mathbf{B}_{\mathcal{S}} = \mathbf{V}, \;\; \mathbf{C}_{\mathcal{S}} = \mathbf{W} \tag{8.34}$$

is a minimal descriptor realization of an interpolant of the data, i.e.,

$$\mathbf{H}(s) = \mathbf{W}(\mathbb{L}_s - s\mathbb{L})^{-1}\mathbf{V} \tag{8.35}$$

is a rational interpolant of the data.
(b) *If the solution is not unique, all solutions of the same McMillan degree are parametrized in terms of*

$$\mathbf{E} = -\mathbb{L}, \; \mathbf{A} = -(\mathbb{L}_s + \mathbf{LKR}), \; \mathbf{B} = \mathbf{V} - \mathbf{LK}, \; \mathbf{C} = \mathbf{W} - \mathbf{KR}, \; \mathbf{D} = \mathbf{K}, \tag{8.36}$$

where the parameter $\mathbf{K} \in \mathbb{C}^{p \times m}$.

Proof. [33] (a) Multiplying equation (8.29) by s and subtracting it from equation (8.31), we get

$$\mathbf{M}(\mathbb{L}_s - s\mathbb{L}) - (\mathbb{L}_s - s\mathbb{L})\mathbf{\Lambda} = (\mathbf{M} - s\mathbf{I})\mathbf{VR} - \mathbf{LW}(\mathbf{\Lambda} - s\mathbf{I}).$$

Multiplying this equation by $\mathbf{e}_i$ on the right and setting $s = \lambda_i$, we obtain

$$(\mathbf{M}-\lambda_i\mathbf{I})(\mathbb{L}_s-\lambda_i\mathbb{L})\mathbf{e}_i=(\mathbf{M}-\lambda_i\mathbf{I})\mathbf{V}\mathbf{r}_i \Rightarrow$$
$$(\mathbb{L}_s-\lambda_i\mathbb{L})\mathbf{e}_i=\mathbf{V}\mathbf{r}_i \Rightarrow \mathbf{W}\mathbf{e}_i=\mathbf{W}(\mathbb{L}_s-\lambda_i\mathbb{L})^{-1}\mathbf{V}\mathbf{r}_i.$$

Therefore $\mathbf{w}_i=\mathbf{H}(\lambda_i)\mathbf{r}_i$. This proves right tangential interpolation. To prove the left tangential interpolation property, we multiply the above equation by $\mathbf{e}_j^T$ on the left and set $s = \mu_j$:

$$\mathbf{e}_j^T(\mathbb{L}_s-\mu_j\mathbb{L})(\mathbf{\Lambda}-\mu_j\mathbf{I})=\mathbf{e}_j^T\mathbf{L}\mathbf{W}(\mathbf{\Lambda}-\mu_j\mathbf{I}) \Rightarrow$$
$$\mathbf{e}_j^T(\mathbb{L}_s-\mu_j\mathbb{L})=\boldsymbol{\ell}_j\mathbf{W} \Rightarrow \mathbf{e}_j^T\mathbf{V}=\boldsymbol{\ell}_j^T\mathbf{W}(\mathbb{L}_s-\mu_j\mathbb{L})^{-1}\mathbf{V}.$$

Therefore, $\mathbf{v}_j^T=\boldsymbol{\ell}_j^T\mathbf{H}(\mu_j)$, which proves the left property.

(b) With $\mathbf{K}\in\mathbb{C}^{p\times m}$, equations (8.29), (8.31) can be rewritten as

$$\mathbf{M}\mathbb{L}-\mathbb{L}\mathbf{\Lambda}=(\mathbf{V}-\mathbf{L}\mathbf{K})\mathbf{R}-\mathbf{L}(\mathbf{W}-\mathbf{K}\mathbf{R}),$$
$$\mathbf{M}(\mathbb{L}_s+\mathbf{L}\mathbf{K}\mathbf{R})-(\mathbb{L}_s+\mathbf{L}\mathbf{K}\mathbf{R})\mathbf{\Lambda}=\mathbf{M}(\mathbf{V}-\mathbf{L}\mathbf{K})\mathbf{R}-\mathbf{L}(\mathbf{W}-\mathbf{K}\mathbf{R})\mathbf{\Lambda}.$$

Repeating the procedure with the new quantities

$$\bar{\mathbb{L}}_s=\mathbb{L}_s+\mathbf{L}\mathbf{K}\mathbf{R},\quad \bar{\mathbf{V}}=\mathbf{V}-\mathbf{L}\mathbf{K},\quad \bar{\mathbf{W}}=\mathbf{W}-\mathbf{K}\mathbf{R}$$

(the Loewner matrix remains unchanged), the desired result follows. □

The case of redundant data

We will now consider the case where more data than absolutely necessary are provided, which is realistic for applications. As shown in [33], in this case, the problem has a solution provided that

$$\text{rank}\,[\xi\,\mathbb{L}-\mathbb{L}_s]=\text{rank}\,[\mathbb{L},\ \mathbb{L}_s]=\text{rank}\begin{bmatrix}\mathbb{L}\\ \mathbb{L}_s\end{bmatrix}=r \tag{8.37}$$

for all $\xi\in\{\lambda_j\}\cup\{\mu_i\}$. Consider, then, the short SVDs

$$[\mathbb{L},\ \mathbb{L}_s]=\mathbf{Y}\Sigma_\ell\tilde{\mathbf{X}}^*,\quad \begin{bmatrix}\mathbb{L}\\ \mathbb{L}_s\end{bmatrix}=\tilde{\mathbf{Y}}\Sigma_r\mathbf{X}^*, \tag{8.38}$$

where $\Sigma_\ell,\Sigma_r\in\mathbb{R}^{r\times r}$, $\mathbf{Y}\in\mathbb{C}^{q\times r}$, $\mathbf{X}\in\mathbb{C}^{k\times r}$.

Theorem 8.21 ([33]). *The quadruple* $(\mathbf{E}_\mathcal{S},\mathbf{A}_\mathcal{S},\mathbf{B}_\mathcal{S},\mathbf{C}_\mathcal{S})$ *of size* $r\times r$, $r\times r$, $r\times m$, $p\times r$, *respectively, given by*

$$\mathbf{E}_\mathcal{S}=-\mathbf{Y}^*\mathbb{L}\mathbf{X},\quad \mathbf{A}_\mathcal{S}=-\mathbf{Y}^*\mathbb{L}_s\mathbf{X},\quad \mathbf{B}_\mathcal{S}=\mathbf{Y}^*\mathbf{V},\quad \mathbf{C}_\mathcal{S}=\mathbf{W}\mathbf{X}, \tag{8.39}$$

is a descriptor realization of an (approximate) interpolant of the data with McMillan degree $\nu=\text{rank}\,\mathbb{L}$.

Next, in the case of exact data (i.e., the data are available with no measurement noise), we list four properties that shed light on the Loewner framework.

Proposition 8.22 ([33]). *From the above construction, we have*

$$\mathbf{Y}\mathbf{Y}^*\mathbb{L} = \mathbb{L}, \;\; \mathbf{Y}\mathbf{Y}^*\mathbb{L}_s = \mathbb{L}_s, \;\; \mathbf{Y}\mathbf{Y}^*\mathbf{V} = \mathbf{V}, \tag{8.40}$$

$$\mathbb{L}\mathbf{X}\mathbf{X}^* = \mathbb{L}, \;\; \mathbb{L}_s\mathbf{X}\mathbf{X}^* = \mathbb{L}_s, \;\; \mathbf{W}\mathbf{X}\mathbf{X}^* = \mathbf{W}. \tag{8.41}$$

Proposition 8.23. *The original pencil* $(\mathbb{L}_s, \mathbb{L})$ *and the projected pencil* $(\mathbf{A}, \mathbf{E})$ *have the same nontrivial eigenvalues.*

Proof. Let $(\mathbf{z}, \lambda)$ be a right eigenpair of $(\mathbb{L}_s, \mathbb{L})$. Then $\mathbb{L}_s\mathbf{z} = \lambda\mathbb{L}\mathbf{z} \Rightarrow$

$$\mathbb{L}_s\mathbf{X}\mathbf{X}^*\mathbf{z} = \lambda\mathbb{L}\mathbf{X}\mathbf{X}^*\mathbf{z} \;\Rightarrow\; \underbrace{\mathbf{Y}^*\mathbb{L}_s\mathbf{X}}_{\mathbf{A}}\mathbf{X}^*\mathbf{z} = \lambda\underbrace{\mathbf{Y}^*\mathbb{L}\mathbf{X}}_{\mathbf{E}}\mathbf{X}^*\mathbf{z}.$$

Thus, $(\mathbf{X}^*\mathbf{z}, \lambda)$ is an eigenpair of $(\mathbf{A}, \mathbf{E})$. Conversely, if $(\mathbf{z}, \lambda)$ is an eigenpair of $(\mathbf{A}, \mathbf{E})$, then $(\mathbf{X}\mathbf{z}, \lambda)$ is an eigenpair of the original pencil $(\mathbb{L}_s, \mathbb{L})$. We reason similarly for left eigenpairs. □

Furthermore, even though the pencil $(\mathbb{L}_s, \mathbb{L})$ is singular, a kind of generalized interpolation property holds.

Proposition 8.24 (Interpolation property). *Let* $\mathbf{z}_i$ *satisfy* $(\lambda_i\mathbb{L} - \mathbb{L}_s)\mathbf{z}_i = \mathbf{V}\mathbf{r}_i$*. It follows that* $\mathbf{z}_i = \mathbf{e}_i + \mathbf{z}_0$ *and* $\mathbf{W}\mathbf{z}_i = \mathbf{w}_i$*, where* $\mathbf{W}\mathbf{z}_0 = 0$.

The next result shows that the projection may indeed be chosen *arbitrarily*, instead of using $\mathbf{X}, \mathbf{Y}$ defined by (8.38).

Proposition 8.25. *Let* $\mathbf{\Phi}$ *and* $\mathbf{\Psi}$ *be such that* $\mathbf{X}^*\mathbf{\Phi}$ *and* $\mathbf{\Psi}^*\mathbf{Y}$ *are square and nonsingular. Then,*

$$(\mathbf{Y}^*\mathbb{L}\mathbf{X},\; \mathbf{Y}^*\mathbb{L}_s\mathbf{X},\; \mathbf{Y}^*\mathbf{V},\; \mathbf{W}\mathbf{X}) \quad \text{and} \quad (\mathbf{\Phi}^*\mathbb{L}\mathbf{\Psi},\; \mathbf{\Phi}^*\mathbb{L}_s\mathbf{\Psi},\; \mathbf{\Phi}^*\mathbf{V},\; \mathbf{W}\mathbf{\Psi})$$

are equivalent minimal descriptor realizations for the same system.

Proof. Given $\mathbf{\Phi} \in \mathbb{R}^{N\times k}$, (8.40) implies $\mathbf{\Phi}^*\mathbf{Y}\mathbf{Y}^*\mathbb{L} = \mathbf{\Phi}^*\mathbb{L}$. Given $\mathbf{\Psi} \in \mathbb{R}^{N\times k}$, (8.41) implies $\mathbb{L}\mathbf{X}\mathbf{X}^*\mathbf{\Psi} = \mathbb{L}\mathbf{\Psi}$. Combining these two relationships, we obtain

$$\underbrace{\mathbf{\Phi}^*\mathbf{Y}}_{\mathbf{T}_1}\mathbf{Y}^*\mathbb{L}\mathbf{X}\underbrace{\mathbf{X}^*\mathbf{\Psi}}_{\mathbf{T}_2} = \mathbf{\Phi}^*\mathbb{L}\mathbf{\Psi} \quad\Rightarrow\quad \mathbf{T}_1\mathbf{Y}^*\mathbb{L}\mathbf{X}\mathbf{T}_2 = \mathbf{\Phi}^*\mathbb{L}\mathbf{\Psi}.$$

Similarly, we have

$$\mathbf{T}_1\mathbf{Y}^*\mathbb{L}_s\mathbf{X}\mathbf{T}_2 = \mathbf{\Phi}^*\mathbb{L}_s\mathbf{\Psi}, \;\; \mathbf{T}_1\mathbf{Y}^*\mathbf{V} = \mathbf{\Phi}^*\mathbf{V}, \;\; \mathbf{W}\mathbf{X}\mathbf{T}_2 = \mathbf{W}\mathbf{\Psi}.$$

Hence, the nonsingularity of $\mathbf{T}_1$ and $\mathbf{T}_2$ implies that the two quadruples are equivalent. □

Remark 8.26. (a) The Loewner approach constructs a descriptor representation $(\mathbb{L}, \mathbb{L}_s, \mathbf{V}, \mathbf{W})$ of an underlying dynamical system exclusively from the data, with no further manipulations involved (i.e., matrix factorizations or inversions). In general, the pencil $(\mathbb{L}_s, \mathbb{L})$ is singular and needs to be projected to a regular pencil $(\mathbf{A}_\aleph, \mathbf{E}_\aleph)$ as in (8.39).

(b) In the Loewner framework, by construction, **D**-terms are absorbed in the other matrices of the realization (see also Remark 8.1). Extracting the **D**-term involves an eigenvalue decomposition of the pencil $(\mathbb{L}_s, \mathbb{L})$ (see Example 8.39 and Example 8.41).
(c) Notice the similarity between the components of formulas (8.25) and (8.39), where $\mathbf{Y} = \mathbf{I}$ and $\mathbf{X} = \mathbb{J}$.

Interpolation property of the projected systems

Given tangential interpolation data, the question arises whether projected quadruples as in (8.39) satisfy interpolation conditions as well. Toward this goal, recall the Sylvester equations (8.29) and (8.31). Adding $\mathbf{M}\mathbb{L}\mathbf{\Lambda}$ to and subtracting it from the second equation and collecting terms in $\mathbf{\Lambda}$ on the right and $\mathbf{M}$ on the left, we obtain

$$(\mathbb{L}_s - \mathbf{M}\mathbb{L} + \mathbf{L}\mathbf{W})\mathbf{\Lambda} - \mathbf{M}(\mathbb{L}_s - \mathbb{L}\mathbf{\Lambda} + \mathbf{V}\mathbf{R}) = 0.$$

Since $(\mathbb{L}_s - \mathbf{M}\mathbb{L} + \mathbf{L}\mathbf{W}) = (\mathbb{L}_s - \mathbb{L}\mathbf{\Lambda} + \mathbf{V}\mathbf{R}) =: \mathbf{Z}$, and assuming that $\mathbf{\Lambda}$ and $\mathbf{M}$ have no common eigenvalues, it follows that $\mathbf{Z} = 0$, and consequently

$$\mathbb{L}_s - \mathbf{M}\mathbb{L} = -\mathbf{L}\mathbf{W}, \quad \mathbb{L}_s - \mathbb{L}\mathbf{\Lambda} = -\mathbf{V}\mathbf{R}. \tag{8.42}$$

After projection, the following quantities (i.e., matrices of reduced dimension) are obtained:

$$\widehat{\mathbb{L}} = \mathbf{Y}^*\mathbb{L}\mathbf{X}, \quad \widehat{\mathbb{L}_s} = \mathbf{Y}^*\mathbb{L}_s\mathbf{X}, \quad \widehat{\mathbf{V}} = \mathbf{Y}^*\mathbf{V}, \quad \widehat{\mathbf{W}} = \mathbf{W}\mathbf{X}, \quad \widehat{\mathbf{L}} = \mathbf{Y}^*\mathbf{L}, \quad \widehat{\mathbf{R}} = \mathbf{R}\mathbf{X}.$$

Because of equations (8.42), the associated $\widehat{\mathbf{\Lambda}}$ and $\widehat{\mathbf{M}}$ must satisfy

$$\widehat{\mathbb{L}_s} - \widehat{\mathbf{M}}\widehat{\mathbb{L}} = -\widehat{\mathbf{L}}\widehat{\mathbf{W}}, \quad \widehat{\mathbb{L}_s} - \widehat{\mathbb{L}}\widehat{\mathbf{\Lambda}} = -\widehat{\mathbf{V}}\widehat{\mathbf{R}},$$

and hence

$$\widehat{\mathbf{M}} = (\mathbf{Y}^*\mathbb{L}_s\mathbf{X} + \mathbf{Y}^*\mathbf{L}\mathbf{W}\mathbf{X})(\mathbf{Y}^*\mathbb{L}\mathbf{X})^{-1}, \quad \widehat{\mathbf{\Lambda}} = (\mathbf{Y}^*\mathbb{L}\mathbf{X})^{-1}(\mathbf{Y}^*\mathbb{L}_s\mathbf{X} + \mathbf{Y}^*\mathbf{V}\mathbf{R}\mathbf{X}).$$

Finally, to recover the Loewner data for the projected system, we need to diagonalize $\widehat{\mathbf{M}}$ and $\widehat{\mathbf{\Lambda}}$. Let the EVD (eigenvalue decomposition) of $\tilde{\mathbf{\Lambda}}$ and $\tilde{\mathbf{M}}$ be $\tilde{\mathbf{\Lambda}}\mathbf{T}_{\Lambda} = \mathbf{T}_{\Lambda}\mathbf{D}_{\Lambda}$ and $\mathbf{T}_{\mathbf{M}}\tilde{\mathbf{M}} = \mathbf{D}_{\mathbf{M}}\mathbf{T}_{\mathbf{M}}$, respectively. Substituting in the above equations, we obtain the corresponding Loewner data for the reduced system:

$$\begin{aligned}
&\widetilde{\mathbb{L}} = \mathbf{T}_{\mathbf{M}}\mathbf{Y}^*\mathbb{L}\mathbf{X}\mathbf{T}_{\Lambda}, \quad \widetilde{\mathbb{L}_s} = \mathbf{T}_{\mathbf{M}}\mathbf{Y}^*\mathbb{L}_s\mathbf{X}\mathbf{T}_{\Lambda},\\
&\widetilde{\mathbf{V}} = \mathbf{T}_{\mathbf{M}}\mathbf{Y}^*\mathbf{V}, \quad \widetilde{\mathbf{W}} = \mathbf{W}\mathbf{X}\mathbf{T}_{\Lambda}, \quad \widetilde{\mathbf{L}} = \mathbf{T}_{\mathbf{M}}\mathbf{Y}^*\mathbf{L}, \quad \widetilde{\mathbf{R}} = \mathbf{R}\mathbf{X}\mathbf{T}_{\Lambda}, \quad \widetilde{\mathbf{\Lambda}} = \mathbf{D}_{\Lambda}, \quad \widetilde{\mathbf{M}} = \mathbf{D}_{\mathbf{M}}.
\end{aligned}$$

Thus, the *right* interpolation points are the generalized eigenvalues of the pencil

$$[(\mathbf{Y}^*\mathbb{L}_s\mathbf{X} + \mathbf{Y}^*\mathbf{V}\mathbf{R}\mathbf{X}), (\mathbf{Y}^*\mathbb{L}\mathbf{X})],$$

while the *left* interpolation points are the generalized eigenvalues of the pencil

$$[(\mathbf{Y}^*\mathbb{L}_s\mathbf{X} + \mathbf{Y}^*\mathbf{L}\mathbf{W}\mathbf{X}), (\mathbf{Y}^*\mathbb{L}\mathbf{X})].$$

Furthermore, the *right* and *left* interpolation values/directions are, respectively,

$$\widetilde{\mathbf{W}} = -(\mathbf{W}\mathbf{X}\mathbf{T}_{\Lambda}), \quad \widetilde{\mathbf{R}} = (\mathbf{R}\mathbf{X}\mathbf{T}_{\Lambda}), \quad \widetilde{\mathbf{V}} = -(\mathbf{T}_{\mathbf{M}}\mathbf{Y}^*\mathbf{V}), \quad \widetilde{\mathbf{L}} = (\mathbf{T}_{\mathbf{M}}\mathbf{Y}^*\mathbf{L}).$$

In the scalar case, the *right* and *left* interpolation values are (in MATLAB notation) $-(\mathbf{W}\mathbf{X}\mathbf{T}_{\Lambda})./(\mathbf{R}\mathbf{X}\mathbf{T}_{\Lambda})$, $-(\mathbf{T}_{\mathbf{M}}\mathbf{Y}^*\mathbf{V})./(\mathbf{T}_{\mathbf{M}}\mathbf{Y}^*\mathbf{L})$, respectively.

Parametrization of all interpolants

Parametrization of all interpolants is accomplished by means of the *generating system matrix* denoted by $\boldsymbol{\Theta}(s)$. The main property of $\boldsymbol{\Theta}$ is that any interpolant of a given set of data can be expressed as a linear combination of its columns. This general framework leads to recursive interpolation (that is, interpolation where the data are not provided all at once) [29, 30]. Interestingly, this framework has a system-theoretic interpretation as the feedback interconnection of linear dynamical systems. For a tutorial introduction to the generating system approach, we refer the reader to Section 4.5.3 of [3] and references therein. For the proofs of the results that follow, we refer to [29, 30].

In the present context, $\boldsymbol{\Theta}$ is explicitly defined in terms of the tangential interpolation data, namely $\boldsymbol{\Lambda}, \mathbf{M}, \mathbf{L}, \mathbf{R}, \mathbf{V}, \mathbf{W}$, defined by (8.26), (8.27), as well as the Loewner matrix $\mathbb{L}$ defined in (8.28). If $q = k$ and $\mathbb{L}$ is invertible, we define the $(p+m)\times(p+m)$ rational matrix $\boldsymbol{\Theta}(s)$ and its inverse $\bar{\boldsymbol{\Theta}}(s)$:

$$\boldsymbol{\Theta}(s) = \begin{bmatrix} \mathbf{I}_p & 0 \\ 0 & \mathbf{I}_m \end{bmatrix} + \begin{bmatrix} \mathbf{W} \\ -\mathbf{R} \end{bmatrix} (s\mathbb{L} - \mathbb{L}\boldsymbol{\Lambda})^{-1} \begin{bmatrix} \mathbf{L} & \mathbf{V} \end{bmatrix} = \begin{bmatrix} \boldsymbol{\Theta}_{11}(s) & \boldsymbol{\Theta}_{12}(s) \\ \boldsymbol{\Theta}_{21}(s) & \boldsymbol{\Theta}_{22}(s) \end{bmatrix},$$

$$\bar{\boldsymbol{\Theta}}(s) = \begin{bmatrix} \mathbf{I}_p & 0 \\ 0 & \mathbf{I}_m \end{bmatrix} + \begin{bmatrix} -\mathbf{W} \\ \mathbf{R} \end{bmatrix} (s\mathbb{L} - \mathbf{M}\mathbb{L})^{-1} \begin{bmatrix} \mathbf{L} & \mathbf{V} \end{bmatrix} = \begin{bmatrix} \bar{\boldsymbol{\Theta}}_{11}(s) & \bar{\boldsymbol{\Theta}}_{12}(s) \\ \bar{\boldsymbol{\Theta}}_{21}(s) & \bar{\boldsymbol{\Theta}}_{22}(s) \end{bmatrix}.$$

It should be mentioned that with appropriate changes, the above equations also hold for the case of interpolation with multiplicities.

First, we notice that the left and right interpolation conditions hold.

Proposition 8.27. *The following relationships are satisfied for $j, i = 1, \ldots, k$:*

$$\begin{bmatrix} \ell_j^T & \mathbf{v}_j^T \end{bmatrix} \boldsymbol{\Theta}(\mu_j) = 0_{1\times(p+m)} \quad \text{and} \quad \bar{\boldsymbol{\Theta}}(\lambda_i) \begin{bmatrix} -\mathbf{w}_i \\ \mathbf{r}_i \end{bmatrix} = 0_{(p+m)\times 1}.$$

Next, we assert that all interpolants can be obtained as linear matrix fractions constructed from the entries of $\boldsymbol{\Theta}$ or $\bar{\boldsymbol{\Theta}}$.

Lemma 8.28. *For any polynomial matrices $\mathbf{S}_1(s)$, $\mathbf{S}_2(s)$, $\bar{\mathbf{S}}_1(s)$, $\bar{\mathbf{S}}_2(s)$ of size $p\times m$, $m\times m$, $p\times p$, $p\times m$, respectively, that satisfy*

$$\mathbf{S}_1(s)\mathbf{S}_2(s)^{-1} = \bar{\mathbf{S}}_1(s)^{-1}\bar{\mathbf{S}}_2(s),$$

the following $p\times m$ rational function is an interpolant:

$$\boldsymbol{\Psi}(s) = \boldsymbol{\Psi}_1(s)\boldsymbol{\Psi}_2(s)^{-1} = [\boldsymbol{\Theta}_{11}(s)\mathbf{S}_1(s) - \boldsymbol{\Theta}_{12}(s)\mathbf{S}_2(s)][-\boldsymbol{\Theta}_{21}(s)\mathbf{S}_1(s) + \boldsymbol{\Theta}_{22}(s)\mathbf{S}_2(s)]^{-1},$$

provided that the denominator is nonsingular for all interpolation points λ_i and μ_j. Similarly, $\boldsymbol{\Psi}$ can also be written as

$$\boldsymbol{\Psi}(s) = \bar{\boldsymbol{\Psi}}_1(s)^{-1}\bar{\boldsymbol{\Psi}}_2(s) = [\bar{\mathbf{S}}_1(s)\bar{\boldsymbol{\Theta}}_{11}(s) + \bar{\mathbf{S}}_2(s)\bar{\boldsymbol{\Theta}}_{21}(s)]^{-1}[\bar{\mathbf{S}}_1(s)\bar{\boldsymbol{\Theta}}_{12}(s) + \bar{\mathbf{S}}_2(s)\bar{\boldsymbol{\Theta}}_{22}(s)].$$

The former is a right coprime factorization and the latter is a left coprime factorization of $\boldsymbol{\Psi}$. For right/left coprimeness of the factors, $\mathbf{S}_1$, $\mathbf{S}_2$ and $\bar{\mathbf{S}}_1$, $\bar{\mathbf{S}}_2$ must be right/left coprime, respectively. Conversely, any interpolant that satisfies the left and right tangential interpolation conditions, can be expressed as above for $\mathbf{S}_1$, $\mathbf{S}_2$, $\bar{\mathbf{S}}_1$, $\bar{\mathbf{S}}_2$ appropriately chosen.

Corollary 8.29. (a) *For* $\mathbf{S}_1(s) = \mathbb{O}$ *and* $\mathbf{S}_2(s) = \mathbf{I}$, *we recover the minimal degree interpolant given by* (8.34): $\Psi(s) = -\Theta_{12}(s)\Theta_{22}^{-1}(s) = \mathbf{C}_\delta(s\mathbf{E}_\delta - \mathbf{A}_\delta)^{-1}\mathbf{B}_\delta$.
(b) *For* $\mathbf{S}_1(s) = \mathbf{K}$ *and* $\mathbf{S}_2(s) = \mathbf{I}$, *we recover all minimal degree interpolants defined by* (8.36): $\Psi(s) = \mathbf{C}(s\mathbf{E} - \mathbf{A})^{-1}\mathbf{B} + \mathbf{D}$.

The Loewner algorithm

We are now ready to formulate a high-level algorithm that produces from given data an approximate model, possibly of reduced dimension.

- **Data:** $(\mu_i, \ell_i, \mathbf{v}_i)$, $i = 1, \ldots, q$ and $(\lambda_j, \mathbf{r}_j, \mathbf{w}_j)$, $j = 1, \ldots, k$, as defined in Section 8.1.3.

Algorithm 8.1 gives the procedure for obtaining (reduced) models.

***ALGORITHM* 8.1. The Loewner algorithm.**

---

1. Build the Loewner matrix pencil $(\mathbb{L}_s, \mathbb{L})$ following (8.30) and (8.28). Let these matrices have size $q \times k$.

2. There are two important positive integers to compute, namely the dimension of minimal (approximate) descriptor realizations and their McMillan degree.

   (a) If condition (8.37) is satisfied, r is the dimension of the resulting minimal descriptor realizations.

   (b) In this case, $\nu = \operatorname{rank} \mathbb{L}$ is the McMillan degree of the corresponding realization.

   Both of these numbers can be computed as *numerical ranks*.

3. Following (8.38), matrices $\mathbf{X} \in \mathbb{C}^{k \times r}$, $\mathbf{Y} \in \mathbb{C}^{q \times r}$ are determined and a descriptor realization is given by (8.39):

$$(\mathbf{E}_\delta, \mathbf{A}_\delta, \mathbf{B}_\delta, \mathbf{C}_\delta) \in \mathbb{C}^{r \times r} \times \mathbb{C}^{r \times r} \times \mathbb{C}^{r \times m} \times \mathbb{C}^{p \times r}.$$

4. If $r = \nu$, the transfer function of the ensuing system is strictly proper rational.

5. If $r > \nu$, we need to decide whether there is a $\mathbf{D}$-term or a polynomial term. Toward this goal we compute the generalized eigenvalues of the pencil $(\mathbf{A}_\delta, \mathbf{E}_\delta)$.

   (a) If there is a multiple eigenvalue at infinity and the corresponding eigenspace has the same dimension, the system has a $\mathbb{D}$-term.

   (b) If there are Jordan blocks at infinity, the transfer function has a polynomial term.

8.2.4 ▪ Further issues

Positive real interpolation in the Loewner framework

The classical (scalar) *positive real (PR) interpolation* problem is as follows. Given pairs of points (λ_i, ϕ_i), $i = 1, \ldots, N$, we seek *PR* functions[31] $\phi(s)$ such that

$$\phi(\lambda_i) = \phi_i, \; i = 1, \ldots, N.$$

If $\mathcal{R}e(\lambda_i) > 0$, the classical condition for solution is that the associated *Pick matrix* $\boldsymbol{\Pi}$ be positive semidefinite:

$$\boldsymbol{\Pi} = \left[\frac{\phi_i^* + \phi_j}{\lambda_i^* + \lambda_j} \right]_{i,j=1,\ldots,N} = \boldsymbol{\Pi}^* \geq 0.$$

Observation. $\boldsymbol{\Pi}$ is a *Loewner matrix* with column array (λ_i, ϕ_i) and row array the *mirror-image* set $(-\lambda_j^*, -\phi_j^*)$. Thus, the interpolation data in the Loewner framework are

$$\mathbf{W} = [\phi_1 \cdots \phi_N], \;\; \mathbf{V} = -\mathbf{W}^*, \;\; \boldsymbol{\Lambda} = \mathrm{diag}[\lambda_1, \ldots, \lambda_N], \;\; \mathbf{M} = -\boldsymbol{\Lambda}^*,$$

and $\mathbf{R} = [1, \ldots, 1] = \mathbf{L}^*$ (the latter quantities do not appear in the classical framework). Consequently, the associated Loewner matrix pencil satisfies the Sylvester (Lyapunov) equations

$$\mathbb{L}\boldsymbol{\Lambda} + \boldsymbol{\Lambda}^*\mathbb{L} = \mathbf{W}^*\mathbf{R} + \mathbf{R}^*\mathbf{W} \quad \text{and} \quad \mathbb{L}_s\boldsymbol{\Lambda} + \boldsymbol{\Lambda}^*\mathbb{L}_s = \mathbf{R}^*\mathbf{W}\boldsymbol{\Lambda} - \boldsymbol{\Lambda}^*\mathbf{W}^*\mathbf{R}.$$

Proposition 8.30 (Realization of PR interpolants). *Assuming that* $\boldsymbol{\Pi} = \mathbb{L} \geq 0$, *a minimal PR interpolant is*

$$\mathbf{H}(s) = \mathbf{W}(s\mathbb{L} - \mathbb{L}_s)^{-1}\mathbf{W}^*.$$

Proof. Making use of the fact that $\mathbb{L} = \mathbb{L}^* > 0$ (Hermitian positive definite) and that $\mathbb{L}_s^* = -\mathbb{L}_s$ (skew Hermitian), we obtain

$$\begin{aligned} \mathbf{H}(s) + \mathbf{H}^*(-s) &= \mathbf{W}(s\mathbb{L} - \mathbb{L}_s)^{-1}\mathbf{W}^* + \mathbf{W}(s^*\mathbb{L} - \mathbb{L}_s)^{-1}\mathbf{W}^* \\ &= \mathbf{W}(s\mathbb{L} - \mathbb{L}_s)^{-1}[(s+s^*)\mathbb{L} - \mathbb{L}_s - \mathbb{L}_s^*](s^*\mathbb{L} - \mathbb{L}_s)^{-1}\mathbf{W}^* \\ &= \underbrace{\mathbf{W}(s\mathbb{L} - \mathbb{L}_s)^{-1}}_{\mathbf{K}(s)} [(s+s^*)\mathbb{L}] \underbrace{(s^*\mathbb{L} - \mathbb{L}_s)^{-1}\mathbf{W}^*}_{\mathbf{K}^*(s^*)} \\ &= (s+s^*)\mathbf{K}(s)\mathbb{L}\mathbf{K}^*(s^*) \geq 0 \quad \text{for} \quad s + s^* \geq 0. \end{aligned}$$

This completes the proof. □

We will now turn our attention to the parametrization of minimal solutions. Following Theorem 8.20(b), from

$$\mathbb{L}_s\boldsymbol{\Lambda} + \boldsymbol{\Lambda}^*\mathbb{L}_s = \mathbf{R}^*\mathbf{W}\boldsymbol{\Lambda} - \boldsymbol{\Lambda}^*\mathbf{W}^*\mathbf{R},$$

[31] A function f of the complex variable s is *positive real* (abbreviated PR) if it is analytic in the (open) right half of the complex plane and, in addition, its real part is nonnegative in the same domain.

we have

$$[\mathbb{L}_s - \mathbf{R}^*\delta\mathbf{R}]\mathbf{\Lambda} + \mathbf{\Lambda}^*[\mathbb{L}_s - \mathbf{R}^*\delta\mathbf{R}] = \mathbf{R}^*[\mathbf{W} - \delta\mathbf{R}]\mathbf{\Lambda} - \mathbf{\Lambda}^*[\mathbf{W}^* + \mathbf{R}^*\delta]\mathbf{R}.$$

Hence, a parametrization of all solutions of degree N is given in terms of the (scalar) parameter δ. The associated realization of these interpolants is

$$\mathbf{H}(s) = \delta + [\mathbf{W} - \delta\mathbf{R}]\,[s\,\mathbb{L} - (\mathbb{L}_s - \mathbf{R}^*\,\delta\,\mathbf{R})]^{-1}[\mathbf{W}^* + \mathbf{R}^*\delta], \quad \delta \geq 0. \tag{8.43}$$

Remark 8.31. Lemma 8.28 can be easily formulated and proved for the case $p = m > 1$, where $\mathbf{R} \in \mathbb{C}^{p\times N}$ contains (right) tangential directions and $\mathbf{W} \in \mathbb{C}^{p\times N}$ the associated (right) tangential values.

Connections with Port–Hamiltonian systems. We will conclude our considerations of the PR realization problem by pointing out that the above realizations are in *Port–Hamiltonian (PH)* form. According to [38] *PH systems* with feed-through terms are described as

$$\begin{aligned} \tfrac{d}{dt}\tilde{\mathbf{x}}(t) &= [\mathbf{J} - \hat{\mathbf{R}}]\,\mathbf{Q}\,\tilde{\mathbf{x}}(t) + [\mathbf{G} - \mathbf{P}]\,\mathbf{u}(t), \\ \mathbf{y}(t) &= [\mathbf{G}^* + \mathbf{P}^*]\,\mathbf{Q}\,\tilde{\mathbf{x}} + [\mathbf{M} + \mathbf{S}]\,\mathbf{u}(t), \end{aligned}$$

where

$$\mathbf{Z} = \begin{bmatrix} \hat{\mathbf{R}} & \mathbf{P} \\ \mathbf{P}^* & \mathbf{S} \end{bmatrix} \geq 0, \ \ \mathbf{J} = -\mathbf{J}^*, \ \ \mathbf{M} = -\mathbf{M}^*.$$

This can be rewritten with $\mathbf{x} = \mathbf{Q}\tilde{\mathbf{x}}$ as

$$\left.\begin{aligned} \mathbf{E}\tfrac{d}{dt}\mathbf{x}(t) &= \mathbf{A}\mathbf{x}(t) + \mathbf{B}\mathbf{u}(t) \\ \mathbf{y}(t) &= \mathbf{C}\mathbf{x}(t) + \mathbf{D}\mathbf{u}(t) \end{aligned}\right\} \ \text{where} \ \left\{ \begin{array}{c|c} \mathbf{E} = \mathbf{Q}^{-1},\ \mathbf{A} = \mathbf{J} - \hat{\mathbf{R}} & \mathbf{B} = \mathbf{G} - \mathbf{P} \\ \hline \mathbf{C} = \mathbf{G}^* + \mathbf{P}^* & \mathbf{D} = \mathbf{M} + \mathbf{S} \end{array} \right. .$$

Corollary 8.32. *It readily follows that the realizations in Proposition* 8.30 *and equation* (8.43) *are in PH form as* $\mathbf{E} = \mathbb{L}$, $\mathbf{J} = \mathbb{L}_s$, $\hat{\mathbf{R}} = \mathbf{R}^*\delta\mathbf{R}$, $\mathbf{G} = \mathbf{W}^*$, $\mathbf{P} = \delta\mathbf{R}^*$, $\mathbf{M} = 0$, $\mathbf{S} = \delta$.

Remark 8.33. PR functions and PH systems are important in model reduction as they represent passive resistor-inductor-capacitor electric circuits (closely related to *interconnect analysis* in VLSI synthesis), spring-mass-damper mechanical systems, etc. See also [4] for the construction of passive models from frequency response data.

Poles and zeros as interpolation points

Suppose that one of the values in (8.27) or (8.26) is infinite or zero. This means that the corresponding interpolation point is a pole or a zero of the interpolant that we wish to construct. Such conditions can be easily accounted for in the Loewner framework by appropriate definition of $\mathbf{r}_i$, $\mathbf{w}_i$ or $\boldsymbol{\ell}_j$, $\mathbf{v}_j$. In the former case (specified poles), either $\mathbf{r}_i = 0$ ($\mathbf{w}_i$ arbitrary) or $\boldsymbol{\ell}_j = 0$ ($\mathbf{v}_j$ arbitrary), in which case λ_i or μ_j is a pole, while in the latter, $\mathbf{w}_i = 0$ ($\mathbf{r}_i$ arbitrary) or $\mathbf{v}_j = 0$ ($\boldsymbol{\ell}_j$ arbitrary), in which case λ_i or μ_j is a zero.

Assume, for instance, that the given interpolation data satisfy $\mathbf{r}_i = 0$ for $i = 1, \ldots, k$, i.e., $\mathbf{R} = 0$. Then, combining (8.28) and (8.30), we obtain $\mathbb{L}_s = \mathbb{L}\mathbf{\Lambda}$. This, in turn, implies that λ_i, $i = 1,\dots,k$, are the eigenvalues of the pencil $(\mathbb{L}_s, \mathbb{L})$ and, assuming that $\mathbb{L}$ is nonsingular, they are the poles of the interpolant

$$\mathbf{H}(s) = \mathbb{W}(\mathbb{L}_s - s\mathbb{L})^{-1}\mathbb{V} = \mathbb{W}\mathbb{L}^{-1}(s\mathbf{I} - \mathbf{\Lambda})^{-1}\mathbb{V} = \sum_{i=1}^{k} \frac{\mathbf{res}_i}{s - \lambda_i},$$

where $\mathbf{res}_i = (\mathbf{w}\mathbb{L}^{-1})_i \mathbf{v}_i^T$ is the residue corresponding to the pole λ_i.

Error expressions

We will now derive an expression for the error when interpolation is performed by means of an inexact barycentric formula (see relations (8.19), (8.20), (8.21)). In particular, let

$$\mathbb{L}\mathbf{c} = \mathbf{e},$$

where $\mathbf{e}$ is not necessarily zero. Using the barycentric formula, we attach to $\mathbf{c} = (\alpha_i)$ the rational function

$$\hat{\mathbf{H}}(s) = \frac{\sum_{j=1}^{q+1} \frac{\alpha_j \mathbf{w}_j}{s - \lambda_j}}{\sum_{j=1}^{q+1} \frac{\alpha_j}{s - \lambda_j}}.$$

It readily follows that $(\lambda_j, \mathbf{w}_j)$ are interpolated. Furthermore, it readily follows that

$$\mathbf{v}_i - \hat{\mathbf{H}}(\mu_i) = \frac{e_i}{\sum_{j=1}^{q+1} \frac{\alpha_j}{\mu_j - \lambda_j}}.$$

Therefore, if $\mathbf{c}$ is the left singular vector corresponding to the smallest singular value of $\mathbb{L}$, say σ_{q+1}, i.e., $\mathbb{L}\mathbf{c} = \sigma_{q+1}\mathbf{x}$, where $(\mathbf{x})_{i,1} = x_i$ is the corresponding left singular vector, this expression yields, for $i = 1,\dots,k$,

$$\mathbf{v}_i - \hat{\mathbf{H}}(\mu_i) = \sigma_{q+1} \frac{x_i}{\sum_{j=1}^{q+1} \frac{\alpha_j}{\mu_j - \lambda_j}} \Rightarrow |\mathbf{v}_i - \hat{\mathbf{H}}(\mu_i)| \leq \frac{\sigma_{q+1}}{|\sum_{j=1}^{q+1} \frac{\alpha_j}{\mu_j - \lambda_j}|}.$$

Error bound. For the case when the full-order transfer function $\mathbf{H}(s)$ that generated the measurements $\mathbf{H}(s_i)$ is known, we want to bound the approximation error $\mathbf{H}(s) - \hat{\mathbf{H}}(s)$ over entire intervals, rather than just at the given points. The error can be bounded by the following formula [27]:

$$|\mathbf{H}(s) - \hat{\mathbf{H}}(s)| \leq \Delta(s) \max_s \left| \frac{d\mathbf{H}(s)}{ds} \right|, \quad \Delta(s) = \frac{\sum_{i=1}^{q+1} |x_i|}{|\delta(s)|}, \quad \delta(s) = \sum_{i=1}^{q+1} \frac{\alpha_j}{s - \lambda_i}$$

for $\lambda_i \in \mathbb{R}$, $s \in [\lambda_1, \lambda_{q+1}]$, $\lambda_1 \leq \lambda_2 \leq \cdots \leq \lambda_{q+1}$, and x_i as defined above. The factor $\Delta(s)$ is given by the barycentric coefficients α_j of $\hat{\mathbf{H}}(s)$, while the derivative is given by $\frac{d\mathbf{H}(s)}{ds} = -\mathbf{C}(s\mathbf{E} - \mathbf{A})^{-1}\mathbf{E}(s\mathbf{E} - \mathbf{A})^{-1}\mathbf{B}$. Usually, this bound correctly predicts interpolation at the Lagrange nodes because the Δ factor equals zero for λ_i. However, in general, this bound is loose over the interval, even though it may capture the general shape of the error.

Real Loewner matrices from complex data

In applications, assuming that the underlying system is real, to obtain solutions containing no complex quantities, it must be assumed that, given a certain set of data, the *complex conjugate data* are also provided. Then $\mathbf{\Lambda}$, $\mathbf{M}$ as well as $\mathbf{W}$, $\mathbf{V}$ must contain complex conjugate data. For this purpose, we need the block diagonal matrix $\mathbf{J}$ with $\frac{1}{\sqrt{2}}\begin{bmatrix} 1 & -\imath \\ 1 & \imath \end{bmatrix}$ repeated on the diagonal.

The goal is to form $\mathbb{L}^{\mathrm{R}}$, $\mathbb{L}_s^{\mathrm{R}}$ with real entries. This can be done by first forming $\mathbb{L}^{\mathrm{C}}$, $\mathbb{L}_s^{\mathrm{C}}$ (with complex entries) and then multiplying by $\mathbf{J}, \mathbf{J}^*$:

$$\mathbb{L}^{\mathrm{R}} = \mathbf{J}^*\mathbb{L}^{\mathrm{C}}\mathbf{J}, \quad \mathbb{L}_s^{\mathrm{R}} = \mathbf{J}^*\mathbb{L}_s^{\mathrm{C}}\mathbf{J}.$$

Alternatively, $\mathbb{L}^{\mathrm{R}}$, $\mathbb{L}_s^{\mathrm{R}}$ can be obtained by solving the Sylvester equations

$$\mathbb{L}^{\mathrm{R}}\mathbf{\Lambda}^{\mathrm{R}} - \mathbf{M}^{\mathrm{R}}\mathbb{L}^{\mathrm{R}} = \mathbf{V}^{\mathrm{R}}\mathbf{R}^{\mathrm{R}} - \mathbf{L}^{\mathrm{R}}\mathbf{W}^{\mathrm{R}}, \quad \mathbb{L}_s^{\mathrm{R}}\mathbf{\Lambda}^{\mathrm{R}} - \mathbf{M}^{\mathrm{R}}\mathbb{L}_s^{\mathrm{R}} = \mathbf{M}^{\mathrm{R}}\mathbf{V}^{\mathrm{R}}\mathbf{R}^{\mathrm{R}} - \mathbf{L}^{\mathrm{R}}\mathbf{W}^{\mathrm{R}}\mathbf{\Lambda}^{\mathrm{R}}, \tag{8.44}$$

where the quantities above are defined in terms of the following *real* blocks:

$$\mathbf{\Lambda}^{\mathrm{R}} = \mathrm{blkdiag}\left[\ldots, \begin{bmatrix} 0 & \lambda_j \\ -\lambda_j & 0 \end{bmatrix}, \ldots\right], \quad \mathbf{M}^{\mathrm{R}} = \mathrm{blkdiag}\left[\ldots, \begin{bmatrix} 0 & \mu_i \\ -\mu_i & 0 \end{bmatrix}, \ldots\right],$$

and, assuming that the left and right directions are real vectors,

$$\mathbf{V}^{\mathrm{R}} = \begin{bmatrix} \vdots \\ \mathrm{Re}(\mathbf{v}_i^T) \\ -\mathrm{Im}(\mathbf{v}_i^T) \\ \vdots \end{bmatrix}, \quad \mathbf{L}^{\mathrm{R}} = \begin{bmatrix} \vdots \\ \ell_i^T \\ 0 \\ \vdots \end{bmatrix},$$

$$\mathbf{W}^{\mathrm{R}} = \left[\ldots, \mathrm{Re}(\mathbf{w}_j),\ \mathrm{Im}(\mathbf{w}_j), \ldots\right], \quad \mathbf{R}_j = \left[\ldots, \mathbf{r}_j,\ 0, \ldots\right],$$

where $\mathrm{Re}(\cdot)$ and $\mathrm{Im}(\cdot)$ denote the real and imaginary parts of $(\cdot)$.

8.2.5 ▪ Illustrative examples

To illustrate the above considerations, we consider several examples.

Example 8.34. We would like to recover the rational function $\mathbf{H}(s) = \frac{1}{s^2+1}$ from the following measurements: $\lambda_1 = 1$, $\lambda_2 = 2$, $\lambda_3 = 3$, $\mu_1 = -1$, $\mu_2 = -2$, $\mu_3 = -3$, $\mathbf{W} = [\frac{1}{2}, \frac{1}{5}, \frac{1}{10}] = \mathbf{V}^T$, and $\mathbf{R} = [1\ 1\ 1] = \mathbf{L}^T$; it follows that

$$\mathbb{L} = \begin{bmatrix} 0 & -\frac{1}{10} & -\frac{1}{10} \\ \frac{1}{10} & 0 & -\frac{1}{50} \\ \frac{1}{10} & \frac{1}{50} & 0 \end{bmatrix}, \quad \mathbb{L}_s = \begin{bmatrix} \frac{1}{2} & \frac{3}{10} & \frac{1}{5} \\ \frac{3}{10} & \frac{1}{5} & \frac{7}{50} \\ \frac{1}{5} & \frac{7}{50} & \frac{1}{10} \end{bmatrix}.$$

As condition (8.37) yields $r = 2$, we need to project to a two-dimensional realization. We (randomly) choose one projection matrix

$$\mathbf{X} = \begin{bmatrix} 5 & -5 \\ 1 & 0 \\ 0 & 1 \end{bmatrix} = \mathbf{Y}^T.$$

Thus, the projected quantities are

$$\hat{\mathbb{L}} = \mathbf{Y}^T\mathbb{L}\mathbf{X} = \begin{bmatrix} 0 & -\frac{51}{50} \\ \frac{51}{50} & 0 \end{bmatrix}, \; \hat{\mathbb{L}}_s = \mathbf{Y}^T\mathbb{L}_s\mathbf{X} = \begin{bmatrix} \frac{157}{10} & -\frac{643}{50} \\ -\frac{643}{50} & \frac{53}{5} \end{bmatrix},$$

$$\tilde{\mathbf{W}} = \mathbf{W}\mathbf{X} = \left[\frac{27}{10}, -\frac{12}{5}\right], \tilde{\mathbf{V}} = \mathbf{Y}^T\mathbf{V} = \begin{bmatrix} \frac{27}{10} \\ -\frac{12}{5} \end{bmatrix},$$

$$\tilde{\mathbf{R}} = \mathbf{R}\mathbf{X} = [6, \; -4], \tilde{\mathbf{L}} = \mathbf{Y}^T\mathbf{L} = \begin{bmatrix} 6 \\ -4 \end{bmatrix}.$$

Therefore, it readily follows that we recover the original rational function:

$$(\mathbf{W}\mathbf{X})\left[(\mathbf{Y}^T\mathbb{L}_s\mathbf{X}) - s(\mathbf{Y}^T\mathbb{L}\mathbf{X})\right]^{-1}(\mathbf{Y}^T\mathbf{V}) = \frac{1}{s^2+1}.$$

Making use of the results of Section 8.2.3, we will determine the new interpolation data, which are compatible with the projected matrices $\hat{\mathbb{L}}$, $\hat{\mathbb{L}}_s$, $\tilde{\mathbf{W}}$, $\tilde{\mathbf{R}}$, $\tilde{\mathbf{V}}$, $\tilde{\mathbf{L}}$. The resulting new right interpolation points are the eigenvalues of the pencil $(\hat{\mathbb{L}}_s + \tilde{\mathbf{V}}\tilde{\mathbf{R}}, \hat{\mathbb{L}})$, while the new left interpolation points are the eigenvalues of the pencil $(\hat{\mathbb{L}}_s + \tilde{\mathbf{L}}\tilde{\mathbf{W}}, \hat{\mathbb{L}})$:

$$[\mathbf{T}, \mathbf{d}] = \text{eig}(\hat{\mathbb{L}}^{-1}(\hat{\mathbb{L}}_s + \tilde{\mathbf{V}}\tilde{\mathbf{R}})) \quad \Rightarrow \quad -(\tilde{\mathbf{W}}\mathbf{T})./(\tilde{\mathbf{R}}\mathbf{T}) = \begin{bmatrix} 0.0686 & 0.9767 \end{bmatrix},$$

$$\text{diag}(\mathbf{d}) = \begin{bmatrix} -3.684 \\ 0.154 \end{bmatrix} \Rightarrow \mathbf{H}(\text{diag}(\mathbf{d})) = \begin{bmatrix} 0.0686 \\ 0.9767 \end{bmatrix},$$

$$[\mathbf{T}_1, \mathbf{d}_1] = \text{eig}((\hat{\mathbb{L}}_s + \tilde{\mathbf{L}}\tilde{\mathbf{W}})(\hat{\mathbb{L}})^{-1}) \Rightarrow -(\mathbf{T}_1^{-1}\tilde{\mathbf{V}})./(\mathbf{T}_1^{-1}\tilde{\mathbf{L}}) = \begin{bmatrix} 0.9767 & 0.0686 \end{bmatrix},$$

$$\text{diag}(\mathbf{d}_1) = \begin{bmatrix} -0.154 \\ 3.684 \end{bmatrix} \Rightarrow \mathbf{H}(\text{diag}(\mathbf{d}_1)) = \begin{bmatrix} 0.9767 \\ 0.0686 \end{bmatrix}.$$

This example illustrates the interpolation property of the projected systems as described in Section 8.2.3. In particular, it shows that, starting from left and right interpolation points $(-1, -2, -3)$ and $(1, 2, 3)$, respectively, the projected system corresponds to left and right interpolation points $(-3.684, 0.154)$, $(3.684, -0.154)$. In other words, when the original left and right triples are compressed, the latter left and right pairs are obtained. ■

Example 8.35. Next, we wish to recover the polynomial $\phi(s) = s^2$ by means of measurements. From

$$\boldsymbol{\Lambda} = \text{diag}(1, \; 2, \; 3), \; \mathbf{M} = \text{diag}(-1, \; -2, \; -3), \; \mathbf{W} = \mathbf{V}^T = [1, \; 4, \; 9],$$

we calculate

$$\mathbb{L} = \begin{bmatrix} 0 & 1 & 2 \\ -1 & 0 & 1 \\ -2 & -1 & 0 \end{bmatrix}, \quad \mathbb{L}_s = \begin{bmatrix} 1 & 3 & 7 \\ 3 & 4 & 7 \\ 7 & 7 & 9 \end{bmatrix}.$$

Since rank $\mathbb{L} = 2$, while (8.37) yields $r = 3$, the minimal descriptor realizations have degree three and McMillan degree two. Therefore, $\mathbf{A} = -\mathbb{L}_s$, $\mathbf{E} = -\mathbb{L}$, $\mathbf{B} = \mathbf{V}$, $\mathbf{C} = \mathbf{W}$

is a desired realization, and the corresponding interpolant is $\phi(s) = \mathbf{W}(\mathbb{L}_s - s\,\mathbb{L})^{-1}\mathbf{V} = s^2$.

We consider two additional points, namely $\lambda_4 = 4$, $\mu_4 = -4$: $\mathbf{\Lambda} = \text{diag}[1,\ 2,\ 3,\ 4]$, $\mathbf{M} = -\mathbf{\Lambda}$. Then, $\mathbf{W} = \mathbf{V}^T = [1,\ 4,\ 9,\ 16]$. The Loewner pencil is updated by means of a new row and a new column:

$$\mathbb{L} = \begin{bmatrix} 0 & 1 & 2 & 3 \\ -1 & 0 & 1 & 2 \\ -2 & -1 & 0 & 1 \\ -3 & -2 & -1 & 0 \end{bmatrix}, \quad \mathbb{L}_s = \begin{bmatrix} 1 & 3 & 7 & 13 \\ 3 & 4 & 7 & 12 \\ 7 & 7 & 9 & 13 \\ 13 & 12 & 13 & 16 \end{bmatrix}.$$

Indeed, the pencil $(\mathbb{L}_s, \mathbb{L})$ has a (generalized) eigenvalue at infinity and a corresponding Jordan chain of length three:

$$\mathbf{v}_0 = \begin{bmatrix} 1 \\ -2 \\ 1 \\ 0 \end{bmatrix}, \quad \mathbf{v}_1 = \begin{bmatrix} -2 \\ 2 \\ 0 \\ 0 \end{bmatrix}, \quad \mathbf{v}_2 = \begin{bmatrix} -2 \\ 4 \\ 0 \\ 0 \end{bmatrix},$$

satisfying $\mathbb{L}\mathbf{v}_0 = \mathbf{0}$, $\mathbb{L}_s\mathbf{v}_0 = \mathbb{L}\mathbf{v}_1$, $\mathbb{L}_s\mathbf{v}_1 = \mathbb{L}\mathbf{v}_2$. Alternatively, one can check this statement by computing the QZ factorization of $(\mathbb{L}_s, \mathbb{L})$ to obtain an upper triangular pencil $(\mathbf{T}_{\mathbb{L}_s}, \mathbf{T}_{\mathbb{L}})$ with diagonal entries as follows:

$\text{diag}(\mathbf{T}_{\mathbb{L}_s})$	$\text{diag}(\mathbf{T}_{\mathbb{L}})$	
$-3.3467\,10^{-8} \ -1.8733\,10^{-1}\imath$	$1.4534\,10^{-8}$	$\leftarrow$ zero eig
$-2.0228\,10^{-7} \ +1.1323\,\imath$	$8.7842\,10^{-8}$	$\leftarrow$ zero eig
$5.3653\,10^{-15}$	$1.5794\,10^{-15}$	
3.9262	0	$\leftarrow$ zero eig

Consequently, the quotients of the first, second, and fourth entries of the diagonal yield the three infinite eigenvalues, while the third entry indicates an undetermined eigenvalue.

We will now project the quadruple $(\mathbb{L}_s, \mathbb{L}, \mathbf{V}, \mathbf{W})$ to get a minimal realization. The projectors are chosen randomly (use the command `round(randn(4,3))` in MATLAB):

$$\hat{\mathbf{X}} = \begin{bmatrix} -2 & 1 & 0 \\ -1 & 0 & -1 \\ 1 & 1 & 3 \\ -1 & -2 & 1 \end{bmatrix}, \quad \hat{\mathbf{Y}}^T = \begin{bmatrix} 1 & 0 & 1 & -1 \\ -1 & 1 & -2 & -2 \\ 0 & 0 & 0 & 1 \end{bmatrix},$$

satisfying the condition of Proposition 8.25. Consequently, the projected quadruple yields a minimal realization of the underlying rational function:

$$\hat{\mathbf{Y}}^T\mathbb{L}\hat{\mathbf{X}} = \begin{bmatrix} -5 & -4 & 11 \\ -19 & 16 & -5 \\ 7 & -4 & -1 \end{bmatrix}, \quad \hat{\mathbf{Y}}^T\mathbb{L}_s\mathbf{X} = \begin{bmatrix} 5 & -22 & 21 \\ 128 & 36 & -154 \\ -41 & -6 & 43 \end{bmatrix},$$

$\hat{\mathbf{Y}}^T\mathbf{V} = [-6,\ 47,\ 16]^T$, $\mathbf{W}\hat{\mathbf{X}} = [-13,\ -22,\ 39]$, and hence

$$\left[\mathbf{W}\hat{\mathbf{X}}\right] \cdot \left[(\hat{\mathbf{Y}}^T\mathbb{L}_s\mathbf{X}) - s \cdot (\hat{\mathbf{Y}}^T\mathbb{L}\mathbf{X})\right]^{-1} \cdot \left[\hat{\mathbf{Y}}^T\mathbf{V}\right] = s^2 = \phi(s).$$

This example illustrates the following aspects. First, since the rank of the Loewner matrix is two, the McMillan degree of the underlying interpolant is two. Furthermore, (8.37) yields the minimal dimension $r = 2$; since the rank of the shifted Loewner matrix is three, according to Section 8.2.3 this means that the interpolant has a polynomial term. Finally, according to the same section, since there is a Jordan chain of length three at infinity, the interpolant is purely polynomial of order two. ∎

Example 8.36. Consider the descriptor system defined by

$$\mathbf{A}=\begin{bmatrix} 1 & 0 & 0 \\ 0 & 1 & 0 \\ 0 & 0 & 0 \end{bmatrix}, \mathbf{E}=\begin{bmatrix} 0 & -1 & 0 \\ 0 & 0 & 0 \\ 0 & 0 & 1 \end{bmatrix}, \mathbf{B}=\begin{bmatrix} 0 & -1 \\ 1 & 0 \\ 0 & 1 \end{bmatrix}, \mathbf{C}=\begin{bmatrix} 1 & 0 & 0 \\ 0 & -1 & 1 \end{bmatrix}.$$

This is a *minimal descriptor realization*, and the transfer function is

$$\mathbf{H}(s)=\begin{bmatrix} s & 1 \\ 1 & \frac{1}{s} \end{bmatrix}.$$

The purpose of this example is to compare matrix interpolation with tangential interpolation. Notice that the McMillan degree of $\mathbf{H}(s)$ is two, while a minimal descriptor realization has degree three. We choose the following right/left data:

$\lambda_1=\imath,\quad \mathbf{w}_1=\mathbf{H}(\lambda_1)=\begin{bmatrix} 1 & -\imath \\ \imath & 1 \end{bmatrix}$	$\mu_1=2\imath,\quad \mathbf{v}_1=\mathbf{H}(\mu_1)=\begin{bmatrix} 1 & -\imath/2 \\ 2\imath & 1 \end{bmatrix}$
$\lambda_2=-\imath,\quad \mathbf{w}_2=\mathbf{H}(\lambda_2)=\begin{bmatrix} 1 & \imath \\ -\imath & 1 \end{bmatrix}$	$\mu_2=-2\imath,\quad \mathbf{v}_2=\mathbf{H}(\mu_2)=\begin{bmatrix} 1 & \imath/2 \\ -2\imath & 1 \end{bmatrix}$
$\lambda_3=1,\quad \mathbf{w}_3=\mathbf{H}(\lambda_3)=\begin{bmatrix} 1 & 1 \\ 1 & 1 \end{bmatrix}$	$\mu_3=-1,\quad \mathbf{v}_3=\mathbf{H}(\mu_3)=\begin{bmatrix} 1 & -1 \\ -1 & 1 \end{bmatrix}$

$$\Rightarrow \mathbf{V}^T=\begin{bmatrix} \mathbf{v}_1 & \mathbf{v}_2 & \mathbf{v}_3 \end{bmatrix}, \quad \mathbf{W}=[\mathbf{w}_1\ \mathbf{w}_2\ \mathbf{w}_3]; \quad \mathbf{R}=[\mathbf{I}_2\ \mathbf{I}_2\ \mathbf{I}_2]=\mathbf{L}^T.$$

It turns out that the associated Loewner and shifted Loewner matrices are

$$\mathbb{L}=\left[\begin{array}{cc|cc|cc} 0 & \frac{1}{2} & 0 & -\frac{1}{2} & 0 & \frac{\imath}{2} \\ 1 & 0 & 1 & 0 & 1 & 0 \\ \hline 0 & -\frac{1}{2} & 0 & \frac{1}{2} & 0 & -\frac{\imath}{2} \\ 1 & 0 & 1 & 0 & 1 & 0 \\ \hline 0 & -\imath & 0 & \imath & 0 & 1 \\ 1 & 0 & 1 & 0 & 1 & 0 \end{array}\right], \mathbb{L}_s=\left[\begin{array}{cc|cc|cc} 1 & 0 & 1 & 0 & 1 & 0 \\ 3\imath & 1 & \imath & 1 & 2\imath+1 & 1 \\ \hline 1 & 0 & 1 & 0 & 1 & 0 \\ -\imath & 1 & -3\imath & 1 & 1-2\imath & 1 \\ \hline 1 & 0 & 1 & 0 & 1 & 0 \\ \imath-1 & 1 & -\imath-1 & 1 & 0 & 1 \end{array}\right].$$

∎

Remark 8.37. The above example shows that *matrix-valued* Loewner pencils are a special case of the general definition of Loewner pencils given by (8.28), (8.30) for appropriate $\mathbf{\Lambda}$, $\mathbf{M}$, $\mathbf{V}$, $\mathbf{L}$, $\mathbf{W}$, $\mathbf{R}$.

It follows that the rank of $\mathbb{L}$ and $\mathbb{L}_s$ is two, while that of $\xi\,\mathbb{L}-\mathbb{L}_s$ is three, for all interpolation points $\xi\in\{\lambda_j\}\cup\{\mu_i\}$. Hence, according to assumption (8.37), we cannot identify the original system from this data. We notice, however, that $\ker(\xi\,\mathbb{L}(1{:}3,1{:}3)-\mathbb{L}_s(1:3,1:3))=0$ for all interpolation points and therefore the corresponding matrices provide the realization

$$-[\mathbf{w}_1,\ \mathbf{w}_2(:,1)](s\mathbb{L}(1{:}3,1{:}3)-\mathbb{L}_s(1{:}3,1{:}3))^{-1}\begin{bmatrix}\mathbf{v}_1\\ \mathbf{v}_2(1,:)\end{bmatrix}=\mathbf{H}(s).$$

Tangential interpolation for Example 8.36. If we choose the right and left directions as

$$\mathbf{R}=\begin{bmatrix}1&0&-1\\0&1&-1\end{bmatrix},\ \mathbf{L}=\begin{bmatrix}1&0\\-1&1\\-1&1\end{bmatrix},$$

the rank of $[\mathbb{L}\ \ \mathbb{L}_s]$ is different from that of $\left[\begin{smallmatrix}\mathbb{L}\\ \mathbb{L}_s\end{smallmatrix}\right]$, and hence condition (8.37) is not satisfied. Therefore, we will choose new left directions

$$\widehat{\mathbf{L}}=\begin{bmatrix}1&0\\-1&1\\0&1\end{bmatrix}\Rightarrow\widehat{\mathbf{V}}=\begin{bmatrix}\mathbf{L}(1,:)\mathbf{V}_1\\ \mathbf{L}(2,:)\mathbf{V}_2\\ \mathbf{L}(3,:)\mathbf{V}_3\end{bmatrix}=\begin{bmatrix}1&-\frac{\imath}{2}\\-2\imath-1&1-\frac{\imath}{2}\\-1&1\end{bmatrix},$$

$$\widehat{\mathbb{L}}=\begin{bmatrix}0&-\frac{1}{2}&-\frac{\imath}{2}\\1&-\frac{1}{2}&-\frac{\imath}{2}-1\\1&0&-1\end{bmatrix},\ \widehat{\mathbb{L}}_s=\begin{bmatrix}1&0&-1\\-\imath-1&1&2\imath-1\\\imath-1&1&-1\end{bmatrix}.$$

In this case,

$$\ker([\widehat{\mathbb{L}}^T,\ \ \widehat{\mathbb{L}}_s^T])=\ker([\widehat{\mathbb{L}},\ \ \widehat{\mathbb{L}}_s])=3,$$

and $\det(s\widehat{\mathbb{L}}-\widehat{\mathbb{L}}_s)=s(2\imath+1)\neq 0$. Therefore, since this determinant is different from zero at the interpolation points, assumption (8.37) is satisfied and indeed $\mathbf{W}(\widehat{\mathbb{L}}_s-s\widehat{\mathbb{L}})^{-1}\widehat{\mathbf{V}}=\mathbf{H}(s)$. Notice also that this realization is complex as no provisions are made to obtain a real realization.

Example 8.38. Here, we will illustrate the relationship between the McMillan degree, the degree of minimal realizations, and the **D**-term:

$$\mathbf{A}=\begin{bmatrix}0&1&0\\0&0&1\\0&0&0\end{bmatrix},\ \mathbf{C}=\begin{bmatrix}1&0&0\\0&1&0\end{bmatrix},\ \mathbf{B}=\mathbf{I}_3,\ \mathbf{D}=\begin{bmatrix}1&1&1\\1&1&1\end{bmatrix}$$

$$\Rightarrow\mathbf{H}(s)=\mathbf{C}(s\mathbf{I}-\mathbf{A})^{-1}\mathbf{B}+\mathbf{D}=\begin{bmatrix}\frac{1}{s}+1&\frac{1}{s^2}+1&\frac{1}{s^3}+1\\1&\frac{1}{s}+1&\frac{1}{s^2}+1\end{bmatrix}.$$

Let the interpolation points be

$$\mathbf{M}=\operatorname{diag}[1,\ 1,\ -\tfrac{1}{2},\ -\tfrac{1}{2},\ -\tfrac{1}{4},\ -\tfrac{1}{4}],$$
$$\Lambda=\operatorname{diag}[\tfrac{1}{2},\ \tfrac{1}{2},\ \tfrac{1}{2},\ -1,\ -1,\ -1,\ 2,\ 2,\ 2].$$

The interpolation values $\mathbf{w}_i = \mathbf{H}(\lambda_i)$, $\mathbf{v}_i = \mathbf{H}(\mu_i)$, $i = 1,2,3$, are

$$\mathbf{w}_1 = \begin{bmatrix} 3 & 5 & 9 \\ 1 & 3 & 5 \end{bmatrix}, \mathbf{w}_2 = \begin{bmatrix} 0 & 2 & 0 \\ 1 & 0 & 2 \end{bmatrix}, \mathbf{w}_3 = \begin{bmatrix} 3/2 & 5/4 & 9/8 \\ 1 & 3/2 & 5/4 \end{bmatrix},$$

$$\mathbf{v}_1 = \begin{bmatrix} 2 & 2 & 2 \\ 1 & 2 & 2 \end{bmatrix}, \mathbf{v}_2 = \begin{bmatrix} -1 & 5 & -7 \\ 1 & -1 & 5 \end{bmatrix}, \mathbf{v}_3 = \begin{bmatrix} -3 & 17 & -63 \\ 1 & -3 & 17 \end{bmatrix}.$$

This corresponds to *matrix interpolation* (the values considered are matrices) as opposed to tangential interpolation (which will follow next):

$$\mathbb{V} = \left[\begin{array}{ccc} 2 & 2 & 2 \\ 1 & 2 & 2 \\ \hline -1 & 5 & -7 \\ 1 & -1 & 5 \\ \hline -3 & 17 & -63 \\ 1 & -3 & 17 \end{array}\right], \quad \mathbb{W} = \left[\begin{array}{ccc|ccc|ccc} 3 & 5 & 9 & 0 & 2 & 0 & \frac{3}{2} & \frac{5}{4} & \frac{9}{8} \\ 1 & 3 & 5 & 1 & 0 & 2 & 1 & \frac{3}{2} & \frac{5}{4} \end{array}\right], \quad \mathbf{R} = [\mathbf{I}_3 \ \mathbf{I}_3 \ \mathbf{I}_3], \ \ \mathbf{L} = \begin{bmatrix} \mathbf{I}_2 \\ \mathbf{I}_2 \\ \mathbf{I}_2 \end{bmatrix}$$

$$\Rightarrow \quad \mathbb{L} = \left[\begin{array}{ccc|ccc|ccc} -2 & -6 & -14 & 1 & 0 & 1 & -\frac{1}{2} & -\frac{3}{4} & -\frac{7}{8} \\ 0 & -2 & -6 & 0 & 1 & 0 & 0 & -\frac{1}{2} & -\frac{3}{4} \\ \hline 4 & 0 & 16 & -2 & 6 & -14 & 1 & -\frac{3}{2} & \frac{13}{4} \\ 0 & 4 & 0 & 0 & -2 & 6 & 0 & 1 & -\frac{3}{2} \\ \hline 8 & -16 & 96 & -4 & 20 & -84 & 2 & -7 & \frac{57}{2} \\ 0 & 8 & -16 & 0 & -4 & 20 & 0 & 2 & -7 \end{array}\right] \in \mathbb{R}^{6\times 9},$$

$$\text{and} \qquad \mathbb{L}_s = \left[\begin{array}{ccc|ccc|ccc} 1 & -1 & -5 & 1 & 2 & 1 & 1 & \frac{1}{2} & \frac{1}{4} \\ 1 & 1 & -1 & 1 & 1 & 2 & 1 & 1 & \frac{1}{2} \\ \hline 1 & 5 & 1 & 1 & -1 & 7 & 1 & 2 & -\frac{1}{2} \\ 1 & 1 & 5 & 1 & 1 & -1 & 1 & 1 & 2 \\ \hline 1 & 9 & -15 & 1 & -3 & 21 & 1 & 3 & -6 \\ 1 & 1 & 9 & 1 & 1 & -3 & 1 & 1 & 3 \end{array}\right] \in \mathbb{R}^{6\times 9}.$$

It readily follows that, while $\operatorname{rank}\mathbb{L} = \operatorname{rank}\mathbb{L}_s = 3$, the rank of $[\mathbb{L};\ \mathbb{L}_s]$ is equal to the rank of $[\mathbb{L},\ \mathbb{L}_s]$, which is equal to four. Furthermore the rank of $\xi\mathbb{L} - \mathbb{L}_s$ is also four, for all $\xi \in \{\mu_i\} \cup \{\lambda_j\}$. Consequently, according to (8.37), the dimension of the minimal realization is $r = 4$. ∎

Tangential interpolation for Example 8.38. If we define the index set $I = [1\,2\,3\,4]$, we get

$$\boldsymbol{\Lambda}(I,I) = \operatorname{diag}\begin{bmatrix} \frac{1}{2} & \frac{1}{2} & \frac{1}{2} & -1 \end{bmatrix}, \ \mathbf{M}(I,I) = \operatorname{diag}\begin{bmatrix} 1 & 1 & -\frac{1}{2} & -\frac{1}{2} \end{bmatrix},$$

$$\mathbb{W}(:,I) = \begin{bmatrix} 3 & 5 & 9 & 0 \\ 1 & 3 & 5 & 1 \end{bmatrix}, \ \mathbb{V}(I,:) = \begin{bmatrix} 2 & 2 & 2 \\ 1 & 2 & 2 \\ -1 & 5 & -7 \\ 1 & -1 & 5 \end{bmatrix},$$

$$\mathbb{L}(I,I) = \begin{bmatrix} -2 & -6 & -14 & 1 \\ 0 & -2 & -6 & 0 \\ 4 & 0 & 16 & -2 \\ 0 & 4 & 0 & 0 \end{bmatrix}, \; \mathbb{L}_s(I,I) = \begin{bmatrix} 1 & -1 & -5 & 1 \\ 1 & 1 & -1 & 1 \\ 1 & 5 & 1 & 1 \\ 1 & 1 & 5 & 1 \end{bmatrix}.$$

Since condition (8.37) is satisfied with $r = 4$ and the rank of the Loewner matrix is three, we recover a minimal descriptor realization with incorporated $\mathbf{D}$-term:

$$\mathbf{W}(:,I)(\mathbb{L}_s(I,I) - s\mathbb{L}(I,I))^{-1}\mathbf{V}(I,:) = \mathbf{H}(s).$$

An alternative way to obtain the interpolant is to project the original matrices by randomly generated projectors:

$$\mathbf{Y}^T = \begin{bmatrix} 0 & 0 & 0 & 0 & 0 & 1 \\ 0 & 1 & 1 & 0 & 0 & 1 \\ 1 & 1 & 1 & -1 & 1 & 1 \\ 0 & 1 & 0 & 0 & 0 & -1 \end{bmatrix}, \; \mathbf{X} = \begin{bmatrix} 0 & 1 & -1 & 1 \\ 0 & 0 & 1 & -1 \\ 1 & 1 & 0 & 0 \\ 1 & -1 & 0 & 1 \\ -1 & -1 & 0 & 0 \\ 0 & 1 & 1 & 1 \\ 0 & 0 & 1 & 1 \\ 0 & 1 & 0 & 0 \\ 1 & 0 & 1 & -1 \end{bmatrix}$$

$$\Rightarrow \mathbf{E}_\mathcal{S} = \begin{bmatrix} -19 & 10 & 21 & 19 \\ -\frac{31}{2} & 3 & \frac{9}{2} & \frac{15}{2} \\ \frac{565}{8} & -\frac{103}{4} & -\frac{687}{8} & -\frac{641}{8} \\ \frac{45}{4} & -\frac{35}{2} & -\frac{95}{4} & -\frac{65}{4} \end{bmatrix}, \; \mathbf{A}_\mathcal{S} = \begin{bmatrix} 12 & 6 & 1 & -4 \\ 14 & 18 & 16 & 5 \\ -\frac{63}{4} & \frac{41}{2} & \frac{153}{4} & \frac{127}{4} \\ -\frac{25}{2} & -5 & \frac{5}{2} & \frac{15}{2} \end{bmatrix},$$

$$\mathbf{B}_\mathcal{S} = \begin{bmatrix} 1 & -3 & 17 \\ 1 & 4 & 12 \\ -1 & 24 & -54 \\ 0 & 5 & -15 \end{bmatrix}, \mathbf{C}_\mathcal{S} = \begin{bmatrix} \frac{65}{8} & \frac{45}{4} & \frac{37}{8} & -\frac{13}{8} \\ \frac{29}{4} & \frac{17}{2} & \frac{25}{4} & \frac{3}{4} \end{bmatrix}$$

$$\Rightarrow \mathbf{C}_\mathcal{S}(\mathbf{A}_\mathcal{S} - s\mathbf{E}_\mathcal{S})^{-1}\mathbf{B}_\mathcal{S} = \mathbf{H}(s).$$

We thus obtain a different (equivalent) descriptor realization with incorporated $\mathbf{D}$-term.

Example 8.39. We consider the transfer function

$$\mathbf{H}(s) = \frac{1}{s(s+2)} \begin{bmatrix} s & 0 \\ 1 & s+2 \end{bmatrix} + \begin{bmatrix} 1 & 2 \\ 0 & 0 \end{bmatrix},$$

which is to be reconstructed from measurements. Let

$$\Lambda = \text{diag}[\imath, \; -\imath, \; 3\imath, \; -3\imath] \quad \text{and} \quad \mathbf{M} = \text{diag}[2\imath, \; -2\imath, \; 4\imath, \; -4\imath].$$

The resulting matrix measurements are

$$\mathbf{w}_1 = \mathbf{H}(\imath) = \begin{bmatrix} \frac{7}{5} - \frac{\imath}{5} & 2 \\ -\frac{1}{5} - \frac{2\imath}{5} & -\imath \end{bmatrix}, \qquad \mathbf{w}_2 = \mathbf{H}(-\imath) = \overline{\mathbf{w}_1},$$

$$\mathbf{w}_3 = \mathbf{H}(3\imath) = \begin{bmatrix} \frac{15}{13} - \frac{3\imath}{13} & 2 \\ -\frac{1}{13} - \frac{2\imath}{39} & -\frac{\imath}{3} \end{bmatrix}, \qquad \mathbf{w}_4 = \mathbf{H}(-3\imath) = \overline{\mathbf{w}_3},$$

$$\mathbf{v}_1 = \mathbf{H}(2\imath) = \begin{bmatrix} \frac{5}{4} - \frac{\imath}{4} & 2 \\ -\frac{1}{8} - \frac{\imath}{8} & -\frac{\imath}{2} \end{bmatrix}, \qquad \mathbf{v}_2 = \mathbf{H}(-2\imath) = \overline{\mathbf{v}_1},$$

$$\mathbf{v}_3 = \mathbf{H}(4\imath) = \begin{bmatrix} \frac{11}{10} - \frac{\imath}{5} & 2 \\ -\frac{1}{20} - \frac{\imath}{40} & -\frac{\imath}{4} \end{bmatrix}, \qquad \mathbf{v}_4 = \mathbf{H}(-4\imath) = \overline{\mathbf{v}_3}.$$

With $\mathbf{R} = \begin{bmatrix} 1 & 1 & 0 & 0 \\ 0 & 0 & 1 & 1 \end{bmatrix} = \mathbf{L}^T$, the following tangential data result:

$$\begin{aligned} &k = 4, \\ &\{\lambda, \mathbf{r}, \mathbf{w}\} = \{(i, \mathbf{e}_1, \mathbf{H}(i)\mathbf{e}_1), (-\imath, \mathbf{e}_1, \mathbf{H}(-\imath)\mathbf{e}_1), (3\imath, \mathbf{e}_2, \mathbf{H}(3\imath)\mathbf{e}_2), (-3\imath, \mathbf{e}_2, \mathbf{H}(-3\imath)\mathbf{e}_2)\}, \\ &q = 4, \\ &\{\mu, \ell, \mathbf{v}\} = \left\{ \begin{array}{l} (2\imath, \mathbf{e}_1{}^T, \mathbf{e}_1{}^T\mathbf{H}(2\imath)), (-2\imath, \mathbf{e}_1{}^T, \mathbf{e}_1{}^T\mathbf{H}(-2\imath)), \\ \qquad (4\imath, \mathbf{e}_2{}^T, \mathbf{e}_2{}^T\mathbf{H}(4\imath)), (-4\imath, \mathbf{e}_2{}^T, \mathbf{e}_2{}^T\mathbf{H}(-4\imath)) \end{array} \right\}. \end{aligned}$$

We obtain the complex matrices $\mathbb{L}^{\mathrm{C}}$, $\mathbb{L}_s^{\mathrm{C}}$, $\mathbb{W}^{\mathrm{C}}$, and $\mathbb{V}^{\mathrm{C}}$. To obtain real matrices, we follow the procedure described in Section 8.2.4. In other words, we multiply by $\mathbf{J} = \mathrm{blkdiag}[\hat{\mathbf{J}}, \hat{\mathbf{J}}]$, where $\hat{\mathbf{J}} = \frac{1}{\sqrt{2}} \begin{bmatrix} 1 & -\imath \\ 1 & \imath \end{bmatrix}$,

$$\mathbb{L}^{\mathrm{R}} = \mathbf{J}^* \mathbb{L}^{\mathrm{C}} \mathbf{J},\ \mathbb{L}_s^{\mathrm{R}} = \mathbf{J}^* \mathbb{L}_s^{\mathrm{C}} \mathbf{J},\ \mathbb{W}^{\mathrm{R}} = \mathbb{W}^{\mathrm{C}} \mathbf{J},\ \mathbb{V}^{\mathrm{R}} = \mathbf{J}^* \mathbb{V}^{\mathrm{C}},$$

to obtain

$$\mathbb{L}^{\mathrm{R}} = \begin{bmatrix} -\frac{1}{5} & \frac{1}{10} & 0 & 0 \\ -\frac{1}{5} & \frac{1}{10} & 0 & 0 \\ \frac{1}{25} & -\frac{1}{50} & 0 & 0 \\ \frac{2}{25} & \frac{21}{100} & 0 & \frac{1}{6} \end{bmatrix}, \qquad \mathbb{L}_s^{\mathrm{R}} = \begin{bmatrix} \frac{12}{5} & -\frac{1}{5} & 4 & 0 \\ \frac{2}{5} & -\frac{1}{5} & 0 & 0 \\ -\frac{2}{25} & \frac{1}{25} & 0 & 0 \\ -\frac{4}{25} & \frac{2}{25} & 0 & 0 \end{bmatrix},$$

$$\mathbb{W}^{\mathrm{R}} = \frac{1}{\sqrt{2}} \begin{bmatrix} \frac{14}{5} & -\frac{2}{5} & 4 & 0 \\ -\frac{2}{5} & -\frac{4}{5} & 0 & -\frac{2}{3} \end{bmatrix}, \qquad \mathbb{V}^{\mathrm{R}} = \frac{1}{\sqrt{2}} \begin{bmatrix} \frac{5}{2} & 4 \\ \frac{1}{2} & 0 \\ -\frac{1}{10} & 0 \\ \frac{1}{20} & \frac{1}{2} \end{bmatrix}.$$

It readily follows that $\operatorname{rank} \mathbb{L}^{\mathrm{R}} = 2$, $\operatorname{rank} \mathbb{L}_s^{\mathrm{R}} = 2$, while the rank of both $[\mathbb{L}^{\mathrm{R}}\ \ \mathbb{L}_s^{\mathrm{R}}]$ and $[\mathbb{L}^{\mathrm{R}};\ \mathbb{L}_s^{\mathrm{R}}]$ is three. Thus, the dimension of minimal descriptor realizations (including the $\mathbf{D}$-term) is three, while the McMillan degree is two. Hence, we have to project to a realization of order three. We (randomly) pick

$$\mathbf{Y}^T = \begin{bmatrix} 0 & 1 & 1 & 0 \\ 0 & -2 & 0 & -1 \\ 1 & 0 & -2 & 1 \end{bmatrix}, \quad \mathbf{X} = \begin{bmatrix} 2 & -1 & 0 \\ 0 & 0 & -1 \\ 0 & 0 & 1 \\ 1 & -2 & -1 \end{bmatrix}.$$

Thus, $\mathbf{E}_\delta = -\mathbf{Y}^T \mathbb{L}^R \mathbf{X}$, $\mathbf{A}_\delta = -\mathbf{Y}^T \mathbb{L}_s^R \mathbf{X}$, $\mathbf{B}_\delta = \mathbf{Y}^T \mathbf{V}^R$, $\mathbf{C}_\delta = \mathbb{W}^R \mathbf{X}$, where

$$\mathbf{E}_\delta = \begin{bmatrix} \frac{8}{25} & -\frac{4}{25} & \frac{2}{25} \\ -\frac{71}{150} & -\frac{1}{75} & -\frac{173}{300} \\ \frac{7}{30} & \frac{2}{15} & \frac{31}{60} \end{bmatrix}, \quad \mathbf{A}_\delta = \begin{bmatrix} -\frac{16}{25} & \frac{8}{25} & -\frac{4}{25} \\ \frac{32}{25} & -\frac{16}{25} & \frac{8}{25} \\ -\frac{24}{5} & \frac{12}{5} & -\frac{21}{5} \end{bmatrix},$$

$$\mathbf{B}_\delta = \frac{1}{\sqrt{2}} \begin{bmatrix} \frac{2}{5} & 0 \\ -\frac{21}{20} & -\frac{1}{2} \\ \frac{11}{4} & \frac{9}{2} \end{bmatrix}, \quad \mathbf{C}_\delta = \frac{1}{\sqrt{2}} \begin{bmatrix} \frac{28}{5} & -\frac{14}{5} & \frac{22}{5} \\ -\frac{22}{15} & \frac{26}{15} & \frac{22}{15} \end{bmatrix}.$$

To extract the **D**-term, we proceed as follows. The pencil $(\mathbf{A}_\delta, \mathbf{E}_\delta)$ has two finite eigenvalues at -2, 0 and an infinite eigenvalue. The matrices of the associated right and left eigenvectors are

$$\mathbf{T}_1 = \begin{bmatrix} -13 & 1 & -7 \\ -14 & 2 & -11 \\ 6 & 0 & 6 \end{bmatrix}, \quad \mathbf{T}_2 = \begin{bmatrix} 1 & 0 & 0 \\ 2 & 1 & 0 \\ 3 & 4 & 4 \end{bmatrix}.$$

Therefore,

$$\mathbf{T}_2 \mathbf{E}_\delta \mathbf{T}_1 = \left[\begin{array}{cc|c} -\frac{36}{25} & 0 & 0 \\ 0 & -\frac{1}{2} & 0 \\ \hline 0 & 0 & 0 \end{array}\right], \quad \mathbf{T}_2 \mathbf{A}_\delta \mathbf{T}_1 = \left[\begin{array}{cc|c} \frac{72}{25} & 0 & 0 \\ 0 & 0 & 0 \\ \hline 0 & 0 & -72 \end{array}\right],$$

$$\mathbf{C}_\delta \mathbf{T}_1 = \frac{1}{\sqrt{2}} \left[\begin{array}{cc|c} -\frac{36}{5} & 0 & 18 \\ \frac{18}{5} & 2 & 0 \end{array}\right], \quad \mathbf{T}_2 \mathbf{B}_\delta = \frac{1}{\sqrt{2}} \left[\begin{array}{cc} \frac{2}{5} & 0 \\ -\frac{1}{4} & -\frac{1}{2} \\ \hline 8 & 16 \end{array}\right].$$

Thus, a minimal realization of order two is

$$\mathbf{E} = -\mathrm{diag}\left[\frac{36}{25}, \ \frac{1}{2}\right], \ \mathbf{A} = \mathrm{diag}\left[\frac{72}{25}, \ 0\right], \ \mathbf{C} = \frac{1}{\sqrt{2}} \begin{bmatrix} -\frac{36}{5} & 0 \\ \frac{18}{5} & 2 \end{bmatrix},$$

$$\mathbf{B} = \frac{1}{\sqrt{2}} \begin{bmatrix} \frac{2}{5} & 0 \\ -\frac{1}{4} & -\frac{1}{2} \end{bmatrix}, \ \mathbf{D} = \frac{1}{2} \begin{bmatrix} 18 \\ 0 \end{bmatrix} \frac{1}{72} [8 \quad 16] = \begin{bmatrix} 1 & 2 \\ 0 & 0 \end{bmatrix}.$$

■

8.3 ▪ Reduced-order modeling from data

We are now ready to address the problem of constructing reduced-order (linear) models from measurements in the Loewner framework. More precisely, we seek to construct models that fit the set of measurements defined in (8.7), (8.8), or (8.9), (8.10).

8.3.1 ▪ Solution in the Loewner framework

We assume that data sets contain samples of the frequency response of an underlying system; for instance, in the electronics industry, measurements of the S- (scattering-)

parameters may be provided. We denote these pairs of measurements by $(\imath\omega_j, \mathbf{S}^{(j)})$, $\omega_j \in \mathbb{R}$, $\mathbf{S}^{(j)} \in \mathbb{C}^{p\times m}$, $j = 1,\ldots,N$ (i.e., $\mathbf{S}^{(j)}$ is the measurement at frequency ω_j); see also Remark 8.2. As we did in Section 8.2.2, we arrange these measurements in two arrays, a *column* and a *row* array:

$$P_c = \{(\lambda_j, \mathbf{W}_j) : j = 1,\ldots,k\}, \quad P_r = \{(\mu_i, \mathbf{V}_i) : i = 1,\ldots,q\},$$

where we have redefined

$$\left.\begin{array}{lll} \lambda_j = \imath\omega_j, & \mathbf{W}_j = \mathbf{S}^{(j)}, & j = 1,\ldots,k \\ \mu_i = \imath\omega_{k+i}, & \mathbf{V}_i = \mathbf{S}^{(k+i)}, & i = 1,\ldots,q \end{array}\right\} \quad \text{and} \quad q+k = N.$$

To obtain a real system, the given set must be closed under conjugation, i.e., in addition to the above, the pairs $(-\imath\omega_j, \overline{\mathbf{S}^{(j)}})$, $j = 1,\ldots,N$, must also belong to the set of measurements. To generate *right and left tangential data sets*, we need the right and left directions $\mathbf{r}_i \in \mathbb{R}^m$, $\boldsymbol{\ell}_j \in \mathbb{R}^p$, which can be chosen either as columns of the identity matrix or as random vectors. Thus, the right data set corresponding to (8.7) is

$$\left\{\lambda_j, \overline{\lambda}_j; \mathbf{r}_j, \mathbf{r}_j; \mathbf{w}_j = \mathbf{W}_j\mathbf{r}_j, \overline{\mathbf{w}}_j = \overline{\mathbf{W}}_j\mathbf{r}_j\right\}, \quad j = 1,\ldots,k,$$

while the left data set corresponding to (8.8) is

$$\left\{\mu_i, \overline{\mu}_i; \boldsymbol{\ell}_i, \boldsymbol{\ell}_i; \mathbf{v}_i^T = \boldsymbol{\ell}_i^T\mathbf{V}_i, \overline{\mathbf{v}}_i^T = \boldsymbol{\ell}_i^T\overline{\mathbf{V}}_i\right\}, \quad i = 1,\ldots,q.$$

The associated Loewner and shifted Loewner matrices constructed as in (8.28), (8.30) have dimension $2q \times 2k$. To ensure real matrix entries, we follow the procedure described in Section 8.2.4. We define

$$\mathbf{J}_r = \text{blkdiag}\left[\hat{\mathbf{J}},\ldots,\hat{\mathbf{J}}\right] \in \mathbb{C}^{2r\times 2r}, \; \hat{\mathbf{J}} = \frac{1}{\sqrt{2}}\begin{bmatrix} 1 & -ß \\ 1 & ß \end{bmatrix}$$

and transform the quantities of the Loewner framework as follows:

$$\begin{array}{lllll} \boldsymbol{\Lambda}^{\mathbf{R}} & = & \mathbf{J}_k^*\boldsymbol{\Lambda}\mathbf{J}_k, & \mathbf{R}^{\mathbf{R}} = \mathbf{R}\mathbf{J}_k, & \mathbf{W}^{\mathbf{R}} = \mathbf{W}\mathbf{J}_k, \\ \mathbf{M}^{\mathbf{R}} & = & \mathbf{J}_q^*\mathbf{M}\mathbf{J}_q, & \mathbf{L}^{\mathbf{R}} = \mathbf{J}_q^*\mathbf{L}, & \mathbf{V}^{\mathbf{R}} = \mathbf{J}_q^*\mathbf{V}, \\ \mathbb{L}^{\mathbf{R}} & = & \mathbf{J}_q^*\mathbb{L}\mathbf{J}_k, & \mathbb{L}_s^{\mathbf{R}} = \mathbf{J}_q^*\mathbb{L}_s\mathbf{J}_k. & \end{array}$$

The transformed quantities have real entries. Then, computing the two SVDs of the Loewner pencil given in (8.38) and provided that condition (8.37) is satisfied, one obtains the projection defined by $\mathbf{X} \in \mathbb{R}^{2k\times r}$, $\mathbf{Y} \in \mathbb{R}^{2q\times r}$ as in (8.40), (8.41). The quantity r is the numerical rank of the corresponding matrices in (8.38) and depends on the application. Thus,

$$\mathbf{E}_{\mathcal{S}} = -\mathbf{Y}^T\mathbb{L}^{\mathbf{R}}\mathbf{X}, \; \mathbf{A}_{\mathcal{S}} = -\mathbf{Y}^T\mathbb{L}_s^{\mathbf{R}}\mathbf{X}, \; \mathbf{B}_{\mathcal{S}} = \mathbf{Y}^T\mathbf{V}^{\mathbf{R}}, \; \mathbf{C}_{\mathcal{S}} = \mathbf{W}^{\mathbf{R}}\mathbf{X},$$

with $\mathbf{D}$ incorporated in the other matrices as an approximate reduced-order descriptor realization of the data of order r.

8.3.2 ▪ Numerical results

The Loewner algorithm described in Section 8.2.3 is applied in two cases. The first consists of an a priori given system, while the second consists of frequency response measurements exclusively. During this procedure we monitor the accuracy and CPU time required to produce an RMO. The accuracy is assessed using

- the normalized H_∞-norm of the error system, defined as

$$H_\infty \text{ error } = \frac{\max_{i=1,\dots,N} \sigma_1\left(\mathbf{H}(\imath\omega_i) - \mathbf{S}^{(i)}\right)}{\max_{i=1,\dots,N} \sigma_1\left(\mathbf{S}^{(i)}\right)}, \tag{8.45}$$

  where $\sigma_1(\cdot)$ denotes the largest singular value of $(\cdot)$, and

- the normalized H_2-norm of the error system

$$(H_2 \text{ error})^2 = \frac{\sum_{i=1}^{N} \left\|\mathbf{H}(\imath\omega_i) - \mathbf{S}^{(i)}\right\|_F^2}{\sum_{i=1}^{N} \left\|\mathbf{S}^{(i)}\right\|_F^2}, \tag{8.46}$$

  where $\|\cdot\|_F^2$ stands for the (square of the) Frobenius norm (namely the sum of the squares of the magnitude of all entries).

The former error measure evaluates the maximum deviation in the singular values of $\mathbf{H}$ on the imaginary axis, while the latter evaluates the error in the magnitude of all entries, proving to be a good estimate of the overall performance of the model. For details on these norms, we refer to [3]. The experiments were performed on a Pentium Dual-Core at 2.2 GHz with 3 GB RAM.

Example 8.40 (Low-order given system). We consider the band-stop filter described in [22]; this system has two ports (i.e., two inputs and two outputs), state-space dimension 14, and a $\mathbf{D}$-term of size 2×2 and full rank; thus, if we incorporate $\mathbf{D}$ in the other matrices as described in Remark 8.1, we obtain a minimal descriptor realization of dimension 16.[32]

We take $k = 608$ samples of the transfer function on the imaginary axis (frequency response measurements) between 10^{-3} and 10^3 rad/sec (Figure 8.2, upper pane). Figure 8.2, lower pane, shows the first 30 normalized singular values of the resulting Loewner and shifted Loewner matrices (the rest are zero). The Loewner matrix has rank 14, while the shifted Loewner matrix has rank 16, so, based on the drop in singular values, we generate models of order 16 with $\mathbf{D} = 0$. Table 8.1 shows the CPU time and the errors for the resulting models. Two algorithms are compared: the complex SVD approach, where the Loewner pencil is left complex, followed by the real SVD approach, where the Loewner pencil is transformed to real form, as explained earlier.
This example shows that the Loewner approach recovers the original system with small error. Furthermore, it also shows that working in real arithmetic is more expensive timewise than working in complex arithmetic. ∎

[32] A band-stop filter is a dynamical system that blocks signals with frequency in a given interval while it lets through, almost unchanged, signals with frequency outside the given interval. Such systems are characterized by a frequency response that is nearly zero in the given interval of frequencies while it is almost equal to one otherwise. The upper pane of Figure 8.2 shows that the system under consideration exhibits such a behavior between the first input and the second output, or vice versa (red curve).

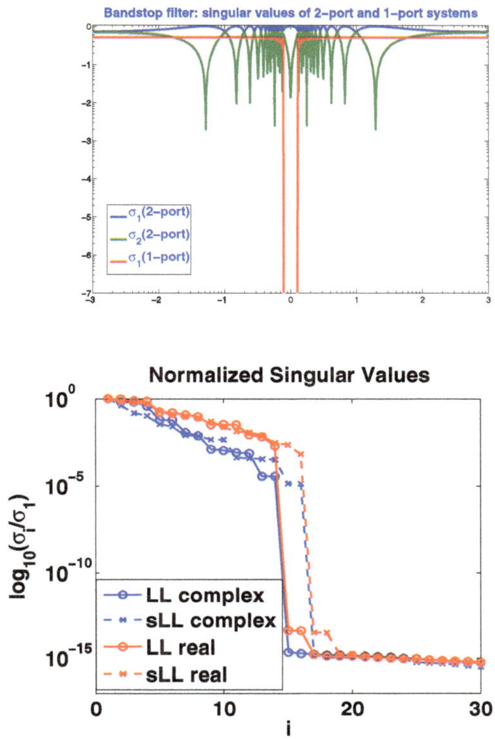

Figure 8.2. *Upper pane: the blue and green curves are the two singular values of the transfer function as a function of frequency; the red curve is the singular value (magnitude) of the* $(1,2)$ *entry of the transfer function, which exhibits band-stop behavior. Lower pane: singular values of the Loewner and the shifted Loewner matrices in both complex (blue curves) and real (red curves) form.*

Table 8.1. *Results for* $k = 608$ *noise-free measurements of an order-*14 *system with* $p = 2$ *ports (two inputs, two outputs).*

Algorithm	CPU (s)	H_∞-error	H_2-error
SVD Complex	0.88	1.3937e–010	2,4269e–011
SVD Real	1.82	1.3146e–012	3.0687e–013

Example 8.41 (Model reduction from measured data). Next, we apply the Loewner framework to a set of measured data containing $k = 100$ S-parameter measurements of an electronic device with $p = 50$ ports (50 inputs and 50 outputs). Measurements were performed using a VNA and were provided by CST AG. The accuracy of the models obtained was assessed by means of the norms (8.45) and (8.46). The frequency of the 100 measurements (100 matrices of size 50×50) ranges between 10 MHz and 1 GHz. For better conditioning, all frequencies were scaled by 10^{-6}.

MIMO case: Figure 8.3 shows the normalized singular values of the Loewner and shifted Loewner matrices constructed using all measurements in the MIMO case. The tangential directions have been chosen as columns of the identity matrix. According to the algorithm of Section 8.2.3, the drop in the singular values suggests that there is an underlying **D**-term. The singular values of the Loewner matrix decay about three

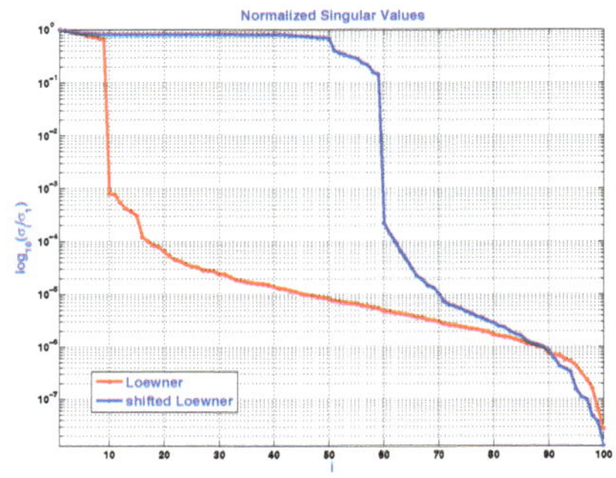

Figure 8.3. *Singular value drop of the Loewner and shifted Loewner matrices.*

Table 8.2. *Results for a device with $p = 50$ ports (50 inputs, 50 outputs).*

Algorithm	CPU time (sec)	$\mathcal{H}_\infty$-error	$\mathcal{H}_2$-error
Loewner ($n = 59$ with $\mathbf{D} = 0$)	0.05	5.3326e–3	4.5556e–4
Loewner ($n = 9$ with $\mathbf{D} \neq 0$)	0.15	6.1575e–3	5.9957e–4

orders of magnitude between the 9th and 10th, while the singular values of the shifted Loewner matrix decay several orders of magnitude between the 59th and 60th. The system can be decomposed into a strictly proper part of order nine plus a direct feed-through term $\mathbf{D}$ of size 50×50 and full rank.

Table 8.2 presents the errors for two models constructed with the Loewner approach: a model of order $n = 59$ obtained by projecting the Loewner pencil onto the subspace of singular vectors associated with the 59 dominant singular values of the Loewner pencil and a model of order $n = 9$ obtained after extracting the $\mathbf{D}$ matrix. The extraction of the $\mathbf{D}$ matrix is realized in two steps. First, the strictly proper part is extracted by projecting the $(\mathbf{A}, \mathbf{E})$ matrix pencil of the model onto the eigenvectors associated with eigenvalues with small magnitude. Subsequently, the $\mathbf{D}$ matrix is determined as the mean over all frequencies of the difference between the measurement and the evaluation of the strictly proper model at each frequency. The model obtained with the Loewner approach is shown in Figure 8.4. It shows small discrepancies in approximating the lowest singular values of the system function (due to the small scale on the y-axis).

It should be mentioned that state-of-the-art approaches are very expensive for such a large number of inputs and outputs. However, our approach is especially suited for such problems since the resulting Loewner matrices only depend on the number of measurements and not the number of ports. Finally, the very low CPU time should be stressed.

SISO case: Finally, we focus on the $(1, 31)$ entry of the S-parameter matrix. As in the MIMO case, the drop in the singular values of the Loewner pencil (see Figure 8.5) shows that the data suggest a model of order four, together with a $\mathbf{D}$-term. Knowing the order of the system, we construct models of order five with $\mathbf{D} = 0$, as well as of order four with $\mathbf{D} \neq 0$. The errors of the constructed models are summarized in Table 8.3. The accuracy for the $\mathbf{D} \neq 0$ case deteriorates slightly because $\mathbf{D}$ was constructed by averaging over all frequencies. ■

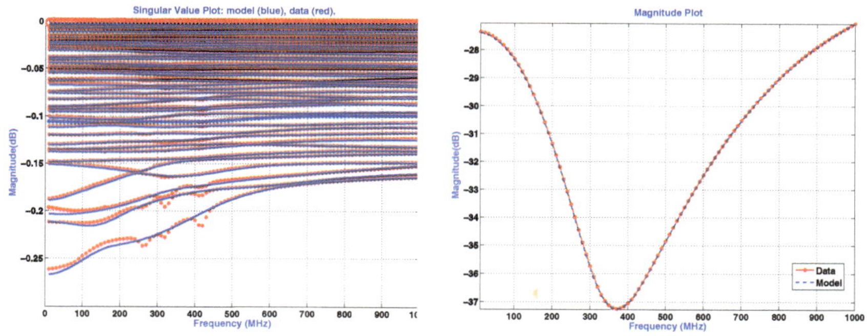

Figure 8.4. *Left pane: fit between the data and the* 50 *singular values (the device has* 50 *ports) of the transfer function of the model constructed from* $k = 100$ *samples. Right pane: data versus model for the* $(1,31)$ *entry.*

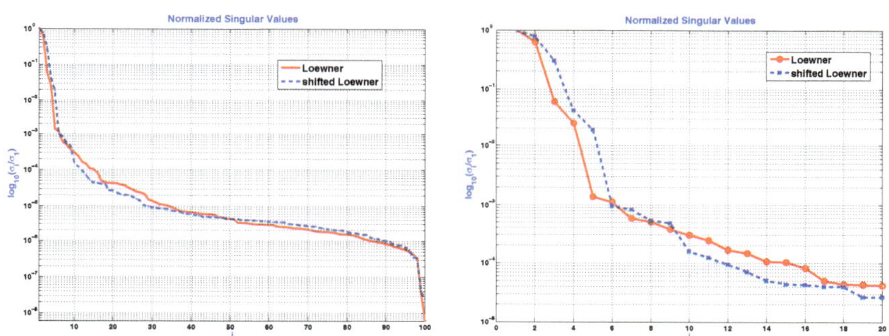

Figure 8.5. *Left pane: the singular value drop of the Loewner matrix pencil for the* $(1,31)$ *entry. Right pane: detail of left pane plot.*

Table 8.3. *Errors for a (scalar) model fitting the* $(1,31)$ *entry of the device with* $p = 50$ *ports (*50 *inputs and* 50 *outputs).*

Algorithm	Order 5, $\mathbf{D}=0$		Order 4, $\mathbf{D}\neq 0$	
	$\mathcal{H}_\infty$-error	$\mathcal{H}_2$-error	$\mathcal{H}_\infty$-error	$\mathcal{H}_2$-error
Loewner	1.5128e−3	8.7226e−4	8.7645e−3	6.4048e−3

8.4 ▪ Summary

In this chapter we have presented a model reduction approach from data, be it measured (e.g., S-parameters) or computed by DNS. The main tool is the *Loewner pencil*, which is constructed exclusively from data. Its determination requires no computations like matrix factorizations or inversions and hence constitutes a natural approach to the problem of data-driven modeling. Furthermore, the singular values of the Loewner pencil provide a trade-off between accuracy of fit and model complexity. Here is a summary of the main features.

- Given input/output data, we can construct, with *no computation*, a singular high-order model in generalized state-space or descriptor form.

- In applications, the singular pencil $(\mathbb{L}_s, \mathbb{L})$ must be reduced to obtain a minimal- or low-order model.
- The approach is a *natural way* to construct full and reduced models because it
  - does not *force* the inversion of $\mathbf{E}$,
  - can deal with many inputs and outputs.
- In this framework, the *singular values of* $\mathbb{L}$, $\mathbb{L}_s$ offer a *trade-off between accuracy of fit and complexity of the reduced system*. The resulting (a posteriori computable) error is proportional to the first neglected singular value of $\mathbb{L}$.
- The Loewner framework has been extended to
  - *multiple-point (Hermite) interpolation* (i.e., values and derivatives at various points are provided) [9, 33];
  - *recursive modeling* (i.e., data become available successively) [29, 30];
  - *linear parametrized systems* [9, 24, 25]; and
  - *bilinear systems* by means of *generalized Loewner pencils* [23].
- The underlying philosophy is therefore to *collect data and extract the desired information.*

Bibliography

[1] P. AMSTUTZ, *Une méthode d'interpolation par des fonctions rationelles*, Annales des Télécommunications, 22 (1967), pp. 62–64.

[2] B. D. O. ANDERSON AND A. C. ANTOULAS, *Rational interpolation and state variable realization*, Linear Algebra and Its Applications, Special Issue on Matrix Problems, 137/138 (1990), pp. 479–509.

[3] A. C. ANTOULAS, *Approximation of Large-Scale Dynamical Systems*, Advances in Design and Control 6, SIAM, Philadelphia, 2005.

[4] ——, *On the construction of passive models from frequency response data*, Automatisierungstechnik, 56 (2008), pp. 447–452.

[5] A. C. ANTOULAS AND B. D. O. ANDERSON, *On the scalar rational interpolation problem*, IMA Journal of Mathematical Control and Information, 3 (1986), pp. 61–88.

[6] ——, *On the problem of stable rational interpolation*, Linear Algebra and Its Applications, 122-124 (1989), pp. 301–329.

[7] ——, *State-space and polynomial approaches to rational interpolation*, in Realization and Modeling in System Theory, M. A. Kasshoek, J. H. van Schuppen, and A. C. M. Ran, Editors, Proceedings of MTNS-89, Vol I, 1990, pp. 73–82.

[8] A. C. ANTOULAS, C. A. BEATTIE, AND S. GUGERCIN, *Interpolatory model reduction of large-scale dynamical systems*, in Efficient Modeling and Control of Large-Scale Systems, J. Mohammadpour and K. Grigoriadis, eds., Springer-Verlag, 2010, pp. 2–58.

[9] A. C. ANTOULAS, A. C. IONITA, AND S. LEFTERIU, *On two-variable rational interpolation*, Linear Algebra and Its Applications, 436 (2012), pp. 2889–2915.

[10] U. BAUR, P. BENNER, AND L. FENG, *Model order reduction for linear and nonlinear systems: A system-theoretic perspective*, Archives of Computational Methods in Engineering, 21 (2014), pp. 331–358.

[11] C. A. BEATTIE AND S. GUGERCIN, *Model reduction by rational interpolation*, in Model Reduction and Approximation: Theory and Algorithms, SIAM, Philadelphia, 2017, pp. 297–334.

[12] V. BELEVITCH, *Interpolation matrices*, Philips Research Reports, 25 (1970), pp. 337–369.

[13] P. BENNER, J. R. LI, AND T. PENZL, *Numerical solution of large-scale Lyapunov equations, Riccati equations, and linear-quadratic optimal control problems*, Numerical Linear Algebra with Applications, 15 (2008), pp. 755–777.

[14] P. BENNER AND A. SCHNEIDER, *Balanced Truncation for Descriptor Systems with Many Terminals*, Max Planck Institute Magdeburg Preprint MPIMD/13-17, 2013.

[15] T. BERGER AND T. REIS, *Controllability of linear differential algebraic systems—A survey*, in Surveys in Differential Algebraic Equations I, A. Ilchmann and T. Reis, eds., Springer-Verlag, 2013, pp. 1–61.

[16] J.-P. BERRUT AND G. KLEIN, *Recent advances in linear barycentric rational interpolation*, Journal of Computational and Applied Mathematics, 259 (2014), pp. 95–107.

[17] J.-P. BERRUT AND L. N. TREFETHEN, *Barycentric Lagrange interpolation*, SIAM Review, 46 (2004), pp. 501–517.

[18] R. BHATIA, *Matrix Analysis*, Springer-Verlag, 1996.

[19] R. BHATIA AND T. SANO, *Loewner matrices and operator convexity*, Mathematische Annalen, 344 (2009), pp. 703–736.

[20] W. F. DONOGHUE JR., *Monotone Matrix Functions and Analytic Continuation*, Springer-Verlag, 1974.

[21] M. FIEDLER, *Hankel and Loewner matrices*, Linear Algebra and Its Applications, 58 (1985), pp. 75–95.

[22] S. GUGERCIN, C. A. BEATTIE, AND A. C. ANTOULAS, *Data-Driven and Interpolatory Model Reduction*, SIAM Series on Computational Science and Engineering, SIAM, Philadelphia, to appear 2018.

[23] A. C. IONITA, *Lagrange Rational Interpolation and Its Applications to Approximation of Large-Scale Dynamical System*, PhD thesis, Rice University, 2013.

[24] A. C. IONITA AND A. C. ANTOULAS, *Data-driven parametrized model reduction in the Loewner framework*, SIAM Journal on Scientific Computing, 36 (2014), pp. A984–A1007.

[25] ——, *Parametrized model order reduction from transfer function measurements*, in Reduced Order Methods for Modeling and Computational Reduction, A. Quarteroni and G. Rozza, eds., Springer-Verlag, 2014, pp. 51–66.

[26] T. KAILATH AND A. H. SAYED, *Dispacement structure: Theory and applications*, SIAM Review, 37 (1995), pp. 297–386.

[27] L. KNOCKAERT, *A simple and accurate algorithm for barycentric rational interpolation*, IEEE Signal Processing Letters, 15 (2008), pp. 154–157.

[28] M. KUIJPER AND J. M. SCHUMACHER, *Realization of autoregressive equations in pencil and descriptor form*, SIAM Journal on Control and Optimization, 28 (1990), pp. 1162–1189.

[29] S. LEFTERIU AND A. C. ANTOULAS, *Modeling multi-port systems from frequency response data via tangential interpolation*, in Proceedings of the 13th IEEE Workshop on Signal Propagation on Interconnects, 2009, pp. 1–4.

[30] ——, *A new approach to modeling multiport systems from frequency-domain data*, IEEE Transactions on Computer-Aided Design, 29 (2010), pp. 14–27.

[31] C. LOEWNER, short biography: `http://en.wikipedia.org/wiki/Charles_Loewner`.

[32] K. LÖWNER, *Über monotone Matrixfunctionen*, Mathematische Zeitschrift, 38 (1934), pp. 177–216.

[33] A. J. MAYO AND A. C. ANTOULAS, *A framework for the solution of the generalized realization problem*, Linear Algebra and Its Applications, 425 (2007), pp. 634–662.

[34] V. MEHRMANN AND T. STYKEL, *Balanced truncation model reduction for large-scale systems in descriptor form*, in Dimension Reduction of Large-Scale Systems, P. Benner, V. Mehrmann, and D. C. Sorensen, eds., Lecture Notes in Computational Science and Engineering, 45, Springer-Verlag, 2005, pp. 83–115.

[35] J. MEINGUET, *On the solubility of the Cauchy interpolation problem*, in Approximation Theory, A. Talbot, ed., Academic Press, 1979, pp. 137–163.

[36] C. SCHNEIDER AND W. WERNER, *Some new aspects of rational interpolation*, Mathematics of Computation, 47 (1985), pp. 285–299.

[37] T. STYKEL, *Gramian based model reduction for descriptor systems*, Mathematics of Control, Signals, and Systems, 16 (2004), pp. 297–319.

[38] A. J. VAN DER SCHAFT, *Port-Hamiltonian differential-algebraic systems*, in Surveys in Differential-Algebraic Equations I, A. Ilchmann and T. Reis, eds., Springer-Verlag, 2013, pp. 173–226.

Chapter 9

Comparison of Methods for Parametric Model Order Reduction of Time-Dependent Problems

Ulrike Baur, Peter Benner, Bernard Haasdonk, Christian Himpe, Immanuel Martini, and Mario Ohlberger[33]

9.1 ▪ Introduction

In this work, we consider parametrized dynamical systems of order n,

$$\begin{aligned} \mathrm{E}(\mathsf{p})\,\dot{\mathbf{x}}(t;\mathsf{p}) &= \mathbf{A}(\mathsf{p})\,\mathbf{x}(t;\mathsf{p}) + \mathbf{B}(\mathsf{p})\,\mathbf{u}(t), \\ \mathbf{y}(t;\mathsf{p}) &= \mathrm{C}(\mathsf{p})\,\mathbf{x}(t;\mathsf{p}), \end{aligned} \tag{9.1}$$

with a parameter (vector) $\mathsf{p} \in \mathbb{R}^d$; $\mathbf{x}(t;\mathsf{p}) \in \mathbb{R}^n$, $\mathbf{y}(t;\mathsf{p}) \in \mathbb{R}^q$, and $\mathbf{u}(t) \in \mathbb{R}^m$; and parameter-dependent system matrices $\mathrm{E}(\mathsf{p})$, $\mathbf{A}(\mathsf{p}) \in \mathbb{R}^{n\times n}$, $\mathbf{B}(\mathsf{p}) \in \mathbb{R}^{n\times m}$, and $\mathrm{C}(\mathsf{p}) \in \mathbb{R}^{q\times n}$.

It is assumed that, for all considered parameter values, $\mathrm{E}(\mathsf{p})$ is invertible and the system is stable, i.e., the eigenvalues of $\mathrm{E}^{-1}(\mathsf{p})\mathbf{A}(\mathsf{p})$ lie in the open left half of the complex plane. In the following, we omit the parameter dependency of the state $\mathbf{x}$ and of the output $\mathbf{y}$ in our notation for a simplified presentation. In many research fields, such as signal processing or control theory, the system is analyzed in the frequency domain. The system's response in the frequency domain is described by a linear mapping called the *transfer function*, which maps the Laplace transform of the inputs to the Laplace transform of the outputs. The parametrized transfer function corresponding to (9.1) for $s \in \overline{\mathbb{C}}_+ := \{s \in \mathbb{C} \mid \mathrm{Re}(s) \geq 0\}$ is defined by

$$\mathbf{H}(s,\mathsf{p}) = \mathrm{C}(\mathsf{p})(s\mathrm{E}(\mathsf{p}) - \mathbf{A}(\mathsf{p}))^{-1}\mathbf{B}(\mathsf{p}). \tag{9.2}$$

[33]This work was supported partially by the German Research Foundation (DFG) grant BE 2174/AN 693, Multivariate Interpolation Methods for Parametric Model Reduction (MIM4PMOR), by the DFG EXC 1003 Cells in Motion—Cluster of Excellence, and by the Center for Developing Mathematics in Interaction, DEMAIN, Münster, Germany. The authors also want to acknowledge the Baden-Württemberg Stiftung gGmbH for funding as well as the DFG for financial support within the Cluster of Excellence in Simulation Technology (EXC 310/1) at the University of Stuttgart.

We would like to thank Lihong Feng, Maria Cruz Varona, Matthias Geuß, and Heiko Peuscher (né Panzer) for providing their code as well as for helpful discussions and Chris Beattie for reading a draft version of this manuscript and giving various suggestions for improvement.

Parametric model order reduction (PMOR) based on projection seeks (full column rank) matrices $\mathbf{V}, \mathbf{W} \in \mathbb{R}^{n\times r}$ with $r \ll n$ such that the output error between the original and the reduced-order system

$$\begin{aligned} \mathbf{W}^T\mathbf{E}(\mathsf{p})\mathbf{V}\dot{\hat{\mathbf{x}}}(t) &= \mathbf{W}^T\mathbf{A}(\mathsf{p})\mathbf{V}\hat{\mathbf{x}}(t)+\mathbf{W}^T\mathbf{B}(\mathsf{p})\mathbf{u}(t), \\ \hat{\mathbf{y}}(t) &= \mathbf{C}(\mathsf{p})\mathbf{V}\hat{\mathbf{x}}(t) \end{aligned} \tag{9.3}$$

or, analogously, in the frequency domain

$$\hat{\mathbf{H}}(s,\mathsf{p}) = \mathbf{C}(\mathsf{p})\mathbf{V}(s\mathbf{W}^T\mathbf{E}(\mathsf{p})\mathbf{V} - \mathbf{W}^T\mathbf{A}(\mathsf{p})\mathbf{V})^{-1}\mathbf{W}^T\mathbf{B}(\mathsf{p}) \tag{9.4}$$

is small and the computational time to simulate (9.1) and (9.2) decreases significantly by using (9.3) or (9.4) instead. The simulation time of a system will be called the *online complexity* since it has to be computed for every new value of p and input **u**. The time required to compute a reduced-order, parametrized model will be called the *offline complexity* of the PMOR method.

The reduction is especially of value (with respect to reduced computational complexity) if the parameter dependency in (9.1) is affine in the system matrices [2, 23], i.e., we have the following matrix representation:

$$\begin{aligned} \mathbf{E}(\mathsf{p}) &= \mathbf{E}_0 + e_1(\mathsf{p})\mathbf{E}_1 + \cdots + e_{P_E}(\mathsf{p})\mathbf{E}_{P_E}, \\ \mathbf{A}(\mathsf{p}) &= \mathbf{A}_0 + f_1(\mathsf{p})\mathbf{A}_1 + \cdots + f_{P_A}(\mathsf{p})\mathbf{A}_{P_A}, \\ \mathbf{B}(\mathsf{p}) &= \mathbf{B}_0 + g_1(\mathsf{p})\mathbf{B}_1 + \cdots + g_{P_B}(\mathsf{p})\mathbf{B}_{P_B}, \\ \mathbf{C}(\mathsf{p}) &= \mathbf{C}_0 + h_1(\mathsf{p})\mathbf{C}_1 + \cdots + h_{P_C}(\mathsf{p})\mathbf{C}_{P_C}, \end{aligned}$$

leading to reduced-order matrices

$$\begin{aligned} \hat{\mathbf{E}}(\mathsf{p}) &:= \mathbf{W}^T\mathbf{E}(\mathsf{p})\mathbf{V} = \mathbf{W}^T\mathbf{E}_0\mathbf{V} + \textstyle\sum_{i=1}^{P_E} e_i(\mathsf{p})\mathbf{W}^T\mathbf{E}_i\mathbf{V}, \\ \hat{\mathbf{A}}(\mathsf{p}) &:= \mathbf{W}^T\mathbf{A}(\mathsf{p})\mathbf{V} = \mathbf{W}^T\mathbf{A}_0\mathbf{V} + \textstyle\sum_{i=1}^{P_A} f_i(\mathsf{p})\mathbf{W}^T\mathbf{A}_i\mathbf{V}, \\ \hat{\mathbf{B}}(\mathsf{p}) &:= \mathbf{W}^T\mathbf{B}(\mathsf{p}) = \mathbf{W}^T\mathbf{B}_0 + \textstyle\sum_{i=1}^{P_B} g_i(\mathsf{p})\mathbf{W}^T\mathbf{B}_i, \\ \hat{\mathbf{C}}(\mathsf{p}) &:= \mathbf{C}(\mathsf{p})\mathbf{V} = \mathbf{C}_0\mathbf{V} + \textstyle\sum_{i=1}^{P_C} h_i(\mathsf{p})\mathbf{C}_i\mathbf{V}. \end{aligned} \tag{9.5}$$

It is assumed that the number of summands P_E, P_A, P_B, P_C is moderate. The reduced parameter-independent matrices $\mathbf{W}^T\mathbf{E}_i\mathbf{V}, \mathbf{W}^T\mathbf{A}_i\mathbf{V} \in \mathbb{R}^{r\times r}$, $\mathbf{W}^T\mathbf{B}_i \in \mathbb{R}^{r\times m}$, and $\mathbf{C}_i\mathbf{V} \in \mathbb{R}^{q\times r}$ can be precomputed. These constant reduced matrices are computed during the offline phase of the method (and thus not for every new value of p).

It is also assumed that the initial state is zero, although the methods in principle can also be used in the case of a varying or parameter-dependent initial state. Some of the methods can easily be modified for the case of singular **E**, while others are still under investigation. Also, not all approaches require stability. We restrict the comparison to the preassigned assumptions to make it applicable to a wide range of different methods.

In the following, we compare several methods for PMOR. All approaches are briefly introduced in Section 9.2, with error measures described in Section 9.3. We apply each method to three benchmarks selected from the MOR Wiki benchmark collection [32]; see Section 9.5 for details. The results and a discussion of the comparison can be found in Section 9.6.

This work complements a recent survey on model reduction methods for parametric systems [6], where several approaches to and aspects of PMOR are discussed in detail but no numerical experiments are included. For a comparison of MOR methods for systems without parameter dependency, see, for instance, [4, 18].

The comparisons described here should be thought of as a first attempt to look at the advantages and disadvantages of various state-of-the-art strategies in PMOR. For a simplified presentation of the results and a better comparability of the approaches, we limit the comparison to SISO systems, which depend affinely on one parameter ($d = 1$). This will be extended to MIMO systems with more than one parameter ($d > 1$) in future work.

9.2 ▪ Methods for PMOR

The computation of the projection matrices $\mathbf{V}$ and $\mathbf{W}$ is of main interest in PMOR and differs very much among the presented approaches. We will briefly describe the methods that will be compared in the following.

9.2.1 ▪ POD and POD-greedy

First, we apply two approaches to compute matrices $\mathbf{V}$ and $\mathbf{W}$ based on proper orthogonal decomposition (POD). POD is a state-space approximation method providing optimal approximation spaces in the mean squared error sense [8, 35, 36]. To be more precise, POD assumes a set of points $\mathbf{x}_1, \ldots, \mathbf{x}_N \in \mathbb{R}^n$, compactly written as a matrix $\mathbf{X} = [\mathbf{x}_1, \ldots, \mathbf{x}_N]$, and computes for arbitrarily chosen $r \in \{1, \ldots, N\}$,

$$POD_r(\mathbf{X}) := \arg\min_{\mathbf{V}} \frac{1}{N} \sum_{i=1}^{N} \left\| \mathbf{x}_i - \mathbf{V}\mathbf{V}^T \mathbf{x}_i \right\|^2$$

over all matrices $\mathbf{V} \in \mathbb{R}^{n \times r}$ that satisfy $\mathbf{V}^T\mathbf{V} = \mathbf{I}$. This can be directly computed by the singular value decomposition (SVD) $\mathbf{V}_1\mathbf{S}\mathbf{V}_2^T \overset{\text{SVD}}{=} \mathbf{X}$ and by taking the r leftmost columns of $\mathbf{V}_1$ as matrix $\mathbf{V}$. These columns are then also denoted as POD modes.

In the following methods, we only address the construction of a matrix $\mathbf{V}$ with orthogonal columns in the setting of parametric systems on a finite time interval $[0, T]$ with a fixed input $\mathbf{u}(\cdot)$. After construction of $\mathbf{V}$, we simply choose $\mathbf{W} = \mathbf{V}$.

I. **Global POD:** We assume we have a set of training parameter samples $\{\mathsf{p}_1, \ldots, \mathsf{p}_K\}$ and a time discretization by choosing $J \in \mathbb{N}$ and setting $\Delta t := T/J$ and $t_i := i\Delta t$, $i = 0, \ldots, J$. These define the set $\{\mathbf{x}_1, \ldots, \mathbf{x}_N\} := \{\mathbf{x}(t_0; \mathsf{p}_1), \ldots, \mathbf{x}(t_J; \mathsf{p}_K)\}$ by computing $K \cdot (J+1)$ solution snapshots of the full model. The POD of this (potentially huge) set of snapshots yields a matrix $\mathbf{V} = POD_r(\mathbf{X})$ that can be used for system projection and promises good average approximation over the time and parameter range if J, K, and r are chosen sufficiently large.

II. **POD-greedy:** The size of the SVD problem of the global POD can be prohibitive, which is circumvented by the following greedy procedure that incrementally constructs a basis $\mathbf{V}$ by several small POD computations: starting with an initial basis $\mathbf{V}$ and the corresponding reduced system, one can detect the single parameter p_j among the parameter samples that is currently worst resolved by the reduced model (e.g., measured by error norms as specified in the next section). For this "worst" sample p_j, the high-dimensional trajectory $\mathbf{x}(\cdot; \mathsf{p}_j)$ is computed and the new information of this trajectory is extracted by computing $\tilde{r} \geq 1$ POD modes of the orthogonal projection error trajectory to the current $\mathbf{V}$. This means we first orthogonalize all trajectory elements and collect them into a matrix, i.e., $\mathbf{X}'$ is chosen as the matrix with columns $\{\mathbf{x}(t_i; \mathsf{p}_j) - \mathbf{V}\mathbf{V}^T\mathbf{x}(t_i; \mathsf{p}_j)\}_{i=0}^{J}$ in arbitrary order. Then, we extract the new POD mode(s) $\mathbf{V}' := POD_{\tilde{r}}(\mathbf{X}')$. This matrix $\mathbf{V}'$ is orthonormal to $\mathbf{V}$ by construction, and we extend the basis $\mathbf{V} := [\mathbf{V}, \mathbf{V}']$. This extension loop is repeated until a desired accuracy is obtained on the training set of parameters or a desired basis size r is reached. The choice $\bar{r} = 1$, i.e., adding a single POD mode in each iteration, promises to result in the most compact basis, while $\bar{r} > 1$ could result in an accelerated basis generation time. This POD-greedy algorithm [22, 37] is standard in reduced basis (RB) methods [10, 23, 25] and has provable quasi-optimal convergence rates [20]. Adaptive techniques of selecting parameters or snapshots for reducing the complexity of the RB generation have been proposed, for example, in [21, 40].

In procedures I and II above, variation of the input $\mathbf{u}$ can be allowed via a parametrization of the input. First, if a finite number n_u of input signals $\mathbf{u}_i$ is to be expected, one can extend the parameter vector by $\bar{\mathsf{p}} = (\mathsf{p}, i)$ and hence directly include the input variation in the POD or POD-greedy procedure by varying $\bar{\mathsf{p}}$ over $\mathbb{R}^d \times \{1,\ldots,n_u\}$.

Second, if a parametric model of the prospective input signals is available, e.g., $\mathbf{u}(t) = \mathbf{u}(t;\mathsf{p}')$ with $\mathsf{p}' \in \mathbb{R}^{d'}$, this can also be directly included in a parametric sampling of the input signal manifold by extending the parameter vector by $\bar{\mathsf{p}} := (\mathsf{p},\mathsf{p}')$ and varying $\bar{\mathsf{p}}$ over $\mathbb{R}^d \times \mathbb{R}^{d'}$.

9.2.2 ▪ Interpolatory methods for PMOR

Next, interpolatory methods for PMOR are considered that are based on well-known techniques for model order reduction (MOR) of deterministic (nonparametric) systems such as moment matching, the iterative rational Krylov algorithm (IRKA) [19], and balanced truncation (BT). These (deterministic) MOR methods are used to reduce the order of the system (9.1) at a certain number of fixed parameter values $\mathsf{p}_1,\ldots,\mathsf{p}_K$ to a local reduced order r', i.e., we apply MOR K times on the systems with transfer functions

$$\mathbf{H}_j(s) := \mathbf{H}(s,\mathsf{p}_j) = \mathbf{C}(\mathsf{p}_j)(s\mathbf{E}(\mathsf{p}_j) - \mathbf{A}(\mathsf{p}_j))^{-1}\mathbf{B}(\mathsf{p}_j), \quad j = 1,\ldots,K.$$

The following (local) quantities are computed and stored after reduction for further use in interpolation-based PMOR:

1. the projection matrices $\mathbf{V}_j$, $\mathbf{W}_j \in \mathbb{R}^{n\times r'}$ for $j = 1,\ldots,K$, and
2. reduced system matrices for $j = 1,\ldots,K$:

$$\begin{aligned} \hat{\mathbf{E}}_j &= \mathbf{W}_j^T\mathbf{E}(\mathsf{p}_j)\mathbf{V}_j \in \mathbb{R}^{r'\times r'}, & \hat{\mathbf{A}}_j &= \mathbf{W}_j^T\mathbf{A}(\mathsf{p}_j)\mathbf{V}_j \in \mathbb{R}^{r'\times r'}, \\ \hat{\mathbf{B}}_j &= \mathbf{W}_j^T\mathbf{B}(\mathsf{p}_j) \in \mathbb{R}^{r'\times m}, & \hat{\mathbf{C}}_j &= \mathbf{C}(\mathsf{p}_j)\mathbf{V}_j \in \mathbb{R}^{q\times r'}. \end{aligned} \tag{9.6}$$

The PMOR approaches considered here use different interpolation strategies that employ some of these quantities to derive a reduced-order system (9.3) to approximate (9.1) over the whole parameter interval.

A short description of the methods follows.

I. The first method is called *PMOR by matrix interpolation* (MatrInt) [33]. A parametrized reduced-order system (9.3) of order r' is obtained by interpolating the locally reduced system matrices (9.6), where

$$\begin{aligned} \hat{\mathbf{E}}(\mathsf{p}) &= \textstyle\sum_{j=1}^K \omega_j(\mathsf{p})\mathbf{M}_j\hat{\mathbf{E}}_j\mathbf{T}_j^{-1}, & \hat{\mathbf{A}}(\mathsf{p}) &= \textstyle\sum_{j=1}^K \omega_j(\mathsf{p})\mathbf{M}_j\hat{\mathbf{A}}_j\mathbf{T}_j^{-1}, \\ \hat{\mathbf{B}}(\mathsf{p}) &= \textstyle\sum_{j=1}^K \omega_j(\mathsf{p})\mathbf{M}_j\hat{\mathbf{B}}_j, & \hat{\mathbf{C}}(\mathsf{p}) &= \textstyle\sum_{j=1}^K \omega_j(\mathsf{p})\hat{\mathbf{C}}_j\mathbf{T}_j^{-1}, \end{aligned} \tag{9.7}$$

with properly chosen transformation matrices $\mathbf{M}_j$, $\mathbf{T}_j \in \mathbb{R}^{r'\times r'}$ and weights ω_j. The transformation matrices are chosen to give a common physical meaning to all reduced-state vectors — $\mathbf{M}_j = (\mathbf{W}_j^T\mathbf{R})^{-1}$, $\mathbf{T}_j = \mathbf{R}^T\mathbf{V}_j$ with $\mathbf{R} \in \mathbb{R}^{n\times r'}$ — obtained from a thin

SVD of $[\omega_1(\mathsf{p})\mathbf{V}_1, \omega_2(\mathsf{p})\mathbf{V}_2, \ldots, \omega_K(\mathsf{p})\mathbf{V}_K]$. Note that a single $\mathbf{R}$ is used to compute the transformation matrices $\mathbf{M}_j$, $\mathbf{T}_j$ for $j = 1, \ldots, K$ and that the computation is part of the online phase since the weights depend on p. Modifications of the approach that avoid this online step are proposed in [1, 33].

Stability of the parametrized reduced-order system can be achieved by a further offline step, which includes the solution of K low-dimensional Lyapunov equations [17]. Transformations based on these solutions make the locally reduced systems contractive.

II. The next approach for PMOR is *transfer function interpolation* (TransFncInt). This is based on interpolation of locally reduced-order models in the frequency domain [3]. Here, reduced-order transfer functions

$$\hat{\mathbf{H}}_j(s) = \hat{\mathbf{C}}_j(s\hat{\mathbf{E}}_j - \hat{\mathbf{A}}_j)^{-1}\hat{\mathbf{B}}_j, \tag{9.8}$$

with $\hat{\mathbf{E}}_j$, $\hat{\mathbf{A}}_j$, $\hat{\mathbf{B}}_j$, $\hat{\mathbf{C}}_j$ computed by (9.6), are taken as "data points" to construct a (parameter-dependent) interpolant. Using polynomial interpolation, the reduced-order transfer function looks like

$$\hat{\mathbf{H}}(s, \mathsf{p}) = \sum_{j=1}^{K} L_j(\mathsf{p})\hat{\mathbf{H}}_j(s), \tag{9.9}$$

where $L_j(\mathsf{p})$ are the Lagrange basis polynomials. A possible realization of order $K \cdot r'$, here described, for example, for a system $\hat{\mathbf{A}}$, $\hat{\mathbf{B}}(\mathsf{p})$, $\hat{\mathbf{C}}$, $\hat{\mathbf{E}}$ with $\mathsf{p} \in \mathbb{R}$, is

$$\hat{\mathbf{H}}(s, \mathsf{p}) = \hat{\mathbf{C}}(s\hat{\mathbf{E}} - \hat{\mathbf{A}})^{-1}\hat{\mathbf{B}}(\mathsf{p}),$$

with

$$\hat{\mathbf{C}} := \begin{bmatrix} \hat{\mathbf{C}}_1, & \ldots, & \hat{\mathbf{C}}_K \end{bmatrix}, \; \hat{\mathbf{B}}_j(\mathsf{p}) := \left(\prod_{i=1, i \neq j}^{K} \frac{\mathsf{p} - \mathsf{p}_i}{\mathsf{p}_j - \mathsf{p}_i} \right) \hat{\mathbf{B}}_j,$$

$$s\hat{\mathbf{E}} - \hat{\mathbf{A}} := \begin{bmatrix} s\hat{\mathbf{E}}_1 - \hat{\mathbf{A}}_1 & & \\ & \ddots & \\ & & s\hat{\mathbf{E}}_K - \hat{\mathbf{A}}_K \end{bmatrix}, \; \hat{\mathbf{B}}(\mathsf{p}) := \begin{bmatrix} \hat{\mathbf{B}}_1(\mathsf{p}) \\ \vdots \\ \hat{\mathbf{B}}_K(\mathsf{p}) \end{bmatrix}.$$

Note that this realization does not allow us to reconstruct the state in the required form $\mathbf{x} \approx \mathbf{V}\hat{\mathbf{x}}$.

TransFncInt also does not provide a reduced-order model (ROM) in parametrized state-space form for more than one parameter (and for other interpolation techniques than polynomial interpolation). For higher-dimensional parameter spaces, piecewise-polynomial interpolation on sparse grids provides an efficient implementation of this method [3]. Other interpolation techniques, such as rational interpolation, can also be used to construct $\hat{\mathbf{H}}(s, \mathsf{p})$; see [5].

Note that the preservation of stability can be guaranteed in the parametrized reduced-order system if BT is taken for the local reduction. BT preserves the stability of the local transfer functions $\hat{\mathbf{H}}_j$ in (9.8), which guarantees that the eigenvalues of the interpolated transfer function $\hat{\mathbf{H}}(s, \mathsf{p})$ also lie in $\mathbb{C}_-$.

III. We further consider an approach called *piecewise $\mathcal{H}_2$ tangential interpolation* (with $\mathcal{H}_2$ optimal frequency points) (PWH2TanInt) [2]. The local projection matrices are computed by IRKA (with local reduced order r') and concatenated,

$$\mathbf{V} = [\mathbf{V}_1, \; \mathbf{V}_2, \ldots, \mathbf{V}_K], \quad \mathbf{W} = [\mathbf{W}_1, \; \mathbf{W}_2, \ldots, \mathbf{W}_K],$$

to obtain (9.4). Thus, the dimension of the reduced-order system is $K \cdot r'$. Note that the number of columns $K \cdot r'$ of $\mathbf{V}$ and $\mathbf{W}$ can further be reduced by an SVD or a rank-revealing QR factorization to ensure that $\mathbf{V}$ and $\mathbf{W}$ have full rank. IRKA computes optimal (frequency) shifts s_i and corresponding tangential directions $\mathbf{b}_{ij}$ and $\mathbf{c}_{ij}$ such that (9.4) matches the p-gradient and p-Hessian of the original system response (9.2) with respect to the parameters:

$$\nabla_{\mathsf{p}} \mathbf{c}_{ij}^T \mathbf{H}(s_i, \mathsf{p}_j)\mathbf{b}_{ij} = \nabla_{\mathsf{p}} \mathbf{c}_{ij}^T \hat{\mathbf{H}}(s_i, \mathsf{p}_j)\mathbf{b}_{ij}, \quad \nabla_{\mathsf{p}}^2 \mathbf{c}_{ij}^T \mathbf{H}(s_i, \mathsf{p}_j)\mathbf{b}_{ij} = \nabla_{\mathsf{p}}^2 \mathbf{c}_{ij}^T \hat{\mathbf{H}}(s_i, \mathsf{p}_j)\mathbf{b}_{ij}$$

for $i = 1, \ldots, r'$, $j = 1, \ldots, K$. Additionally, the usual tangential interpolation properties hold:

$$\mathbf{H}(s_i, \mathsf{p}_j)\mathbf{b}_{ij} = \hat{\mathbf{H}}(s_i, \mathsf{p}_j)\mathbf{b}_{ij}, \quad \mathbf{c}_{ij}^T \mathbf{H}(s_i, \mathsf{p}_j) = \mathbf{c}_{ij}^T \hat{\mathbf{H}}(s_i, \mathsf{p}_j).$$

IV. The generalization of moment-matching MOR called multi-(parameter) moment matching (MultiPMomMtch) or multivariate Padé approximation was first considered in [9, 38]. Improvements that avoid explicit moment matching can be found in [11–13, 29]. This method is based on a multivariate Taylor expansion with expansion points in frequency and parameter space. We denote the frequency expansion points by $s_1, \ldots, s_L$ in the following. It ensures the following moment matching:

$$\frac{\partial^k}{\partial s^k} \frac{\partial^l}{\partial \mathsf{p}^l} \mathbf{H}(s_i, \mathsf{p}_j) = \frac{\partial^k}{\partial s^k} \frac{\partial^l}{\partial \mathsf{p}^l} \hat{\mathbf{H}}(s_i, \mathsf{p}_j)$$

for $i = 1, \ldots, L$, $j = 1, \ldots, K$, $k = 0, \ldots, M$, and $l = 0, \ldots, M$ and in all directions (in contrast to tangential interpolation). The dimension of the reduced-order system is $K \cdot L \cdot \frac{M}{2} \cdot (1+M)$ (for SISO systems). This dimension might get reduced by truncating linearly dependent columns during the (repeated modified Gram–Schmidt) orthogonalization process [13].

9.2.3 ▪ Empirical cross-Gramian

The empirical cross-Gramian (emWX) is a snapshot-based method to compute the cross-Gramian, which is applicable to square systems.[34] This method requires an invertible matrix $\mathbf{E}$; thus, it can be described in terms of linear control systems (9.1) with $\mathbf{E}^{-1}\mathbf{A} \to \mathbf{A}$, $\mathbf{E} \to \mathbf{I}$, and $\mathbf{E}^{-1}\mathbf{B} \to \mathbf{B}$. The cross-Gramian [14] is defined as

$$\mathbf{W}_X := \int_0^\infty e^{\mathbf{A}t}\mathbf{B}\mathbf{C}e^{\mathbf{A}t}\,dt. \tag{9.10}$$

For a system with a symmetric transfer function, which includes all SISO systems, the absolute values of the cross-Gramian's eigenvalues equal the Hankel singular values on which BT is based. Instead of computing the cross-Gramian by solving a Sylvester matrix equation or through the empirical cross-Gramian from [24], we choose a controllability-based approach here. To compute the cross-Gramian, the underlying linear control system's (9.1) vector field is augmented by the negative adjoint vector field and the output functional is augmented by the adjoint output functional:

$$\begin{pmatrix} \dot{\mathbf{x}} \\ \dot{\bar{\mathbf{x}}} \end{pmatrix} = \begin{pmatrix} \mathbf{A} & 0 \\ 0 & \mathbf{A}^T \end{pmatrix} \begin{pmatrix} \mathbf{x} \\ \bar{\mathbf{x}} \end{pmatrix} + \begin{pmatrix} \mathbf{B} \\ \mathbf{C}^T \end{pmatrix} \begin{pmatrix} \mathbf{u} \\ \bar{\mathbf{u}} \end{pmatrix},$$

$$\begin{pmatrix} \mathbf{y} \\ \bar{\mathbf{y}} \end{pmatrix} = \begin{pmatrix} \mathbf{C} & \mathbf{B}^T \end{pmatrix} \begin{pmatrix} \mathbf{x} \\ \bar{\mathbf{x}} \end{pmatrix},$$

[34] Systems with the same number of inputs and outputs.

with $\bar{\mathbf{x}}_0$ and $\bar{\mathbf{u}}$ chosen appropriately; a simple choice is $\bar{\mathbf{x}}_0 = \mathbf{x}_0$, $\bar{\mathbf{u}} = \mathbf{u}$. The controllability Gramian of this augmented system, $\widehat{\mathbf{W}}_C$, has the form [15]

$$\widehat{\mathbf{W}}_C = \begin{pmatrix} \mathbf{W}_C & \mathbf{W}_X \\ \mathbf{W}_X^T & \mathbf{W}_O \end{pmatrix}, \tag{9.11}$$

which contains the cross-Gramian $\mathbf{W}_X$ as its upper right block.

To compute this version of the cross Gramian, the empirical controllability Gramian is employed. Empirical Gramians are computed using impulse-response snapshots. Hence, to compute the empirical cross-Gramian using the relation from (9.11) amounts to computing an empirical controllability Gramian from [27, 28]. Since only the upper right block of the augmented system's controllability Gramian is required to obtain the empirical cross-Gramian, it is sufficient to compute $\mathbf{W}_X$ using snapshots $\mathbf{X} := [\mathbf{x}(t_0), \ldots, \mathbf{x}(t_J)] \in \mathbb{R}^{n \times J}$ and adjoint snapshots $\bar{\mathbf{X}} := [\bar{\mathbf{x}}(t_0), \ldots, \bar{\mathbf{x}}(t_J)] \in \mathbb{R}^{n \times J}$, given by

$$\mathbf{x}(t) = \int_0^t e^{\mathbf{A}\tau} \mathbf{B} \mathbf{u}(\tau) d\tau, \quad \bar{\mathbf{x}}(t) = \int_0^t e^{\mathbf{A}^T \tau} \mathbf{C}^T \mathbf{u}(\tau) d\tau.$$

Then, the empirical cross-Gramian is computed by

$$\mathbf{W}_X \approx \Delta t \, \mathbf{X} \bar{\mathbf{X}}^T,$$

which corresponds to a discrete evaluation of (9.10). The snapshots can be obtained by solving $\mathbf{x}(t)$ and $\bar{\mathbf{x}}(t)$ at discrete times $t_0, \ldots, t_J$, for instance using a Runge–Kutta method. This method is closely related to POD as described in Section 9.2.1 and to balanced POD from [39]; see also Chapters 1 and 6.

For a parametrized system (9.1), the parameter space is discretized and the mean empirical cross-Gramian $\overline{\mathbf{W}}_X$ is computed over all points p_j, $j = 1, \ldots, K$, in the discretized parameter space, using the snapshots $\mathbf{X}(\mathsf{p}_j) := [\mathbf{x}(t_0; \mathsf{p}_j), \ldots, \mathbf{x}(t_J; \mathsf{p}_j)]$ and adjoint snapshots $\bar{\mathbf{X}}(\mathsf{p}_j) := [\bar{\mathbf{x}}(t_0; \mathsf{p}_j), \ldots, \bar{\mathbf{x}}(t_J; \mathsf{p}_j)]$, with discrete times $t_0, \ldots, t_J$ and $\Delta t = |t_i - t_j|$, $i \neq j$:

$$\overline{\mathbf{W}}_X = \frac{1}{K} \sum_{j=1}^{K} \mathbf{W}_X(\mathsf{p}_j) \approx \frac{1}{K} \sum_{j=1}^{K} \Delta t \, \mathbf{X}(\mathsf{p}_j) \bar{\mathbf{X}}(\mathsf{p}_j)^T =: \widetilde{\mathbf{W}}_X.$$

An SVD of the empirical cross Gramian provides an approximate balancing projection $\mathbf{V}_1$:

$$\widetilde{\mathbf{W}}_X \overset{\text{SVD}}{=} \mathbf{V}_1 \mathbf{S} \mathbf{V}_2^T.$$

With the Galerkin projection $\mathbf{V}_1$, the ROM is constructed similar to the POD method. The columns of $\mathbf{V}_1$ that correspond to the lowest entries of the diagonal matrix $\mathbf{S}$ are truncated, which amounts to taking the r leftmost columns of $\mathbf{V}_1$ as projection matrix $\mathbf{V}$. After choosing $\mathbf{W} = \mathbf{V}$, the reduced system is given by (9.5).

Remark 9.1. Note that the interpolatory approaches MatrInt and TransFncInt are the only approaches in the comparison that can *efficiently* deal with general parametrizations (instead of the considered affine dependency of (9.5)). The reduced representations in (9.7) and in (9.9) allow us to evaluate the reduced-order system at any parameter value without solving the original system. This leads to a low online complexity.

9.3 ▪ Performance measures

PMOR seeks a reduced-order, parametrized representation of the original system such that the output error between the original and the reduced-order system, i.e., $||\mathbf{y}-\hat{\mathbf{y}}||$, is small in some norm. This demand can be satisfied by a good approximation of (9.2) by a reduced-order, parametrized transfer function (9.4), i.e., forcing $\hat{\mathbf{H}}(s,\mathsf{p}) \approx \mathbf{H}(s,\mathsf{p})$ over a wide frequency range for s and a wide parameter range for p. Alternatively, a good approximation of $\mathbf{x}(t)$ by $\mathbf{V}\hat{\mathbf{x}}(t)$ will also ensure a small output error.

The choice of an appropriate error measure depends on the application. In this work, we consider many different error norms to get good insight into the qualities of the presented methods.

All error measures have in common that they require a discretization of the parameter space as a first step. To this end, we introduce the parameter test grid $\mathsf{p}_1,\dots,\mathsf{p}_{\bar{K}}$ with $\bar{K} > K$ and compute the errors for every parameter grid point p_j, $j=1,\dots,\bar{K}$. We denote the corresponding local quantities by $\mathbf{H}_j := \mathbf{H}(\cdot,\mathsf{p}_j)$, $\hat{\mathbf{H}}_j := \hat{\mathbf{H}}(\cdot,\mathsf{p}_j)$, $\mathbf{y}_j := \mathbf{y}(\cdot;\mathsf{p}_j)$, $\mathbf{x}_j := \mathbf{x}(\cdot;\mathsf{p}_j)$.

The state and the output errors are computed in the time domain for a finite time interval $[0,T]$. We drive the original and the reduced-order system with the same input $\mathbf{u}(\cdot)$. The time interval is discretized by $J+1$ time points $t_0,\dots,t_J$.

The error in state space is estimated by the $\mathscr{L}_2$-norm for square-integrable functions:

$$\begin{aligned} \mathrm{e}(\mathbf{x}_j,\hat{\mathbf{x}}_j)_{\mathscr{L}_2} &:= ||\mathbf{x}_j-\mathbf{V}\hat{\mathbf{x}}_j||_{\mathscr{L}_2([0,T])} = \sqrt{\textstyle\int_0^T ||\mathbf{x}_j(t)-\mathbf{V}\hat{\mathbf{x}}_j(t)||_2^2\,dt} \\ &\approx \sqrt{\Delta t \textstyle\sum_{i=0}^{J} ||(\mathbf{x}_j(t_i)-\mathbf{V}\hat{\mathbf{x}}_j(t_i))||_2^2}. \end{aligned}$$

The output error is computed in two different norms:

$$\begin{aligned} \mathrm{e}(\mathbf{y}_j,\hat{\mathbf{y}}_j)_{\mathscr{L}_2} &:= ||\mathbf{y}_j-\hat{\mathbf{y}}_j||_{\mathscr{L}_2([0,T])} = \sqrt{\textstyle\int_0^T ||\mathbf{y}_j(t)-\hat{\mathbf{y}}_j(t)||_2^2\,dt} \\ &\approx \sqrt{\Delta t \textstyle\sum_{i=0}^{J} ||(\mathbf{y}_j(t_i)-\hat{\mathbf{y}}_j(t_i))||_2^2} \end{aligned}$$

and the $\mathscr{L}_\infty$-norm

$$\mathrm{e}(\mathbf{y}_j,\hat{\mathbf{y}}_j)_{\mathscr{L}_\infty} := ||\mathbf{y}_j-\hat{\mathbf{y}}_j||_{\mathscr{L}_\infty([0,T])} \approx \max_{i=0,\dots,J} ||\mathbf{y}_j(t_i)-\hat{\mathbf{y}}_j(t_i)||_2.$$

These errors are computed as relative errors with denominators $||\mathbf{x}_j||_{\mathscr{L}_2}$, $||\mathbf{y}_j||_{\mathscr{L}_2}$, and $||\mathbf{y}_j||_{\mathscr{L}_\infty}$, respectively.

The frequency response error can be computed by the Hardy $\mathscr{H}_2$-norm and the $\mathscr{H}_\infty$-norm, which provide useful error measures for many classes of input signals.

To compute the $\mathscr{H}_\infty$-norm, we additionally require a fine grid of frequency points $\omega_1,\dots,\omega_{\bar{L}} \in \mathbb{R}$ ($\bar{L} > L$) to compute an estimate of

$$\begin{aligned} \mathrm{e}(\mathbf{H}_j,\hat{\mathbf{H}}_j)_{\mathscr{H}_\infty} &:= ||\mathbf{H}_j-\hat{\mathbf{H}}_j||_{\mathscr{H}_\infty} = \sup_{\omega\in\mathbb{R}} \bar{\sigma}(\mathbf{H}_j(\imath\omega)-\hat{\mathbf{H}}_j(\imath\omega)) \\ &\approx \max_{1\le i\le \bar{L}} \bar{\sigma}(\mathbf{H}_j(\imath\omega_i)-\hat{\mathbf{H}}_j(\imath\omega_i)), \end{aligned}$$

with $\imath = \sqrt{-1}$ and $\bar{\sigma}(\mathbf{H}_j(\imath\omega))$ as the largest singular value of the $q \times m$ matrix $\mathbf{H}_j(\imath\omega)$. The error in the $\mathscr{H}_\infty$-norm is computed as a scaled error, where the maximum of all $||\mathbf{H}_j(\imath\omega_i)-\hat{\mathbf{H}}_j(\imath\omega_i)||_2$ divided by $\max_{1\le l\le \bar{L}} ||\mathbf{H}_j(\imath\omega_l)||_2$ for $i=1,\dots,\bar{L}$ is taken.

The error in the $\mathcal{H}_2$-norm is computed without sampling the frequency space by

$$\begin{aligned} e(\mathbf{H}_j,\hat{\mathbf{H}}_j)_{\mathcal{H}_2} &:= ||\mathbf{H}_j-\hat{\mathbf{H}}_j||_{\mathcal{H}_2} \\ &= \sqrt{\frac{1}{2\pi}\int_{-\infty}^{\infty} \operatorname{trace}\left[(\mathbf{H}_j(\imath\omega)-\hat{\mathbf{H}}_j(\imath\omega))^*(\mathbf{H}_j(\imath\omega)-\hat{\mathbf{H}}_j(\imath\omega))\right] d\omega} \\ &= \sqrt{\operatorname{trace}\left[[\mathbf{C}_j,\hat{\mathbf{C}}_j]\mathbf{P}[\mathbf{C}_j,\hat{\mathbf{C}}_j]^T\right]}. \end{aligned}$$

$\mathbf{P}$ is the controllability Gramian of the error system and is obtained as solution of a Lyapunov equation associated with the error system [41, Section 4.6]. In Section 9.6, the $\mathcal{H}_2$-error is computed as relative error with denominator $||\mathbf{H}_j||^2_{\mathcal{H}_2}$. Estimates for a combined error norm in the frequency (or time) and parameter domains can be obtained by taking the maximum over the parameter test grid, i.e.,

$$e(\mathbf{H},\hat{\mathbf{H}})_{\mathcal{H}_\infty\otimes\mathcal{L}_\infty} := ||\mathbf{H}-\hat{\mathbf{H}}||_{\mathcal{H}_\infty\otimes\mathcal{L}_\infty} \approx \max_{j=1,\dots,\bar{K}} ||\mathbf{H}_j-\hat{\mathbf{H}}_j||_{\mathcal{H}_\infty}.$$

These combined errors are listed in the tables of the corresponding benchmarks in Section 9.6. The following short notation is used in the tables:

$$\begin{aligned} e(\mathbf{x},\hat{\mathbf{x}})_{\mathcal{L}_2} &:= e(\mathbf{x},\hat{\mathbf{x}})_{\mathcal{L}_2\otimes\mathcal{L}_\infty}, \\ e(\mathbf{y},\hat{\mathbf{y}})_{\mathcal{L}_2} &:= e(\mathbf{y},\hat{\mathbf{y}})_{\mathcal{L}_2\otimes\mathcal{L}_\infty}, \\ e(\mathbf{H},\hat{\mathbf{H}})_{\mathcal{H}_\infty} &:= e(\mathbf{H},\hat{\mathbf{H}})_{\mathcal{H}_\infty\otimes\mathcal{L}_\infty}, \\ e(\mathbf{H},\hat{\mathbf{H}})_{\mathcal{H}_2} &:= e(\mathbf{H},\hat{\mathbf{H}})_{\mathcal{H}_2\otimes\mathcal{L}_\infty}. \end{aligned}$$

For $\mathbf{u}\in\mathcal{L}_2([0,\infty)\to\mathbb{R}^m)$, the bounds

$$||\mathbf{y}_j-\hat{\mathbf{y}}_j||_{\mathcal{L}_2([0,\infty))} \le ||\mathbf{H}_j-\hat{\mathbf{H}}_j||_{\mathcal{H}_\infty} ||\mathbf{u}||_{\mathcal{L}_2([0,\infty))}$$

and

$$||\mathbf{y}_j-\hat{\mathbf{y}}_j||_{\mathcal{L}_\infty([0,\infty))} \le ||\mathbf{H}_j-\hat{\mathbf{H}}_j||_{\mathcal{H}_2} ||\mathbf{u}||_{\mathcal{L}_2([0,\infty))}$$

connect the time domain errors with the error measures in the frequency domain.

The complexity of methods for PMOR can be divided into two parts: operations in the offline phase, where the original system size is reduced, and the online complexity, which describes the costs for computing simulations for a new parameter using the reduced-order, parametrized model. The online complexity can be considered for the transient and for the frequency response.

The number of simulation runs z in an application times the online costs of the reduced system plus the offline costs should be smaller than z times the online costs of the full-order system to justify the use of PMOR. This will be measured by the break-even quantity in Section 9.6.

9.4 ▪ Expectations

Before showing the numerical results, we provide some formal comparison and discussion of the methods.

The POD method is based solely on state-space simulation. Thus, it is expected to perform well for time domain error measures, especially with respect to the state error. The POD-greedy method is a variant of the POD method that not only requires fewer large dense matrix operations but also can give more accurate results. Note that by construction, POD minimizes the mean error over the training parameter set, whereas POD-greedy minimizes the maximal error. A good approximation of the state leads to an equally good approximation of the output, so that we also expect the output errors to be quite small. In contrast, we cannot guess anything about the quality of the frequency space approximation. Regarding the computational effort in the offline phase, the POD-greedy algorithm is expected to be more expensive than a global POD as it requires many iterations of solving small PODs. However, the online computational times should not show much of a difference because the reduced orders are equal.

Interpolatory methods that are based on a combination of local reduction at a set of parameter sampling points and interpolation, i.e., transfer function interpolation (TransFncInt) and matrix interpolation (MatrInt), are expected to produce smaller errors at the selected parameter values. The error between the parameter points depends on the interpolation technique, whereas the error at the parameter sampling points depends on the applied MOR method. A good approximation of the transfer function with respect to the $\mathcal{H}_\infty$-norm and of the output with respect to the $\mathcal{L}_2$-norm is expected by combining TransFncInt with BT. PWH2TanInt is an approach for rational interpolation that computes an ROM that matches the gradient (with respect to the parameters) of the original transfer function at the parameter sampling points. The ROM is locally $\mathcal{H}_2$-optimal if IRKA is applied to compute the local projection matrices $\mathbf{V}_j$ and $\mathbf{W}_j$ by determining optimal frequency expansion points (and tangent directions). Thus, this method is anticipated to produce small $\mathcal{H}_2$-errors. The offline costs of these interpolatory methods depend heavily on the costs for the underlying MOR approach (times the number of parameter sampling points).

The quality of the reduced-order system obtained by MultiPMomMtch depends on the initial choice of frequency and parameter expansion points and on the chosen order of matched moments. The selection of all of these quantities is largely ad hoc and not automated in this work. The accuracy in all error measures cannot be predicted a priori.

Since we use state-space simulations of the system and the adjoint system, the empirical cross-Gramian should perform similarly to the POD methods with respect to the state-space norms. Yet because we incorporate both controllability and observability information, we expect better accuracy in the frequency-space norms.

The online complexity of all methods, except MatrInt, is comparable since the computed reduced-order systems are of the same dimension. MatrInt and TransFncInt have an additional online step: the computation of transformation matrices and the interpolation of the system matrices in MatrInt and the construction of a state-space realization in the implementation of TransFncInt considered. Note also that MatrInt computes a system that is much smaller than all other considered approaches and is therefore less expensive during the simulation.

9.5 ▪ Benchmarks

We consider three examples from the benchmark collection [32] and apply the methods introduced in Section 9.2. All systems are SISO systems with an (affine) parameter dependency on a single parameter.

9.5.1 ▪ Synthetic system[35]

A synthetic parametrized system of order n can be constructed as

$$H(s,\mathsf{p}) = \sum_{i=1}^{n} \frac{\mathbf{r}_i}{s-\sigma_i} = \mathbf{C}(s\mathbf{I}-\mathbf{A}_0-\mathsf{p}\mathbf{A}_1)^{-1}\mathbf{B},$$

where $\mathbf{r}_i$ and σ_i are the residues and poles of the transfer function H. The parameter p scales the real part of the system poles $\sigma_i = \mathsf{p}a_i + \imath b_i$. The smaller p is, the closer the poles are to the imaginary axis and the more amplitude of the frequency response changes; see Figures 9.1 and 9.2. Also, the decay of the Hankel singular values is influenced by p. In our particular setting, we took the following entries in the system matrices:

$$\mathbf{A}_0 = \begin{bmatrix} 0 & b_1 & & & 0 \\ -b_1 & 0 & & & \\ & & \ddots & & \\ & & & 0 & b_k \\ 0 & & & -b_k & 0 \end{bmatrix}, \quad \mathbf{A}_1 = \begin{bmatrix} a_1 & & & & 0 \\ & a_1 & & & \\ & & \ddots & & \\ & & & a_k & \\ 0 & & & & a_k \end{bmatrix},$$

$$\mathbf{B} = [2,0,2,0,\ldots]^T, \quad \mathbf{C} = [1,0,1,0,\ldots],$$

with $b_i \in [10,\ 10^3]$; $a_i \in [-10^3,\ -10]$ for $i = 1,\ldots,k$; $k = 500$; and an original system size $n = 1000$. The parameter range is chosen as $\mathsf{p} \in [0.1,\ 1]$. The frequency response of the system over the whole parameter interval is shown in Figure 9.1.

The motivation for this benchmark comes from the simple construction, which allows a flexible choice of n and of the system poles and residues. For further details, see the MOR Wiki page of the benchmark [32].

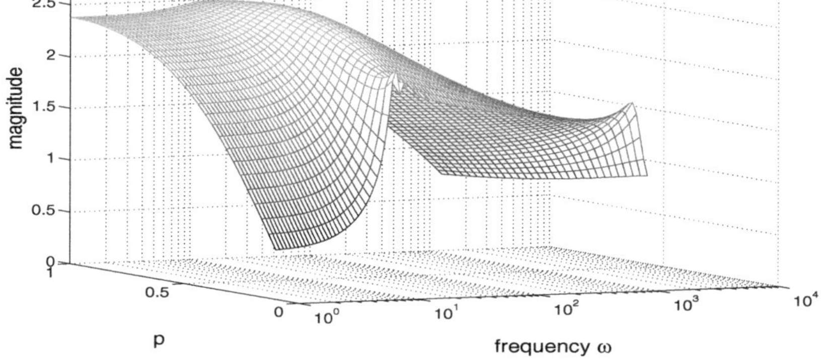

Figure 9.1. *Frequency response of the synthetic system.*

[35]http://www.modelreduction.org/index.php/Synthetic_parametric_model.

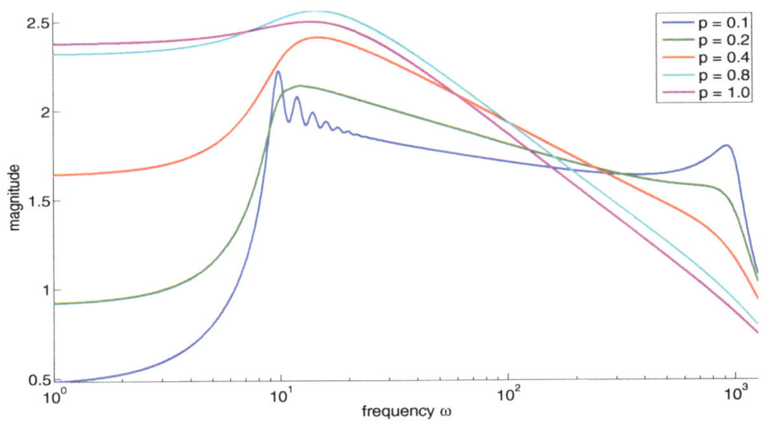

Figure 9.2. *Frequency response of the synthetic system for some parameter values.*

9.5.2 ▪ Microthruster unit[36]

The second benchmark is a real-world example from microsystems technology. This parametric model was originally presented in the Oberwolfach Model Reduction Benchmark Collection [26] under the name "Boundary Condition Independent Thermal Model" [34] and is also listed in the MOR Wiki [32]. The model describes the thermal conduction in a semiconductor chip, where flexibility in specifying the boundary conditions allows the simulation of temperature changes in the environment. This allows independent designers to observe how the surrounding influences the temperature distribution in the chip. The thermal problem is modeled as homogeneous heat diffusion with heat exchange occurring at three device interfaces modeled with convection boundary conditions. These conditions introduce film coefficients describing the heat exchange on the three device interfaces. We assume two film coefficients as being fixed at 10^4, to stay within our specified setting, and one parameter (the film coefficient at the top) as variable within the range $[1, 10^4]$.

Discretization leads to a system of ordinary differential equations (ODEs)

$$\mathbf{E}\dot{\mathbf{x}}(t) = (\mathbf{A}_0 + \mathsf{p}\mathbf{A}_1)\mathbf{x}(t) + \mathbf{B}u(t), \;\; y(t) = \mathbf{C}\mathbf{x}(t),$$

where $\mathbf{E} \in \mathbb{R}^{4257\times4257}$ and $\mathbf{A}_0 \in \mathbb{R}^{4257\times4257}$ are system matrices, $\mathbf{A}_1 \in \mathbb{R}^{4257\times4257}$ is a diagonal matrix arising from the discretization of the convection boundary condition on the top interface, and $\mathbf{B} \in \mathbb{R}^{4257}$ is a constant load vector. Originally, the system had seven outputs. We take a sum over all rows to obtain a single output with $\mathbf{C} \in \mathbb{R}^{1\times4257}$. The frequency response of the system is shown in Figure 9.3 for p varying in $[1, 10^4]$ and for five selected values of p in Figure 9.4.

9.5.3 ▪ Anemometer[37]

An anemometer is a flow-sensing device that consists of a heater and temperature sensors placed both before and after the heater, either directly in the flow or in its vicinity.

[36]http://www.modelreduction.org/index.php/Microthruster_Unit.
[37]http://www.modelreduction.org/index.php/Anemometer.

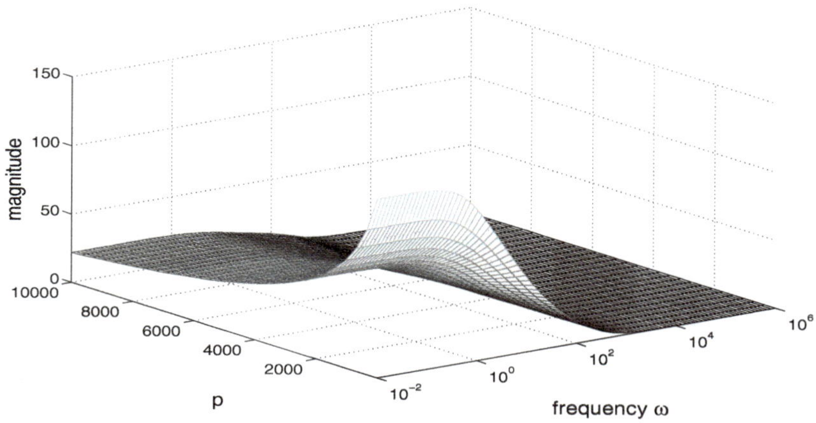

Figure 9.3. *Frequency response of the microthruster.*

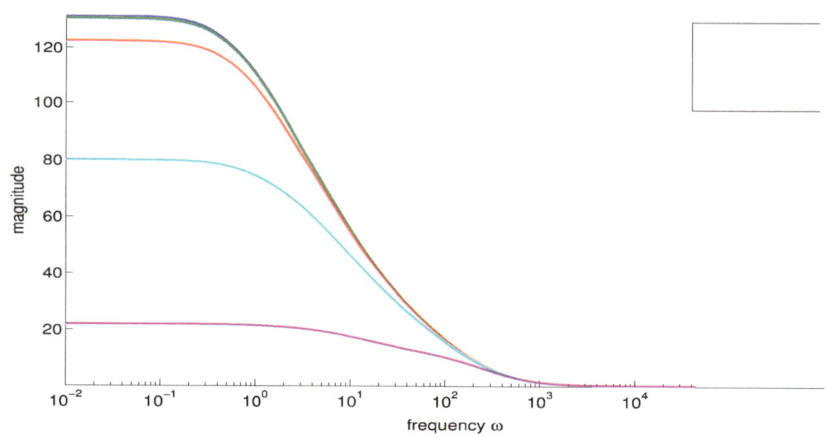

Figure 9.4. *Frequency response of the microthruster for some parameter values.*

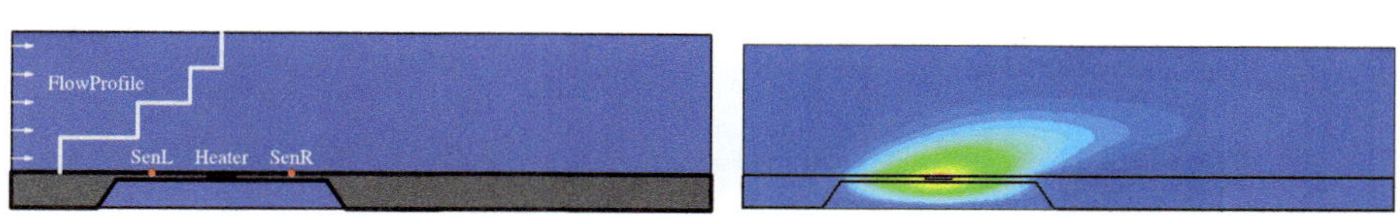

Figure 9.5. *2D model of an anemometer courtesy of* [31]. *Left: schematics. Right: calculated temperature profile.*

With no flow, the heat dissipates symmetrically into the fluid. This symmetry is disturbed if a flow is applied to the fluid, which leads to convection of the temperature field and therefore to a difference between the temperature sensors (Figure 9.5) from which the fluid velocity can be determined.

The physical model can be expressed by the convection-diffusion partial differential equation (PDE) [30]

$$\rho c \frac{\partial T}{\partial t} = \nabla \cdot (\varkappa \nabla T) - \rho c v \nabla T + \dot{q}, \tag{9.12}$$

where ρ denotes the mass density, c is the specific heat, $\varkappa$ is the thermal conductivity, v is the fluid velocity, T is the temperature, and $\dot{q}$ is the heat flow into the system caused by the heater.

The model (9.12) is discretized in space using the finite element (FE) method with triangular elements. The order of the discretized system is $n = 29,008$ after applying (zero) Dirichlet boundary conditions. The n-dimensional ODE system has the following transfer function:

$$H(s,\mathsf{p}) = \mathbf{C}(s\mathbf{E} - \mathbf{A}_0 - \mathsf{p}\mathbf{A}_1)^{-1}\mathbf{B},$$

with the fluid velocity $\mathsf{p}(= v)$ as scalar parameter. Here, $\mathbf{E}$ is the heat capacitance matrix, $\mathbf{A}_0$ describes the thermal conduction, and $\mathbf{A}_1$ contains the convection terms, which are given by a cascaded flow profile. $\mathbf{B}$ is the load vector, which characterizes the spatial heat distribution into the fluid introduced by the heater. The initial conditions are set to zero. The dependency of the frequency response on the parameter $\mathsf{p} \in [0, 1]$, i.e., $\bar{\sigma}(H(\imath\omega, \mathsf{p}))$, can be seen in Figure 9.6 and in Figure 9.7 for some parameter points in $[0, 1]$. For more information about the system, see [31].

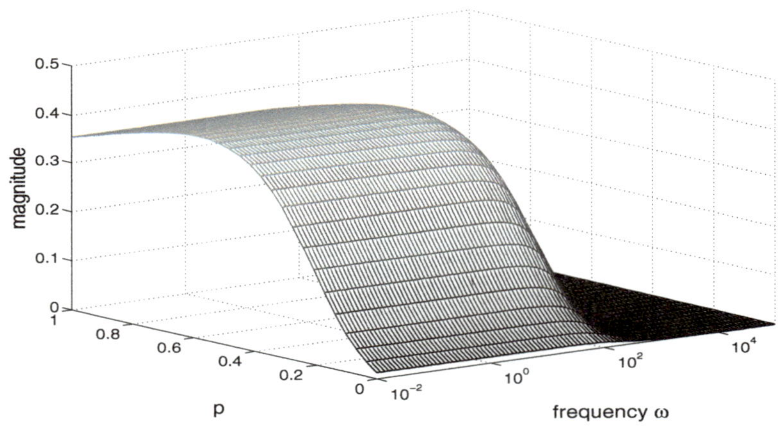

Figure 9.6. *Frequency response of the anemometer.*

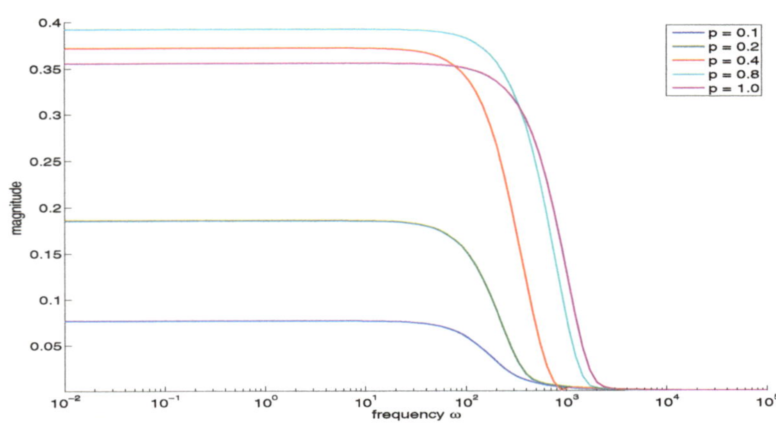

Figure 9.7. *Frequency response of the anemometer for some parameter values.*

9.6 ▪ Numerical results

All methods use the same initial sampling of the parameter space, i.e., the same choice of $\{\mathsf{p}_1, \ldots, \mathsf{p}_K\}$, to compute either snapshots (POD, emWX) or local reduced quantities (in interpolatory approaches). Furthermore, the reduced order in the parametrized reduced-order system is either fixed by r (in POD, emWX) or determined by $K \cdot r'$, where r' is the local reduced order in the interpolatory approaches. To maintain comparability, we set $r = K \cdot r'$ for POD and emWX. The final reduced order in MatrInt is r', which leads to smaller online costs. Note that in our experiments we also tried higher orders than r' for MatrInt; this led to better approximation in the vicinity of the reference points (because of an improvement in the *local* reduced models) but deteriorated the global (overall) approximation quality because of higher errors *between* the reference points. This is probably because with increasing reduced order, the local subspaces at different reference points tend to become more and more different, such that the interpolation mechanism based on compatibility transformation progressively fails. Earlier results confirm that MatrInt typically works better for lightly damped systems, whose distinct dominant dynamics are captured at all reference points and facilitate the interpolation due to the comparable local subspaces.

All methods, except TransFncInt and MatrInt, are computed with one-sided (Galerkin) projections to promote stability in the reduced-order system. Note that there is a loss in fidelity from using one-sided projections. TransFncInt used in combination with BT (i.e., with two-sided projections for computing the local reduced-order systems) guarantees the preservation of stability of the resulting parametrized reduced-order system [3]. Thus, this approach still uses two-sided (Petrov–Galerkin) projections. MatrInt uses a two-sided IRKA method followed by solving a low-order Lyapunov equation for each parameter sampling point to preserve stability [16, 17].

All error measures as described in Section 9.3 are computed and compared in the following, using an error grid with $\bar{K} = 100$ (with $K < \bar{K}$) parameter points. The pointwise (in p) errors are plotted in the corresponding figures in the following subsections. The maximum values of these errors (on the error grid) are listed in Tables 9.1, 9.2, and 9.3. The time domain grid for all time domain errors equals the time discretization $t_i = i\Delta t$, $i = 0, \ldots, J$, in the snapshot-based PMOR approaches. A unit impulse is applied as input to all systems to compute the snapshots as well as the time domain error measures. To compute the frequency response, $\bar{L} = 100$ frequency points are taken.

The benchmarks were computed on an octa-core CPU (AMD FX-9590) with 32 GB memory (DDR3-2133) by MATLAB R2014a (Intel MKL) on Ubuntu 14.04 running Linux Kernel 3.19 using four pinned threads.

9.6.1 ▪ Synthetic system

All PMOR approaches introduced in Section 9.2 are applied to the synthetic system with $n = 1000$ from Section 9.5.1 to compute parametrized reduced-order systems. To this end, an initial discretization of the parameter interval $[0.1, 1]$ is required. We choose an equidistant grid with four sampling points to compute snapshots as well as to compute local reduced quantities for interpolation. In the interpolatory approaches, the local reduced order is $r' = 25$. This leads to a reduced order of $r = 100$ in all approaches except MatrInt, where a summation over the reduced system matrices leads to an overall reduced dimension of $r = 25$. The reduced order in all approaches based on snapshots is also set to 100. Here, the reduced-order systems computed by POD and POD-greedy are of slightly smaller dimension; see Table 9.1. This is due to a smaller numerical rank of the snapshot matrix using a tolerance of 1.e−7.

The state is computed using the backward Euler method on the time interval [0, 1] with 1000 constant time steps of size $\Delta t = 0.001$.

The frequency domain error measures are computed on $[\imath\, 1,\ \imath\, 10^{3.1})$, the domain where the frequency responses differ along the parameter interval; see Figure 9.2. In MultiPMomMtch an initial discretization of the considered frequency range is also required. We took 16 logarithmically scaled frequency expansion points $s_1, \ldots, s_L$. The highest number of moments, included in the projection matrix $\mathbf{V}$, is chosen as $M = 2$.

The results can be found in Table 9.1 and in Figures 9.8–9.13. A discussion of all results can be found in Section 9.6.4.

Table 9.1. *Synthetic results for all PMOR methods considered (relative errors).*

Method	r	$e(\mathbf{x},\hat{\mathbf{x}})_{\mathcal{L}_2}$	$e(y,\hat{y})_{\mathcal{L}_2}$	$e(H,\hat{H})_{\mathcal{H}_\infty}$	$e(H,\hat{H})_{\mathcal{H}_2}$
POD	89	1.5e-10	7.7e-14	4.4e-2	1.2e-2
POD-greedy	87	2.7e-10	1.6e-13	4.9e-2	1.4e-2
MatrInt	25	6.4e-1	5.8e-3	2.2e-1	1.4e-1
TransFncInt	100	-	8.1e-6	5.3e-2	1.4e-2
PWH2TanInt	100	1.3e-5	1.2e-10	1.1e-2	6.5e-3
MultiPMomMtch	100	7.7e-4	2.4e-6	3.2e-2	2.1e-2
emWX	100	7.0e-8	3.5e-9	9.2e-2	5.8e-2

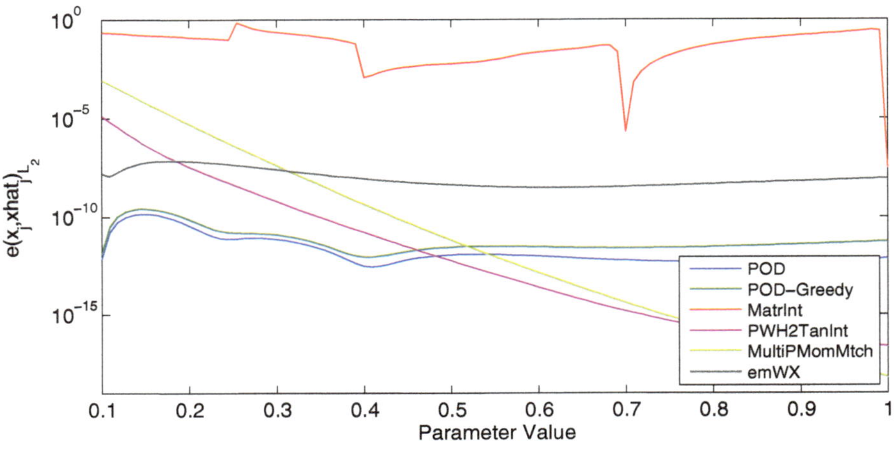

Figure 9.8. *Relative $\mathcal{L}_2$ state error for the synthetic system.*

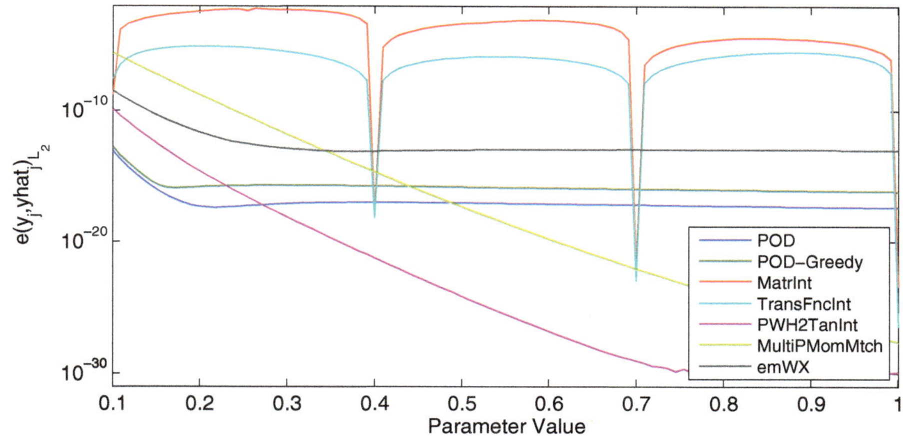

Figure 9.9. *Relative $\mathcal{L}_2$ output error for the synthetic system.*

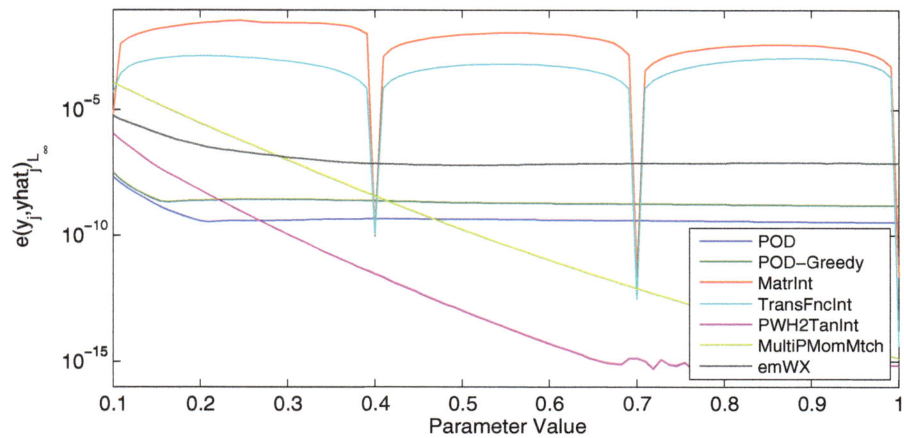

Figure 9.10. *Relative $\mathcal{L}_\infty$ output error for the synthetic system.*

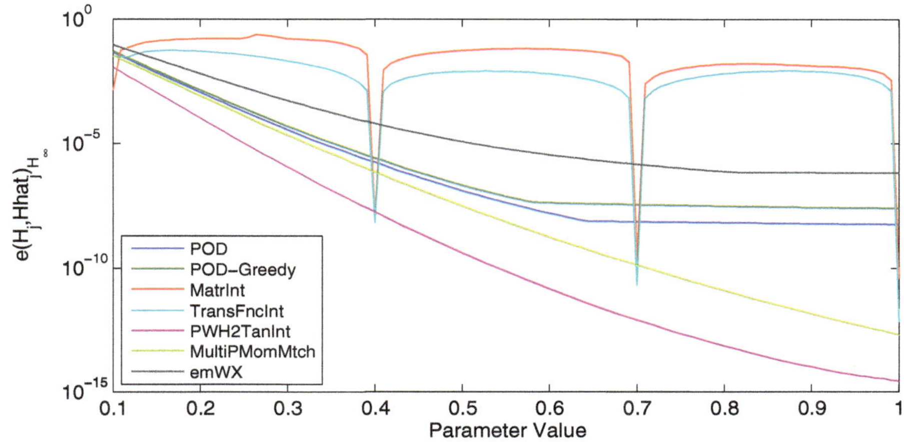

Figure 9.11. *Scaled $\mathcal{H}_\infty$-error for the synthetic system.*

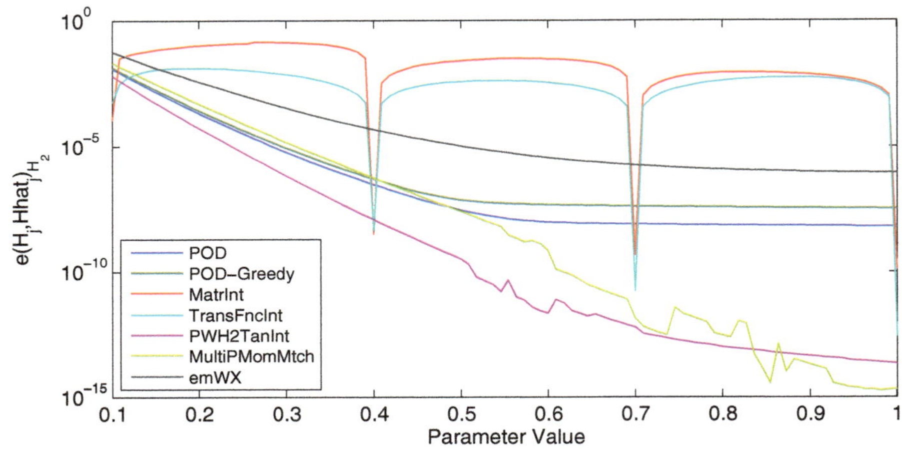

Figure 9.12. *Relative $\mathcal{H}_2$-error for the synthetic system.*

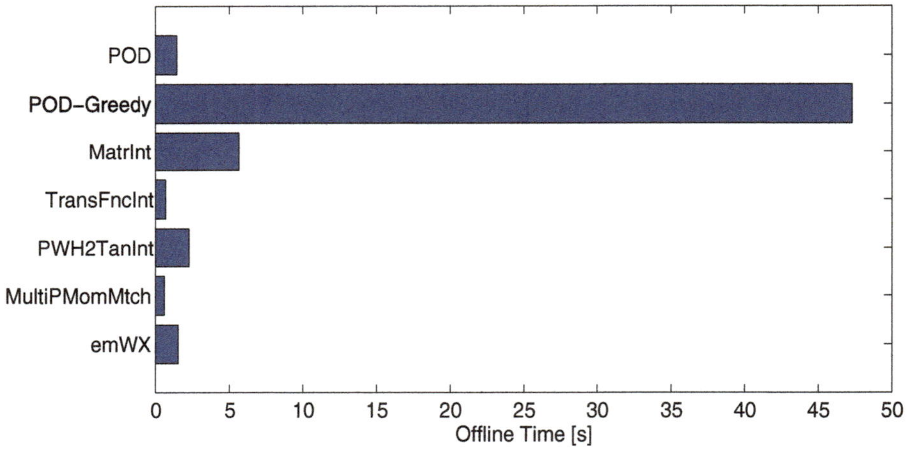

Figure 9.13. *Offline times for the synthetic system.*

9.6.2 ▪ Microthruster unit

The film coefficient p at the top interface is in the interval [1, 10000]. This parameter interval is discretized with 10 parameter sampling points to compute snapshots, locally reduced-order systems for interpolation, or Krylov subspaces for projection. The local reduced order is set to $r' = 10$. This gives a reduced order of $r = 100$ (MatrInt $r = 10$) in the parametrized reduced-order system. We consider a frequency range of $[\imath\ 0.001, \imath\ 10^6]$ and select four points in the interval as expansion points in MultiPMomMtch. The highest order of matched moments is prescribed by $M = 2$.

Trajectories and time domain errors are computed in the time horizon [0, 20] with constant time steps of size $\Delta t = 0.1$ (200 time points) using the backward Euler method.

See Table 9.2 and Figures 9.14–9.19 for the results achieved.

Table 9.2. *Microthruster results for all PMOR methods considered (relative errors).*

Method	r	$e(\mathbf{x},\hat{\mathbf{x}})_{\mathscr{L}_2}$	$e(y,\hat{y})_{\mathscr{L}_2}$	$e(H,\hat{H})_{\mathscr{H}_\infty}$	$e(H,\hat{H})_{\mathscr{H}_2}$
POD	100	1.4e–19	1.5e–19	2.9e–3	9.2e–2
POD-greedy	100	1.7e–19	1.7e–19	1.5e–3	8.6e–2
MatrInt	10	4.1e–2	1.4e–2	2.4e–1	9.0e–2
TransFncInt	100	-	8.7e–6	9.2e–3	4.1e–2
PWH2TanInt	100	4.8e–9	8.0e–14	2.0e–6	2.3e–2
MultiPMomMtch	100	3.2e–4	7.5e–5	2.9e–2	3.5e–2
emWX	100	4.7e–7	1.8e–8	1.3e–2	3.4e–1

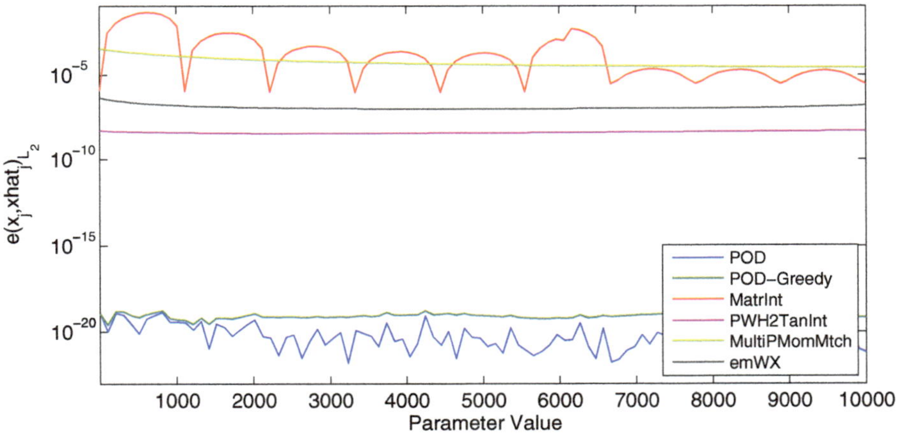

Figure 9.14. *Relative $\mathscr{L}_2$ state error for the microthruster.*

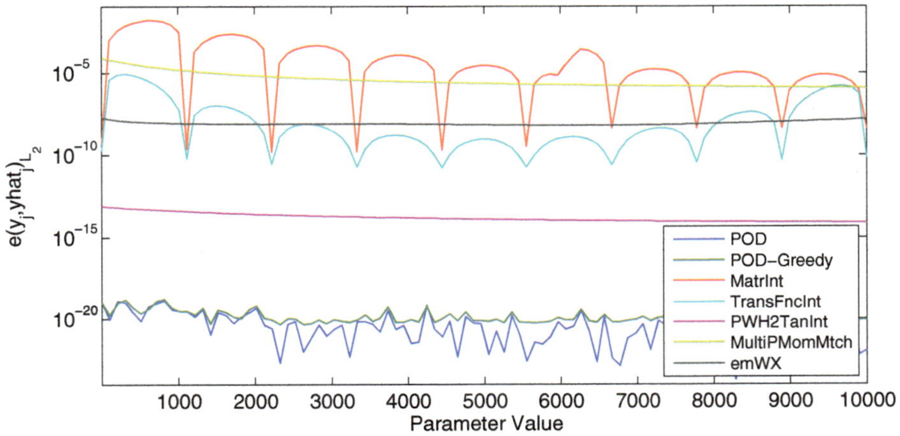

Figure 9.15. *Relative $\mathscr{L}_2$ output error for the microthruster.*

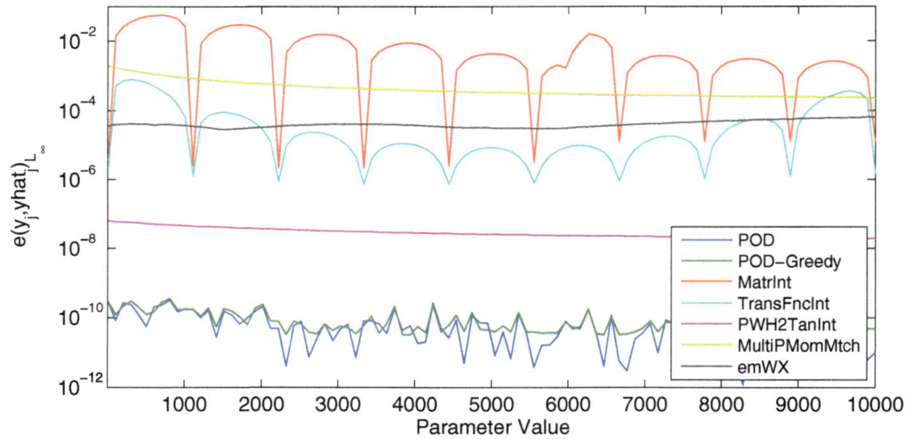

Figure 9.16. *Relative $\mathscr{L}_\infty$ output error for the microthruster.*

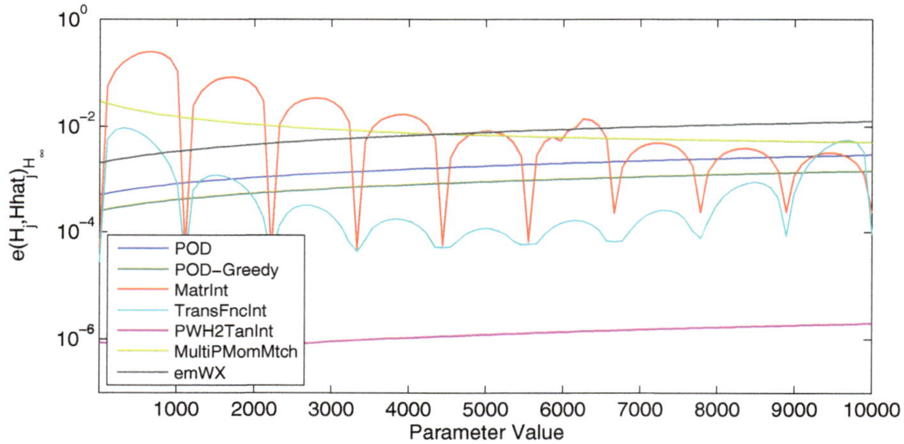

Figure 9.17. *Scaled $\mathscr{H}_\infty$-error for the microthruster.*

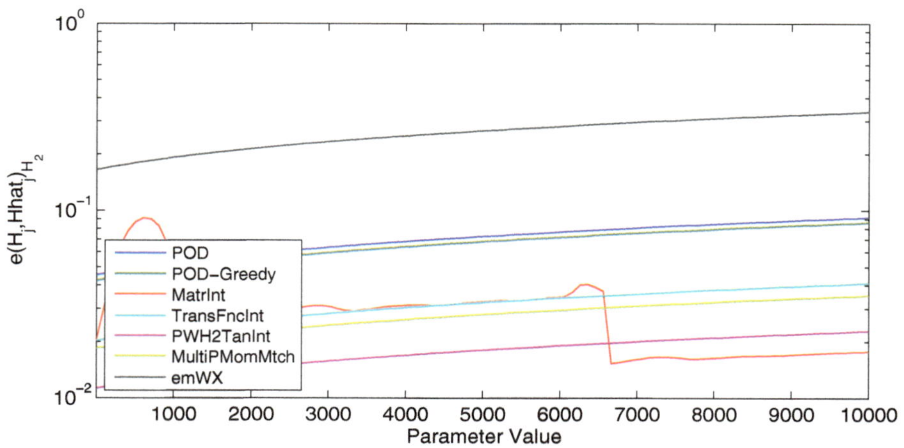

Figure 9.18. *Relative $\mathscr{H}_2$-error for the microthruster.*

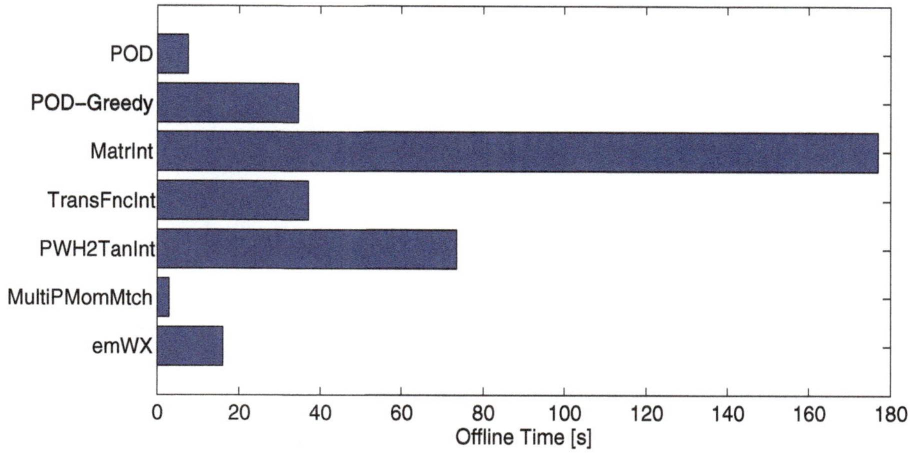

Figure 9.19. *Offline time for the microthruster.*

9.6.3 ▪ Anemometer

We consider the discretized convection-diffusion equation (9.12) from Section 9.5.3, i.e., a sparse ODE system of order $n = 29,008$ with one parameter $\mathsf{p} \in [0, 1]$ that influences the convection. Sixteen parameter sampling points are taken from this interval. The local reduced order is set to $r' = 10$, which leads to $r = 160$ for all approaches except MatrInt (with $r = 10$). The state and output errors are computed for the time horizon $[0, 0.05]$ with 50 time points. As in the microthruster example, the trajectories are computed using the backward Euler method with constant step size of 0.001.

The errors in the frequency domain are computed in the interval $[\imath\, 0.01, \imath\, 10^5]$. A choice of 10 frequency expansion points and $M = 2$ is taken for MultiPMomMtch. This gives a dimension of 480 in the projection matrix $\mathbf{V}$, which is truncated to $r = 160$ to get a comparable online time.

Table 9.3. *Anemometer results for all PMOR methods considered (relative errors).*

Method	r	$e(\mathbf{x},\hat{\mathbf{x}})_{\mathscr{L}_2}$	$e(y,\hat{y})_{\mathscr{L}_2}$	$e(H,\hat{H})_{\mathscr{H}_\infty}$	$e(H,\hat{H})_{\mathscr{H}_2}$
POD	160	1.6e−14	3.0e−11	9.3e−2	3.0e+0
POD-greedy	160	1.4e−14	8.8e−11	3.7e−2	1.2e+0
MatrInt	10	4.9e−3	5.5e−1	8.9e−1	7.6e−1
TransFncInt	160	-	8.2e−3	2.3e−1	3.5e−1
PWH2TanInt	160	2.5e−6	2.9e−10	6.0e−5	2.5e−3
MultiPMomMtch	160	2.3e−7	2.9e−6	3.8e−1	1.3e+1
emWX	160	8.0e−7	1.6e−2	9.8e−1	8.5e+0

9.6.4 ▪ Discussion of the results

The error measures from Section 9.3 are computed for all the PMOR methods considered here as applied to the benchmarks described in Section 9.5.

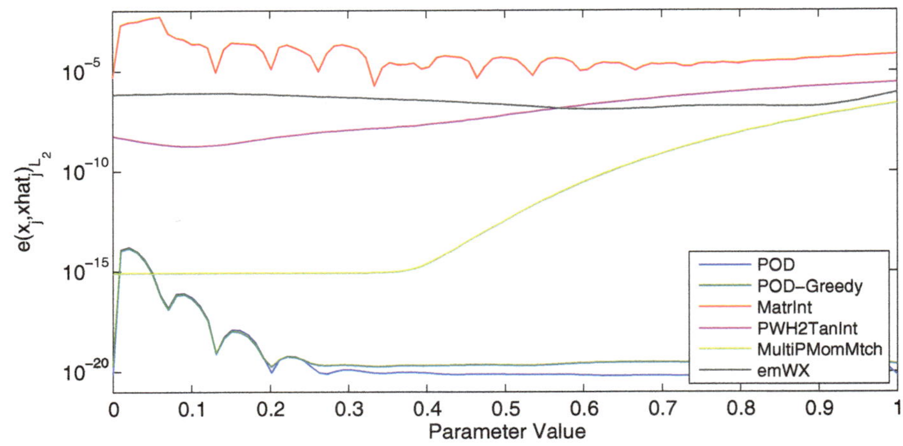

Figure 9.20. *Relative $\mathscr{L}_2$ state error for the anemometer.*

Time domain errors

Figures 9.8, 9.14, and 9.20 show the pointwise $\mathscr{L}_2$-errors in state space. Note that state-space errors for TransFncInt cannot be computed because of the absence of a lifting map that maps the reduced state back to the original (full-order) state space for the realization considered.

It can be seen that the results for POD and POD-greedy are consistent with the expectations. The errors appear to be uniform and very small throughout the parameter interval for all benchmarks considered. The state-space errors computed by emWX are also uniformly distributed but larger than the POD errors.

For the synthetic system in Figure 9.8, it can be seen that the PWH2TanInt and MultiPMomMtch errors are decreasing in the parameter interval. This can be explained by the shape of the frequency response in Figures 9.1 and 9.2. The frequency response slightly oscillates for $\mathsf{p} = 0.1$, which makes the approximation at the beginning of the parameter interval much more difficult. This also explains the error decay in the frequency domain for most of the methods considered. MatrInt fails to reproduce the state.

The state-space errors of MultiPMomMtch, PWH2TanInt, and emWX for the microthruster in Figure 9.14 are very uniform over p with average size of 4.e–9 (PWH2TanInt), 6.e–5 (MultiPMomMtch), and 1.e–7 (emWX). The average accuracy in the POD methods is about 1.e–20. For MatrInt, the errors are of moderate size (about 1.e–6) at the parameter sampling points but worse between them.

PWH2TanInt and emWX approximate the state uniformly well with a small average error size of 4.e–7 for the anemometer example; see Figure 9.20. Here, the state-space error is smaller (about 8.e–16) in MultiPMomMtch for $p \in [0, 0.4]$ and is increasing to 2.e–7 in $[0.4, 1]$. The state approximation computed by MatrInt is worse but still of moderate size of 2.6e–4 on average.

The output errors for the synthetic benchmark of the methods POD, POD-greedy, PWH2TanInt, MultiPMomMtch, and emWX are according to the state-space errors discussed above; see Figures 9.9 and 9.10. The results obtained by TransFncInt and MatrInt show the expected wave-like behavior with minima at the parameter sampling points and larger errors between them.

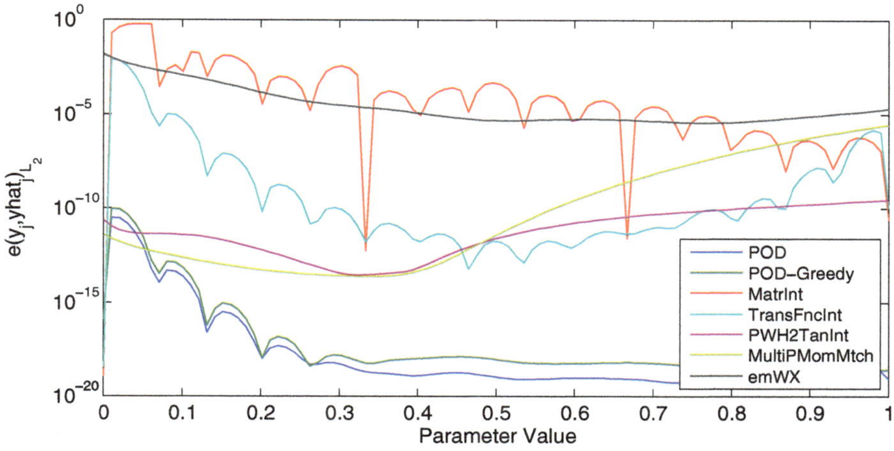

Figure 9.21. *Relative $\mathscr{L}_2$ output error for the anemometer.*

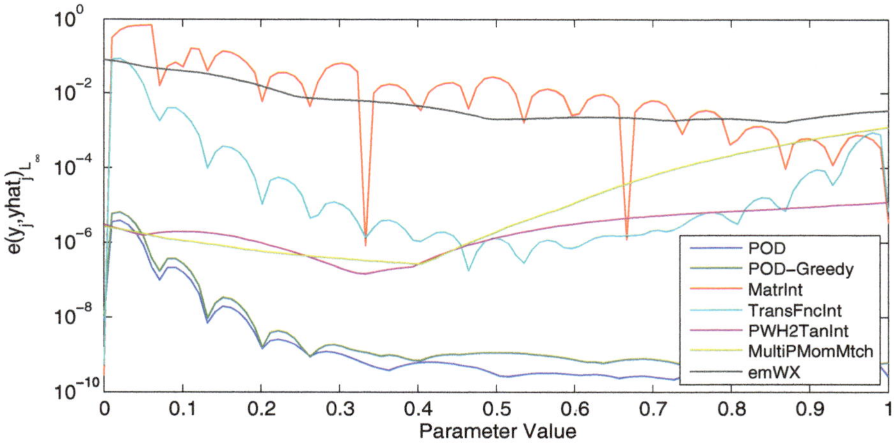

Figure 9.22. *Relative $\mathscr{L}_\infty$ output error for the anemometer.*

In the other two benchmarks, the POD methods yield better approximations than all other approaches; see Figures 9.15, 9.16, 9.21, and 9.22. The $\mathscr{L}_2$ output errors for PWH2TanInt, MultiPMomMtch, and emWX are uniformly distributed over the interval for the microthruster benchmark in Figure 9.15 but of different size on average: 2.e–14 (PWH2TanInt), 6.e–6 (MultiPMomMtch), and 8.e–9 (emWX). The errors obtained by TransFncInt are of medium size. The error curve is wave-like with average size of 5.e–7. The average in MatrInt is 1.e–5.

MultiPMomMtch approximates the output in the anemometer benchmark much better than MatrInt and with a similar quality to PWH2TanInt. The errors obtained by TransFncInt are of medium size. The errors for emWX are larger for the anemometer example. In most of the cases, we observe that the approximation qualities for the states and the outputs are related. An exception are the PWH2TanInt and emWX methods for the anemometer. Although the $\mathscr{L}_2$ state errors of the two methods are comparable, the output errors are several orders of magnitude lower for PWH2TanInt; see Figure 9.22.

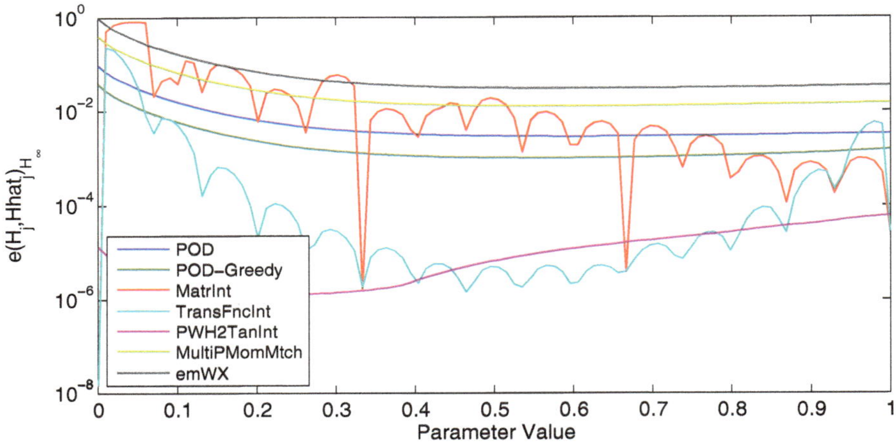

Figure 9.23. *Scaled $\mathcal{H}_\infty$-error for the anemometer.*

Frequency domain errors

The $\mathcal{H}_\infty$-errors in POD, POD-greedy, and emWX appear to be uniform in p and lie, on average, between 9.e–4 and 8.e–2 for all examples. The frequency domain errors of MatrInt and TransFncInt show the expected behavior for all examples. The error curves have minima at the parameter sampling points, showing waves between them. The errors are smaller in TransFncInt, especially in the anemometer and microthruster examples. Both interpolatory approaches produce larger errors between the sampling points for the synthetic system. These results can also be explained by the frequency response of the system given in Figures 9.1 and 9.2.

PWH2TanInt computes the smallest $\mathcal{H}_\infty$-errors for all benchmarks considered. The error is about 3.e–4 for the synthetic benchmark, 1.e–6 for the microthruster example, and 1.e–5 for the anemometer on average. It is nearly exponentially decreasing for the synthetic benchmark, with a maximum error of 1.e–2 for $p = 0.1$ in Figure 9.11, and is nearly uniform over the parameter interval for the other two benchmarks; see Figures 9.17 and 9.23. The $\mathcal{H}_\infty$-error in MultiPMomMtch is also small and exponentially decreasing in the synthetic benchmark. For the microthruster and the anemometer examples, the errors obtained by MultiPMomMtch are uniformly distributed.

With an average size of 9.e–3, it is of similar approximation quality to emWX for the microthruster benchmark. For the anemometer example, MultiPMomMtch produces a uniformly distributed larger error of about 3.e–2; the average error size in emWX is 8.e–2.

The $\mathcal{H}_2$-errors are plotted in Figures 9.12, 9.18, and 9.24. The error curves in Figure 9.12 show a very similar behavior to the $\mathcal{H}_\infty$-errors for the synthetic system (even with smaller maximum values; see Table 9.1) in Figure 9.11. The errors for the microthruster in Figure 9.18 are of comparable size between 1.e–2 and 3.e–1 in all approaches and again of similar shape to the $\mathcal{H}_\infty$-error curves for the anemometer. Here, POD, POD-greedy, MultiPMomMtch, and emWX produce errors larger than one at the beginning of the parameter interval.

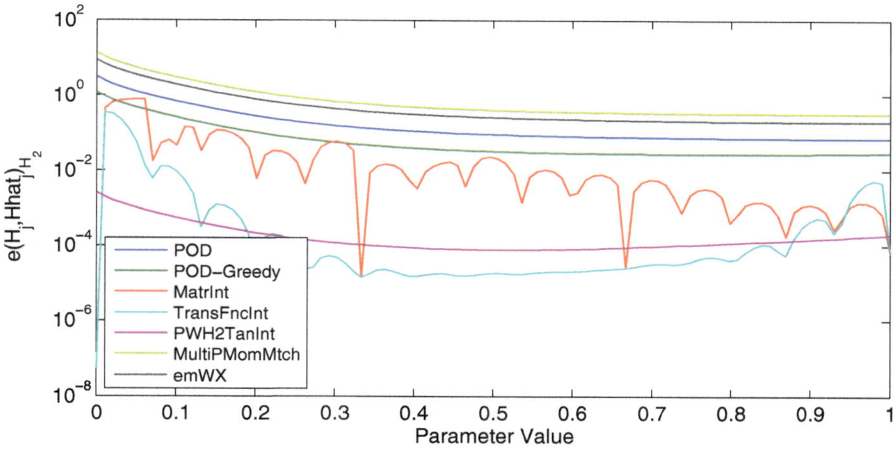

Figure 9.24. *Relative $\mathcal{H}_2$-error for the anemometer.*

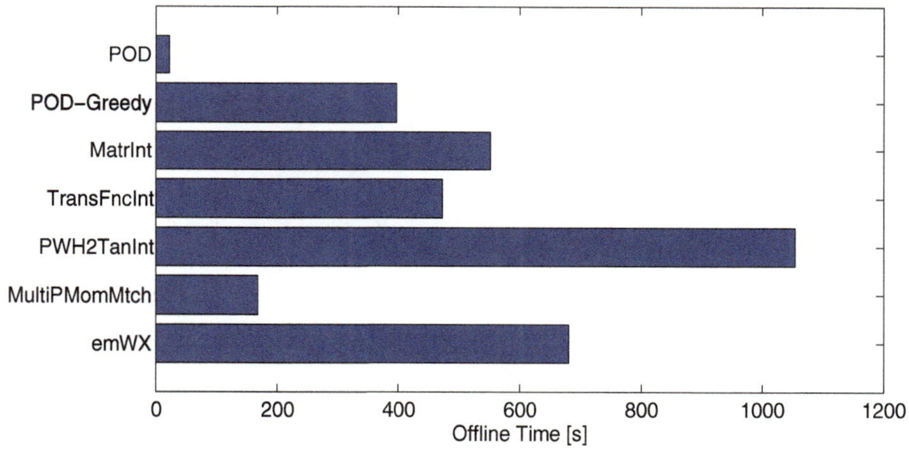

Figure 9.25. *Offline times for the anemometer.*

Computational time

The offline computational costs for the (small) synthetic system in Figure 9.13 are low for all methods except POD-greedy. Since we considered only four training parameters for the synthetic system, the number of iterations (and so the offline time) in the POD-greedy approach could be reduced drastically by choosing more POD modes in each iteration.

For the medium-size benchmark microthruster in Figure 9.19, POD and MultiPMomMtch are efficient with respect to the offline time. POD-greedy, TransFncInt, PWH2TanInt, and emWX have slightly higher offline costs. The main complexity of the offline phase in the interpolatory approaches comes from applying K times a deterministic MOR method on $\mathbf{H}(s, \mathsf{p}_j)$. MatrInt is more expensive in the offline phase because of additional steps for stability preservation like solving Lyapunov equations.

It can be seen in Figure 9.25 that for the anemometer, an example with larger original system size, again POD and MultiPMomMtch have the lowest offline times. The snapshot-based methods benefit from an efficient implementation of the backward Euler method. A single LU decomposition per parameter value can be used to solve the linear systems because of the choice of a constant time-step size. MultiPMomMtch has very low offline costs since it only requires a few factorizations of (sparse) matrices followed by a number of forward/backward solves. The anemometer offline times for POD-greedy, MatrInt, TransFncInt, and emWX are higher but nearly half the times for PWH2TanInt. In PWH2TanInt, the computation of an SVD of $\mathbf{V} \in \mathbb{R}^{n \times Kr'}$ leads to the highest offline time. The computational cost for emWX arises from three sources: first, the computation of snapshots; second, the Gramian matrix assembly; and, last, the SVD of the cross-Gramian, which is also the dominant component of the computational time for large-scale systems.

The averaged (over the parameter interval) online times for the computation of the transient and the frequency response with corresponding break-even quantities are shown in Tables 9.4–9.6 for the three benchmarks. The break-even quantity of a PMOR approach is the number of online simulations (in frequency or in time) such that offline plus online time of the reduced-order system is smaller than the simulation time (again in frequency or in time) of the original system. Note that the transient online costs are obtained by solving the system at $J+1$ time points. The number of time points differs among the examples (1000 in the synthetic benchmark, 200 for the microthruster, and 50 for the anemometer). This is not the case in the computation of the frequency response, where the transfer function is evaluated 200 times for all examples.

Table 9.4. *Simulation times [sec] in the synthetic benchmark for all PMOR methods considered.*

		Transient	
Method	Offline	Online	Break-even
original	-	0.0497	-
POD	1.39	0.0296	70
POD-greedy	46.09	0.0291	2237
MatrInt	5.48	0.0191	180
TransFncInt	0.59	0.0343	39
PWH2TanInt	2.30	0.0324	134
MultiPMomMtch	0.54	0.0328	33
emWX	1.43	0.0327	85

The simulation (online) time in the frequency domain is not computed for the small synthetic benchmark in Table 9.4. Here, it requires more time to compute the frequency response of the reduced system than to simulate the full (but sparse) original system in the frequency domain. This is different for the larger benchmarks, where the online complexity in the frequency domain and the corresponding break-even quantities can be found in Tables 9.5 and 9.6.

The online times of all PMOR approaches, i.e., the simulation times of the computed reduced-order systems, are comparable when the reduced dimensions are equal. This is the case for all approaches considered except MatrInt. The reduced order of

Table 9.5. *Simulation times [sec] in the microthruster benchmark for all PMOR methods considered.*

		Transient		Frequency	
Method	Offline	Online	Break-even	Online	Break-even
original	-	0.27	-	3.37	-
POD	8.81	0.0079	34	0.084	3
POD-greedy	32.85	0.0078	125	0.085	11
MatrInt	173.70	0.0057	654	0.0076	52
TransFncInt	35.78	0.0097	137	0.082	11
PWH2TanInt	72.14	0.0078	274	0.086	22
MultiPMomMtch	2.80	0.0080	11	0.084	1
emWX	17.86	0.0075	68	0.081	6

the system computed by MatrInt is much smaller. This results in a smaller online simulation time for systems of smaller dimension. For larger systems, the additional online steps in MatrInt (and TransFncInt) are visible in the (transient) online times; see Section 9.4 for details.

Note that the transient online costs of the original system benefit from the constant time step in the ODE solver. Here, PMOR only pays off for simulations in the time domain when the transient response has to be computed many times. This would have been different (with smaller break-even quantities) when the time steps could vary. For the small synthetic benchmark, POD, TransFncInt, MultiPMomMtch, and emWX might be useful (depending on the application) with break-even quantities below 100. POD-greedy in this scenario should obviously be run with $r' > 1$ to considerably reduce the offline time.

For simulations in the frequency domain, all break-even quantities are in the range from 1 to 52 (see Tables 9.5 and 9.6), such that all PMOR approaches pay off.

Table 9.6. *Simulation times [sec] in the anemometer benchmark for all PMOR methods considered.*

		Transient		Frequency	
Method	Offline	Online	Break-even	Online	Break-even
original	-	1.25	-	31.52	-
POD	20.56	0.0055	17	0.18	1
POD-greedy	375.18	0.0057	302	0.18	16
MatrInt	541.07	0.015	439	0.019	18
TransFncInt	463.37	0.0084	374	0.15	16
PWH2TanInt	1015.93	0.0056	817	0.18	33
MultiPMomMtch	182.98	0.0056	148	0.18	6
emWX	681.28	0.0055	548	0.18	31

9.7 ▪ Conclusions

POD seems to give the best results for state-space approximations. However, it may not be feasible if the number of training parameters or the dimension of the state is too large. In this case, POD-greedy should be preferred.

PWH2TanInt computes the best approximations in the frequency domain and also provides good results with respect to the time domain error measures. However, it requires larger offline times when applied to large-scale systems like the anemometer. So it will depend on the application (on the number of computed simulations) whether a reduction by PWH2TanInt pays off for larger systems. TransFncInt performs well for almost all error measures and benchmarks considered. The errors in the frequency domain are small for the anemometer and the microthruster, as expected. TransFncInt has problems when applied to transfer functions with peaks, as in the synthetic system. MultiPMomMtch is very efficient with respect to the offline time. However, it requires good tuning and therefore knowledge about sensitive regions in frequency and parameter space. Much smaller errors could be obtained by a more sophisticated choice of expansion points. The results obtained by applying MatrInt do not fit exactly in the framework of this comparison. This is due to a much smaller reduced order, which, not surprisingly, leads to larger errors. MatrInt is well adapted to systems with dominant eigenmodes since mode veering and crossing can be recognized by the approach. Furthermore, MatrInt can be applied to any kind of parameter dependency, even if it is not analytically given. MatrInt could therefore not fully prove its strengths in this work, as it did not exploit the affine parameter dependency of the benchmark examples considered here. The performance could be improved in the future by a better choice of transformation matrices.

emWX exhibits nearly constant behavior over the tested parameter space for most of the different error norms and benchmarks. However, the expected better match in the frequency space norms than with the POD methods was not achieved. Improvements to the accuracy (especially in the frequency domain errors) can be achieved in several ways: including more snapshots, using two-sided projections, utilizing an enhanced decomposition algorithm to obtain the projection matrices, or using a higher-order integrator for the snapshots. This would also improve the accuracy in the other snapshot-based approaches, POD and POD-greedy.

This chapter was conceived as a first attempt to compare the state-of-the-art approaches on three chosen benchmarks. The implemented framework can be used in the future to compare the approaches on other benchmarks (e.g., lightly damped systems), where the methods can perform differently. An extension of the experiments to multiparametric systems is certainly necessary to judge the ability of the various PMOR approaches to deal with the curse of dimensionality. Future work could also include a more challenging computation of the time domain error measures by using different inputs for snapshot and error computation by extending the time interval.

Bibliography

[1] D. Amsallem and C. Farhat, *An online method for interpolating linear parametric reduced-order models*, SIAM J. Sci. Comput., 33 (2011), pp. 2169–2198.

[2] U. Baur, C. A. Beattie, P. Benner, and S. Gugercin, *Interpolatory projection methods for parameterized model reduction*, SIAM J. Sci. Comput., 33 (2011), pp. 2489–2518.

[3] U. BAUR AND P. BENNER, *Modellreduktion für parametrisierte Systeme durch balanciertes Abschneiden und Interpolation (Model Reduction for Parametric Systems Using Balanced Truncation and Interpolation)*, at-Automatisierungstechnik, 57 (2009), pp. 411–420.

[4] U. BAUR, P. BENNER, AND L. FENG, *Model order reduction for linear and non-linear systems: A system-theoretic perspective*, Arch. Comput. Methods Eng., 21 (2014), pp. 331–358.

[5] U. BAUR, P. BENNER, A. GREINER, J. G. KORVINK, J. LIENEMANN, AND C. MOOSMANN, *Parameter preserving model reduction for MEMS applications*, Math. Comput. Model. Dyn. Syst., 17 (2011), pp. 297–317.

[6] P. BENNER, S. GUGERCIN, AND K. WILLCOX, *A survey of projection-based model reduction methods for parametric dynamical systems*, SIAM Review, 57 (2015), pp. 483–531.

[7] P. BENNER, V. MEHRMANN, AND D. C. SORENSEN, *Dimension Reduction of Large-Scale Systems*, vol. 45 of Lect. Notes Comput. Sci. Eng., Springer-Verlag, Berlin/Heidelberg, 2005.

[8] G. BERKOOZ, P. HOLMES, AND J. L. LUMLEY, *The proper orthogonal decomposition in the analysis of turbulent flows*, Ann. Rev. Fluid Mech., 25 (1993), pp. 539–575.

[9] L. DANIEL, O. C. SIONG, L. S. CHAY, K. H. LEE, AND J. WHITE, *A multiparameter moment-matching model-reduction approach for generating geometrically parameterized interconnect performance models*, IEEE Trans. Comput.-Aided Design Integr. Circuits Systems, 23 (2004), pp. 678–693.

[10] J. L. EFTANG, D. J. KNEZEVIC, AND A. T. PATERA, *An hp certified reduced basis method for parametrized parabolic partial differential equations*, Math. Comput. Model. Dyn. Systems, 17 (2011), pp. 395–422.

[11] O. FARLE, V. HILL, P. INGELSTRÖM, AND R. DYCZIJ-EDLINGER, *Multi-parameter polynomial order reduction of linear finite element models*, Math. Comput. Model. Dyn. Systems, 14 (2008), pp. 421–434.

[12] L. FENG AND P. BENNER, *Parametrische Modellreduktion durch impliziten Momentenabgleich*, in Tagungsband GMA-FA 1.30 “Modellbildung, Identifizierung und Simulation in der Automatisierungstechnik,” Workshop in Anif, B. Lohmann and A. Kugi, eds., 2007, pp. 34–47.

[13] ——, *A robust algorithm for parametric model order reduction based on implicit moment matching*, in Reduced Order Methods for Modeling and Computational Reduction, *MS&A Series*, vol. 9, Springer-Verlag, Berlin, Heidelberg, New York, 2014, Chapter 6, pp. 159–186.

[14] K. V. FERNANDO AND H. NICHOLSON, *On the structure of balanced and other principal representations of SISO systems*, IEEE Trans. Automat. Control, 28 (1984), pp. 228–231.

[15] K. V. FERNANDO AND H. NICHOLSON, *On the cross-Gramian for symmetric MIMO systems*, IEEE Trans. Circuits Systems, 32 (1985), pp. 487–489.

[16] M. GEUSS, H. PANZER, A. WIRTZ, AND B. LOHMANN, *A general framework for parametric model order reduction by matrix interpolation*, in Workshop on Model Reduction of Parametrized Systems II (MoRePaS II), 2012.

[17] M. GEUSS, H. PANZER, T. WOLF, AND B. LOHMANN, *Stability preservation for parametric model order reduction by matrix interpolation*, in Proc. European Control Conf. ECC 2014, Strasbourg, 2014, pp. 1098–1103.

[18] S. GUGERCIN AND A. C. ANTOULAS, *A comparative study of 7 algorithms for model reduction*, in Proc. 39th IEEE Conference on Decision and Control, Sydney, Australia, vol. 3, 2000, pp. 2367–2372.

[19] S. GUGERCIN, A. C. ANTOULAS, AND C. A. BEATTIE, *$\mathcal{H}_2$ model reduction for large-scale dynamical systems*, SIAM J. Matrix Anal. Appl., 30 (2008), pp. 609–638.

[20] B. HAASDONK, *Convergence rates of the POD-Greedy method*, ESAIM: Math. Model. Numer. Anal., 47 (2013), pp. 859–873.

[21] B. HAASDONK, M. DIHLMANN, AND M. OHLBERGER, *A training set and multiple basis generation approach for parametrized model reduction based on adaptive grids in parameter space*, Math. Comput. Model. Dyn. Systems, 17 (2011), pp. 423–442.

[22] B. HAASDONK AND M. OHLBERGER, *Reduced basis method for finite volume approximations of parametrized linear evolution equations*, ESAIM: Math. Model. Numer. Anal., 42 (2008), pp. 277–302.

[23] ———, *Efficient reduced models and a-posteriori error estimation for parametrized dynamical systems by offline/online decomposition*, Math. Comput. Model. Dyn. Systems, 17 (2011), pp. 145–161.

[24] C. HIMPE AND M. OHLBERGER, *Cross-Gramian based combined state and parameter reduction for large-scale control systems*, Math. Prob. Eng., 2014 (2014), pp. 1–13.

[25] D. J. KNEZEVIC AND A. T. PATERA, *A certified reduced basis method for the Fokker–Planck equation of dilute polymeric fluids: FENE dumbbells in extensional flow*, SIAM J. Sci. Comput., 32 (2010), pp. 793–817.

[26] J. KORVINK AND E. RUDNYI, *Oberwolfach benchmark collection*, Chapter 11 (pages 311–315) of [7].

[27] S. LALL, J. E. MARSDEN, AND S. GLAVAŠKI, *Empirical model reduction of controlled nonlinear systems*, in Proc. IFAC World Congress, vol. F, 1999, pp. 473–478.

[28] S. LALL, J. E. MARSDEN, AND S. GLAVAŠKI, *A subspace approach to balanced truncation for model reduction of nonlinear control systems*, Internat. J. Robust and Nonlinear Cont., 12 (2002), pp. 519–535.

[29] Y. T. LI, Z. BAI, Y. SU, AND X. ZENG, *Parameterized model order reduction via a two-directional Arnoldi process*, in Proc. 2007 IEEE/ACM International Conference on Computer-Aided Design, IEEE Press, Piscataway, NJ, 2007, pp. 868–873.

[30] C. MOOSMANN, E. B. RUDNYI, A. GREINER, AND J. G. KORVINK, *Model order reduction for linear convective thermal flow*, in THERMINIC 2004, Sophia Antipolis, France, 2004, pp. 317–321.

[31] C. MOOSMANN, E. B. RUDNYI, A. GREINER, J. G. KORVINK, AND M. HORNUNG, *Parameter preserving model order reduction of a flow meter*, in 2005 NSTI Nanotech, Nanotechnology Conference and Trade Show, pp. 8–12.

[32] *MOR Wiki—Model Order Reduction Wiki*, `http://www.modelreduction.org`, 2014.

[33] H. PANZER, J. MOHRING, R. EID, AND B. LOHMANN, *Parametric model order reduction by matrix interpolation*, at-Automatisierungstechnik, 58 (2010), pp. 475–484.

[34] E. RUDNYI AND J. KORVINK, *Boundary condition independent thermal model*, Chapter 17 (pages 345–348) of [7].

[35] L. SIROVICH, *Turbulence and the dynamics of coherent structures. I. Coherent structures*, Quart. Appl. Math., 45 (1987), pp. 561–571.

[36] S. VOLKWEIN, *Model Reduction Using Proper Orthogonal Decomposition*, lecture notes, University of Konstanz, 2013.

[37] S. WALDHERR AND B. HAASDONK, *Efficient parametric analysis of the chemical master equation through model order reduction*, BMC Systems Biology, 6 (2012), p. 81.

[38] D. S. WEILE, E. MICHIELSSEN, E. GRIMME, AND K. GALLIVAN, *A method for generating rational interpolant reduced order models of two-parameter linear systems*, Appl. Math. Lett., 12 (1999), pp. 93–102.

[39] K. WILLCOX AND J. PERAIRE, *Balanced model reduction via the proper orthogonal decomposition*, AIAA J., 40 (2002), pp. 2323–2330.

[40] Y. ZHANG, L. FENG, S. LI, AND P. BENNER, *Accelerating PDE constrained optimization by the reduced basis method: Application to batch chromatography*, Internat. J. Numer. Methods Eng., 104 (2015), pp. 983–1007.

[41] K. ZHOU, J. C. DOYLE, AND K. GLOVER, *Robust and Optimal Control*, Prentice-Hall, Upper Saddle River, NJ, 1996.

Index